MOLECULAR BIOLOGY

MOLECULAR BIOLOGY

CHANNARAYAPPA

Professor and Head, Department of Biotechnology
MS Ramaiah Institute of Technology (MSRIT)
Bengaluru, Karnataka

Universities Press

UNIVERSITIES PRESS (INDIA) PRIVATE LIMITED

Registered Office
3-6-747/1/A & 3-6-754/1, Himayatnagar, Hyderabad 500 029, Telangana, India
info@universitiespress.com; www.universitiespress.com

Distributed by
Orient Blackswan Private Limited

Registered Office
3-6-752 Himayatnagar, Hyderabad 500 029, Telangana, India

Other offices
Bangalore / Bhopal / Chennai / Ernakulam / Guwahati / Hyderabad
Jaipur / Kolkata / Lucknow / Mumbai / New Delhi / Noida / Patna

First published 2015

ISBN 978 81 7371 946 2

Cover and book design

Typeset in CG Times 11/13.3 *by*
OSDATA, Hyderabad

Printed at

B.B. Press, Noida

Published by
Universities Press (India) Private Limited
3-6-747/1/A & 3-6-754/1, Himayatnagar,
Hyderabad 500 029, Telangana, India.

Preface

Molecular biology is one of the basic and rapidly growing fields in the biological sciences. It has widespread applications covering genetic research, drug delivery systems, stem cell therapy and cancer treatment. Recent developments in the subject have confirmed that the functioning of biological systems is based on the flow of genetic information from the DNA. Information contained in the nucleotides of the DNA is passed on to the RNA and ultimately affects the cellular proteins. Thus, control is exercised only by modifying the various stages of protein synthesis. Therefore, studying how the cellular mechanisms are regulated through the different elements of transfer of genetic information is the right approach in molecular biology. Various books that are available do not focus on this aspect adequately. Hence, this book is an attempt to bridge this gap.

It gives a comprehensive overview of the subject and current information has been included. It is written in a clear and simple language and incorporates several student-friendly features. There are numerous illustrations and tables that will enable the readers to grasp the topics easily. The information obtained from various sources has been gratefully acknowledged. Each chapter begins with Learning Objectives and includes Key Points and Self-assessment Questions. The Further Reading section guides the students towards advanced discussion of the topics. It is hoped that the book will be a valuable textbook to students of biotechnology, genetics and other courses which have molecular biology as a component.

I would like to thank a number of people for their support while writing this book. My wife, Bhagya, son Shrikara C and daughter Shrinidhi C have given unstinting encouragement which inspired me to put in long hours to complete the book. I acknowledge with deep gratitude the suggestions and encouragement given by Dr MR Jayaram, Chairman, Gokula Education Foundation (GEF) Bengaluru, Mr MR Seetharam, Vice-Chairman GEF and Director of MS Ramaiah Institute of Technology (MSRIT), Bengaluru, Mr MR Kodhandaram, Trustee, GEF, Dr SY Kulkarni, Principal, MSRIT, Dr M Udayakumar, Professor (Retired) Department of Crop Physiology (UAS), Mr and Mrs Venkataswamy (Retired Estate Officer) UAS, Bengaluru. I would like to thank the publishers, Universities Press, especially Ms Ramaa Kishore and Dr Gita S Dattatri, for their excellent editorial assistance and fine production of this book.

Channarayappa
crayappa@rediffmail.com

Contents

CHAPTER 1

Introduction to Molecular Biology

LEARNING OBJECTIVES

After reading this chapter, you will be able to:

- Appreciate the important milestones in the development of molecular biology
- Understand the state-of-the-art instruments used in molecular biology
- List some of the sophisticated techniques used in molecular biology

1.1 INTRODUCTION

Molecular biology is concerned with understanding the mechanisms responsible for transmission and expression of the genetic information that governs cell structure and function. Its central theme consists of studying the interactions between the various systems of a cell, including the interrelationship of DNA, RNA, protein synthesis and also learning how these interactions are regulated. Interestingly, all cells have common basic properties and this unifies cell biology particularly at the molecular level. Such unity has facilitated the researchers to choose simple organisms as models for many fundamental experiments, with the expectation that similar molecular mechanisms are operative in diverse organisms. The information generated initially in bacterial studies and bacterial viruses supplemented the basic information for studies in other complex organisms.

Advances in research in molecular cell biology depend on the development of new instruments and techniques to study cell structure and function. Thus, availability of experimental tools and techniques is critical and biologists need to learn the appropriate technologies for collecting data and annotating the molecular events.

1.2 HISTORY OF MOLECULAR BIOLOGY

Molecular biology truly began in the 1930s with the convergence of various, previously distinct biological disciplines: biochemistry, genetics, microbiology and virology. With the hope of understanding life at its most fundamental level, numerous physicists and chemists also took an interest in what would become molecular biology. In its modern sense, molecular biology attempts to explain the phenomena of life starting from the molecular properties that generate them to techniques used to understand and manipulate the genetic material.

Thomas Hunt Morgan and his colleagues studied the fruit fly, *Drosophila*, as a model organism to establish the relationship between the gene and the chromosomes in heredity (Morgan 1926). A former student of Morgan, HJ Müller (1926), recognized the 'gene as a basis of life', and so set out to investigate its structure. Müller discovered the mutagenic effect of X-rays on *Drosophila*, and used this phenomenon as a tool to explore the size and nature of the gene. However, Müller could explain the more fundamental properties of genes and their actions only in a limited manner. In 1940, George Beadle and Edward Tatum demonstrated the existence of a precise relationship between genes and proteins. In the course of their experiments connecting genetics with biochemistry, they switched from *Drosophila* to the fungus, *Neurospora*. In 1944, Oswald Avery, working at the Rockefeller Institute at New York, demonstrated that genes are made up of DNA and Alfred Hershey and Martha Chase (1952) confirmed that the genetic material of the bacteriophage, the virus which infects bacteria, is made up of DNA. In 1953, James Watson and Francis Crick discovered the double helical structure of the DNA molecule. In 1961, Francois Jacob and Jacques Monod hypothesized the existence of an intermediary between DNA and its protein products, which they called messenger RNA (mRNA). Between 1961 and 1965, the relationship between the information contained in DNA and the structure of proteins was determined: the existence of a genetic code, which creates a correspondence between the succession of nucleotides in the DNA sequence and the series of amino acids in proteins was confirmed. The chief discoveries of molecular biology, however, took place in a very short period of only about 25 years. The new and more sophisticated technologies of genetic engineering could permit the isolation and characterization of genes, in particular, those of highly complex organisms. These studies revealed the structure and function of the macromolecules. A milestone in that process was the work of Linus Pauling, which for the first time linked the specific genetic mutation in patients with sickle cell disease, demonstrated change in an individual protein, the haemoglobin in the erythrocytes of heterozygous or homozygous individuals.

1.3 SOME MODEL ORGANISMS USED IN MOLECULAR BIOLOGY

Since the fundamental properties of all cells have been conserved during evolution, the information derived from the study of one type of cell is generally applicable to other cells. This has great implications in molecular biology where we use different types of cells and organisms as experimental models. The features of a few cell types that make them particularly advantageous as experimental models are discussed briefly in the following sections.

1.3.1 Virus

The simplest life form (biological agent) is, of course, the virus (**Fig. 1.1**). It consists of an inner core of nucleic acid (DNA or RNA) enclosed in a protein coat (see Chapter 2). The key to the virus's survival strategy

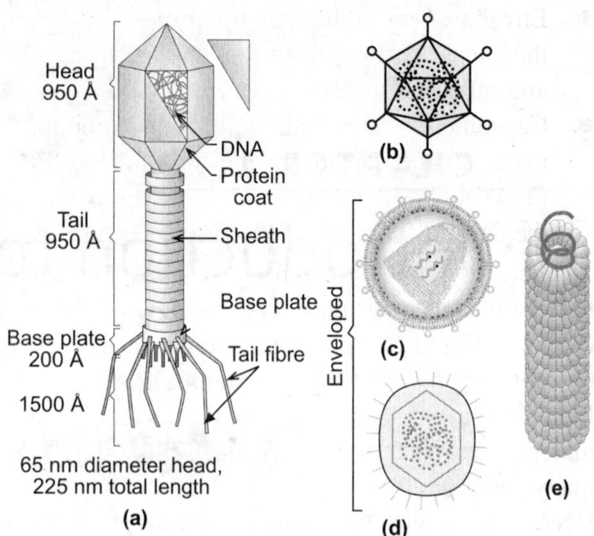

Fig. 1.1 A few basic types of viral structures (a) Icosahedral tailed phage T4 (b) Adenovirus (c) HIV (d) Herpes simplex virus (HSV) (e) Tobacco mosaic virus (TMV)

is its highly parasitic nature. The existence of viruses was discovered in 1852 by the Russian botanist DI Ivanovsky in an infectious extract from tobacco plants. In 1898, the Dutchman Beijerink coined the name 'virus' (Latin for poison). Viruses, in general, have the following characteristics:

- They are the smallest agents having biological activities.
- They do not have a cellular structure and are particulate in nature.
- They reproduce only by invading living cells, where they are all parasitic, and are obligate endoparasites, meaning, they can only live parasitically inside other cells.
- Viruses have a simple structure, consisting of a small piece of nucleic acid (DNA or RNA), surrounded by a protein or lipoprotein coat.

Viruses range in size of 20–300 nm and cannot be seen even under a light microscope. It is believed that viruses might have originated when a piece of genetic material from prokaryotic or eukaryotic cells escaped and retained the potential to replicate on getting back into the host cell. The structure of viruses consists of the following:

- Core—the genetic material, either DNA or RNA, which may be single or double stranded
- Capsid—protective coat of protein subunits surrounding the core
- Nucleocapsid—the combined structure formed by the core and capsid

- Envelope—an additional lipoprotein layer around the capsid derived from the host cell (seen in HIV and influenza viruses)
- Capsomeres—identical repeating subunits that make up the capsid. In the tobacco mosaic virus (TMV), 2,130 identical protein capsomeres form the capsid.

The viruses that attack bacteria form a group called bacteriophages or phages. The lifecycle of the phage is shown in **Fig. 1.2**. Phages that complete their lifecycle by rupturing the host cell after multiplication are called lytic. However, some other phages insert their genetic material into the host DNA and remain dormant for many generations. Each time the host divides, the phage DNA is also copied. This dormant phase of the phage is called the prophage. Eventually, when conditions turn favourable, it is activated again and completes its lifecycle by causing death of the host cell in the usual way. Such phages are called lysogenic.

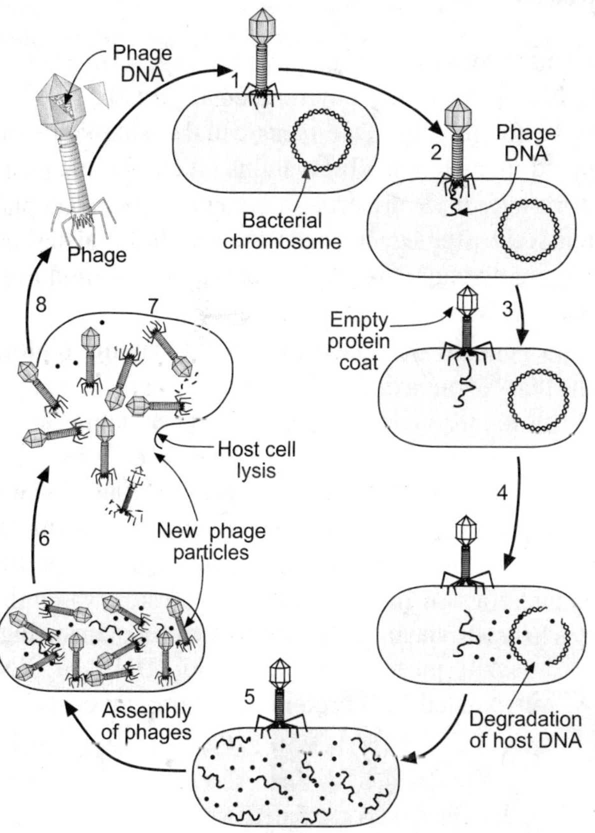

Fig. 1.2 Lifecycle of a bacteriophage 1) Phage approaches the bacterium and the tail fibres fit into receptor sites on cell surface 2) Tail fibres bend to anchor the pins and the base plate to the cell surface; tail sheath contract forcing the hollow to spike into the cell with the help of lysozyme in the base plate; DNA is injected into the cell 3) Phage DNA codes for production of phage enzymes 4) Phage enzymes de-activate the host DNA 5) Phage DNA replicates and codes for new coat proteins 6) New phage particles are released and more lysozyme is made 7) Cell lyzes and 200–1,000 phages released 8) Phage infect further bacteria

1.3.2 Escherichia coli

Prokaryotic cells like bacteria (**Fig. 1.3**) have become an ideal model organism because they are very small and can be easily cultured. The most thoroughly investigated species of bacteria is *E. coli*. Much of the present-day understanding of molecular biology—DNA replication, the genetic code, gene expression, and protein synthesis—has been derived from studies on the bacterium. The small size of its genome (4.6 mega base pairs (Mbp) and 4,000 genes) provides obvious advantages for genetic analysis. It grows rapidly (divides every 20 minutes) under normal laboratory conditions. Importantly, a clonal population of *E. coli*,

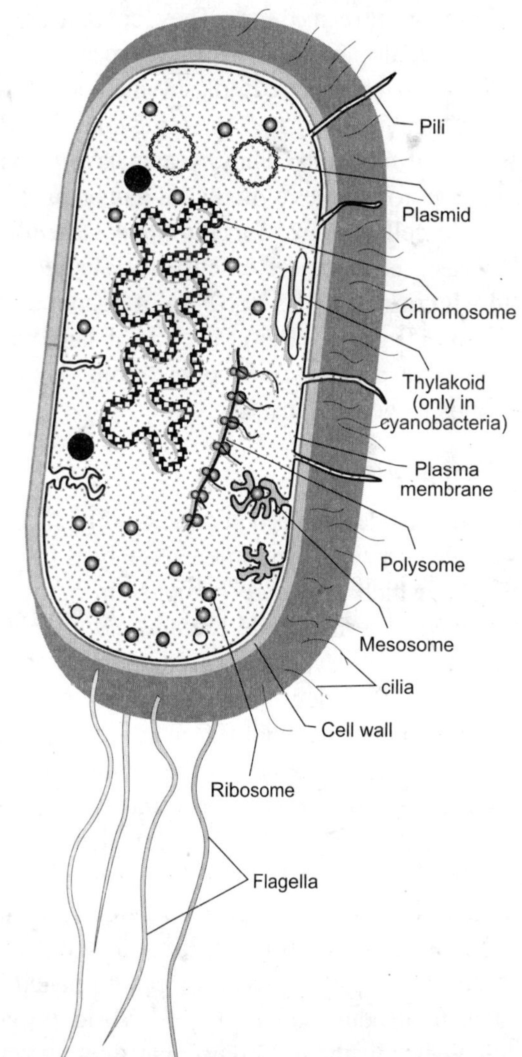

Fig. 1.3 A typical prokaryotic cell

in which all cells are derived from division of a single cell of origin, can be readily isolated as a colony grown on semisolid agar-containing medium. Since, bacterial colonies containing as many as 10^8 cells can develop overnight, selecting genetic variants of an *E. coli* strain, for example, mutants that are resistant to an antibiotic or transformed cell after transfection with a vector containing a gene of interest, is easy and rapid. Bacteria also have an ability to carry out all the biosynthetic reactions using very simple defined medium, which make them extremely useful in elucidating the biochemical pathways that are fundamental to understanding the cellular functions (see **Chapter 2**).

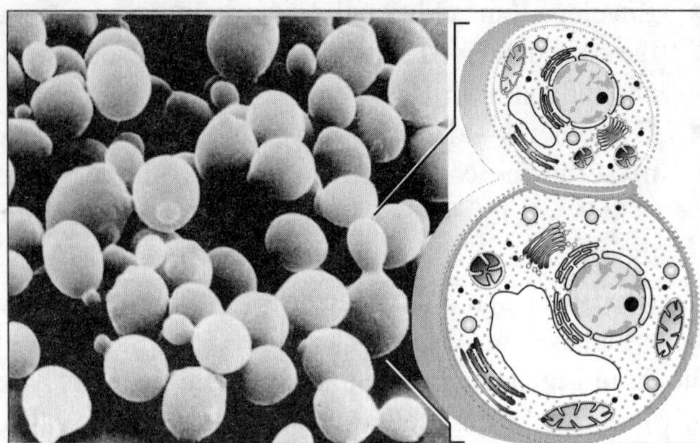

Fig. 1.4 Model organism yeast: A single cellular eukaryote. Left: Scanning electron micrograph of yeast. Right: Various parts of the yeast cell (***Source:*** modified from: 'http://www.google.co.in/imgres?imgurl=http://www.wadsworth.org/databank/hirez/wongp1.gif)

1.3.3 Yeasts

Yeasts are single celled and the simplest eukaryotic systems. They have a number of advantages similar to *E. coli*. They can be grown rapidly in the laboratory with a simple culture medium. They contain discrete nucleus with multiple linear chromosomes packaged into chromatin, and their cytoplasm includes a wide range of intracellular organelles (e.g., mitochondria) and cytoskeletal filaments. Thus, yeasts have provided a crucial model for studies of many fundamental aspects of eukaryotic cell biology. *Saccharomyces cerevisiae*, (baker's yeast) is the most extensively studied yeast (**Fig. 1.4**). It consists of 12×10^6 bp of genomic DNA and about 6,000 genes. The existence of haploid and diploid cells in *S. cerevisiae* populations facilitates various kinds of genetic analysis.

1.3.4 Nematode

To understand the developmental process of multicellular organisms we carry out experimental analysis of plants and animals, which are more complex in nature and highly evolved. The nematode *Caenorhabditis elegans* (**Fig. 1.5**) possesses several notable features that make it one of the most widely used models for studies of animal development and cell differentiation. *C. elegans* has a very rapid lifecycle. It can be cultured on petri dishes and fed a simple diet of bacteria. It grows well at a range of temperatures. At 25 °Ç, fertilized embryos complete development in 12 hours and hatch into free-living animals capable of complex behaviour. The adult hermaphrodite can produce up to 300 self-progeny over the course of about 4 days, or can be mated with rare males to produce up to 1,000 hybrid progeny. The adult lives for about 15 days, but under stressful conditions, the first juvenile or L1 stage animal can enter an alternative developmental stage called dauer, which is resistant to environmental stresses and can survive for months. Due to its cellular simplicity, the detailed microscopic observations on embryonic origin and lineage of all the cells in the adult worm has been extensively studied. Genetic studies have also been made in mutants that carry mutations for important genes.

The genome of *C. elegans* (~100 mbp) is larger than that of unicellular eukaryotes and is simpler and more manageable than of most animals. Its genome sequence has been determined, revealing that it contains approximately 19,000 genes. *C. elegans* has a simple body plan. The worm can be conveniently grown and subjected to genetic manipulations in the laboratory. Interestingly, similar genes have also been shown to work in complex animals

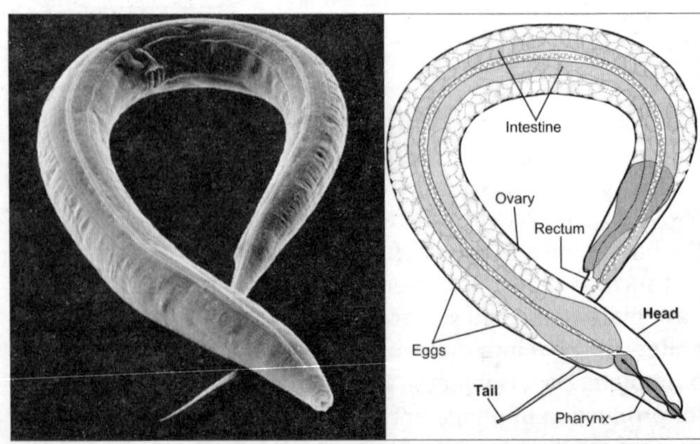

Fig. 1.5 Main parts of the nematode *C. elegans* (***Source:*** modified from http://www.wormatlas.org/hermaphrodite/introduction/mainframe.htm)

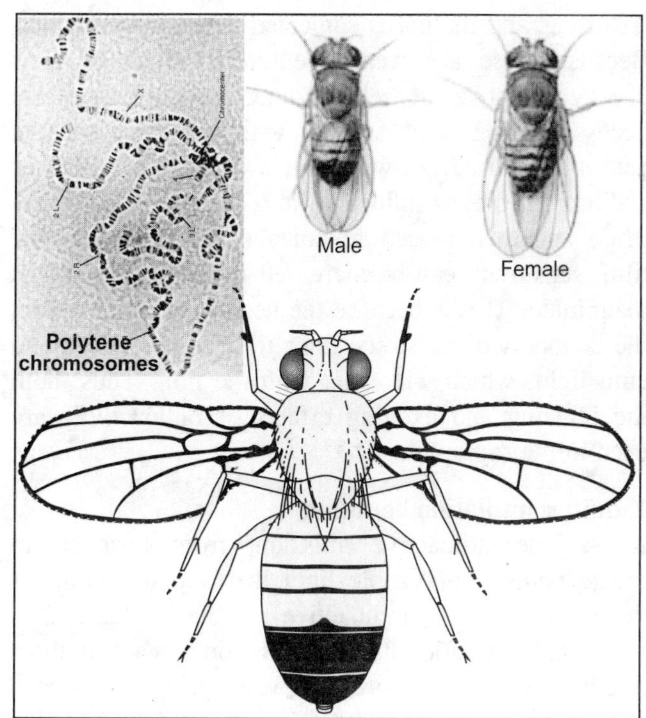

Fig. 1.6 *Drosophila melanogaster*

(including humans) making *C. elegans* an important model for studies of animal development.

1.3.5 Fruit fly: Drosophila melanogaster

The genome of *Drosophila* (**Fig. 1.6**) contains only about 14,000 genes. Fruit fly can be easily maintained and bred in the laboratory, and its short reproductive cycle (about 2 weeks) makes it a very useful organism for genetic experiments. Many fundamental concepts of genetics, such as the relationship between genes and chromosomes, were derived from studies on *Drosophila*. Many genes that control development and differentiation have been identified; and current methods of molecular biology have allowed the functions of these genes to be analyzed in detail. These studies also led to advancement in understanding the molecular mechanisms that govern animal development, particularly with respect to formation of the body plan of complex multicellular organisms.

1.3.6 Arabidopsis thaliana: A model plant

Understanding the biology of plants has significant economic importance as well as intellectual interest. Since the size and complexity of the plant nuclear genome varies enormously, an optimal model for studies of development should be a relatively simple organism. It is also known that in many organisms only a small percentage of the genome actually encodes proteins and a vast majority of the DNA is non-coding and apparently functionless or with unknown function. Much of this non-coding DNA in the nuclear genome comprises of repetitive DNA. Molecular studies, however, require an organism with the smallest genome containing no or lesser non-coding sequences.

The small flowering plant, *Arabidopsis thaliana* meets these criteria and is, therefore, widely used as a model to study the molecular biology of plants. It is a small, dicotyledonous weed that has become the model species for most aspects of plant biology. It has a short lifecycle and produces large numbers of offspring, making it ideally suited for genetic and mutational analysis. It is also easily transformed and has one of the smallest genomes in higher plants (125 Mbp). *A. thaliana* was, therefore, chosen to be the first plant to have its complete genome sequenced. These studies have led to the identification of genes involved in various aspects of plant development.

1.4 TOOLS OF MOLECULAR BIOLOGY

Since the mid-1970s, molecular genetics has seen an explosive growth and a number of techniques have been developed for the analysis and manipulation of macromolecules, particularly DNA. Some of the important laboratory techniques used in molecular biology are described in the following sections.

1.4.1 Use of radioisotopes

Radioactive tracers have been used for decades as they are easy to detect and have a high resolution. A tracer is a substance that reveals its presence in one way or another and thus can be localized or monitored during the course of an experiment. Depending on the type of substance and the experiment, a tracer can be fluorescently labelled, spin labelled, density labelled, or radioactively labelled. In each case, a labelled group enables a molecule to be detected without affecting the specificity of its interactions.

Autoradiography

Autoradiography is used to detect the sites containing radioisotopes. If a photographic emulsion is brought into close contact with a radioactive source, the particles emitted by the source leave tiny, black silver grains in the emulsion after photographic development (**Fig. 1.7**). The radioactive emissions from the bands of DNA expose the film, just as visible light would. To enhance the sensitivity of autoradiography, one can use an intensifying screen. This is a screen coated with a compound that fluoresces when it is excited by ß-rays at low temperature. When these high-energy electrons strike the screen, they cause fluorescence, which is detected by the film.

The emulsion is applied to the section on the slide or grid as a very thin overlying layer. The specimen is taken in a light-proof container to allow the emulsion to be exposed by the emissions. The longer the exposure to the emulsion before development, the greater the number of silver grains formed. When the developed slide or grid is examined in the microscope, the location of the silver grains in the layer of emulsion just above the tissue indicates the location of the radioactivity in the underlying cells.

Fluorography

In order to resolve the separated macromolecules in gel electrophoresis, macromolecules are labelled with

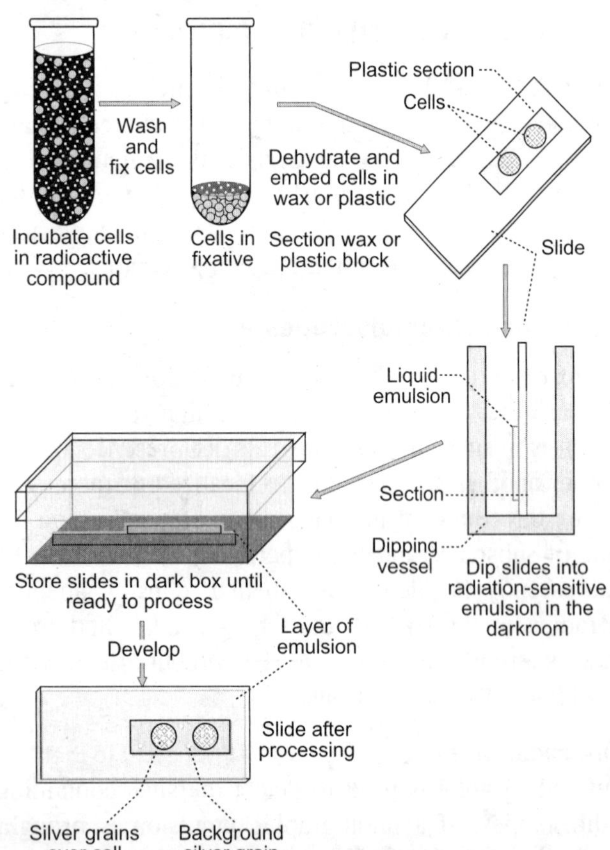

Fig. 1.7 Step-by-step procedure for the preparation of an autoradiograph

^{3}H or ^{14}C and the gel is subjected to autoradiography. Because these are weak ß-emitters, much of their energy is lost in the gel and long exposure times are necessary even when sources with very high specific activity are used. However, if a fluor (e.g., PPO or sodium silicate) is infiltrated into the gel, and the gel dried and then placed in contact with a pre-flashed film, sensitivity can be increased by several orders of magnitude. This is because the negatrons emitted from the isotope will cause the fluor to become excited and emit light, which will react with the film. Thus, both the ionizing and exciting effects of radioactivity are used in fluorography.

Liquid scintillation counting

It uses the radioactive emissions from a sample to create photons of visible light that a photomultiplier tube can detect. A radioactive sample is placed in a vial with scintillation fluid. This liquid contains a fluor, a compound that fluoresces when it is bombarded with radioactivity. In effect it converts the invisible radioactivity into visible light. A vial in a dark chamber with a photomultiplier tube detects the light resulting from the radioactive emissions exciting the fluor. The instrument counts these bursts of light, or scintillations, and records them as counts per minutes (cpm).

1.4.2 Subcellular fractionation

In order to study the detailed functions and structure of organelles, it is very important to isolate them in a form that can be used for biochemical studies. This is usually accomplished by differential centrifugation—a technique developed by Albert Claude, Christian de Duve and their colleagues in the 1940s and 1950s to separate the components of cells on the basis of their size and density.

The first step in subcellular fractionation is the disruption of the plasma membrane without destroying the internal components of the cell through methods like sonication, grinding in a mechanical homogenizer, or treatment with a high-speed blender. These techniques principally break the plasma membrane and the endoplasmic reticulum into small fragments while leaving other components intact (called lyzate or homogenate). The broken cells are then fractionated into their components by a series of centrifugations in an ultracentrifuge, which rotates samples at very high speeds to produce forces up to 500,000 times greater than gravity (i.e., >500,000 g). This force causes cell components to move towards the bottom of the centrifuge tube and form a pellet with the largest and heaviest structures sedimenting most rapidly (**Fig. 1.8**). The fractions obtained from differential centrifugation are enriched but not pure. A degree of purification can be achieved by density-gradient centrifugation, in which organelles are separated by sedimentation through a gradient of a dense substance, such as sucrose. The particles of different sizes sediment through the gradient at different rates and they move as discrete bands.

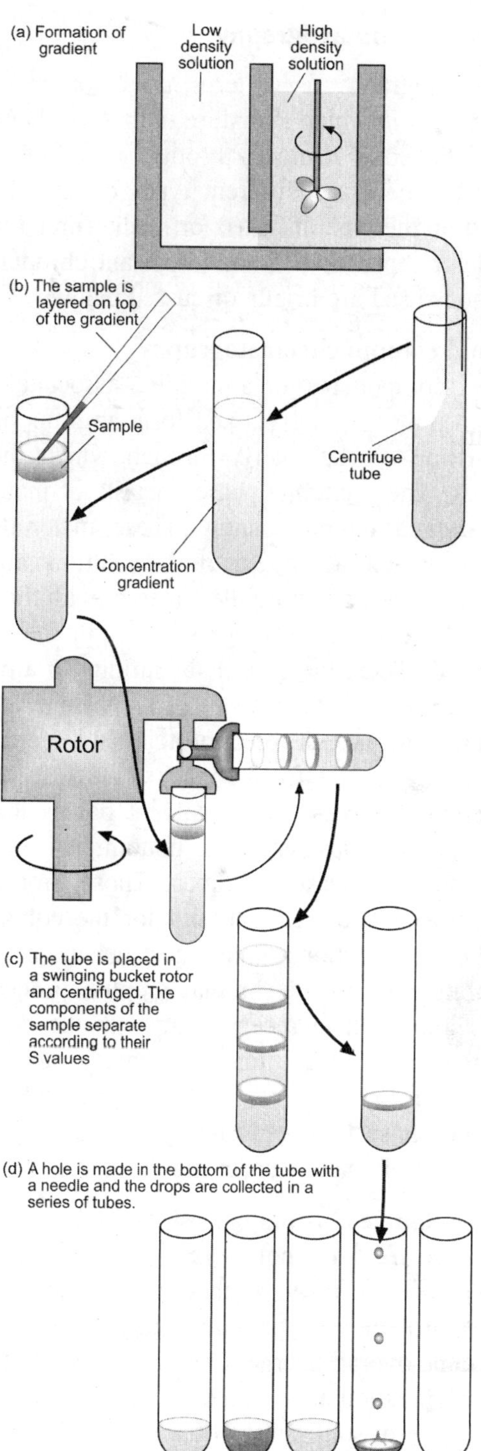

Fig. 1.8 Purification of subcellular fraction by density gradient equilibrium centrifugation; the medium is composed of a continuous density gradient and the different components sediment until they reach a level equal to their own density, where they form bands.

1.4.3 Isolation, purification and fractionation of proteins

Biosynthesis and properties of many different proteins are discussed in detail in Chapter 13. However, isolation of a particular protein in pure form is very important for further characterisation. Purification of a protein is generally performed by the step-wise removal of contaminants. Protein purification is often difficult, especially when two proteins have similar properties, such as overall charge, but very different molecular size or shape. Consequently, the complete purification of a given protein usually requires the use of successive techniques that take advantage of differences in the properties of the proteins being separated. If the protein is an enzyme, its catalytic activity may be used as an assay to monitor purification. Alternatively, immunologic, electrophoretic, electron microscopic or other assays can also be used.

Selective precipitation

This technique takes advantage of differences in solubility of proteins by selectively precipitating the desired protein. The solubility properties of a protein are determined largely by the distribution of hydrophilic and hydrophobic side chains on its surface. The salt most commonly employed for selective protein precipitation is ammonium sulphate, which is extremely soluble in water and has high ionic strength. Purification is achieved by gradually adding a solution of saturated ammonium sulphate to the crude protein extract. As addition of salt continues, the precipitation of proteins increases. The protein is collected and categorized according to the concentration of the salt solution at which it is formed (fraction). Once the protein is precipitated at a particular salt concentration, other proteins are left behind in solution (called contaminants). Thus, the protein being sought can be obtained by re-dissolving a precipitate collected at a specific salt concentration.

1.4.4 Chromatography

Chromatography is a term used for a variety of techniques in which a mixture of dissolved components is fractioned as it moves through some type of porous matrix. There are different types of chromatography based on the type of matrix or media (immobile phase) used for separation. Some important chromatographic methods used are briefly discussed below.

Liquid column chromatography

Here, components in a mixture associate with any one of a moving solvent and form an immobile phase, consisting of the matrix through which the solvent moves. The immobile phase consists of materials that are packed into a column. These materials contain sites to which the proteins in the solution can bind. As individual protein molecules interact with the materials of the matrix, their progress through the column is retarded. Thus, the greater the affinity of a protein for the matrix material, the slower its passage through the column. Because different proteins in the mixture have different affinities for the matrix, they are retarded to different degrees. As the solvent passes through the column and drips out at the bottom, it is collected as fractions in a series of tubes. Those proteins in the mixture with the least affinity for the column appear in the first fractions to emerge from the column. The resolution of many chromatographic procedures has been improved in recent years with the development of high performance liquid chromatography (HPLC), in which long, narrow columns are used, and the mobile phase is forced through a tightly packed non-compressible matrix under high pressure.

Ion exchange chromatography (IEC)

Proteins are long polymers of amino acids and are polyvalent electrolytes; therefore, it is unlikely that many proteins in a partially purified preparation have the same overall charge. Ionic charge is used as a basis for purification in a variety of techniques, including IEC. The overall charge of a protein is a summation of all the individual charges of its component amino acids. As the pH of the medium is lowered, negatively charged groups get neutralised and positively charged groups increase. The opposite occurs as the pH is increased. At a certain pH for each protein, the total number of negative charges equals the total number of positive charges. This is called the isoelectric point (pI), at which the protein is neutral. The pI of most proteins is below pH 7.

IEC depends on the ionic bonding of proteins to an inert matrix material, such as cellulose, containing covalently linked charged groups. Two of the most commonly employed IEC matrix are diethylaminoethyl (DEAE) cellulose and carboxymethyl (CM) cellulose. DEAE cellulose is positively charged and, therefore, binds negatively charged molecules; it is an anion exchanger. In contrast, CM cellulose is negatively charged and acts as a cation exchanger. The resin is packed into a column, and the protein solution is allowed to percolate through the column in a buffer whose composition promotes the binding of some or all of the proteins to the resin. Proteins are bound to the resin reversibly and can be displaced by increasing the ionic strength of the buffer and/or changing its pH. Proteins are eluted from the column in the order of the least strongly bound to the most strongly bound.

Gel filtration chromatography

Gel filtration separates proteins primarily on the basis of their effective size (hydrodynamic radius). Like IEC, the separation material consists of tiny beads that are packed into a column through which the protein solution slowly passes. The materials used in gel filtration are composed of cross-linked polysaccharides (dextrans or agarose) of different porosity, which allow proteins to diffuse in and out of the beads. For example, if a mixture contains 120, 250 and 75 kDa proteins, one way the protein might be purified is to pass the mixture through a column of Sephadex G-150 beads, which allows entry to globular proteins that are less than about 200 kDa. When the protein mixture passes through the column bed, the 250 kDa-protein is eluted as soon as the pre-existing solvent in the column (the bed volume) has moved out. In contrast, the other two proteins can diffuse into the interstices within the beads and are retarded in their passage through the column. These proteins move down its length but at different rates. Among those proteins that enter the beads, smaller species are retarded to a greater extent. Consequently, the 125 kDa-protein is eluted in a purified state, while the 75 kDa-protein remains in the column.

Affinity chromatography

It takes advantage of the unique structural properties of a protein, allowing one protein species to be specifically withdrawn from the solution. Proteins interact with specific substances: enzymes with substrates, receptors with ligands, antigens with antibodies and so forth. Each of these types of proteins can be removed from solution by passing a mixture of proteins through a column in which the specific interacting molecule is covalently linked to an inert, immobilized material. If, for example, an impure preparation of an acetylcholine receptor is passed through a column containing agarose beads to which an acetylcholine analogue is attached, the receptor binds specifically to the beads as long as the conditions in the column are suitable to promote the interaction. Once all the contaminating proteins have passed through the column and are collected at the bottom, the acetylcholine receptor molecules can be displaced from their binding sites on the matrix by changing the ionic composition and/or pH of the solvent in the column. Thus, unlike the other chromatographic procedures that separate proteins on the basis of size or charge, affinity chromatography can achieve a near-total purification of the desired molecule in a single step.

1.4.5 Mass spectrometry

Mass spectrometers are analytical instruments used primarily to measure the mass of molecules, determine their chemical formulae and molecular structure, and identify unknown substances. They accomplish these feats by converting the substances in a sample into positively charged, gaseous ions, which are accelerated through a curved tube towards a negatively charged plate. As the ions pass through the tube, they are subjected to a magnetic field that causes them to separate from one another according to their molecular mass or according to their mass-to-charge (m/z) ratio. The ions strike an electronic detector located at the end of the tube. Smaller ions travel faster and strike the detector more rapidly. The input to the detector is converted into a series of peaks of ascending m/z ratio.

Proteins are generally digested with trypsin and the resulting peptides gently ionised and made gaseous by one of the two procedures. Matrix-assisted laser desorption ionisation (MALDI) or electrospray ionisation (ESI). In MALDI, the protein sample is applied as a part of a crystalline matrix, which is irradiated by a laser pulse. The energy of the laser excites the matrix and the absorbed energy converts the peptides into gaseous ions. In ESI, an electric potential is applied to a peptide solution, causing the peptides to ionize and the liquid to spray as a fine mist of charged particles that enter the spectrometer. Because it acts on molecules in solution, ESI is well suited to ionize peptides prepared from proteins fractionated by liquid chromatography.

After the molecular mass of the peptides in the sample has been determined, the complete protein can be identified by a database search. If the protein is not identified clearly, one or more of the peptides generated by tryptic digestion can be fragmented in a second step and subjected to another round of mass spectrometry. This two-step procedure (called tandem MS or MS/MS) yields the amino acid sequence of the peptide(s) and hence the unmistakable identification of the protein. MS/MS is so powerful that complex mixtures of hundreds of unknown proteins can be digested and subjected to mass spectrometry and the identity of each of them in the mixture determined simultaneously.

1.4.6 X-ray crystallography

X-ray crystallography (or X-ray diffraction) uses protein crystals, which are bombarded with a fine beam of X-rays. The radiation that is scattered (diffracted) by the electrons of the protein's atom strikes an electron-sensitive detector placed behind the crystal. The diffraction pattern produced by a crystal is determined by the structure within the protein. The positions and intensities of the reflections, such as those on the photographic plate can be related mathematically to the electron densities within the proteins.The resolution obtained by X-ray diffraction depends on the number of spots that are analyzed.

Over the years, X-ray diffraction technology has improved greatly. In most current studies:

- intense, highly focused X-ray beams are generated by synchrotrons, which are high-energy particle accelerators that produce X-rays as a by-product, and
- highly sensitive electronic detectors (CCDs) provide a digital readout of the diffraction data.

Use of these instruments in conjunction with increasingly powerful computers allows researchers to collect and analyze sufficient data to determine the tertiary structure of most proteins in a matter of hours. It is ideally suited for determining the structure of soluble proteins that lend themselves to crystallization, but cannot be very challenging for the study of complex multi-subunit structures, such as ribosomes or proteasomes, or for membrane proteins, for which it is difficult to obtain the 3D crystals required for analysis.

1.4.7 Gel electrophoresis

Gel electrophoresis is one of the principal tools of molecular biology and a powerful method for separating nucleic acids (DNA and RNA) and proteins according to their size. DNA and RNA molecules are highly charged near neutral pH because the phosphate group in each nucleotide contributes one negative charge. As a result, nucleic acids move towards the positive electrode during gel electrophoresis. Smaller molecules move through the gel matrix more readily, so that the molecules of different length, such as restriction fragments, separate. Because the gel matrix restricts random diffusion of the molecules, those of varying lengths separate into 'bands' whose width equals that of the well into which the original DNA mixture was placed. The resolving power of gel electrophoresis is so great that single-stranded DNA molecules up to about 500 nucleotides long can be separated if they differ in length by just one nucleotide. This high resolution is a critical aspect of the procedures for sequencing DNA and other molecular studies.

Separated molecules on gels can be visualized by different techniques. If the DNA is not radio-labelled, the gel is incubated in a solution containing the fluorescent dye, ethidium bromide, which binds to DNA by intercalating between the base pairs. This binding concentrates the ethidium in the DNA and also increases its intrinsic fluorescence. As a result, when the gel is illuminated with ultraviolet light, the regions of the gel containing DNA fluoresce brightly. Radioactively labelled DNA can be visualized by autoradiography of the gel. In this case, the gel is laid against a sheet of photographic film in the dark, exposing the film at the positions where labelled DNA is present. When the film is developed, a photographic image of the DNA is observed.

In molecular genetics, electrophoresis is used to sequence nucleotides and to assign various loci to particular chromosomes. In population genetics, electrophoresis has made estimation of the variability in natural populations possible. In biochemical genetics, electrophoretic technique can be used to study enzyme pathways.

1.4.8 Blotting techniques

Blotting methods are extensively used to detect DNA, and RNA sequences and proteins. The three common methods that are described below are Southern blotting, Northern blotting and Western blotting.

Southern blotting

This technique, named after the biologist Edwin M Southern, can identify specific DNA sequence in a complex mixture of restriction fragments or within a mixture of many chromosomal DNA fragments. Prior to a Southern blotting experiment, a gene of interest or a fragment of a gene is cloned. This cloned DNA is labelled (e.g., radiolabelled or fluorescent dye) *in vitro*, and then the labelled DNA is used as a probe to detect the presence of the gene within a mixture of many fragments. The DNA to be analyzed (genomic, cloned or PCR amplified) is digested completely with a restriction enzyme(s). The complex mixture of fragments is subjected to gel electrophoresis to separate the fragments according to size. However, many different fragments are of exactly the same length, and these do not separate from each other.

An important variable in the Southern blot procedure is the temperature and ionic strength of the hybridisation and wash steps. If the experiment is done at a very high temperature (65 °C) or at high salt (NaCl) concentrations, then the probe DNA and chromosomal fragment must be highly complementary. This condition is called high stringency. Conditions of high stringency are used to detect a match between the cloned gene and a chromosomal DNA fragment. At lower stringency, probes can hybridize genes other than those cloned that are similar but not identical.

Northern blotting

Northern blotting is used to detect a particular RNA in a mixture of RNAs. An RNA sample, often the total cellular RNA, is denatured by treatment with an agent (e.g., formaldehyde) that prevents hydrogen bonding between base pairs, ensuring that all the RNA molecules have an unfolded, linear conformation. The individual RNAs then are separated according to size by gel electrophoresis and transferred to a nitrocellulose filter to which the extended denatured RNAs adhere. The filter is then exposed to a labelled DNA probe and subjected to autoradiography. Because the amount of a specific RNA in a sample can be estimated from a Northern blot, the procedure is widely used to compare the amounts of a particular mRNA in cells under different conditions.

Western blotting

Western blotting is used to locate a specific protein in a given cell type or at a particular stage of development. After the proteins have been extracted from cells, they are loaded onto a gel that separates them by molecular weight. The proteins within the gel are blotted onto a nitrocellulose (or nylon) filter. In the following step, a specific protein of interest is detected on a filter paper by using specific antibodies as the probe. A second antibody, specific to the first and labelled with a marker, usually fluorescent, locates the first antibody. Fluorescence of the second antibody would locate the presence of the first antibodies and thus indicate which of the clones is expressing the particular gene.

1.4.9 DNA microarray

This technique provides a radically different approach to current laboratory strategies because large-scale analysis and quantification of genes and gene expression can be achieved simultaneously.

A DNA microarray is a collection of spots attached to a solid support such as a microscopic slide or membrane or gel matrix (**Fig. 1.9**). Each spot in a microarray contains one or more DNA molecules. In a microarray it is possible to put down a large number (hundreds and thousands) of very small (100 μ diameter) spots in a small area on a single slide (about 1.2 cm × 1.2 cm). Currently, the arrays are synthesised on or off the glass and require complex fabrication methods similar to that used in producing microchips. They can also be spotted by robotic ultra-fine microarray deposition instruments that dispense volumes as low as 30 pl. Alternatively on-chip fabrication as used by Affymetrix builds up layers of nucleosides using a process called photolithography. Here, wafer-thin masks with holes allow photoactivation of specific deoxy nucleotide triphosphates dNTPs, which are linked at specific regions on the chip. The whole process allows layers of oligonucleotides to be built up, with the position of each nucleotide being well defined. The arrays may represent a variety of nucleic acid materials like mRNA produced in a particular cell type, termed cDNA expression arrays, or coding and regulatory regions of a particular gene or group of genes.

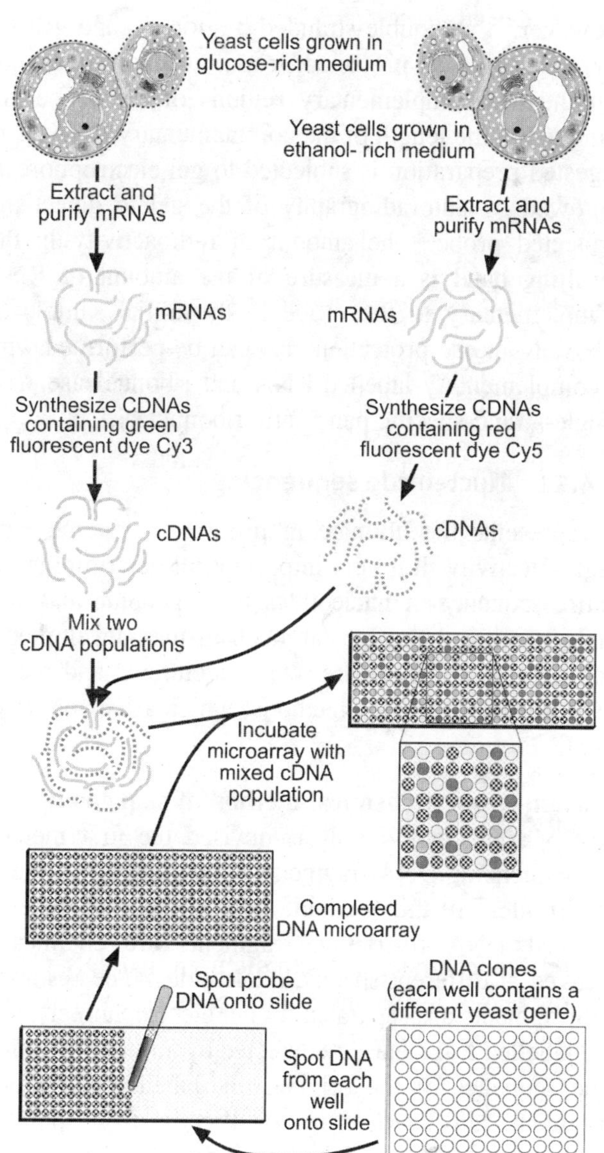

Fig. 1.9 Production of DNA microarrays and their use in monitoring gene expression

1.4.10 Nuclease protection assay

Another technique different from the previous ones, used for detecting and quantitating specific RNA molecules employs endonucleases that digest single-stranded but not double-stranded nucleic acids. This method was originally designated using endonuclease S1, an enzyme from the mould *Aspergillus oryzae* that digests single-stranded RNA only. A labelled DNA strand, or probe, complementary to RNA of interest is prepared from a cloned DNA. A source of RNA such as the total polyadenylated RNA isolated from a particular tissue or type of cultured cell, is incubated from a particular tissue or type of cultured cell with a high concentration of the labelled DNA probe under conditions so that RNA complementary to the probe hybridizes to it. The preparation then is treated with endonuclease S1, which digests all the unhybridized RNA and probe molecules.

However, the double-stranded region in the RNA–DNA hybrids is protected from nuclease digestion, but the non-complementary regions of the probe are trimmed to the length of the complementary RNA. The digested preparation is subjected to gel electrophoresis followed by autoradiography of the gel to detect the protected probe. The amount of radioactivity in the resulting band is a measure of the amount of RNA complementary to the probe in the initial sample of RNA. Nuclease protection also can be performed with a complementary labelled RNA and ribonuclease A, a single-strand-specific pancreatic ribonuclease.

1.4.11 Nucleotide sequencing

It represents the ultimate in probing a genome with high selectivity. It is very important for determining the entire sequence of nucleotides for a genome and also understanding the molecular mechanism of development and regulation. There are two conventional and recent PCR-based sequencing methods, which are extensively used.

Maxam–Gilbert chemical method of sequencing
AM Maxam and W Gilbert devised the first method for sequencing DNA fragments containing up to ~500 nucleotides. In this method, four samples of an end-labelled DNA restriction fragments are chemically cleaved at different specific nucleotides. The resulting subfragments are separated by gel electrophoresis, and the labelled fragments are detected by autoradiography. The sequence of the original end-labelled restriction fragment can be determined directly from parallel electrophoretograms of the four samples.

Sanger method
This method is also called dideoxy sequencing because it involves use of 2', 3'-dideoxynucleoside triphosphates (ddNTPs), which lack a 3'-hydroxyl group. In this method, the single-stranded DNA to be sequenced serves as the template strand for *in vitro* DNA synthesis. A synthetic 5'-end-labelled oligonucleotide is used as the primer. The separated fragments then are detected by autoradiography. The sequence of the original DNA template strand can be read directly from the resulting autoradiogram.

Shotgun sequencing of a bacterial genome
In this method, the complete sequence of the genome can be assembled. A bacterial genome is first sheared into many random fragments with an average size of 1 kb. These pieces of genomic DNA are cloned into a plasmid recombinant DNA vector. DNA is prepared from individual recombinant DNA colonies and separately sequenced on sequenators using the dideoxy method. By using this sequencing strategy, the complete sequence of the genome can be assembled.

1.4.12 Locating genes of interest

A number of disease loci have been identified and located to certain chromosomes by using *in situ* mapping techniques such as FISH. A breast cancer gene provides a good example. Other genes that have been found this way include those for cystic fibrosis and Huntington's disease. Finding the gene may be simple, but the procedure for isolating it is very complex. Searching begins by looking at the pedigrees of families, segregating the disease and then trying to correlate the occurrence of the disease with a particular RFLP or microsatellite marker. Once the gene has been localized to a particular region of a particular chromosome, then with a genomic library, the target gene can be found in the neighbourhood of the marker by chromosome walking. For known genes, the sequence and protein product can be determined.

A number of other techniques that have been used to identify the potential gene-encoding sequence are shown in **Table 1.1.**

1.4.13 Polymerase chain reaction (PCR)

One of the most important developments in applied molecular genetics was the invention of DNA amplification strategy called the polymerase chain reaction (PCR). It can be a good alternative to cloning and it mimics nature's way of replicating DNA. PCR is a rapid procedure for *in vitro* enzymatic amplification of a specific segment of DNA. It is capable of increasing the amount of a target DNA sequence in a sample by synthesizing many copies of the segment. This method of amplifying rare sequences from a mixture has vastly increased the sensitivity of procedures used in human genetics. For example, the ß-globin gene in a small sample of DNA isolated from an individual can be specifically amplified by PCR to determine whether the person is a carrier of the mutant sickle-cell allele. Amplified DNA can also be used for sequencing, which may reveal if the mutant allele is present in the sample or not. A typical PCR is carried out in discrete cycles and after n cycles, the target DNA is amplified 2^n times.

A target DNA sequence to be amplified is chosen first, but it is not necessary to know the entire nucleotide sequence of the target DNA but two small stretches of DNA of known sequence that flank this target segment must be found and used to design two synthetic oligonucleotide primers. These two oligonucleotides are used as primers for a series of reactions that are catalyzed by a DNA polymerase. These oligonucleotides typically have different sequences and are complementary to sequences that lie on opposite strands of the template DNA and flank the segment of DNA that needs to be amplified. In a typical PCR set up, the template DNA is first denatured by heating (94 °C) in the presence of a large molar excess of each of the two oligonucleotides and the four dNTPs. The mixture is then cooled to a temperature (55°C) that allows the oligonucleotide primers to anneal to their target sequences, after which the annealed primers are extended (72°C) with DNA polymerase (*Taq* polymerase). The cycle of denaturation, annealing and DNA synthesis is then repeated many times. Because the products of one round of amplification serve as templates for the next, each successive cycle essentially doubles the amount of the desired DNA product. It is this fact that accounts for the exponential increase in PCR products with the number of cycles.

Applications of PCR

PCR amplification is being used to carry out molecular cloning and analysis of DNA in the following ways:

- Generation of specific sequences of cloned double-stranded DNA for use as probes
- Generation of probes specific for uncloned genes by selective amplification of particular segments of cDNA
- Generation of libraries of cDNA from small amounts of mRNA
- Generation of large amounts of DNA for sequencing
- Analysis of mutations
- Chromosome crawling/walking
- Identification of unknown genes, called differential display
- Identification of viral agents in the early stage of disease establishment (e.g., HIV)
- Amplification of ancient or aged DNA.

Site-directed mutagenesis using PCR: PCR can be used to introduce a specific mutation into a specific base pair or pairs in a cloned gene. The primer is designed in such a way that it contains the mutant base pair (either deleted or added) in the place of the normal base pair(s). The primer modification must not be at the 3'-end. This mutant primer is used with the plasmid containing the cloned gene. PCR is used to synthesize the entire plasmid and after several rounds of replication, the majority of plasmid molecules could be expected to contain the mutation. This mixture of mutant and wild-type molecules is transformed into *E. coli* and the plasmids in individual colonies are

Table 1.1 Techniques used to determine putative gene-encoding sequences

Identification technique	*Expected results*
Zoo blotting (cross-hybridization)	Evolutionary conservation of DNA sequences that suggest functional significance
Homology searching	Gene database searching to gene family-related sequences
Identification of open reading frames/ promoters/splice sites/ ribosome binding site	Consensus sequences
Northern blot hybridization	Detection of mRNA
Exon trapping technique	Artificial RNA splicing assay
Expressed sequence tags (ESTs)	cDNAs amplified by PCR that represent a part of a gene
Promoter trapping	A gene whose promoter is trapped can be identified.
Transposon tagging	New genes can be identified
Differential display	Detection of newly expressed or inactivated genes between, for example, normal and a diseased tissues

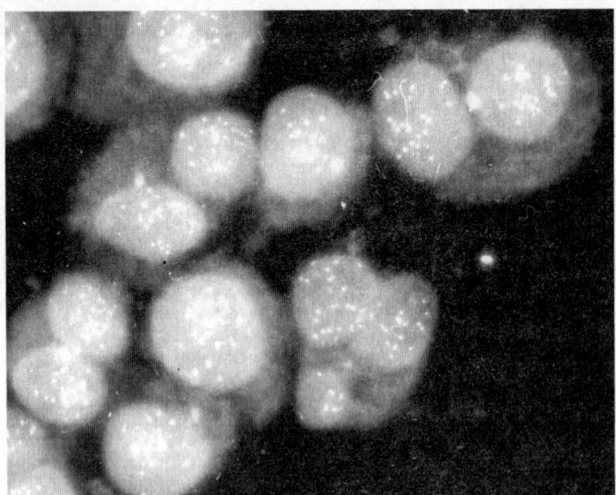

Fig. 1.10 Immunostaining of kinetochore proteins present in a centromere by human anti-kinetochore antibodies and FITC labelled secondary antibodies (***Source:*** Channarayappa 1992)

tested for the presence of the mutation by determining the DNA sequence of the cloned gene (**Fig. 1.10**).

1.5 RECOMBINANT DNA TECHNOLOGY

Recombinant DNA technology involves the joining together of DNA segments derived from biologically different sources. The discovery of a few important enzymes provided tremendous impetus for further refinement of the technique and also permitted DNA cloning (**Table 1.2**). One group of enzymes, called restriction enzymes, cut the DNA at specific sequences of a few base pairs (recognition sequences), generating a reproducible set of fragments. The other important group, called DNA ligases, can insert DNA restriction fragments into replicating DNA molecules producing recombinant DNA. These DNA molecules can then be introduced into appropriate cells and all the descendants from a single such cell, called a clone, carry the same recombinant DNA molecule. Once a clone of cells bearing a desired segment of DNA is isolated, vectors and other physical means (e.g., microinjection, electroporation, gene gun), can transmit these new sequences into host cells where they can be propagated and amplified unlimited number of times.

Genetic material from organisms of quite diverse origins (across the kingdoms) with little homology of DNA sequence can now be introduced into the host cell for propagation and expression. Recombinant DNA technology generally involves several steps as shown below (**Fig. 1.11**):

- Isolation of DNA is an essential step in molecular biology experiments. Depending upon the quality and quantity of DNA, nature of experiment and the organism from which DNA should be isolated, one can choose appropriate DNA isolation protocol.
- Generating a series of fragments of DNA from different sources containing gene sequences of interest and also codes for proteins of interest; can also be synthesised by chemical means
- Isolation and purification of plasmid DNA for cloning
- Generation of DNA fragments for genetic manipulation
- Joining (splicing) these segments into a composite molecule that is able to replicate—usually a bacterial plasmid called a vector
- Transforming bacterial cells with the recombinant molecule so that they are first replicated (cloned) and then expressed
- Selection of cells that carry the desired recombinant DNA molecules; selective propagation of cells containing the vector.
- Extraction and purification of the cloned DNA, and used for various molecular studies and transformation.

1.5.1 Cloning

Cloning is the creation of an organism i.e., the exact genotype or an identical copy of another. This includes the same sequence of DNA molecule, blood type, genetic abilities and so on.

Types of cloning

Some important types of cloning are as follows:

Molecular cloning: This includes recombinant DNA technology or DNA cloning. The recombinant technology is used to create an animal that has the same nuclear DNA as the original animal that it is cloned from. Genetic material from the diploid nucleus is moved to an enucleated egg of the same species. This process is known as 'somatic cell nuclear transfer'. The newly created egg cell is supplied with chemicals and electric current to kick-start division. When the embryo is developed to a certain degree, it is transferred into the uterus of a female host. For example, Dolly was obtained by reproductive cloning.

Therapeutic cloning: It is also known as 'embryo cloning' because it deals with the production of human

Table 1.2 Important enzymes and their use in recombinant DNA technology

Enzyme	*Properties and applications*
Klenow fragment (part of *E. coli* DNA polymerase I)	Lacks 5'→3' exonuclease activity, used for general DNA synthesis purposes, e.g., labelling by random priming, end-filling, second-strand cDNA synthesis, primer extension
T4 DNA polymerase	Active 3'→5' exonuclease activity, used for replacement labelling, generating blunt-ended DNA
T7 DNA polymerase	Rapid and highly processive, modified version, sequenase, lacks 5'→3' exonuclease activity and is used for DNA sequencing
Thermostable DNA polymerases (e.g., *Taq* DNA polymerase)	Functions at elevated temperatures, used for PCR and applications where DNA secondary structure is difficult to determine
Reverse transcriptase	RNA-dependent DNA polymerase activity, used for first-strand cDNA synthesis, RNA sequencing
RNA polymerase (T7, T3, SP6)	DNA-dependent RNA polymerase with strong promoter-specificity, used for *in vitro* transcription, RNA labelling
Restriction endonucleases	Sequence-specific DNA endonucleases, over 1,000 commercially available; used for clone mapping and preparing DNA fragments
S1 nuclease, mung bean nuclease	Single-strand-specific DNA endonuclease, used for mapping DNA–RNA hybrids, cleaving single-stranded DNA
RnaseA	General RNA endonuclease, used to remove contaminating RNA from DNA
Dnase I	General DNA endonuclease, used to introduce nicks into DNA and remove contaminating DNA from RNA
T4 DNA ligase	Generates phosphodiester bonds, used for ligation of cohesive and blunt termini
Calf intestinal alkaline phosphatase	Removes 5' phosphate groups from DNA and RNA, used to block self-ligation and for 5' end-labelling
T4 polynucleotide kinase	Adds 5' phosphate groups to DNA and RNA, used for 5' end-labelling
Terminal deoxynucleotidyl transferase	Template-independent DNA polymerase, used for 3' end-labelling, tailing

embryos. The ultimate goal is not to create a new, whole human being, but to harvest stem cells in order to pursue research. Stem cells can be used to create any other cell types in the human body. Using these stem cells scientists are optimistic that we can create replacement cells for humans, in order to treat various diseases or for organ transplants.

Cloning involves the following steps.

Isolation of DNA: For cloning, a gene can be isolated from its native site within a chromosome and inserted into a smaller segment of DNA known as a vector. When it is introduced into a living cell, a vector can replicate and produce many identical copies of the inserted gene. Therefore, to clone a particular gene, either the chromosomal DNA or the vector can be a source. Besides natural sources like plasmids and viruses, vectors can also be constructed artificially. It is very desirable that the vector should have a selectable marker, such as a gene for an antibiotic resistance determinant. Vectors must contain restriction sites or DNA sequences that are recognized and cleaved by restriction endonucleases. These enzymatically cleaved sites provide places for the cloned gene to be inserted into the vector.

Cutting, modifying and joining DNA molecules: DNA can be cleaved randomly by a number of mechanical, chemical or enzymatic methods, the extent of the treatment governing the average size of the fragment produced. However, the only way to generate precise and defined fragments is to use restriction enzymes. The restriction enzymes bind to a specific base sequence and then cleave the DNA backbone. Currently, several hundred different restriction enzymes are used.

Digestion of DNA by restriction enzymes: Genomic DNA or long DNA fragments are cut into suitable sizes leaving fragments with specific sequences at the ends. This is very conveniently achieved by using restriction endonucleases. In 1978, Arber, Smith and Nathans won the Nobel prize in physiology and medicine for their pioneering work in the study of restriction endonucleases. Restriction enzymes are bacterial

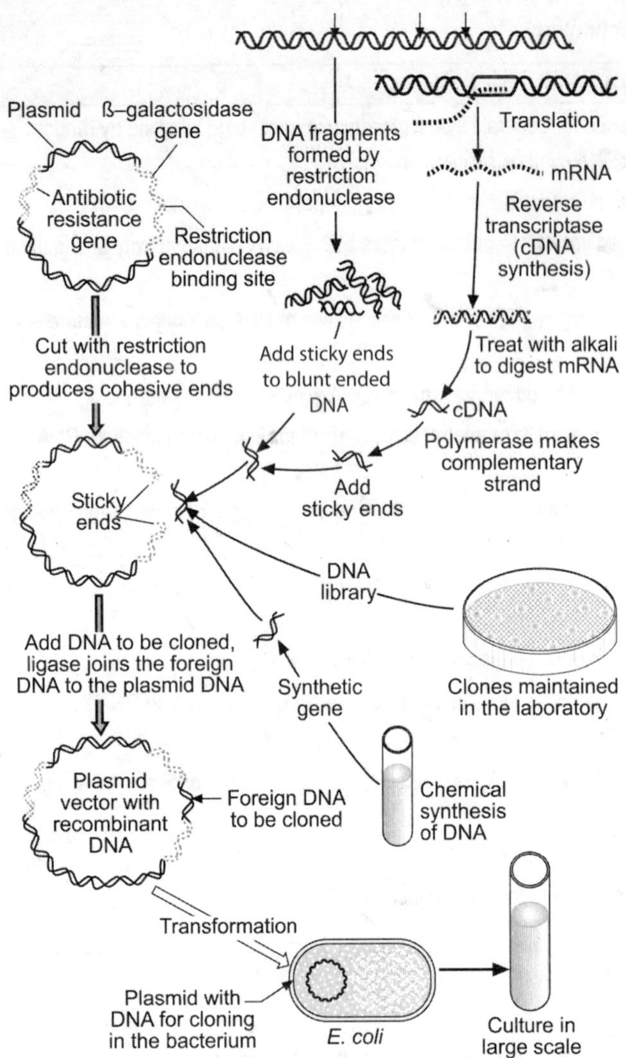

Fig. 1.11 General scheme for recombinant cloning

enzymes that recognize specific 4–8 bp sequences, called restriction sites, and then cleave both DNA strands at this specific site. Their function in the bacterial cell is to restrict (destroy) foreign DNA that might enter the cell. The enzyme recognizes the foreign DNA and cuts it at several sites along the molecule. Another enzyme, called a modification enzyme, protects the bacterium's own DNA from cleavage by modifying it at or near each potential group to one or two bases, usually within the restriction site. When a methyl group is present there, the restriction endonuclease cannot cut the DNA. Each bacterium has its unique restriction enzymes and each enzyme recognises only one type of sequence.

Three types of restriction endonucleases (type I, II and III) are known. Type II endonucleases cleave at specific sites they recognize, leading to predictable cleavage patterns; therefore, they are used in molecular biology.

Restriction enzymes generate DNA fragments with 'sticky' ends or blunt ends: Several hundred restriction enzymes have been identified, which can generate fragments that have a single-stranded 'tail' (sticky ends) at both ends. The tails on the fragments generated at a given restriction site are complementary to those on all other fragments generated by the same restriction enzyme. At room temperature, these sticky ends can transiently base pair with those on other DNA fragments generated by the same restriction enzyme, regardless of the source of the DNA and permit DNA from widely differing species to be ligated, forming chimeric molecules.

Some restriction enzymes cleave both DNA strands exactly in the middle of the recognition site. These restriction enzymes generate DNA restriction fragments with 'blunt' ('flush') ends in which all of the nucleotides at the fragment ends are base-paired to nucleotides in the complementary strand. In addition to ligating complementary sticky ends, the DNA ligase from bacteriophage T4 can ligate any two blunt DNA ends. However, blunt-end ligation requires a higher DNA concentration than ligation of sticky ends.

DNA ligase covalently links restriction fragments: DNA ligase can catalyze the formation of a 3'→5' phosphodiester bond between the 3'-hydroxyl end of one restriction fragment strand and the 5'-phosphate end of another restriction fragment strand when the sticky ends are transiently base-paired. When DNA ligase and ATP are added to a solution containing restriction fragments with sticky ends, the fragments become covalently linked.

Restriction fragments can be directly spliced into vectors: Restriction enzymes conveniently generate DNA fragments and DNA ligase can covalently link them. These properties can be used to insert foreign DNA into a suitable vector *in vitro*. For example, *E. coli* plasmid vectors can be constructed with a polylinker, or **multiple cloning site sequence (MCS)**, a synthetic sequence that contains one copy of several different restriction sites. When such a vector is treated with a restriction enzyme that recognizes a recognition sequence in the polylinker, it is cut. In the presence of DNA ligase, DNA fragments produced with the same restriction enzyme are inserted into the plasmid.

The reaction conditions and the ratio of the DNA fragments to be inserted to cut vectors are chosen so as to achieve insertion of one restriction fragment per plasmid vector. The recombinant plasmids produced *in vitro* can be used to transform antibiotic-sensitive *E. coli* cells. All of the cells in each antibiotic-resistant clone, that remain after selection contain plasmids with the same inserted DNA fragment; but different clones carry different fragments.

Modification of cut ends: The simplest strategy for cloning is thus to cut the donor and vector DNA with the same restriction enzyme and join them with DNA ligase. However, this allows the vector to close up without an insert, generating a high background of non-recombinant vectors. There are two procedures, which prevent vector self-ligation.

- The open vector can be dephosphorylated using the enzyme, alkaline phosphatase. The 5' phosphate groups at each end of the linearized vector are required for ligation and if these are removed, the vector cannot re-clone unless 5' phosphate groups are provided by a bridge of donor DNA.
- The vector and donor can be prepared using a pair of restriction enzymes which generate incompatible ends. Additional advantage of this strategy is that the orientation of the insert can be predicted (**directional cloning**).

General purpose cloning vectors: The success of cloning systems in achieving the isolation of many copies of a particular DNA sequence greatly depend on the kind of vector used to carry the DNA insert into cells. To be useful, a vector must have three properties:

- It must be able to replicate.
- There must be some way to introduce vector DNA into a cell, and
- There must be some means of detecting its presence, preferably by a plating test in petri dishes.

Naturally occurring Plasmids and bacteriophages make poor vectors because they lack cloning versatility, and accommodate less cloned DNA. They need to be modified by inserting cloning sites and retaining only the essential sequences for replication to facilitate insertion of bigger exogenous DNA.

DNA transfer to cloning host: Once a recombinant vector has been constructed *in vitro*, it must be introduced into host cells for cloning. *E. coli* is the major host for general cloning purposes, but this bacterium is not naturally competent to take up DNA from the surrounding medium. Competence can be induced by chemical treatment, such as incubation in the presence of divalent cations or a brief heat shock. The uptake of plasmid DNA by bacterium is termed transformation and that of naked phage DNA is called transfection.

Electroporation is an alternative technique where DNA enters the cells through pores created by transient high voltage. This is also a highly efficient method for introducing DNA into cells and can generate up to 10^9 colonies per μg of vector DNA. Transduction is also an effective method for transferring DNA by using phage and cosmids.

The vector is packaged in phage head precursors, tails and packaging proteins. The bacterial cultures are infected with the packaged clones. DNA enters the cells through the normal phage infection route and up to 10^6 colonies (**cosmids**) or plaques (**lambda vectors**) can be generated per μg vector DNA.

Recovery of cloned DNA: After transfer of recombinant DNA to host cells, they are cultured for a short while to allow recovery and then plated out to form colonies or plaques under the appropriate selective regime. Usually, each colony or plaque represents a clone of identical cells or phages and can be picked into liquid medium for a second round of cloning. Plating out is the process which fractionates the heterogeneous population of recombinant vectors into isolated homogenous clones, and indicates that on average, each host cell takes a single recombinant molecule.

The cloned DNA can be recovered by lysis followed by selective precipitation, sedimentation and dialysis. However, these processes are laborious, expensive and time consuming. A technique of purification of DNA by adsorption to glass beads or to resin in spin columns is simpler and a number of kits are available commercially.

1.5.2 Isolation of DNA sequences from complex DNA

The cloning techniques described above allow any DNA sequence to be inserted into a vector and cloned to facilitate further analysis and manipulation. Under the circumstances where the source DNA is not complex, or where it is highly enriched for a particular sequence, it may be possible to isolate the desired donor DNA fragment directly, and insert it into a vector for cloning. This is applicable to genomic DNA fragments from small genomes, previously obtained clones, PCR products and cDNAs representing superabundant messages in particular tissues.

However, in most cases, the source of a target sequence is complex. It is, therefore, necessary to construct a DNA library, a representative collection of all DNA fragments from a particular source cloned in vectors.

A DNA sequence of a single gene size is only a very small part of a genome; therefore, isolating a DNA fragment containing a particular gene often involves a number of steps. First, a DNA library should be constructed which must contain all the DNA fragments derived from a genome. Secondly, the DNA fragment containing the gene of interest should be identified by taking advantage of the key properties that distinguish it from other DNA fragments. Location of DNA sequences of a gene is a critical step for the isolation of genes.

Genes (a sequence of nucleotides, code for a polypeptide) are composed of stretches of nucleotide sequences in a DNA; to isolate genes one must isolate DNA. Isolation of specific genes from total DNA of eukaryotes is not easy and many properties of genes are used to advantage to isolate the genes. To increase the probability of success in cloning a particular gene from such a highly complex genome requires enriching the DNA sequence before cloning into a vector. A combination of several methods could yield an enrichment of several thousand fold.

1.5.3 Construction of DNA libraries

Sometimes it is desirable to make clones of all the genes from an organism and subsequently fish out the clone of interest. A large group of clones that contains all of the pieces of a chromosomal or mRNA on individual vector molecules is known as a **DNA library**. There are two main types of DNA libraries: genomic library and cDNA library.

Genomic library: It represents a collection of plasmid-containing fragments that is large enough so that each segment of the entire genomic DNA is represented at least once. The library size (number of clones) required varies depending on the size of the genome and vector used to clone. A general method for producing a gene library is called 'shotgun' cloning.

cDNA library: In multicellular organisms, a set of genes are transcribed into mRNA, which are unique to a cell type or tissue. For example, mRNAs encoding globin proteins are found only in erythrocyte precursor cells called reticulocytes. Likewise, the mRNA encoding albumin, the major protein in the serum, is produced only in liver cells where albumin is synthesized. The specific DNA sequences expressed as mRNAs in a particular cell type can be cloned by synthesizing DNA copies of the mRNA and then cloning the DNA copies in plasmid or bacteriophage λ vectors. DNA copies of mRNAs are called **complementary DNAs (cDNAs)**. Clones of such DNA copies of mRNAs are called cDNA clones. In addition to representing only the sequences expressed as mRNAs in a particular cell type, cDNA clones are free from introns present in genomic DNA of the respective cDNAs. A large collection of cDNA copies of all the mRNAs in a cell type is called a cDNA library. These libraries can be exploited to isolate differentially expressed genes by difference cloning.

Identification of specific clones in a recombinant library: After obtaining a genomic or cDNA library, a specific clone of interest must be identified. The most common method involves screening a library by hybridisation with radioactively labelled DNA or RNA probes. In an alternative method, called **expression cloning**, a specific clone in a library of cloned DNA is identified based on some property of its encoded protein.

1.5.4 Expression cloning

It is one of the most basic techniques of molecular biology to study protein function. Many proteins with interesting or useful functions are normally expressed at very low concentrations. The first step in obtaining large amounts of low-abundance proteins is to purify a sufficient quantity of the protein to allow determination of a portion of its amino acid sequence. A synthetic oligonucleotide probe is developed based on the determined amino acid sequence. This probe is used to identify a cDNA encoding the protein of interest. Once the desired cDNA is cloned, large amounts of the encoded protein often can be synthesized in engineered *E. coli* cells. DNA libraries can also be screened for the properties of a specific protein encoded in the cloned DNA.

Analyzing and sequencing cloned DNA: The complete characterization of a cloned DNA requires determination of its nucleotide sequence from which the amino acid sequence of the encoded protein can be deduced. The sequence of genomic DNA includes introns as well as exons and regions that control gene expression by determining the type of cell in which the encoded protein is expressed as well as the time in development and the amount of protein produced. The presence of replication origins and sequences that interact with proteins in chromosomes can also be determined.

1.5.5 Applications of genetic engineering

There are many and diverse applications of recombinant DNA technology some of which are listed below.

- Studies of regulation by subcloning
- Construction of industrially important bacteria
- Designing products suitable for industries and consumer preferences (designer organisms)
- Genetic engineering of plants
- Production of drugs
- Synthetic vaccines
- Gene therapy
- Solving environmental problems by genetically modified bacteria

Mutagenesis of cloned DNAs

In classical genetic studies, altered phenotypes of mutant organisms helped in identifying the genes and determining their functions. The isolation of genes by recombinant DNA technology, however, has opened a new approach to mutagenesis. It is now possible to introduce any desired alteration into a cloned gene and study the effect of the mutation on gene function (called **reverse genetics**). The ability to introduce specific mutations very precisely into cloned DNAs (*in vitro* mutagenesis) has proven to be a powerful tool in studying gene expression and function. Newly synthesized DNA molecules containing the mutation can then be isolated and characterized.

Introducing mutations directly into cellular genes: In addition to introducing mutations into cloned genes, it is often desirable to deactivate or inhibit the normal gene already in the chromosomal DNAs. One efficient approach is based on the ability of a cloned gene introduced into a cell to undergo homologous recombination with its chromosomal copy (**Fig. 1.12**). Here, the cloned gene replaces the normal allele, and so mutations introduced into the cloned gene *in vitro* become incorporated into the chromosomal copy of the gene.

1.6 PROTEOMICS

Recent successes in genome sequencing has provided the portrait of the world of protein. Now it has been suggested that the human genome contains about 30,000 genes, each of which can potentially give rise to a variety of different proteins. However, to date, only a fraction of these molecules has been characterized. The entire inventory of proteins that is produced by an organism (e.g., human, plants) is known as its **proteome**.

Proteomics is defined as the identification of the full set of proteins produced by a cell or tissue under a particular set of conditions. It also addresses a number of questions related to particular proteins. This includes the specific activity of the protein *in vitro* and *in vivo*, the 3D structure, its role in the developmental stages of an organism, its location in the cell, protein modifications, stability and amount of protein, interaction with other

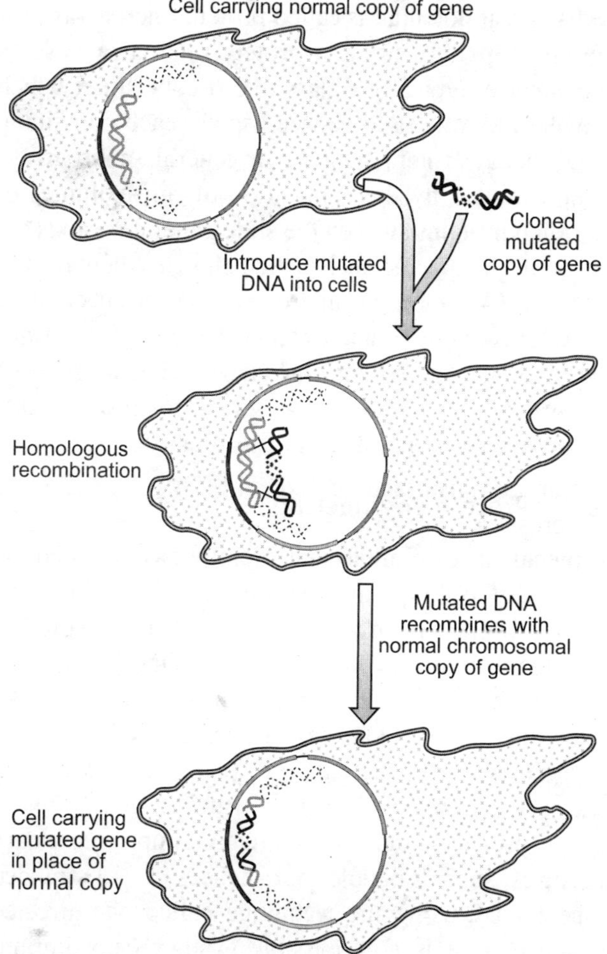

Fig. 1.12 Gene inactivation by homologous recombination

proteins and so on. Therefore, the tools of proteomics provide a snapshot of the cell's full repertoire of proteins.

Proteomics is based on three principal methods: two-dimensional gel electrophoresis for protein separation, mass spectrometry for precise determination of the molecular weight and identity of a protein and bioinformatics for coding protein sequences.

1.6.1 Protein microarray technology

Studying the genome at the protein level has become the core of many research areas, and is the next step in understanding gene function. There are so many proteins with complex relationships within an organism; a high-throughput systematic approach for protein analysis is inevitable. Researchers are working hard to develop techniques that allow protein function to be determined on a large scale, rather than one protein at a time. Though several new technologies have been developed, one technique that gives very high output and has great potential is called protein microarrays (or **protein chips**). In brief, a protein microarray uses a solid surface, typically a glass microscope slide, which is covered by microscopic-sized spots, each containing an individual protein sample. In general, microarrays are constructed by the application of tiny volumes of individual proteins sites on the slide generating an array of proteins to specific such as that shown schematically in **Fig. 1.13**. Once the microarray has been created, the proteins that it contains can be screened for various types of activities. Protein chips have the potential to be used in clinical laboratories to screen for proteins that are characteristic of particular disorders.

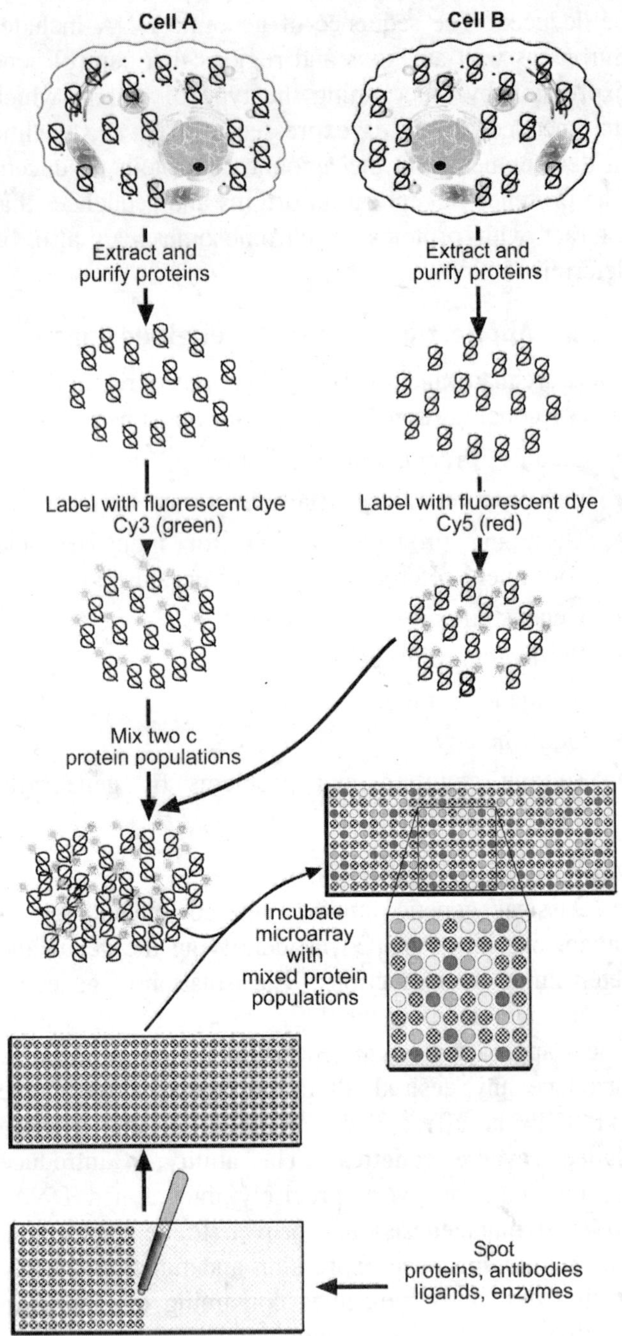

Fig. 1.13 Schematic representation of protein microarrays experiment; Cy5-labelled proteins of cell A mixed with Cy3-labelled proteins of cell B and incubated to monitor gene expression

1.6.2 Protein engineering

Recent advances in molecular biology have created the opportunity to design and mass-produce novel proteins that are different from those made by living organisms. It is possible with current recombinant DNA techniques to create an artificial gene that can be used in the production of a protein having any desired sequence of amino acids. In the past few years, there has been great success in synthesizing artificial genes whose peptide products fold into relatively simple secondary structures, such as bundles of α-helices or 2–3 stranded ß sheets. By using the technique called site-directed mutagenesis, it is now possible to alter (substitution, deletion, addition, duplication and so on) a single nucleotide in a gene. This technique is versatile because any protein can be tailored to increase or decrease its activities.

Another important application of protein engineering is the development of new drugs. Drug companies have access to chemical libraries that contain hundreds of

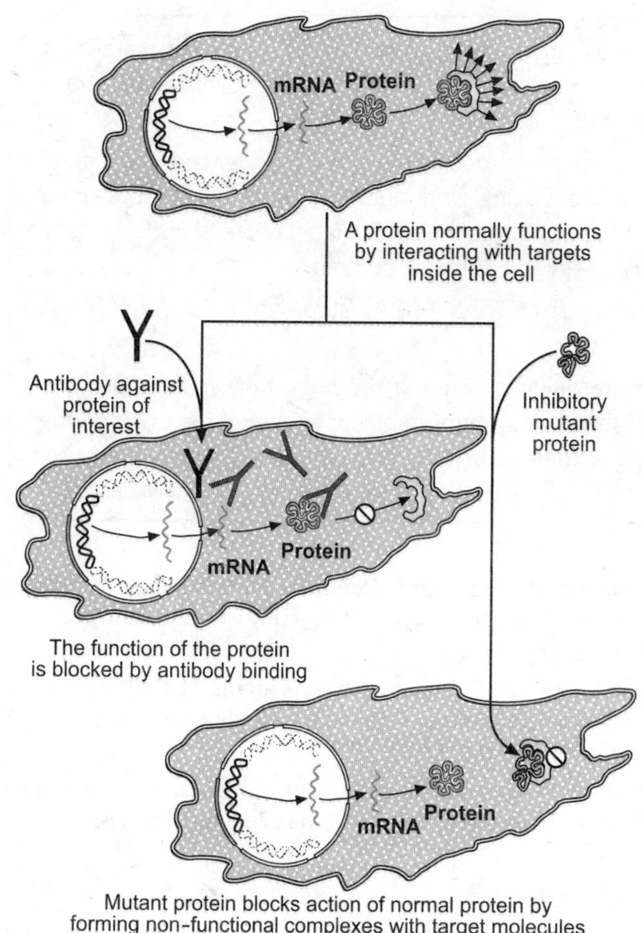

Fig. 1.14 Direct inhibition of protein function by micro-injection of antibodies

thousands of different organic compounds that have been either isolated from cells or synthesized in the laboratory. One way to search for potential drugs is to expose the protein in question to combinations of these compounds and determine which compounds, if any are capable of binding with high affinity. Alternatively, tertiary structures of proteins have been determined using computers, molecules whose shape and size might allow them to fit into the apparent cracks and crevices of the protein, rendering it inactive.

Like deactivating a gene or inducing degradation of mRNA, it is possible to interfere directly into the function of proteins within cells. One simple approach is by microinjecting antibodies into the cell that blocks the activity of the protein against which they are directed (**Fig. 1.14**). The alternative approach is, some mutant proteins are known to interfere with the function of their normal counterparts when they are expressed within the same cell and it works by competing with the normal protein for binding to its target molecule. If cloned DNAs encoding such mutant proteins (called dominant inhibitory mutants) are introduced into cells by gene transfer, they block normal gene function and the effects can be studied.

KEY POINTS

- Molecular biology is the study of the structure and functions of biological systems at the molecular level. It is fundamentally concerned with understanding the mechanisms responsible for transmission and expression of the genetic information that governs cell structure and function.
- Research in molecular cell biology like most experimental sciences, depends on the laboratory methods that can be used to study cell structure and function.
- The successes of molecular biology are derived from the development of new techniques: X-ray diffraction, electron microscopy, ultra-centrifugation and electrophoresis.
- Several different types of cells and organisms are used as experimental models to study various aspects of cell and molecular biology. Some of the organisms useful as experimental models are: viruses, *E. coli*, yeast, nematodes, *Drosophila melanogaster* and *Arabidopsis thaliana*.
- Some of the important techniques used in molecular biology are: radioisotopes, centrifuge, various chromatography techniques, mass spectrometry, X-ray crystallography, gel electrophoresis, blotting techniques, nucleotide and protein sequencing techniques, PCR, immunotechniques and recently DNA microarray technology.

- The advancement in recombinant DNA technology provided the techniques to manipulate genetic material in various organisms. These new methodologies enabled molecular biologists to construct novel combinations of genetic material artificially in the laboratory and then extended to the field.
- To date, millions of base pairs of DNA (genomic DNA) from many species have been sequenced. These data are either obtained directly from large-scale genome-sequencing work and/or contributed by individual scientists. All these data are now stored in various GenBanks. With the help of highly processive computers, it is possible to search the sequences in a GenBank using several different programs.
- If a new sequence is needed to be characterized for its structural and functional roles, it is now possible to search for related sequences and the organisms.
- Applications of recombinant DNA based technology are diverse. Some of the important applications are: studies of regulation by subcloning, construction of industrially important bacteria, genetic engineering of plants, production of drugs, synthetic vaccines, gene therapy and solving environmental problems by genetically modified bacteria.

Self-assessment Questions

1. Which are the model organisms commonly used in molecular biology and why?
2. Explain the use of radioisotopes in molecular biology.
3. Briefly describe autoradiography, fluorography and liquid scintillation count.
4. What is centrifugation? Discuss the various types of centrifugation.
5. Briefly discuss different types of commonly used chromatographic techniques.
6. Explain the principles of mass spectrometry and X-ray crystallography.
7. Describe with a diagram the functioning of gel electrophoresis.
8. Write in detail on the applications of PCR in molecular biology.
9. Explain the different steps involved in recombinant DNA technology.
10. What is meant by digestion by restriction enzymes?
11. How do you construct different DNA libraries?
12. Explain the different techniques used to screen recombinant molecules.
13. What is proteomics?
14. How do you engineer recombinant proteins?

FURTHER READING

Textbooks

Alberts Bruce, A Johnson, J Lewis, M Raff, K Roberts and P Walter (2002). *Molecular Biology of the Cell*, 4th ed. New York: Garland.

Brown TA (2002). *Genomes*, 2nd ed. Oxford: BIOS Scientific.

Channarayappa (2009). *Cell Biology*. Hyderabad: Universities Press.

Channarayappa (2006). *Molecular Biotechnology: Principles and Practices.* Hyderabad: Universities Press and CRC Press worldwide.

Sambrook J, EF Fritsch and T Maniatis (1989). *Molecular Cloning: A Laboratory Manual.* 2nd ed. New York: Cold Spring Harbor Laboratory Press.

Walker JM and EB Gingold (1993). *Molecular Biology and Biotechnology.* New Delhi: Panima Education Book Agency.

References

Avery Oswald T, CM MacLeod and M McCarty (1944). Studies on the chemical nature of the substance inducing transformation of pneumococcal types. *J. Exp. Med.* 79: 137–58.

Baltimore D (1970). Viral RNA-dependent DNA polymerase. *Nature* 226: 1209–11.

Barrell BG, MA Gillian and CA Hutchison III (1976). Overlapping genes in bacteriophage PhiX174. *Nature.* 264:34–41.

Benzer S (1968). Genes and behaviour. *Eng. Sci.* 32: 50–52.

Cairns J, GS Stent and JD Watson (eds) (1966). Phage and the origins of molecular biology. Cold Spring Harbour Laboratory Press

Caruthers M (1985). Gene synthesis machines: DNA chemistry and its uses. *Science* 230:281–86.

Channarayappa (1990). Cytogenetic analysis of clastogens and aneuploidogens by immunofluorescent and micronucleus methods, PhD thesis submitted to the West Virginia University, Morgantown, West Virginia, USA. W microfilm 3365.

Channarayappa (1993). Resistance mechanism of *Lycopersicon* species to the vector *Bemisia tabaci* and ultrastructural changes in leaf curl virus infected tomato leaves, PhD thesis submitted to the University of Agricultural Sciences, GKVK, Bangalore.

Chu TH, I Martinez, P Olson and R Dornburg (1995). Highly efficient eukaryotic gene expression vectors for peptide secretion. *Biotechniques* 18:890–99.

Cohen S (1984). The biochemical origins of molecular biology. *Tr. Biochem. Sci.* 9: 334–36.

Crick F (1970). Central dogma of molecular biology. *Nature.* 227: 561–63.

Gao HG, JK Chen, J Stewart, Channarayappa, WZ Whong and T Ong (1997). Distribution of *p53* and *K-ras* mutations in human lung cancer tissues. *Carcinogenesis* 18(3):473–78.

Hershey AD and M Chase (1952). Independent functions of viral protein and nucleic acid in growth of bacteriophage. *J. Gen. Physio.* 36: 39–56.

Jacob F and J Monod (1961). Genetic regulatory mechanisms in the synthesis of proteins. *J. Mol. Bio.* 3:318–56.

Jordan E and FS Collins (1996). Human genome project: A march of genetic maps. *Nature.* 380: 111–12.

Maxam AM and W Gilbert (1977). A new method for sequencing DNA. *PNAS* (USA) 74:560–64.

Melton DW (1994). Gene targeting in the mouse. *Bioessays* 16:633–38.

Muller HJ (1926) 1929). The gene as the basis of life. Proceedings of the International Congress of Plant Science 1:897–921.

Nirenberg MW and JH Matthaei (1961). The dependence of cell-free protein synthesis in *E. coli* upon naturally occurring or synthetic polyribonucleotides. *PNAS* (USA) 47: 1588–1602.

Pauling L and Robert B Corey (1953). A proposed structure for the nucleic acids. *PNAS* (USA) 39: 84–97.

Perriman RJ and WL Gerlach (1990). Manipulating gene expression by ribozyme technology. *Curr. Opin. Biotech.* 1: 86–91.

Sanger F (1988). Sequences, sequences and sequences. *Ann. Rev. Biochem.* 57:1–28.

Van Brunt J (1990). Amplifying genes: PCR and its alternatives. *Biotechnology* 8:291–93.

Watson JD and FC Crick (1953a). A structure for Deoxyribose Nucleic Acid. *Nature* 171: 737–38.

CHAPTER 2

The Cell: Structure and Functions

LEARNING OBJECTIVES

After reading this chapter, you will be able to:

- Understand that the cell is the basic unit of life
- List the structural and functional properties of cells
- Appreciate the various components of cells and their functions

2.1 INTRODUCTION

Understanding the structure and functions of cells is important as it contributes to our knowledge of the architecture of various life forms. Cells have DNA as their genetic material, are surrounded by plasma membranes, and use the same basic mechanisms for energy metabolism. However, many organisms, such as bacteria, amoeba and yeast, are made up of single cells that are capable of living as independent organisms, whereas more complex organisms are composed of collections of cells that function in a coordinated manner, with different cells specialised to perform particular tasks. The fundamental similarities between cells of different organisms provide a unifying theme to cell biology.

A living cell is a complex, multifunctional unit. Even the simplest of cells performs a number of different functions. Cells are divided into two basic types: **prokaryotic** and **eukaryotic**. The former type consist of bacteria and blue-green algae. The latter are found in protozoa, unicellular algae, fungi and higher plants and animals.

According to the **Cell theory of biology**:

- All organisms are composed of cells.
- The cell is the structural unit of life—units smaller than cells cannot survive.
- Cells arise by division of pre-existing cells—there is no spontaneous generation. Cells can be cultured to produce more cells: *in vitro* and *in vivo*.

2.2 GENERAL PROPERTIES OF CELLS

Cells exhibit great variety in shape, size and function. Despite this, cells share some common properties. Some of these are given beow:

Cells are complex and highly organised: They contain numerous internal structures called **organelles** which may or may not be surrounded by membranes.

Cells contain a genetic blueprint: Genes are instructions for cells to create specific proteins. Genetic information is stored in the triplet nucleotides called **codons**. The genetic code is universal in all organisms while methods of synthesis might vary. Smooth functioning of the metabolic pathways is possible as long as there is error-free transfer of information. Errors in the genetic material are called **mutations**.

Cells arise from the division of other cells: Cell division transfers the genes from the parent to the daughter cells. There are two main types of cell division: **mitosis** and **meiosis**. In mitosis, the genetic complement of each daughter cell is identical to the other and to the mother cell. This is called **asexual reproduction**. In meiosis, the genetic complement of each daughter cell is reduced by half. Daughter cells inherit cytoplasm and organelles from the mother cell by cytokinesis.

Cells acquire and utilise energy: Plant cells perform photosynthesis to convert light energy and CO_2 into chemical energy (ATP and glucose). Most cells release energy stored in organic compounds by converting them to release ATP and CO_2. This process is called respiration and ATP is known as the energy currency of the cell.

Cells can perform a variety of chemical reactions: The chemical reactions of the cell can consist of building simple organic molecules into complex molecules (**anabolism**) and breaking down complex molecules to release energy (**catabolism**). Together these reactions constitute **metabolism.**

Cells can engage in various mechanical activities: Cells can move, and within the cells organelles can also move. Cells can respond to stimuli such as light (**phototaxis**), chemicals (**chemotaxis**) and touch.

Cells can regulate activities: Cells control DNA synthesis and cell division. Cells make specific proteins only when needed by gene regulation and can turn metabolic pathways off or on.

2.2.1 Types of cells

The two basic types of cells are prokaryotic and eukaryotic.

Prokaryotes: These are small (5–10 μm) and lack membrane-bound organelles. They are considered the earliest type of cells.

Eukaryotes: They have evolved from prokaryotes by endosymbiotic association of two or more prokaryotes. They contain membrane-bound organelles.

Both the cell types contain the following structures:

- Plasma membrane which separates the cell from its external environment
- Cytoplasm which constitutes the fluid inside the cell
- Nuclear material which contains the genetic information stored as DNA.

Table 2.1 Major differences between plant and animal cells

	Plant cell	*Animal cell*
Cellulose cell wall	Present	Absent
Vacuoles	Usually large	Absent or very small
Plastids	Present	Absent
Golgi bodies	Rudimentary in the form of dictyosomes	Well developed
Centriole	Absent	Present

The differences between plant and animal cells are shown in **Table 2.1.**

2.3 VIRUSES

Viruses are non-cellular microscopic particles. They exist as parasites of prokaryotic and eukaryotic cells. They cannot reproduce in free-living conditions. A virus contains a small amount of nucleic acid surrounded by one or more protective shells (**capsid**). Genetic material of viruses may be either DNA or RNA, but not both. The DNA (**Fig. 2.1**) or RNA (**Fig. 2.2**) genomes may be single-stranded or double-stranded and may be linear

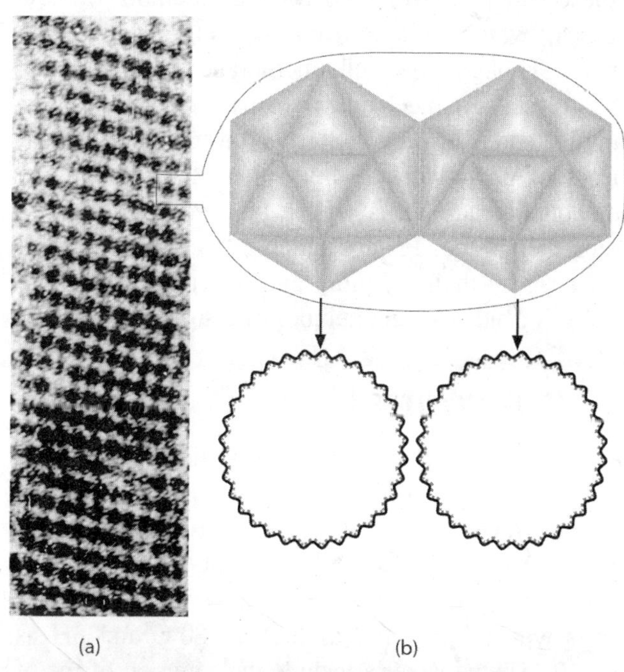

Fig. 2.1 (a) TEM showing aggregates of series of geminivirus infected in tomato leaf cell (b) Geminate twin particles and single-stranded DNA molecules (***Source:*** Channarayappa 1992)

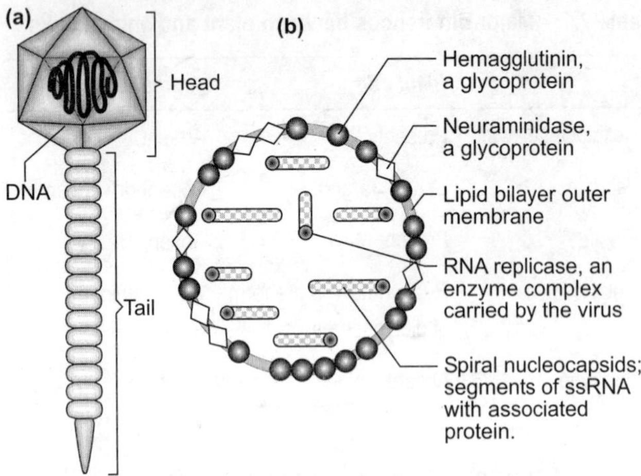

Fig. 2.2 Structure of viruses (a) A phage lambda chromosome wrapped around a protein core in the head (b) Structure of the influenza virion

or circular. Viral genome may contain very few to less than 100 genes. The virus that attacks the bacteria is known as **bacteriophage** or **phage**. Viroids are very small, naked RNA molecules without any protein or other type of coat. They do not contain genes coding for proteins.

Viruses enter the cells they infect and release their genetic material. The viral genes direct the synthesis of the components required for the assembly of new particles, which escape from the cell to propagate further. Though the cell membrane is a barrier, there are several mechanisms to pass through it. For example, some viruses can enter through the normal process of endocytosis, while others like the human immunodeficiency virus (HIV) fuses with the host's plasma membrane. Phages attach to the rigid cell wall of bacteria by their tail fibre and inject the DNA into the cell. Viroids cause a number of diseases in plants.

2.4 PROKARYOTES

Prokaryotes, which include all varieties of bacteria, are divided into two main groups: **archaebacteria** and **eubacteria**, which diverged early in evolution. Many archaebacteria live in extreme environments. For example, thermoacidophiles live in hot sulphur springs with temperature as high as 80°C and pH as low as 2. The eubacteria include the common forms of bacteria. The largest and most complex prokaryotes are the **cyanobacteria**, which can carry out photosynthesis (**Fig. 2.3**).

Perhaps the most obvious structural characteristic of bacteria is their small size. For example, *Escherichia coli*, an 'average-sized' bacterium with average cell length of ca. 1 μm has a cell volume of approximately 1–2 μm^3. This corresponds to a wet mass of ca. 1 pg, assuming that the cell consists mostly of water. The dry mass of a single cell can be estimated as 20% of the wet mass, amounting to 0.2 pg. Small size is extremely important because it allows for a large surface area-to-volume ratio to enable rapid uptake and intracellular distribution of nutrients and excretion of wastes. Otherwise, the diffusion of nutrients and waste products across the bacterial cell membrane limits the metabolic rate making the cell less fit for survival.

2.4.1 Cellular architecture

Bacteria contain a well-developed cell structure with unique features, which is responsible for many of their biological properties. Because of the simplicity of bacteria relative to larger organisms and the ease with which they can be manipulated experimentally, the cell structure of bacteria has been well studied, revealing many biochemical principles that have been subsequently applied to other organisms. Structurally, a bacterial cell (**Fig. 2.4**) has three architectural regions: appendages in the form of flagella and pili (or **fimbriae**), a cell envelope consisting of a capsule, cell

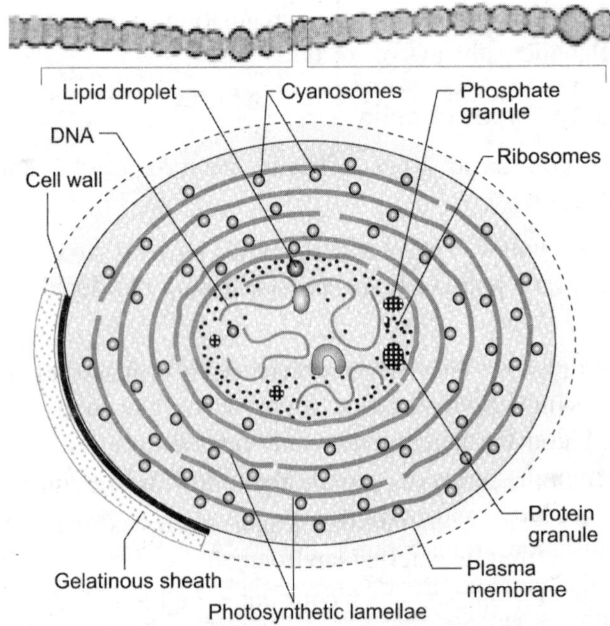

Fig. 2.3 A prokaryotic cell of cyanobacteria (electron microscopic view)

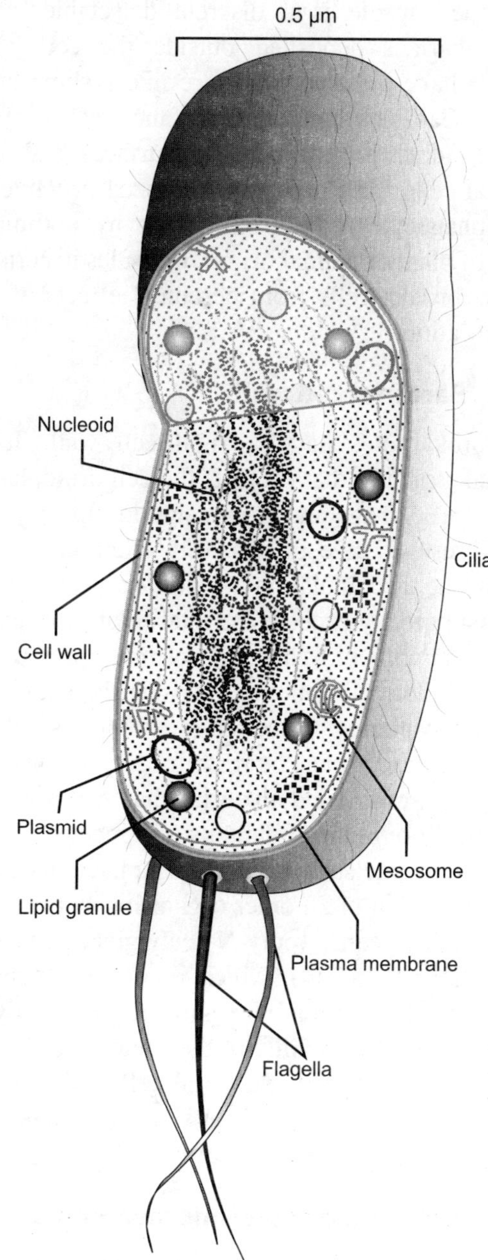

Fig. 2.4 A prokaryotic cell: *Escherichia coli*

wall and plasma membrane, and a cytoplasmic region that contains DNA and ribosomes.

The bacterial chromosome is not enclosed in a membrane-bound nucleus. This means that the transfer of cellular information through the processes of translation, transcription and DNA replication occur in the same compartment and there can be interaction with other cytoplasmic structures, most notably ribosomes. The bacterial chromosome is not packaged using histones to form chromatin as in eukaryotes, but it exists as a highly compact supercoiled structure called **nucleioid** (**Fig. 2.5**). The naked circular DNA is structurally identical to the DNA found in a eukaryotic cell's nucleus, except that instead of being like a long string, with two ends, it is formed into a closed circle. A single chromosome in a cell carries all the essential genes for the cell. Along with chromosomal DNA, most bacteria also contain small independent pieces of DNA called plasmids that often encode for traits that are advantageous but not essential.

2.4.2 Appendages

Flagella are filamentous protein structures attached to the cell surface that provide the swimming movement for most motile prokaryotes. Prokaryotic flagella are very thin with a diameter of about 20 nm, well below the resolving power of a light microscope. The flagellar filament is propelled by a motor apparatus in the plasma membrane which allows the cell to swim (**Fig. 2.6**).

The occurrence of tactic behaviour provides evidence for the ecological (survival) advantage of flagella in bacteria and other prokaryotes.

Fimbriae and pili are interchangeable terms used to refer to short, hair-like structures on the surface of

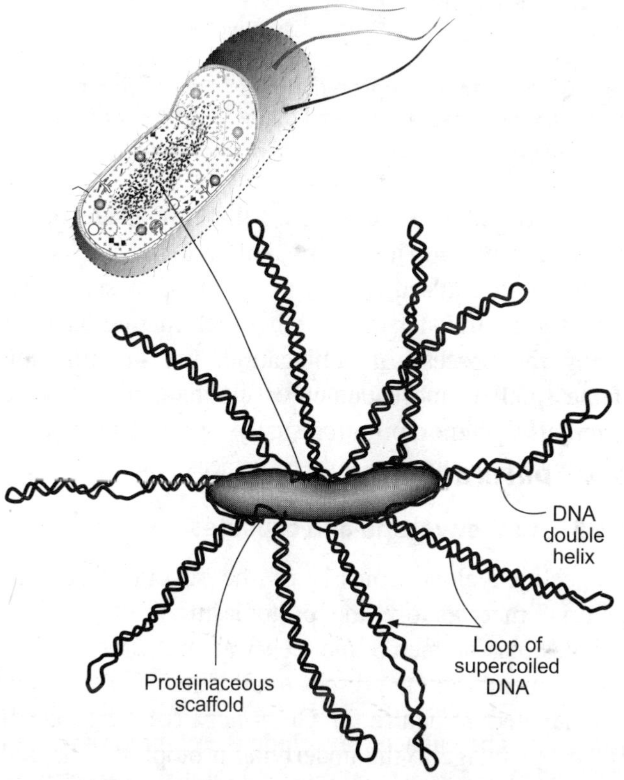

Fig. 2.5 Organisation of DNA (nucleoid) and proteins in the bacterial chromosome

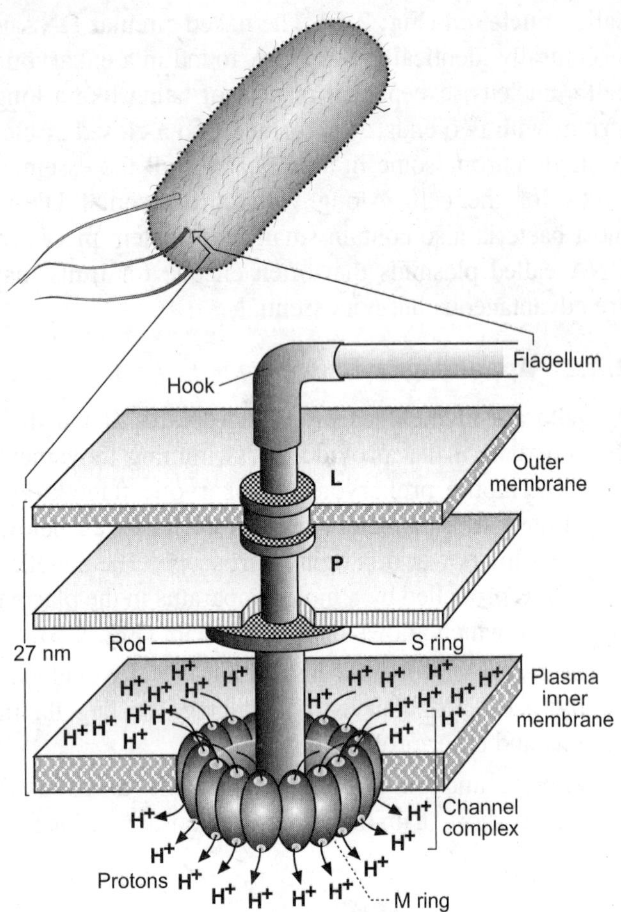

Fig. 2.6 A schematic representation of the flagellar motor of a prokaryotic cell; the overall diameter of the motor is about 30 nm

prokaryotic cells. Like flagella, they are made up of proteins. They are however, shorter than flagella. In *E. coli*, a specialised type of pilus, the F or sex pilus, mediates the transfer of DNA between mating bacteria during the process of conjugation, but the function of the smaller, more numerous common pili is quite different. Common pili are usually involved in specific adherence (attachment) to free surfaces.

2.4.3 Cell envelope and capsules

The cell envelope consists of the several layers of material that enclose the protoplasm of the cell. It includes plasma membrane, cell wall and capsule. The plasma membrane is present in all cells. The cell wall is a layered structure in Gram-negative bacteria. It prevents damage to the underlying protoplast. The cell capsule is made up of a polysaccharide and hence is also called **glycocalyx**.

A true capsule is a discrete detectable layer of polysaccharides deposited outside the cell wall. It can also have a matrix structure like a slime layer or biofilm. Capsules have several functions. They often mediate adherence of cells to surfaces and protect bacterial cells from predatory protozoa or white blood cells (phagocytes), or from attack by antimicrobial agents of plant or animal origin. Capsules in certain soil bacteria protect cells from perennial effects of drying or desiccation.

2.4.4 Bacterial cell wall

Most prokaryotes have a rigid cell wall. It is an essential structure that protects the cell protoplast from mechanical damage and from osmotic rupture or lysis. Bacteria usually live in relatively dilute environments so that the accumulation of solutes inside the cell cytoplasm greatly exceeds the total solute concentration in the environment. Thus, the osmotic pressure from inside the plasma membrane may be the equivalent of 10–25 atmospheres. Since the membrane is a delicate, plastic structure it must be restrained by an outside wall made of porous, rigid material that has high tensile strength. Murein, the ubiquitous component of bacterial cell walls, is one such material. The bacterial cell wall differs from that of all other organisms by the presence of peptidoglycan (poly-N-acetylglucosamine and N-acetylmuramic acid), which is located immediately outside the cytoplasmic membrane. Peptidoglycan is responsible for the rigidity of the bacterial cell wall and gives shape to the cell. Bacterial cell walls are of two main structural types based on staining characteristics: Gram-positive and Gram-negative.

2.4.5 Bacterial cytoplasmic membrane

The cytoplasmic membrane (or plasma membrane) of a bacterial cell is a delicate structure that encloses the cell cytoplasm (or protoplasm). It contains 40% phospholipid and 60% protein. The phospholipids are amphoteric molecules, meaning they have a water-soluble hydrophilic region (the glycerol 'head') attached to two insoluble hydrophobic fatty-acid 'tails'. In water, such molecules naturally form the molecular bilayer characteristic of membranes. Dispersed throughout the bilayer are various structural and enzymatic proteins, which carry out most of the functions of the membrane. Such an arrangement is called the **fluid mosaic membrane**. Unlike eukaryotes, bacterial membranes generally do not contain sterols but have fatty acids.

The relative proportions of these fatty acids can be modulated by the bacterium to maintain the optimum fluidity of the membrane (e.g., following temperature change).

As a phospholipid bilayer, the lipid portion of the outer membrane is impermeable to charged molecules. However, channels called porins are present in the outer membrane, which facilitate passive transport of many ions, sugars and amino acids across the outer membrane. These molecules are, therefore, present in the periplasm, the region between the cytoplasmic and outer membranes. The periplasm is believed to exist as a gel-like state rather than a liquid due to its high concentration of proteins and peptidoglycan. **Table 2.2** lists the important functions of the plasma membrane.

2.4.6 Cytoplasm

The cytoplasm is the fluid that fills the cell. It is mostly water, but contains a large assortment of molecular and structural components. The cytoplasm of bacterial cells consists of an aqueous solution of three groups of molecules:

- Macromolecules such as proteins (enzymes), DNA, mRNA and tRNA
- Small molecules that are energy sources, precursors of macromolecules, metabolites or vitamins
- Various inorganic ions and cofactors.

The distinct granular appearance of prokaryotic cytoplasm is due to the ribosomes which are the sites of protein synthesis in living organisms. The ribosomes of prokaryotes are smaller than those of eukaryotes. The cytoplasmic constituents of bacterial cells invariably include the prokaryotic chromosome (nucleoid), ribosomes and several hundred proteins and enzymes. The chromosome is typically one large circular molecule of DNA, more or less free in the cytoplasm. Prokaryotes sometimes possess smaller extrachromosomal pieces of DNA called **plasmids**.

Table 2.2 Some important functions of the prokaryotic plasma membrane

- Provides osmotic or permeability barrier
- Location of transport systems for specific solutes (nutrients and ions)
- Energy-generating functions, involving respiratory and photosynthetic electron transport systems, establishment of proton motive force and ATP-synthesising ATPase
- Synthesis of membrane lipids (including lipopolysaccharide in Gram-negative cells)
- Synthesis of murein (peptidoglycan)
- Coordination of DNA replication and segregation during formation of septum in cell division
- Location of specialised enzyme systems, such as for CO_2 or nitrogen fixation

2.4.7 Inclusions

Inclusions are distinct granules that may occupy a substantial part of the cytoplasm. They are usually storehouses of reserve materials. For example, they could contain glycogen (a polymer of glucose) or polybetahydroxybutyric acid (a type of fat) granules. Some inclusion bodies are actually membranous vesicles which contain photosynthetic pigments or enzymes. A bacterial structure sometimes observed as an inclusion is actually a type of dormant cell, called an **endospore**.

Endospores are the best known bacterial adaptation for long-term stress. They exhibit no signs of life and are called cryptobiotic. They are highly resistant to environmental stresses such as high temperature, irradiation, strong acids, disinfectants and so on. They retain viability indefinitely and under appropriate environmental conditions, they germinate back into vegetative cells. Hence, endospore formation is a mechanism of survival rather than a mechanism of reproduction.

2.5 EUKARYOTES

Eukaryotic cells are found in animals (**Fig. 2.7**), plants (**Fig. 2.8**), fungi and protists (**Fig. 2.9**), and there are structural differences among these four groups. Eukaryotic cells also have many common structures and similar functions. They are much larger and more complex. They are characterised by the presence of a prominent nucleus, a number of cytoplasmic organelles, and a cytoskeleton. The largest and most prominent organelle of eukaryotic cells is the nucleus, with a diameter of approximately 5 μm. Effective study of cell structure uses an electron microscope, which is capable of much higher resolution than a light microscope.

2.5.1 Plasma membrane

Plasma membrane encloses the cell and represents its outer living boundary. On the exterior of this membrane is present an 'extracellular matrix' in animal

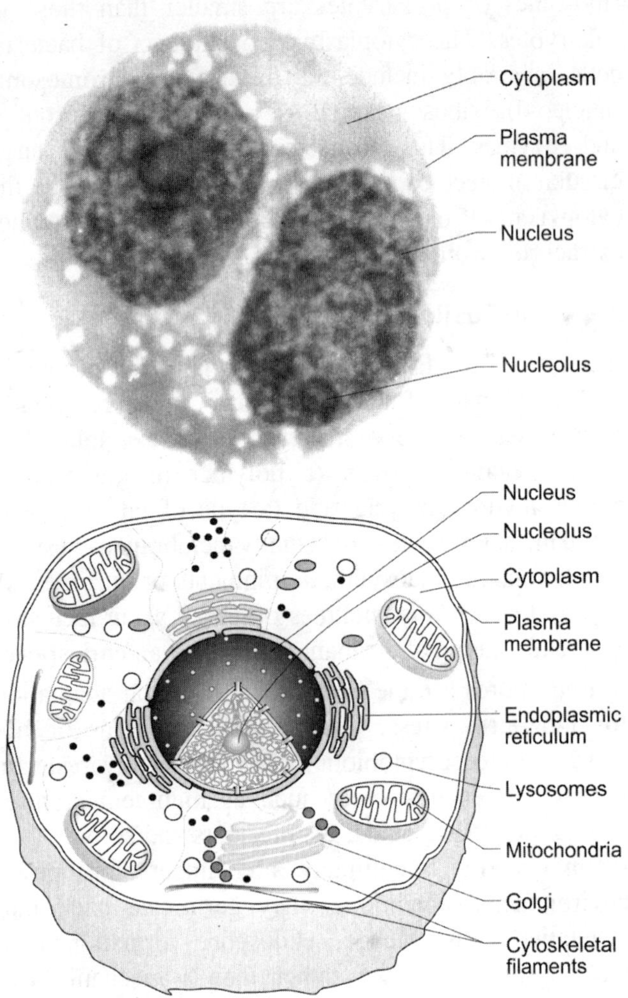

Fig. 2.7 (a) Photograph showing human lymphocyte with two nuclei (b) Drawing of a typical animal cell and its parts (***Source:*** Channarayappa 1990)

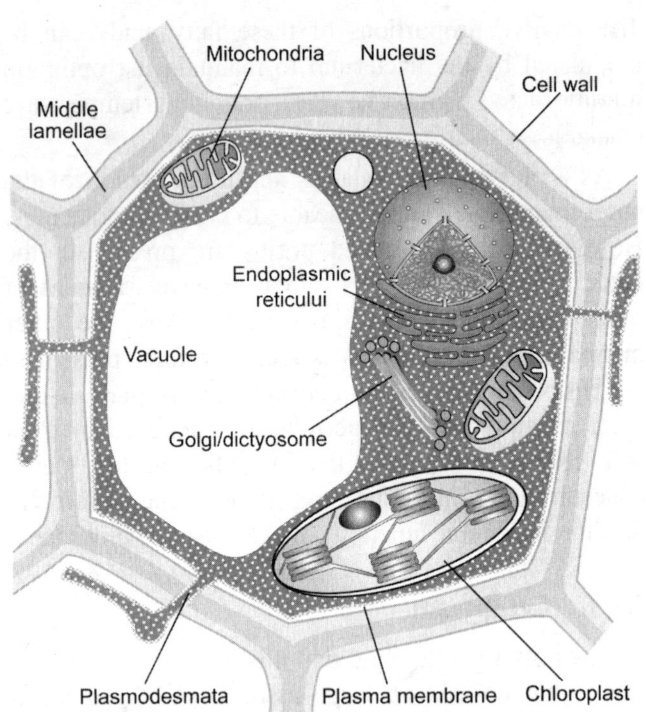

Fig. 2.8 Diagrammatic illustration of the structure a plant cell

cells and cell wall in plants. In all cells, the plasma membrane contains a variety of proteins, which help in cell recognition, transport of material in either direction or in sensing the signals received from the outer environment. Some membrane proteins are also involved in establishing and maintaining ion gradients across the membranes to facilitate the synthesis of ATP, movement of solutes or ions across the membrane and transmission of signals in all kinds of cells including the nerve and muscle cells.

The chemical composition of the plasma membrane includes lipids (28%–79%), proteins (20%–70%), oligosaccharides (1%–5%), and water (20% of total weight). However, their relative proportions greatly vary depending on the cell type, function, metabolic condition and membrane type. The main lipid components of membranes are phospholipids, cholesterol, and glycolipids. The major types of membrane-phospholipids are phosphatidyl serine, phosphatidyl choline and sphingomyelin, all of which are neutral phospholipids which tend to pack tightly in the membrane bilayer.

2.5.2 Nucleus

The nucleus is the most distinguishing feature of the eukaryotic cell. It is the largest organelle seen clearly when the cell is not dividing (**Fig. 2.10**). It is a highly specialised organelle that serves as the centre for information processing and control inside the cell. This organelle has two major functions: it stores the cell's hereditary material or DNA, and it coordinates the cell's activities, which include growth, intermediary metabolism, protein synthesis, repairs and reproduction. Generally each cell contains only one nucleus, but there are exceptions, where some specialised cells or organisms can have more than one. The nuclear envelope, a double-layered membrane, separates the contents of the nucleus from the cytoplasm. The envelope has holes called nuclear pores and is also attached to a network of tubules and sacs, called the endoplasmic reticulum. The semi-fluid matrix found inside the nucleus is called

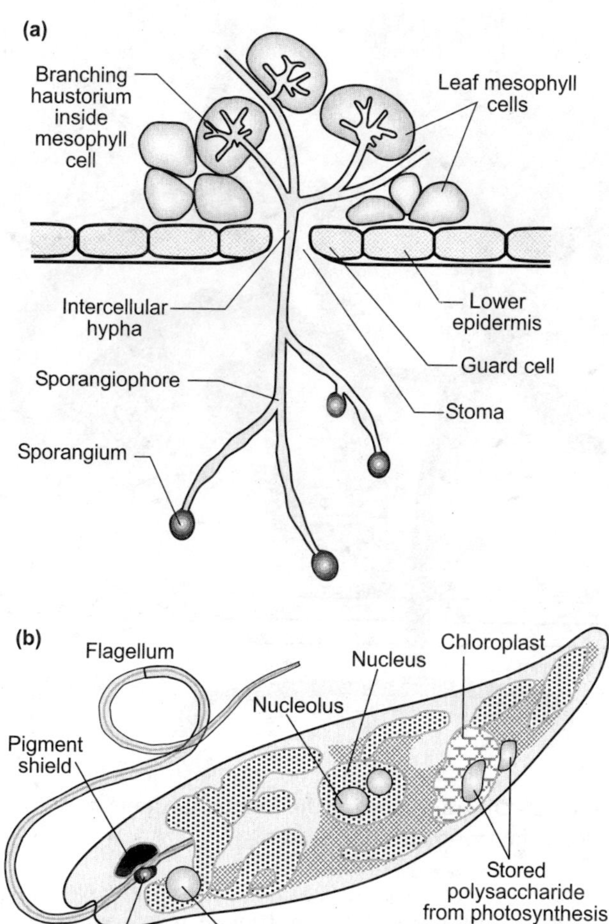

Fig. 2.9 Lower eukaryotes (a) Fungi (*Phytophthora infestans*) growing in a diseased potato leaf, with sporangiophores emerging from the underside of the leaf (b) A protozoan, *Euglena*

nucleoplasm and contains most of the nuclear material, the chromatin. The nucleus also contains one or more nucleoli and a variety of other smaller components, such as Cajal bodies, **GEMS** (Gemini of coiled bodies) and interchromatin granule clusters.

Nucleoli: It is a membrane-less organelle within the nucleus that manufactures ribosomes which are essential for protein synthesis. It has DNA, RNA and proteins, stains uniformly and deeply and looks like a large dark spot within the nucleus (**Fig. 2.11**). After a cell divides, a nucleolus is formed when chromosomes are brought together into nucleolar organising regions. During cell division, the nucleolus disappears. Some studies suggest that the nucleolus may be involved with cellular ageing and, therefore, may affect the senescence of an organism.

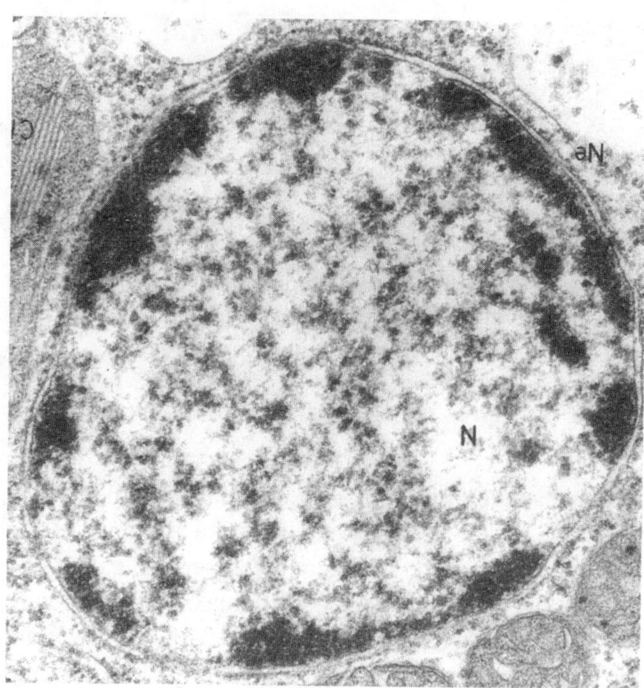

Fig. 2.10 TEM showing nucleus in tomato plant leaf cell (***Source:*** Channarayappa 1992)

Chromatin: These are the fibrillar structures that form a network and condense to form distinct bodies called chromosomes during cell division. The chromosomes are composed of DNA and histones. The number, shape and arrangement of genes on chromosomes is characteristic of each species. Each gene is a set of instructions for the construction of a specific protein. Packed inside the nucleus of every human cell is nearly

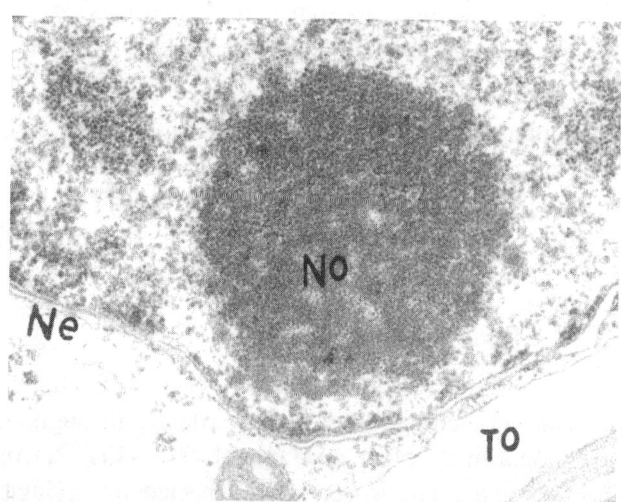

Fig. 2.11 TEM showing nucleolus (a dense dark spot in the nucleus) from tomato leaf cell (***Source:*** Channarayappa 1992)

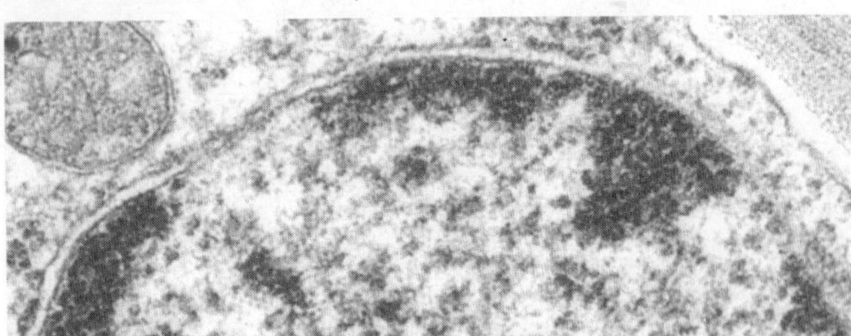

Fig. 2.12 TEM of nuclear envelope, showing (arrows) two lipid bilayers and perinuclear space (***Source:*** Channarayappa 1992)

6 feet of DNA, which is divided into 46 individual molecules.

Nuclear envelope: It is a double-layered membrane (**Fig. 2.12**) that normally encloses the contents of the nucleus. The space between the layers is called the perinuclear space and appears to connect with the rough endoplasmic reticulum. The nuclear pores regulate the passage of molecules between the nucleus and the cytoplasm. The inner surface has a protein lining called the nuclear lamina, which binds to chromatin and other nuclear components. During mitosis, or cell division, the nuclear envelope disintegrates, but re-forms after the daughter cells have formed and the chromatin begins to unravel and disperse.

Nuclear pores: These regulate the selective passage of molecules between the nucleus and cytoplasm. Nucleotides which are the building blocks for building DNA and RNA are allowed into the nucleus as also energy molecules needed for building genetic material. The nuclear pores are important for ribosomes to leave the nucleus. Though ribosomes are constructed in the nucleus, they must move to the cytoplasm in order to function.

Nucleoplasm: It fills the interior of the nucleus. The fluid is mostly water with a high concentration of nucleotides, which are used to make DNA and RNA, and the group of enzymes, which control the DNA and RNA construction reactions.

2.5.3 Endoplasmic reticulum

The endoplasmic reticulum (**ER**) is a network of flattened sacs and branching tubules that extends throughout the cytoplasm in plant and animal cells (**Fig. 2.13**). The sacs and tubules are interconnected by a single continuous membrane so that the organelle has only one large, highly convoluted lumen. The ER membrane allows molecules to be selectively transferred between the lumen and cytoplasm, and since it is connected to the nuclear envelope, it acts as the pipeline between the nucleus and cytoplasm too. There are two basic kinds of endoplasmic reticulum: **rough** and **smooth**. The surface of rough ER (rER) is covered with ribosomes. It is involved in the production and processing of proteins that will be exported, or secreted, from the cell.

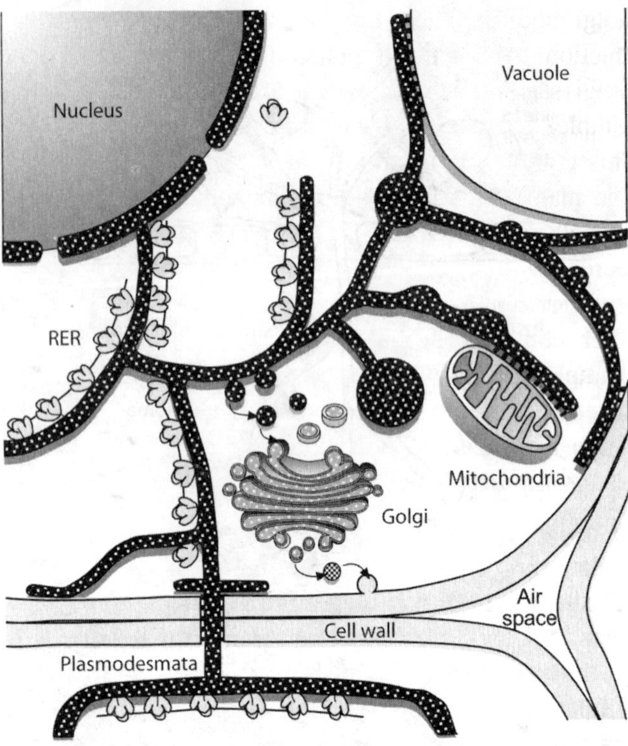

Fig. 2.13 Diagram showing the functional domains of the endoplasmic reticulum (ER) system

The ER has several functions. It helps to compartmentalise the cell, and serves as routes for the transport of materials from one part of the cell to another. It is associated with the synthesis of lipids and proteins. It also manufactures, processes and transports a wide variety of other biochemical compounds for use inside and outside the cell. Proteins that are targeted to remain in the lumen are known as **ER-resident proteins**. These special proteins, which are necessary for the ER to carry out its normal functions, contain a specialised retention signal consisting of a specific sequence of amino acids that enables them to be retained by the organelle.

2.5.4 Golgi bodies

The Golgi apparatus or complex is found universally in both plant and animal cells. Each cell contains many

Golgi bodies. Their number varies according to its function. Animal cells generally contain about 10–20 Golgi stacks per cell, which are linked into a single complex by tubular connections between the cisternae. This complex is usually located close to the nucleus. The main function of the Golgi bodies is to package the cell products into small vesicles—microbodies and secretory vesicles (**Fig. 2.14**). Microbodies contain useful enzymes that are packaged and removed from other contents of the cell. Lysosomes are the best example of microbodies. Secretory vesicles contain materials that are moved or secreted out of the cell. These materials are not waste products. In fact, they are usually hormones and pheromones.

Lysosomes: These are small, spherical organelles contained by a single-layer membrane, though their size and shape vary to some extent. They serve as digestion compartments and perform intracellular digestion by the process called autophagy. They break down cellular waste products, fats, carbohydrates, proteins, and other macromolecules into simple compounds (**Fig. 2.15**).

Peroxisomes: They contain enzymes that help to remove toxic substances, especially hydrogen peroxide. The peroxisomes in liver cells detoxify alcohol and other harmful compounds by oxidation. Others initiate the production of phospholipids, which are typically used in the formation of membranes. A number of metabolic disorders have been attributed to molecular defects in peroxisomes.

In plants, they play different roles depending on their location. During germination, they help in mobilising lipids in the growing seed; in leaves of C_3 plants they help in photorespiration and in nodules of leguminous roots, they are involved in the conversion of fixed nitrogen into nitrogen-rich organic compounds (**Fig. 2.16**).

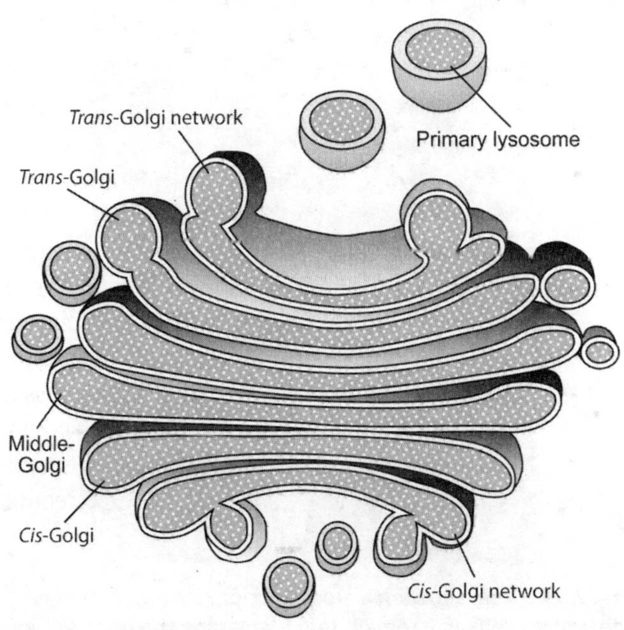

Fig. 2.14 Golgi apparatus

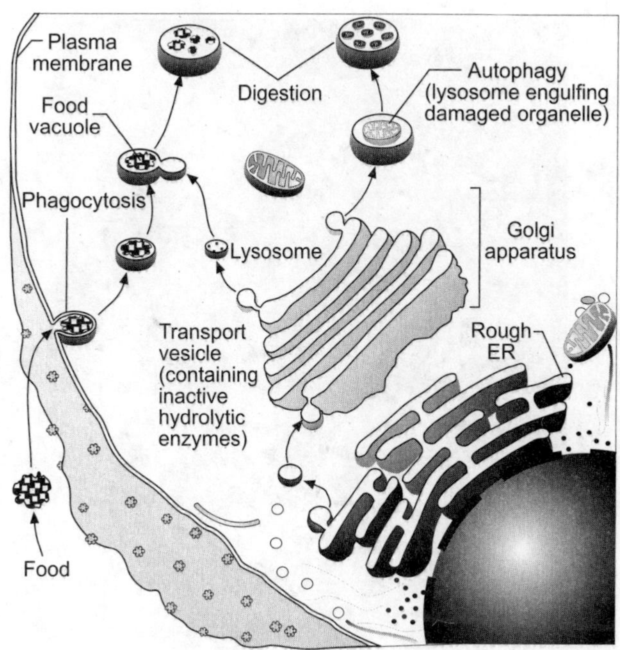

Fig. 2.15 Formation and function of lysosomes

2.5.7 Ribosomes

Eukaryotic ribosomes are similar in many aspects to prokaryotic ones but are larger and more complex. These tiny organelles are composed of approximately 60% ribosomal RNA (rRNA) and 40% protein. Even though they are called organelles, they are much smaller and are not bound by a membrane. A cell may have several thousands to a few million ribosomes. They are mainly bound to the ER and nuclear envelope, and also freely scattered in the cytoplasm. Ribosomes serve as the machinery for production of proteins. They are, therefore, most abundant in cells that are active in protein synthesis. Proteins synthesised by free ribosomes are retained in the cell while those produced by bound ribosomes are transported out of the cell.

In eukaryotes, the rRNA in ribosomes is organised into four strands (28S, 18S, 5.8S and 5S). They are produced and assembled in the nucleolus. Ribosomal

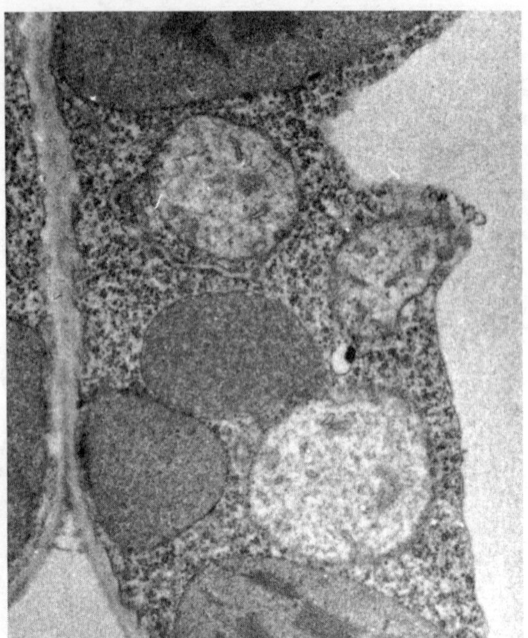

Fig. 2.16 TEM showing a group of peroxisomes in tomato leaf cell (*Source:* Channarayappa 1992)

proteins enter the nucleolus and combine with the four rRNA strands to create the two ribosomal subunits (small and large) that make up the completed ribosome. The ribosome units leave the nucleus through the nuclear pores and blend into the cytoplasm during protein synthesis. They separate when there is no synthesis of proteins. The eukaryotic ribosomes are 80S and are comprised of 40s and 60s subunits.

Protein synthesis requires the assistance of two other kinds of RNA molecules in addition to rRNA: the mRNA and transfer RNA (tRNA). Once the protein backbone is formed, the ribosome releases the protein. Ribosomes are very efficient organelles. A single ribosome in a eukaryotic cell can add two amino acids to a protein chain every second. In prokaryotes, ribosomes can work even faster (adding 20 amino acids per second). Ribosomes can also be found inside mitochondria and chloroplasts of plants. These are similar to those present in the cells of bacteria and blue-green algae.

2.5.8 Centrioles

They are found only in animal cells and are located near the nucleus in the **centrosome**, a granular mass that serves as an organising centre for microtubules. Within the centrosome, the pair of centrioles are positioned at right angles to each other. Each centriole is made of nine bundles of microtubules (three per bundle) arranged in a ring. Centrioles play a notable role in cell division. As mitosis progresses, the centrosome divides and moves one set of centrioles to each of the new microtubule-organising centres. From these new centres, microtubules radiate in star-shaped clusters known as **asters**. As the asters move to opposing poles, the microtubules, with the help of the centrioles, become organised into a spindle-shaped formation that spans the cell. These spindle fibres act as guides for the alignment of the chromosomes as they separate later during the process.

2.5.9 Vacuole

There are different types of vacuoles found in the cytoplasm. Examples are food vacuoles, contractile vacuoles and the specialised central vacuoles which are found only in plant cells (**Fig. 2.17**). The central vacuole mostly contains water and some impurities—mineral or protein—so that the water concentration is always less than 100%. When the cell is surrounded by sufficient water, osmosis causes the central vacuole to swell and press against the inside of the cell wall. This

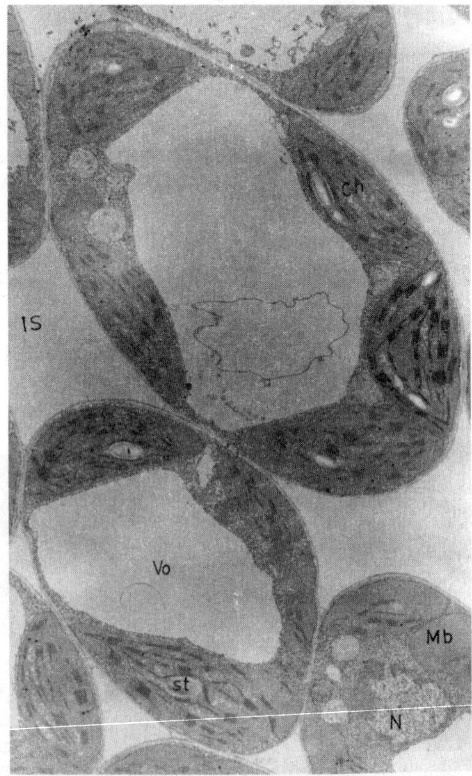

Fig. 2.17 TEM showing a cross-sectional view of spongy mesophyll cells in a tomato leaf displaying the large volume occupied by the central vacuoles (V) (*Source:* Channarayappa 1992)

phenomenon in the leaf tissue maintains the turgidity of the lamina so that maximum sunlight can be trapped during photosynthesis.

2.5.13 Cell wall

The cell wall is a well-organised structure and is essential for plant development and functioning. It contains the primary wall, secondary wall and middle lamella. The structure of the cell wall has the following features:

- It is the non-living, outermost layer present in all plant cells.
- It is made up of cellulose but may also contain other substances such as pectin and lignin.
- It is not homogeneous but consists of fine threads or fibres called microfibrils.
- It may be thin (1 μ) and transparent or thick.

The cell wall performs the following functions:

- It protects the delicate inner parts of the cell.
- Being rigid, it gives shape and mechanical strength to the cell.
- It acts as a physical barrier against infection.
- It protects against environmental stress.
- It provides rigidity and flexibility to the cell.
- It freely allows the passage of water and other chemicals into and out of the cells.
- The walls of two adjacent cells are firmly cemented by the middle lamella made of calcium pectate.
- There are breaks in the primary wall of the adjacent cells through which cytoplasm of one cell remains connected with the other. These cytoplasmic strands which connect one cell to the other are known as **plasmodesmata** (**Fig. 2.18**).

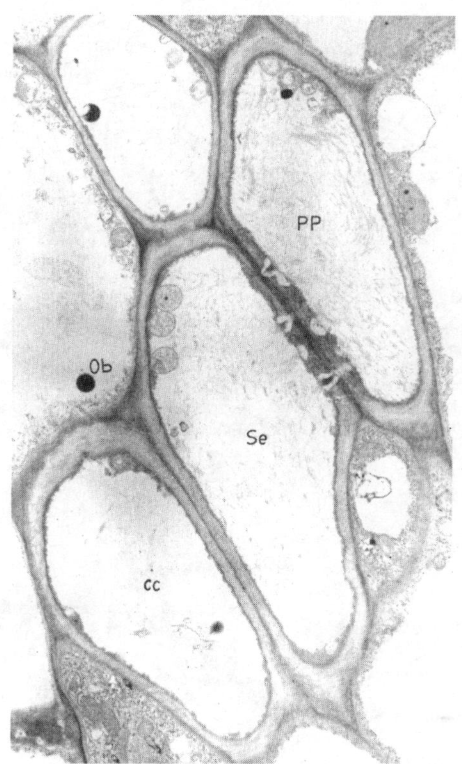

Fig. 2.18 TEM showing a plasmodesmata connecting two tomato leaf cells (***Source:*** Channarayappa 1992)

2.5.14 Mitochondria

Mitochondria (singular: *mitochondrion*) are rod-shaped organelles that are considered to be the powerhouse of the cell. Plant mitochondria in cells fixed for electron microscopy commonly appear spherical to ellipsoid, generally approximately 1 μm thick and 1–3 μm long. Its main function is to convert oxygen and nutrients into adenosine triphosphate (ATP). Mitochondria also supply various compounds, including organic acids and amino acids that are used as building blocks in synthetic reactions elsewhere in the cell. ATP is the chemical energy 'currency' of the cell that powers the cell's metabolic activities. This process is called **aerobic respiration** and is the reason why animals breathe oxygen. Without mitochondria, higher eukaryotes cannot exist, because their cells would then only be able to obtain energy from anaerobic respiration (in the absence of oxygen), a process much less efficient than the aerobic pathway. In fact, mitochondria enable cells to produce 15 times more ATP than they could otherwise and complex animals, like humans, need large amounts of energy in order to survive.

The mitochondrion has two specialised membranes, which divide the organelle into a narrow inter-membrane space and a much larger internal matrix, each of which contains highly specialised proteins (**Fig. 2.19**). The outer membrane of a mitochondrion contains many channels formed by the protein porin and acts like a sieve, filtering out molecules that are too big. Similarly, the inner membrane, which is highly convoluted, forms foldings called **cristae** through which selective passage of molecules is allowed. The transport proteins present on the inner membrane allow only the materials essential to the matrix. Together, the various compartments of a mitochondrion are able to work in harmony to generate ATP in a complex multistep process.

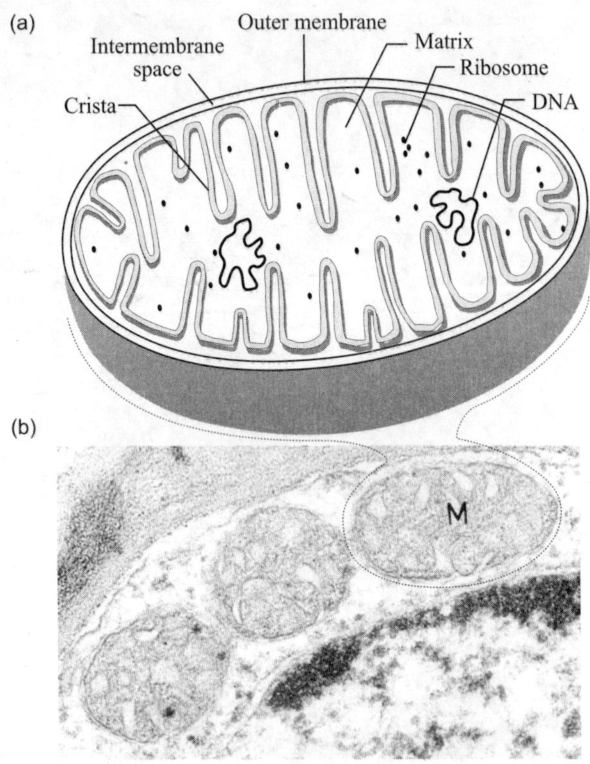

Fig. 2.19 (a) A typical eukaryotic mitochondrion (b) TEM showing distribution of mitochondria in tomato plant leaf cells (***Source:*** Channarayappa 1992)

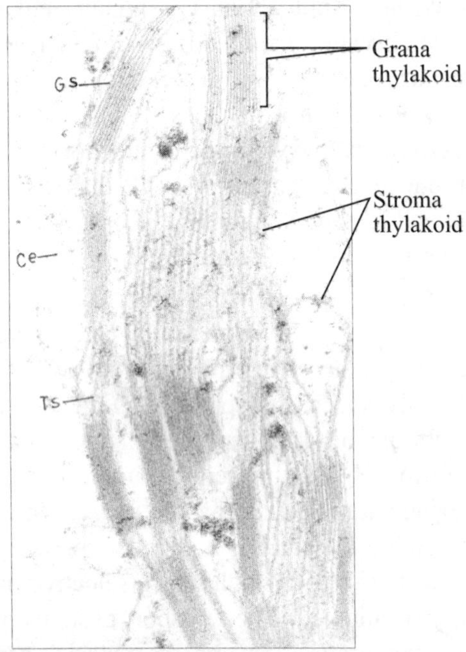

Fig. 2.21 TEM of stacked grana and unstacked stroma thylakoids in tomato leaf chloroplast (***Source:*** Channarayappa 1992)

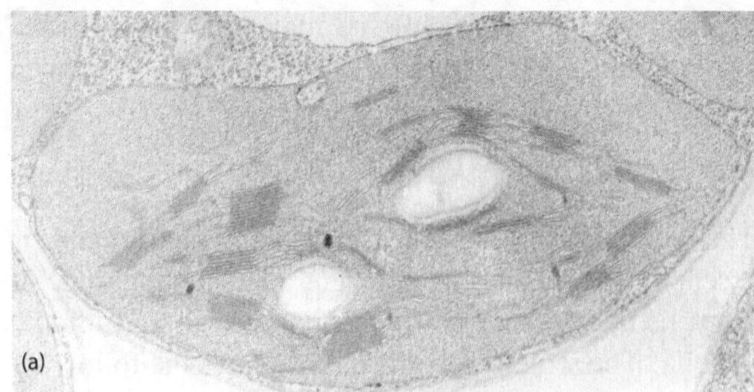

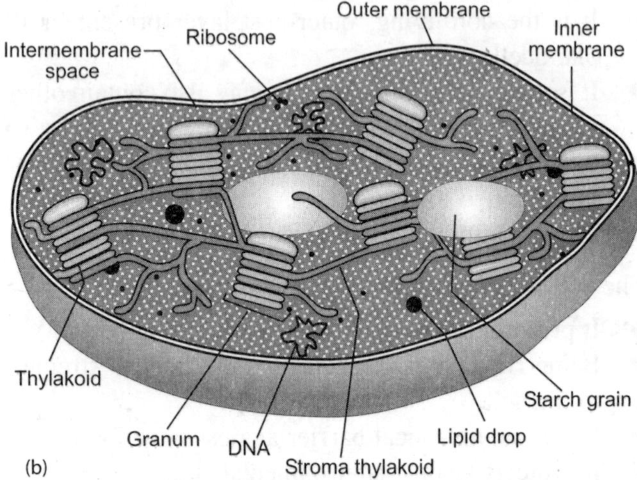

Fig. 2.20 (a) TEM of tomato leaf cell chloroplast showing two starch grains and thylakoid system (b) Labelled parts of a chloroplast (***Source:*** Channarayappa 1992)

2.5.15 Plastids

Plastids (Greek, *plastikos* 'moulded') are major organelles found only in plant and algal cells. There are several kinds of plastids and they can convert from one form to another depending upon environmental conditions. They are responsible for photosynthesis, for the storage of a wide variety of products, and for the synthesis of key molecules required for the basic architecture and functioning of plant cells. The best known of these plastids is the chloroplast (**Fig. 2.20**), which performs the function of photosynthesis. Chloroplasts have a double membrane like mitochondria. It is formed into many thylakoid structures (**Fig. 2.21**) as in prokaryotic cells.

Photosynthesis occurs in the chloroplasts of the cells of green plants. The pathways for harvesting light energy and for transport of electrons occur in the thylakoid membranes.

KEY POINTS

- The smallest living part of an organism is the cell.
- Cells exhibit a wide variety of shapes and sizes and their number varies between organisms.
- All cells are surrounded by plasma membrane, use DNA as their genetic material and follow the same basic mechanisms for energy metabolism.
- Viruses and viroids are non-cellular microscopic particles. They exist as parasites of prokaryotic and eukaryotic cells. They cannot reproduce in free-living conditions.
- The simplest cellular forms are the prokaryotes. They include bacteria, which are divided into two groups: the archaebacteria and eubacteria.
- Bacteria contain a well-developed cell structure, which is responsible for many of their unique biological properties. Their cytoplasm may not have internal membranes but contains a number of characteristic cell structures.
- The most recognisable extracellular bacterial cell structures are flagella attached to the cell surface, that provides the swimming movement for most motile prokaryotes.
- Eukaryotic cells are found in animals, plants, fungi and protists. There are structural differences among these groups.
- Eukaryotic cells are typically characterised by the presence of a prominent nucleus, a variety of cytoplasmic organelles (mitochondria and chloroplasts) and a cytoskeleton.
- The plasma membrane encloses the cell and represents its outer living boundary. On the exterior of this membrane is an 'extracellular matrix' in animal cells and cell wall in plants.
- The nucleus is a highly specialised organelle that serves as the information processing and administrative centre of the cell.
- The endoplasmic reticulum (ER) is a network of flattened sacs that extends throughout the cytoplasm. It has several functions including transport of materials and synthesis of lipids and proteins.
- The Golgi apparatus is found in both plant and animal cells. It also helps in the distribution of chemical products in the cell.
- Lysosomes, peroxisomes and glyoxisomes are microbodies which contain enzymes for removing toxic substances from the cell and for degradation of non-functional organelles.
- Vacuoles are found in the cytoplasm of plant cells.
- In vascular plants, symplasmic transport occurs via plasmodesmata.
- Plastids are major organelles found only in plant and algal cells and they are the sites for photosynthesis.

Self-assessment Questions

1. Explain briefly the general properties of cells.
2. Describe the structure of a prokaryotic cell and its components with the help of a diagram.
3. What are the functions of the appendages in a bacterial cell?
4. How is the bacterial cell wall organised?
5. Explain the structure and functions of the cytoplasmic membrane in bacteria.
6. Describe the structure and components of a eukaryotic cell.
7. Explain the cytoskeletal architecture of a eukaryotic cell.
8. Explain the bioenergetics of mitochondria and chloroplast in eukaryotic cells.
9. Describe the structure of the plant cell wall.

FURTHER READING

Textbooks

Buchanan BB, W Gruissem and RL Jones (2004). *Biochemistry and Molecular Biology of Plants.* IK International Pvt. Ltd.

Channarayappa (2010). *Cell Biology.* Hyderabad: Universities Press (India) Private Limited.

Channarayappa (2006). *Molecular Biotechnology: Principles and Practices.* Hyderabad: Universities Press (India) Pvt. Ltd., and CRC Press worldwide.

Cooper GM and RE Hausman (2004). *The Cell: A Molecular Approach,* 3rd ed. Washington DC: ASM Press.

Gupta PK (1999). *Cell and Molecular Biology.* Meerut: Rastogi Publications.

Madigan MT and JM Martinko (2005). *Brock Biology of Microorganisms,* 11th ed. New Jersey: Pearson Prentice-Hall.

Twyman RM (1998). *Advanced Molecular Biology,* New Delhi: Viva books Pvt. Ltd.

References

Berg JM, JL Tymoczko and L Stryer (2006). *Biochemistry,* 5th ed. New York: Freeman.

Cano RJ and MK Borucki (1995). Revival and identification of bacterial spores in 25- to 40-million-year-old Dominican amber. *Science* 19:268(5213): 1060–64.

Channarayappa (1992). Resistance mechanism of *Lycopersicon* species to the vector *Bemisia tabaci* and ultrastructural changes in leaf curl virus infected tomato plants. PhD thesis submitted to the University of Agricultural Sciences, GKVK, Bangalore.

Channarayappa (1990). Cytogenetic analysis of clastogens and aneuploidogens by immunofluorescent and micronucleus methods. PhD thesis submitted to the West Virginia University, Morgantown, USA.

Eyster KM (2007). The membrane and lipids as integral participants in signal transduction. *Adv. Physiol. Edu.* 31: 5–16.

Fischman J (1995). Have 25-million-year-old bacteria returned to life? *Science* 19: 268: 977.

Gitai Z (2005). The new bacterial cell biology: Moving parts and subcellular architecture. *Cell* 120(5): 577–86.

Parkes RJ (2000). A case of bacterial immortality? *Nature* 19:407(6806): 844–45.

Shih YL and L Rothfield (2006). The bacterial cytoskeleton. *Microbiol. Mol. Biol. Rev.* 70(3): 729–54.

Vreeland RH, WD Rosenzweig and DW Powers (2000). Isolation of a 250 million-year-old halotolerant bacterium from a primary salt crystal. *Nature* 19: 407(6806): 897–900.

CHAPTER 3

Chemistry of Cells

LEARNING OBJECTIVES

After reading this chapter, you will be able to:

- Know about the molecular composition of various cell components
- Understand the different kinds of biomolecules and their structures
- List the important metabolic reactions and their role in cellular functions

3.1 INTRODUCTION

The living cell is like a symphony of thousands of chemical reactions all accurately timed and sequentially coordinated to perform the functions necessary for life. Only six elements, viz., carbon, hydrogen, oxygen, nitrogen, phosphorus, and sulphur (i.e., CHNOPS) make up about 98% of the mass of all living organisms.

Individual atoms are organised into molecules which have different structural groups. Different combinations of atoms in a molecule can create extraordinary diversity. The central region of an atom is dense and contains positively charged protons and neutrons with neutral charge. The outer part of the atom is less dense, and holds negatively charged electrons, which orbit around the nucleus. The first orbit accommodates two electrons and the remaining orbits can hold eight electrons each. Only protons and neutrons (together known as baryons) contribute to the total atomic mass because the electrons have neglible mass which is therefore ignored. The atom will normally contain equal number of protons and electrons so that its overall atomic charge is neutral. Atoms can become charged only when they gain or lose electrons. In this state, the atom is known as an ion.

3.2 ELEMENTS OF ATOMIC STRUCTURE

About 25 of the 92 natural elements are known to be essential to life, but four of these—carbon (C), oxygen (O), hydrogen (H), and nitrogen (N)—make up 96% of living matter. Phosphorus (P), sulphur (S), calcium (Ca), potassium (K) and a few other elements account for most of the remaining 4% of an organism's weight. Trace elements are necessary in minute quantities only.

The smallest unit of matter is the atom. The stable forms of subatomic particles are neutrons, protons and electrons. Neutrons and protons are packed together tightly to form a dense core or the nucleus. The electron is the smallest subatomic particle which moves about the nucleus at nearly the speed of light. Protons and electrons are electrically charged. Each electron has one unit of negative charge, while each proton has one positive charge. A neutron, as its name implies, is neutral in charge. Atoms of different elements differ in their number of subatomic particles (**Fig. 3.1**).

The electrons of an atom are arranged around the nucleus in orbits. Their distance from the nucleus depends upon the amount of energy they have. Those with the most energy can travel the farthest away, while those with less are drawn closer. These particles, as they position themselves around the centre of the atom,

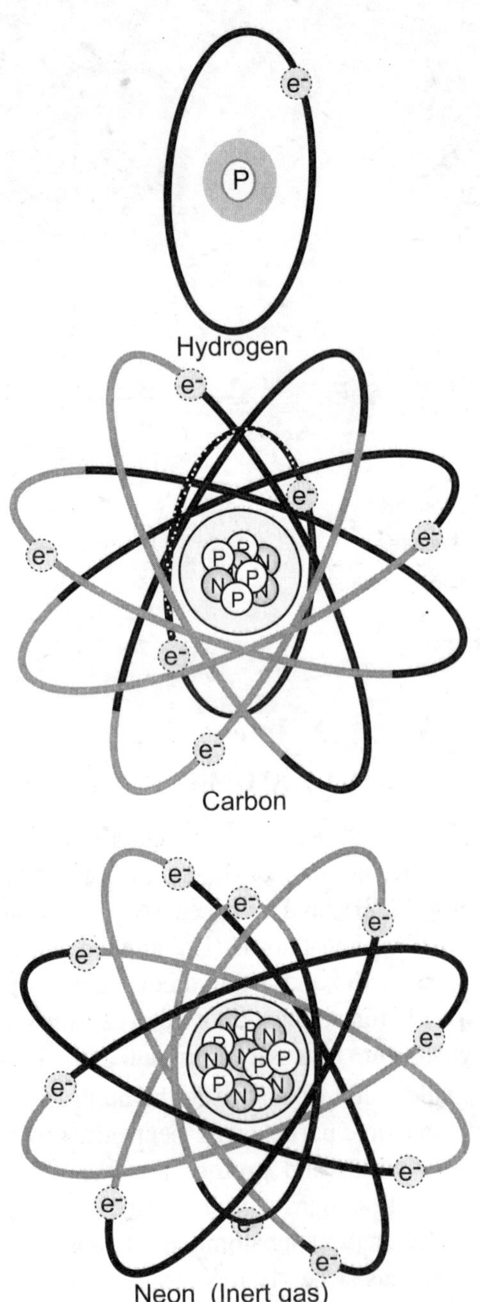

Fig. 3.1 Structure of atoms

have a tendency to pair up and form energy levels, or electron shells. Each level can contain a certain number of electrons. If a level fills up and another electron is added, the lone electron becomes part of a new level that can accommodate more electrons. The three-dimensional space where an electron is found 90% of the time is called an **orbital**. No more than two electrons can occupy the first orbital (shell, *s* orbital). The second electron shell (*2s*) can hold eight electrons, two in each of the four orbitals. The other three orbitals, called *p* orbitals, are dumb bell-shaped, each oriented at right angles to the other two. The chemical behaviour of an atom is determined by its electron configuration. That is, the distribution of electrons in the atom's electron shells. Electrons are generally affected by three tendencies:

- Electrons seem to pair.
- They are negatively charged and so are subject to the electrostatic attraction–repulsion rules.
- Atoms tend to complete their outermost electron energy level.

These factors are the basis for the vast variety of chemical reactions and chemical bonds that occur in nature. As atoms transfer electrons to fill their outermost or valence electron shells and then react because of their unbalanced electrostatic charges, unique structures as simple as a water molecule or as complex as a DNA strand can form.

Elements can be distinguished on the basis of their **atomic number**. It refers to the number of protons in an element. Hydrogen, for example, has only one proton in its nucleus and so has an atomic number of 1. Another property of an atom is the **atomic mass**, or **atomic weight**. The atomic mass of an atom is obtained by adding the number of protons and neutrons that it has. Theoretically, it should always be an integer. But there are different forms of certain elements like hydrogen, oxygen, carbon, uranium and so on, containing the same number of protons but different number of neutrons. These forms are called **isotopes**. When the atomic mass for an element is calculated, then, the mass of each isotope is evaluated depending on the prevalence of each isotope in nature. From this data, a weighted average for the atomic mass of each element is calculated, and this is what is given in the periodic table of elements. Hence, you would find atomic masses expressed in fractions.

3.3 TYPES OF CHEMICAL FORCES

What are the forces or structures that hold the atoms together in a biomolecule? Atoms can react with each other in many ways, forming stable or weak interactions (**bonding**). These bonds take many forms, but the following are the most important in biological systems (**Fig. 3.2**).

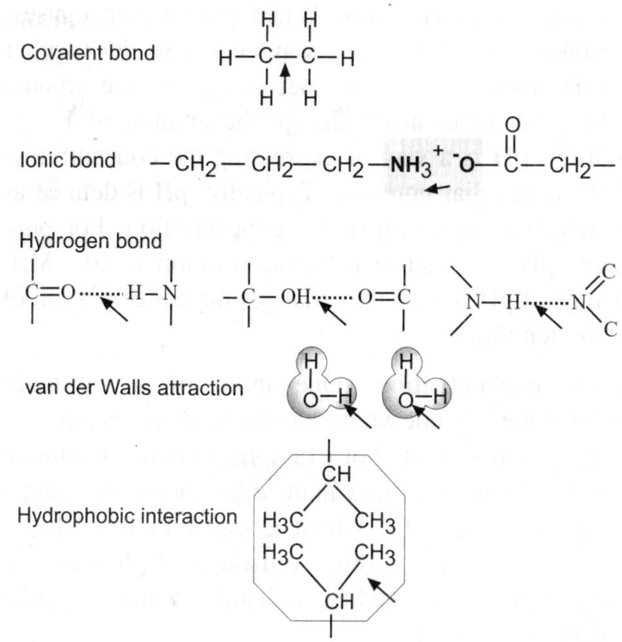

Fig. 3.2 Major chemical bonds and interactions (arrows)

3.3.1 Ionic bonds

Atoms with partially filled orbitals are generally unstable. They gain stability either by gaining or losing electrons, or by sharing them with another atom to compensate the imbalance. For example, when sodium (Na) atoms and chlorine (Cl) atoms are mixed, they rapidly convert into sodium chloride (NaCl). Na with just one electron in its outer orbital readily shares with Cl, which requires just one electron to complete its outer orbital. Both gain stability by forming the NaCl molecule. Substances like NaCl are called ionic compounds, since they contain charged ions. The ions are held in a regular arrangement within a crystal by the electrical attraction between opposite charges, which are called **ionic bonds**. Actually millions or billions of alternating atoms organise in three dimensions. Biomolecules contain a number of groups with positive and negative charges, like carboxyl (-COO^-) and amino (-NH_3^+) groups, which can form ionic bonds. Usually, ionic bonds are quite strong in the absence of water. In hydrophobic environments (such as inside a folded protein), ionic bonds can become quite strong (50 kcal/mole or more). Ionic bonds allow many biological molecules to interact, form folded shapes that are less strong than covalent but moderately stable.

3.3.2 Covalent bonds

Covalent bonds are strong bonds (50–100 kcal/mole) because atoms share the electron pairs (e.g., CH_4, H_2O) equally, instead of gaining or losing electrons as in the ionic bonds. In case of the water molecule, oxygen and hydrogen share one pair of electrons in each bond. Molecules are primarily held together by covalent bonds, which are the strongest and require significant energy input to dissociate.

Covalent bonds are formed when at least one of the constituent atoms prefers to share the electron and thus has an intermediate electronegativity. Carbon is one such atom and hence it almost exclusively forms covalent bonds. However, not all covalent bonds are the same. For example, if a carbon atom shares electrons with another carbon atom, usually the share of each is equal. On the other hand, if a carbon atom shares an electron with an oxygen atom, the latter has a greater tendency to accept an electron and therefore, the electron is not equally shared but spends more of its time around the oxygen atom. Thus, a partial negative charge is associated with the oxygen atom and there must also be an equal and opposite partial positive charge associated with the carbon atom. Such a bond is said to be **polar**. Similarly, in the water molecule, the oxygen atom has a weakly negative charge and the hydrogen atom has a weakly positive charge, creating a bipolar condition. In biological molecules, local charges are particularly noticeable where there are oxygen–hydrogen, nitrogen–hydrogen, and carbon–oxygen or carbon–nitrogen bonds. The weak electrical attractions between local charges are called **hydrogen bonds** (about 3–5 kcal/mole).

Covalent bonds can be single (H–H), double (O=O or O=C=O) or triple (N≡N) bonds. The number of bonds between the atoms depends on the valence.

3.3.3 van der Waals bonds

The van der Waals bonds are the weakest of all bonds (1–2 kcal/mole). They are due to non-specific attractive force between atoms, which come close to each other. As any two atoms approach each other, electron clouds begin to overlap and create a situation known as **induced dipole**. One electron 'pushes' another to the opposite side of its atom, creating a momentary electron deficit. Therefore, a slight positive charge arises. This situation oscillates back and forth between the two atoms, creating a weak attractive force ~1 kcal/mole. When many atoms are involved, the individual forces add up and then the force can be very strong.

In this type of bonding, there are no permanent charges associated with the participating groups. The

forms, some energy is lost as heat. Chemical substances have potential energy within the structures of their molecules, some of which may release as heat during chemical reactions.

3.5.1 Oxidation–reduction reaction

When organic materials like coal, gas, oil and wood are used to release heat energy, it is an oxidation. According to the chemical definition, burning is the loss of electrons from a material. During the reaction, oxygen replaces the other atoms attached to the carbon in the cellulose. The most stable end product consists of two oxygen molecules bound to one carbon, or carbon dioxide. Carbon dioxide cannot react further with oxygen, and thus it cannot be oxidised. The oxygen atoms, on the other hand, have been reduced. Whenever there is an oxidation reaction, there must be an accompanying reduction. Oxidation gives off energy; reduction gains energy.

Cells use the same process to derive energy from the environment—by oxidising glucose to release energy as heat. Some of this released energy will be captured by adenosine triphosphate (ATP), which acts as an energy intermediate.

3.5.2 Phosphorylation

Many biological molecules have phosphoryl groups attached to their carbon skeletons. Some molecules readily give them up to other molecules. They are said to have a high phosphoryl group transfer potential. In many reactions in which a molecule gives up its phosphoryl group, free energy is released. These molecules are, therefore, considered to be energy stores. The most important molecule of this nature is ATP. If the phosphate group at the end of the chain is broken, adenosine diphosphate (ADP) and the inorganic ion orthophosphate (HPO_3^{2-}) are the end products. This is a hydrolysis reaction. If two phosphates are removed as pyrophosphate, $HP_2O_7^{3-}$ (symbolised as PPi), then adenosine monophosphate (AMP) is formed. Both these reactions have a similar, highly negative, ΔG value.

3.5.3 Catalysis

In biological systems, for two atoms or compounds to react, they must first be brought into close proximity and be put into the appropriate positions for the reaction. Activation energy is the amount of energy needed to commence the reaction. Cells use certain chemicals as a 'scaffold' to line up the reacting compounds and to create a conducive situation for the reaction to start. This process is called **catalysis**. A **catalyst** causes a reaction to occur more readily by lowering the activation energy. Although the catalyst may participate in the reaction, it is never permanently changed by the reaction. Thus, a catalyst can be used over and over again to catalyse the same reaction.

Enzymes

Enzymes are the biological catalysts and they are always protein molecules. Enzymes catalyse almost every chemical reaction that occurs in the cells. Consider a reaction converting a substrate (S) into a product (P). S must first be converted to the transition state (S^t), in which it exists for about 10^{-14} to 10^{-13} of a second and the molecule is distorted to an electronic conformation which readily converts to P. Thus, a negative free energy change is an essential condition for the overall reaction S→P to occur. Enzyme specificity is critical for many reasons:

- It allows only the appropriate reactants to come together, so that improper or toxic products are not made due to mistakes.
- The rate of a particular reaction can be controlled by the activity of the catalyst. Cells control enzymes either by adjusting their amount or more often, by altering their ability to bind or transform reactants.

3.6 TYPES OF BIOMOLECULES FOUND IN CELLS

There are different types of molecules found in cells (**Fig. 3.5**). Molecules like water, H_2O (70% of a typical cell) are small. These are called **micromolecules**. On the other hand, a large molecule like that of table sugar, $C_{12}H_{22}O_{11}$ (sucrose) has 45 atoms. Fats, proteins and their derivatives and nucleic acids form the building blocks of cellular constituents. These may contain tens of thousands of atoms. These gigantic molecules are called **macromolecules**. Lipids are the major group of biological macromolecules and they are insoluble in water.

The four major types of macromolecules found in living cells are carbohydrates (polysaccharides), lipids, proteins and nucleic acids (**Table 3.1**), which are made of smaller, repeating subunits called monomers. The monomers are not always identical but they always have similar chemical structures. They undergo polymerisation to form large, complex molecules called polymers.

Table 3.1 Four major types of macromolecules found in living cells

Macromolecule	*Elements*	*Monomer*	*Polymer*	*Example*
Carbohydrate	C, H, O	Simple sugars	Polysaccharide	Starch
Lipids	C, H, O	Fatty acids, glycerol	Lipid	Fats, oils, waxes
Proteins	C, H, O, N, S	Amino acids	Polypeptides	Insulin
Nucleic acids	C, H, O, N, P	Nucleotides	Nucleic acids	DNA/RNA

The fundamental building blocks of living cells are the biomolecules. They are carbon compounds naturally produced by living organisms. They are divided into three classes according to their size—small, medium and large. Monosaccharides, amino acids, glycerol, fatty acids and nucleotides are the building blocks of the larger biomolecules. Monosaccharides join together to form **disaccharides** and **polysaccharides** (**starch**, **cellulose** and **glycogen**). Amino acids link in specific sequences to form proteins. Glycerol and fatty acids join to form lipids and with **nucleotides** they make up the **nucleic acids** (**DNA, RNA**). Lipids act as the building blocks for cell membranes by organising into a double layer; these bilipid membranes perform many vital functions for cells. The 13 vitamins are another group of biomolecules but are chemically unrelated. Besides the polymeric biomolecules, numerous organic molecules are absorbed by the living systems.

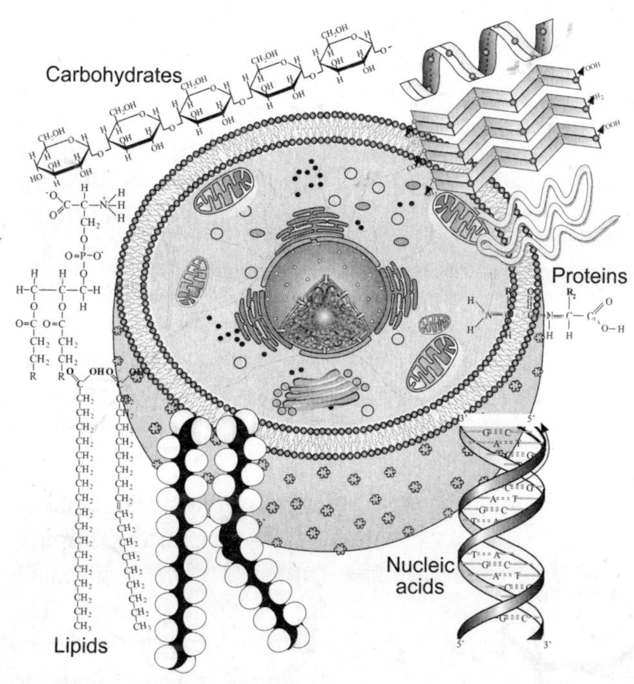

Fig. 3.5 The four major biomolecules that build the cell

3.6.1 Carbohydrates

Carbohydrates are compounds that contain carbon (C), hydrogen (H) and oxygen (O). They are classified as **monosaccharides**, **oligosaccharides** or **polysaccharides**. They are often built as long chains by stringing together smaller units. The simplest carbohydrates also contain either an aldehyde moiety (polyhydroxyaldehydes) or a ketone moiety (polyhydroxyketones). From two to ten monosaccharide units, linked by glycosidic bonds, make up an oligosaccharide. Polysaccharides are much larger molecules, containing hundreds of monosaccharide units. The presence of hydroxyl groups allows carbohydrates to interact with the aqueous environment and to participate in hydrogen bonding, both within and between chains. They combine with lipids to form glycolipids or with proteins to form glycoproteins. Derivatives of carbohydrates can contain nitrogen, phosphates and sulphur compounds.

Chemical structure of carbohydrates

The predominant carbohydrates encountered in human cells are structurally related to the aldotriose glyceraldehyde and to the ketotriose dihydroxyacetone. All carbohydrates contain at least one asymmetric (chiral) carbon and are, therefore, optically active. They exist in either of the two conformations, depending upon the orientation of the hydroxyl group about the asymmetric carbon farthest from the carbonyl. With a few exceptions, most carbohydrates that are of physiological significance exist in the D-conformation. The mirror-image conformations, called enantiomers, are in the L-conformation (**Fig. 3.6**).

Monosaccharides: These are simple sugars. Examples of monosaccharides are the trioses (glyceraldehyde or dihydroxyacetone), pentoses (ribose and deoxyribose) and hexoses (glucose, fructose, and galactose). The monosaccharides commonly found in humans are classified according to the number of carbons they contain in their backbone structures (**Table 3.2**). The major monosaccharides contain four to six carbon atoms.

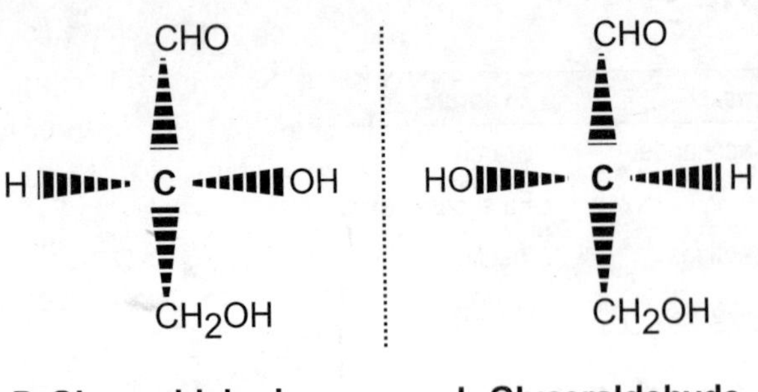

Fig. 3.6 D– and L– forms of glyceraldehyde

Disaccharides: They are formed by joining two monosaccharides. Several physiologically important disaccharides include sucrose, maltose and lactose.

Table 3.2 Classification of carbohydrates

Number of Carbons	*Category*	*Examples*
3	Triose	Glyceraldehyde, dihydroxyacetone
4	Tetrose	Erythrose
5	Pentose	Ribose, ribulose, xylulose
6	Hexose	Glucose, galactose, mannose, fructose
7	Heptose	Sedoheptulose
9	Nonose	Neuraminic acid also called sialic acid
Many	Polyose	Starch, amylase, cellulose, pectinose, hemicellulose

(a) Hollow helix

(b) Twisted ribbon

(c) ß-glucose cellulose is a polymer of glucose

ß-N-acetylglucosamine: chitin is its polymer

(d) α -1,6 linkage

Fig. 3.7 Polysaccharide structure (a) Amylose (b) Cellulose (c) Chitin (d) Amylopectin and glycogen

Covalent bonds between the anomeric hydroxyl of a cyclic sugar and the hydroxyl of a second sugar (or another alcohol-containing compound) are termed glycosidic bonds, and the resultant molecules are glycosides.

Monosaccharides and disaccharides are sweet, water-soluble and crystalline. Sucrose in sugarcane and sugar beet is composed of glucose and fructose through an α-(1,2) β-glycosidic bond. Lactose is found exclusively in the milk of mammals and consists of galactose and glucose in a β-(1,4) glycosidic bond (**Fig. 3.7**). Maltose is the major degradation product of starch and is composed of two glucose monomers in an α-(1,4) glycosidic bond.

Trisaccharides: Several biologically important trisaccharides ($C_{18}H_{36}O_{18}$) are also present in the cells. For example, raffinose is one such trisaccharide with monosaccharide units (D-glucose, D-fructose and D-galactose). Other trisaccharides are mannotriose, rabinose, rhamminose, gentianose and melezitose. Sugars of higher order are also found in cells. Sugars containing 2–9 sugar residues are often collectively termed as oligosaccharides.

Polysaccharides: They are polymerised monosaccharides (with more than ten units), complex, non sweet carbohydrates. Examples are starch, cellulose and glycogen. They are generally large and often have a complex, branched structure. They are insoluble in water and do not form crystals (**Fig. 3.7**). Most of the carbohydrates found in nature occur in the form of high-molecular-weight polymers. When polysaccharides are composed of the same monosaccharide units, they are called homopolysaccharides (or homoglycans) while those containing more than one type of monosaccharide are termed heteropolysaccharides (or heteroglycans). The number of possible structures of a polysaccharide involving the same monosaccharide units is large because each monosaccharide has several free hydroxyl groups that can allow condensation with another monosaccharide. For example, three hexoses can join in different ways to give a large number of trisaccharides. Therefore, it is extremely difficult to determine the structure of a polysaccharide.

Glycogen: It is the major form of stored carbohydrate in animals. It is mainly found in liver and muscles. This crucial molecule is a homopolymer of glucose in α-(1, 4) linkage and it is also highly branched, with α-(1, 6) branch linkages at every eighth to tenth residue. Glycogen has a compact structure that results from the coiling of the polymer chains. This allows a large amount of energy to be stored for a given volume, with little effect on cellular osmolarity (**Fig. 3.7**).

Starch: Starch is the major form of stored carbohydrate in plant cells. It is a polymer of D-glucopyranose units linked by α-1,4 linkages although occasionally α-1,6 linkages are also found (as in amylopectin) giving it a branched structure. Its structure is identical to that of glycogen, except for a much lower degree of branching (about every 20–30 residues). Unbranched starch is called amylose; branched starch is called amylopectin. Starch consists of amylose and amylopectin in the ratio of 1:4. If starch is mixed in hot water, only amylose is dissolved and not the amylopectin.

Cellulose: It is the most important constituent of the cell wall (**Fig. 3.8**). It is insoluble in most solvents and thus differs from starch and glycogen. It is a linear polymer of glucose with β-1,4 glycosidic bonds, where a number of linear chains are held together to form microfibrillae. A large number of chains lie parallel and are linked by hydrogen bonds. Cellulose can be hydrolysed by cellulase found in some snails, termites and the bacteria found in the rumen of cattle and other ruminants.

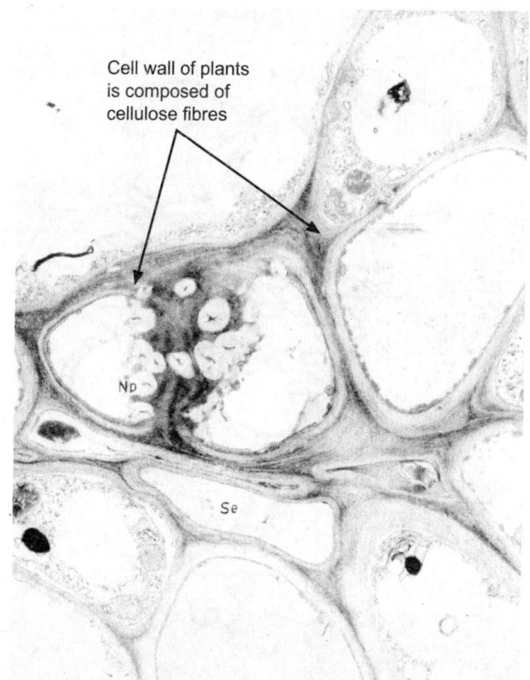

Fig. 3.8 TEM of tomato plant cells showing a thick cell wall; the cell wall is mainly composed of numerous cellulose fibres. (***Source:*** Channarayappa 1992)

Functions of carbohydrates

The main functions of carbohydrates are as follows:

- They are an excellent source of energy for the various activities of the cells. Glucose enters the glycolysis pathway in the cytoplasm and then the citric acid cycle in mitochondria thus generating large amounts of ATP.
- Some carbohydrates may have a structural function. For example, cellulose allows the plants to stand erect and gives toughness to the wood. Sugar polymers like starch and glycogen make up the stored forms of energy. Starch is found in plant products such as potatoes and glycogen is found in animals.
- Carbohydrates are essential for cells to communicate with each other, especially by glycoproteins and glycolipids.
- They also help cells adhere to each other and the extracellular material (ECM) surrounding the cells. The capsule of some bacteria consisting of polysaccharide provides protection for a cell in diverse environments.
- The ability of the body to defend itself against invading microbes and the removal of foreign material from the body (such as the capture of dust and pollen by the mucus in our nose and throat) is also dependent on the properties of carbohydrates.

3.6.2 Proteins

Like carbohydrates, proteins are also composed of smaller units. The monomers that make up proteins are called amino acids. Amino acids are molecules that contain both amino and carboxylic acid functional groups (**Fig. 3.9**). They are the building blocks of long polymer chains. With two to ten amino acids such chains are called peptides, with 10–100 they are often called polypeptides, and longer chains are known as proteins. Proteins of all types have many structural and functional roles in organisms. There are 20 amino acids encoded by the standard **genetic code**, but there are more than 500 natural amino acids which have been identified in the cytosol. When amino acids other than the set of 20 are observed in proteins, this is usually

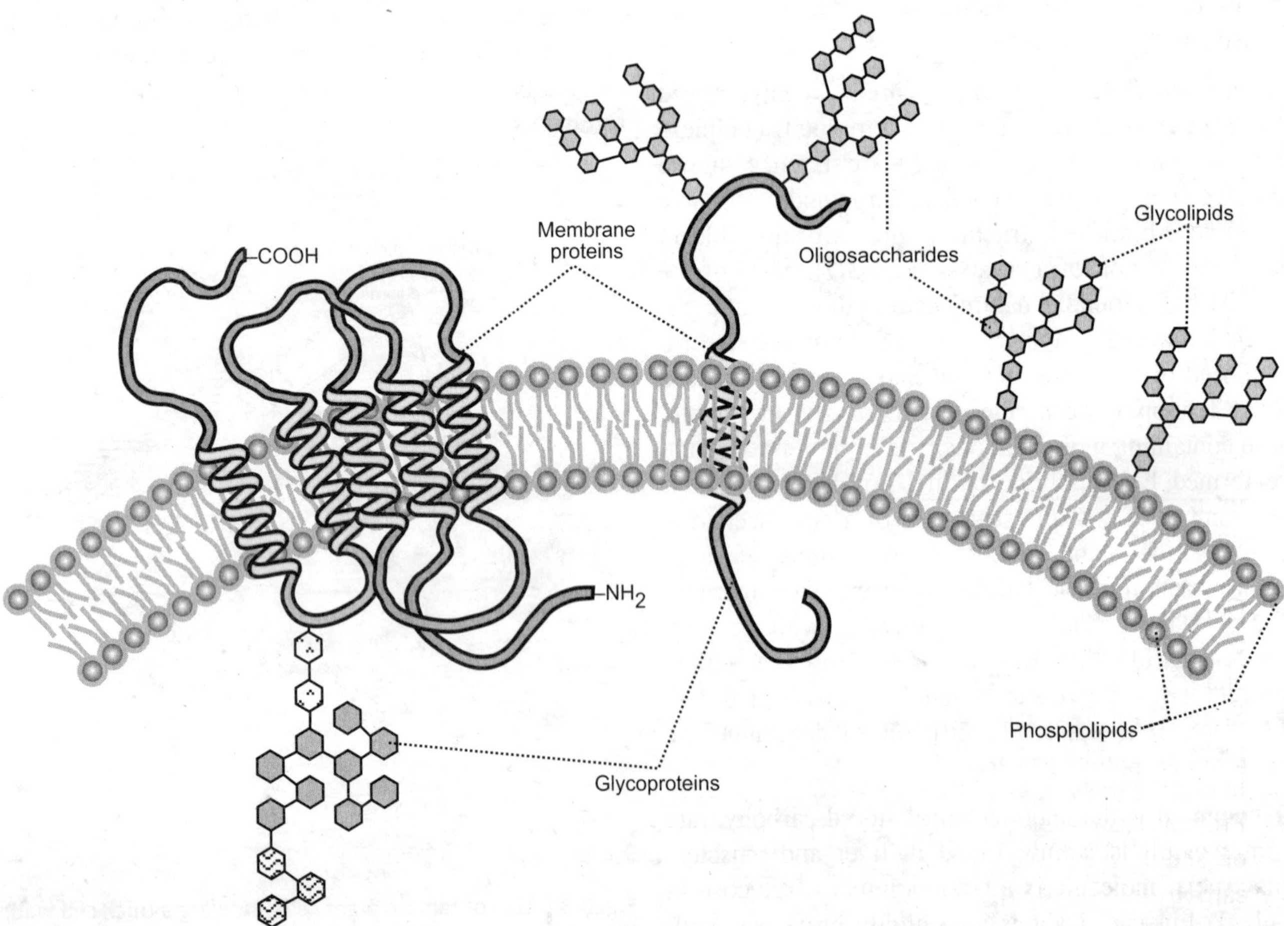

Fig. 3.9 Glycoproteins and glycolipids

the result of modification after translation (protein synthesis). Only two amino acids other than the standard 20 are known to be incorporated into proteins during translation, in certain organisms: selenocysteine is incorporated into some proteins at a UGA codon, which is normally a stop codon. Pyrrolysine is incorporated into some proteins at a UAG—another stop codon. For instance, these are used to produce methane in enzymes called methanogens. Besides those used in protein synthesis, other biologically important amino acids include carnitine (used in lipid transport within a cell), ornithine, GABA and taurine.

Functions of proteins

Proteins are very important molecules as they are involved in virtually all cellular functions. Each protein within the body has a specific function. Some proteins are involved in structural support, while others are involved in bodily movement, or in defence against germs. Some of the important functions of proteins are as follows:

- They form many of the structural features of the body including hair, nails and muscles. In silkworm, the cocoon is built largely by proteins of fibroin and sericins. Proteins are a major structural component of cells and cell membranes.
- They aid in transporting materials across cellular membranes. An example would be the uptake of glucose into cells from the bloodstream.
- They act as biological catalysts. A large group of proteins, known as enzymes, are able to speed up chemical reactions that are necessary for cells to work properly. For example, there are numerous enzymes that are involved in breaking down the food we eat and making the nutrients available.
- Antibodies are specialised proteins involved in defending the body from antigens (foreign invaders). They immobilise the antigens so that they can be destroyed by white blood cells.
- Contractile proteins are responsible for muscle contraction and movement. Examples include actin and myosin.
- Storage proteins are stores of amino acids. Examples include ovalbumin and casein. Ovalbumin is found in egg white and casein is milk-based.
- Proteins can also provide energy in the case of shortage of main energy molecules such as carbohydrates and fats.
- Proteins are often responsible for maintaining contact between adjacent cells and between cells and their local environment. A good example would be the cell–cell interactions that keep the cells in our skin bound together. These interactions are dependent on proteins from neighbouring cells binding tightly to each other.
- Proteins work to control the activity of cells, including decisions regarding cell division.
- Many hormones—signals that travel through the body to change the behaviour of cells and organs—are composed of proteins. For example, insulin, a small protein hormone regulates the uptake of glucose from the bloodstream.

Levels of protein structure

Differences in the three-dimensional structures of proteins correspond to the huge diversity of their functions. Protein structure is organised hierarchically from the so-called primary structure to the quaternary structure at four levels (**Fig. 3.10**). Higher-level structures are called motifs and domains. The different conformations are also caused by the different sequences of amino acid residues. The **primary structure** is the sequence of residues in the polypeptide chain. **Secondary structure** is a local regularly occurring structure in proteins and is mainly formed through hydrogen bonds between the backbone atoms. Motifs and domains are combinations of secondary structures. The former consist only of a few secondary structures. They could, but not always, have a function. A domain is more complex. It is usually defined as a modular functional unit folding independently and has a definite function.

Tertiary structure refers to the three-dimensional structure of a protein and describes the packing of alpha helices, beta sheets, and random coils with respect to each other for the whole polypeptide chain. The tertiary structure is stabilised by various types of bonds.

Quaternary structure is possible if there is more than one polypeptide chain present in a complex protein.

Production of therapeutic proteins

Many proteins are made by biofarming of genetically modified bacteria. For example, the human gene for a protein is inserted into the bacteria's DNA along with its section that regulates expression, either carrying a plasmid or the bacterial chromosome. However, prokaryotic cells cannot make some human proteins.

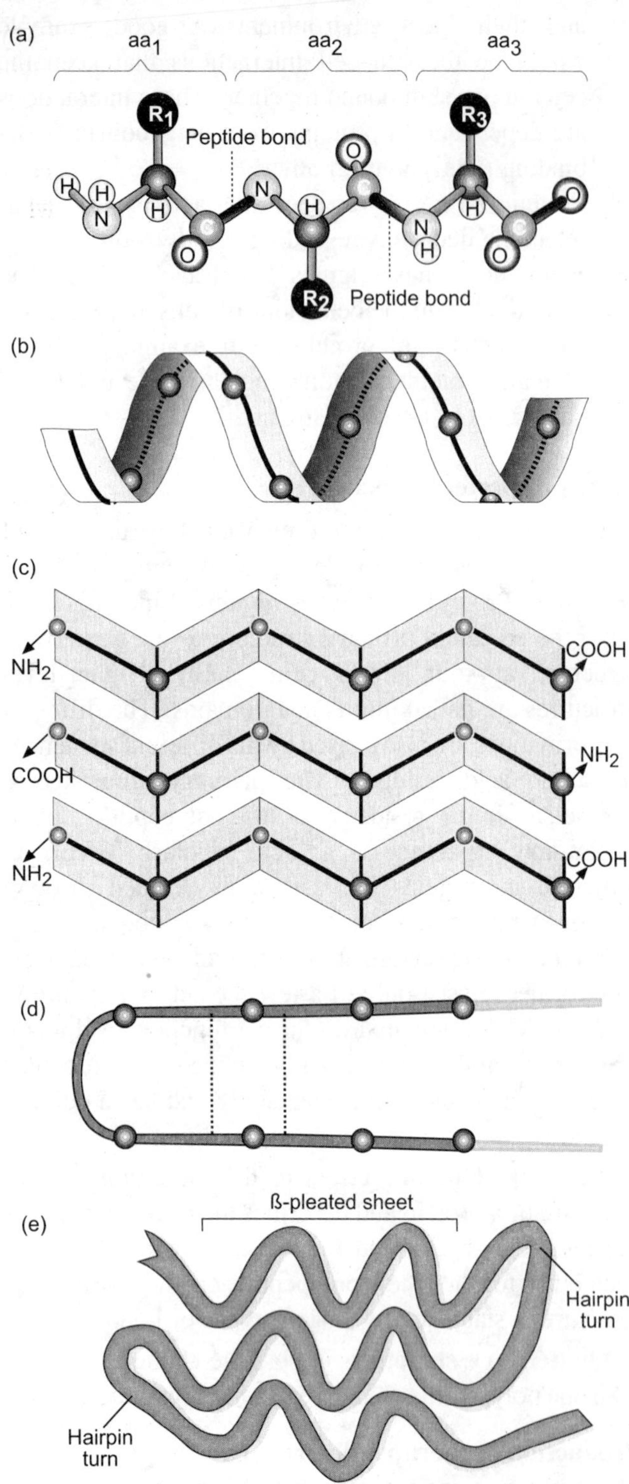

Fig. 3.10 Protein secondary structures (each amino acid residue is shown) (a) Primary structure (b) Alpha helix (c) ß-pleated sheet (d) Hairpin turns (e) Hairpin turns allow a single polypeptide to form a ß-pleated sheet

Therefore, alternatively, genetically modified yeast, plants and animals are used for their production on a commercial scale. Genetically modified sheep now produce a specific human blood clotting factor. The inserted human gene is expressed in the mammary glands and the protein is extracted from ewe's milk. Similarly, specific antibodies can be obtained from the egg white of genetically modified chicken. With the development of recombinant DNA technology, it is now possible to produce new proteins of any particular sequence of amino acids (**designer protein**). By knowing the amino acid sequence of the gene, a base sequence can be calculated, and a new gene can be synthesised in the laboratory. Then, it can be inserted into the DNA of a suitable vector or host cell chromosome and activated to be expressed, leading to the production of a new protein.

3.6.3 Lipids

They are biomolecules such as fats, oils, waxes and steroid hormones. Regardless of their structure, location or function in a cell or body, all lipids share common features that enable them to be grouped together, i.e., they do not dissolve in water (i.e., they are hydrophobic). Lipids are soluble in organic solvents such as chloroform, ether and benzene. Like the carbohydrates, they are composed primarily of carbon, hydrogen and oxygen. The hydrophobic nature of the lipids dictates many of their uses in biological systems. Fats are a good source of stored energy while oils and waxes are used to form protective layers on our skin preventing infection. Some lipids like the steroid hormones are important regulators of cell activity.

Lipids are chiefly fatty acid esters, and are the basic building blocks of biological membranes. Most lipids consist of a polar or hydrophilic head (typically glycerol) and one to three non-polar or hydrophobic fatty acid tails. Fatty acids consist of unbranched chains of carbon atoms that are connected by single bonds alone (**saturated** fatty acids) or by both single and double bonds (**unsaturated** fatty acids). The chains are usually 14–24 carbon groups long, but it is always an even number. For lipids present in biological membranes, the hydrophilic head belongs to one of the following three classes:

- **Glycolipids**—whose heads contain an oligosaccharide with 1–15 saccharide residues (**Fig. 3.9**)

- **Phospholipids**—whose heads contain a positively charged group that is linked to the tail by a negatively charged phosphate group
- **Sterols**—whose heads contain a planar steroid ring, for example, cholesterol.

Other lipids include prostaglandins and leukotrienes, which are both 20-carbon fatty acyl units synthesised from arachidonic acid. They are also known as fatty acids. However, all lipids are not esters of fatty acids. For example, in cerebrosides, the fatty acid's moiety is an amide and in many other lipids there are ether linkages (e.g., plasmogens).

Functions of lipids

The lipids of physiological importance for humans have the following biological functions:

- They serve as structural components of biological membranes. Cells are surrounded by a thin layer of lipids. This layer is made up of a special type of lipid that has both hydrophobic and hydrophilic properties. The hydrophilic ends of these molecules face the water-filled environment inside the cells and the watery environment outside the cells. A hydrophobic region exists inside the two layers. The membrane that surrounds the cells is rich in proteins and other lipids such as cholesterol. The lipid bilayer is impermeable to most substances. Water and some other small molecules can freely pass through the membrane while other molecules must be actively transported via protein channels embedded in the membrane.
- They provide energy reserves, predominantly in the form of triacylglycerols, giving 9.7 kcal/g, that is obtained from carbohydrates.
- Both lipids and lipid derivatives serve as vitamins (e.g., vitamins A, D, E and K) and hormones.
- They provide essential fatty acids such as linoleic and linolenic acids. Some of these fatty acids are precursors for biosynthesis of prostaglandins, which are essential for maintaining several functions of the body.
- Lipophilic bile acids aid in solubility of lipids.

Table 3.3 lists some of the fatty acids that play a major role in normal physiological functions.

3.6.4 Nucleic acids

Nucleic acids (NA) are polymers of nucleotides. When nucleobases attach to a pentose sugar, they form nucleosides which are phosphorylated by specific kinases in the cell to produce nucleotides (see **Chapter 7**). There are two main types of nucleic acids: deoxyribonucleic acid (DNA) and ribonucleic acid (RNA). They are composed of monomer subunits, which are often referred to by the single letter abbreviations A, C, G, T and U. One of the differences between DNA and RNA is the subset of nucleotides used to build the polymers. DNA contains A, C, G and T while RNA contains A, C, G and U. DNA is composed of two long strings (polymers) of nucleotides twisted around each other to form the spiral or helical structure. The nucleotide containing adenine (A) always pairs with the nucleotide containing thymine (T). Likewise, guanine

Table 3.3 Physiologically important fatty acids

Numerical Symbol	*Common name*	*Structure*	*Features*
14:0	Myristic acid	$CH_3(CH_2)_{12}COOH$	Often found attached to the N-terminal of plasma membrane-associated cytoplasmic proteins
16:0	Palmitic acid	$CH_3(CH_2)_{14}COOH$	End product of mammalian fatty acid synthesis
$16:1^{\Delta 9}$	Palmitoleic acid	$CH_3(CH_2)_5C{=}C(CH_2)_7COOH$	
18:0	Stearic acid	$CH_3(CH_2)_{16}COOH$	
$18:1^{\Delta 9}$	Oleic acid	$CH_3(CH_2)_7C{=}C(CH_2)_7COOH$	
$18:2^{\Delta 9,12}$	Linoleic acid	$CH_3(CH_2)_4C{=}CCH_2C{=}C(CH_2)_7COOH$	Essential fatty acid
$18:3^{\Delta 9,12,15}$	Linolenic acid	$CH_3CH_2C{=}CCH_2C{=}CCH_2C{=}C(CH_2)_7COOH$	Essential fatty acid
$20:4^{\Delta 5,8,11,14}$	Arachidonic acid	$CH_3(CH_2)_3(CH_2C{=}C)_4(CH_2)_3COOH$	Precursor for eicosanoid synthesis

(G) always pairs with cytosine (C). The polymers that form DNA can be extremely long, reaching millions of nucleotides per molecule.

All of the nucleated cells in the human body have the same DNA content regardless of their function. The difference is which parts of the DNA are being used in any given cell. For example, the cells that make up the liver contain the same DNA as the cells that make up muscles. The dramatically different activities of these two cell types depends on the portions of DNA that are active in the cells. Changes in the sequence of DNA can lead to alterations in cell behaviour. Unregulated growth, as seen in cancer, is usually the result of changes in the structure of DNA caused by mutations.

As a class, nucleotides may be considered one of the most important metabolites of the cell. Nucleic acids are also required for numerous other important functions within the cell. These functions include:

- Serving as energy stores for future use in phosphate transfer reactions; these reactions are predominantly carried out by ATPs.
- Forming a portion of several important co-enzymes such as NAD^+, $NADP^+$, FAD and co-enzyme A.
- Serving as mediators of numerous important cellular processes such as second messengers in signal transduction events; the predominant second messenger is cyclic-AMP (cAMP), a cyclic derivative of AMP formed from ATP.
- Controlling numerous enzymatic reactions through allosteric effects on enzyme activity.
- Serving as activated intermediates in numerous biosynthetic reactions; these activated intermediates include S-adenosylmethionine (S-AdoMet) involved in methyl transfer reactions as well as the many sugar-coupled nucleotides involved in the synthesis of glycogen and glycoprotein.

3.7 COMBINATIONS OF BIOMOLECULES

Biomolecules, however, work together to perform specific functions and to build important structural features of the cells. For example, in the plasma membrane, in addition to the lipid bilayer, there are numerous proteins and sugars. Many of the proteins that are important in the development and/or detection of cancer are glycoproteins. Often many biomolecules

KEY POINTS

- An atom is the smallest particle of matter. The six elements—carbon, hydrogen, oxygen, nitrogen, phosphorus and sulphur—make up about 98% of the mass of all living organisms.
- Molecules are made up of atoms which are held together by forming stable interactions called bonds (ionic, covalent, hydrogen, van der Waals, hydrophobic). These bonds take many forms and are the most important in biological systems.
- All life processes occur in water (aqueous environment).
- The molecules are always in motion, randomly colliding with each other. Weak bonds or temporary bonds stabilise certain associations and allow useful reactions to occur; enzymes speed up reactions, antibodies bind to antigens and so on.
- Energy is the fundamental principle in a wide range of phenomena. Energy is always conserved and converted from one form to another. For example, potential energy is released to give kinetic energy to drive the reactions. Cells oxidise glucose to release energy as heat and some of this released energy will be captured by the molecule ATP, which acts as an energy intermediate.
- Oxidation gives off energy whereas reduction does the opposite. Activation energy refers to the energy necessary to start a chemical reaction.
- Enzymes are the biological catalysts and they are always protein molecules. Enzymes catalyse almost every chemical reaction that occurs in the cells: they can join the reactants in such a way that a reaction between them becomes more favourable.
- Enzyme specificity is critical for two reasons: first, it ensures that only the proper reactants are brought together and secondly, the rate of a particular reaction can be controlled by the activity of the catalyst.

- Cells control enzymes sometimes by adjusting their amount or more often by changing their ability to bind or transform reactants.
- The major molecules present in the cell are organic—based on a skeleton of covalently bonded carbon atoms.
- The four types of macromolecules found in living cells are carbohydrates (polysaccharides), lipids, proteins, and nucleic acids, which are made of smaller, repeating subunits called monomers.

join together to form functional structures in cells. Some of these complex structures are discussed in other chapters.

Self-assessment Questions

1. Discuss briefly the different types of chemical forces between the atoms.
2. What are the properties that make water a unique molecule?
3. Describe the role of oxidation–reduction reaction in the cells.
4. Discuss the major biomolecules present in cells.
5. Explain how infinite number of different proteins can be made using only 20 amino acids.
6. Explain the important role of enzymes in cells.
7. What are the main functions of lipids in cells?
8. Describe the functions of nucleic acids.

FURTHER READING

Textbooks

Berg JM, JL Tymoczko and L Stryer (2006). *Biochemistry*, 5th ed. New York: Freeman.

Channarayappa (2010). *Cell Biology*. Hyderabad: Universities Press (India) Pvt. Ltd.

Cooper GM and RE Hausman (2004). *The Cell: A Molecular Approach*, 3rd edition. Washington DC: ASM Press.

Gupta PK (1999). *Cell and Molecular Biology*. Meerut: Rastogi Publications.

Spiller GA (ed.) (2006). *Handbook of Lipids in Human Nutrition*. Boca Raton: CRC Press.

Twyman RM (1998). *Advanced Molecular Biology*. New Delhi: Viva books Pvt. Ltd.

References

Channarayappa (2006). *Molecular Biotechnology: Principles and Practices*. Hyderabad: Universities Press (India) Pvt. Ltd and CRC Press worldwide.

Channarayappa (1992). Resistance mechanism of *Lycopersicon* species to the vector *Bemisia tabaci* and ultrastructural changes in leaf curl virus-infected tomato plants. PhD thesis submitted to the University of Agricultural Sciences, GKVK, Bangalore.

Channarayappa (1990). Cytogenetic analysis of clastogens and aneuploidogens by immunofluorescent and micronucleus methods. PhD thesis submitted to the West Virginia University, Morgantown, USA.

Eyster KM (2007). The membrane and lipids as integral participants in signal transduction. *Adv. Physiol. Edu.* 31: 5–16.

Fahy E, S Subramaniam, HA Brown *et al.* (2005). A comprehensive classification system for lipids. *J. Lipid Res.* 46 (5): 839–61.

Hölzl G and P Dörmann (2007). Structure and function of glycoglycerolipids in plants and bacteria. *Prog. Lipid Res.* 46: 225–43.

Mathews CK, KE van Hode and KG Ahern KG (2000). *Biochemistry*, 3rd edition, Redwood City CA: Benjamin Cummings.

Sutton J (1998). *Biology*. London: Macmillan Foundations.

CHAPTER 4

Basic Rules of Inheritance

LEARNING OBJECTIVES

After reading this chapter, you will be able to:

- Understand the concept of heredity and inheritance pattern of genetic traits
- List the exceptions to the Mendelian laws of inheritance and heritability

4.1 INTRODUCTION

Understanding the basic principles and the laws governing heredity is very important as it is closely related to the several processes studied in molecular biology. **Heredity** is the cause of the similarities between individuals through the process by which all living beings produce offspring. It involves the transmission of information that specifies a particular pattern of growth and organisation from parents to offspring. In contrast, **variation** is the cause of the differences between individuals. Thus, the science of **genetics** attempts to explain the mechanism and the basis for both similarities and differences between related individuals. It also explains the variation found in all forms of life. The structure of the gene and its expression is the fundamental aspect of molecular biology as well as genetics. Molecular genetics also covers the area of development. All organisms begin life as a single cell, but in higher forms, this single cell gives rise to the very complex multicellular organism containing numerous cell types.

4.2 MENDELIAN INHERITANCE

Gregor Johann Mendel, an Austrian monk, laid the foundations for the basic rules in modern genetics, in his monastery in Brno in erstwhile Czechoslovakia. Mendel's research uncovered the basic rules governing inheritance through breeding experiments in the garden pea, *Pisum sativum*. He derived his rules of heredity and variation from the results of crosses between pure breeding, contrasting varieties of the garden pea and crosses involving hybrid plants (**Fig. 4.1**). Although he worked exclusively with one plant species, his conclusions are applicable to all sexually reproducing eukaryotes, including those which cannot be studied in the same manner (e.g., humans).

Mendel's principles of inheritance deduced from experiments in the garden pea can be summarised as follows.

- The hereditary characters are determined by factors (occur in pairs), which Mendel called as Formbildungelementen (form-building elements).
- Different traits were specified by different forms of each gene (i.e., different **alleles**).
- When two phenotypically different alleles combine in the same individual (heterozygote condition, Aa), one trait exhibits dominance over the other allele of the pair: the allele expressed (**dominant** allele, A) contributes to the phenotype, whereas, the other allele (**recessive** allele, a) does not. When dominant and recessive alleles are present together (Aa), the recessive allele was masked by

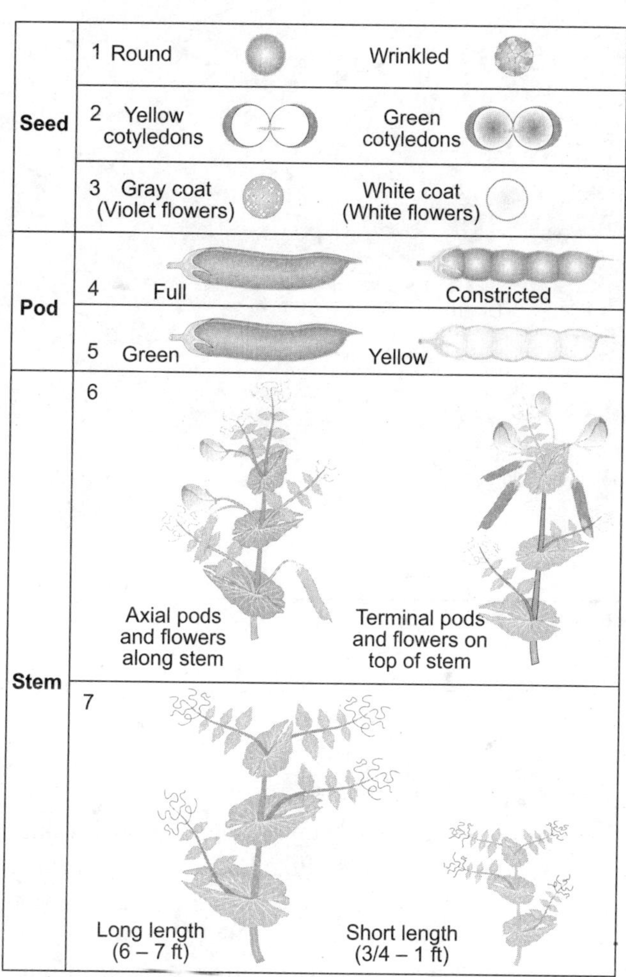

Fig. 4.1 The seven characteristiics of garden pea described by Mendel

the dominant one. Recessive genes can only be expressed in homozygous (aa) individuals.

- The genetic factors (genes) do not blend, but remain discrete (particulate) as they are transmitted to the next generation.
- During sexual reproduction (meiosis), pairs of alleles segregate equally, so that equivalent number of gametes carrying each allele is formed. This separation of alleles formed the basis of Mendel's **law of segregation**.
- The segregation of the pair of alleles for one trait had no effect on the segregation of alleles for another trait. This finding formed the basis of Mendel's **law of independent assortment**.

4.2.1 Monohybrid inheritance

Five of Mendel's principles can be inferred from the **one-point cross** (one factor cross), where a single gene locus is isolated for study. A contrasting pure line produces hybrid progeny and establishes the **principle of dominance**. For example, when pure lines of tall (TT) and dwarf (tt) plants are crossed (**Fig. 4.2**), the F1 plants all show tall (Tt), indicating that tall character is dominant over dwarf. A **pure line** breeds true for a particular trait when self-crossed (or inbred). It can be inferred that the pure line contains only one type of allele; this means all the progeny are homozygous at the locus of interest. A cross between contrasting pure lines thus produces a generation of uniform hybrids, where each individual is heterozygous, carrying one allele from each line. This is the **first filial generation (F1 generation)**.

Mechanism of the monohybrid cross

It is clear that the factors responsible for transmission of traits from parent to offspring and further development of that trait at an appropriate time and place in a particular environment are the genes, which occur in two or more structural forms, called alleles. Because the gametes constitute the only link in sexually reproducing organisms between parent and offspring, it is clear that genes must be transmitted from generation to generation *via* this gametic bridge. The sperm and egg in this kind of cross do make equal genetic contribution to the next generation.

Law of segregation: An F1 self-cross (self-fertilisation) or, where this is not possible, an intercross between F1 individuals can be termed a monohybrid cross because the participants are heterozygous at one particular locus. Mendel went on to mate the first-generation plants (e.g., cross of smooth × wrinkled) with one another (**Fig. 4.3**). Three-quarters of the resulting second-generation plants (3:1) still produced smooth seeds, but the seeds of the remaining quarter were again wrinkled. The trait that could not be seen in the first generation had reappeared in the second. This is known as **monohybrid ratio**, and would be expected to arise only if there was equal segregation of alleles at meiosis. Such a cross demonstrates the **principle of equal segregation**, which has long been known as **Mendel's first law—the principle of segregation**. It states that hereditary patterns are determined by factors (genes) that occur in pairs in an individual but segregate from each other in the formation of sex cells (gametes) so that any one gamete receives only one or the other of the paired alleles. The double number of genes is then restored in the progeny. Comparable breeding experiments showed that other characteristics, such as

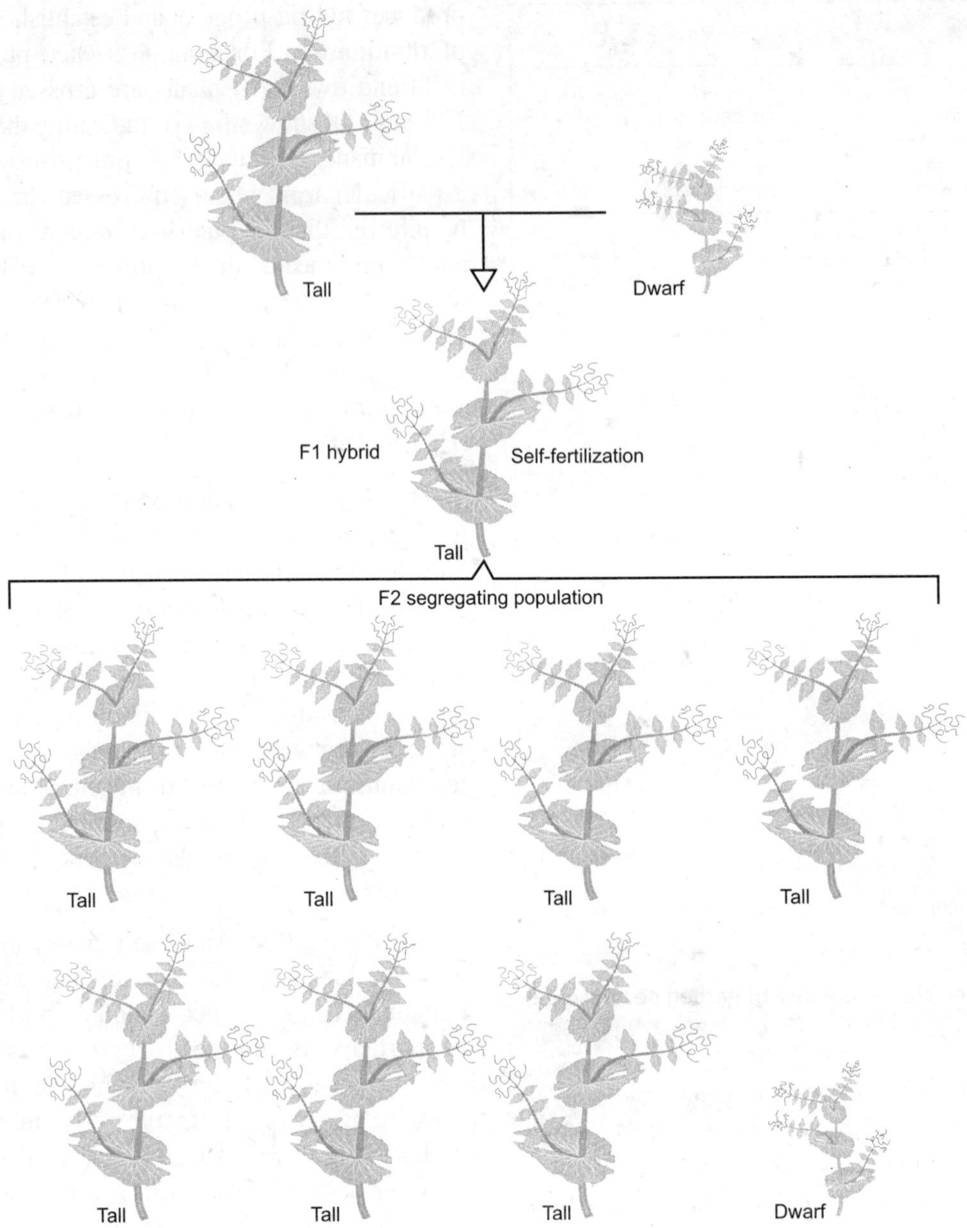

Fig. 4.2 Results of monohybrid crosses for the tallness trait in pea

seed colour, plant height and so on showed the same trend.

4.2.2 Dihybrid inheritance

Mendel also studied the inheritance of two different characteristics simultaneously; he found that they could be inherited independently of one another (the principle of independent assortment). Mating plants that produced smooth, yellow seeds (doubly homozygous) with those that yielded wrinkled, green seeds (doubly homozygous) produced plants in the first generation that made only smooth, yellow seeds (doubly heterozygous). The F1 individuals in this instance are referred to as dihybrid individuals. If these are selfed or intercrossed (a **dihybrid cross**) the F2 generation shows a 9:3:3:1 ratio of the four possible phenotypes. This is termed the **dihybrid ratio** and is derived as shown in **Fig. 4.4**. The plants of the second generation not only yielded peas with the two initial combinations, but also produced some peas with the novel combinations of wrinkled, yellow and smooth, green. The dihybrid ratio could only arise if the segregation of one pair of alleles had no effect on that of the other, i.e., the two-allele pairs show independent assortment at meiosis. From these

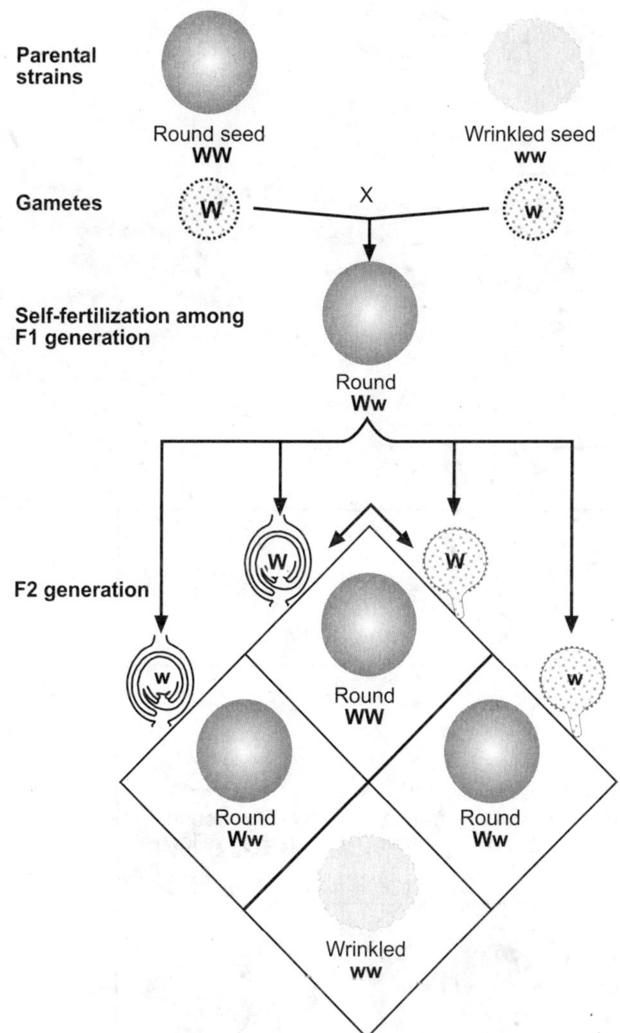

Fig. 4.3 Inheritance of dominant and recessive genes; the results of monohybrid cross showing segregation of a gene for seed shape (W/w = round/wrinkled).

results Mendel deduced that hereditary traits are carried and transmitted to the progeny as discrete units. The principle of independent assortment has become known as **Mendel's second law**.

Mendel's experiments also led him to conclude that each individual carries two units for a given characteristic, but passes just one copy to each progeny. Furthermore, some variants of a particular trait are dominant over others. If they are inherited together, the dominant character is expressed, while the other, recessive variant is not seen.

As mentioned earlier, outward appearance cannot reveal whether an individual who shows a dominant trait is homozygous or heterozygous for that characteristic (e.g., a round seed pea may be homozygous or heterozygous for seed colour). A backcross (a cross involving a filial generation and one of its parents) confirms that the F1 generation is heterozygous, for both single and multiple gene loci. If the F1 generation is crossed to homozygous parent carrying the recessive allele (test crossing), the 1:1 ratio, in two locus crosses, ratio of 1:1:1:1 (**Fig. 4.5**) and in three locus, ratio of 1:1:1:1:1:1:1:1 of phenotypes in the first backcross generation confirms the F1 genotype. This type of experimental analysis demonstrates the power of genetic crosses involving a test stock (which carries recessive alleles at all loci under study) to determine unknown genotypes and a similar principle can be used in genetic mapping. The reappearance of the recessive phenotype in the F2 generation confirms that the pair of alleles remain particulate during transmission and are neither displaced nor blended in the heterozygote (hybrid).

4.3 DEVIATIONS FROM THE MENDELIAN LAWS

Some of the important features of deviations are described below.

Deviation from the monohybrid ratio

Most characteristics generally follow the Mendelian inheritance patterns. But experiments have shown that specific deviations do occur in the monohybrid segregation ratio. The results of such deviations from monohybrid ratios have demonstrated that all but one of Mendel's six principles may be broken, a fact that suggests that Mendel enjoyed a degree of luck in his choice of seven characters which did not suffer from any of the complications discussed below.

Parental non-equivalence: A fundamental conclusion revealed from Mendel's experiments was that for every locus, one allele is contributed by each parent. Thus, the results of reciprocal crosses are equivalent and form the basis for autosomal inheritance patterns in pedigrees. Parental equivalence reflects that diploid eukaryotic cells carry two sets of chromosomes, one from each parent. However, there are two important exceptions: **sex-linked inheritance** (due to structural hemizygosity), and **monoallelic expression** (due to functional hemizygosity).

Sex-linked inheritance: The chromosomes which determine the sex of the organism are called sex chromosomes, and the genes on them are said to be sex-linked. Sex chromosomes in animals are

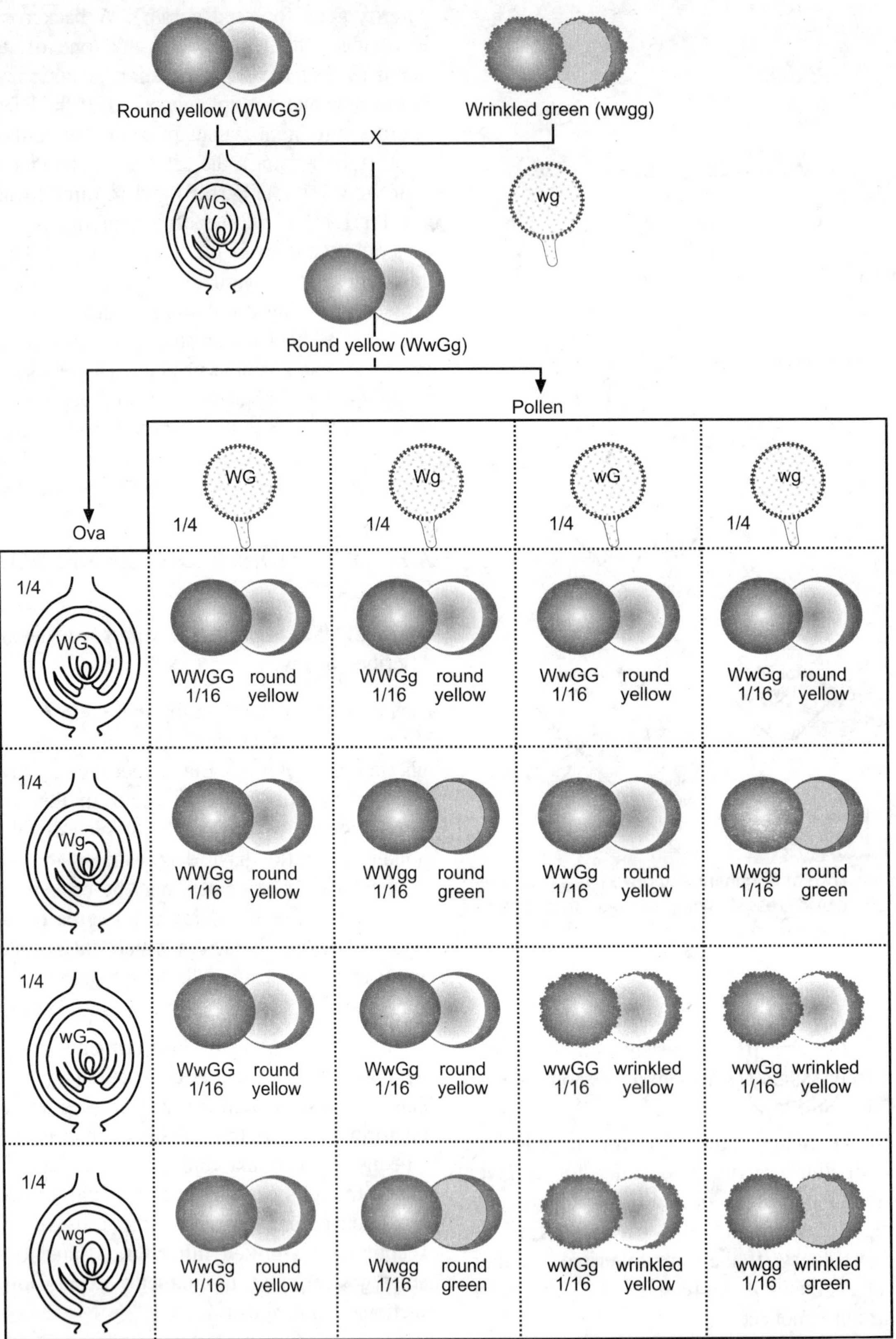

Fig. 4.4 The results of dihybrid cross showing independent assortment of the two pairs of alleles resulting in the 9:3:3:1 phenotypic ratio

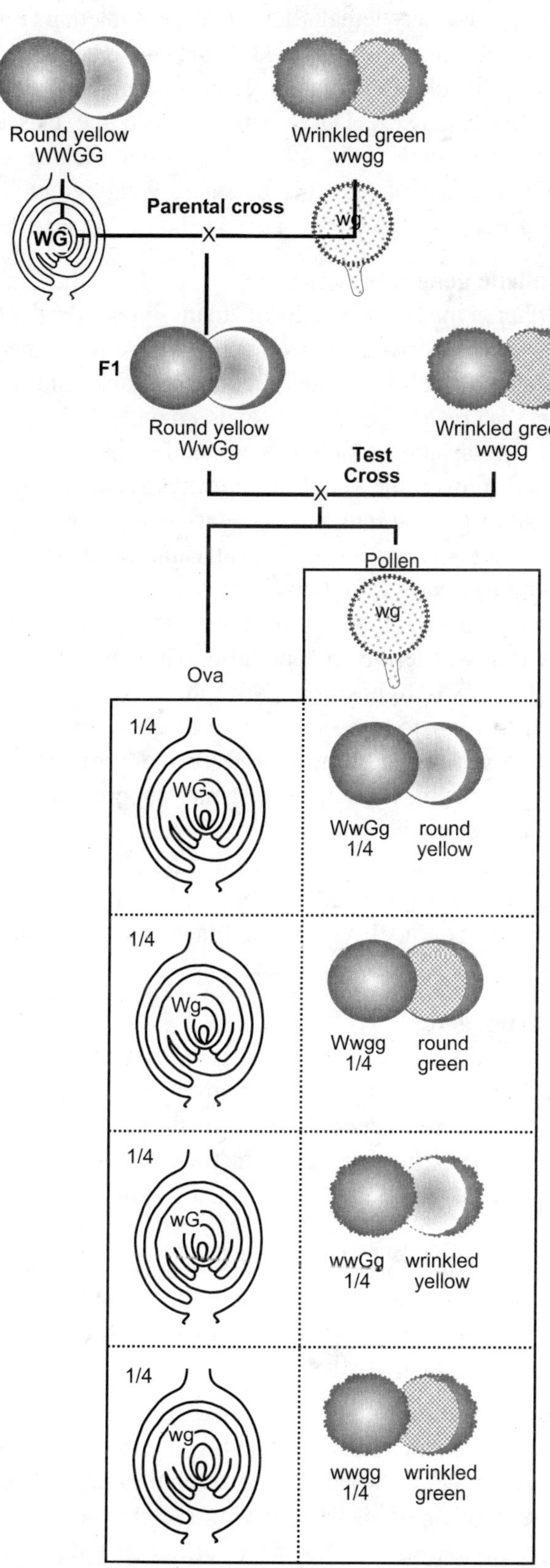

Fig. 4.5 The results of the dihybrid test cross showing two pairs of alleles resulting in the 1:1:1:1 phenotypic ratio

asymmetrically distributed between the sexes. In mammals, females are homogametic (carry two copies of the X-chromosome), whereas males are heterogametic (carry one X-chromosome and one Y-chromosome). In X- and Y-chromosomes, there are two short regions of homology, the major and minor pseudoautosomal regions, which facilitate pairing during meiosis. The major pseudoautosomal region is the site of an obligatory cross-over, and genes located there are inherited in a normal autosomal fashion (*pseudoautosomal inheritance*). Other genes on sex chromosomes are described as sex-linked because their expression depends upon how the sex chromosomes (X and Y) are distributed.

The X-linked genes show phenotypic sex-specificity, whereas for autosomally transmitted traits the segregation ratios are sex-independent. Further, because males inherit their X-chromosome only from their mothers and transmit it only to their daughters, the sex-phenotype relationship alternates in each generation, a phenomenon termed *criss-cross inheritance* and is the major characteristic used to distinguish X-linked inheritance patterns in human pedigrees. In crosses and pedigrees, Y-linked genes will be identical because the characteristics they control express only in males and are passed solely through the sons (**holandric**).

Monoallelic expression: Autosomal genes are inherited from both parents, but only one allele becomes active. This is termed as mono-allelic expression, and the locus is functionally, but not structurally, hemizygous. There are two types of monoallelic expression: parental imprinting, where the gene inherited from one parent is specifically repressed and random inactivation, where the parental allele to be repressed is chosen randomly. (see **Chapter 16**).

Maternal effect and maternal inheritance

Generally, inheritance of characters is through nuclear genes. However, under several circumstances reciprocal crosses are non-equivalent. An example is the maternal effect, where the phenotype of an individual depends entirely on the genotype of the mother. Genes that show the maternal effect inherit in a normal Mendelian fashion, but the phenotype is not observed until the following generation and thus depends on the (equivalent) contributions of the embryo's maternal grandparents. Reciprocal crosses carried out in the grandparental generation would thus be equivalent with respect to the embryonic phenotype. In this way, the maternal effect differs from maternal inheritance which is a form of non-Mendelian inheritance where genes are transmitted solely through the female line because they are located in organelle genome in the cytoplasm, rather than in the nuclear genome. Maternally inherited genes are not specifically linked to development (i.e., they are expressed throughout the lifetime of an individual) and there is no male contribution. Thus, reciprocal crosses in all generations would be non-equivalent.

Distortion of segregation ratios

The mechanism of equal segregation is the strongest of Mendel's rules and it is derived from the observation that contributing alleles are represented equally in the progeny of a monohybrid cross. However, there are several ways in which equal representation is altered, resulting in distortion of the Mendelian ratios. Such distortions are grouped into two major classes described as below:

Distortion of segregation: It occurs before fertilisation, due to a disproportionate representation of gametes carrying each allele and is termed **meiotic drive**. Generally, two types (genetic and chromosomal drive) of drive mechanisms occur predominantly in the different sexes. Genetic drive usually occurs in males and involves selective inactivation of the sperm of a particular haplotype. Two loci are involved in this type of system, a *trans*-acting driver or distorter and a *cis*-acting target. Genetic drive is uncommon in females because they produce far fewer gametes and would be placed at a selective disadvantage by large-scale gamete inactivation. Drive in females often occurs earlier through the process of chromosomal drive, where the property of a given bivalent at meiosis causes it to adopt a particular orientation in the spindle and thus undergo preferential segregation into either the egg or the polar body. Chromosomal drive would not work in males because of identical meiotic products.

Lethal alleles: If distortion occurs after fertilisation, it reflects differing viabilities of zygotes with alternative genotypes. In its most extreme form, distortion results in the total absence of a particular genotype, indicating the presence of lethal alleles, whose effects are manifest early in development. If lethal gene is dominant, it results in the recovery of only a homozygous recessive genotype. In contrast, the presence of a recessive lethal generates a characteristic 2:1 segregation ratio of dominant homozygotes to heterozygotes. Lethal alleles usually represent the loss of function of an essential gene product, thus leaky lethal alleles may be sublethal. For example, in corn, tall (D-) and dwarf (dd) phenotypes are known in addition to the green and albino condition. The cross DdGg × DdGg will produce the classic 9:3:3:1 phenotypic ratio of seedlings, but because of the lethal effect of gg, this becomes 9 tall green: 3 dwarf green.

Non-allelic gene interactions

The phenotype is a result of many gene products expressed in a given environment. The environment includes not only external factors but also internal factors such as hormones and enzymes. Several genes are usually required to specify the enzymes involved in even the simplest pathways. One common example of genetic interaction occurs whenever two or more genes specify enzymes that catalyse steps in a common pathway. Mutations in any one of these genes produce metabolic blocks. Originally, this type of gene interaction was termed as **epistatic**. The gene or locus suppressed was **hypostatic**. Now the term 'epistasis' is used to refer to almost any type of gene interaction that involves the masking of one of the gene effects. Epistasis involves interallelic gene suppression, or the masking effect that one gene locus has upon the expression of another. Consequently, the classical phenotypic ratio of 9:3:3:1 observed in the progeny of dihybrid parents becomes modified by epistasis into ratios of various combinations.

Modifying genes

These are a group of genes which either enhance and/or reduce the phenotypic effect produced by a major gene. Such genes generally have small and cumulative effect on the expression of that major gene. As a result, a continuous variation is generated in the phenotype governed by a single major gene; this converts the qualitative character into a quantitative one. Many modifying genes do not have any known effect of their own, but some others are known to affect the same or some other characteristics.

Major and minor genes: In mice, rats, guinea pigs and rabbits a kind of white spotting is produced by a recessive gene, s, when it is in the homozygous state (ss). A number of such modifying factors, designated as s_1, s_2, s_3 and so on influence (enhance or reduce) the expression of this spotting gene. Their effect is small but cumulative; as a result, mice with little or no white spot (full colour) to those with little or no coloured spot (full white) can be produced by these modifying genes. The inheritance pattern of this trait often exhibits the typical features of a quantitative trait, e.g., length of the corolla in *Nicotiana*. The modifying factors produce some spotting even in the absence of the spotting gene s, but their effect is more pronounced when it is present. Obviously, the spotting gene s is a major gene controlling spotting, while the modifying genes are minor ones affecting this trait. Most quantitative features of crop plants may be controlled in a similar fashion: a few major genes having additive effect with a few to several minor genes controlling them. In many other cases also, the modifying genes are known to increase as well as decrease the degree of expression of genes for self-incompatibility, disease resistance, cytoplasmic male sterility and so on.

Epistasis

A gene is said to be epistatic when its presence suppresses the effect of a gene at another locus. Epistatic genes are sometimes called '**inhibiting genes**' because of their effect on the other genes. They are said to be hypostatic. Epistasis may be dominant or recessive i.e., one or two specific alleles may be required at the epistatic locus for the effect to occur. The dihybrid ratios would be 12:3:1 and 9:3:4 respectively. Epistasis is often observed in genetic pathways or hierarchies where there is an order of gene action, early-acting genes being epistatic to later-acting ones. **Phenotypic modification** is similar to epistasis except that the modifying allele alters rather than masks the phenotype of the downstream genes.

Fur colour in mice is controlled by a pair of genes occupying different loci. The epistatic gene determines the presence of colour and has two alleles, coloured (dominant) and albino (white) (recessive). The hypostatic gene determines the nature of the colour and its alleles are agouti (grey) (dominant) and black (recessive). The mice may have agouti or black fur depending upon their genotypes, but this will only appear if accompanied by the allele for coloured fur. The albino condition appears in mice that are homozygous recessive for colour even if the alleles for agouti and black fur are present. Three possible phenotypes can occur and they are agouti, black and albino. A variety of phenotypic ratios can be obtained depending on the genotypes of the mating pair.

Redundancy and dosage: Alleles of identical genes which arise by duplication may interact with each other in a manner which superficially appears allelic. However, they may be separated by recombination, showing that they occupy discrete loci, and such interaction is termed pseudo-allelic. The dihybrid ratio would be 15:4. Multiple redundancies are often seen in transgenic organisms where many copies of a transgene have integrated. Transgenic animals may show copy-number-dependent gene expression (i.e., levels of gene product correspond to copy number), or there may be interaction between repetitive copies resulting in homology-dependent gene silencing.

Suppression: An allele at one locus (the suppressor allele) prevents the expression of a specific (and usually deleterious) allele at a second locus by compensating in some way for its effect. The term is usually used to describe an interaction where a mutation compensates for a second mutation found at a different locus and restores the wild-type phenotype (second-site mutations). Suppression may be dominant or recessive, depending upon whether one or two alleles are required at the suppressor locus. The dihybrid ratios would be 12:3:1 and 9:3:4 respectively. Epistasis differs from suppression in that the former is gene-specific but not allele-specific (i.e., epistasis prevents the expression of all alleles at a particular locus), whereas the latter is gene-specific and allele-specific (i.e., suppression compensates for the effect of a particular allele at the second locus).

Synergistic enhancement: This is similar to suppression mechanism except that the enhancer allele specifically increases the effect of a second mutation, instead of suppressing it. As with suppressor alleles, enhancer alleles may be dominant or recessive. Synergistic enhancement differs from additive effects because the enhancer locus alone does not contribute to the phenotype associated with the target (enhanced) locus, i.e., if the target allele is not present, the enhancer locus has no effect on the phenotype. Enhancer alleles should not be confused with enhancer regulatory elements.

Pleiotropy: Generally, one gene affects a single characteristic. But often, many genes are known to affect more than one character (interconnected). Such genes are known as pleiotropic genes and the condition is termed as pleiotropy. An example of a pleiotropic gene in human beings is the recessive gene s (denoted as Hb^s), which produces sickle cell anemia in the ss homozygotes. More than 50% of the individuals homozygous for this gene (ss) die before the age of 20 years. This gene carries a defect that causes substitution of a valine for a glutamic acid at position 6 of the ß-chain of hemoglobin. This mutant hemoglobin remains in solution as long as oxygen concentration is high. However, at lower oxygen concentrations, filamentous aggregates of the sickle cell hemoglobin precipitates causing the red blood cells to assume the characteristic sickle shape. In addition to this primary effect, it also affects a number of other characteristics in individuals homozygous for this gene. These include, among other things, 'tower skull', dilation of heart and heart failure, poor physical development, impaired mental function, pneumonia due to lung damage, rheumatism, paralysis, kidney damage and failure.

A number of other recessive genes produce marked and often detrimental pleiotropic effects. Genes showing pleiotropy produce a single polypeptide just like other non-pleiotropic genes. But their polypeptides govern the biochemical reaction which is basic to many developmental events.

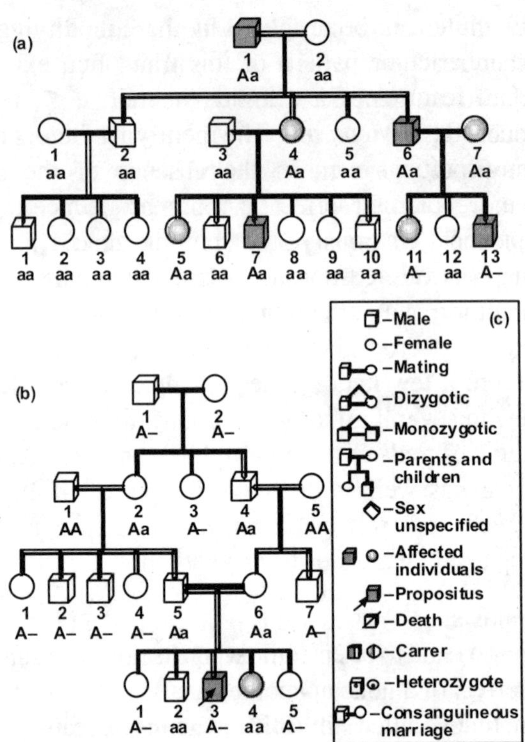

Fig. 4.6 A pedigree involving a rare phenotype (a) Controlled by recessive allele a (b) Controlled by a dominant allele A (c) Symbols used in human pedigree analysis

Pedigree analysis

Mendel in his crossing experiments used selectively made crosses and then analysed a large number of offspring. However, when studying larger animals and human traits, it is not ethical to control parental crosses. Instead, we must rely on the information that is contained within family trees. This type of approach is known as a **pedigree analysis (Fig. 4.6)** and is aimed to determine the type of inheritance pattern that a gene will follow. In pedigree analysis, the oldest generation is at the top of the pedigree, and the most recent generation is at the bottom. Vertical lines connect each succeeding generation. A male (square) and female (circle) that produce, one or more offspring are directly connected by a horizontal line. A vertical line connects parents with their offspring. The siblings (brothers and sisters) are denoted by two or more individuals projecting from the same horizontal line. When a pedigree involves the transmission of a human trait or disease, affected individuals are depicted by filled symbols that distinguish them from unaffected individuals (unfilled).

Pedigree analysis is largely used to determine the inheritance pattern of human genetic diseases. It helps to discern the dominant/recessive relationship between two alleles. If the disease follows a simple Mendelian pattern of inheritance and is caused by a recessive allele, an individual must inherit two copies of the abnormal allele to exhibit the disease. The recessive pattern of inheritance makes two important predictions. First, two heterozygous normal individuals on an average produce 1/4th of their offspring affected; secondly, all the offspring of two affected individuals will be completely affected. A dominant trait, however, predicts that affected individuals inherit the gene from at least one affected parent.

Gene interactions

It is very common to say that a gene (say H or L) produces a given characteristic (e.g., plant height). Actually, there is no gene that produces a characteristic by itself. Genes generally produce their phenotypic effects by directing the synthesis of specific proteins or RNA, which can act mostly as enzyme or structural components. The specific function of individual proteins affects that function within living cells. Molecular genetic approach helps us in understanding the relationship between a protein's function and its effect on the phenotype. Most of the phenotypic expressions involve a series of metabolic reactions. For example, the production of hydrocyanic acid (HCN) from a substrate in clovers involves two biochemical reactions. The first reaction converts the substrate into a cyanogenic glucoside. The enzyme catalysing this reaction is produced by gene L. The release of HCN from the cyanogenic glucoside is accomplished by another reaction, which is controlled by the enzyme produced by gene H. Clearly, neither gene L nor H produces HCN by itself; two genes act in series on a substrate to convert into HCN in clovers. A similar but more complicated situation may occur for most biochemical reaction pathways. The developments of most phenotypes are quite complicated and it is often difficult to resolve them by simple reactions.

4.4 CHROMOSOMAL BASIS OF INHERITANCE

On the basis of the evidence suggested by de Vries, Correns and Tschermak (1902), Sutton and Boveri proposed that chromosomes were the carriers of Mendel's factors, the so-called **chromosome theory of heredity**. According to this theory, a pair of homologous chromosomes carries each pair of factors, with each chromosome carrying one of the 'factors'. The term 'gene' was coined by Johannsen, in 1900, to replace 'factors' and alternative forms were termed as alleles, which influence phenotypic expression. Alleles are the alternative forms in which a gene may exist and they occupy the same loci on homologous chromosomes.

Linkage groups and chromosomes

Morgan's research with *Drosophila* established that the majority of phenotypic characteristics were transmitted together in four groups and these were called **linkage groups**. It was observed that the number of linkage groups corresponded to the number of pairs of chromosomes. Studies on other organisms produced similar results. Breeding experiments using a variety of organisms revealed that some linkage groups were larger than others. Examination of chromosomes in these organisms showed that they varied in length.

4.4.1 Biochemical genetics

Studies on many characteristics demonstrate that genes control the formation of enzymes and proteins that control the steps in biochemical pathways. In most of the cases, dominant alleles exercise control. Recessive alleles often produce non-functioning enzymes that cannot catalyse specific steps. The heterozygote dominant is usually normal because one allele produces a functional enzyme, which may be sufficient. The study of the relationship between genes and enzymes is generally called biochemical genetics because it involves the genetic control of biochemical pathways.

Beadle and Tatum isolated mutants with nutritional requirements that defined steps in a biochemical pathway and they used mutants to work out the steps in the biosynthesis of niacin in the pink bread mould, *Neurospora crassa*. They could also ascertain the substances accumulated in the mutants. For example, if a biochemical pathway is blocked at a certain point, then the substrate at that point cannot convert into the next product and it builds up in the cell. They concluded from their studies that one gene controls the production of one enzyme, and they proposed the **one gene one enzyme hypothesis**.

4.4.2 Gene pool

Genetic disease can result from the inheritance of mutant genes, which produce abnormal structure or function. Familial hypercholesterolemia, cystic fibrosis, sickle cell anaemia, haemophilia and muscular dystrophy are single gene diseases. Polygenic diseases result from several defective genes, each of which by itself has little effect. Examples are diabetes, hypertension and cancer. Chromosomal diseases result from addition or deletion of whole or portions of chromosomes during meiosis. An example is the Down syndrome or trisomy 21 in which the egg has an extra copy of chromosome 21.

Cancer is a genetic disorder that is not generally inherited. It arises from mutations in the somatic cells and results in the failure of the control system that regulates cell division. The result is uncontrolled growth. Dominant cancer-producing genes, called **oncogenes**, code for abnormal forms of cell surface receptors that bind growth factors and produce a continuous growth signal. Recessive cancer-producing genes, called **tumour suppressor genes**, fail to produce proteins that inhibit various steps in cell replication. Abnormal replication of cells forms a tissue mass called tumour. If these cells remain localised it is called a benign tumour and if they invade the surrounding tissue they are said to be malignant. Cancers that develop in epithelial cells are called carcinomas, ones in muscle cells are called sarcomas and ones in white blood cells are called lymphomas. Lung, colon, and breast are the organs most commonly affected. Incidence of cancer increases with age due to the accumulation of defective mutations. Mutagens that increase the probability of cancerous transformation of a cell are called carcinogens.

4.5 QUANTITATIVE INHERITANCE

The genetic laws formulated from early studies in garden pea were based on obvious differences in the general appearance of phenotypes, such as those found between short and tall plants, green and yellow seeds and so on. These phenotypic characteristics have always been discontinuous. Such traits are called qualitative and are controlled by a single or very few genes. However, not all inherited traits are expressed in this discontinuous fashion. Critical evaluation of these characteristics revealed that even Mendel's own pea plants, short or tall were not uniformly small or large. Instead gradations could be observed from short-to-tall. These quantitative variations were obviously difficult to analyse genetically, since their phenotypic range appeared to be continuous, without distinct steps that could be accounted for by distinct genes. In fact, most measurable characteristics, called quantitative characteristics or **complex traits** (historically referred to as **metric traits**), follow a normal distribution curve ranging from low-to-high values but without sharp separation in values between the many classes within this range.

Many characteristics show continuous variation i.e., phenotypes are measured in quantitative terms and cannot be placed into discrete traits. The phenotypes often show a normal distribution about a mean value. Such quantitative characteristics are inherited in a complex manner; the genotype cannot be deduced from phenotype and no simple rules of heredity can be used to predict the outcome of a cross. The inheritance of such characteristics is studied using statistical methods (**biometrics**).

4.5.1 The polygenic theory of quantitative inheritance

Since quantitative characteristics are unsuitable for Mendelian analysis, they were believed to be controlled by fundamentally different factors. The polygenic theory of Fisher established a common basis for the transmission of Mendelian and quantitative characteristics. The theory proposed that these characteristics may be controlled by a large number of segregating loci (**polygenes**), each of which contributes in a small but additive manner. The locations on chromosomes where these genes reside are called **quantitative trait loci** (**QTLs**). In the simplest form of the model, each contributing locus is di-allelic and one allele contributes to the phenotype whilst the other has no effect. The value of each allele is the same and the phenotype thus depends upon the number of segregating loci. As the number of loci increases, so does the number of phenotypes and the distribution approaches that of a normal curve. This is because, in the middle range of phenotypic values, many different genotypes will generate the same phenotype (**heterogeneity of locus**).

Significance of polygenic inheritance

The concept of additive traits is important because it demonstrates that Mendelian rules of inheritance cannot explain traits that have a continuous distribution. Additive traits are also of practical interest. Many products (plant and animal) like milk production and fruit and vegetable yield exhibit polygenic inheritance. In addition, many human traits, such as height and IQ, appear to be polygenic, although with substantial environmental influences. Biometricians used statistical techniques to study traits characterised by continuous variation and claimed that single discrete genes were not responsible for the observed inheritance patterns. The biometric school founded by F Galton and K Pearson demonstrated that many quantitative traits, such as height were indeed inherited quantitatively.

4.5.2 Components of phenotypic variance

The variance observed between individuals when they are measured for most quantitative traits may be caused by genetic differences between individuals as well as by environmental causes and occasionally by interaction between genotype and environment. Interaction occurs when the genotypes act differently in different environments, so that accurate quantitative predictions cannot be made by considering genotype and environment alone. The breakdown of observed variation into its components is commonly done by statistical methods. For our purposes, we may consider that the observed phenotypic variance, called V_P, is composed of three main sources of variation: genotypic, environmental and interaction.

$$V_P = V_G + V_E + V_{GE}$$

The environmental variance (V_E) may include such obviously variable factors as growth in different localities or in different years, or even less obvious ones, such as different maternal environments, different order in birth sequences and differences between conditions in one part of the body and the other. It includes all variation not caused by genetic differences. Genotype–environmental interactions (V_{GE}) are those in which the relative performance of different genotypes varies with the environments in which the genotypes are placed.

4.5.3 Heritability

It refers to the amount of phenotypic variation within a group of individuals that is due to genetic factors. If all the phenotypic variation in a group were due to genetic variation, the heritability would have a value of 1. If all the variation were due to environmental effects, the heritability would equal 0. Breeders want to improve the yields of their crops to the greatest degree they can. They must choose the parents of the next generation on the basis of the current generation's yields. The genetic variance ($V_A + V_D$) estimated above is of considerable importance in evaluating and predicting the degree of resemblance between one generation and another. Such predictions, of course, depend greatly upon the amount of environmental variance. The accuracy of prediction would be low, for example, for a trait whose measurements show little genetic variance but a relatively high degree of environmental variance. In such instances, the parental environment may differ sufficiently from that of the offspring so that the observed phenotype of the offspring depends little on the measurements of the parents. For parents to transmit characteristics to their offspring in some predictable degree, it is obvious that environmental variance should be low and genetic variance high, i.e., the trait should be only slightly affected by environmental differences but greatly affected by genetic differences.

KEY POINTS

- Modern genetics began with the experiments of an Augustinian monk named Gregor Mendel who discovered the fundamental principles of inheritance. Later, Jacob and Monod discovered the operon method of gene regulation. Genetic analysis provides essential information about the functions of genes and their products *in vivo*.
- Mendel's research helped him derive his rules of heredity and variation from the results of crosses between pure breeding, contrasting varieties of the garden pea *Pisum sativum* and crosses involving hybrid plants.
- Mendel's principles of inheritance cover the following mechanisms: dominance–recessiveness, law of segregation and law of independent assortment.
- Most characteristics follow in general the Mendelian pattern of inheritance, but there are many deviations to this. Some of the exceptions include parental equivalence, linkage groups, sex-linked inheritance, mono-allelic expression, maternal inheritance, allelic variation and gene interactions, distortion of segregation ratios, gender influence and gene conversions.
- Studies on many characteristics demonstrated that genes control the formation of enzymes and proteins that control the steps in biochemical pathways. George Beadle and Edward Tatum did pioneering work on gene control in the production of enzymes in the biochemical pathways in *Neurospora crassa.*
- Alleles are variants of the same gene. Genetic disease can result from the inheritance of mutant genes which produce abnormal structure or function.
- Cancer is a genetic disorder that is not generally inherited. It occurs due to mutations in the somatic cells.
- Many characteristics show continuous variation i.e., phenotypes are measured in quantitative terms and cannot be placed into discrete traits. Such quantitative characteristics are inherited in a complex manner.
- Quantitative differences between genotypes for polygenic traits may strongly be affected by the environment.
- The study of a quantitative trait in a large population generally reveals that very few individuals possess the extreme phenotypes and progressively more individuals are found nearer the average value for that population (normal distribution).
- The term heritability refers to the amount of phenotypic variation within a group of individuals that is due to genetic factors. If all the phenotypic variation in a group were due to genetic variation, the heritability would have a value of 1 and if it is due to environmental effects, the heritability would equal 0.

Self-assessment Questions

1. What are Mendel's principles of inheritance deduced from the crossing experiments in pea?
2. Briefly explain the deviations from Mendelian principles.
3. Describe sex-linked inheritance with suitable examples.
4. Discuss the different types of genetic interactions.
5. How are genes modified?
6. What is pedigree analysis?
7. Discuss briefly the chromosomal basis of inheritance.
8. Describe the one gene one enzyme hypothesis.
9. What is quantitative inheritance?

FURTHER READING

Textbooks

Brooker RJ (1999). *Genetics: Analysis and Principles.* BenjaminCummins–Addison Wesley Longman Inc.

Falconer DS and Mackay TFC (1996). *Introduction to Quantitative Genetics.* 4th ed. Harlow: Longman Group.

Hadorn E (1961). *Developmental Genetics and Lethal Factors.* New York: John Wiley.

Strickberger MW (1999). *Genetics.* 3rd ed. New Delhi: Prentice-Hall of India

Tamarin RH (2002). *Principles of Genetics.* 7th ed. New Delhi: Tata McGraw-Hill.

Twyman RM (1998). *Advanced Molecular Biology: A Concise Reference.* BIOS Scientific Publishers.

References

Beadle G and E Tatum (1941). Genetic control of biochemical reactions in *Neurospora. PNAS,* USA 27:499–506.

Becker WA (1975). *Manual of Quantitative Genetics.* Pullman WA: Student Book Corp.

Briggs RW (1966). Recognition and classification of some genetic traits in maize. *Jour. Hered.* 57:35–42.

Fuchs JA, JD Smith and LS Bird (1972). Genetic basis for an 11:5 dihybrid ratio observed in *Gossypium hirsutum. Jour. Hered.* 63:300–304.

Keightley P (1996). A metabolic basis for dominance and recessivity. *Genetics* 143:621–25.

Mendel G (1866). Versuch uber Pflanzen–Hybriden (Experiments in Plant Hybridization. Reprinted in Peters, JA (ed.) 1959, *Classic Papers in Genetics.* Englewood Cliffs, NJ: Prentice Hall.

Novitski E and S Blixt (1978). Mendel, linkage and synteny. *Bioscience* 28:34–35.

Punnett RC (1923). *Heredity in Poultry.* New York: Macmillan.

Singh BD (2000). *Plant Breeding: Principles and Methods,* 6th ed. New Delhi: Kalyani Publishers.

Wolf U (1995). The genetic contribution of phenotype. *Hum. Genet.* 95:1127–1148.

Wright S (1963). Genetic interaction. In *Methodology in Mammalian Genetics,* WJ Burdette (ed.), San Francisco: Holden–Day. 159–92.

CHAPTER 5

Nucleic Acids

LEARNING OBJECTIVES

After reading this chapter, you will be able to:

- Understand the chemical composition of DNA and RNA
- Outline the structure of the nucleic acids
- Appreciate the role of DNA as the genetic material in cellular organisms

5.1 INTRODUCTION

By the end of the 19th century, scientists had recognised DNA as the genetic material and a major constituent of chromosomes of most organisms. The developments in molecular biology have their genesis in the discovery that DNA lies behind all the cellular activities. With the development of molecular techniques, the role of nucleic acids in storage and transfer of genetic information has been well elucidated. However, DNA serves several other biological functions as well. It serves also as a replicable template for the copying of cells, as a scaffold for the organisation of proteins and as a polyfunctional switch for the control of gene expression. Recently, biologists also used DNA structure as a starting point for the design of many new molecules with functions other than those listed for naturally occurring DNA. In almost all cellular organisms, DNA encodes all the information needed to specify the structure of every protein in the cell. Acellular organisms (viruses) use either DNA or RNA for this purpose. However, RNA is an absolute requirement for the decoding of information (expression) from the genome in all organisms and no case has been reported so far where protein is synthesised *in vivo* directly from the DNA.

5.2 MOLECULAR COMPOSITION OF NUCLEIC ACIDS

Nucleic acids are high molecular weight polymeric compounds containing repeating units of nucleotide subunits. On complete hydrolysis, nucleic acids yield three components: nitrogenous bases (pyrimidine and purine bases), a pentose sugar component and phosphoric acid. Partial hydrolysis yields compounds known as nucleosides and nucleotides. Nucleic acids in the nucleotides contain either a deoxyribose sugar (called deoxyribonucleic acid, DNA) or ribose sugar (called ribonucleic acid, RNA). Both DNA and RNA are built from two purine- and two pyrimidine-containing molecules (**Table 5.1**).

Nitrogenous bases

Nucleic acids are biological molecules that possess heterocyclic nitrogenous bases as principal components of their structure. The term base in this context simply denotes that the group is, chemically speaking, basic (alkaline; since NH_2 and $>C=O$ in the bases are good proton acceptors). The bases of DNA or RNA are heterocyclic aromatic rings, with a variety of substituents (**Fig. 5.1**). The bases of nucleic acids are derivatives of two parent compounds, either pyrimidine or purine. Minor amounts of other bases are also found in nucleic acids. All the bases are flat or planar

Table 5.1 Components of nucleic acids

Components	*DNA*	*RNA*
Bases:		
Purines	Adenine (A)	Adenine (A)
	Guanine (G)	Guanine (G)
Pyrimidines	Cytosine (C)	Cytosine (C)
	Thymine (T)	Uracil (U)
Sugars	D-2-deoxyribofuranose	D-ribofuranose
Acids	Phosphoric acid	Phosphoric acid

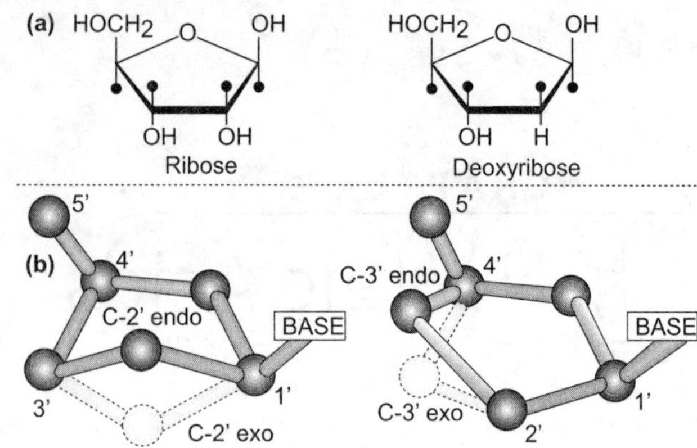

Fig. 5.2 The chemical structures of sugars and phosphate group (a) Ribose and deoxyribose sugars (b) Ribofuranose rings in nucleotides exist in four different puckered conformations

molecules and this allows them to stack like a tower of plates. In an aqueous environment, this stacking is enhanced by hydrophobic interactions, between the rings of adjacent bases. These interactions play a major role in stabilising the overall structure of nucleic acids.

Tautomerism of bases: The common structure of oxygen-containing bases at the physiological pH is in the *keto* (or lactam) form. However, there is equilibrium between this form and the *enol* (or lactim) form. The pyrimidine bases can undergo *keto–enol* tautomerism and as in the pyrimidines, purine bases can also exist in two tautomeric forms. Other naturally occurring purine derivatives include hypoxanthine, xanthine and uric acid (see **Chapter 19**).

Pentose sugars of nucleic acids

Pentoses are five-carbon sugars. RNA contains the pentose ß-D-ribose, while DNA contains ß-2-deoxy-D-ribose. In both instances, the pentose is in the five-member ring form, known as **furanose**: D-ribofuranose for RNA and 2-deoxy-D-ribofuranose for DNA (**Fig. 5.2a**). Some RNAs, notably ribosomal RNAs, contain very small amounts of 2-O-methyl ribose. Following IUPAC-IUB recommendations, when these ribofuranoses are found in nucleotides, their atoms are numbered as 1' (one prime), 2' (two-prime), and so on, merely to distinguish them from the numbering of the bases. The minor but unique difference of a hydroxyl group at the 2'-position has far-reaching effects on the secondary structures available to RNA and DNA, as well as their relative susceptibilities to chemical and enzymatic hydrolysis. The base of a nucleotide joined covalently (at N-1 of pyrimidines and N-9 of purines) in an N-ß-glycosyl bond to the 1' carbon of the pentose and the phosphate is esterified to the 5' carbon. The pentose ring is not planar but occurs in one of a variety of conformations generally described as 'puckered' (**Fig. 5.2b**).

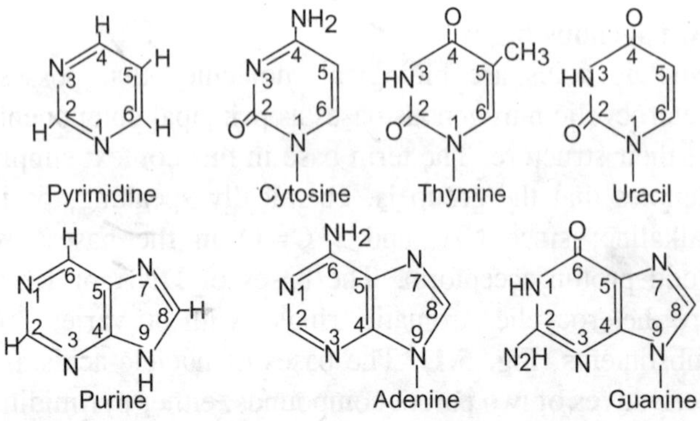

Fig. 5.1 The chemical structures of nucleobases

Nucleosides

Nucleosides are compounds formed when a purine or pyrimidine base is linked to a sugar *via* a glycosidic bond. Glycosidic bonds, by definition, involve the carbonyl carbon atom of the sugar, which in cyclic structures is joined to the ring O atom. The bases are attached covalently to the 1'-position of a pentose sugar ring to form a nucleoside (**Fig. 5.3**). These bonds can be either α or ß, depending on their orientation relative to the aromatic carbon atom. Glycosidic bonds in nucleosides and nucleotides are always of the ß-configuration, as represented in **Fig. 5.4.**

The base can exist in two distinct orientations about the N-glycosidic bond. These conformations are identified as, *syn* and *anti* (**Fig. 5.4**).

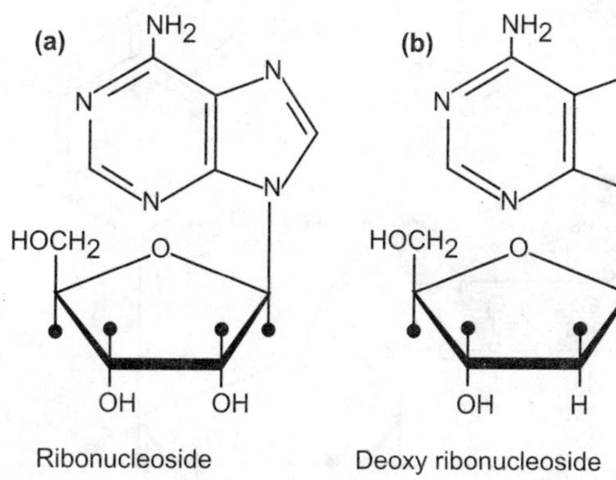

(a) Ribonucleoside (b) Deoxy ribonucleoside

Fig. 5.3 The chemical structure of nucleosides (a) Adenosine and (b) Deoxyadenosine

Nucleotides

Nucleosides are found in the cell primarily in their phosphorylated form and are termed nucleotides. A nucleotide is a nucleoside with one or more phosphate groups bound covalently to the nucleoside (i.e., they are all phosphoric esters of the nucleosides). Since the ribonucleosides have three free hydroxyl groups on the sugar ring, three possible ribonucleoside monophosphates can be formed. Adenosine, for example, can give rise to three monophosphates (adenylic acids), adenosine 5'-phosphate, adenosine 3'-phosphate and adenosine 2'-phosphate. In the same way guanosine, cytidine and uridine can give rise to three guanosine monophosphates (guanylic acids), three cytidine monophosphates (cytidylic acids) and three uridine monophosphates (uridylic acids), respectively.

A nucleotide results when phosphoric acid is esterified to a sugar -OH group of the 5'-carbon atom of a nucleoside. If the sugar is deoxyribose, then the compounds are termed deoxyribonucleotides (**Fig. 5.5**). The majority of the monomeric nucleotides in the cells are nucleotides having 5'-phosphate groups and up to three phosphates may be attached to the 5' position, to form, for example, deoxyadenosine 5'-triphosphate, commonly abbreviated to dATP.

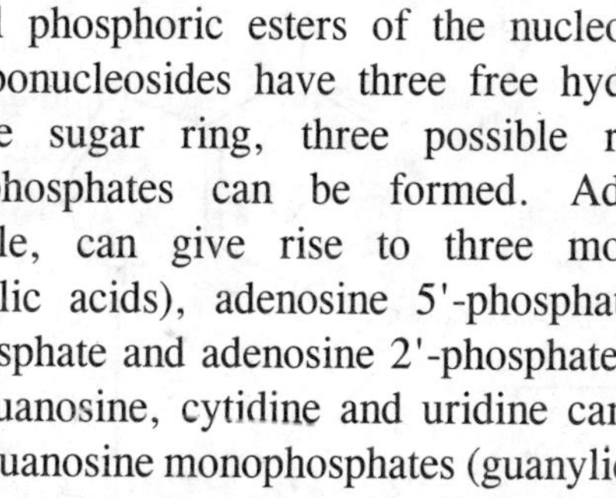

syn-adenosine *anti*-adenosine

Fig. 5.4 The base can exist in two distinct orientations *syn* and *anti* about the N-glycosidic bond— the two distinct orientations of adenosine base

In the course of DNA or RNA synthesis, two phosphates split off as pyrophosphate to leave one phosphate per nucleotide incorporated into the nucleic acid chain. The repeat unit of a DNA or RNA chain is, hence, a nucleotide. Nucleotides play a number of roles:

- They participate as essential intermediates in all aspects of cellular metabolism.
- Polynucleotides contain the elements of heredity and act as agents for genetic information.
- They are a part of several important co-enzymes such as NAD^+, $NADP^+$ FAD and co-enzyme A.
- They control numerous enzymatic reactions through allosteric effects on enzyme activity.

They serve as activated intermediates in biosynthetic reactions. One such activated intermediate is S-adenosylmethionine (S-AdoMet) which is involved in methyl transfer reactions. Further, many sugar-coupled nucleotides are involved in glycogen and glycoprotein synthesis.

Complete hydrolysis of nucleic acids releases nitrogenous bases, a five-carbon sugar and phosphoric acid in equal amounts.

Nucleoside triphosphates (NTP) have high group transfer potential: NTPs are indispensable agents in metabolism because the phosphoric anhydride bonds they possess are a prime source of chemical energy for biological activity. ATP has been termed as the energy currency of the cell. GTP is the major energy source for protein synthesis; CTP is an essential metabolite in phospholipid synthesis and UTP forms activated intermediates with sugars that go on to serve as substrates in the biosynthesis of complex carbohydrates and polysaccharides. The evolution of metabolism has led to the dedication of one of these four NTPs to each of the major branches of metabolism.

Nucleotides function as regulatory molecules: Some nucleotides have various regulatory roles in cells. Cells respond to environment by receiving signals from hormones or external chemical signals. The interaction of these extracellular chemical signals (first messenger) with receptors on the cell surface often leads to the production of second messengers inside the cell, which in turn lead to adaptive changes in the cell. Often the second messenger is a nucleotide and the most common nucleotide that acts as a second messenger is adenosine

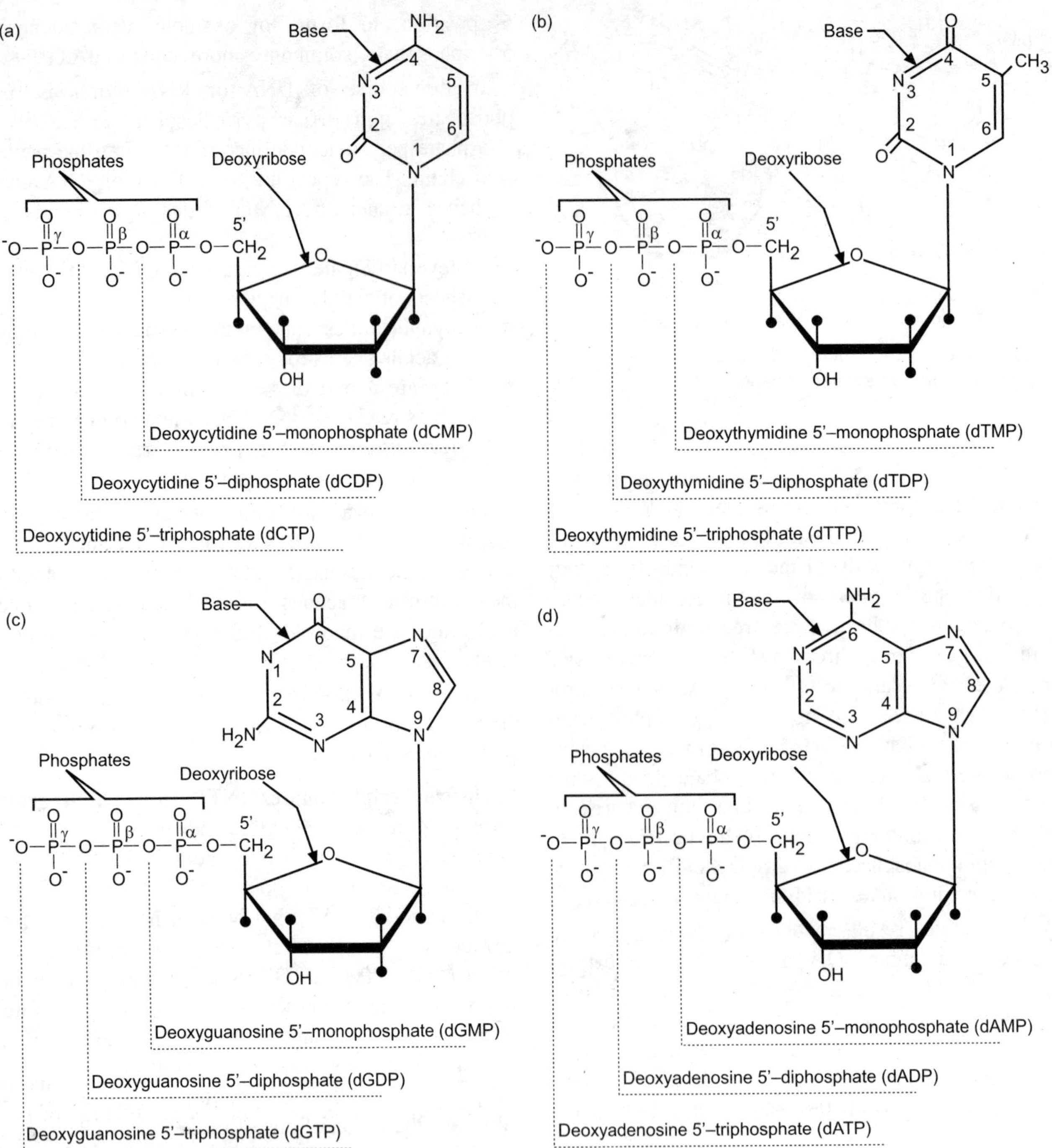

Fig. 5.5 Chemical structures of (a) and (b) Pyrimidine and (c) and (d) Deoxynucleotides

3', 5'-cyclic monophosphate (cyclic AMP or cAMP), a derivative of ATP (**Fig. 5.6a**). Cyclic AMP is formed in a reaction catalysed by adenylate cyclase—an enzyme associated with the inner face of the plasma membrane.

Guanosine 3', 5'-cyclic monophosphate (cGMP) (**Fig. 5.6b**) occurs in many cells and also has regulatory functions. Formation of cGMP occurs in response to receptor-mediated signals similar to those for activation of adenylate cyclase. However, in this case, it is guanylate cyclase that is coupled to the receptor. The most important signal transduction cascade mediated by the coupled cGMP is that of photoreception. The nucleotide **ppGpp** (**Fig 5.6c**) is another regulatory molecule produced in bacteria, in response to the

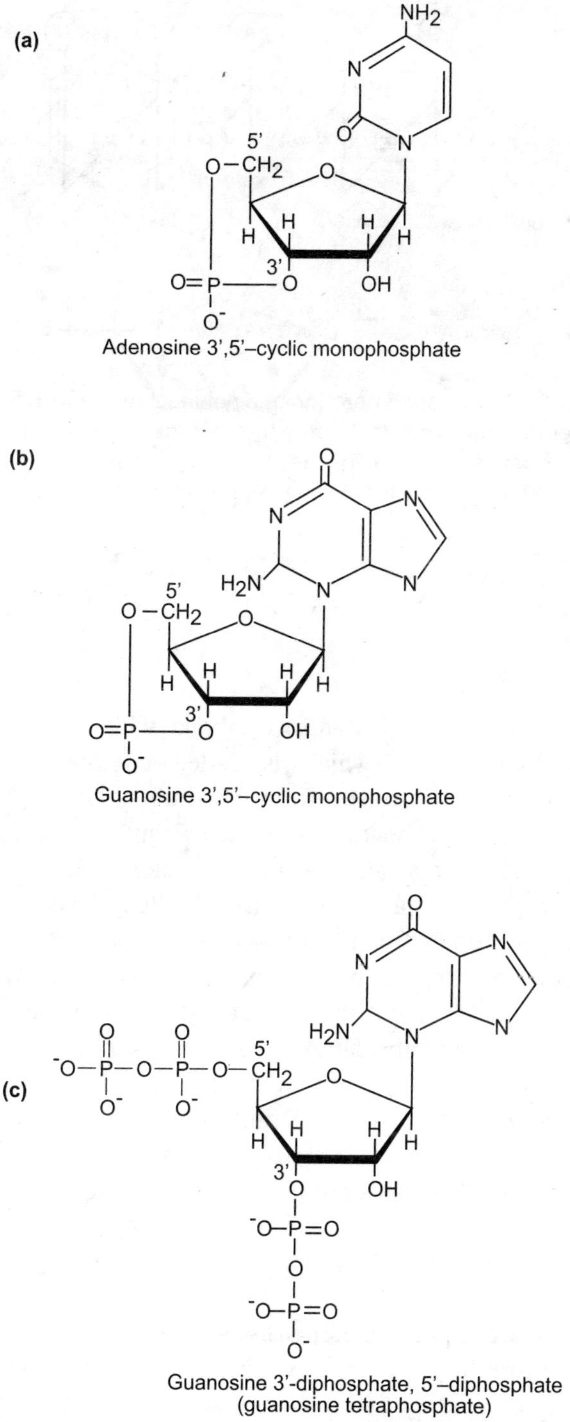

(a)

Adenosine 3',5'–cyclic monophosphate

(b)

Guanosine 3',5'–cyclic monophosphate

(c)

Guanosine 3'-diphosphate, 5'–diphosphate (guanosine tetraphosphate)

Fig. 5.6 The three regulatory nucleotides (a) Cyclic AMP (cAMP) (b) Cyclic GMP (cGMP) (c) Guanosine tetraphosphate (ppGpp)

slowing down of protein synthesis during amino acid starvation. This nucleotide inhibits the synthesis of the rRNA and tRNA molecules needed for protein synthesis, preventing the unnecessary production of nucleic acids.

Synthetic nucleotide analogues

A large number of nucleotide analogues are chemically synthesised. These nucleotide analogues can inhibit specific enzymatic activities and interfere with the synthesis of DNA. They disrupt the normal replication process by interfering with the formation of correct Watson Crick base-pairing as well as terminating the chain. Some of the synthetic analogues and their uses are given below:

- 6-mercaptopurine, 5-fluorouracil, 5-iodo-2'-deoxyuridine and 6-thioguanine are used as chemotherapeutic agents.
- They have been targeted for use as antiviral agents. Several analogues such as AZT (azidothymidine) and ddI (dideoxyinosine) are used to interfere with the replication of HIV.
- Several purine analogues are used to treat gout. The most common is allopurinol, which resembles hypoxanthine.
- Nucleotide analogues are used after organ transplantation in order to suppress the immune system and reduce the likelihood of transplant rejection by the host.

5.3 THE PRIMARY STRUCTURE OF POLYNUCLEOTIDES

Polynucleotides (nucleic acids) are polymers made up of hundreds, thousands or millions of nucleotides coupled together by phosphodiester linkages, i.e., bonds joining 5' and 3' carbons of adjacent nucleotide sugars. They are either polydeoxyribonucleotides (DNA) or polyribonucleotides (RNA). Polynucleotides are formed by the condensation of two or more nucleotides. The condensation occurs between the alcohol and a 5'-phosphate of one nucleotide and the 3'-hydroxyl of a second, with the elimination of H_2O, forming a phosphodiester bond. The primary structure of DNA and RNA proceeds in the 5'→3' direction. The sequence of bases read in the 5'→3' direction constitutes the primary structure of an oligonucleotide or nucleic acid. The common representation of the primary structure of DNA or RNA is to write the nucleotide sequences from left to right synonymous with the 5'→3' direction: 5'-pCpApTpG-3'.

Phosphodiester bonds

Nucleic acids are linear polymers of nucleotides linked 3' to 5' by phosphodiester bridges. They are formed as 5'-nucleoside monophosphates and are joined into a polymer by the covalent linkage of a phosphate group between the 5'-hydroxyl of one sugar molecule and the 3'-hydroxyl of the next (**Fig. 5.7**). This kind of bond or linkage is called a **phosphodiester** bond (since the phosphate is chemically in the form of a **diester**). The backbones of both DNA and RNA are hydrophilic. Any nucleic acid chain of whatever length (unless it is circular) has a free 5'-end, which may or may not have any attached phosphate groups and a free 3'-end, which is most likely to be a free hydroxyl group. At neutral pH, each phosphate group has a single negative charge. This is why nucleic acids are termed acids; they are the anions of strong acids. Nucleic acids are, thus, highly charged polymers.

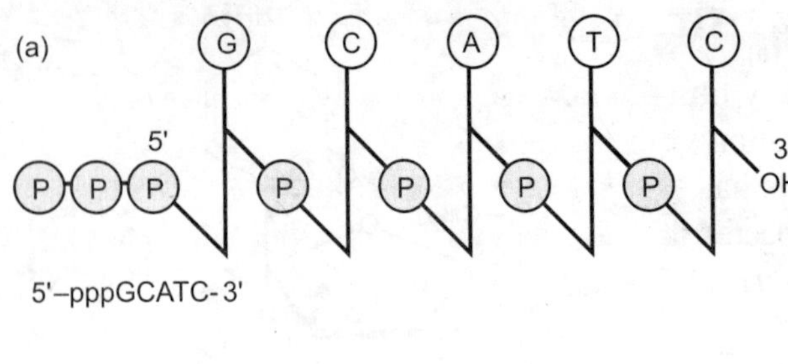

Fig. 5.8 A shortened notation of a polynucleotide with five nucleotides drawn 5' to 3', from left to right. In most cases, nucleic acid sequences are presented in a format that does not include symbols for the sugar–phosphate backbone, as in (c)

Base sequence

The only variation that occurs in the chemical structure of nucleic acids is the nature of the base at each nucleotide position. The order in which the bases follow one another along the chain, called base sequence (or nucleotide sequence) is characteristic of a nucleic acid. These bases are not a part of the sugar–phosphate backbone but serve as distinctive side chains in the polymer. The representation of polynucleotide chains by complete formulae is clumsy and it has become customary to use the schematic system illustrated in **Fig. 5.8**. Conventionally the repeating nucleotides of DNA or RNA are represented by a single letter A, G, C, T, or U.

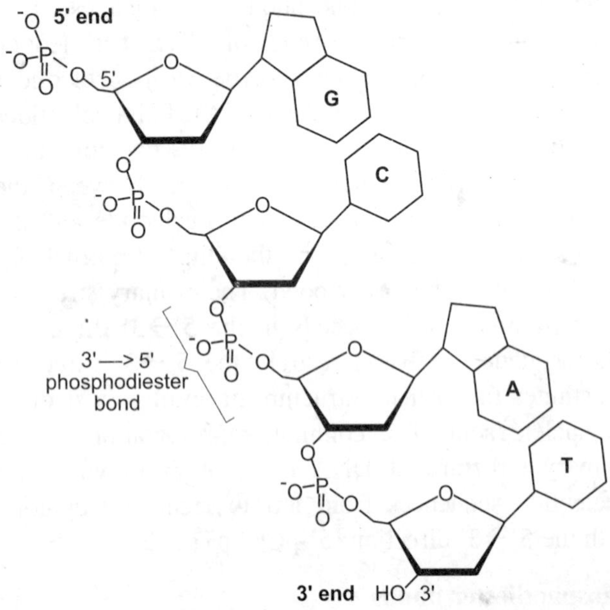

Fig. 5.7 Phosphodiester bonds and the covalent structure of a DNA strand

Base pairing

A base on one strand forms a bond with a base on the other strand, which is called complementary base-pairing. Purines form hydrogen bonds to pyrimidines—A bonding only to T and C bonding only to G. This arrangement of two nucleotides joined together across the double helix is called a **base pair (bp)**. In a double helix, the two strands are also held together by forces generated by the hydrophobic effect and pi stacking, but these forces are not affected by the sequence of the DNA. Since the hydrogen bonds are not covalent, they can be broken and rejoined relatively easily. The two strands of DNA in a double helix can be separated apart like a zipper, either by a mechanical force or high temperature. As a result of these complementarities, all the information in the dsDNA helix is duplicated on each strand, which is vital in DNA replication. This reversible and specific interaction between complementary base pairs is critical for all the functions of DNA in the cell. The bases A and T are held by two hydrogen bonds, whereas G and C pair is held by three hydrogen bonds. The GC base pair, therefore, is stronger than the AT base pair. As a result, it is both the percentage of GC bp and the overall length of a DNA double helix that determine the strength of the association between the two strands of DNA. In the laboratory, the strength of this interaction can be measured by finding the temperature required to break the hydrogen bonds, their melting temperature (called **T_m value**).

5.4 THE SECONDARY STRUCTURE OF DNA

Many observations made on deoxyribonucleic acids in solution suggested that DNA molecules extracted from cells had structural characteristics different from those expected in molecules having only the primary structure described above. This suggests that native DNA has a secondary structure. Based on some of the following facts, Crick and Watson in 1953, proposed a model, which explained the various properties of DNA.

5.4.1 The DNA double helix

Another important discovery that shed more light on the structure of DNA was the result obtained by X-ray diffraction studies of the DNA molecule. X-ray diffraction was used extensively in the study of the molecular architecture of DNA by Astbury (1947), Franklin and Gosling (1953), and later by Wilkins and Alex Stokes, who estimated the diameter and pitch of the helix. The molecules in a crystal are arranged in an orderly way, so that when a beam of X-rays is aimed at the crystal, the beam scatters in an orderly fashion. The scatter pattern can be recorded on photographic film or computer-controlled devices. The nature of this pattern depends upon the structure of the crystal. They showed that DNA produces a characteristic x-ray diffraction pattern, which shows the helical nature of DNA molecule. Using the information obtained by X-ray diffraction studies and chemical observations, Watson and Crick (1953) deduced the basic features of the three-dimensional structure of the DNA molecule (**DNA double helix model**). Using scale models, they were able to show that the two strands form the familiar double helix with the flat base pairs and stacked between intertwined sugar–phosphate chains (**Fig. 5.9a**), as one above the other like a pile of plates that resembles a spiral staircase. The negatively charged sugar–phosphate backbones of the molecules are on the outside, and the planar bases of each strand stack one above the other in the centre of the helix. Between the backbone strands run the **major** and **minor grooves**, which also follow a helical path. The two strands are joined non-covalently by hydrogen bonding between the bases to form base pairs. There are around 10 base pairs per turn in the form of a right-handed helix with two polynucleotide chains wound around the same axis (**Fig. 5.9b**). Another feature of DNA structure is the existence of polarity in each strand, that is, one end of a DNA strand has a 5' phosphate and the other end has a 3' hydroxyl group. The two strands are oriented in opposite directions (**antiparallel**) in terms of their 5'→3' direction and, most crucially, the two strands are **complementary** (but not identical) in terms of sequence.

The structures of the bases and the constraints of the DNA backbone dictate that the bases form hydrogen bonds with each other as purine–pyrimidine pairs, which have very similar geometry and dimensions (**Fig. 5.10**). Guanine pairs with cytosine (by three H bonds) and adenine pairs with thymine (by two H bonds). Hence, any sequence can be accommodated within a regular dsDNA structure. The sequence of one strand uniquely specifies the sequence of the other. This has implications for the mechanism of copying (replication) of DNA and the transcription of a DNA sequence into RNA. The double helical structure of DNA is very stable, and *in vivo* the DNA strands only unwind and separate in a localised manner during replication and transcription.

5.4.2 The dynamic structure of DNA

An additional and separate phenomenon is the dynamic secondary structure of DNA. The double helix is not a totally fixed or rigid molecule but undergoes considerable internal deformation in a continuous manner. This aspect of DNA secondary structure is best studied by ^{3}H-exchange experiments and NMR results. The ^{3}H-exchange requires that the base pair is opened and the probability of finding a particular base pair open is only 10^{-5}, but each base pair will breathe about 100 times a second. A number of studies indicated that small segments of the double helix can swing apart ('breathe') and protrude into the external medium in a manner closely dependent on the environment of the DNA molecule and supercoiled molecules show an enhanced tendency to form single-stranded regions. The breathing of a base pair is determined by its stability of and its nearest neighbour and sequences.

5.4.3 Alternate forms of DNA double helix

Depending on the conditions chosen to produce the DNA fibres, they can have a variety of possible secondary structures. Considerable rotation is possible around a number of bonds in the sugar–phosphate backbone, and thermal fluctuation can produce bending, stretching and unpairing of the strands. Many significant deviations from the Watson–Crick DNA structures (B-form) are found in cellular DNA,

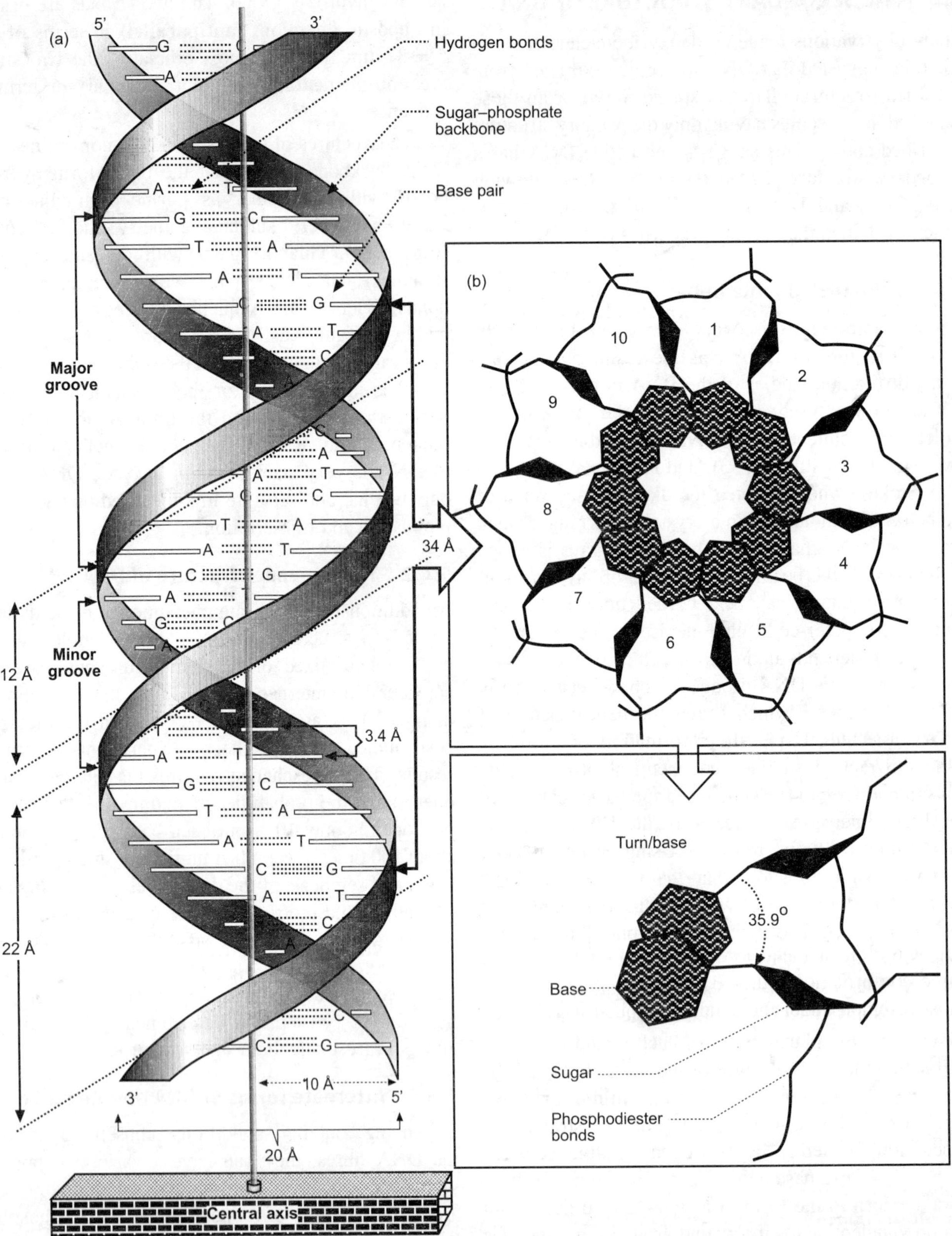

Fig. 5.9 A schematic view of the three-dimensional structure of DNA double helix (a) Geometry of the helix (b) Diagram of one of the strands of the helix, viewed down the central axis

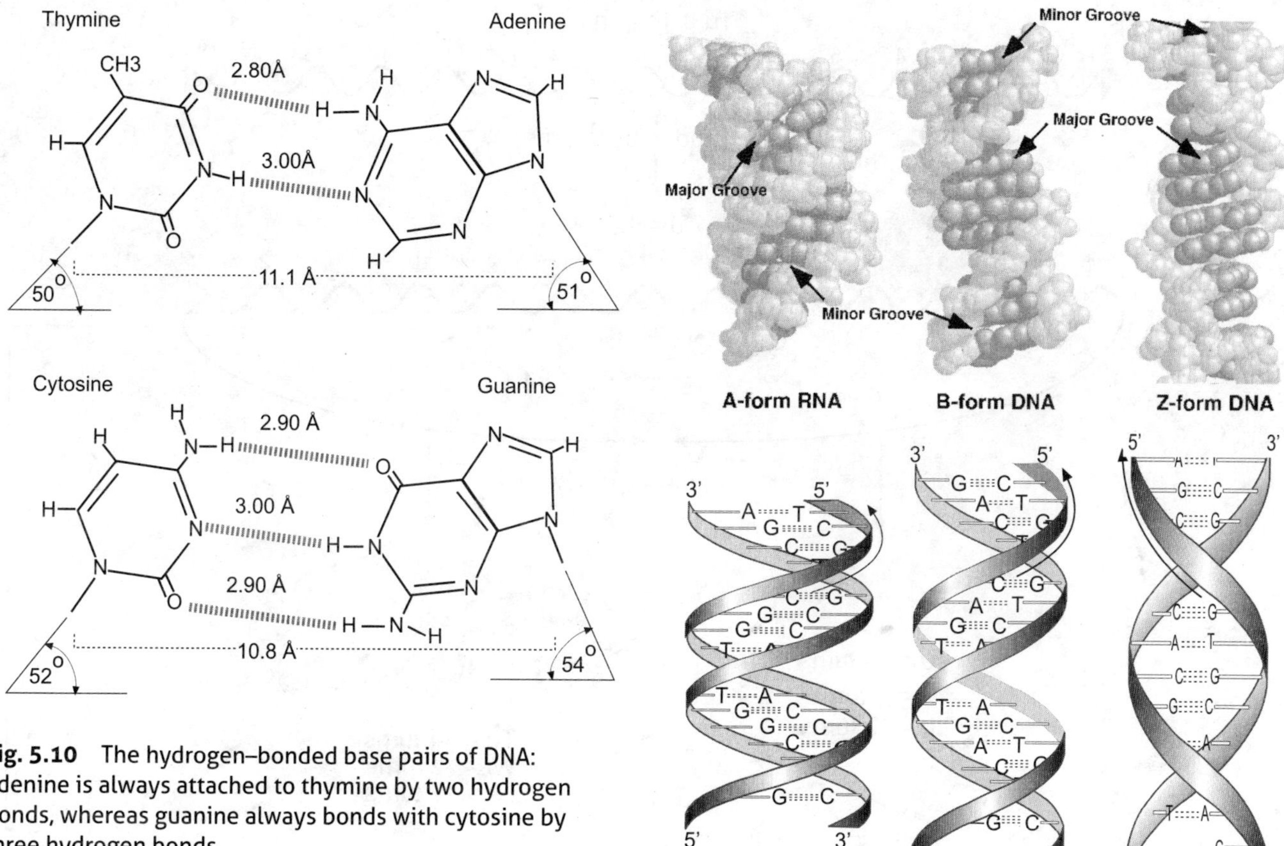

Fig. 5.10 The hydrogen–bonded base pairs of DNA: adenine is always attached to thymine by two hydrogen bonds, whereas guanine always bonds with cytosine by three hydrogen bonds

Fig. 5.11 The A-, B- and Z-forms of the DNA double helix; the alternative pyrimidine–purine sequence of Z–DNA is the key to the 'left–handedness' of the Z-helix

and some or all of these may play important roles in DNA metabolism. The conformations so far identified are: A-DNA, B-DNA, C-DNA, D-DNA, E-DNA, H-DNA, L-DNA, Z-DNA and P-forms. However, only A-DNA, B-DNA and Z-DNA (**Fig. 5.11**) are found to exist in nature. The conformation adopted by the DNA depends on the sequence of the DNA, the amount and direction of supercoiling, chemical modifications of the bases and also solution conditions, such as the concentration of metal ions and polyamines. Of these three conformations, the 'B' form described above is most common under the conditions found in cells. The other two alternative double-helical forms of DNA differ in their geometry and dimensions.

DNA Topology

Topological characteristics of DNA and specifically DNA supercoiling, influence all major DNA transactions in living cells. DNA supercoiling induces the formation of unusual secondary structure by specific DNA repeats which can also affect DNA functioning. DNA supercoiling refers to the over- or under-winding of a DNA strand, and is an expression of the strain on that strand (**Fig. 5.12**). Supercoiling is important in a number of biological processes, such as compacting DNA. Additionally, certain enzymes such as topoisomerases are able to change DNA topology to facilitate functions such as DNA replication or transcription. Mathematical expressions are used to describe supercoiling by comparing different coiled states to the relaxed B-form DNA. As a general rule, the DNA of most organisms is negatively supercoiled.

5.5 THE TERTIARY STRUCTURE OF NUCLEIC ACIDS

The tertiary structures of nucleic acids reflect interactions that contribute to the three-dimensional shape. This includes interactions between different single strands and secondary structure elements and topological properties of nucleic acids. In DNA, tertiary interactions involve single strands interacting with duplexes or among duplexes, resulting in the

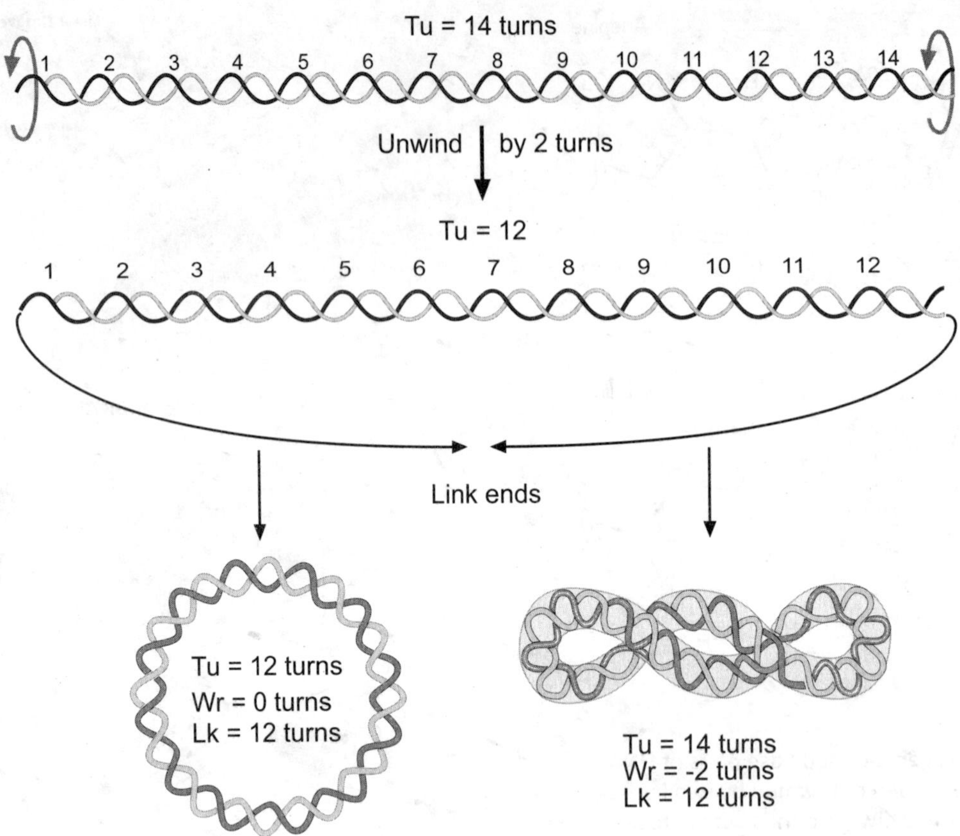

Fig. 5.12 DNA super coiling

formation of triple and quadruple strand structures. Guanine can form **base tetrads** (**Fig. 5.13**), and DNA-containing runs of guanosine residues can form quadruplex structures, which may be the precursors to telomeres. If a chromosome lacked telomeres, it would become shorter each time it was replicated. These specialised chromosome quadruplex structures protect the DNA ends from exonucleases and stop the DNA repair systems in the cell from treating them as damage to be corrected (see **Chapter 8**). These structures are often stabilised by chelation of a metal ion in the centre of each four-base unit. In addition to these stacked structures, telomeres also form large loop structures called **telomere loops** or **T-loops**. Here, the single-stranded DNA curls around in a circle stabilised by telomere-binding proteins.

Quaternary structure and other properties: In many structures, nucleic acids interact in *trans* (i.e., the ribosome and spliceosome), and this may be considered a quaternary level of nucleic acid structure. Nucleic acids also interact with a large number of proteins (e.g., genome structural proteins, transcription factors, enzymes, splicing factors). Many of these proteins have a significant effect on DNA or RNA conformation.

5.6 MODIFICATIONS OF THE REGULATORY BASES

The expression of genes is influenced by modifications of the bases in DNA. In humans, the most common base modification is cytosine methylation to produce 5-methylcytosine. This modification reduces gene expression and is important in X-chromosome inactivation. The level of methylation varies between organisms, with *C. elegans* lacking cytosine methylation, while vertebrates show high levels, with up to 1% of their DNA being 5-methylcytosine. The 5-methylcytosine, abnormally, due to spontaneous deamination of 5-methylcytosine converts into thymine. Therefore, methylated cytosines can be mutation hotspots. Other base modifications include adenine methylation in bacteria and the glycosylation of uracil to produce the 'J-base' in kinetoplastids. Various chemical mutagens can modify the bases in different

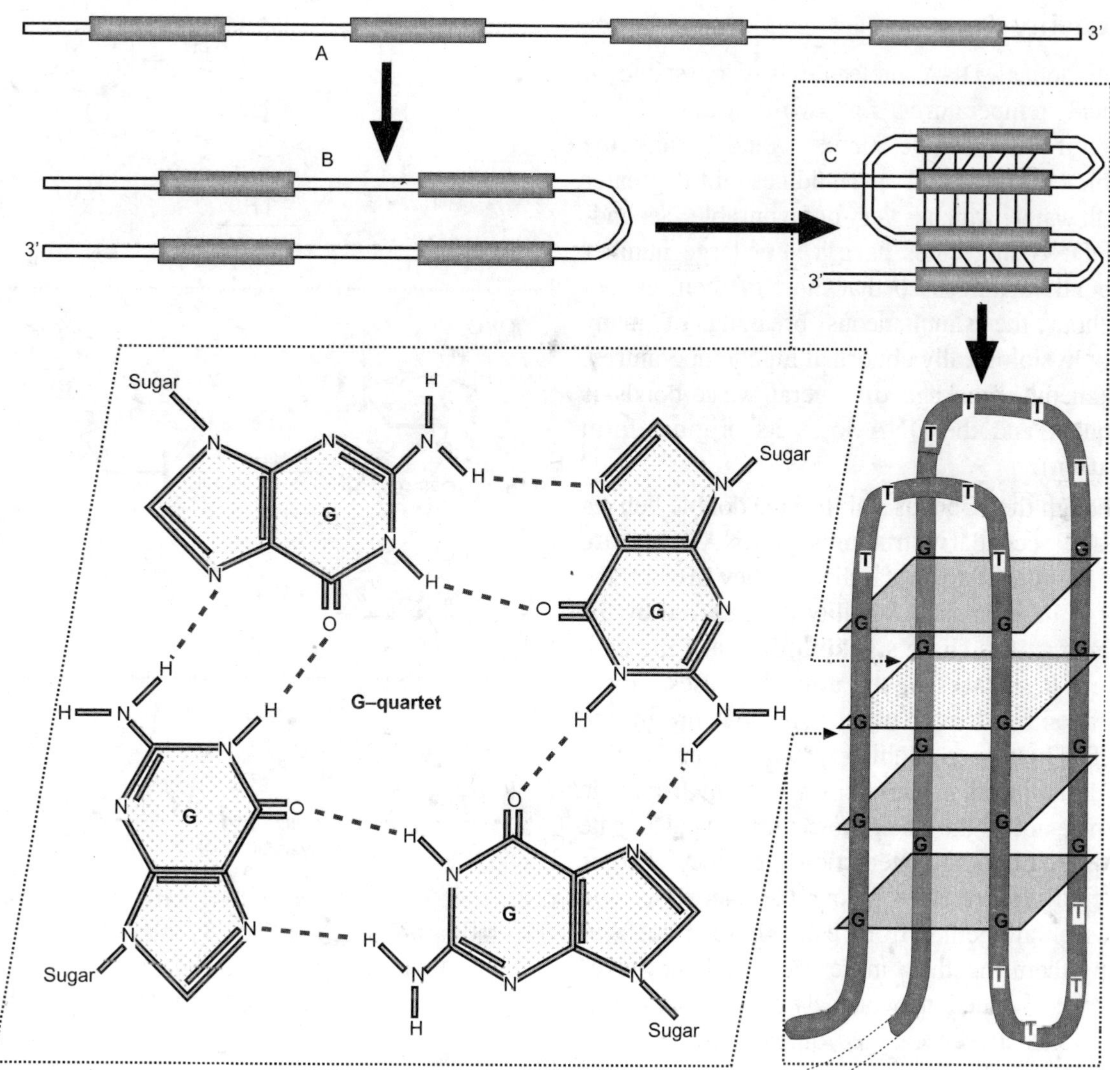

Fig. 5.13 A model of how the G-rich strand could fold upon itself to form two or four-stranded structures; in step A, repetitive sequences replicate a number of times. In step B, the DNA strand folds to form loop, and step C, the first loop again folds to form four-stranded structure. The G–quartet structure formed by hydrogen bonding between four guanine bases present in a single DNA strand folded back upon itself is shown at two levels.

ways (see Chapter **Molecular mutagenesis and DNA repair**).

Interaction with proteins

Most functions of the DNA depend on interaction with proteins. These protein interactions can be specific or otherwise, where the protein can only bind to a particular DNA sequence. Enzymes can also bind to DNA and of these, the polymerases that copy the DNA base sequence in transcription (see **Chapter 9**) and DNA replication (see **Chapter 8**) are particularly important. A distinct group of DNA-binding proteins are the single-stranded DNA-binding (SSB) proteins that specifically bind single-stranded DNA. The binding of SSB proteins seems to stabilise ssDNA and protect it from forming stem loops or being degraded by nucleases. Various classes of transcription factors are unique proteins that specifically bind particular DNA sequences. These proteins control gene expression (see Chapters **14** and **15**).

5.7 PHYSICAL AND CHEMICAL PROPERTIES OF NUCLEIC ACIDS

The major physical and chemical properties are discussed in the following sections.

5.7.1 Stability

The double-helical DNA molecule is very stable at physiological temperatures for two reasons. First, disruption of the double helix would bring the hydrophobic purines and pyrimidines into greater contact with water, which is very unfavourable. Second, individual DNA molecules have a very large number of weak bonds arranged so that most of them cannot break without the simultaneous breaking of many others. At physiologically abnormal high temperatures, the simultaneous breakage of several weak bonds is more frequent and the DNA loses its original form (**denaturation**).

Even though the H-bonds stabilise the double helixes of DNA and secondary structures of RNA, they are not solely responsible for stability as they are for the specificity of base pairing. Stability is conferred by the stacking interactions (**base stacking**) between the base pairs. The flat surfaces of the aromatic bases cannot form hydrogen bonds with water when they are in free solution. The hydrogen bonding network of bulk water becomes destabilised in the vicinity of a hydrophobic surface, since not all the water molecules can participate in full hydrogen bonding interactions, and they become more ordered. Hence, it is favourable energy-wise to exclude water altogether from pairs of such surfaces by stacking them together; more water ends up in the bulk hydrogen-bonded network. This also maximises the interaction between charged dipoles on the bases. Even in ssDNA, the bases have a tendency to stack on top of each other, but this stacking is maximised in dsDNA, and the hydrophobic effect ensures that this is the most energy-wise favourable arrangement. The charged phosphate groups on the exterior of the molecule are able to react with the dipoles of water (**electrostatic interactions**), further contributing to the stability of the DNA duplex.

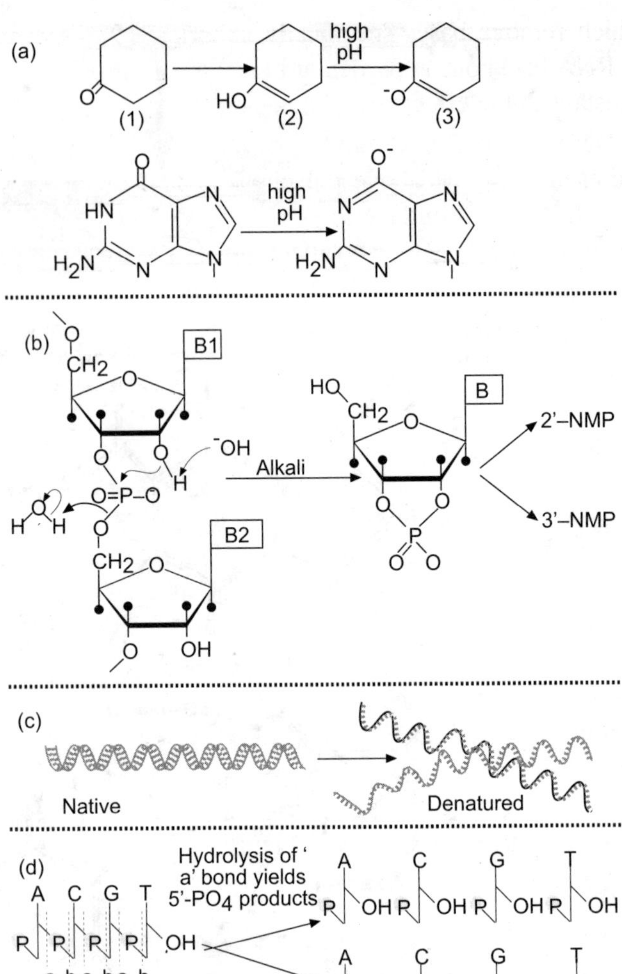

Fig. 5.14 The denaturation of DNA at high pH (a) alkali shifts the tautomeric ratio to the enolate form (b) the tautomeric shift of deoxyguanosine (c) the denaturation of double-helical DNA (d) cleavage in polynucleotide chains: 'a' cleavage yields 5'-phosphate products, whereas 'b' cleavage gives 3' phosphate products

5.7.2 Hydrolysis

Hydrolysis of nucleic acids is affected by the following factors.

Effect of acid and alkali on nucleic acids

The majority of reactions that lead to hydrolysis of nucleic acids break the bonds in the polynucleotide backbone. Such reactions are important because they can be used to manipulate these molecules for analysis of nucleic acids (**Fig. 5.14**).

In strong acids and at elevated temperatures, for example, perchloric acid ($HClO_4$) at more than 100°C, both DNA and RNA (except tRNA) are hydrolysed completely to their constituents (bases, pentose sugar and phosphate). In more dilute mineral acid, for example at pH 3–4, the most easily hydrolysed bonds are selectively broken. Gentle treatment of DNA with 1 mM HCl leads to selective removal of its purine bases by cleavage of the purine glycoside bonds. More complex chemical treatments have been developed which remove bases specifically and cleave the DNA or RNA backbone at particular bases. RNA is relatively resistant to the effects of dilute acids.

DNA is not susceptible to alkaline hydrolysis. Increasing the pH above the physiological range (pH 7–8) has more subtle effects on DNA structure. On the other hand, RNA is alkali-labile and is readily hydrolysed by dilute NaOH. Cleavage is a random process in RNA and the ultimate product will be a mixture of nucleosides. Even at neutral pH, RNA is much more susceptible to hydrolysis than DNA.

Nucleases

The enzymes that act on nucleic acids are called nucleases. They hydrolyse nucleic acids. They exhibit chemical specificity and are classified as either **deoxyribonucleases (DNases)** or **ribonucleases (RNase)**. DNases act either only on single-stranded DNA or only on double-stranded or on both. Further some nucleases act only at the end of a nucleic acid, removing a single nucleotide at a time. These are called **exonucleases**, and may also be specific for the 3' or 5' end of the strand (**Fig. 5.15**). Many nucleases act within the strand and are called **endonucleases**.

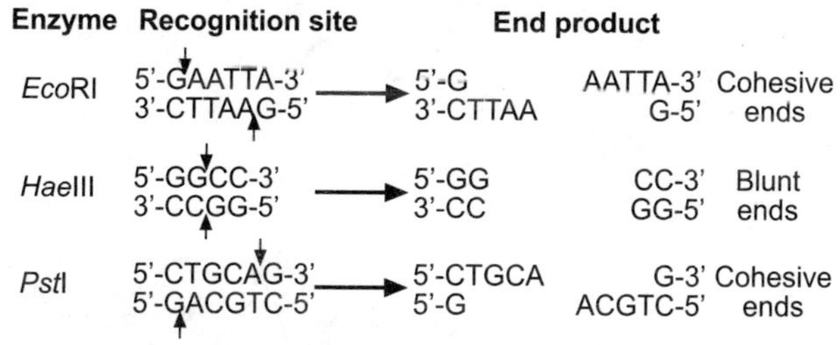

Fig. 5.15 Restriction endonucleases; a few commonly used restriction endonucleases and source organisms, extracted respectively from *E. coli* (EcoRI), *Haemophilus aegyptus* (HaeIII) and *Providencia stuarti* (PstI), as well as the places where the cleavage takes place are indicated by arrows

Viscosity and density

The physical properties of DNA are strongly influenced by many factors. Genomic DNA is very thin, long and has a high axial ratio and is around 2 nm in diameter, and may have a length ranging from μm (viruses) to several centimetres (in eukaryotes). DNA is a relatively stiff molecule and therefore, DNA solutions have a high viscosity.

5.8 OPTICAL AND THERMAL PROPERTIES OF NUCLEIC ACIDS

The important optical and thermal properties of nucleic acids are discussed in the following sections.

5.8.1 Ultraviolet absorption

The density and nature of biomolecules in solution can be estimated by their inherent thermal and optical properties. Nucleic acids absorb ultraviolet (UV) light due to the conjugated aromatic nature of the bases (purine and pyrimidine rings). The sugar–phosphate backbone does not contribute appreciably to absorption. The wavelength of maximum absorption of light by both DNA and RNA is 260 nm (λ_{max}), which is conveniently distinct from the λ_{max} of protein (280 nm). By using these optical properties, quantity, purity and nature of nucleic acids can be measured.

Hyperchromacity: Even though the λ_{max} for DNA or RNA bases is constant, the extinction coefficient depends on the environment of the bases. The absorption at 260 nm (A260) is greatest for isolated nucleotides, intermediate for ssDNA or RNA and least for dsDNA. This effect may be due to the fixing of the bases in a hydrophobic environment by stacking. The classical term for this change in absorbance is hyperchromacity. Double-stranded DNA is hypochromic (optically less coloured) relative to ssDNA. Alternatively, ssDNA may be said to be hyperchromic when compared with dsDNA.

5.8.2 Denaturation

The double-stranded helical structure of DNA is maintained by hydrogen bonds between the bases of complementary pairs. High temperature, high pH and a number of chemical agents can break hydrogen bonds and bring about separation of the strands. When the helical structure of DNA is disrupted and the strands are separated, the DNA molecules are said to be denatured. The reverse process of denaturation is renaturation, in which the denatured DNA will re-form the duplex structure if the denaturing conditions are removed. Renaturation is dependent on both concentration and time (**Fig. 5.16**). The process of denaturation and renaturation can be observed conveniently by the increase in hyperchromacity as double-stranded nucleic acids are converted to single strands or vice versa.

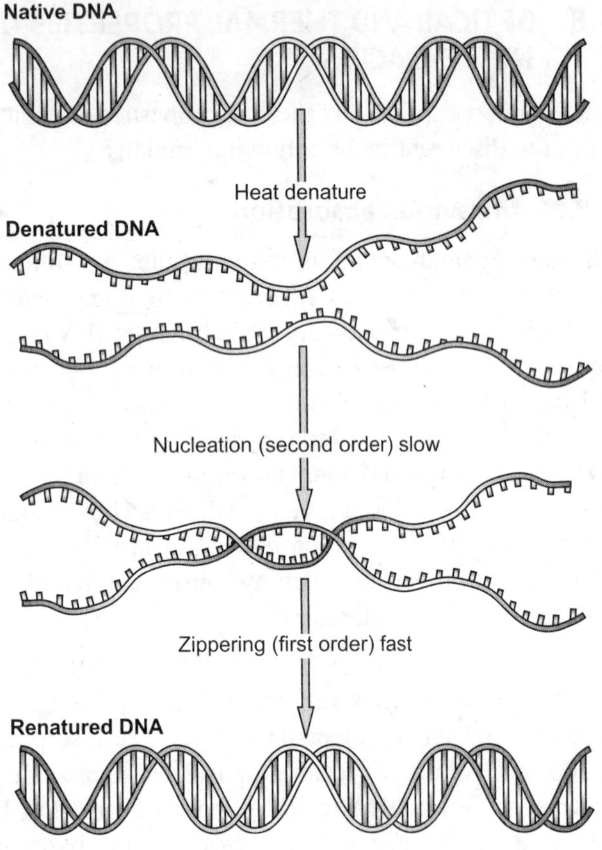

Fig. 5.16 Steps in the thermal denaturation and renaturation of DNA. The nucleation phase of the reaction is a second-order process depending on sequence alignment of the two strands. This process takes place slowly because it takes time for complementary sequences to encounter one another in solution and then align themselves in register. Once the sequences are aligned, the strands zipper up quickly

Thermal denaturation

The thermal behaviour of dsDNA and RNA are very different. When duplex DNA melts, the hydrogen bonds (non-covalent forces) break and the bases unstack with the consequence that the absorption at 260 nm rises by about 20%–30%. As the temperature is increased, the absorbance of an RNA sample gradually and erratically increases as the stacking of the bases in double-stranded regions reduces. Shorter regions of base-pairing will denature since they are more thermally mobile. In contrast, the thermal denaturation, or melting of dsDNA is cooperative. The denaturation of the ends of the molecule and of the more mobile AT-rich internal regions, will destabilise adjacent regions of the helix. This leads to a progressive and concerted melting of the whole structure at a well-defined temperature corresponding to midpoint of the smooth transition, which is known as the **melting temperature (T_m)**. Originally, denaturation was always accomplished by heating a DNA solution; so a graph of a varying property as a function of temperature is called a **melting curve** (**Fig. 5.17**). In the case of DNA, the simplest way to detect denaturation is to monitor the ability of DNA in a solution to absorb UV-light whose wavelength is 260 nm (bases absorb maximum at 260 nm light strongly).

Chemical denaturation

Heating is not the only way to denature DNA. Chemical agents that destabilise hydrogen bonds, such as alkaline solutions, concentrated solutions of formamide, urea, dimethyl sulphoxide (DMSO), or high pH are also denaturants. Most common ones are urea (H_2NCONH_2) and formamide ($HCONH_2$).

5.8.3 Renaturation

Renaturation (reuniting the separated DNA strands) of DNA requires base pair matching. Separated strands of DNA will re-associate when appropriate conditions (mainly temperature and salt concentrations) are achieved. Renaturation has proved to be a valuable tool in molecular biology because it helps in determining the genetic relatedness between different organisms.

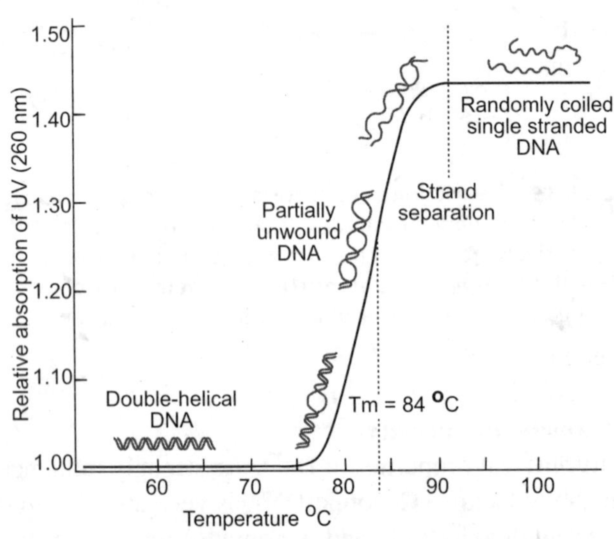

Fig. 5.17 A melting curve of DNA showing T_m and possible molecular conformations for various degrees of melting. This temperature curve is of T4 DNA showing marked change in UV absorption that occurs at approximately 84°C when the DNA changes from double helical form to single strands.

Requirements for renaturation

The parameters that affect the degree to which ssDNA re-associates correctly are given below.

- The salt concentration must be high (>0.25 M) to neutralise the negative charges of the phosphate groups, which would otherwise cause the complementary strands to repel each other.
- The temperature must be high enough to disrupt hydrogen bonds that form at random between short sequences of bases within the same strand, but not too high. The optimal temperature for renaturation is 20°C–25°C below the value of T_m. A temperature about 20 °C below T_m is usually optimal.
- The concentration of DNA in the solution is also important.

The longer the time allowed, more annealing will occur.

5.9 SPECIAL FEATURES OF DNA

Classical experiments of Mendel and others have showed that traits inherit in an unaltered way as they pass from parent to offspring. This implies that living organisms contain a genetic material that governs an individual's traits which has transferred during the process of reproduction. To fulfil its role, the genetic material must meet several criteria.

- The genetic material must contain the information necessary to construct an entire organism.
- It should replicate accurately during cell growth and duplication.
- Its structure must be sufficiently stable so that the heritable changes (mutations) occur only very rarely.
- It should have the potential to carry out all the necessary biological activities.
- It should transmit information to every cell.

DNA fulfills all these requirements completely and is certainly a most suitable candidate for the role of genetic material. In early 1900s, other geneticists consistently reported the above properties in DNA. It has many structural, physical and chemical properties that make it highly suitable as genetic material.

The absence of an enslaved nucleophile (the 2'-hydroxyl group) at each phosphodiester linkage makes DNA ~100,000-fold more stable than RNA under physiological conditions. RNA is vulnerable to 'spontaneous' hydrolysis. The 2'-OH group on RNA has the ability to catalyse cleavage of the adjacent phosphoester bond, and this reaction is accelerated by extreme conditions (e.g., high pH, high temperatures, and the presence of certain divalent cations such as Zn^{2+}). Similarly, DNA phosphodiester bonds are 100-fold more resistant to hydrolytic degradation than the peptide bonds of proteins.

DNA is double stranded and the complementary base pairing of nucleotides makes it stable and an ideal molecule for storage and transfer of information.

Complementary base pairing between two strands of DNA prevents folding into secondary structures (hairpins, cruciforms, loops) and formation of intricate active sites that may have self-cleaving (autocatalytic) activities. Complementary base pairing between purine and pyrimidine (A:T and G:C) makes DNA linear and uniform all along the molecule. Therefore, it facilitates proper organisation and efficient packing of DNA in the nucleus.

The double-stranded nature of DNA is also very useful in DNA repair. When nucleotides are damaged or patches of DNA sequences in a strand are lost, DNA repair enzymes use the other complementary strand as template for repairing the gaps in a damaged strand; therefore, accurate repair is possible.

Since DNA contains thymidine nucleotide instead of uracil as seen in RNA, it helps repair enzymes to identify precisely the uracil nucleotides, which are formed by the process of deamination. If DNA had uracil instead of thymine like RNA, the repair enzyme would fail to discriminate between uracil formed by deamination and the normal ones, which would increase mutations (C→A transversion mutations).

5.10 STRUCTURE AND FUNCTIONS OF RNA

Most genetic systems studied to date contain genes made of DNA. But some viruses, including several phages, plant and animal viruses and all viroids have RNA genes. In addition to genomic RNA, almost all cells use RNA as a structural molecule involved in many functions. For each function, a specific type of RNA is required and each differs in length of chain and secondary and tertiary structures. Cells contain up to eight times more of RNA when compared to that DNA. The structure of RNA is similar to that of DNA but there are a number of important differences. As mentioned before, in RNA, ribose replaces 2'-deoxyribose and the base thymine is replaced by another base, uracil. In addition, RNA molecules normally exist as a single polynucleotide strand and do not form a double helix.

Major classes of RNA

RNA has a number of important biological functions, and on this basis, it is classifed into the following types: messenger RNA (**mRNA**), ribosomal RNA (**rRNA**), and transfer RNA (**tRNA**). Eukaryotic cells contain an additional minor type RNA, e.g., small nuclear RNA (snRNA).

5.11 EVIDENCE OF DNA AS THE CARRIER OF GENETIC INFORMATION

Biological organisms need to store and transfer information. The genetic instructions for building and running a cell need to be retained intact, used and passed on to future cells at cell division. The information molecules of the cell are the nucleic acids. In almost all organisms, the nucleic acids play the role of genetic material and contain the information that specifies the inherited characteristics of each. In the following sections, we see how experimental evidence supports this hypothesis.

Search for genetic material

By the beginning of the 20th century, scientists began speculating as to what kind of molecules could be stable and heritable permanently, yet be capable of permanent sudden change (mutant forms) that provide the basis for evolution. Until the 1940s, there appeared to be no direct way to attack the chemical essence of the gene. Some early studies in microorganisms played an important role in developing an experimental strategy to identify the genetic material. In the early 1940s, research on *Neurospora*, spearheaded by Beadle and Tatum, generated evidence to support the hypothesis of Archibald E Garrod that genes work by controlling the synthesis of specific enzymes (the one gene–one enzyme hypothesis).

How does genetic information depend on DNA?

In higher organisms, the quantity of DNA contained in the nuclei of diploid, somatic cells (having $2n$ chromosomes), is twice that contained in the nuclei of haploid, germinal cells, called the gametes (having n chromosomes). A correlation can be observed between the genetic relationship of certain species and the analogy of base composition of their deoxyribonucleic acids or even the degree of homology of the nucleotide sequences (measured by hybridisation technique).

5.11.1 Transformation experiments in bacteria

The development of our understanding that DNA is the genetic material, began with an observation in pathogenic bacteria responsible for causing human pneumonia. The bacteria *Streptococcus pneumoniae* (or *Diplococcus pneumoniae* or *Pneumococcus*) which cause certain types of pneumonia, occurs in two major forms. The first is **smooth (S)**, whose cells secrete a covering capsule of polysaccharide materials, causing its colonies on agar to be smooth and shiny. The smooth (S) form is virulent and produces septicemia in mice. The second type is **rough (R)**, lacks a capsule and the agar colonies have a rough, rather dull surface. This form is non-virulent. The capsules are classified as types I, II, III and IV and the polysaccharides on each type is unique and elicits formation of antibodies specific to it. When cultured separately, both R and S colonies transmit their respective characteristics to the offspring except for rare mutations. Whatever the type of bacteria, it is highly sensitive to heat and if the temperature is raised sufficiently high, it is killed and can no longer divide.

Griffith, in 1928, in one of the seminal sets of experiments showed that heat-killed bacteria on one type could have a hereditary influence on bacteria of another type. In one experiment, he injected dead bacteria of IIIS type into mice with live bacteria that ordinarily produced non-virulent rough colonies (type IIR) and obtained virulent live cultures which were of the IIIS variety. Since the injection of heat-killed type IIIS bacteria on its own could not result in any live bacterial cultures, type II could have mutated or the transformation into type IIIS must have occurred through the transfer of some active substance (**Fig. 5.18**). Work in the next 15 years showed that Griffith's results could be duplicated by mixing different types of heat-killed and live bacterial strains in mice (*in vivo*) as well as by mixing them in test tubes (*in vitro*).

DNA as the transforming principle

In 1945, Oswald Avery, Colin MacLeod and Maclyn McCarty showed that DNA is indeed the transforming principle. They sequentially purified the cell-free extract, and discovered that removal of protein, polysaccharides, lipids, or RNA from the extracts did

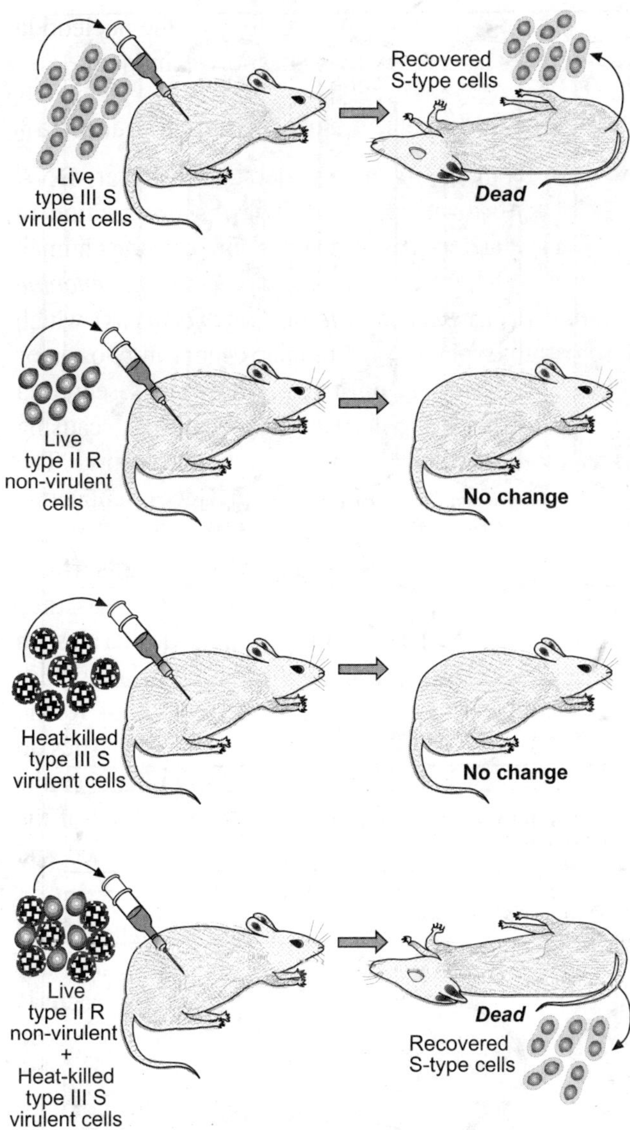

Fig. 5.18 Griffith's transformation experiment with Streptococcus. S-type cells will kill mice but heat-killed S type do not. When heat-killed S type cells are injected with live non-pathogenic R-type cells, S-type cells are recovered from dead mice

not significantly diminish their ability to bring about transformation. First, they removed the protein from the extract with organic solvents. By the end of the purification procedure, they had obtained a relatively pure, viscous material that had the physical and chemical properties of DNA. This 'purified DNA' was then added to a live culture of R cells. When the cells were plated on agar and allowed to form colonies, some of the colonies that grew were S type (**Fig. 5.19**). Next, they subjected it to digestion with various enzymes. The proteases, trypsin and chymotrypsin, which destroy protein, had no effect on transformation. Neither did ribonuclease digestion, which degrades RNA. However, when they digested with enzyme deoxyribonuclease (DNase), which cleaves DNA, the virulent cell extract lost its transforming ability.

Since 1944 numerous transforming experiments have been carried out with a variety of bacteria as well as with yeasts and some tissue-cultured cells of mammals. In all cases, the properties of the transforming principle have been found to be those of DNA. Also, labelling of transforming DNA with radioactive phosphorus, results in the incorporation of this radioactivity into recipient bacteria in direct proportion to the number of transformants. The pure DNA nature of the transforming principle in these experiments thus provided strong evidence that DNA is the genetic material.

RNA as genetic material

Tobacco mosaic virus (TMV) is one of the many viruses using RNA instead of DNA as its sole nucleic acid. Structurally, RNA is enclosed in a protein coat (envelope) that aids in the infection of plant cells and protects the RNA from enzymes such as ribonuclease. The protein is specific for a particular strain, and there are different TMV strains with different protein envelopes. Through biochemical methods, Fraenkel-Conrat and others (1955) found that the protein could be separated from the RNA, but RNA or protein by itself was not infective. Only when the protein recombined ('reconstituted') with isolated RNA was there viral infectivity and new viral particles were produced. The question was whether chemically isolated TMV RNA could be an infective agent on its own and thereby produce new viral particles. Experiments by Gierer and Schran in 1956 showed that, this was indeed the case. Pure TMV RNA rubbed into tobacco leaves caused the synthesis of new TMV virus and, as expected, such synthesis could be prevented by digestion of RNA by ribonuclease. Furthermore, experiments by Fraenkel-Conrat and Singer showed that reconstituted TMV, containing RNA from one strain and protein from another, always led to the synthesis of viral particles whose protein was determined by the parental RNA rather than the parental protein (**Fig. 5.20**).

Although this evidence related to the hereditary function of RNA, it also confirmed that RNA is a nucleic acid rather than protein.

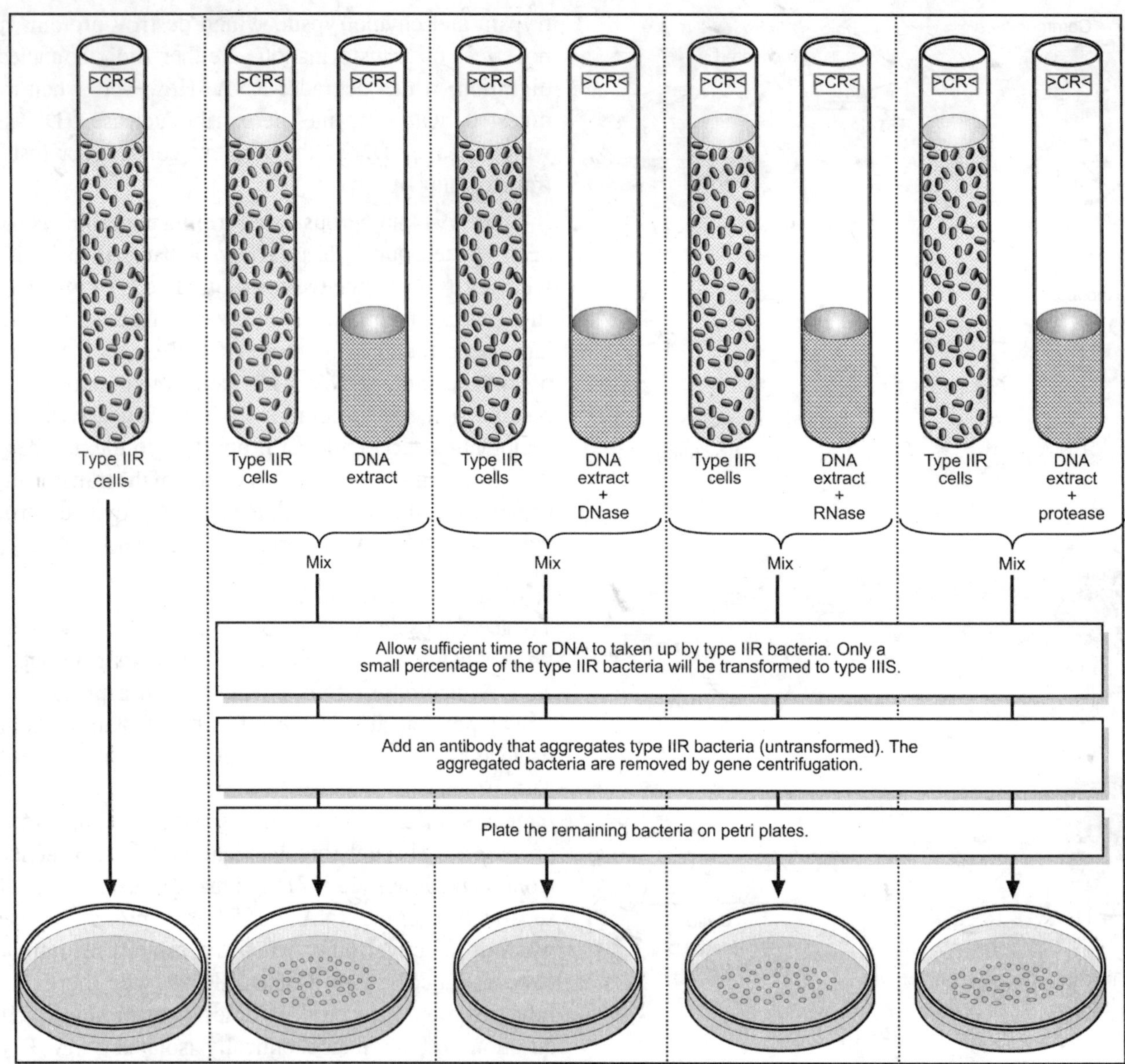

Fig. 5.19 Experimental protocol used by Avery MacLeod and McCarty to identify the transforming principle

5.12 THE CENTRAL DOGMA

The above experiment points to the existence of a genetic material, nucleic acid that is transmitted between generations and found in all living organisms. As will be discussed in the next chapters, molecular studies that analyse the replication of nucleic acids support very strongly the structure proposed by Watson and Crick. However, in addition to its own physical replication and transmission, this material also has significant effects on the biological activity of the cell, and therefore on the structural and functional traits of the organism. In fact, an understanding of the exact relationship between genetic material, its cellular products (proteins), and its consequent effects, only began to develop once molecular studies could be combined with a genetic analysis of how biological traits are transmitted.

It may be sufficient to point out at this stage that central to our understanding of the relationship between genetic material and protein is the assumption that the 3D structure of a protein (i.e., its form, shape and subsequent function) is determined by the linear sequence of amino acids constituting it. This linear sequence of bases in a nucleic acid formed by means of a protein-synthesising apparatus requires three different kinds of RNA. In brief, the genetic material, through the process of transcription produces a molecule of

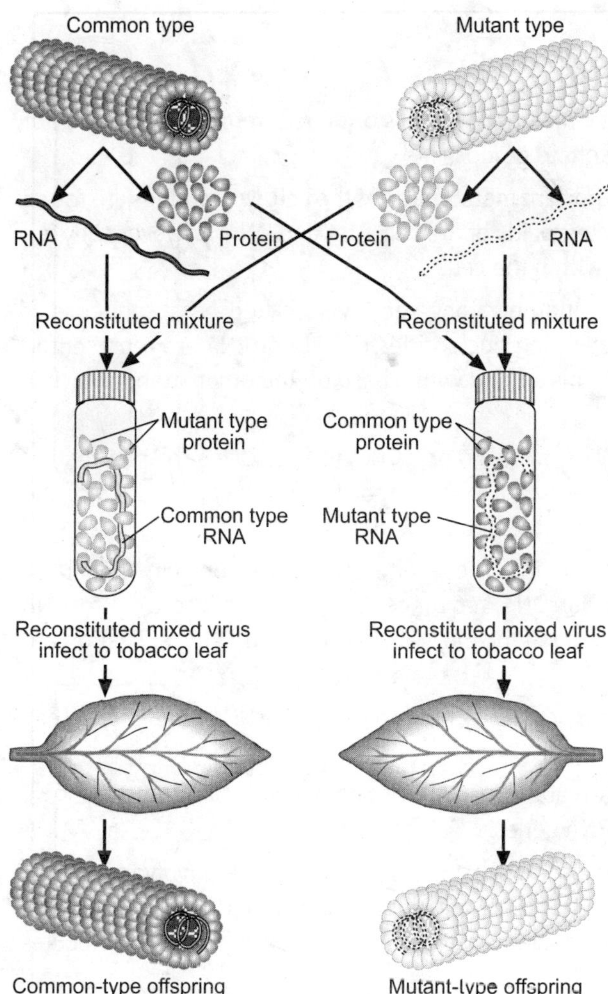

Fig. 5.20 Reconstitution experiment demonstrating that RNA, not the protein component of the virus, controls inheritance

messenger RNA, which is, base-for-base, a complement to the bases on one of its strands. Through the mediation of ribosomes, which contain ribosomal RNA and protein, a sequence of bases in the messenger RNA is then 'translated' into a sequence of amino acids. This translation follows the 'triplet' rule that a sequence of three messenger RNA bases designates one of the 20 different kinds of amino acids.

During this process of translation, no physical material is actually inserted by the messenger RNA into the protein. The mRNA only designates the linear position in which each amino acid is to be placed by special molecules of transfer RNA that bring the amino acids to the messenger. With the aid of the

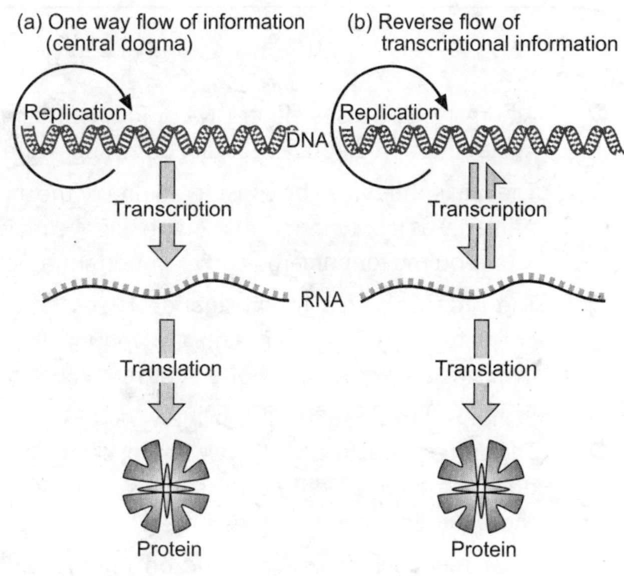

Fig. 5.21 Central dogma. (a) One-way flow of information (b) Reversed flow of transcriptional information

ribosome and various enzymes, the amino acids are then connected in sequence through peptide linkages. Thus, a polypeptide chain of amino acids is formed in which the precise position of each component has been ultimately designated by the genetic material.

In 1958, Crick proposed that this process moves only in the direction of nucleic acid-to-protein and not the other way round. This concept, which became known as **'the central dogma'** states that once sequential information has passed into protein, it cannot be transferred back to nucleic acids again (**Fig. 5.21**). Only nucleic acids, through complementary base pairing, can determine the base sequences of other nucleic acids, and nucleic acids can only determine the amino acid sequences of proteins. This is not to say that proteins cannot affect nucleic acids or cause errors in nucleic acid replication. It only means that a long chain of amino acids in a protein is not translated either into a specific long chain of bases in a nucleic acid or into a similar long-chained sequence of amino acids in another protein. Thus, changes in amino acids, caused perhaps by somatic 'experiences' do not lead to directly correlated nucleotide sequence changes in the genetic material responsible for synthesising these proteins. There is no mechanism for the inheritance of such 'acquired characters'.

KEY POINTS

- All organisms posses either DNA or RNA as their genetic material, which is obtained from polymerisation of nucleotides. Nucleotides contain base, sugar and phosphate groups.
- The nucleotides may be considered one of the most important metabolites of the cell. Nucleotides are found primarily as monomeric units comprising the major nucleic acids of the cell, RNA and DNA. However, they are also required for numerous other important functions within the cell.
- The order in which the nucleotide bases follow one another along the chain, called base sequence is characteristic of a nucleic acid. Conventionally, the repeating nucleotides of DNA or RNA are represented by a single letter A, G, C, T or U. A base on one strand forms a bond with a base on the other strand, which is called complementary base pairing.
- The correct structure of DNA was proposed by James Watson and Francis Crick (1953). Using the information obtained by X-ray diffraction studies and chemical observations, they deduced the basic features of the three-dimensional structure of the DNA molecule (DNA double helix model).
- Most functions of DNA depend on interaction with proteins. These interactions can be non-specific or specific, where the protein can only bind to a particular DNA sequence. Enzymes can also bind to DNA and of these, the polymerases that copy the DNA base sequence in transcription and DNA replication are particularly important.
- Transcription factors are the other unique proteins that specifically bind to particular DNA sequences. They are involved in the control of gene expression.
- Most genetic systems studied to date contain genes made of DNA. But some viruses, including several phages, plant and animal viruses and all viroids have RNA genes.
- In addition to genomic RNA in viruses, almost all cells use RNA as a structural molecule involved in many functions. The structure of RNA is similar to that of DNA but there are a number of important differences: in RNA, ribose replaces 2′-deoxyribose and the base thymine is replaced by another base, uracil. RNA molecules normally exist as a single polynucleotide strand.
- In 1958, Crick proposed that informational process goes only in the direction of nucleic acid to protein but not in reverse. This concept, which became known as 'the central dogma' essentially means that once sequential information has passed into protein, it cannot be transferred back to nucleic acids again.

Self-assessment questions

1. What are nucleic acids? What are the types commonly found in cells?
2. List the different components of nucleic acids.
3. Differentiate between ribonucleotides and deoxyribonucleotides.
4. Besides being a component of nucleic acids, what are the other roles of nucleotides?
5. What properties of nucleotides make the two DNA strands undergo base pairing?
6. What makes DNA the more stable genetic material?
7. What is thermal denaturation and renaturation of DNA?
8. Briefly explain the different types of RNA and their functions in a cell.
9. Explain in detail the experiments to prove that DNA is the genetic material.
10. What is the central dogma in microbiology?

FURTHER READING

Textbooks

Adams RLP, JT Knowler and DP Leader (1992). *The Biochemistry of the Nucleic Acids*, 11th ed. London: Chapman and Hall.

Callandine CR and HR Drew (1992). *Understanding DNA: The Molecule and How it Works*. Academic Press.

Gesteland RF, TR Cech and JF Atkins (eds) (1999). *The RNA World*, 2nd ed. New York: Cold Spring Harbor Laboratory Press.

Saenger W (1984). *Principles of Nucleic Acid Structure*. New York: Springer Verlag.

Smith and Wood (1991). *Molecular and Cell Biochemistry, Biological Molecules*. London: Chapman and Hall.

Weaver RF (1999). *Molecular Biology*. McGraw-Hill.

References

Avery OT, CM MacLeod and M McCarty (1944). Studies on the chemical nature of the substance inducting transformation of pneumococcal types: Induction of transformation by a deoxyribonucleic acid fraction isolated from pneumococcus type III. *J. of Exp. Med.* 79:137–58.

Baker TA and SH Wickner (1992). Genetics and enzymology of DNA replication in *E. coli. Annu. Rev. Genet.*, 26: 447–77.

Bauer WR, FHC Crick and JH White (1980). Supercoiled DNA. *Scientific American,* 243:118–35.

Chambers DA, KBM Reid and RI Cohen (1994). DNA: The double helix and the biomedical revolution at 40 years. *FASEB J.* 8:1219–26.

Dickerson RE (1983). The DNA helix and how it is read. *Scientific American.* 249(6):86–102.

Dickerson RE et al. (1989). Definitions and nomenclature of nucleic acid structure parameters. *EMBO J.*, 8:1–5.

Eaton BE and WA Pieken (1995). Ribonucleosides and RNA. *Annu. Rev. Biochem.*, 64:837–65.

Felsenfeld G (1985). DNA. *Scientific American.* 253(4):44–55.

Korenberg A and TA Baker (1992). *DNA Replication.* New York: WH Freeman.

Gellert GH (1981). DNA topoisomerases. *Annu. Rev. Biochem.*, 50:879–910.

Lebrun A and R Lavern (1997). Unusual DNA conformations. *Curr. Op. Struct. Biol.*, 7:348–55.

Rich AR, A Nordheim and HJ Wang (1984). The chemistry and biology of left-handed Z–DNA. *Ann. Rev. Biochem.*, 53: 791–846.

Roca J (1995). The mechanisms of DNA toposiomerases. *Trends Biochem. Sci.*, 20:156–60.

Wang JC (2002). Cellular roles of DNA topoisomerases: A molecular perspective. *Nat. Rev. Mol. Cell Biol.*, 3:430–40.

Watson JD and FHC Crick (1953). Molecular structure of nucleic acids. *Nature* 171:737–38.

Wells RD (1988). Unusual DNA structures. *J. Biol. Chem.* 263:1095–98.

Zimmerman SB (1982). The three-dimensional structure of DNA. *Ann. Rev. Biochem.*, 51:395–427.

CHAPTER 6

Structure and Organisation of Genome

LEARNING OBJECTIVES

After reading this chapter, you will be able to:

- Understand the structure and evolution of genes
- Discuss the organisation of genetic material
- List the structural and functional aspects of chromosomes

6.1 INTRODUCTION

The concept of gene is important because its physical and chemical nature is the basis for all the biological activities. Genes are like information bullets that can fire when metabolism of a cell has to be activated. It can be defined as a hereditary unit of a DNA sequence that occupies a specific location on a chromosome and contains information to produce a functional product such as protein and RNAs. Genes also contain regulatory sequences that dictate them to express only in certain conditions. In the classical sense, a gene is known to occupy distinct chromosomal regions. Since a gene is responsible for some phenotype, it must perform a function in each cell in which it is active. Since genes can undergo mutation, a permanent change in the organisation of the genetic material is also possible.

The total complement of genes in an organism or cell is known as its **genome**. In prokaryotes, the vast majority of genes are located on a single chromosome of circular DNA, while eukaryotes usually have multiple individual linear DNA helices packed into dense DNA–protein complexes called **chromosomes**. The genomes of most eukaryotes are larger and more complex. The basic structural unit of the eukaryotic chromosome is massive in relation to its component macromolecules. Genetic material (both DNA and RNA) occurs in various forms in different cell types. Eukaryotic cells contain DNA in the nucleus and in other organelles (mitochondria and plastids).

6.2 NATURE OF THE GENE

Earlier studies revealed that genes were discrete factors that were retained throughout the life of an organism and then passed on to each of its progeny. Later, these hereditary factors were shown to reside on chromosomes and contain DNA. The classical principles of genetics deduced by Mendel in 1865 were on the basis of the results of breeding experiments carried out with peas. Later, geneticists showed the correlation between genes and chromosomes and gene linkages.

Seymour Benzer postulated the three functional units (**cistron, recon** and **muton**) to correlate the physiological aspects of the gene. Beadle and Tatum (1940) suggested that genes direct the synthesis of proteins that act as enzymes and this is called one gene–one enzyme hypothesis. This was later developed into the more specific form of one gene–one polypeptide, and was supported by data from *Neurospora*, plants, animals and humans. In most cases, there is a gene–product correspondence and the gene is usually collinear with its product. However, in higher eukaryotic genomes, there is often a complex relationship between the gene and its product. Almost all the higher eukaryotic genes contain intervening sequences (**introns**), which are not represented in the final product; therefore, they are not a part of its final function (**Fig. 6.1**). Most genes are discrete non-overlapping units, i.e., they do not share information with other genes. The open reading frames may be transcribed from opposite strands and translated in different directions.

6.2.1 Structural components of a gene

In molecular terms, a gene can be defined as the sequence of nucleic acids that is necessary for the synthesis of a functional product (polypeptide or RNA molecule). According to this definition, a gene includes, besides the nucleotides encoding the amino acid sequence of a protein, referred to as the coding gene, a functional RNA such as rRNA or tRNA. A gene also includes the DNA sequences required to make a particular RNA transcript.

In eukaryotic genes, such regions for control of transcription (known as **enhancers**) may be located 50 kb or more away from the coding region. Other critical non-coding regions in a eukaryotic genes are the DNA sequences that specify signals for 3' cleavage and polyadenylation, and for splicing of the primary RNA transcripts. Cells and organelles may contain more than one copy of their genome. Each gene is located at a particular position along the chromosome, termed the locus, whilst the alternate form of this gene is termed as the allele. Each gene is present in two allelic forms: identical (**homozygous**) or variant (**heterozygous**). The occurrence of different alleles at the same site in the genome is termed **polymorphism**.

Sequence analysis of thousands of genes from eubacteria has revealed that the vast majority contains uninterrupted protein-coding regions. However, a few exceptions have been identified with a few genes that contain intervening sequences. In a bacterial chromosome, genes encoding enzymes involved in related functions are generally located next to each other. For example, the three genes encoding the enzymes required for lactose metabolism map are in one contiguous stretch on the *E. coli* genome. The

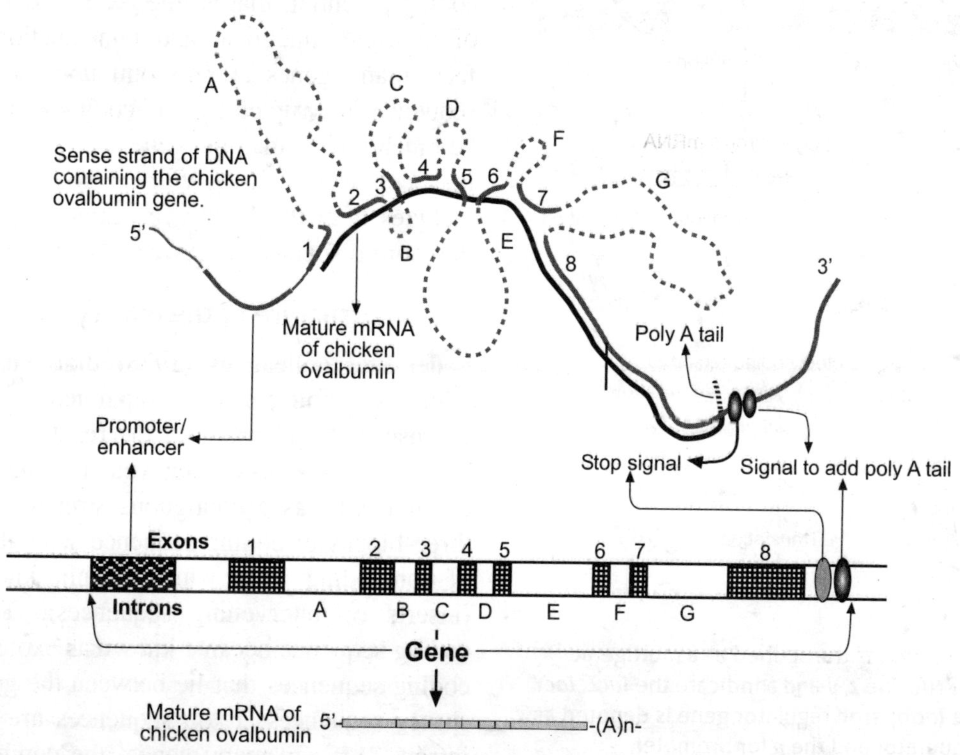

Fig. 6.1 The mapping of introns in the chicken ovalbumin gene; electron micrograph prepared after hybridising mRNA (bold line) with the chicken ovalbumin gene; the bold lines (1 to 8) represent 8 exons and loops (A to G, dashed lines) introns

Table 6.1 **Sequences in gene structure**

Prokaryotic gene	*Eukaryotic gene*
Recognition region (~50 bp)	Recognition region (~50 bp)
Transcription initiation site	Transcription initiation site
5' untranslated region	5' untranslated region
Translation initiation	Translation initiation
Coding region	Coding region consists of alternating exon/intron, splice donor and acceptor sites
Translation stop site	Translation stop site
3' untranslated region	3' untranslated region
Translation stop site	Polyadenylation signal
	Translation stop site

genes in this cluster comprise a single transcription unit referred to as an **operon**. Ribosomes initiate translation at the beginning of each of the genes in this mRNA, producing the three polypeptides required for lactose metabolism. Since the *lac* mRNA encodes several polypeptides, it is said to be a polycistronic mRNA (**Fig. 6.2**). The differences in the structure of gene

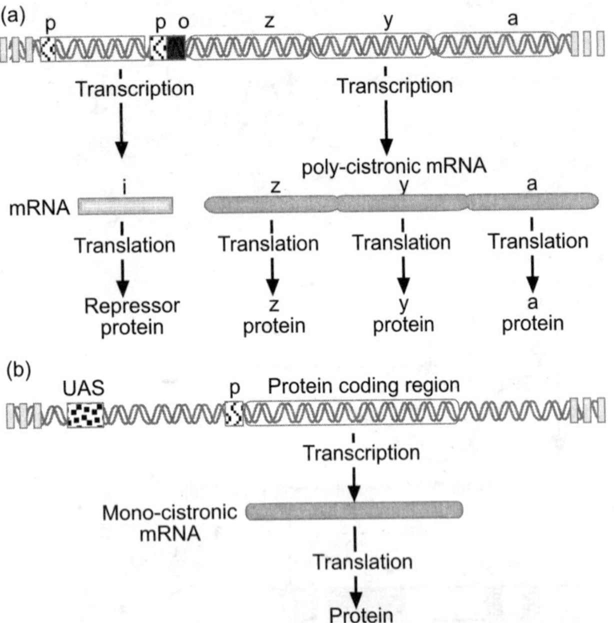

Fig. 6.2 The *lac* operon is transcribed as a multigenic (polycistronic) mRNA. The z, y and a indicate the *lac*Z, *lac*Y and *lac*A loci. The *lac* operon regulator gene is denoted as *i*; the *o* stands for operator and the *p* for promoter.
(b) Monocistronic eukaryotic gene having protein coding gene and promoter of its own; UAS: upstream activating sequences

sequences in prokaryotes and eukaryotes are listed in **Table 6.1**.

Although most genes are transcribed into mRNAs that encode proteins, clearly some RNAs, including tRNAs and rRNAs, do not encode proteins. Many other RNA molecules described in later chapters also do not encode proteins.

6.3 ORGANISATION OF GENES ON CHROMOSOMES

In *E. coli,* most of the genes are present as a single copy although there are seven sets of the genes that encode the ribosomal RNAs. Many of the genes in prokaryotes occur in operons (groups of genes of related functions), which may, for instance, encode all the enzymes in a metabolic pathway. The genes of an operon are united by common controls of their expression and tend to be transcribed from common promoters into polycistronic mRNAs. Prokaryotic DNA is used efficiently with only short regions of non-coding DNA between genes.

The mammalian genome with approximately 3×10^9 bp has over 600 times the coding potential of *E. coli* but the order of complexity of about 50 times is sufficient (**Fig. 6.3**). Eukaryotic DNA has excess coding potential, that is, the genes occur in the pattern of multiple functional and non-functional copies. In fact, many genes are discontinuous with the coding sequence interrupted by non-coding elements that may be much longer than the gene itself. Much of the DNA in a cell is repetitive with no known function. This includes both the highly repetitive satellite DNA and the more dispersed repetitive elements.

6.3.1 Structure of the eukaryotic gene

Leder and colleagues (1978) digested mouse DNA with restriction enzymes, separated the fragments on agarose gels and analysed the resultant with Southern blotting. Their results showed that the ß-globin gene did not exist as a contiguous stretch of DNA but as three blocks of coding sequence separated by sections of non-coding DNA which became known as introns (inserts or intervening sequences). The section of coding sequence became known as exons. But the non-coding sequences that lie between the genes are called spacer sequences. Exon sequences are conserved but introns vary. In many genes, the combined length of the introns exceeds that of the exons. Thus, the chicken ovalbumin gene contains 7,564 bp whereas the mRNA

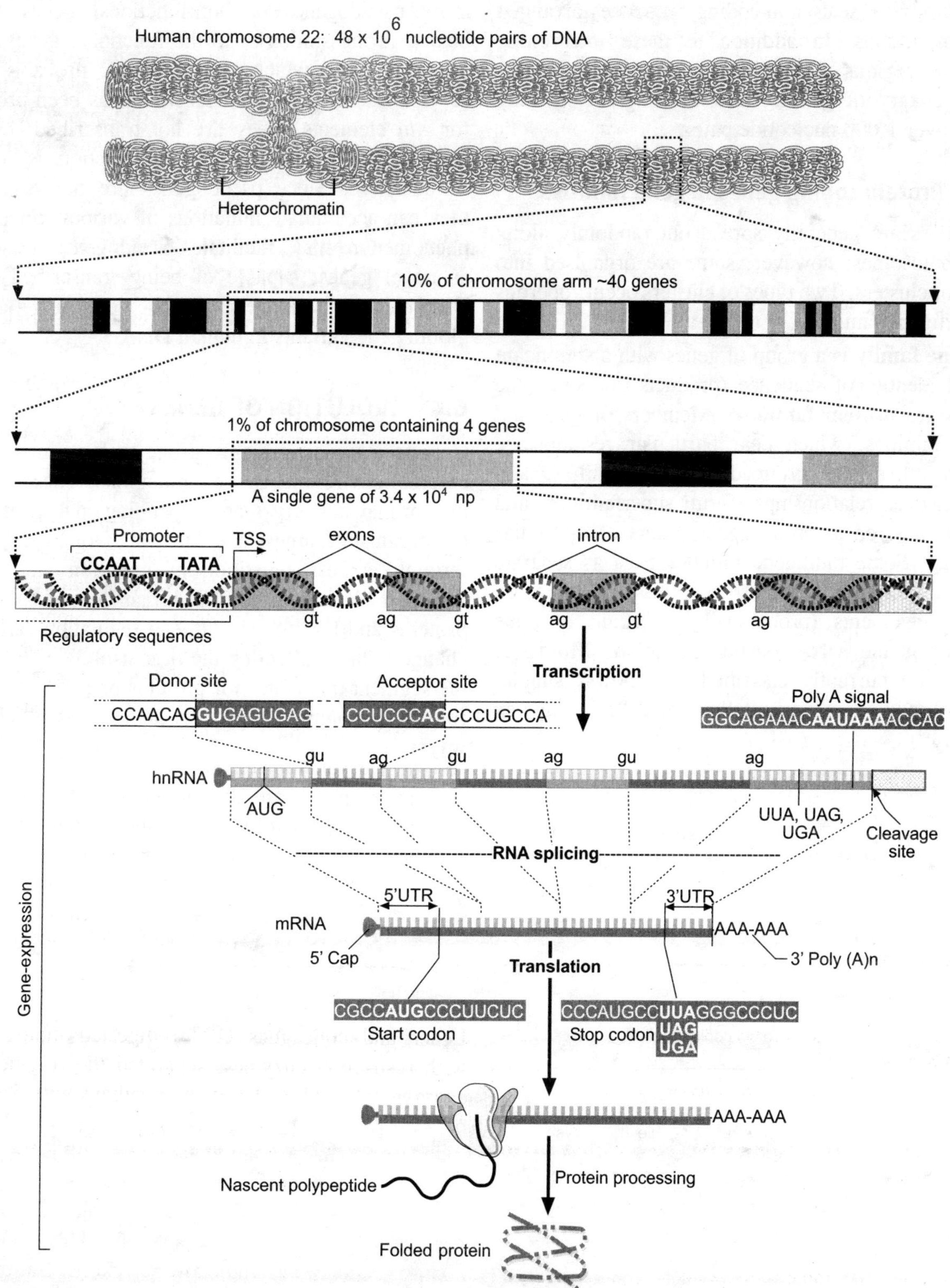

Fig. 6.3 A schematic diagram of the organisation of genes on human chromosome 22, one of the smallest human chromosomes; expanded portions of the chromosome from highest (a portion in chromatid) to lowest (a gene sequence) structural form

is only 1,872 nucleotides long. The difference of 5,692 nucleotides represents non-coding sequences arranged in seven introns. In addition to these intervening sequences, regions of DNA that have control functions precede eukaryotic genes. These flanking regions may contain over 1,000 nucleotide pairs.

6.3.2 Protein coding genes in gene families

Most genes are generally spread out randomly along the chromosomes; however, some are organised into groups or clusters. Two types of clusters occur: operons and multigene families.

A **gene family** is a group of genes with a significant level of identity of sequence (products of these gene families are **protein families**). Members of gene and protein families, which bear particular resemblance to each other, are grouped into subfamilies, and more distant relationships form superfamilies and megafamilies representing higher orders of molecular taxonomy. Some multigene families exist as separate clusters on different chromosomes; this probably arose by rearrangements (probably by breaking of gene clusters) of the DNA during evolution. Multigene families are currently classified as follows: simple, complex and developmentally controlled complex (**Fig. 6.4**).

Pseudogenes: Gene families often include one or more pseudogenes or non-functional genes. These bear a strong similarity to the functional genes in the sequence from which they evolved, probably by a reverse transcriptase mechanism as has been proposed for *Alu* elements. They are not transcribed, because some element required for transcription is missing or defective. Since pseudogenes are not expressed, they can accumulate mutations of various kinds. This phenomenon may facilitate pseudogenes to evolve as novel genes, capable of being transcribed again. Examples of pseudogenes can be seen in α– and ß–globin gene variants in human DNA.

6.4 EVOLUTION OF GENES

Genetic material is highly prone to both internal and external environments, i.e., mutations. These changes may or may not perpetuate, depending on their effect on the organism's fitness and survival. Organisms evolve through endless repetition of mutation and natural selection. A segment of DNA that does not code for proteins and has no role in gene regulation can freely change without affecting the organism as compared to the segment that codes for proteins or has a regulatory role. The evolutionary process can take place in several ways:

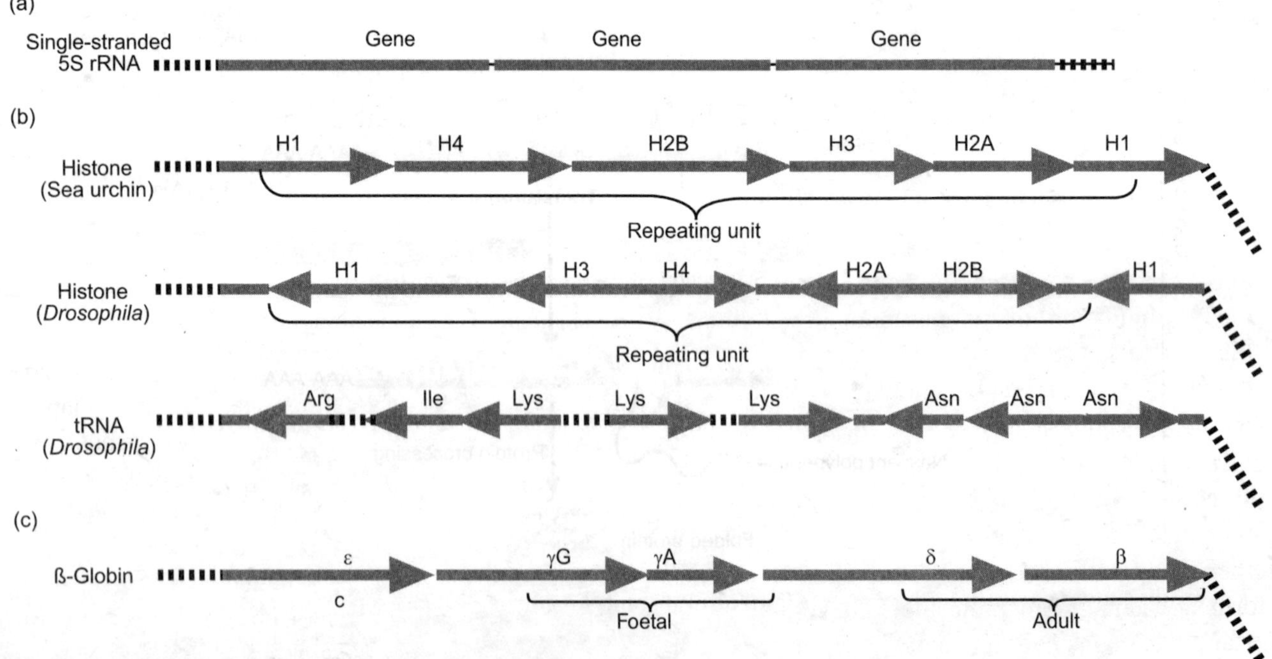

Fig. 6.4 Five examples of gene families; the arrows indicate the direction of transcription
(a) Simple multigene family (b) Complex multigene families (c) Developmentally controlled complex multigene family

Intragenic mutation: An existing gene can undergo mutations within its DNA sequence.

Gene duplication: An existing gene can be duplicated into a pair of closely related genes within a cell.

DNA rearrangement/segment shuffling: Two or more genes can combine to make a hybrid gene consisting of DNA segments that originally belonged to separate genes.

Horizontal (intercellular) transfer: A fragment of the genome can be transferred from one cell of an organism to the cell of another. On the other hand, in vertical transfer of genetic material, it passes from parent to progeny.

Relationship between gene families may be orthologous or paralogous. In the former, genes which have diverged by speciation, perform the same functions in different species (e.g., ß-globin in humans and mouse). In the latter, genes have duplicated and diverged within a genome. That is, such genes are members of multigene families (human α- and ß-globin genes). Both the types are together called homologues.

6.4.1 Gene duplication and divergence

Increasing biological complexity requires increasing number of functions ultimately specified by genes. Thus, the origin of new functions must reflect the creation of new genes, or diversification of the use of pre-existing ones. Both processes seem to have occurred during evolution to increase the total gene numbers with increase in biological complexity. New genes can arise from three sources:

- Through import from other genomes
- Through intrinsic duplication of pre-existing genes
- Spontaneously from random, non-coding DNA by the accumulation of point mutations. However, this source can be discounted because it is unlikely based on probability calculations.

The gene family that provides clear evidence of repeated duplication events is the ß-like globin gene family (**Fig. 6.5**). Two identical ß-globin polypeptides combine with two identical α-globin polypeptides (another gene family) and with four small haeme groups to form a haemoglobin molecule. The ß-like globin gene family contains five functional genes designated ß, δ, ${}^{A}\gamma$, ${}^{G}\gamma$ and ε; the encoded polypeptides are similarly designated. All the haemoglobins formed from the different ß-like globins carry oxygen in the blood, but they exhibit somewhat different properties. It is thought that an initial duplication gave rise to the prototypes of the two families, the α and β families. The human α-globin family (on chromosome 16) contains α and ζ genes, and the ß-globin family (on chromosome 11) contains ß, γ, δ and ε genes. The best understood mechanism in which tandem copies of a gene can be generated is homologous reciprocal recombination at meiosis, but involving unequal crossing-over. Many other gene families (e.g., actins, keratins, protein kinases, transcription factors, tubulins, immunoglobulins, variable regions, transplantation antigens and so on) could have also evolved by a similar mechanism.

Tandemly repeated genes

The genes for the rRNAs, each type of tRNA, and one family of histone proteins occur in invertebrates and some vertebrates as tandemly repeated arrays. These are different from the duplicated genes of gene families as they encode identical or nearly identical proteins or functional RNAs. Most often copies of a sequence appear one after another, in a head-to-tail fashion, over a long stretch of DNA. All these copies are exactly or almost-exactly, like all the others. However, the non-transcribed spacer regions between the transcribed regions can vary. Arrays of tandem repeated histone DNA are somewhat more complex; however, each histone gene, too, has multiple identical copies. The tandemly repeated rRNA, tRNA and histone genes are needed to meet the greater cellular demand for their transcripts. Most of the RNA in a cell consists of rRNA and tRNA.

Synteny

In classical genetics, synteny describes the physical co-localisation of genetic loci on the same chromosome within an individual or species. Biologists usually refer to synteny as the conservation of blocks of order within two sets of chromosomes that are being compared with each other. Genomic sequencing and mapping have enabled comparison of the general structures of genomes of many different species. The general finding is that organisms of relatively recent divergence show similar blocks of genes in the same relative positions in the genome. This situation is called synteny, translated roughly as possessing common chromosome sequences. For example, many of the genes of humans are syntenic with those of other mammals—not only apes but also cows, mice, and so on. Study of synteny can show how the genome is cut and pasted in the course of evolution.

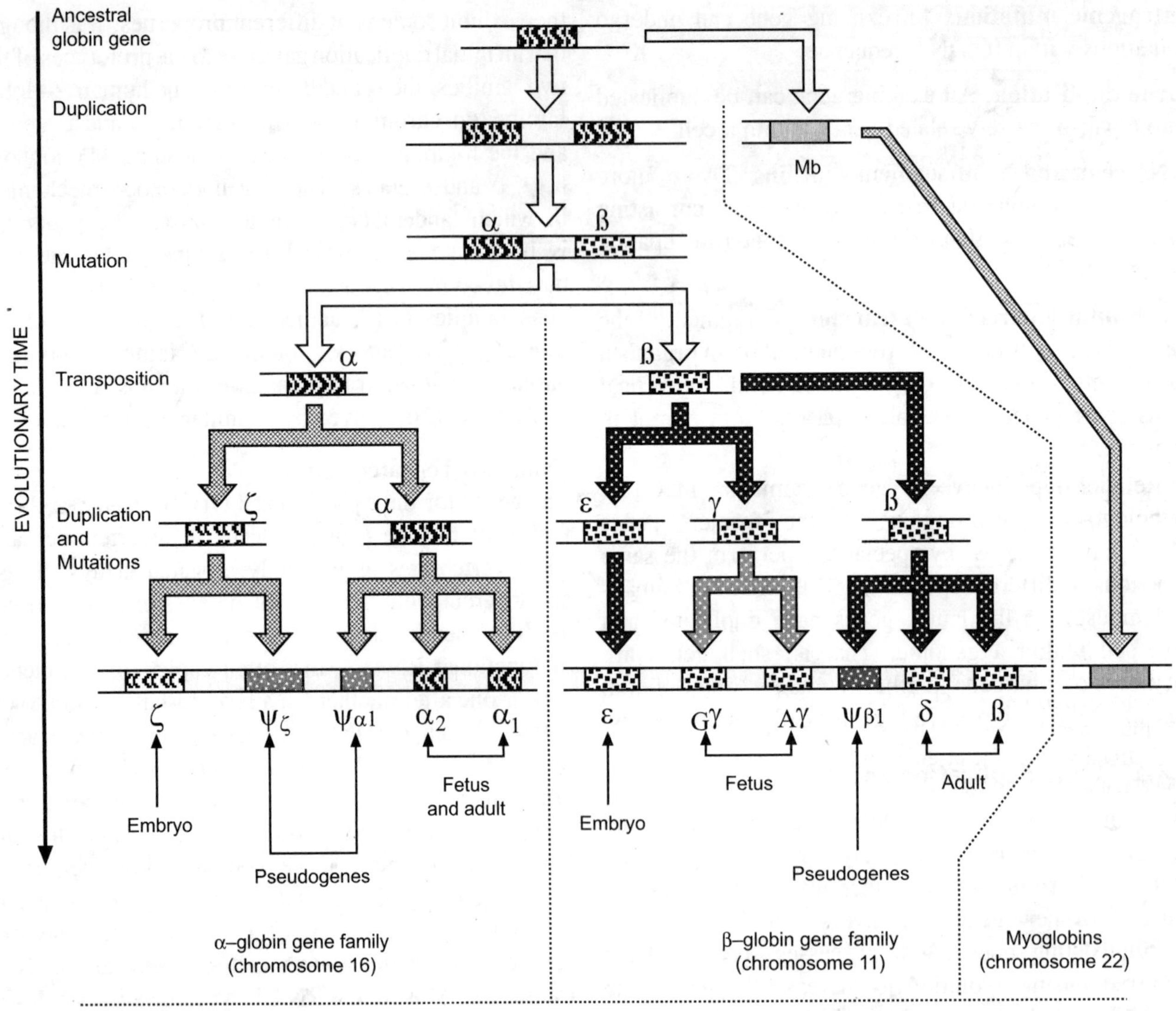

Fig. 6.5 Evolution of the globin genes; families of related genes for α-globin and β-globin, haemoglobin subunits and myoglobins

Many human diseases caused by genetic predisposition have synteny with other mammals and the degree of overlap is indicative of how well a treatment for one will affect the other.

6.5 REPEATING DNA

The coding regions of genes are interrupted by intron sequences, which contribute substantially (25% in humans) to the large size of the genome in higher eukaryotes. Besides the duplicated protein-coding genes and the tandemly repeated genes encoding rRNAs, tRNAs and histones, eukaryotic cells contain multiple copies of other DNA sequences in the genome. These are generally referred to as repeating DNA. Many of these sequences are quite short and occur as tandem repeats of similar or identical copies of a few different types of sequences; others are much longer and are interspersed at many places in the genome.

6.5.1 Satellite DNA

It is non-coding and tandemly repetitive. Further analysis has divided these repetitive DNA into several categories (**Fig. 6.6**). These long stretches of DNA with very short, simple sequences like $(ATAAACT)_n$ can often be separated from the major portion of the DNA by sedimentation to equilibrium in density

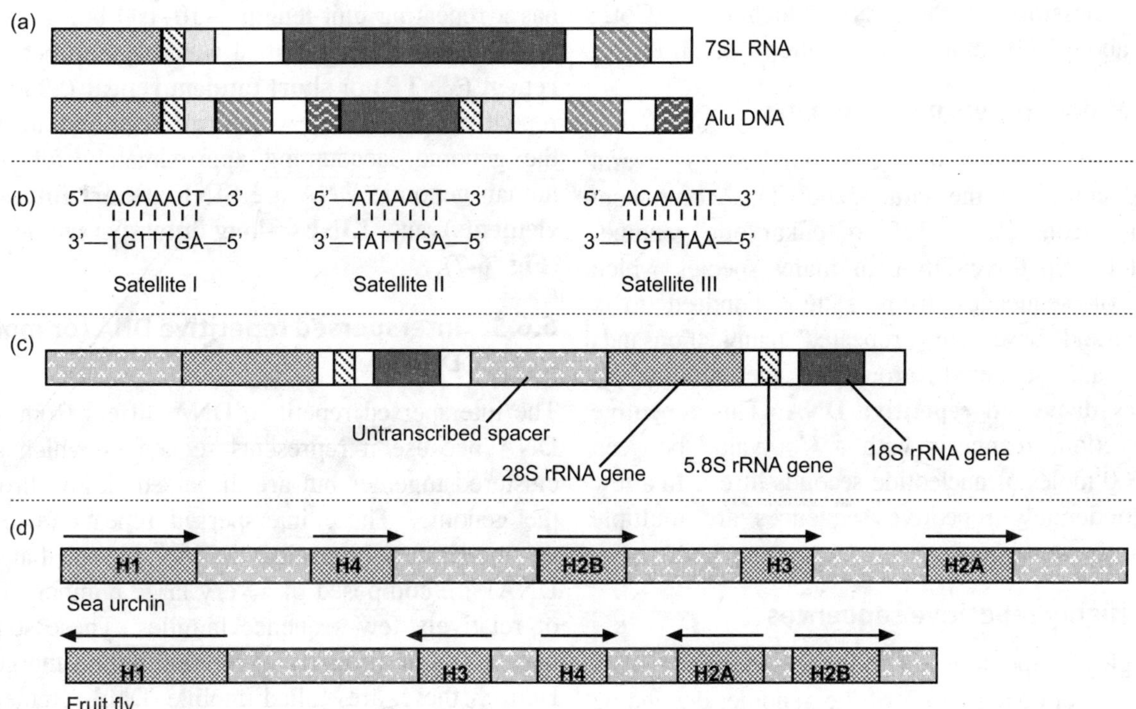

Fig. 6.6 The genomes of higher eukaryotes contain many repetitive DNA sequences (a) *Alu* sequences in human DNA are similar to portions of 7SL RNA, an abundant cytoplasmic RNA. (b) The sedimentation pattern of fragmented *Drosophila* DNA exhibits a main band and three less dense satellite bands consisting of highly repetitive heptanucleotide sequences. (c) Organisation of the genes for 40S precursor of 18S, 5.8S and 28S rRNAs in *Xenopus* (d) Maps of the histone gene cluster in a sea urchin and fruit fly

gradients. Repetitive DNA sequences that are AT-rich have lower density compared to DNA with average composition while GC-rich sequences are denser. Thus, in a density-gradient experiment, the reiterated, simple-sequence fragments form satellite bands about the main-band DNA. Therefore, banding clusters of repeated sequences are referred to as satellite DNAs. Often they may occur as multiple copies scattered throughout the genome (interspersed), such as *Alu* elements. One estimate suggests that approximately 50% of the mammalian DNA consists of highly repetitive sequences (10%–20% satellite DNA), some of which are repeated more than 10^5 to 10^6 times.

6.5.2 *Alu* elements

These sequences are a kind of repeated DNA sequences scattered throughout the genome. They do not code for proteins but their true function remains mysterious. However, some of these sequences may represent control elements of some kind. These sequences exist in hundreds and thousands of copies in the human genome, with each copy about 300 bp. Their name reflects the common existence of a single site for the restriction endonuclease *Alu*I in most members of this class. The *Alu* sequences transcribe into RNA inefficiently, but they are not translated.

6.6 CLASSIFICATION OF EUKARYOTIC DNA

The three major classes of eukaryotic genomic DNA are described below.

6.6.1 Unique or single-copy sequence

They are found only once or a few times within the genome and are usually sequences for genes. This is usually the major component and typically comprises about 30% to 75% of the chromosomal DNA in most organisms. This fraction of genomic DNA is the slowest to re-associate on a Cot curve, and corresponds to the coding regions of genes, which occur in only one or a few copies per haploid genome. In *E. coli* genome, virtually all the DNA has a unique sequence since it consists predominantly of more or less contiguous single-copy genes. In human cells, unique DNA is generally classified as that DNA which has a $Cot_{1/2}$ value of about 1,000 moles of nucleotide seconds $litre^{-1}$.

6.6.2 Middle repetitive sequences

They are those found in a few hundred to several thousand copies of the same gene. This component constitutes from 1% to 30% of eukaryotic genome. Moderately repetitive DNA in many species which consists of sequences from a few hundred to a few thousand base pairs repeated many thousands of times and scattered throughout the genome, is known as **dispersed repetitive DNA**. This repetitive DNA fraction, reanneals with a Cot value between 100–1,000 moles of nucleotide seconds $litre^{-1}$. In a few cases, moderately repetitive sequences are multiple copies of the same gene.

6.6.3 Highly repetitive sequences

The highly repetitive or simple-sequence DNA constitutes from 5% to 45% of the genome, depending on the species. The sequences in this class are typically from 5 bp to 300 bp per repeat and are duplicated as many as 10^6 or more times per genome. Since they consist of repeats of such short sequences, they have non-average G + C content. For example, one type of simple-sequence repeat in *Drosophila* consists of tandem repeats of the seven nucleotides unit, ACAAACT. Such DNA can be separated from the rest of the cellular DNA by equilibrium density-gradient centrifugation.

6.6.4 Minisatellite and microsatellite

A microsatellite generally has the length of the repeating unit as less than 10 bp, and the minisatellite has a repeating unit length ~10–100 bp. These types of sequences are also called **variable number tandem repeat (VNTR)** or **short tandem repeat (STR)**. Other repetitive sequences, which are scattered throughout the genome, accounting approximately for 45% of human genomic DNA are **SINEs (short interspersed elements)** and **LINEs (long interspersed elements)** (**Fig. 6.7**).

6.6.5 Interspersed repetitive DNA (or mobile DNA)

The interspersed repetitive DNA differs from satellite DNA because it represents sequences which are not clustered together but are dispersed singly throughout the genome. These interspersed repeats (also called as moderately repeated DNA, intermediate-repeat DNA) are composed of a very large number of copies of relatively few sequence families. These sequences have a unique property of moving within the genome. Hence, these are called mobile DNA elements (or transposable elements). The process by which these sequences are copied and inserted into a new site in the genome is called **transposition**. These mobile elements are basically molecular symbionts that, in most cases, appear to have no specific function in the biology of their host organisms, but exist only to maintain themselves; for this reason they are also called **selfish DNA** (see **Chapter 18**).

6.7 ORGANISATION OF GENOMES

Genetic material (both DNA and RNA) occurs in various forms in different cell types (**Fig. 6.8**). The length of a typical bacterial chromosome is about 10^3

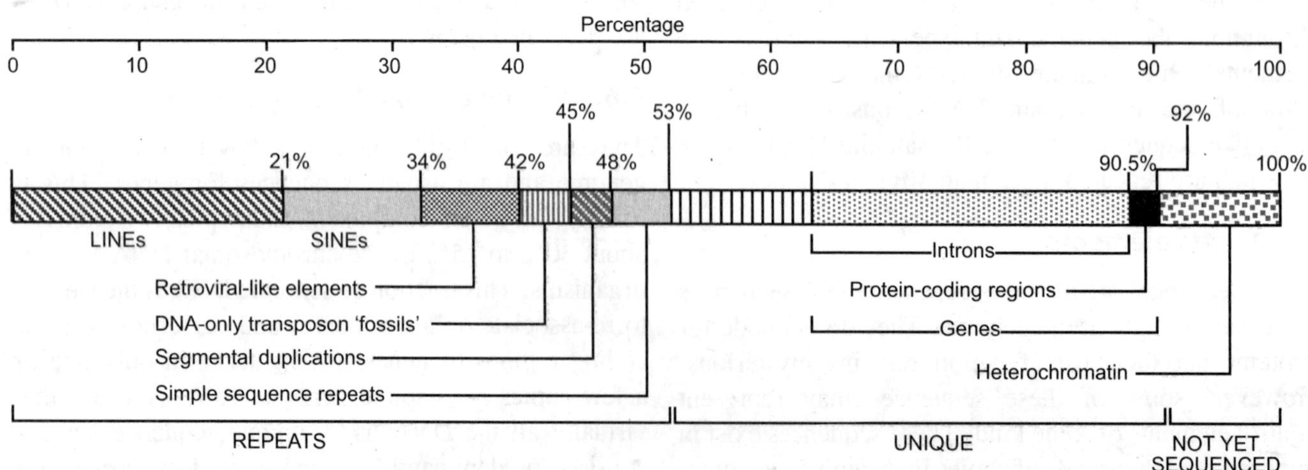

Fig. 6.7 The nucleotide sequence content of the human genome

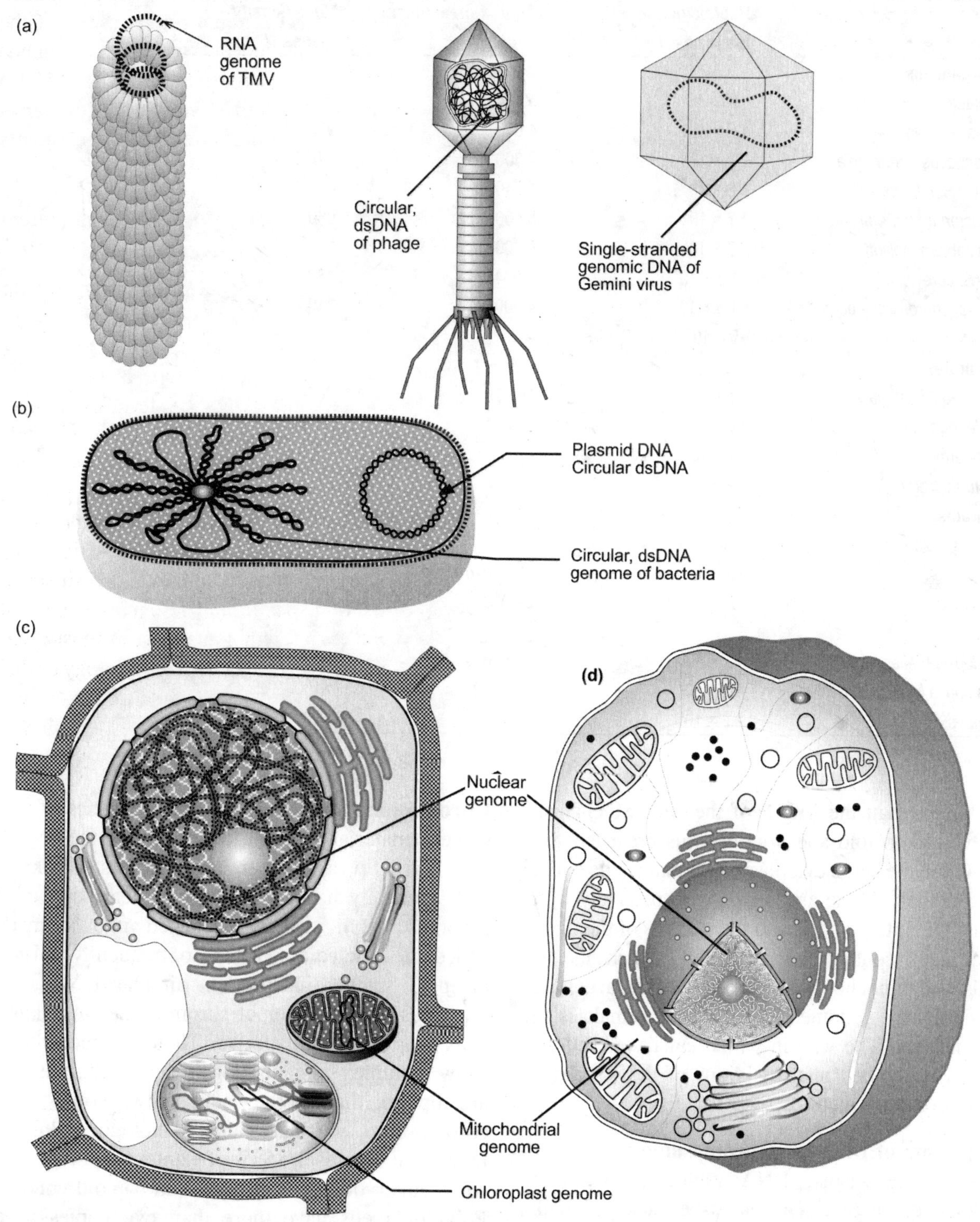

Fig. 6.8 Schematic view of the structure and organisation of genomic DNA of (a) viral genomes (b) the *E. coli* chromosome with a plasmid DNA (c) plant nuclear chromosome with mitochondrial and chloroplast genomes (d) animal nuclear chromosome with mitochondrial genomes

Table 6.2 Genome size, number of genes and gene density of various organisms

Organism	*Haploid genome size (base pairs)*	*Approximate number of genes*	*Gene density (genes/10^6 bp)*
Microorganisms			
Mycoplasma	$<10^6$	500	860
Bacteria (*E. coli*)	4.6×10^6	4,000	950
Streptococcus pneumonia	2.2×10^6	2,300	1,060
Bacteria (*H. influenzae*)	1.8×10^6	1,740	-
Agrobacterium tumefaciens	7.7×10^6	5,400	960
Sinorhizobium meliloti	7.7×10^6	6,200	930
Yeast (*S. cerevisiae*)	$\sim 1.3 \times 10^7$	6,034	480
Schizosaccharomyces pombe	$\sim 1.3 \times 10^7$	4,900	410
Moulds	$\sim 5 \times 10^7$	-	-
Invertebrates			
Caenorhabditis elegans	$\sim 8 \times 10^7$	19,099	200
D. melanogaster	1.8×10^8	13,061	80
Echinoderm	$>4 \times 10^8$	-	-
Locusta migratoria	5×10^9	-	-
Vertebrates			
Birds/amphibia	$\sim 10^9$		-
Homo sapiens	$\sim 3 \times 10^9$	25,000–45,000	9.3
Mus musculus	$\sim 2.5 \times 10^9$	29,000	12
Plants			
Arabidopsis thaliana	1.2×10^8	25,500	200
Oryza sativa (rice)	4.3×10^6	>45,000	>100
Zea mays	2.2×10^9	>45,000	>20

times longer than the length of the cell. Mechanisms have evolved to fold and organise this great length of DNA within the cell's cytoplasm. All plant cells contain three genomes. These are localised respectively in the nucleus, the mitochondria and the plastids of the plants. Animal cells contain two genomes, which includes nucleus and mitochondrial genomes. The largest and most significant genome in plants and animals is that of the nucleus. The way in which eukaryotic DNA is packaged in the cell nucleus is one of the wonders of macromolecular structures.

Organisation of DNA into chromosomes

In cellular organisms, DNA molecules are large. This long length of DNA has to be packaged within a cell by forming a highly organised structure called chromosome, composed of DNA and a variety of associated proteins. A chromosome is a discrete DNA molecule, which carries essential genetic information, together with any associated proteins.

Chromosome number is specific to species

Genetic material in a cell, depending on the organism, may exist in a number of segments. Prokaryotic cells generally have only one complete copy of their chromosome(s), that is packaged into a structure called the **nucleoid.** They also frequently carry one or more smaller independent circular DNAs, called **plasmids**. The number of chromosomes in eukaryotic cells varies between species. Their chromosomes also show complex organisation. Most eukaryotic cells are diploid; that is, they contain two copies of each chromosome, which are called homologues. But a good number of organisms (50% of plants) are polyploids and some eukaryotic cells also exist in haploid conditions. Polyploid cells have more than two copies of each chromosome; in some cases, there can be hundreds or even in thousands.

Size of the genome

Table 6.2 lists the amount of DNA in terms of base pairs in different organisms. Apparently, there is a relationship between increase in the DNA content and an increase in the complexity of organisms. However, the genome size of many eukaryotes does not appear to be related to genetic complexity. For example, the genomes of salamanders and lilies contain more than ten times the amount of DNA found in the human genome; yet these organisms are clearly not ten times more advanced when compared to humans.

6.8 TYPES OF GENOMES

6.8.1 Viral genomes

Viruses are small infectious particles containing nucleic acids as their genetic material, packaged into a capsid of proteins. The genome can be DNA or RNA, but not both. In some cases, it is single-stranded (ss), whereas in others it is double-stranded (ds). Depending on the virus, the genome can be linear or circular. Viral genomes vary in size from a few thousands to more than one hundred thousand nucleotides.

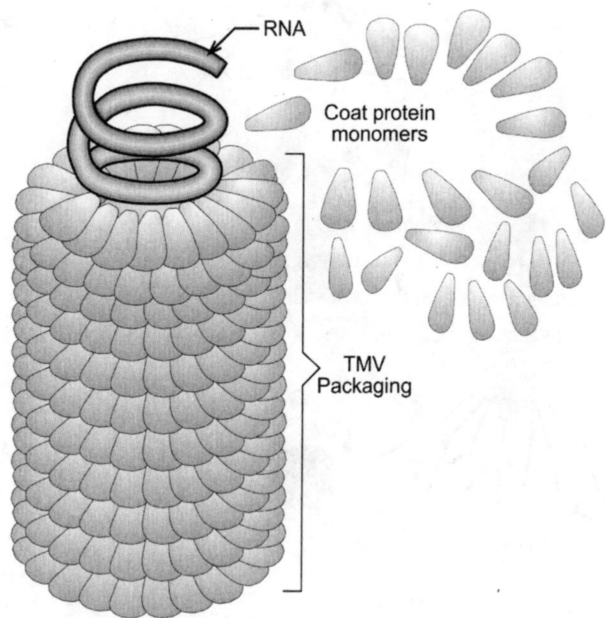

Fig. 6.9 Packaging of the single-stranded RNA of TMV genome; association with the monomeric protein subunits generated the spiral arrangement of the coat around the genomic RNA

Packaging of viral genomes

In an infected cell, the lifecycle of the virus eventually leads to the synthesis of viral nucleic acids and proteins. Newly synthesised viral chromosomes and capsid proteins must then come together and assemble to make a mature virus particle. Viruses with a simple structure may assemble themselves. Some viruses, however, do not do so. Hence, for correct assembly they require the help of proteins that are not found in the mature particle. This process is called **directed assemble**. The viral capsid, could be symmetrical or quasi-symmetrical structure assembled from only one or a few types of protein monomers. The capsid can have a rod-like or spherical shape.

In single-stranded RNA viruses, for example, the tobacco mosaic virus (TMV), the capsid assembles around the genomic RNA. The RNA becomes coiled inside the protein shell (**Fig. 6.9**). The RNA–protein interaction determines the position of RNA within the capsid.

In DNA viruses, the capsids (mostly spherical) are first assembled from a small set of proteins. Initially the capsid has a protein core, but subsequently the capsid takes a distinct shape as it loses its protein core and becomes empty. Finally, it acquires DNA and expands in size while maintaining its shape (**Fig. 6.10**). In the λ phage, the genome occurs as a concatameric molecule in which several genomes are joined end to end. The ends of genomes are marked by sequences called ***cos* sites**.

6.8.2 Prokaryotic chromosomes

The DNA is packaged into a region of the cell known as the nucleoid or folded chromosome (**Fig. 6.11**), which is composed of a single circular DNA molecule and a few associated proteins. The nucleoid region has a very high concentration of DNA, and contain all the proteins associated with DNA, such as polymerases, repressors and others. It does not condense prior to cell division but it is organised into a series of looped domains which reflect functional as well as structural order.

In bacterial cells without plasmids, all genes are located on a single large, circular chromosome. The mass of the DNA in an *E. coli* cell consists of 4.6 million base pairs and the 1 mm-long DNA molecule is contained within cells that are only 2 μm long and about 0.5–1 μm wide. A free DNA molecule of this size would form a random coil about 1,000 times the volume of an *E. coli* cell. The folded genome of *E. coli* has about 100 loops, folds or domains of ~40 kb (13μm) of DNA. Several mechanisms operate to compact *E. coli* chromosomal DNA sufficiently to fit inside the bacterial cell. In addition, numerous small protein molecules

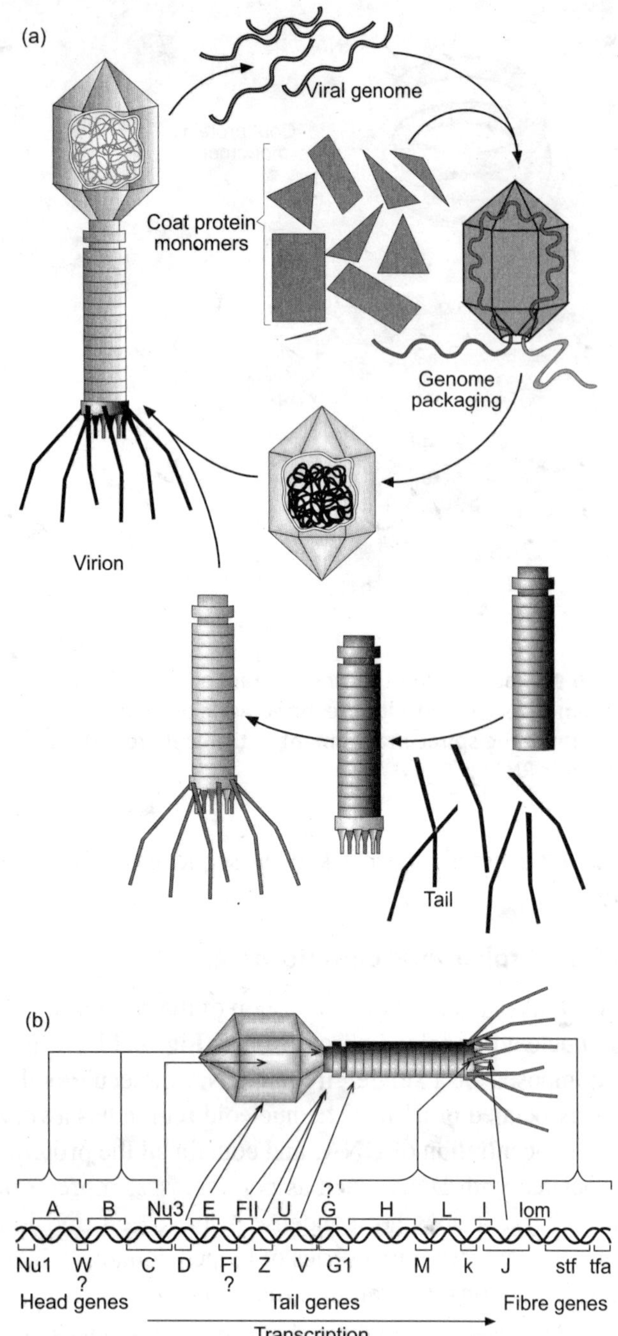

Fig. 6.10 (a) Organisation in bacteriophage and packaging of viral components (b) Genes and components of tailed phages; the main components of the lambdoid phage family and the conserved order of the genes for these components is shown. Heads (or capsids), tails and fibres are composed of several proteins. Component genes and their protein products include: Nu1, terminase small subunit; A, terminase large subunit; W, head stabilisation; B, portal protein; C, capsid component; Nu3, scaffold protein; D, head-DNA stabilisation; E, major head protein; FI, DNA maturation; FII, head stabilisation; Z–H, tail components; M, L, K and I, tail assembly proteins; J, tail tip, host specificity; lom, outer membrane protein; stf and tfa, tail fibre proteins

associate with the chromosomal DNA, causing it to fold into a more compact structure (**Fig. 6.12**). The most abundant of these proteins, H-NS, is a dimer of a 17.6 kDa polypeptide, which binds DNA tightly and compacts it considerably. Protein HU, a small basic dimeric protein binds DNA non-specifically by wrapping the DNA around the protein.

DNA supercoiling

Since the DNA double helices already coil around each other, the formation of additional coils (whether in the 'tightening' or 'loosening' direction) due to the twisting forces is referred to as supercoiling. Both underwinding and overwinding of the DNA double helix can induce supercoiling of the helix. Since B-DNA is a right-handed helix, underwinding is a left-handed twisting and overwinding is a right-handed twist. The formation of short bubbles or hairpin structures of unpaired, single-stranded DNA from severe negative supercoiling can be further stabilised by cruciform structures (**Fig. 6.13**). They occur in regions of palindromic (inverted repeat) DNA. Such cruciform structures serve as 'docking sites' for proteins involved in DNA function. Supercoiling thus favours large, local deviations from the standard conformation.

6.8.3 Organellar genomes

In eukaryotes, in addition to the main nucleus, cell organelles like mitochondria and chloroplasts also replicate autonomously. Most eukaryotes contain mitochondria, whereas plants in addition to mitochondria also carry chloroplasts. These two organelles have genomes of their own and they resemble that of bacteria and not the nuclear one. The origins of these organelles have been postulated that these were originally bacteria which were taken up by primitive eukaryotic cells by an endosymbiotic process.

Mitochondrial genome

Mitochondrial DNA (mtDNA) is a single double-stranded circular molecule (2.1 Mb, 16,569 bp). There are several copies in each mitochondrion and there are many mitochondria in each eukaryotic cell. Mitochondria originated by endosymbiosis of a prokaryotic cell early in evolution. Mitochondrial DNA is similar to prokaryotic DNA (**Fig. 6.14a**). There are no histones or any other protein associated with mtDNA. The genes contain no introns. Because it is in a highly oxidising environment it has a much higher rate of mutations than nuclear DNA. The genes in

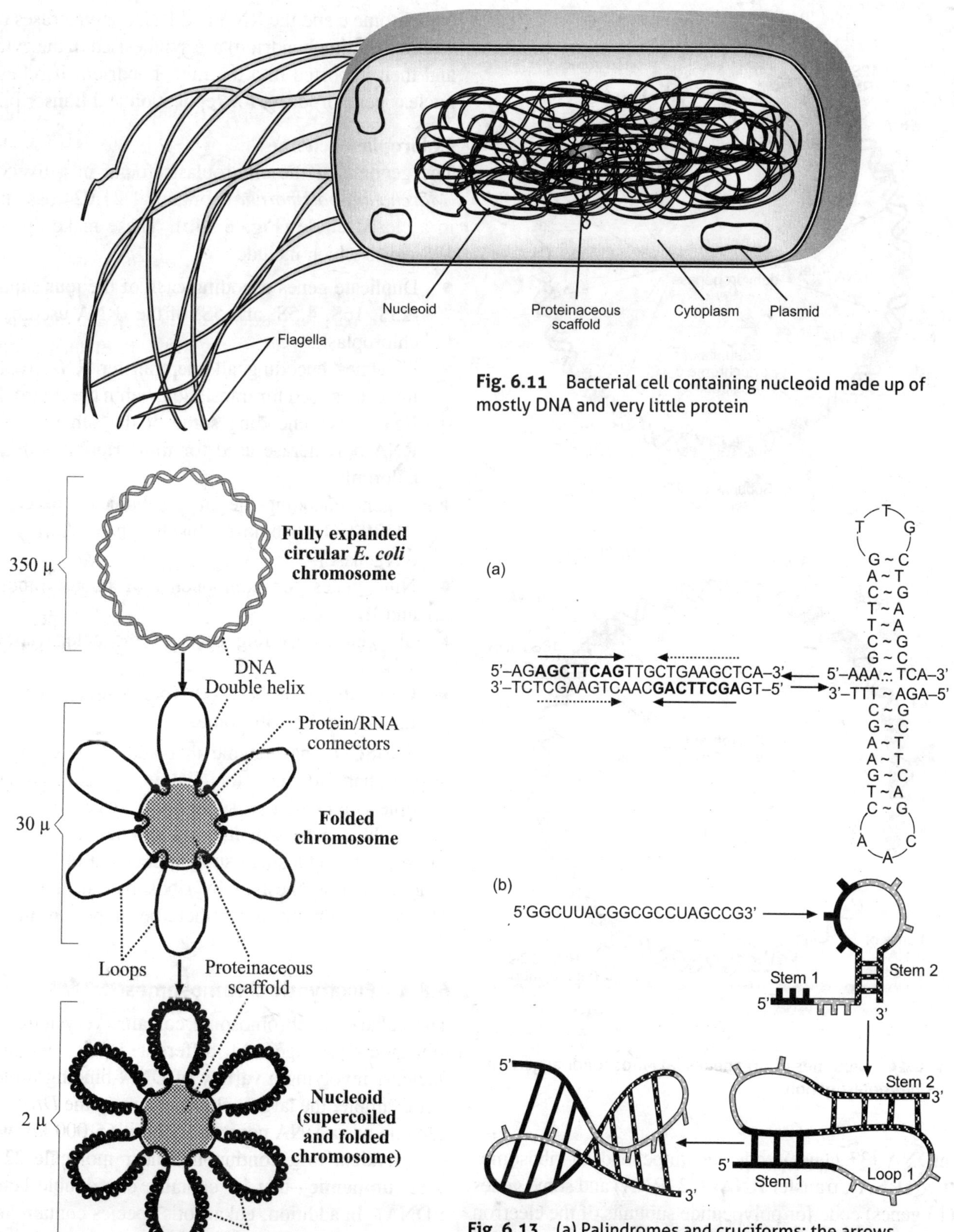

Fig. 6.11 Bacterial cell containing nucleoid made up of mostly DNA and very little protein

Fig. 6.12 Diagrammatic representation of the organisation of DNA and proteins in bacterial chromosome; only six loops in place of the normal 100 loops are shown

Fig. 6.13 (a) Palindromes and cruciforms; the arrows indicate the palindromic sequence which can fold back on itself to form the cruciform structure. (b) A pseudoknot in which stems from different regions align to form a continuous duplex molecule

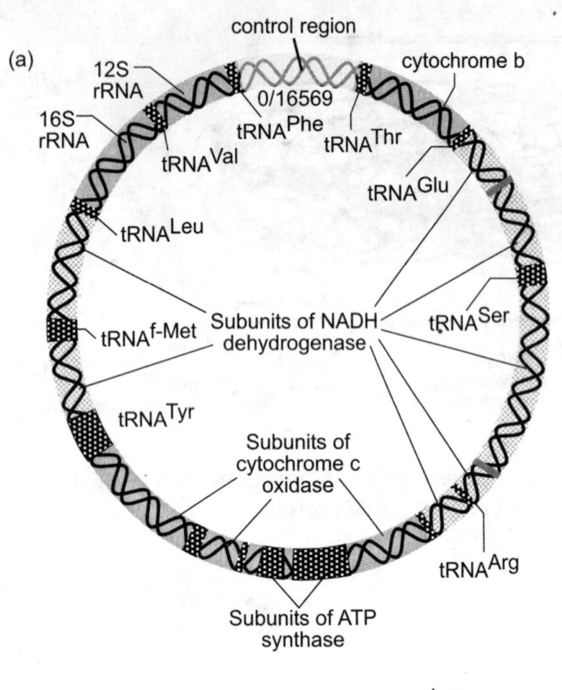

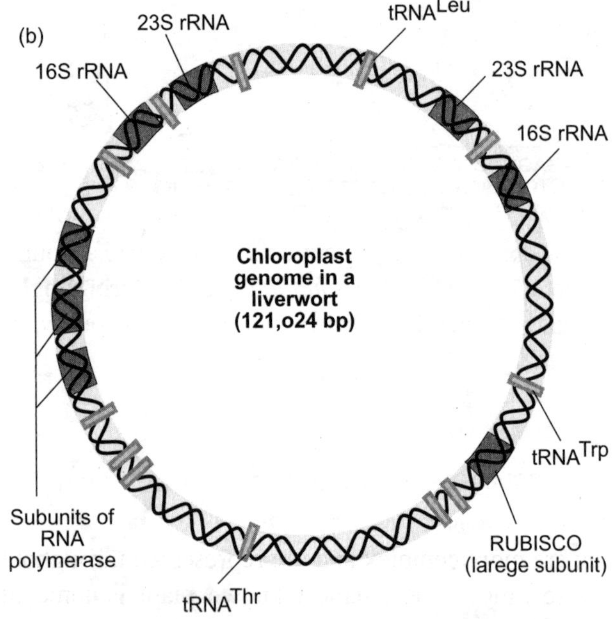

Fig. 6.14 Genomes of organelles (a) Mitochondrial genome (b) Chloroplast genome

mtDNA (37 genes) code for mitochondrial ribosomes (two rRNA), transfer RNAs (22 tRNA) and some genes (13 genes) code for polypeptide subunits of the electron transport chain embedded in the inner membrane and are common to all mitochondria. All gene products are used within the mitochondrion, but proteins encoded by nuclear genes are also used. These proteins (e.g., cytochrome c and the RNA and DNA polymerases used within the mitochondrion) are synthesised in the cytosol and then imported into the mitochondrion. It relies on nuclear gene products for replication and transcription.

Chloroplast genome

The genome of the chloroplasts found in a liverwort (*Marchantia polymorpha*) contains 1,21,024 base pairs in a closed circle (**Fig. 6.14b**). These make up some 128 genes which include:

- Duplicate genes encoding each of the four subunits (23S, 16S, 4.5S, and 5S) of the rRNA used by the chloroplast.
- 37 genes encoding all the transfer RNA (tRNA) molecules used for translation within the chloroplast.
- Four genes encoding some of the subunits of the RNA polymerase used for transcription within the chloroplast.
- A gene encoding the large subunit of the enzyme RUBISCO (ribulose bisphosphate carboxylase oxygenase).
- Nine genes for components of photosynthesis I and II.
- Six genes encoding parts of the chloroplastATP synthase.
- Genes for 19 of the ~60 proteins used to construct the chloroplast ribosome.

RUBISCO, the enzyme that adds CO_2 to ribulose bisphosphate to start the Calvin cycle, consists of multiple copies of two subunits: a large one encoded in the chloroplast genome and synthesised within the chloroplast, and a small subunit encoded in the nuclear genome and synthesised by ribosomes in the cytosol. The small subunit must then be imported into the chloroplast.

6.8.4 Eukaryotic chromosomes

The eukaryotic chromosome contains very long DNA molecules packaged in different levels within the nucleus, involving a variety of DNA-binding proteins. For example, the largest chromosome in the *Drosophila* genome has a DNA content of about 65,000 kb, which is equivalent to a continuous linear molecule 22 mm long (**uninemic**—that is, contains one double helix of a DNA). In addition, eukaryotic species contain one or more sets of chromosomes and each set is composed of several different linear chromosomes. Humans, for example, have two sets of chromosomes each with 23 chromosomes.

Yeast genome

The simplest eukaryotic genome (1.2×10^7 base pairs of DNA) is found in yeast, *Saccharomyces cerevisiae* (**Fig. 6.15**). The yeast genome contains about 6,000 genes, including 5,885 predicted protein-encoding sequences, 140 rRNA genes, 275 tRNA genes and 40 genes encoding small nuclear RNAs involved in RNA processing. Yeasts, thus, have a high density of protein-coding sequences, similar to bacterial genomes, covering about 70% of the total yeast DNA.

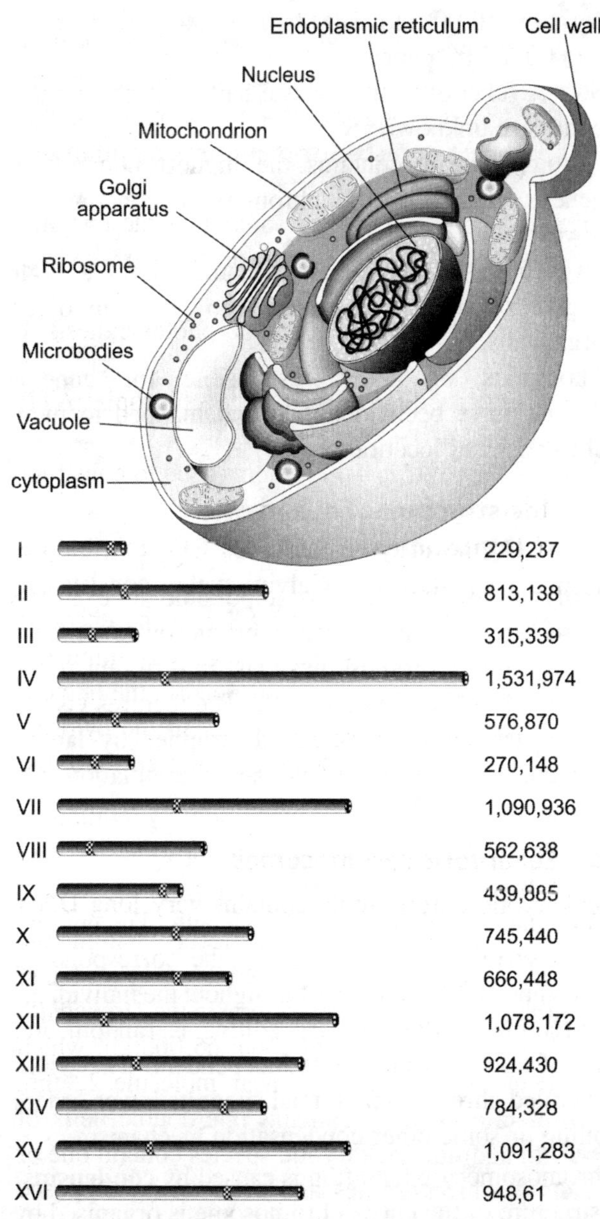

Fig. 6.15 Yeast (*Saccharomyces cerevisiae*) genome has 16 chromosomes and 6,702 genes

Genomes of *Caenorhabditis elegans* and *Drosophila melanogaster*

The genomes of *C. elegans* (nematode) and *Drosophila* (fruit fly) are relatively simple and intermediate in size.

The *C. elegans* genome is 97×10^6 base pairs and contains about 19,000 predicted protein-coding sequences—approximately three times that of the yeast. Its genes span about 5 kb and contain an average of five introns. Approximately 40% of the *C. elegans* proteins have similarity to known proteins of other organisms. Although the adult contains only 959 somatic cells, it has all the specialised cell types found in more complicated animals (digestive system, reproductive organs, nervous system and so on). The complete pattern of cell division leading to development in *C. elegans* has been described; many of these genes have already been found to be related to genes involved in controlling the proliferation and differentiation of mammalian cells.

In *Drosophila*, a special tool for genetic analysis is provided by the giant **polytene chromosome** found in the salivary gland tissue in the larvae. The banding pattern of polytene chromosomes thus provides a high-resolution physical map of the *Drosophila* genome. There are approximately 180×10^6 base pairs, of which about one-third is heterochromatin. The *Drosophila* genome contains approximately 13,600 genes and an average of four introns. In total, protein-coding sequence accounts for about 13% of the *Drosophila* genome. Surprisingly, a complex animal like *Drosophila* has only about twice the number of unique genes found in yeast, which appears to be a much simpler organism. This suggests that the complexity of multicellular organisms is not simply related to a greater number of genes.

Plant genomes

They are more complex and are represented throughout the size range. The smallest known plant genome of 7×10^7 bp belongs to *Arabidopsis thaliana* and one of the largest is a member of the lily family, *Fritillaria assyriaca*, with 1×10^{11} bp. Rice, maize and wheat fall between those two, having genome sizes of 5×10^8, 7.6×10^9 and 1.6×10^{10} bp respectively.

Genome sequencing of *Arabidopsis thaliana* is a major event in plant biology. Sequence analysis of the *Arabidopsis* genome indicates that it contains approximately 26,000 protein-coding genes. This is almost the same number of genes as in most plants. However, many of these genes appear to have resulted from duplication of large segments of the genome, which is about 60% of the genome. Therefore, the number of distinct protein-coding genes in *Arabidopsis* is estimated to be about 15,000 (i.e., 25% of the genome).

Human genome

It has approximately 3×10^9 base pairs of DNA and extended length of DNA is about 1 m. The human genome is distributed among 24 chromosomes (22 autosomes and the 2 sex chromosomes), each containing between 45 and 280 Mb of DNA (**Fig. 6.16**). The draft of human genome sequences was produced by two independent researchers and published in 2001. Highlights from the Human Genome Project are:

- The human genome consists of about 3.1 billion bp and so far 2.85 billion bp have been fully sequenced.

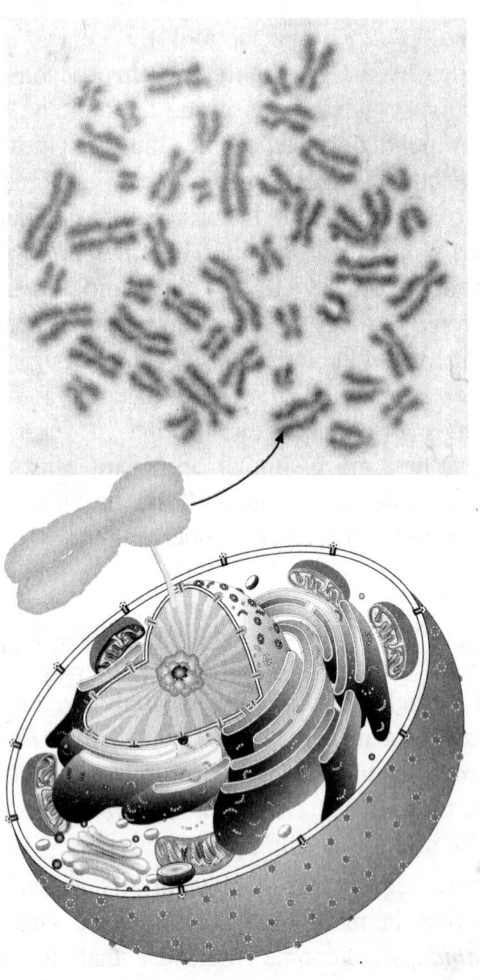

Fig. 6.16 Animal cell (a) A photomicrograph of human metaphase chromosomes (b) The 3D structure of animal cell (*Source:* Channarayappa 1990)

- More than 99.9% of genome sequence is the same between individuals of all nationalities and backgrounds.
- Interestingly, less than 2% of the genome codes for genes.
- Surprisingly, the majority of our DNA is non-protein-coding and repetitive DNA sequences account for at least 50% of the non-coding DNA.
- The genome contains ~ 20,000–25,000 protein-coding genes.
- Many human genes are capable of making more than one protein, allowing human cells to make perhaps 80,000 to 1,00,000 proteins from only 20,000–25,000 genes.
- So far, functions for over one-half of all the human genes are unknown.
- Chromosome 1 contains the highest number of genes, whereas the Y chromosome contains the fewest genes.
- Over 50% of the human genome shows a high degree of sequence similarity to genes in other organisms.
- Thousands of genes responsible for genetic disease have been identified and mapped to their chromosomal locations.

6.8.5 The structure of eukaryotic chromosomes

In nearly all organisms it is impossible to examine the structure of individual chromosomes except during division. Chromosomes exist in different states throughout the lifecycle of the cell. The features visible in a compact chromosome are determined by lateral replication, coiling and longitudinal differentiation.

During mitosis, due to contraction by coiling, a metaphase chromatid or anaphase chromosome becomes more or less a solid cylinder usually wide and of constant diameter along most of its length. The number of gyres seems to be constant for the corresponding chromosome of different cells throughout the individual. The direction of the standard coiling is random and coiling of sister chromatids is independent. The width of the coiled chromosome thread suggests lower orders of coiling or some other condensation mechanism.

Chromosome condensation is caused by **condensins**. The structure of the entire chromosome is organised by interactions with proteins of the **structural maintenance of chromosome** (**SMC**) family (**Fig. 6.17**). SMC are ATPases that fall into two functional groups:

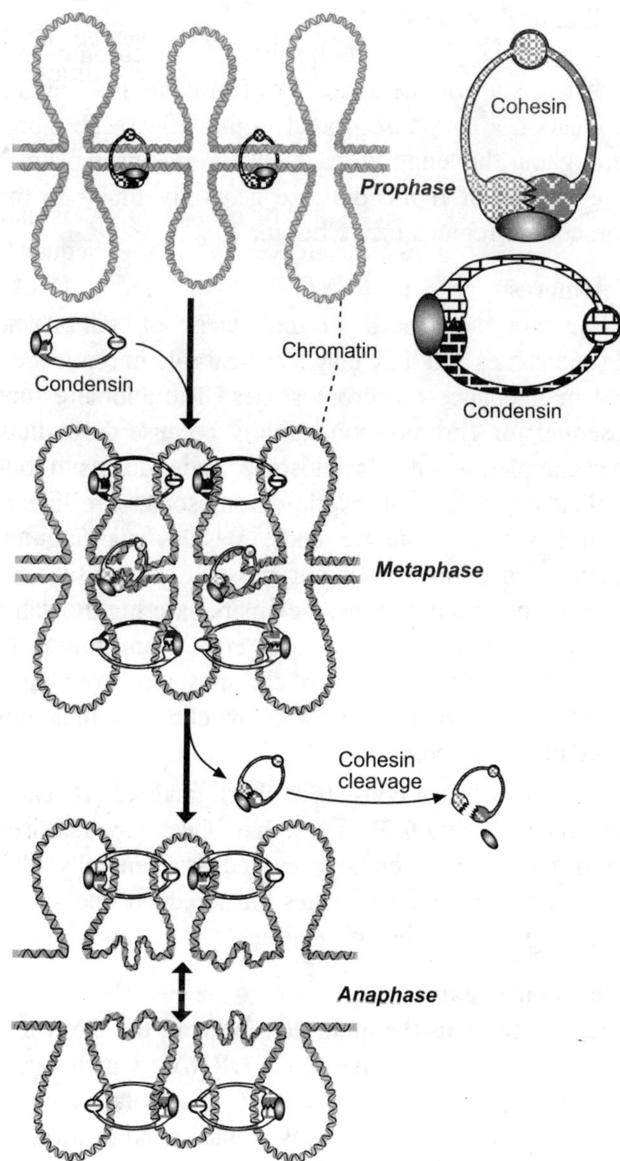

Fig. 6.17 A model for the structure of cohesins and condensins

Condensins are involved with the control of the overall structure and are responsible for the condensation into compact chromosomes at mitosis.

Cohesins are concerned with connections between sister chromatids that must be released at mitosis.

Longitudinal differentiation

In stained preparations, as seen in the light microscope, chromosomes appear to possess a few definitive morphological characteristics by which individual members of a set may be distinguished from each other. The important features are: (i) number (ii) length (iii) location of centromere and relative arm lengths and (iv) presence or absence of satellites.

Chromosome number: Every species has a definite number of somatic and gametic chromosomes. Somatic chromosome number is the number of chromosomes found in somatic cells (e.g., meristematic tissues) and is denoted by $2n$. The two sets (or copies) of a chromosome are ordinarily identical in morphology, gene content and gene order and they are known as homologous chromosomes. On the other hand, gametic chromosome number is precisely one-half of the somatic number (written as n) and denotes the number of chromosomes found in the gametes of a species.

Length (or size) of chromosome: The DNA in each chromosome is believed to be a single linear molecule, which can be up to several centimetres long. All DNA must be packaged into the nucleus, so in their most highly condensed forms, the chromosomes have an enormously high DNA concentration of perhaps 200 mg/ml. More than 50% of the mass of chromatin is protein (**Fig. 6.18**).

Centromere: It is a specialised region of the chromosome (**Fig. 6.19**) that plays a critical role in ensuring the correct distribution of duplicated chromosomes to daughter cells during mitosis. It serves both as the site of association of sister chromatids and as the attachment site for microtubules of the mitotic spindle. The centromere is essential for segregation because it provides a site for the assembly of kinetochores and facilitates attachment of microtubules. In the absence of a centromere, the replicated chromosomes segregate randomly, leading to frequent loss of duplication of chromosomes. If present in multiple copies, centromeres can cause a single chromosome to be pulled into both daughter cells, leading to chromosome breakage.

6.9 MOLECULAR NATURE OF CHROMOSOME STRUCTURE

The primary role of the chromosome is to provide a framework which allows each linear segment of the genome to replicate and segregate efficiently. If either of these processes fails, there is a resulting chromosome imbalance in the daughter cells. Three specific *cis*-acting sites are required for stable chromosome maintenance: an **origin of replication**, a **centromere** and **telomere**.

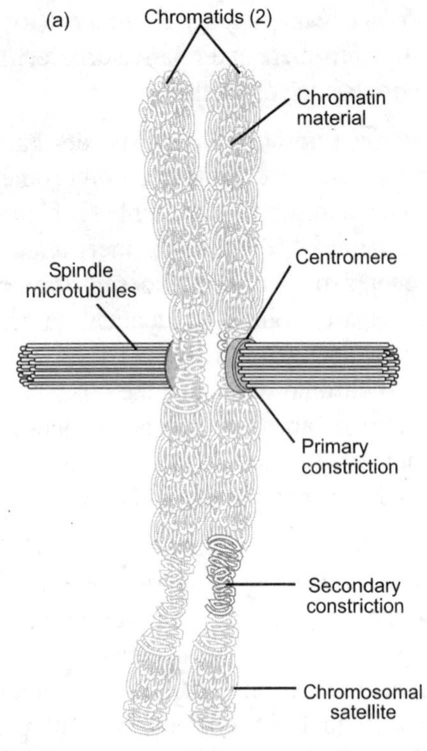

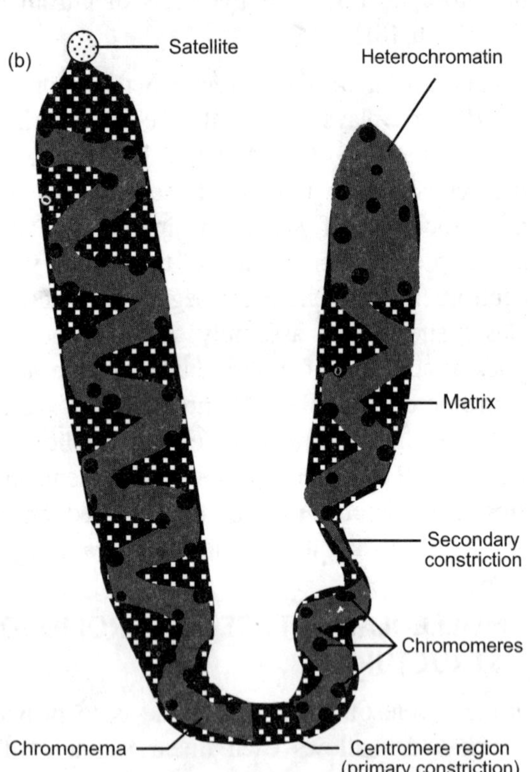

Fig. 6.18 (a) A generalised structure of a eukaryotic metaphase chromosome showing the two chromatids connected at the centromere region (b) A classical illustration of a generalised mitotic chromosome showing major regional differences often described under the light microscope

Origin of replication

It is essential for the replication of the chromosome as it is a site for the assembly of proteins that initiate replication. They are found some 30–40 kb apart throughout the length of each eukaryotic chromosome. The origins of replication are generally found in the non-coding regions (see **Chapter 8**).

Telomeres

These are sequences at the ends of eukaryotic chromosomes and they play a critical role in replication and maintenance of chromosomes. The telomeres are essential for chromosome stability because they allow the completion of chromosome ends during DNA replication. Since eukaryotic chromosomes are linear, each has two telomeric ends. Besides marking the termination of the linear chromosome they also have several specific functions. Telomeres are highly stable and prevent the chromosomal ends from breaking. They also prevent the ends of chromosomes from being degraded by exonuclease and allow chromosomal ends to be properly replicated.

Telomeric DNA consists of short, tandemly repeated sequences (**Table 6.3**). These have been characterised from a number of eukaryotes and are generally GC-rich, with guanidine residues clustered on one strand and cytidine residues on the other.

Nucleolar organiser

This site contains the multiple repeats of the DNA unit coding for the ribosomal RNA (rRNA), which forms a major component of nucleoli. Nucleolar material disperses entirely at the end of prophase and re-forms at each nucleolar organiser (NO). At late telophase, two or more organisers are closed and the nucleoli may fuse later in interphase.

Table 6.3 Telomere end DNA sequences

Organism	*Telomeric repeat sequences*
Yeast	
Saccharomyces cerevisiae	$G_{1-3}T$
Schizosaccharomyces pombe	$G_{2-5}TTAC$
Protozoans	
Tetrahymena	GGGGTTT
Dictyostelium	$G_{1-8}A$
Plants	
Arabidopsis	AGGGTTT
Mammals	
Human	AGGGTT

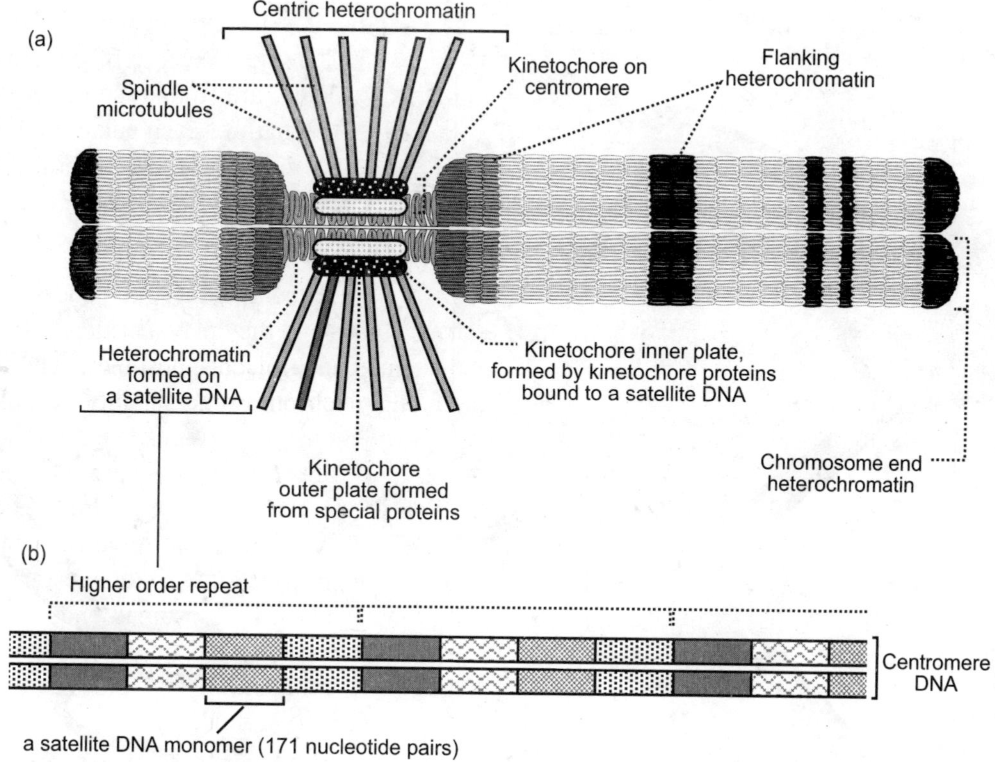

Fig. 6.19 The structure of a human centromere (a) The structure and different components of an entire chromosome (b) The organisation of α satellite DNA sequences, which are repeated many thousands of times at a centromere

6.10 STRUCTURE OF CHROMATIN

Eukaryotic chromosomes in the interphase are generally referred to as chromatin to describe the nucleoprotein fibres in the nucleus. Light-microscope studies in the 1930s distinguished between two types of chromatin in the interphase nuclei of many higher eukaryotic cells. The extent of chromatin condensation varies during the lifecycle of the cell and plays an important role in regulating gene expression. In the interphase cells, the highly condensed form is called **heterochromatin** and the less condensed form is called **euchromatin** (**Fig. 6.20**).

Heterochromatin comprises the dense, readily-stained areas of the nucleus or chromosome and was believed to represent inactive chromatin which was not undergoing transcription. In a typical mammalian cell, approximately 10% of the genome is packaged into heterochromatin. Although it is present throughout the chromosome (**Fig. 6.21**), it is concentrated in specific regions, including the centromeres and telomeres. Analysis of chromosomes of eukaryotic cells has shown

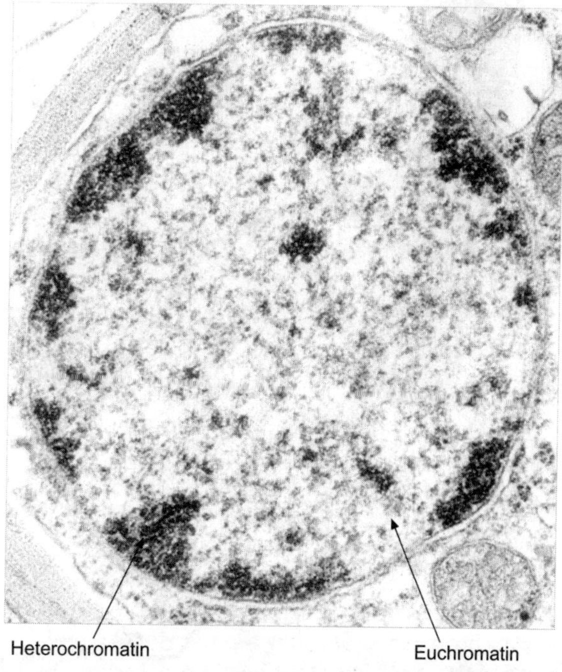

Fig. 6.20 TEM photomicrograph of tomato leaf nucleus showing heterochromatin (densely stained patches) mostly located at peripheral region and euchromatin (lightly stained areas) (***Source:*** Channarayappa 1992)

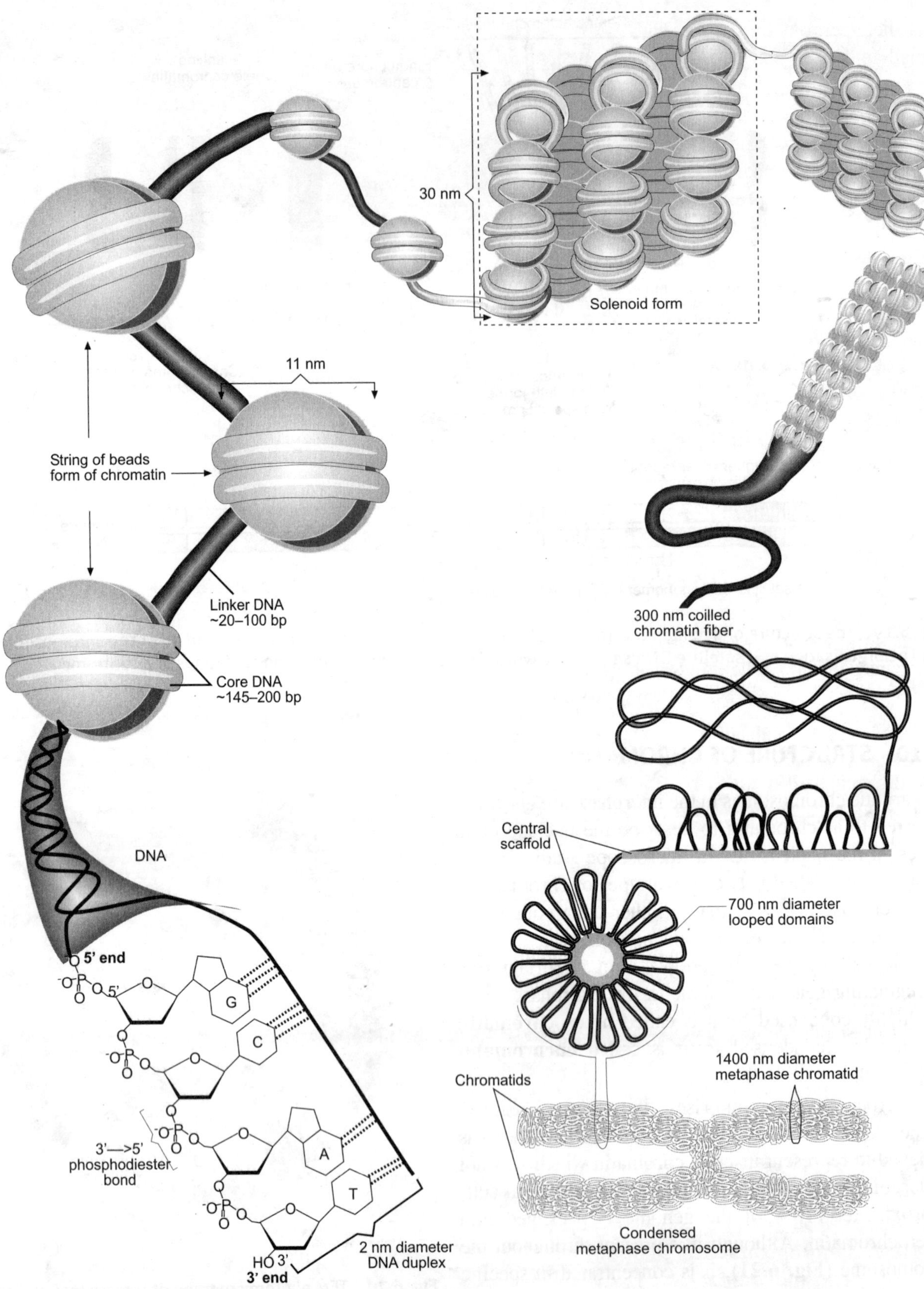

Fig. 6.21 Schematic view of the structure of nucleosome and various stages in the condensation of DNA and chromatin in forming the metaphase chromosome

that heterochromatin contains DNA and protein, with small amounts of chromosomal RNA.

Chromatin is deactivated by the addition of proteins to the nucleosomal fibre. The inactivation may be due to a variety of effects, including condensation of chromatin to make it inaccessible to the apparatus needed for gene expression or the addition of proteins that directly inhibit transcription. Two systems that have been characterised at the molecular level involve HP1 in mammals and the SIR complex in yeast, which have similar general mechanism. RAP1 has the crucial role of identifying the DNA sequences at which heterochromatin forms. It recruits SIR3/SIR4 and they interact directly with the histones H3/H4. Once SIR3/SIR4 has bound to histones H3/H4, the complex may polymerise further and spread along the chromatin fibre.

Most DNA condensed into heterochromatin does not contain known coding sequences. Rather, it may contain functional regions. Bulk of the constitutive heterochromatin may be found in the centromeric region or at other places along a chromosome. The distal arm of the Y chromosome in male mammals contains constitutive heterochromatin.

The **facultative heterochromatin** (at times condensed) is specifically inactivated during certain phases of an organism's life or in certain types of differentiated cells. An example of facultative heterochromatin can be seen by comparing male and female cells of a mammal. At about the 16th day of embryonic development in human females, one of the two X chromosomes in a cell is heterochromatinised, partially or wholly and becomes inactive genetically. This X chromosome remains heterochromatic for the lifetime of the cell and its descendants. The cells of males have a tiny Y chromosome and a much larger X chromosome. Although cells of females contain two X chromosomes, only one of them is transcriptionally active. The other X chromosome remains condensed as a heterochromatic clump called **Barr body** after the researcher who discovered it in 1949. Heterochromatin also plays an important role in protecting the genome from being overtaken by 'parasitic' mobile elements.

6.10.1 Cytological features of normal chromosomes

When chromatin is extracted from nuclei and examined under the electron microscope, at low salt concentration, isolated chromatin resembles beads on a string. Chromatin in the extended form is a thin filament of DNA connecting the bead-like structures termed nucleosomes which are about 10 nm in diameter and are the primary structural units of chromatin. If chromatin is isolated at physiological salt concentration (~0.15 M KCl), it assumes a more condensed fibre-like form, 30 nm in diameter. Individual nucleosomes can be isolated by nuclease digestion of extracted chromatin, because the DNA component of nucleosomes is much less susceptible to digestion than is the linker DNA connecting the nucleosome.

6.10.2 Higher-order structure of chromatin

The chromosomes in most eukaryotic cells are readily visible during mitosis, when they coil up to form highly condensed structures. In fact, a typical interphase chromosome reduces its length by a factor of ten and produces a dramatic change in the appearance of chromosomes. Eukaryotic chromatin must be compacted to fit within the cell.

Packaging of chromatid fibres

Since the nucleosome has a width of only 110 Å (11 nm), and metaphase chromosomes appear to be made of a fibre having diameter of about 1,400 nm (**Fig. 6.22**), several additional levels of chromatin compaction lead to the metaphase chromosome.

Nucleosome: First-order packaging involves the winding of DNA around a core complex called nucleosome, whose diameter is 11 nm.

Solenoid: Nucleosomes also associate to form a second order of packaging spontaneously. The nucleosomes and the DNA strands linking them are packed closely to produce a helix with a diameter of 30 nm containing about six nucleosomes per turn (reduced by a factor of 7). This is known as the solenoid fibre or hollow contact helix of nucleosomes. Woodcock and colleagues have proposed a variable zig-zag ribbon structure for the 30-nm fibre.

Extended chromosome: Chromosomes are further compacted by folding the solenoid fibre into a series of radial loop domains. Worcel produced the first evidence in support of this model in 1976 when they subjected *Drosophila* chromatin to mild digestion with DNase I and measured the sedimentation coefficients of the digested chromatin. They found that the coefficients decreased gradually with digestion and

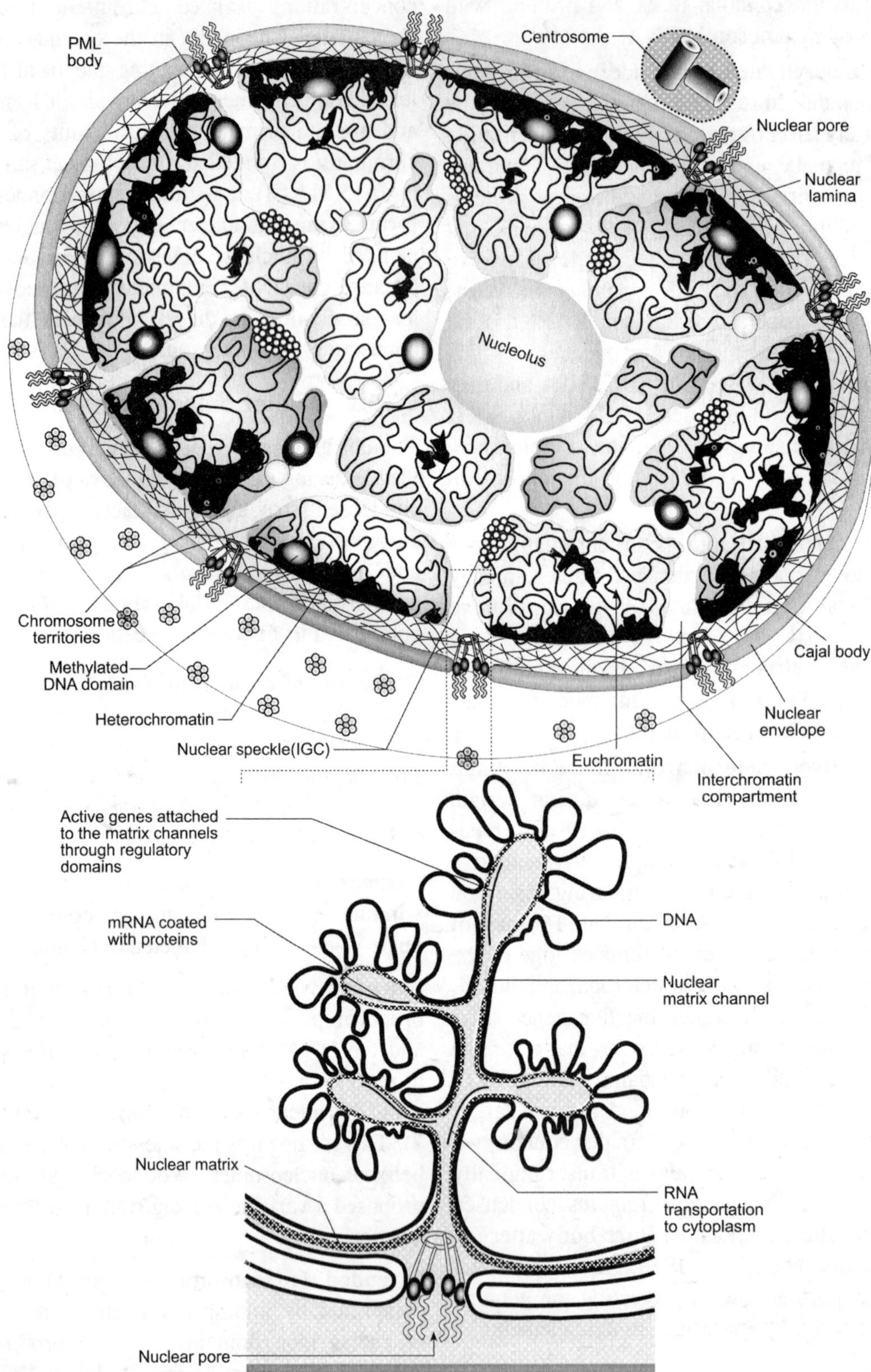

Fig. 6.22 Organisation of chromosomes in the nucleus (a) Individual chromosomes occupy distinct areas of the nucleus called chromosome territories; presence of active chromatin along the nuclear periphery and channels. Speculative model of interphase chromatin organisation, envisioning the nuclear matrix as a series in internal channel (***Source:*** adapted from various sources).

reached a plateau. Worcel calculated that each loop in a *Drosophila* chromosome contains about 85 kb. The size of the loops in the extended chromosome is about 300 nm. Loops of DNA are anchored to a central scaffold in the skeleton of the chromosome.

Compressed chromosome: The size of the loops in the compressed chromosome is about 700 nm. These loops are stabilised by proteins and not by RNA. The loops are constrained by interaction with a protein complex known as the **nuclear scaffold** (also known **nuclear matrix**). The DNA sequences that bind to the nuclear scaffold are known as **scaffold-attached regions** (**SARs**) or **matrix-attached regions** (**MARs**). In addition, the matrix contains an ill-defined fibrous protein network and is also associated with the nuclear pore–lamina complex to which the chromosomes attach.

Metaphase chromosome: It is the highly condensed form of a chromosome. The diameter of the metaphase chromosome is about 1,400 nm.

6.11 SPECIAL CHROMOSOME STRUCTURES

All the members of a species and all the cells in an organism generally have the same type and number of chromosomes. However, in some organisms, certain tissues contain chromosomes, which differ significantly from normal chromosomes in terms of either morphology or function. Such chromosomes are referred to as **special chromosomes**.

6.11.1 Accessory chromosomes

Some species carry extra chromosomes, often appearing to be composed entirely of heterochromatin, which have no effect on the phenotype and often vary between populations, within a given population or even between cells in an individual. These structures, which are variously referred to as **accessory chromosomes**, **satellite chromosomes**, **supernumerary chromosomes**, or in plants, **B-chromosomes** (to distinguish them from the invariant and essential A-chromosomes) often do not take part in mitosis and thus segregate randomly. They may be regarded as giant linear plasmids.

6.11.2 Lamp brush chromosomes

In amphibian oocytes (immature), the diplotene stage of meiosis may last for several months in developing oocytes and during this time the chromosomes are visible under the phase contrast microscope while still undergoing transcription. Such chromosomes have been most extensively studied in the oocytes of *Triturus* (newt) and *Xenopus* (toad). At this stage the chromosomes are paired, the two homologues being held together by chiasmata, and can be seen to extrude long uncoiled lateral loops of transcriptionally active chromatin from an axis of densely packed chromomeres. Each member of the pair consists of two chromatids, i.e., two DNA duplexes (four chromatids). The DNA duplexes run the length of the chromosome, being attached to the axial thread in a series of loops. As there are two chromatids, the loops are paired (**Fig. 6.23**) and the structure has the appearance of a cleaning brush. Hence the name, lamp brush chromosomes. The loops are the site of transcriptional activity and the DNA in the loops may be only partly condensed into nucleosomes. The loops are made more visible by the accumulated RNA. As the RNA polymerase molecules move around a loop, the length of the RNA transcript increases, producing the so-called **Christmas tree effect**.

Undisputedly, loops represent the active sites for gene action (transcription), and the function of lamp brush chromosomes is to produce the large amount of proteins and RNAs stored in eggs. In all eukaryotes, fertilisation of a mature egg triggers a dramatic increase in protein synthesis, but it is not generally accompanied by RNA synthesis.

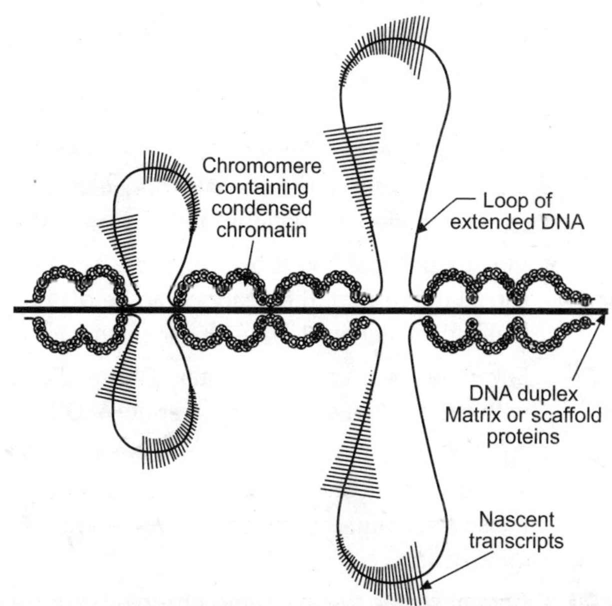

Fig. 6.23 Diagrammatic illustration of part of a lampbrush chromosome showing the Christmas tree effect produced by the nascent transcripts

6.11.3 Polytene chromosomes

In larval secretory tissues, gut epithelium, Malphigian tubules and some fat bodies of dipterans (flies, mosquitoes), as well as in the salivary glands of *Drosophila* we can see giant chromosomes (**Fig. 6.22**). In *Drosophila,* homologous chromosomes pair to form the giant chromosome (**somatic pairing**) so that the total number of giant chromosomes in a $2n$ nucleus is only n. The giant chromosomes radiate as five long and one short arm from a deeply staining and more or less amorphous structure called chromocentre.

Many studies have clearly suggested that each giant chromosome consists of numerous strands, each strand representing one chromatid. Therefore, these chromosomes are also known as **polytene chromosomes**. They are formed by repeated replication of DNA strands without separation of chromatid strands (called **endoreduplication**) and cytoplasmic cell division.

KEY POINTS

- The concept of the gene is fundamental because its physical and chemical nature is the basis for all the biological activities.
- Most genes are discrete non-overlapping units. Overlapping genes are independently regulated but utilise some of the same sequence.
- A gene is an entire nucleic acid sequence that is necessary for the synthesis of a segment of DNA that is expressed to yield a functional product, which may be either an RNA or polypeptide.
- In contrast to polycistronic mRNAs of prokaryotes, most eukaryotic transcription units produce mRNAs that encode only one protein.
- One of the most striking features of eukaryotic DNA is the great quantity (30 to more than 90%) of its non-coding DNA (called 'junk' or 'parasitic' or 'selfish'), which is present in each cell.
- The complex multigene families generally consist of a cluster of several related but not identical genes, each transcribed independently and separated by a spacer. There are several types of complex multigene families.
- Repetitive sequences, which are scattered throughout the genome, accounting approximately for 45% of human genomic DNA are SINEs (short interspersed elements) and LINEs (long interspersed elements). These sequences have a unique property of moving within a genome. Hence, these sequences are called mobile DNA elements (or transposable elements).
- Genetic material (DNA and RNA) occurs in various forms in different cell types. Therefore, DNA has to be packaged within a cell in such a way as to allow programmed access to the encoded information. This is accomplished by forming a highly organised structure called chromosome.
- Prokaryotic cells generally have only one complete copy of their chromosome(s) that is packaged into a structure called the nucleoid and carry one or smaller independent circular DNAs called plasmids.
- Viruses are small infectious particles that contain nucleic acids as their genetic material, surrounded by a capsid of proteins. The genome can be DNA or RNA, and it can be single-stranded (ss) or double-stranded (ds) and the genome can be linear or circular.
- Eukaryotic chromosomes contain DNA molecules of enormous length and with many levels of packaging within the nucleus, involving a variety of DNA-binding proteins.
- The centromere is a specialised region of the chromosome that plays a critical role in ensuring the correct distribution of duplicated chromosomes to daughter cells during mitosis. It is also termed as primary constriction. In some chromosomes, a second constriction is also present, which is known as secondary constriction.
- Chromosomes contain many conserved sequences, called the origin of replication, which is essential for the replication of the chromosome. The sequences at the ends of eukaryotic chromosomes, called telomeres, play a critical role in replication and maintenance of chromosomes.

➲ Heterochromatin comprises the dense, readily-stained areas of the nucleus or chromosome and it represents inactive chromatin not undergoing transcription.

Self-assessment questions

1. Define a gene.
2. Distinguish between coding and non-coding genes.
3. Distinguish between the gene structure of prokaryotes and eukaryotes.
4. Discuss the origin of gene families.
5. Discuss briefly the various mechanisms of gene evolution.
6. What is the role of repetitive DNA in different organisms?
7. How does the size of the genome affect the complexity in an organism?
8. What are the different levels of DNA organisation in eukaryotes?
9. Explain in detail the importance of centromere and telomeres.
10. Define heterochromatin and euchromatin and describe their role in the regulation of gene expression.
11. What are nucleosomes and what is their role in a cell?
12. Discuss with the help of suitable diagrams the different types of special chromosomes.

FURTHER READING

Textbooks

Alberts B, A Johnson, J Lewis, M Raff, K Roberts and P Walter (2002). *Molecular Biology of The Cell*. 4th ed. Garland Science.

Karp G (2005). *Cell and Molecular Biology.* John Wiley and Sons.

Lewin B (2004). *Genes VIII.* Pearson Prentice-Hall.

Singer M and P Berg (1991). *Genes and Genomes: A Changing Perspective.* University Science Books.

Therman E and M Susman (1993). *Human Chromosomes: Structure, Behaviour and Effects*, 3rd ed. Berlin: Springer.

Watson JD, NH Hopkins, JF Roberts, JA Steitz and AM Weiner (1987). *Molecular Biology of the Gene.* 4th ed. Benjamin/ Cummings Publishing Company.

Wolfe A (1999). *Chromatin: Structure and Function.* 3rd ed. New York: Academic Press.

References

Baker TA and SH Wickner (1992). Genetics and enzymology of DNA replication in *E. coli*. *Annu. Rev. Genet.* 26:447–77.

Beerman W and V Clever (1964). Chromosome puffs. *Scientific American* 210:50.

Benzer S (1962). The fine structure of the gene. *Scientific American* 206:70–84.

Bickmore WA and AT Sumner (1989). Mammalian chromosome banding. *Trends Genet.* 5:144–78.

Britten RJ (1996). DNA sequence insertion and evolutionary variation in gene regulation. *PNAS USA* 93:9374–77.

Channarayappa (1990). Cytogenetic analysis of clastogens and aneuploidogens by immunofluorescent and micronucleus methods. PhD thesis submitted to the West Virginia University, Morgantown, West Virginia, USA.

Cooke J, MA Nowak, M Boerlijst and J Maynard Smith (1997). Evolutionary origins and maintenance of redundant gene expression during metazoan development. *Trends Genet.* 13:360–63.

Davidson JN and ML Peterson (1997). Origin of genes encoding multi-enzyme proteins in eukaryotes. *Trends Genet.* 13:281–86.

Eichler EE and D Sankoff (2003). Structural dynamics of eukaryotic chromosome evolution. *Science* 301:793–97.

Felsenfeld G and M Groudine (2003). Controlling the double helix. *Nature* 421:448–53.

Feschotte C, N Jinag and SR Wessler (2002). Plant transposable elements: Where genetics meets genomics. *Nat. Rev. Genet.* 3:329–41.

Gray YH (2000). It takes two transposons to tango: Transposable-element-mediated chromosomal rearrangements. *Trends Genet.* 16:461–68.

Greider CW (1996). Telomere length regulation. *Annu. Rev. Biochem.* 65:337–66.

Hartzog GA and F Winston (1997). Nucleosomes and transcription: Recent lessons from genetics. *Curr. Opin. Genet. Dev.* 7:192–98.

Heinkoff S, EA Greene, S Pietrokovski, P Bork, TK Attwood and L Hood (1997). Gene families: The taxonomy of protein paralogs and chimeras. *Science* 278:609–16.

Hirano T (2000). Chromosome cohesion, condensation and separation. *Annu. Rev. Biochem.* 69:115–44.

Holland PWH, J Garcia-Fernandez, NA Williams and A Sidow (1994). Gene duplications and the origins of vertebrate development. *Development* (suppl.): 125–33.

Igo-Kemenes T, W Horz and HG Zachau (1982). Chromatin. *Ann. Rev. Biochem.* 51:89.

ISCN (1995), *An International System for Human Genetic Nomenclature* (F Mittelman ed.). Basel Karger.

Kazazian HH Jr (1999), An estimated frequency of endogenous insertional mutations in humans. *Nature Genet.* 22:130.

Kornberg RD and A Klug (1981). The nucleosome. *Scientific American* 244:52.

Marsden M and UK Laemmli (1979). Metaphase chromosome structure: evidence for a radial loop model. *Cell* 17:849–58.

Rhodes D (1997). Chromatin structure: The nucleosome core all wrapped up. *Nature* 389:231–33.

Smith S and T de Lange (1997). TRF1: A mammalian telomeric protein. *Trends Genet.* 13:21–27.

von Holt C (1985). Histones in perspective. *BioEssays* 3:120–26.

Wade PA, D Pruss and AP Wolffe (1997). Histone acetylation: Chromatin in action. *Trends Biochem. Sci.* 22:128–32.

Wells RD (1988). Unusual DNA structures. *J. Biol. Chem.* 263:1095–98.

Zhimulev IF (1998). Morphology and structure of polytene chromosomes. *Adv. Genet.* 37:1–567.

CHAPTER 7

The Cell Cycle

LEARNING OBJECTIVES

After reading this chapter, you will be able to:

- List the stages and components of the cell cycle
- Understand the mechanism of regulation of cell cycle
- Appreciate the biological significance of cell cycle regulation
- Explain the activity of telomerase in eukaryotes

7.1 INTRODUCTION

The continuity of life depends on cell growth—an increase in size or mass of a developing/living system. It is an irreversible process and the most fundamental characteristic of all living organisms. Since, growth is multifactorial, it undergoes three growth patterns: **auxetic growth**—an increase in cell mass or auxesis; **multiplicative growth**—an increase in cell number due to cell division and **accretionary growth**—growth due to accumulation of extracellular products. Reproduction is an orderly sequence of events in which an organism replicates its genetic material (see **Chapter 8**), and then divides into two at the end of each cycle of cell division. This process is also called the **cell cycle**. It comprises two alternating events: **cell division**, during which a cell replicates its genome and partitions it to daughter cells, and **cell growth** during which cells increase their mass by synthesising proteins, membrane lipids and other essential components (**Fig. 7.1**). The newly formed daughter cells, further grow and divide, giving rise to a new cell population or different groups of cell populations. Cells divide by duplicating their chromosomes and segregate each duplicate copy of chromosome and essential organelles into daughter cells.

In unicellular species, such as bacteria and yeast, each cell division produces a completely a new organism. In multicellular species, a long and complex sequence of cell division is required to produce a functioning organism. Although the details of the cell cycle vary between organisms and at different times in its life, certain characteristics are universal. The most basic functions of the cell cycle are to duplicate accurately the vast amount of DNA in the chromosomes and then segregate the copies precisely into two genetically identical daughter cells. In addition to regulating the timing of cell division, control mechanisms also prevent transmission of incompletely replicated or damaged genomes to daughter cells. To prevent errors during cell division, all organisms utilise molecular mechanisms to monitor the specific steps of cell division through checkpoints.

Regulation of cell division is, therefore, very critical for the normal development of multicellular organisms. Loss of control ultimately leads to cell death or cancer. During the last few decades, significant progress has been made in our understanding of the regulatory control of step-wise progression of cell division at the different stages of mitotis and meiosis.

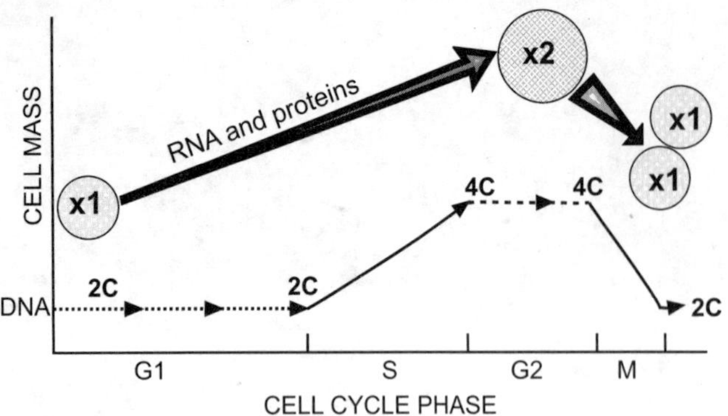

Fig. 7.1 Phases of the cell cycle; the graph compares the accumulation of continuous components with the discontinuous synthesis of DNA

Biochemical and genetic studies have revealed that cell replication is primarily controlled by regulating the timing of two critical events, i.e., nuclear DNA synthesis and the initiation of mitosis. A small group of protein kinases are the master controllers of these events, which phosphorylate at specific regulatory sites, activating some and inhibiting others to coordinate their activities. The regulatory subunits are called **cyclins** because their concentrations change in phase with the cell cycle. The catalytic subunits are called **cyclin-dependent kinases (Cdks)** because they have no kinase activity unless they are associated with a cyclin. Each catalytic subunit can associate with different cyclins, and the associated cyclin determines which proteins to phosphorylate by the Cdk–cyclin complex (**Fig. 7.2**).

Defects in cell cycle regulation are a common cause of the abnormal proliferation of cancer cells. Therefore, studies of cell cycle and cancer have become closely interconnected.

7.2 CELL CYCLE IN EUKARYOTES

In multicellular organisms, different cells have different capacities to grow and divide. These can be grouped into three categories:

Terminally differentiated cells: These include nerve cells, muscle cells, or red blood cells, tracheas and sieve elements in plants. They are highly specialised and lack the ability to divide and remain in that state until they die.

Quiescent pluripotent cells: These are cells that normally do not divide but can be induced to begin DNA synthesis and divide when given an appropriate stimulus. For example, WBCs can be induced to divide after exposure to an appropriate mitogenic compound. Liver cells can be induced to proliferate by the surgical removal of a part of the liver.

Stem cells: These cells normally possess a relatively high level of mitotic activity and are subject to continuous renewal. Spermatogonia that give rise to the male gametes, haematopoietic stem cells that give rise to red and white blood cells, and epithelial cells that line the body cavities and the body surface are some examples.

The cell cycle consists of a **nuclear** or **chromosomal cycle** and a **cytoplasmic** or **cell division cycle** (doubling of cytoplasmic components, which in eukaryotes include the organelles). In eukaryotes, the two major events of the chromosomal cycle—replication and mitosis—are controlled so that they can never occur simultaneously. Conversely, in bacteria the analogous processes, replication and partition, are coordinated so that partially replicated chromosomes can segregate during rapid growth (**Fig. 7.3**).

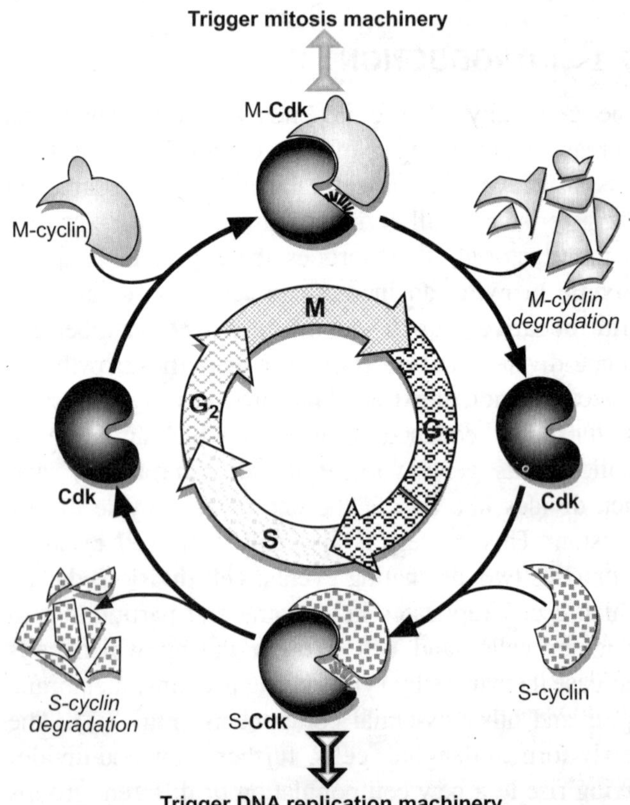

Fig. 7.2 Diagram showing the core of the control system of the cell cycle; only S-cyclin and M-cyclin and their interaction with a single Cdk are shown

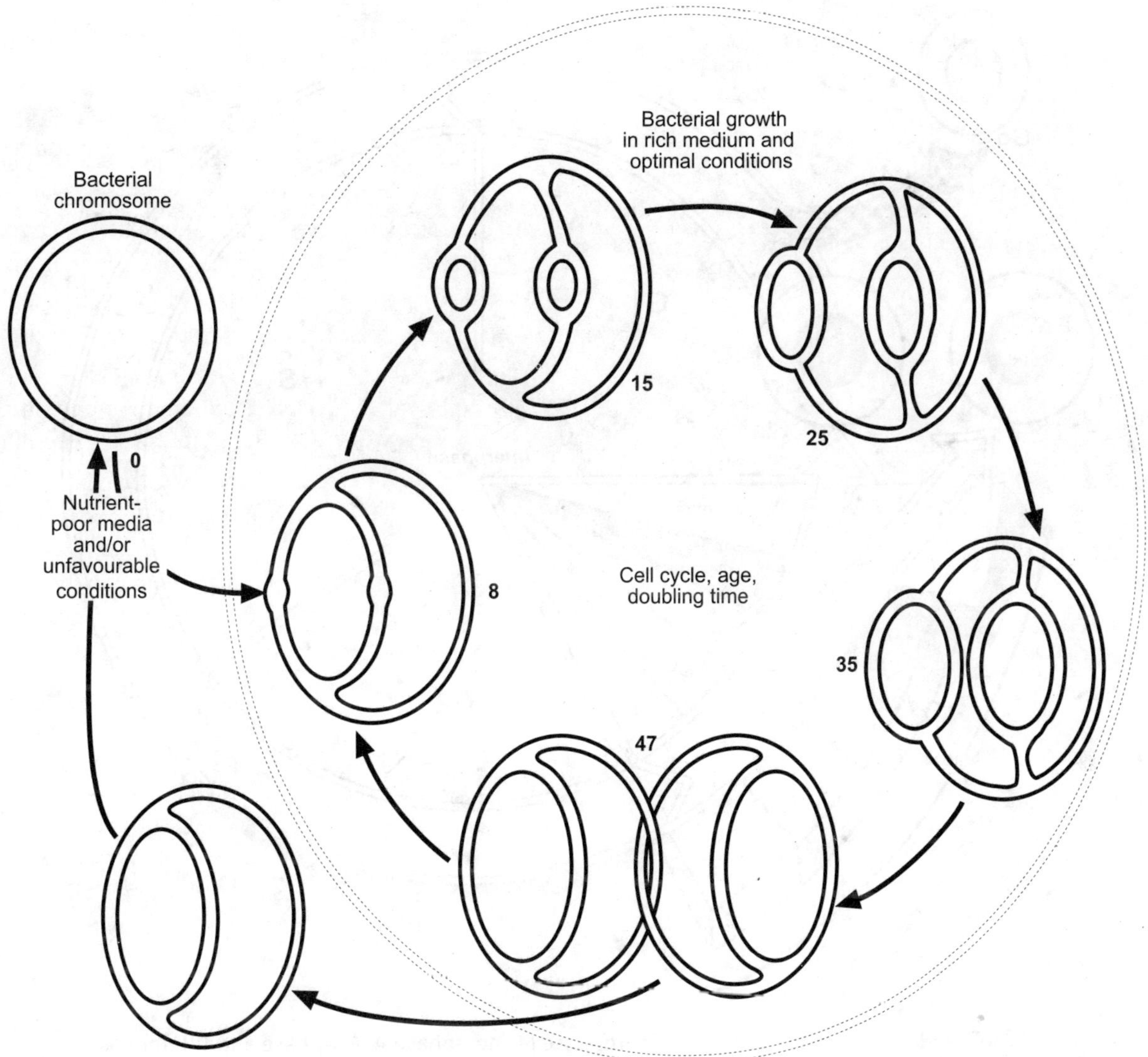

Fig. 7.3 DNA synthesis rates and chromosome configuration in *E. coli* as a function of cell cycle age (about 50 minutes) and cell doubling time (about 20 minutes); the numbers 8, 15, 25, 35 and 47 are time in minutes showing the overlapping of DNA replication cycles. The overlapping of DNA replication in optimal culture conditions reduces the average doubling time to around 20 minutes. If initiation of replication occurs only once per cell division, then doubling time will be around 50 minutes

7.2.1 Phases of the eukaryotic cell cycle

The standard eukaryotic cell cycle is divided into discrete phases, which proceed in a particular order, whereas the stages of the bacterial cell cycle may overlap. The interphase (between divisions) usually is the longest part of a cell cycle. The standard eukaryotic cell cycle is divided into discrete non-overlapping phases (**Fig. 7.4**). The duration of these cell cycle phases varies considerably in different kinds of cells. For example, rapidly dividing human cells progress through the full cell cycle in about 24 hours: mitosis takes ≈30 min; G_1 9 hours; the S phase 10 hours and G_2 4.5 hours. In contrast, in rapidly growing yeast cells, the full cycle takes only ≈90 min. Even shorter cell cycles (<30 min) occur in early embryo cells immediately after fertilisation. However, in shorter cell cycles, complete cell growth does not take place; instead, these early embryonic cell cycles rapidly divide the egg cytoplasm into smaller cells. There is no G_1 or G_2 phase, and DNA replication occurs very rapidly with very short S phase alternating with M phases.

In contrast to the rapid proliferation of embryonic cells, some cells in adult animals cease division altogether (e.g., nerve cells) and many other cells

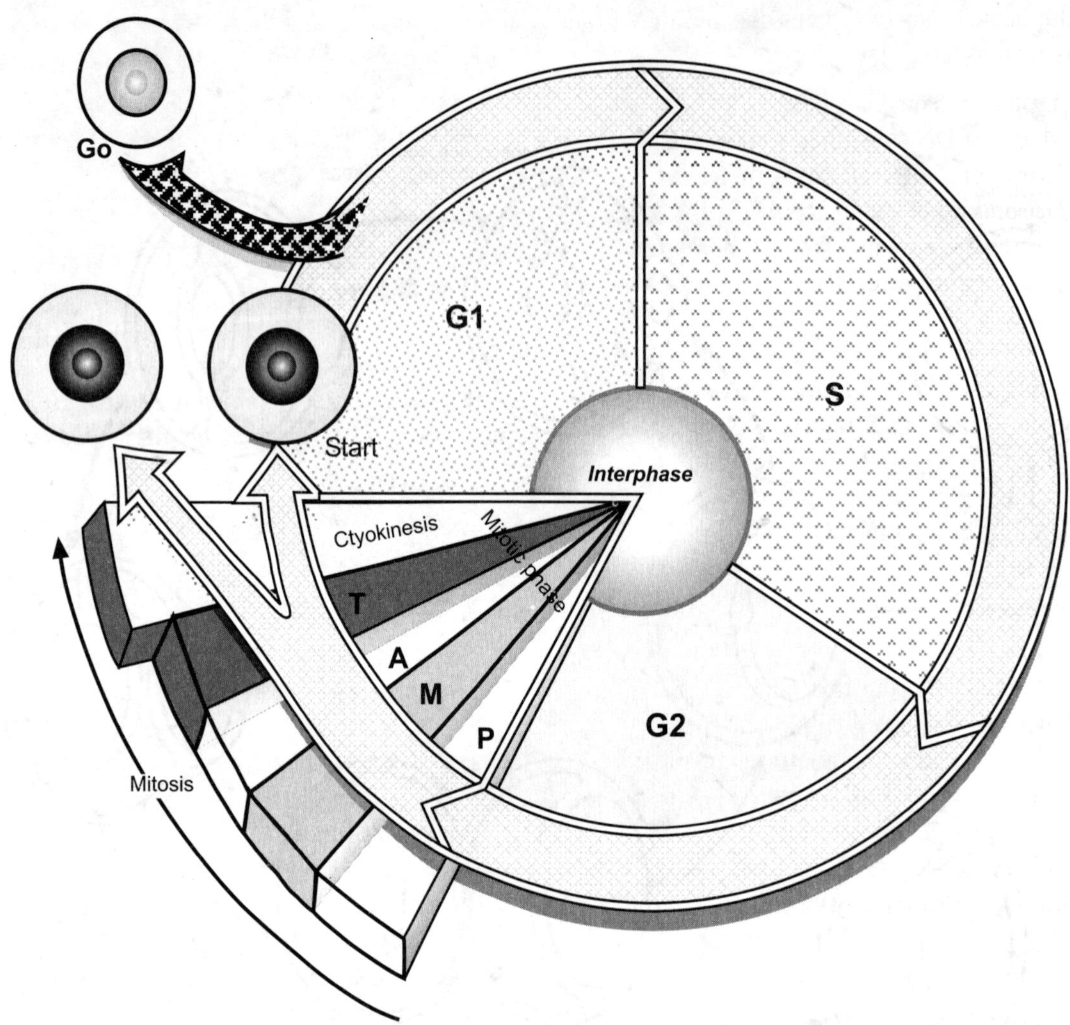

Fig. 7.4 Phases of the standard eukaryotic cell cycle; P: Prophase M: Metaphase A: Anaphase and T: Telophase

divide only occasionally, as needed to replace cells that have been lost (e.g., skin fibroblasts, liver cells). This kind of cells exit G_1 to enter a quiescent stage of the cycle called G_0, where they remain metabolically active but no longer proliferative unless called on to do so by appropriate extracellular signals.

Analysis of the cell cycle requires identification of cells at the different stages. Cells in S phase can be easily identified because they incorporate radioactive thymidine, which is exclusively incorporated into the DNA. Variations of such cell labelling experiments can also be used to determine the length of different stages of the cell cycle. Cells at different stages of the cell cycle can also be distinguished by estimating their DNA contents (which range from 2–4 nm). Experimentally, cellular DNA content can be determined by incubation of cells with a fluorescent dye that binds to DNA, followed by analysis of the intensity of fluorescence in a flow cytometer or fluorescence-activated cell sorter. On biochemical and cytological observations, the entire cell cycle can be broadly divided into **interphase** and **mitotic (M) phase**. Howard and Pelc (1953) have divided cell cycle into following four phases or stages.

Interphase: The process of crossing from one phase of the cell cycle to the next is a cell cycle transition. Transitions in the cell cycle clock are regulated by a family of protein kinases known as cyclin-dependent kinases (Cdks). During interphase, the nuclear envelope remains intact. Chromosomes appear in the form of diffused, long, coiled and indistinctly visible chromatin fibres. DNA duplicates by replication, a large amount of rRNA is synthesised and ribosomal proteins are accumulated in the nucleolus. A pair of centrioles duplicates into two pairs and net membrane biosynthesis increases.

Gap 1 (G_1) phase: The G_1 phase is called the first gap phase, since no DNA synthesis takes place during this stage. However, large amounts of RNAs (rRNA, tRNA, mRNA and other small RNAs), proteins (e.g., regulatory proteins, enzymes) and membranes are synthesised. G_1 begins after the G_0 phase or before the onset of DNA replication. During this phase, the chromatin is fully extended. During the late G_1 stage, the cell commits itself to initiating DNA synthesis and prepares for the following S interval. Nuclei and cells grow throughout the G_1 period. At this point, the cell cycle can continue or permanently stop the division and become reversibly quiescent (G_0). This is an essential phase as the cell prepares for DNA replication that occurs in the S phase. It takes about 30%–50% of the total time of the cell cycle.

Synthesis (S) phase: During this phase, replication of DNA is initiated and synthesis of histone proteins takes place. Thus, at the end of S phase, DNA content is doubled and each chromosome has two DNA molecules (i.e., duplicate set of genes). Replication of DNA and synthesis of all associated proteins must be completed prior to mitotic division. The cell cycle cannot proceed without DNA replication. This stage occupies roughly 35%–45% of the cell cycle.

Gap 2 (G_2) phase: This stage is the second gap between the completion of DNA replication and the onset of mitosis. It is the final stage of the interphase and is a period of preparation for condensation of chromosomes and mitosis. No DNA synthesis and very little RNA synthesis occur but protein synthesis continues. The cell in this phase contains twice the number of chromosomes as a normal diploid cell. This stage may occupy 10%–20% of the cell cycle. As the G_2 phase draws to a close, M phase begins by condensation of chromatin and disappearance of the nuclear membrane.

Post-mitotic stage

Post-mitotic (quiescent or G_0) cells in multicellular organisms can 'exit' the cell cycle and remain without proliferating further for days, weeks, or in some cases (e.g., nerve cells and eye lens cells) even the lifetime of the organism. Most post-mitotic cells in vertebrates exit the cell cycle in $G_1$and enter a phase called G_0. Within the G_0 phase, a cell is regarded as being quiescent or in a non-dividing state. The cell does not undergo any DNA replication and there is very little RNA synthesis. A cell may be induced to enter the subsequent G_1 phase when the quiescent cell is challenged with growth factors and hormones. This re-entry is highly regulated, thereby providing control of cell proliferation.

Mitotic (M) phase

Mitosis (Greek, *mitos* = thread) occurs in the somatic cells and it is meant for the multiplication of cell number during embryogenesis and blastogenesis of plants and animals. It occupies less than 10% of the cell cycle. It consists of nuclear division commonly followed by cytoplasmic division and is divided into prophase, metaphase, anaphase and telophase (**Fig. 7.5**). Many of the steps of meiosis closely resemble corresponding steps in mitosis.

7.2.2 Meiosis

Meiosis can be broadly divided into meiosis I and meiosis II. Since meiosis I is very long and complex it is subdivided conveniently into different phases. After the interphase, the first karyokinesis stage consists of Prophase I, Metaphase I, Anaphase I and Telophase I. Prophase I passes through leptotene, zygotene, pachytene, diplotene and diakinesis stages. At the end of the first telophase, cytokinesis starts and then the interphase is reached. During the second karyokinesis, Prophase II, Metaphase II, Anaphase II and Telophase II stages are observed.

Meiosis, like mitosis, is preceded by the replication of chromosomes (**Fig. 7.6**). During prophase I of meiosis, the duplicated chromosomes pair with their homologues through a process called **synapsis**. The four closely associated chromatids are called bivalents or tetrads. Exchange of genetic material, or crossing over, occurs between homologous (non-sister) chromatids; the results of this event are visible in the microscope by the appearance of X-shaped regions called **chiasmata**. At metaphase-I, bivalents consisting of homologous pairs of chromosomes, align on the metaphase plate. The homologous pairs are held together by chiasmata. At anaphase I of meiosis, centromeres do not divide and sister chromatids do not separate, as they do in mitosis. Rather, the sister chromatids of each chromosome go to the same pole of the cell. Meiosis I separates homologous pairs of chromosomes, not sister chromatids of individual chromosomes. The second meiotic division, meiosis II is virtually identical in mechanism to mitosis, separating the sister chromatids.

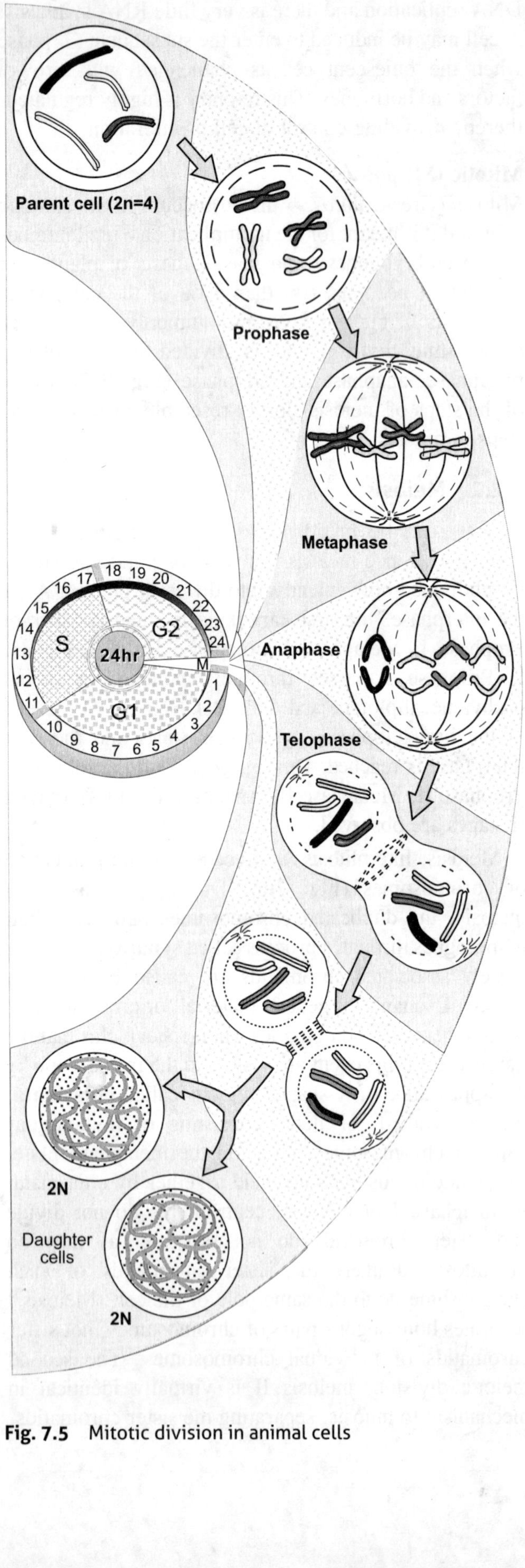

Fig. 7.5 Mitotic division in animal cells

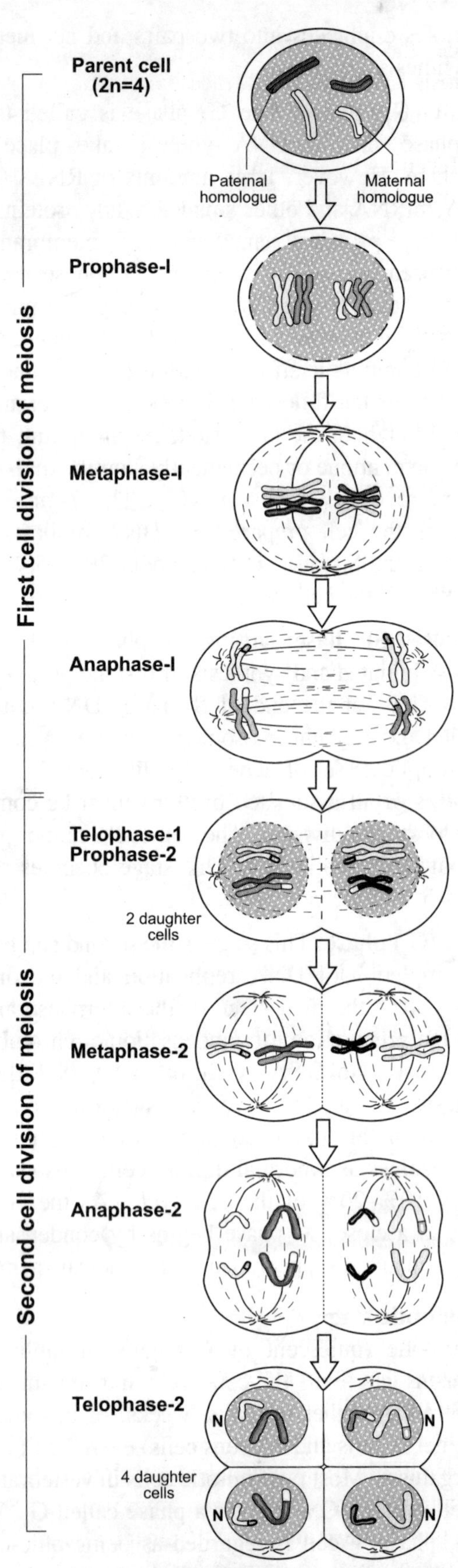

Fig. 7.6 Meiosis in animal cells

Since the chromosomes do not replicate between meiosis I and meiosis II, the final outcome of meiosis is a halving of the number of chromosomes per cell.

Differences between mitosis and meiosis

Mitosis and meiosis differ in the following manner:

- The chromosome number in a cell is reduced by half in meiosis but not in mitosis. The genetic consequences of this difference are important.
- Mitosis produces diploid daughter cells genetically identical to their parent cell and to each other; meiosis produces haploid cells that differ genetically from their parent cell and from each other.
- Although meiosis involves two nuclear divisions, the events that are unique to meiosis all occur during the first division: meiosis I.
- Neither synapses nor crossing over occurs during mitosis.

Variations in the standard cell cycle

There could be cell cycles in which one or both gap phases are omitted, or where either the S phase or the M phase is omitted, leading to halving or doubling of the DNA content, respectively. In addition, a cell may be **arrested** (indefinitely or permanently delayed) at any stage of the cell cycle, as occurs during oocyte maturation and in post-mitotic cells such as neurons. Animal cells may withdraw from the cell cycle altogether, entering a quiescent state termed G_0, where both growth and division are repressed.

7.3 REGULATION OF CELL DIVISION

Regulation of cell division is generally of two types. One is the control of cell metabolism and enlargement. A matured neuron, for example, does not grow in mass or size and does not divide. Most adult differentiated cells in humans neither grow nor divide, making only enough biomolecules (mainly proteins) to replace what is lost. The lack of net growth (in size, mass, or protein content) despite availability of abundant nutrients, imply that cells have precise mechanisms that regulate inappropriate cell enlargement. The second regulation is that of coordinated enlargement and division. If cells doubled in mass and size slightly faster than they divide, they would become larger and larger with each cycle. In contrast, if they double in mass and size more slowly than they divided, they would shrink. Generally, most cell types maintain the same average size over many generations and there must be some mechanisms that bridge division to growth.

Once cell commits to divide, within a cell division cycle there are two points at which a decision may be taken on whether to proceed through another cell cycle or not.

Commitment to chromosome replication: For most cell types, this mechanism functions at a control point in G_1. Once the G_1 phase cells grow sufficiently and meet various other conditions, they become irreversibly committed to division. This 'point of no return' is called the **commitment point** or **start** (in yeast), and is the main control switch in eukaryotic cell cycle.

Commitment to mitotic division occurs at the end of G_2: If the cell does not divide at this point, it remains in the condition of having twice the normal complement of chromosomes.

7.4 REGULATION OF CELL CYCLE PROGRESSION

Regulation of cell cycle has been studied extensively in cultured mammalian cells and in amphibian eggs. There are two complementary approaches used to characterise and isolate the regulatory components of the cell cycle. They are heterokaryon approach and study of conditional mutations in yeasts. In the former approach, nuclei at different stages of the chromosome cycle are joined in a common cytoplasm and their behaviour was observed. The results of using cultured fibroblasts synchronised at different cell cycle stages are shown in **Table 7.1**. The ability of M-phase cells to induce mitosis in any interphase nucleus provided early evidence for the existence of an **M-phase promoting factor**. Similar results were obtained in *Xenopus* nuclear transplantation studies—interphase nuclei formed spindles when injected into eggs arrested at the metaphase of meiosis I and cytoplasm from these eggs could induce meiosis in oocytes arrested at G_2. The large size of the egg was exploited to purify the substance responsible, which was called **maturation promoting factor (MPF)** and is identical to the M-phase promoting factor. The second approach has been to exploit the versatility of yeast genetics to isolate conditional mutants for cell cycle functions. Numerous cell division cycle (cdc) mutants have been identified which are blocked at various stages of the cell cycle. Another class of the so called **wee mutants** (small)

Table 7.1 Heterokaryon experiment to investigate regulatory factors controlling the cell cycle

Fusion	*Result*	*Conclusion*
S × G_1	Both nuclei replicate	S nucleus contains an S-phase promoting factor
S × G_2	S-phase cell completes replication; G_2-phase nucleus waits for S-phase nucleus to complete replication and then both cells enter the M phase	G_2 nucleus cannot respond to S-phase activator (a re-replication block); S-phase activator is also an inhibitor of mitosis
M × G_1, S or G_2	Interphase nucleus enters precocious mitosis (regardless of state of chromosome replication)	M nucleus contains an M-phase promoting factor
G_1 × G_2	Neither nucleus undergoes replication or mitosis	Both S-phase and M-phase activators are present transiently

allows precocious transition of cell size checkpoints, which are smaller in size than wild-type cells. Many of the genes identified from cdc mutants control processes such as DNA replication, repair and mating.

7.4.1 Molecular basis of cell cycle regulation

The sequential stages of the cell cycle reflect alternative states of phosphorylation for key proteins, which mediate the different cell cycle events. The cell cycle transitions represent switches in those phosphorylation states. The G_1–S transition involves the phosphorylation of proteins required for DNA replication, whilst the G_2–M transition involves the phosphorylation of proteins required for mitosis. Recent evidence revealed that certain key regulatory proteins function as signals in the commitment steps. It is now clear that the cell division cycle in a majority of eukaryotes is controlled by three major types of proteins:

- Cyclin-dependent kinases (Cdks)
- Cyclins
- Anaphase-promoting complex (APC).

A mitosis-promoting factor (MPF) was first discovered in *Xenopus* (Masui 1971). When MPF was injected into *Xenopus* oocytes, it stimulated the oocyte to enter the M phase. MPF protein contains at least two components: (1) Proteins called **cyclins** (**Table 7.2**) that undergo cycles of synthesis and accumulation during G_1 and G_2 and degradation during M phase. (2) A start- and M-phase-specific protein kinase called pp34 (pp for phosphoprotein), which is the product of the *cdc2* gene of *S. pombe* and the *cdc28* gene of *S. cerevisiae*. One class of cyclins, **M-cyclins**, is involved in the M-phase commitment and the second class, **G_1-cyclins**, is involved in the start decision. Cell cycle transitions are characterised by the burst of Cdk activity, which cause sudden switches in the phosphorylation states of target proteins responsible for cell cycle events. Sudden spike of kinase activity is not regulated by cyclin synthesis and degradation alone, as cyclins are accumulated gradually in the cell and only the mitotic cyclins are degraded by rapid, targeted proteolysis.

7.4.2 Regulation of Cdk-cyclin activity

Cdks phosphorylate target proteins, but are also regulated by phosphorylation. The yeast cdc28 and cdc2 Cdks are phosphorylated on two key residues, Tyr-15 and Thr-161. Phosphorylated Thr-161 is required for kinase activity, whereas phosphorylated T-15 is

Table 7.2 Principal cyclins and cyclin-dependent kinases (CDKs) active at each stage of the cell cycle

Stage	*S. cerevisiae* (Budding yeast)		*S. pombe* (Fission yeast)		*Mammals*	
	Cyclins	CDKs	Cyclins	CDKs	Cyclins	Cdks
G_1→S	CLN1-3	CDC28 (Cdk1)	Cig1-2	Cdc2	D1-3	CDK2 (p33), CDK4, CDK5
					E Class	CDK2
S-Phase	CLN5,CLB5 CLN6	CDC28 (Cdk1)	Cig2	Cdc2	A Class	CDK2
G_2/M-phase	CLB1-4	CDC28	Cdc13	Cdc2	A and B1, B2	CDK1 (CDC2)

inhibitory and dominant to Thr-161 phosphorylation. The principal determinant of Cdk–cyclin activity in yeast is, thus, the state of phosphorylation of Tyr-15, and some of the upstream regulatory components have also been identified. In *S. pombe*, wee1 is a tyrosine kinase, which phosphorylates cdc2 at Tyr-15 and thus deactivates it. Wee1 activity is antagonised by cdc25 phosphatase, which removes phosphate groups from the same substrate. Both wee1 and cdc25 are themselves regulated by intrinsic and extrinsic signals, and this is believed to be the basis of the G2-M checkpoint in this species. The decision to proceed with mitosis or arrest in G_2, thus, reflects the relative levels of these opposing activities.

There are three phosphorylation sites on mammalian Cdc2 (Cdk1), Thr-14 and Tyr-15, both of which are dominant inhibitors when phosphorylated, and Thr-161, whose phosphorylation is required for kinase activity. An enzyme identified in mammals is responsible for Thr-161 phosphorylation. Remarkably, this turns out to be a yet another Cdk–cyclin complex comprising Cdk7 and cyclin H; it is known as **Cdk-activating kinase** (**CAK**) (q.v. TFIIH). Cdk–cyclin complexes are also regulated by a number of inhibitory proteins described as cyclin-dependent kinase inhibitors of **CKIs** (e.g., Sic1, Far1, Rum1, p15, p16, p18, p19, p21, p27, p57). Two families of Cdk-cyclin inhibitors are found in animals. One blocks all Cdk-cyclin activities and the other specifically inhibits D-cyclin complexes containing Cdk4 and Cdk6 (substrate is the retinoblastoma protein). These regulatory mechanisms are discussed in more detail below.

Subcellular localisation of cell cycle regulators

Cells are highly compartmentalised in which regulatory molecules can be coupled to or separated from their target molecules. Subcellular localisation is a dynamic process in which cell cycle regulators are moved into different compartments at different stages of the cell cycle. For example, cyclin B1, one of the major mitotic cyclin in animal cells, shuttles between the nucleus and cytoplasm until G_2, when it accumulates in the nucleus just prior to the onset of mitosis. Nuclear accumulation of cyclin B1 is facilitated by phosphorylation of a cluster of serine residues that reside in its nuclear export signal. Phosphorylation presumably blocks subsequent export of the cyclin back into the cytoplasm.

7.4.3 Coordination of cell mass and DNA duplication

The continuous progression of the cell division cycle is regulated by both extracellular and intracellular signals that monitor and coordinate the various processes that take place during different cell cycle phases. First, the cell must replicate every DNA sequence of its entire genome once and only once before it begins to divide—each and every time. This is ensured by a **licensing factor**, which enters the nuclei during M phase and is most likely destroyed after DNA replication is initiated. Secondly, it must increase its mass to about twice as much as that at the beginning of G_1 stage so that the two daughter cells produced after division are comparable in mass to the parent cell. The control on mass, however, is not absolute and is related to the growth rate. To achieve the proper cell mass and replication, cells must coordinate during each cell cycle. Cells achieve this feat (coordination) generally by two **control points** or **checkpoints**, which occur late in G_1 and progress from G_1 to S. These regulatory points were first defined by studies in budding yeast (*S. cerevisiae*). In yeast, the first checkpoint called **START** is located in G_1 phase and is a highly regulated event (**Fig. 7.7**). At this point of cell cycle, a cell monitors its internal (e.g., cell size and mass) and external environments (e.g., availability of the nutrients) and accordingly decides whether to

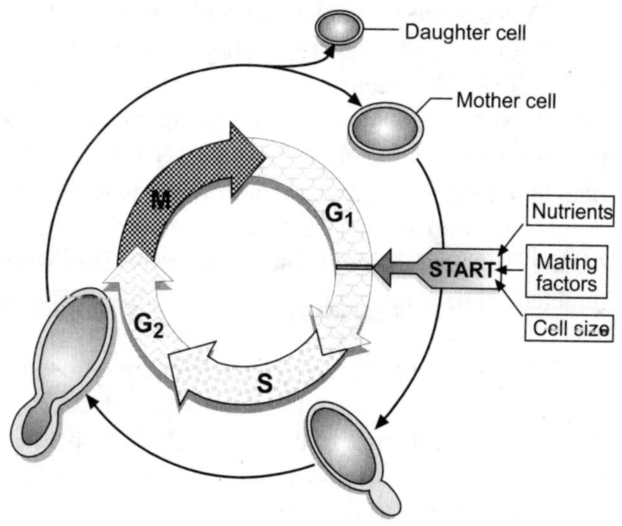

Fig. 7.7 Regulation of the cell cycle of budding yeast; in *S. cerevisiae*, cell cycle is regulated primarily at a point in late G_1 called START

enter S phase or not. If commited to enter S phase, DNA replication begins after a lag period and it will not stop before it is completed. Thus, START represents a decision point at which the cell determines whether cell is ready to support progression through the rest of the division cycle. Polypeptide factors that signal yeast mating also arrest the cell cycle at START, allowing haploid yeast cells to fuse with one another instead of progression to S phase.

The second checkpoint is at the end of G_2 phase. At this point, doubling of cell mass, completeness of DNA replication and absence of DNA damage are evaluated. This appears to be the major control point in haploid organisms. But some yeast can use either G_1 or G_2 as the main control point depending on the nutritional conditions. For example, in fission yeast (*S. pombe*) the cell cycle is primarily regulated at G_2 (transition from G_2 to M) and is the principal point at which cell size and nutrient availability are monitored. In higher animals, the example of cell cycle control at G_2 is provided by oocytes. Vertebrate oocytes can remain arrested in G_2 for long periods (several decades in humans) until their progression to M phase is triggered by hormonal stimulation.

7.4.4 The restriction of replication to once per cell cycle

The cell fusion studies have shown that the S-phase nucleus induces the G_1 nucleus to replicate but not the G_2 or M nuclei. In all cases, replication occurs only once (**Fig. 7.8**). The **G_2 checkpoint** prevents the initiation of mitosis prior to the completion of S phase, ensuring the prevention of incompletely replicated DNA distributed to daughter cells. Once a segment of DNA is replicated in S phase, control mechanisms prevent re-initiation of DNA replication until the cell cycle has been completed. The molecular mechanism that restricts DNA replication to once per cell cycle involves the action of a family of proteins called **minichromosome maintenance** (MCM) complex (composed of cdc46p, cdc47p, cdc54p, mcm2p, mcm3p and mcm6p) that bind to replication origins together with the **origin replication complex** (ORC) proteins.

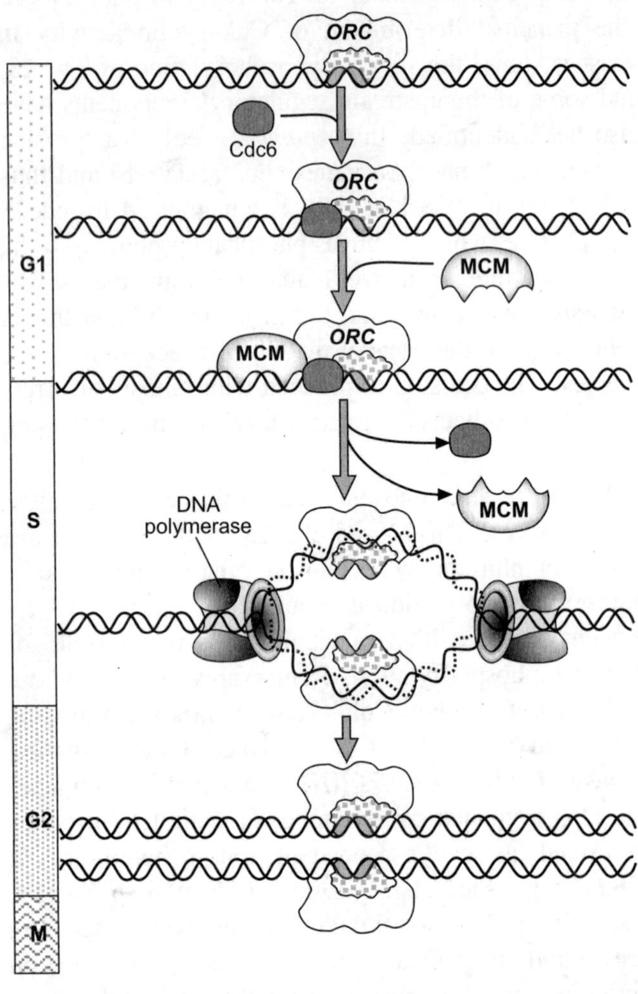

Fig. 7.8 Restriction of DNA replication once per cycle by MCM proteins that bind to the ORC

7.4.5 Unreplicated DNA prevents entry into mitosis

It would be disastrous for a cell to enter mitosis before replicating its entire genomic DNA. If a cell enters mitosis before DNA is replicated fully, it can pass on the damaged or incomplete sets of chromosomes to its daughter cells. This disaster is avoided in most cells by a **DNA replication checkpoint** mechanism, which ensures that the initiation of mitosis cannot occur until the last nucleotide in the genome has been copied (**Fig. 7.9**). The unknown sensor mechanisms detect either the unreplicated DNA or the corresponding unfinished replication forks and send a negative signal to the cell-cycle control system, blocking the activation of M-Cdk (**Fig. 7.10**). The other targets of the negative checkpoint signal are the enzymes that control M-Cdk activation. The negative signal activates a protein kinase that inhibits the cdc25 protein phosphatase. As a result, M-Cdk remains phosphorylated and inactive until DNA replication is complete.

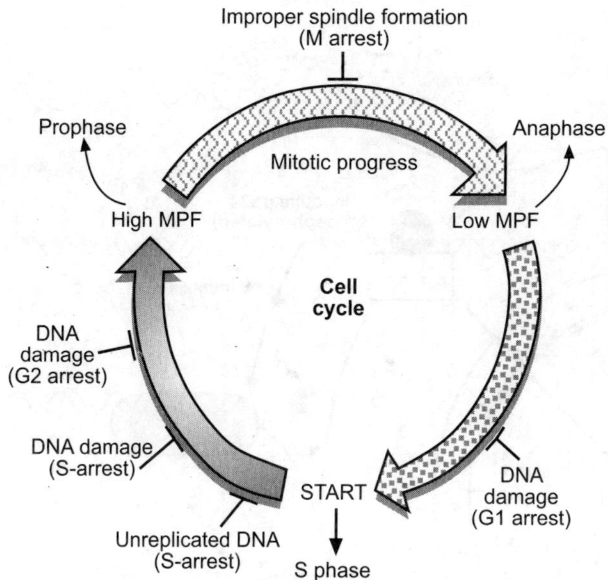

Fig. 7.9 Stages at which checkpoint controls can arrest passage through the cell cycle. MPF: Maturation promoting factor

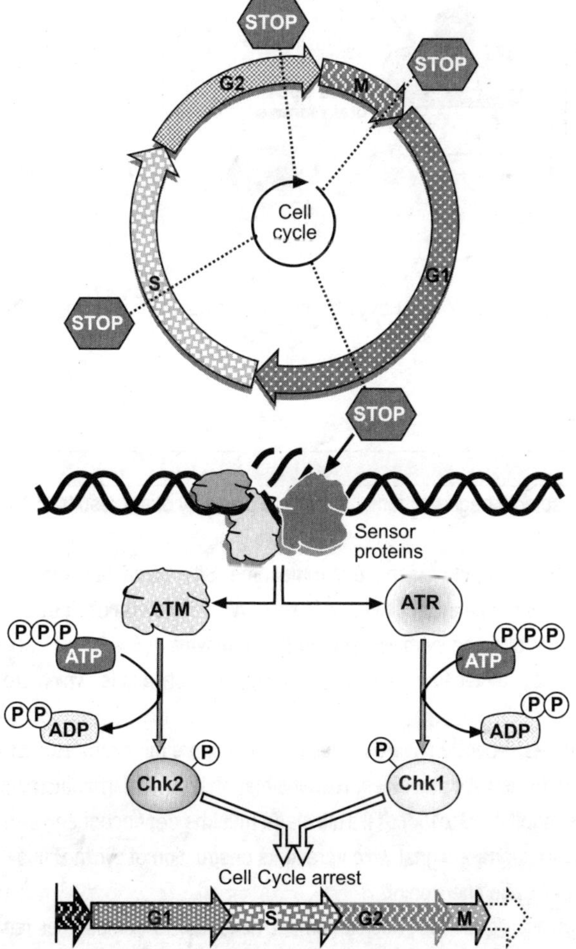

Fig. 7.10 Diagram showing arrest of cell cycle at the G_1, S and G_2 checkpoints

7.4.6 Separation of duplicated chromosomes

One of the most remarkable features of cell-cycle control is that a single protein kinase, M-Cdk (or M-phase CDK–cyclin kinase complex) is able to bring about all the diverse and complex rearrangements that occur in the early stages of mitosis (**Fig. 7.11**). The M-Cdk complex phosphorylates target proteins mediating the structural changes occurring during mitosis. These include condensation of chromatin, assembly of the spindle apparatus, reorganisation of the nucleus and assembly of the cytokinetic machinery. One critical but indirect target of the M-Cdk is the **anaphase-promoting complex** (**APC**) or **cyclosome** (APC/C), a destruction box-dependent ubiquitin ligase, which facilitates the sudden degradation of mitotic cyclins. Other APC targets include proteins required for maintaining the association of sister chromatids, possibly kinetochore components (**Fig. 7.12**) or perhaps a more generally distributed factor like *S. cerevisiae* PDS1. Another important target of M-phase kinase is the myosin light chain, which in its phosphorylated state is prevented from binding to actin and forming the cytoskeletal ring required for cytokinesis. Degradation of cyclin B is dependent upon chromatin kinetochore attachment and inhibited by unattached kinetochores. The inactivation of M-phase kinase allows M-phase kinase substrates such as myosin to be dephosphorylated, thus facilitating the cytoskeletal organisation, which precludes cytokinesis, the reconstruction of the nuclear lamina and decondensation of chromatin. The cell ensures the correct order of events in mitosis, as for the cell cycle as a whole, by organising them into a dependent series (**Table 7.3**).

7.4.7 Separation of sister chromatids

After M-Cdk complex has triggered the complex rearrangements that occur in early mitosis, the cell cycle reaches its culmination with the separation of the **sister chromatids** (**SCs**) at the transition between metaphase and anaphase (**Fig. 7.13**). Although M-Cdk activity plays a key role in this event, an entirely different enzyme complex, anaphase-promoting complex (APC), initiates separation of SCs. The APC is a highly regulated ubiquitin ligase that promotes the destruction of several mitotic regulatory proteins. The attachment of the two sister chromatids to opposite poles of the kinetochore spindles in early mitosis results in forces tending to full the two chromatids apart. These pulling forces initially resisted because the SCs were

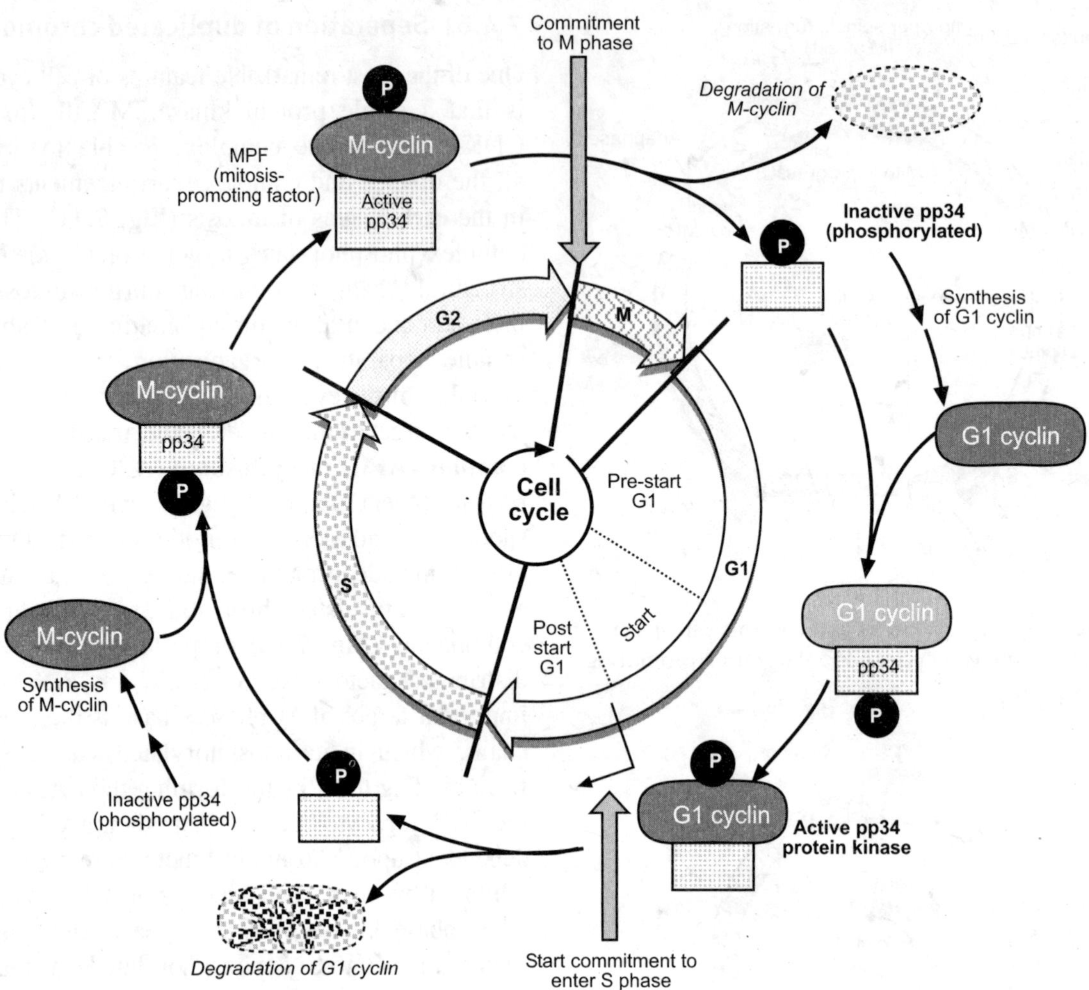

Fig. 7.11 Diagram showing some of the factors that regulate the progression of eukaryotic cells through the mitotic cell cycle

Table 7.3 Important stages of mitotic phase with cellular events and molecular mechanisms

Stage	*Critical events*
Prophase	M-phase kinase activated. Chromatin condenses and mitotic spindle begins to form. These are probably direct results of phosphorylation of target proteins by M-phase kinase.
Prometaphase	In metazoans, breakdown of the nuclear envelope allows mitotic spindle to reach the chromosome. Some spindle microtubules attach to kinetochores. The nuclear reorganisation is also directly regulated by M-phase kinase. When all chromosomes become attached to the spindle, APC is activated by cdc20/Slp1 following inhibition of Mad/Bub pathway.
Metaphase	Cyclin A degraded. Chromosomes guided to metaphase plate held by tension between opposing poles of spindle, which are attached to sister kinetochores.
Anaphase	Cyclin B degraded; M-phase kinase inactive. Paired kinetochores separate towards poles as kinetochore microtubules shorten. Poles also move apart as polar microtubules repel each other. Proteolytic machinery responsible for cyclin B degradation probably also targets a protein which binds sister chromatids together (Scc1/Mcd1). This stage must be dependent on complete alignment of chromosomes. The nature of the early metaphase signal which prevents destruction of cyclin B and precocious segregation is unknown, but constitutes a major cell cycle checkpoint.
Telophase	Substrates of M-phase kinase dephosphorylated (including myosin light chain, nuclear lamins, histones). Chromosomes reach poles, nuclear membrane re-forms, chromatin decondenses and contractile ring assembles.
Cytokinesis	Contractile ring completed. The cell constricts around the remains of mitotic spindle and cleaves into two daughter cells.

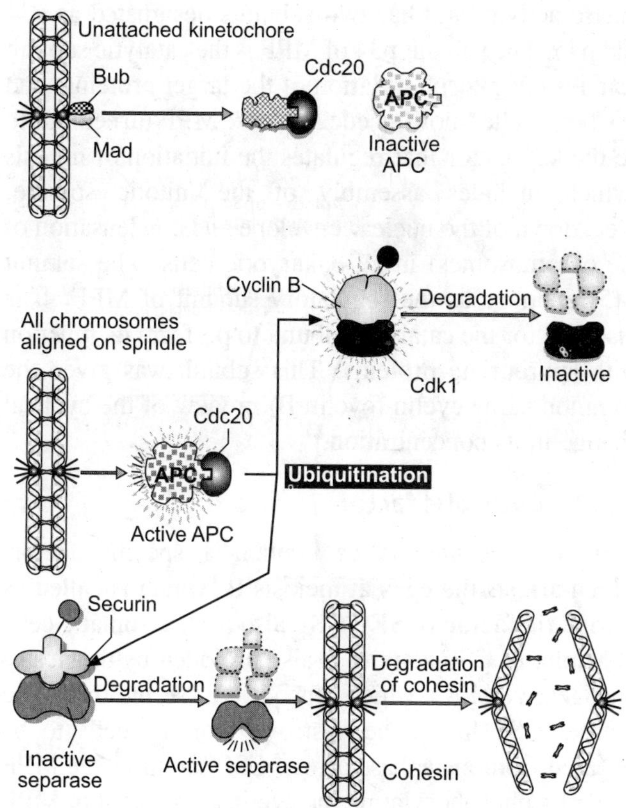

Fig. 7.12 The spindle assembly checkpoint; initiation of anaphase is mediated by activation of the APC ubiquitin ligase. Kinetochores unattached to spindles lead to the assembly of a complex of Mad/Bub proteins in which Mad proteins are activated and prevent APC activation by inhibiting cdc 20. Once all chromosomes are aligned on the spindle, the Mad/Bub complex dissociates and leads to activation of the APC. The APC ubiquitinates securin leading to the activation of separase which degrades a subunit of cohesion breaking the link between sister chromatids and initiating anaphase.

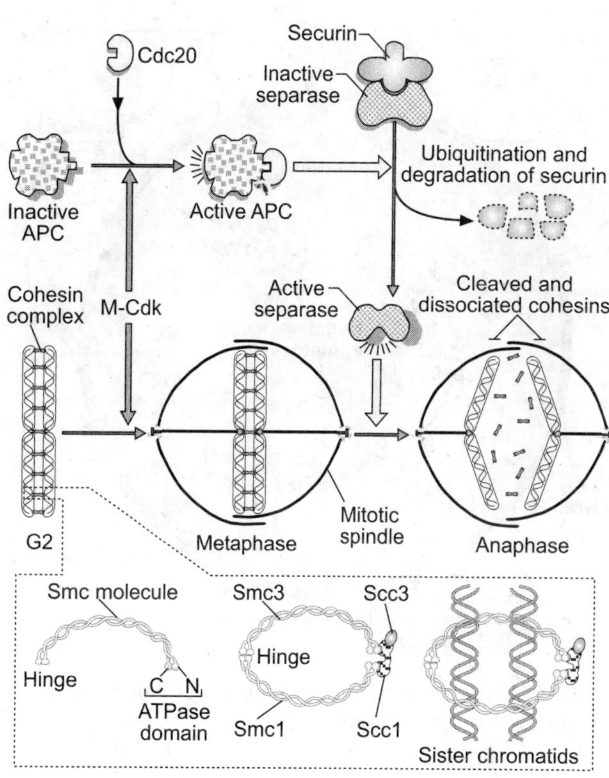

Fig. 7.13 Activation of separation of the sister chromatids

bound tightly together by the cohesin complex proteins deposited along the chromosomes. Anaphase begins with a sudden disruption of the cohesion between SCs, which allows them to separate and move to opposite poles of the spindle. Separation of the SCs requires activation of the APC enzyme complex, suggesting that proteolysis is central to the process. The relevant target of the APC is the protein **securin**. Before anaphase, securin binds to and inhibits the activity of a protease called **separase**. The destruction of securin at the end of metaphase releases separase, which is then free to cleave one of the subunits of the cohesin complex. Immediately, the cohesin complex falls away from the chromosomes and the SCs separate.

7.4.8 Control of entry into and exit from mitosis

The first evidence that *trans*-acting factors regulate the cell cycle came from cell fusion experiments with cultured mammalian cells. When interphase cells in the G_1, S or G_2 stage of the cell cycle were fused to cells in mitosis, their nuclear envelopes broke down and their chromosomes condensed. This finding indicates that some *trans*-acting component(s) in the cytoplasm of the mitotic cells forced interphase nuclei to undergo many of the processes associated with mitosis. Similarly, when cells in G_1 were fused to cells in the S phase and the fused cells exposed to radiolabelled thymidine, the label was incorporated into the DNA of the G_1 nucleus, indicating that DNA synthesis began in the G_1 nucleus shortly after fusion. However, when cells in G_2 were fused to S-phase cells, no incorporation of labelled thymidine occurred in the G_2 nuclei. This *trans*-acting factor in an S-phase cell can enter the nucleus of a G1 cell and stimulate DNA synthesis, but these factors cannot induce DNA synthesis in a G2 nucleus. These and other results suggest that cell cycle is regulated both positively and negatively (**Fig. 7.14**).

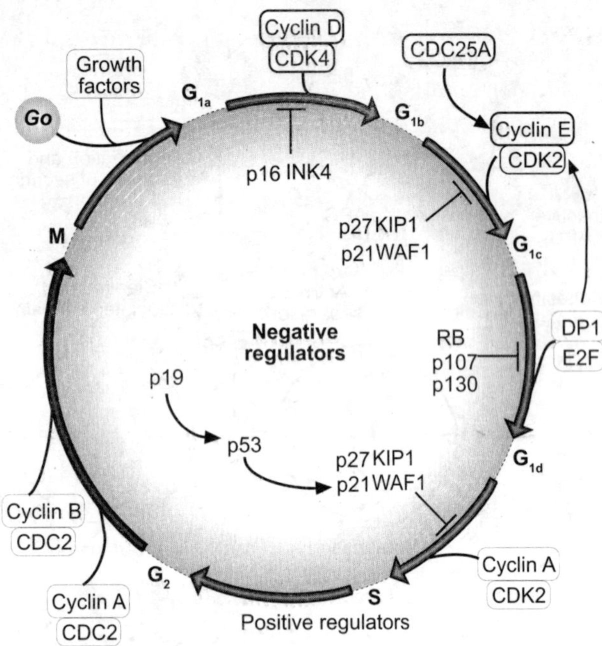

Fig. 7.14 Diagram showing the positive and negative regulators of the mitotic cell cycle

7.5 CONTROL IN MULTICELLULAR ORGANISMS

In unicellular organisms, the single major factor that restricts cell growth is the nutritional limitation. Therefore, cell size is the major cue for regulation of cell division. However, in multicellular eukaryotes with specialised cell types and functions, control necessarily becomes more complex. Although cell size signals are also used, they are insufficient to direct the complex patterns and rates of cell division. Further control of cell cycle is closely linked with the regulation of cell expansion, differentiation and cell death.

7.5.1 Oocyte maturation factors

A breakthrough in identification of the factor that promotes mitosis came from the experiments conducted on oocyte development in the frog, *Xenopus laevis* (**Fig. 7.15**). The process of oocyte maturation, from G_2-arrested oocyte to the egg arrested in metaphase of meiosis II, can be studied *in vitro* by surgically removing G_2-arrested oocytes from the ovary of an adult female frog and treating them with progesterone. This experiment permitted identification of a factor in egg cytoplasm, named maturation-promoting factor (MPF) that can stimulate maturation of oocytes *in vitro* in the absence of progesterone. MPF is a dimer with protein kinase activity and has two subunits designated as p34 and p45. The subunit p34 of MPF is the catalytic subunit (carries out phosphorylation of the target proteins) and has been called now as cdc2. Later, MPF turned out to be the key factor that regulates the initiation of mitosis (which includes assembly of the mitotic spindle, breakdown of the nuclear envelope and condensation of the chromosomes) in all eukaryotic cells. The subunit p45 functions as the regulatory subunit of MPF. It is essential for the catalytic subunit to perform its function with appropriate proteins. This subunit was given the common name cyclin (cyclin B) in view of the cyclical change in its concentration.

7.5.2 Cytosolic factor

Eggs of *Xenopus laevis* contain a specific factor, which arrests the eggs at meiosis II, which is called as **cytostatic factor** (**CSF**). CSF also arrests somatic cells in M phase. CSF activity is also provided by the proto-oncogene *c-mos*, which encodes a serine/threonine kinase. CSF has to be destroyed for the cells to be released from metaphase arrest. CSF stabilises the MPF either by phosphorylating the cyclin component of MPF or the protease responsible for cyclin degradation.

7.5.3 Synthesis and degradation of cyclin B

Experiments with inhibitors revealed that new protein synthesis is required for the increase in MPF during

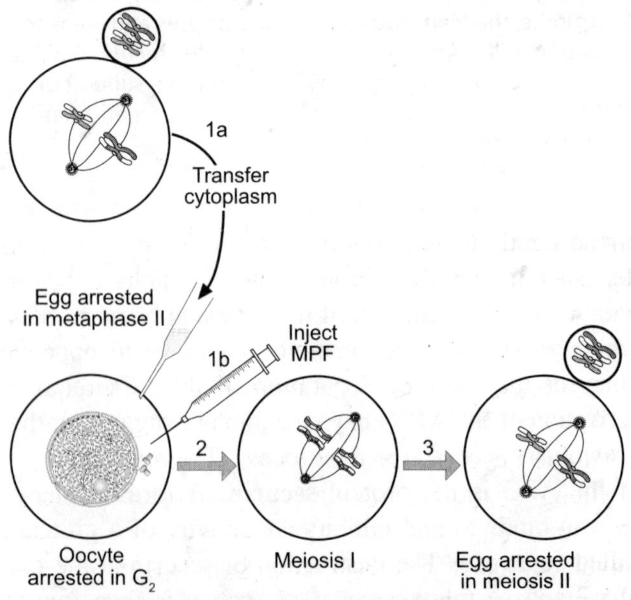

Fig. 7.15 Experimental demonstration of meiotic maturation in arrested *Xenopus* eggs due to a diffusible factor

the mitotic phase in early frog embryos. This clue stimulated the researchers to look for a protein whose synthesis followed the time course of MPF activity in early embryos, reasoning that such a protein might be a component of MPF. In synchronously fertilised sea urchin eggs, a protein was isolated, which peaked in intensity at each mitosis, fell abruptly during anaphase of each mitosis and then slowly accumulated during the next interphase to peak at the beginning of the following mitosis. This protein has been called cyclin. Many such mitotic cyclins have now been found in all eukaryotic cells analysed. The critical cyclin abruptly destroyed at anaphase in early sea urchin embryos is called **cyclin B**.

MPF protein kinase activity, assayed by its phosphorylation of histone H1, was found to rise and fall in synchrony with the concentration of cyclin B. The early events of mitosis such as chromosome condensation and breakdown of nuclear envelope occurred when MPF protein kinase activity reached its highest levels in parallel with the rise in cyclin B concentration. Addition of cycloheximide, an inhibitor of protein synthesis prevented cyclin B synthesis and the rise in MPF activity. These results demonstrate that cyclin B is a crucial protein whose synthesis is required to regulate MPF activity.

7.5.4 Proteasome degradation of polyubiquitinated cyclin

Protein degradation plays an important role in the regulation of several cell cycle pathways. This is facilitated by the protein ubiquitin (activated by enzymes E1, E2 and E3), which forms a 26S proteasome with the target proteins to be degraded (see **Chapter 13**). Several mitotic cyclins from various eukaryotes have shown that all the encoded proteins contain a homologous sequence near the N-terminus called the destruction box. In intact cells, cyclin degradation begins at the onset of anaphase, the period of mitosis when sister chromatids are separated and pulled toward opposite spindle poles.

Mitotic cyclins provide just one example of polyubiquitination, a widely occurring process for marking proteins for rapid degradation. Polyubiquitinated proteins are degraded rapidly by multiprotein complexes of proteolytic enzymes called proteasomes. Different recognition proteins regulate the degradation of different classes of proteins, but the same type of ubiquitin ligase catalyses addition of ubiquitin to all substrates. In mitotic cyclins, the recognition protein interacts with the destruction box, and ubiquitin is added to lysine residues on the C-terminal side of the destruction box. According to the recent model (**Fig. 7.16**), the signals that trigger degradation of cyclin B during anaphase explains, when MPF protein kinase activity reaches a high level, it phosphorylates and thereby activates the cyclin destruction box recognition protein. Polyubiquitination of cyclin B then occurs, leading to the degradation of cyclin B. Since cyclin B is an essential subunit of MPF, its destruction causes inactivation of the MPF protein kinase activity.

7.5.5 Phosphorylation of nuclear proteins

Entry into mitosis is a consequence of the phosphorylation of specific proteins by the protein kinase activity of MPF. The common feature in substrates that are phosphorylated by M phase kinase is the presence of the duo Ser-Pro, flanked by basic residues (often in the form of Ser-Pro-X-Lys). Potential substrates, based upon the ability of M phase kinase to prepare the targets *in vitro*, include H1 histone (involved in chromatid condensation), nucleolin (involved in the arrest of ribosome synthesis), lamins (required for nuclear envelope breakdown) and other structural and enzymatic activities. For example, early in mitosis, MPF phosphorylates specific serine residues in all three nuclear lamins (A, B and C), causing depolymerisation of the lamin intermediate filaments. The resulting phosphorylated lamin A and C dimers are released into solution (**Fig. 7.17**), whereas the phosphorylated lamin B dimers remain associated with the nuclear membrane *via* their isoprenyl anchor. Depolymerisation of the nuclear lamins leads to disintegration of the nuclear lamina mesh and contributes to the breakdown of the nuclear envelope into small vesicles.

7.5.6 Dephosphorylation and late mitotic stages

Inactivation of MPF coincides with the late stages of mitosis (anaphase and telophase) in *Xenopus*. During these stages, sister chromatids separate and move to opposite spindle poles, microtubule dynamics return to interphase conditions, the chromosomes decondense, the nuclear envelope re-forms, the endoplasmic reticulum and Golgi complex are remodelled and cytokinesis occurs. Some of these processes are triggered by protein degradation and others by dephosphorylation.

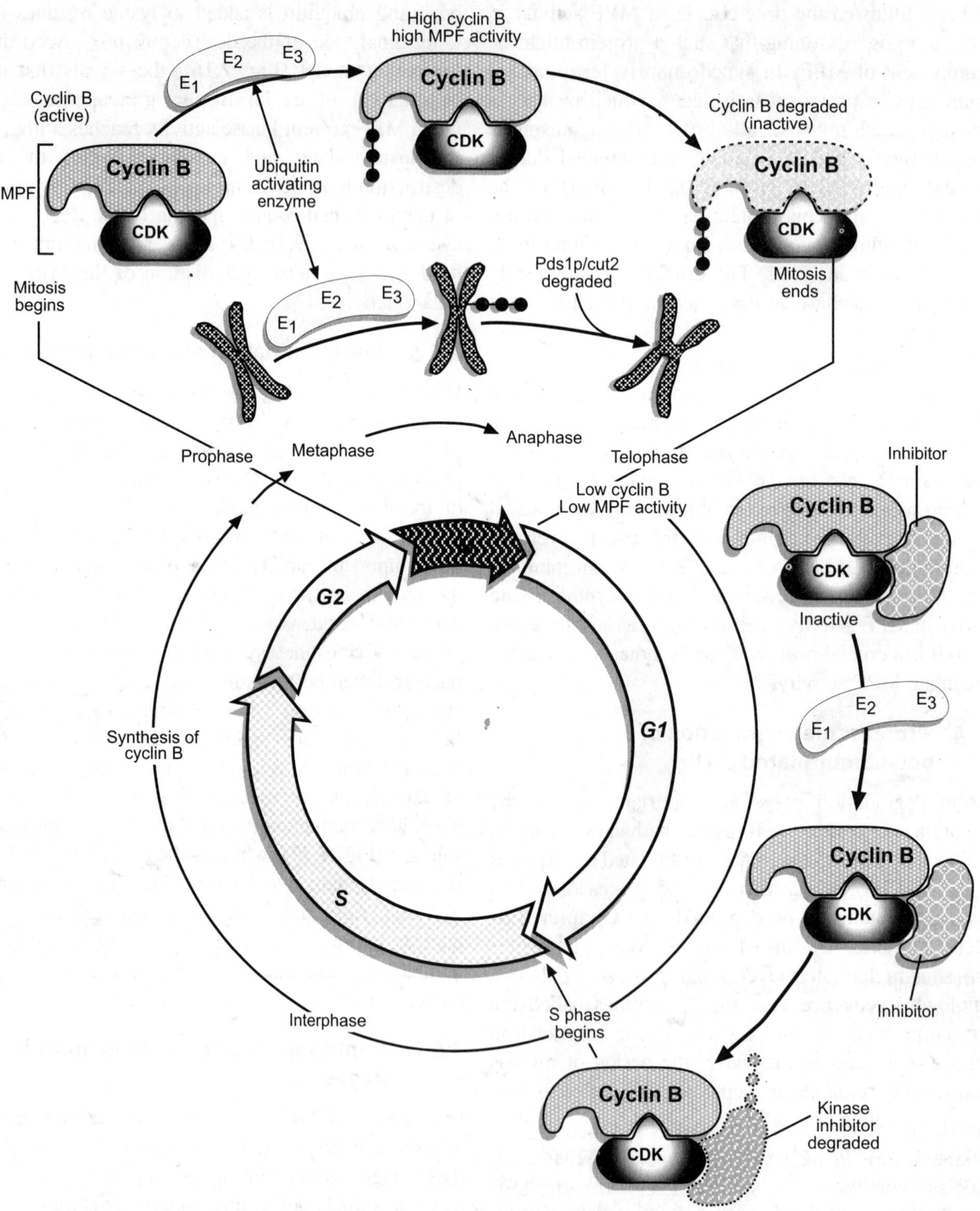

Fig. 7.16 Proposed model for regulation of mitotic cyclin levels in cells; ubiquitination of cell cycle proteins with the help of three enzymes (E_1, E_2 and E_3) is essential for degradation

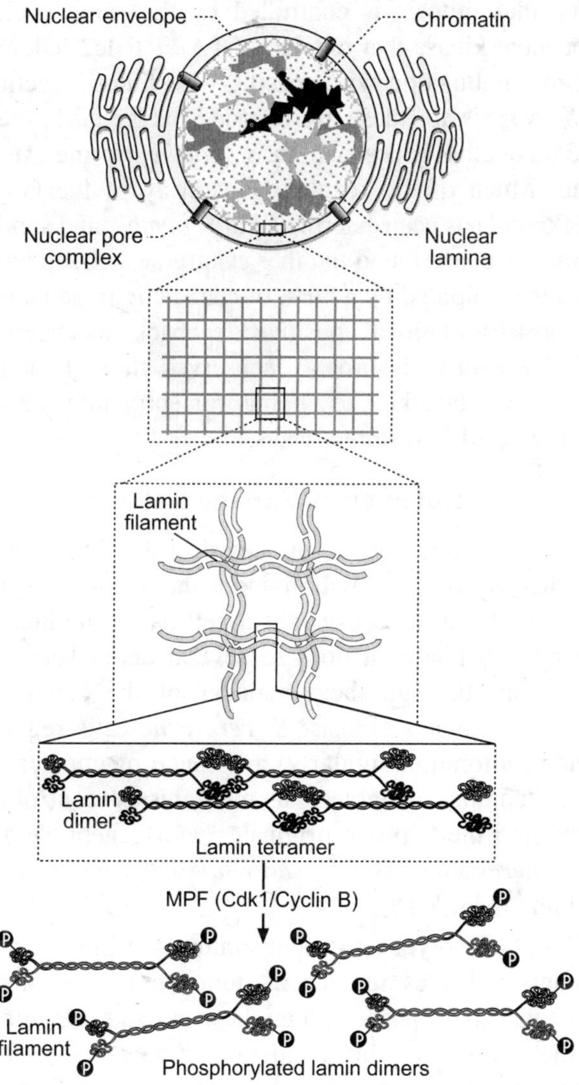

Fig. 7.17 Schematic representation of depolymerisation of the nuclear lamina due to the phosphorylation of lamin proteins by MPF

Experiments with RNase-treated *Xenopus* egg extracts revealed that polyubiquitination and subsequent degradation of proteins other than cyclin B is required to initiate anaphase. Quite possibly, these substrates include the proteins that link sister chromatids at the centromere. Each sister chromatid is attached through its kinetochore to microtubules, whose opposite ends associate with one of the spindle poles. At metaphase, the spindle is in a state of tension with forces pulling the two kinetochores towards the opposite spindle poles balanced by forces pushing the spindle poles apart. Degradation of the protein cross-bridges that link the sister chromatids at the centromeres would leave the poleward forces on the kinetochores unopposed, resulting in the poleward movement of chromatids that defines anaphase.

Phosphorylation of nuclear lamins results in their depolymerisation, leading to breakdown of the nuclear envelope, which is a crucial event of early mitosis. Removal of these phosphates coincides with lamin re-polymerisation and re-formation of the nuclear lamina associated with the nuclei of the daughter cells later in mitosis. Re-assembly of the nuclear envelope occurs as depicted in **Fig. 7.18**.

Cytokinesis, the final step in mitosis, is also dependent on phosphatase action. As MPF activity rises early in mitosis, it phosphorylates the myosin light chain, thereby inhibiting the ATPase activity of myosin and its association with actin filaments. As MPF is deactivated toward the end of anaphase due to the degradation of cyclin B, protein phosphatases dephosphorylate myosin light chain. As a result, the contractile machinery is activated, the cleavage furrow can form, and cytokinesis proceeds. This regulatory mechanism assures that cytokinesis does not occur until anaphase is completed.

7.5.7 MPF and regulation of mitosis

In multicellular organisms, entry into mitosis is carefully regulated, in order to properly coordinate the cell division with cell growth. *Xenopus* wee1 tyrosine activity is high and cdc2 phosphatase activity is low during interphase, holding the MPF assembled from *Xenopus* cdc2 and newly synthesised cyclin B in the inactive state with the cdc2 Tyr-15 phosphorylated. As mitosis is initiated, wee1 activity diminishes and cdc2 activity increases so that MPF is converted into its active form. As a result, although cyclin B is the only protein whose synthesis is required for the cycle in early *Xenopus* embryos, the activities of other proteins, including wee1 and cdc25, must be properly regulated.

MPF activity can also be regulated by controlling transcription of the genes encoding the proteins that regulate MPF activity. For example, after the initial rapid synchronous cell divisions of the early *Drosophila* embryo, all the mRNAs are degraded, and the cells become arrested in G_2. This arrest occurs because of the *Drosophila* homologue of cdc2, called String, is degraded and its synthesis ceases. Because of the resulting decrease in String phosphatase activity, MPF is maintained in its inhibited state. In yeast,

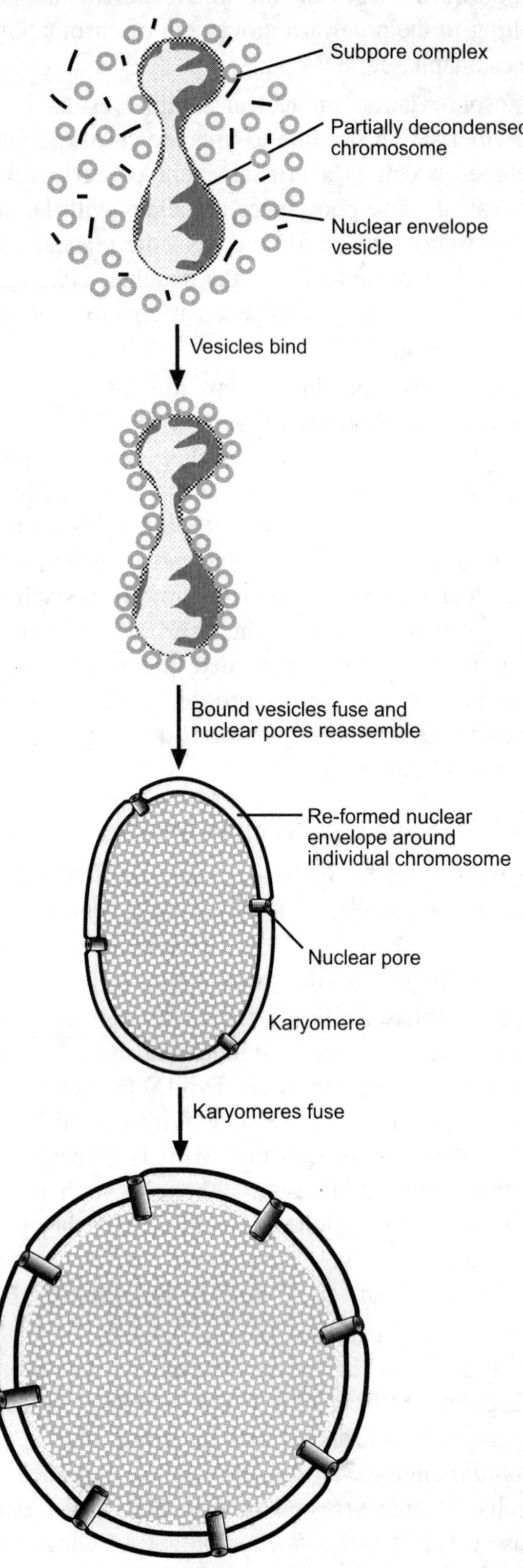

Fig. 7.18 Assembly of the nuclear envelope during telophase

entry into mitosis is controlled by the same cyclin-dependent kinase that regulates START (cdc2, cdc28), but in combination with specific with mitotic cyclins. In *S. cerevisiae*, these are encoded by the clb1, clb2, clb3 and clb4 genes and in *S. pombe* by the *cdc13* gene. Much of the regulatory pathway of the G_2–M checkpoint has been determined in *S. pombe*, and similar components are found in other eukaryotes. Essentially, mitosis is initiated by a burst of cdc2 activity generated by post-translational positive feedback mechanism. The mechanism is broadly conserved throughout the eukaryotes, but the checkpoints for entry into mitosis are regulated in different ways.

7.5.8 Control of entry into the S phase

In most vertebrate cells, the key decision determining whether or not a cell will divide is the one to enter the S phase. In most cases, once a cell has committed to enter the S phase, it does so several hours later and progresses through the remainder of the cell cycle until it completes mitosis. *S. cerevisiae* cells regulate their proliferation similarly, and much of our current understanding of molecular mechanisms controlling entry into the S phase originates with genetic studies of *S. cerevisiae*. The *S. cerevisiae* cells replicate by budding (**Fig. 7.19**).

The phenotype of temperature-sensitive cdc28 mutants indicates that cdc28 function is critical for entry into the S phase. When these mutants are shifted to the non-permissible temperature, they behave like wild-type cells suddenly deprived of nutrients. That is, cells that have grown large enough to pass START at the time of the temperature shift continue through the cell cycle normally and undergo mitosis, whereas those that are too small to have passed START when shifted to the non-permissible temperature do not enter the S phase even though there are plentiful nutrients.

S phase-promoting factors

The MPF is known to contain two subunits: a cyclin-dependent protein kinase (cdk) and mitotic cyclin for activating the catalytic subunit. By analogy, it seemed likely that *S. cerevisiae* contains an **S phase-promoting factor** (**SPF**) that phosphorylates and regulates proteins required for DNA synthesis. Like MPF, SPF was proposed to be a heterodimer composed of cdc28 and a cyclin that acts in G_1. The complexes formed between cdc28 and the three G_1 cyclins (*cln1*, *cln2*, and *cln3*) in

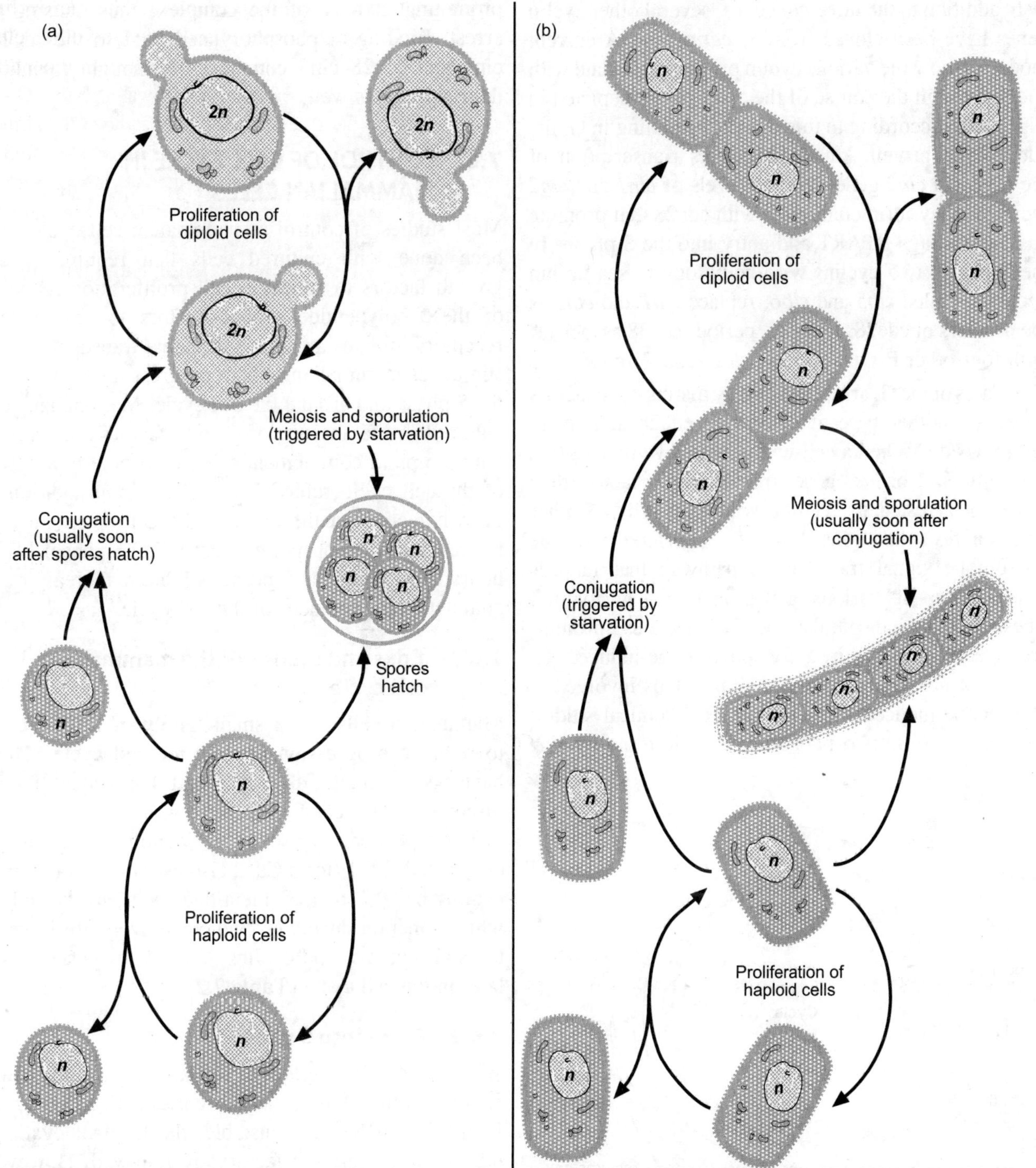

Fig. 7.19 Lifecycle of (a) Budding yeast (*Saccharomyces cerevisiae*) (b) Fission yeast (*Schizosaccharomyces pombe*); under nutrient-rich conditions, budding yeast proliferates as diploid cells while fission yeast proliferates as haploid cells

S. cerevisiae have protein kinase activity and constitute the hypothesised S phase-promoting factors (SPFs).

In addition to the three *cln* genes, several other cyclin genes have been cloned from *S. cerevisiae*. A current model for how the various cyclin proteins associate with cdc28 through the course of the cell cycle is depicted in **Fig. 7.20.** According to this model, beginning in G_1 the cdc28–cln3 protein kinase stimulates transcription of the *cln1* and *cln2* genes. As the levels of *cln1* and *cln2* increase, they form complexes with cdc28 that promote passage through START and entry into the S phase. In the S phase, two cyclins with homology to sea urchin cyclin B, called *clb5* and *clb6*, replace *cln1* and *cln2* as the partners of cdc28. In the G_2 period, cdc28 associates with four other B-type cyclins, *clbs 1, 2, 3* and *4*.

Studies of the G_1 arrest induced by mating-type factors revealed another type of control of cdc28 activity in *S. cerevisiae*. When a cell with a recessive mutation in either the *far*1 or *fus*3 genes treated with α factor, they were not arrested in G1 like wild-type cells. Neither *far*1 nor *fus*3 mutations, however, interfered with the pheromone signal transduction pathway that induces transcription of various genes involved in mating. These findings indicate that the Far1 and Fus3 proteins are specifically required for pheromone-induced G_1 arrest but not for any of the multiple steps involved in pheromone-induced transcription. Biochemical studies showed that Fus3 phosphorylates Far1 in response to α factor. Phosphorylated Far1 can bind to the cdc28–cln2 complex. This binding is thought to inhibit the S phase-promoting activity of the complex, thus causing G_1 arrest. Binding of phosphorylated Far1 to the cdc28–cln1 and cdc28–cln3 complexes is thought to inhibit their activity as well.

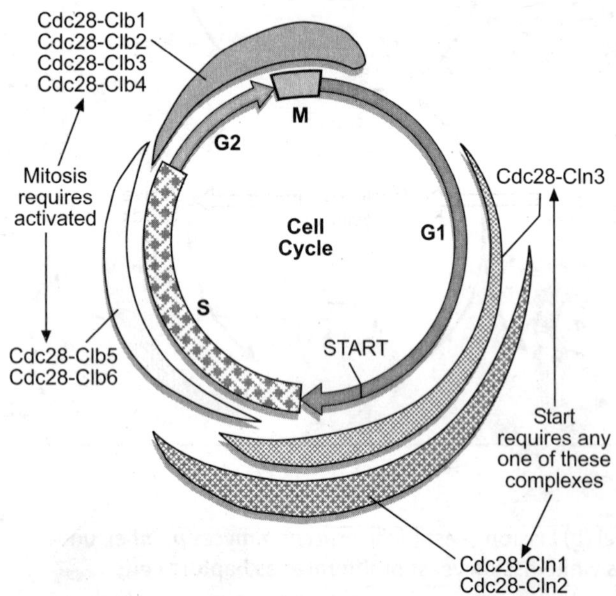

Fig. 7.20 Model showing the action of Cdc 28–cyclin complex in *S. cerevisiae*

7.6 CONTROL OF CELL CYCLE IN MAMMALIAN CELLS

Most studies of control of mammalian cell cycle have been done with cultured cells that require certain growth factors to stimulate cell proliferation. Binding of these polypeptide growth factors to cell surface receptors initiates a cascade of signal transduction that stimulates mammalian cells to progress through G_1 into the S phase and around the cell cycle. Mammalian cells cultured in the absence of growth factors are arrested with a diploid complement of chromosome in a period of the cell cycle referred to as G_0. Upon addition of growth factors to the culture, the cells pass through the restriction point (point of entry into S phase) 14–16 hours later, enter the S phase 6–8 hours thereafter and complete the remainder of the cell cycle.

7.6.1 Cdks and cyclins of the mammalian cell cycle

Mammalian cells use a small family of related Cdks to regulate progression through the cell cycle. These have been named Cdk1, 2, 3 and 4 in order of their discovery. The cdc2 or $p34^{cdc2}$ was the first human cyclin-dependent kinase to be identified and thus is designated Cdk1 (term Cdc2 also used). Several human cyclins have also been identified including D and E, which function during G_1 and collectively are referred to as G_1 cyclins, and cyclins A and B, which function later in the cell cycle (**Table 7.2**).

7.6.2 Entry into S phase

When mammalian cells in G_0 are stimulated with growth factors, cyclin D appears before cyclin E. Since cyclin D and its mRNA are unstable, the level of cyclin D falls when the growth factors are removed. However, in cells that are continuously proliferating because of constant exposure to growth factors, the cyclin D level remains high, since cyclin D mRNA is synthesised and translated into protein continuously (**Fig. 7.20**).

In continuously proliferating mammalian cells maintained in media containing growth factors, cyclin

E levels oscillate, reaching peak levels just before the S phase is initiated. The Cdk2–cyclin E complex is the protein kinase principally responsible for activating DNA initiation factors and inducing entry into the S phase. When growth factors are removed from cultured cells, transcription of the genes encoding cdk2, cdk4 and cyclins D and E ceases, and the levels of all these proteins fall rapidly. Consequently, in G_0 cells, the proteins currently thought to promote passage through the restriction point and to compose the S phase-promoting factors are not present. This certainly explains why cells in G_0 are arrested and do not enter the S phase.

7.6.3 Passage through S phase and G_2

Soon after mammalian cells enter the S phase, Cdk2 becomes associated with cyclin A (**Fig. 7.21**). Synthesis of cyclin A begins as cells approach the $G_1 \rightarrow S$ transition and the protein is immediately transported into the nucleus. Disruption of cyclin A function inhibits DNA synthesis in mammalian cells, suggesting that the cdk2–cyclin A complex is crucial for progression through the S phase. However, cyclin B is first synthesised during the S phase and increases in concentration as cells proceed through G_2, peaking in early mitosis and dropping after anaphase, as in *Xenopus* egg extracts. In human cells, cyclin B first accumulates in the cytoplasm and then breaks down early in mitosis. Consequently, MPF function may be controlled in part by regulation of the nuclear transport of cyclin B. The principal mammalian Cdk in G_2 is cdc2 (Cdk1).

7.6.4 Transcription factor E_2F

Mammalian genes encoding many of the proteins involved in DNA and deoxynucleotide synthesis are induced as cells through the $G_1 \rightarrow S$ transition. A family of transcription factors, together called E_2F, is required for transcription of several of these genes. Binding of two proteins, one called **retinoblastoma protein (Rb)** and a related protein called p107, to E_2F inhibits its ability to activate transcription. Rb is generally identified as a tumour suppressor gene in which loss of function causes tumours. In quiescent cells, or during the first part of G_1, Rb is bound to the transcription factor E_2F.

The Rb protein was identified in children with an inherited eye cancer, known as retinoblastoma. The loss of both copies of the Rb gene leads to excessive cell proliferation in the immature retina, suggesting that the Rb protein is particularly important for restraining the rate of cell division in the developing retina. However, the complete loss of Rb does not immediately cause increased proliferation of other cell types, in part because Hct1 and p27 provide assistance in control of

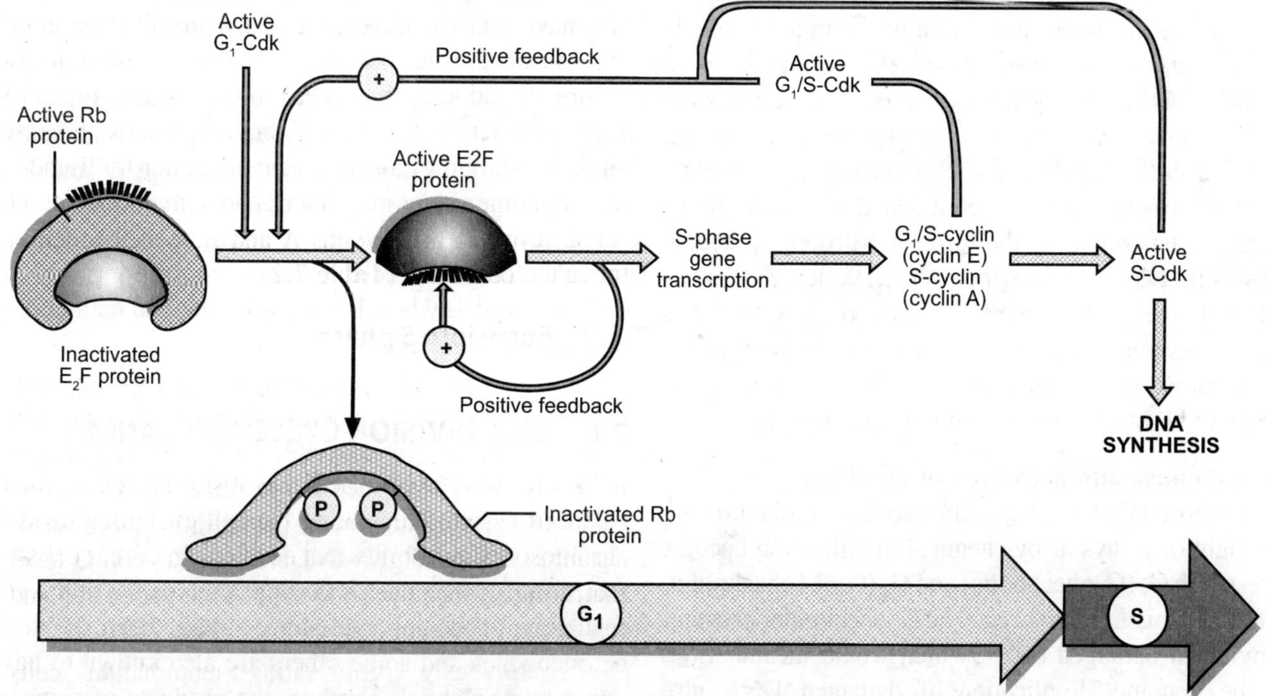

Fig. 7.21 Mechanisms controlling initiation of S phase in animal cells

G_1, and in part because other cell types contain Rb-related proteins that provide back up support in the absence of Rb.

7.7 CHECKPOINTS IN CELL CYCLE

Defects in the mitotic spindle

Inhibitors of microtubule polymerisation, which cause cell arrest during mitosis, commonly are used to prepare cells for karyotyping. If microtubules are inhibited chemically (e.g., treating with vincristine or colchicine), the cells enter mitosis and arrest of cells with condensed chromosomes occurs. As time progresses, a large fraction of the cells in a culture become arrested at metaphase, thus permitting determination of the size, shape and number of mitotic chromosomes (karyotyping). A checkpoint control senses that the mitotic spindle has not assembled properly and prevents activation of the polyubiquitination system that normally leads to degradation of mitotic cyclins and onset of anaphase. As a result, MPF activity remains high, chromosomes remain condensed, and the nuclear envelope does not re-form.

The spindle assembly checkpoint is known to be involved in the improvement of the fidelity of chromosome segregation by delaying anaphase until all chromosomes are properly aligned on the mitotic spindle. The spindle checkpoint provides time for surveillance mechanisms to monitor the mitotic spindle connecting the chromosomes to the two poles with the help of their kinetochores. Thus, kinetochores are not just attachment sites for the spindle, but are also involved in the regulation of chromosome movements. Kinetochores that are not yet attached to spindle fibres release a wait signal in the form of a protein to keep a brake on cell division. When all the kinetochores are attached, the protein signal is somehow disabled and mitosis resumes. Kinesin, a motor protein that moves the kinetochore to the end of the spindle fibre, also seems to be involved in the spindle checkpoint.

DNA damage and activation of checkpoints

Cells whose DNA has been damaged by irradiation with UV light or γ-rays or by chemical modification become arrested in G_1 (G_1 checkpoint) and G_2 (G_2–M checkpoint) until the damage is repaired. The G_1 checkpoint prevents copying of damaged bases, which would fix mutations in the genome. Replication of damaged DNA also causes chromosomal rearrangements at high frequency although the mechanisms are not yet fully understood. Arrest in G_2 allows DNA double-stranded breaks to be replicated before mitosis. If a double-stranded break is not repaired, the broken distal portion of the damaged chromosome is not properly segregated because it is not physically linked to a centromere, which is pulled toward a spindle pole during anaphase.

This response to DNA damage, leading to the arrest of cell division is facilitated by regulation of cdk–cyclin complexes, using three groups of checkpoint proteins.

Sensor proteins—recognise the events that trigger the pathway.

Transducer proteins—activated by sensor proteins and signal the next group by phosphorylation.

Effector proteins—activated by transducer kinases and execute the actions.

The checkpoints sense the DNA damage and transduce an inhibitory signal (**Fig. 7.22**). Four genes (*rad*9, *rad*17, *rad*24 and *mec*3) sense this DNA damage and are used for arrest of cell cycle at G_1 and G_2. The cdc13 rad9 double mutants fail to cause arrest suggesting that rad9p should be sensor or transducer of DNA damage signal. One of the important effectors of the response in mammalian cells is the protein kinase ATM (**a**taxia **t**elangiectasia **m**utated). Loss of ATM function is responsible for the human disease **ataxia telangiectasia**. Patients with this disease are not only abnormally sensitive to DNA damaging agents, but also have many cellular defects that result from errors that occur during normal cell division, due to the failure of inducing the checkpoints. Normally, ATM exists in inactive dimer form, but DNA damage causes phosphorylation of a serine causing dimer to dissociate into monomers (the monomer is an active kinase). The ATM protein is involved in detecting DNA damage (mainly double-stranded breaks), interrupting the cell cycle (with p53) when damage is found, and maintaining normal telomere length.

7.8 CELL DIVISION CYCLE AND CANCER

It is now well established that disturbances in finely balanced systems regulating the cell cycle may lead to chromosomal instability that develops in cancer cells. In fact, some of the genes encoding components of the key regulators of the cell cycle have already been shown to be oncogenes and some others are also known to have oncogenic potential. Further, the products of some of the Cdk and cyclin genes can interact with the products

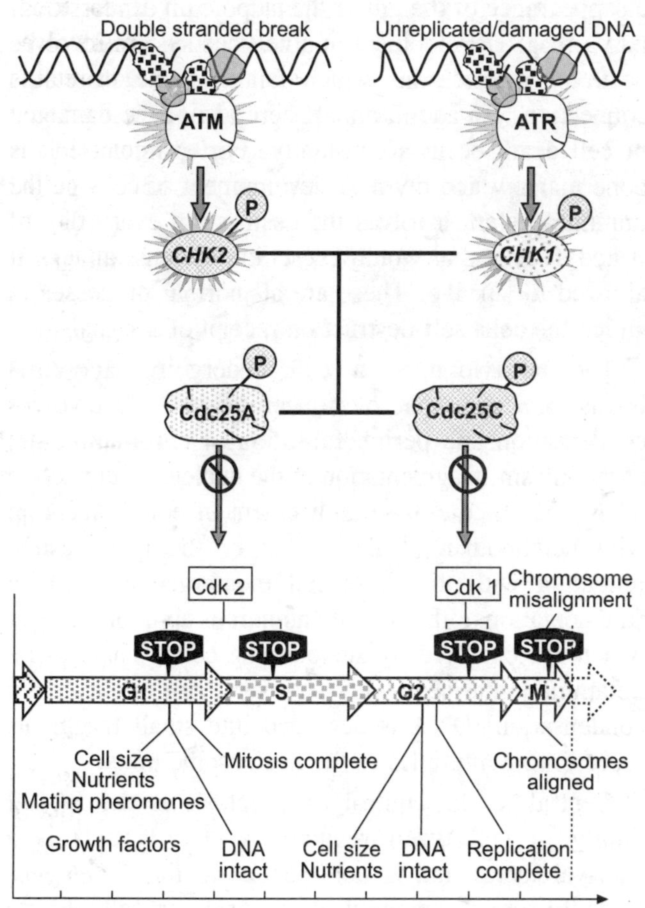

Fig. 7.22 Checkpoints at which specific protein kinases are activated or deactivated in the eukaryotic cell cycle

of tumour suppressor genes such as p53, which in its mutant form is responsible for more than 50% of human cancers. Intrinsic defects in the key regulators of cell division cycle may also cause cancer. For instance, cyclin D1 is overexpressed in most breast cancers, and mice designated to lack D1 have recently been shown to be completely resistant to breast cancers induced due to *neu* and *ras* oncogenes. Similarly, cyclin E, cyclin A and Cdk4 were found to be over expressed in cancer cell lines. Loss of activity of some genes encoding cell cycle inhibitors like *p53* and *Rb* (retinoblastoma) may also lead to cancer. All these and other similar cell cycle genes will be the targets for cancer therapy in future. Therefore, the discovery of the key regulators of the cell cycle has opened new avenues of research in the field of tumour diagnostics and cancer therapy.

The loss of function of some cell cycle inhibitors may also lead to cancer. These inhibitors include:

- p53, which blocks the activity of cdk2, and other Cdks,
- an inhibitor blocking specifically the cdk4,
- an inhibitor mediating TGF-ß's inhibitory effects (see **Chapter 20**).

Diploid chromosomes equally segregate into each daughter cell during cell division. Some times, a defect in spindle assembly checkpoint has also been found to cause aneuploidy associated with tumours. One such defect is a mutation in *hBUB1* (human BUB1), a gene involved in spindle assembly checkpoint, so that the anaphase separation will be disrupted causing aneuploidy. In aneuploidy, if the missing chromosome(s) carry tumour suppressor genes(s), its absence may lead (**Fig. 7.23**) to uncontrolled cell division (see **Chapter 20**).

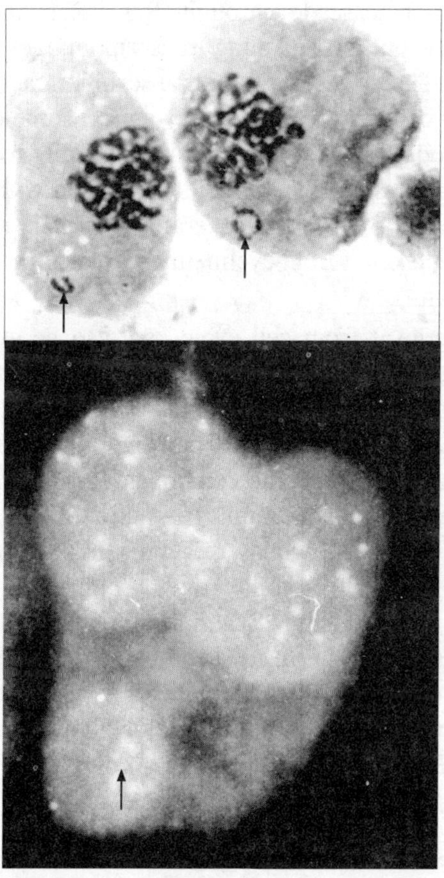

Fig. 7.23 Agents can damage formation of spindles and consequently chromosomes may lag behind during anaphase and end up in the cytoplasm as excluded DNA fragments or micronucleus. (Top) Kinetochores in both the main nucleus and micronuclei (Bottom) Lagging chromosomes, shown by arrow not included in the nucleus (***Source:*** Channarayappa 1990)

7.9 SPECIAL CELL CYCLE SYSTEMS IN ANIMALS

Exit from the cell cycle

As discussed earlier, the withdrawal of growth factors from animal cells at a critical period during G_1 causes them to cease growth and exit from the cell cycle, entering a quiescent state termed G_0. This may be a transient response to the withdrawal of growth factors, a permanent aspect of differentiation, or regulated during development to control the final size of a growing structure. The absence of growth factors stimulates exit from the cell cycle because growth factors are required for the synthesis of the D cyclins; these are very unstable proteins replenished only by new transcription early in G_1. It is this transcription which is activated by growth factor signaling. Growth factors are ligands for receptor tyrosine kinases, which initiate a signal transduction cascade through Ras, Raf and MAP kinases eventually activating transcription factors in the nucleus. The targets for these transcription factors are a set of so-called **immediate early genes**, some of which also encode transcriptional regulators likely to activate D-cyclin transcription. G_0 cells stimulated with growth factors enter the S phase several hours later, reflecting the time taken for D-cyclin mRNA to appear in the cytoplasm.

Withdrawal from cell cycle can also be stimulated by growth inhibitors such as TGF-ß, contact inhibition or loss of substrate. These diverse signals act through a family of small cyclin-dependent kinase inhibitors (CKIs) targeting specific components of the cell cycle machinery. The p16 family of inhibitors specifically inhibits CDK-D cyclins by preventing both their assembly and their ability to phosphorylate the Rb1 protein. The p21 and p27 families inhibit all Cdk–cyclin activities and thus arrest the cell cycle by inhibiting the activity of Cdk–cyclin D complexes by preventing their activation by CAK and also by blocking the activity of cyclin E and cyclin A-containing complexes, and the PCNA/δ-DNA polymerase complex itself.

Apoptosis

Apoptosis means programmed cell death, in response to various intrinsic and external signals. It has an important role in development and is also a defence against chemical insult and infection. In embryonic development, it is essential that at the appropriate stage certain cells be removed. Noticeable examples are the disappearance of the tail of the tadpole on development into a frog. Development of a nervous system involves death of the neurons which fail to make profitable connections. In adult animals here also, a vast amount of cell death occurrs constantly. For example, in the bone marrow and thymus, development of cells of the immune system involves the destruction every day of B and T cells that would cause autoimmune attacks if allowed to survive. These are all normal processes in which the cells self-destruct on receipt of a signal.

The behaviour of a cell undergoing apoptosis is distinct from that of a necrotic cell. It involves condensation and peripheralisation of chromatin, loss of cytoplasm, fragmentation of the nucleus, compaction of organelles, the dis-establishment of communication with neighbouring cells, fusion of the endoplasmic reticulum with the outer cell membrane and finally fragmentation of the cell into numerous apoptotic bodies which are engulfed by surrounding cells. There is no inflammation as seen with necrosis. Prior to chromatin condensation, DNA is degraded into small fragments representing multiples of the nucleosome units.

Central to the control of apoptosis is the BCL-2 family of cell death regulators. BCL-2 itself is a survival factor and is homologous to the *C. elegans* CED-9 protein expressed in all surviving cells during development. Other family members, e.g., BAX, BAD and BAK are stimulators of apoptosis. The proteins form dimers and competition between them, reflects the family of cysteine proteases related to the interleukin-1ß converting enzyme (**ICE proteases**). A number of such enzymes are synthesised by mammalian cells as inactive zymogens and are activated in a proteolytic cascade, culminating in the activation of proteins such as poly(ADP)ribose polymerase and nuclear lamins, the direct effectors of apoptotic cell behaviour.

Mitochondria are important players in the initiation of apoptosis. An early event in apoptosis is the appearance of small holes in the outer mitochondrial membrane. This is due to proteins, which allow the release of cytochrome c into the cytoplasm. The synthesis of these proteins, however, requires p53. Release of cytochrome c is a death sentence. Cytochrome c is a low-molecular-weight protein, only loosely attached on the inner mitochondrial membrane. It has been highly conserved over a billion years of evolution and is a component of every mitochondrion. Cytochrome c, when released into the cytoplasm, mediates oligomerisation of a cytoplasmic adaptor protein called **apaf-1** (apoptotic protease-mediating factor). This clusters together procaspases, and activates caspase-9. Once an active caspase is formed, it brings about a proteolytic cascade, which rapidly activates all the procaspases of the cell. Caspases target more than 300 proteins for cleavage. One important effect is to activate cellular DNases by destroying its inhibitor protein, and this takes the cell's DNA apart. The end result is that the constituents of the cell are destroyed; the cell shrinks in size and is removed by phagocytic white cells.

7.10 REGULATION OF MEIOTIC CELL CYCLE

During meiosis, two important events take place, namely pairing of homologous chromosomes and recombination (exchange of chromosome segments). Sometimes a distinction has been made between presynaptic chromosome pairing and synaptic pairing. The presynaptical pairing implies that alignment of homologous chromosomes and the synaptic pairing is due to synaptonemal complex formation. The special behaviour of the chromosomes in meiosis I requires some special controls. Nonetheless, passage through the cell cycle in meiosis I (as well as meiosis II, which is essentially a mitotic division) uses many of the same players, e.g., MPF and APC (**Fig. 7.24**).

In yeast, synapsis between homologous chromosomes, accompanies the formation of synaptinemal complex (SC) in early meiotic prophase I. But recent studies have indicated that initiation of recombination precedes initiation of synapsis. In contrast to the situation in yeast, in *Drosophila*, synapsis is independent of exchange of chromosome segments and can occur even in the absence of recombination. It is, thus, obvious that the molecular events in meiosis may differ in different organisms, particularly between organisms with small genomes and those with complex bigger genomes.

7.11 REGULATION OF BACTERIAL CELL CYCLE

The simple prokaryote *E. coli* divides by binary fission and can, in a rich medium, attain doubling times of 20 minutes or less. *E. coli* exhibits patterns of DNA synthesis during the cell cycle which can be quite complex and which differ substantially from replication timing in eukaryotic cells. Coordination of DNA synthesis and cell division in *E. coli* depends on a

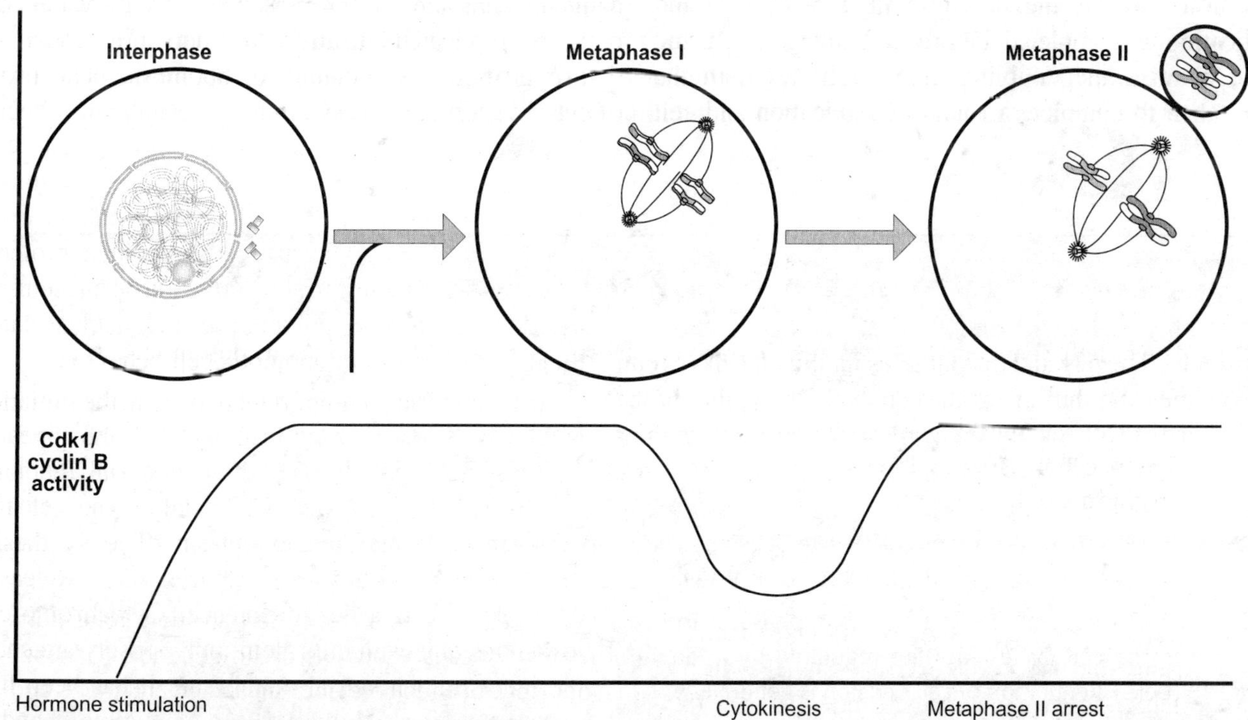

Fig. 7.24 Activity of Cdk/cyclin B(MPF) during meiosis; hormonal stimulation of diplotene oocytes activates Cdk/cyclin B, allows progression of metaphase I. At the end of meiosis I, activity of Cdk/cyclin B rises again and remains high during the arrest of metaphase II

combination of regulatory and kinetic properties. Under many growth conditions, the time for replication of the *E. coli* chromosome is approximately constant at a value around 40 minutes. During DNA synthesis, the rate of advance of a replication fork is also constant. Hence, the rate of DNA synthesis is directly proportional to the number of active replication forks. Cell division control is coupled with DNA synthesis.

The Helmstetter–Cooper model (or I + C + D model) divides the bacterial chromosome cycle into three phases, the interval phase (I), the chromosome replication phase (C) and the division phase (D). DNA replication occurs during the C-phase; its duration is fixed (about 40 min in *E. coli*), reflecting the time taken to replicate the whole chromosome. The D phase begins when replication is complete, and culminates in cell division. The duration of the D phase is also fixed (about 20 min in *E. coli*), and can be regarded as the time required to synthesise the cellular components required for cell division. The minimum duration of the chromosome cycle in *E. coli*, is thus, close to an hour. Since C + D are fixed, any change in the cell doubling time must reflect a change in the duration of I, the interval between successive initiations of replication. The doubling time of *E. coli* can vary from 20 min to 3 h. During slow growth, I > C + D and replication is completed before cell division. During rapid growth, the doubling time is shorter than the time taken to complete a round of replication and cell division (I > C + D). Therefore, during rapid growth daughter cells inherit chromosomes which are already partially replicated (**multiforked chromosomes**) so that replication can be completed before the next round of cell division (**Fig. 7.25**).

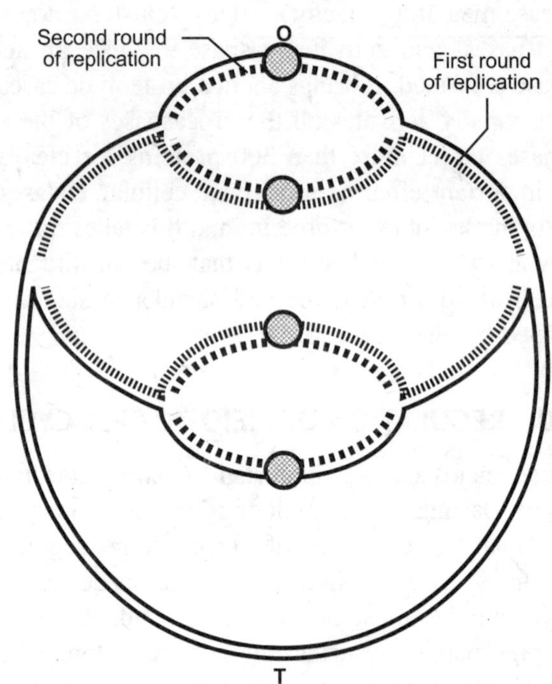

Fig. 7.25 Re-initiation of bacterial replication at new origins (O) before completion of the first round of replication T:Terminus

KEY POINTS

- The orderly sequence of events in which a cell replicates its genetic material and divides into two is called the cell cycle.
- In unicellular species, such as bacteria and yeast, each cell division produces a complete new organism. In multicellular species, long and complex sequence of cell divisions is required to produce a functioning organism.
- The standard eukaryotic cell cycle is divided into discrete phases, which proceed in a particular order, whereas the stages of the bacterial cell cycle may overlap.
- The process of crossing from one phase of the cell cycle to the next is a cell cycle transition. Transitions in the cell cycle clock are regulated by a family of protein kinases known as cyclin-dependent kinases (Cdks). Regulation of cell division is generally of two types: control of cell metabolism and enlargement, and coordinated enlargement and division.
- The sequential stages of the cell cycle reflect alternative states of phosphorylation for key proteins, which mediate the different cell cycle events. Transitions represent switches in those phosphorylation states.

- Cell cycle transitions are characterised by bursts of Cdk activity, which cause sudden switches in the phosphorylation states of target proteins responsible for cell cycle events.
- Cells are highly compartmentalised in which regulatory molecules can either be united with or separated from their target molecules.
- Subcellualr localisation is a dynamic process in which cell cycle regulators are moved into different compartments at different stages of the cell cycle. However, the continuous progression of cell division is regulated by both extracellular and intracellular signals that monitor and coordinate the various processes that take place during different cell cycle phases.
- MPF protein kinase activity induces the onset of mitosis, and that inactivation of MPF protein kinase activity by the degradation of cyclin B is required for the late stages of mitosis.
- In multicellular organisms, entry into mitosis is carefully regulated, in order to properly coordinate the cell division with cell growth.
- In most vertebrate cells, the key decision determining whether or not a cell will divide is the decision to enter the S phase. Mammalian cells use a small family of related Cdks to regulate progression through the cell cycle.
- A number of regulatory systems have evolved, which can sense the progress of the cell cycle and can interrupt subsequent stages in the event of failure. These regulatory mechanisms are termed cell cycle checkpoints and represent intrinsic signalling systems of cell cycle control.
- Disturbances in finely balanced systems regulating the cell cycle may lead to chromosomal instability that develops in cancer cells. Apoptosis is programmed cell death, a form of cell death initiated by the cell itself in response to various intrinsic and external signals.
- In embryonic development, it is essential that at the appropriate stage certain cells be removed.

Self-assessment Questions

1. Describe briefly the different stages of the cell cycle.
2. Describe the role of cyclins and Cdks in the regulation of cell cycle.
3. How does regulation of the cell cycle play a role in DNA repair?
4. What are the main checkpoints in the cell cycle?
5. Explain the mechanism of exit of a cell from the cell cycle.
6. How does regulation of the cell cycle influence apoptosis?
7. Briefly explain the regulation of meiotic cell cycle.
8. Discuss briefly about the regulation of bacterial cell division.

FURTHER READING

Books

Alberts B, D Bary, J Lewis, M Raff, K Roberts and JD Watson (1994). *Molecular Biology of the Cell*, 3rd ed. New York: Garland Publishing.

Lewin B (2000). *Genes VII*. Oxford University Press.

Lodish H, A Berk, P Matsudaira, CA Kaiser, M Krieger, MP Scott, SL Zipursky and J Darnell (2004). *Molecular Cell Biology*, 5th ed. New York: WH Freeman.

Murray AW and T Hunt (1993). *The Cell Cycle: An Introduction*. New York:WH Freeman.

Hutchison C and DM Glover (eds) (1995). *Cell Cycle Control: A Practical Approach*. Oxford University Press.

References

Blow JJ and P Nurse (1990). A cdc2-like protein is involved in the initiation of DNA replication in *Xenopus* egg extracts. *Cell* 62:855–62.

Breeden L and K Nasmyth (1987). Cell cycle control of the yeast ho gene: *Cis*- and *trans*-acting regulators. *Cell* 48: 389–97.

Chong JPJ, P Thommes and JJ Blow (1996). The role of mcm/p1 proteins in the licensing of DNA replication. *TIBS* 21: 102–108.

Donachie WD (1993). The cell cycle of *E. coli*. *Annu. Rev. Microbiol.* 47:199–230.

Dynlacht BD (1997). Regulation of transcription by proteins that control the cell cycle. *Nature* 389:149–52.

Finley D and V Chau (1991). Ubiquitination. *Ann. Rev. Cell Biol.* 7:25–69.

Forsberg SL and P Nurse (1991). Cell cycle regulation in the yeasts *S. cerevisiae* and *S. pombe. Ann. Rev. Cell. Biol.* 7: 227–58.

Hartwell LH and TA Weinert (1989). Checkpoints: Controls that ensure the order of cell cycle events. *Science* 246: 629–34.

Hinds PW, S Mittnacht, V Dulic, A Arnold and SI Reed (1992). Regulation of retinoblastoma protein functions by ectopic expression of human cyclins. *Cell* 70:993–1008.

Hochstrasser M (1992). Ubiquitin and intracellular protein degradation. *Curr. Opin. Cell Biol.* 4:1024–31.

Hollingsworth RE Jr, CE Hensey and WH Lee (1993). Retinoblastoma protein and the cell cycle. *Curr. Opin. Genet. Dev.* 3:55–62.

Nasmyth K (1996). At the heart of the budding yeast cell cycle. *Trends Genet.* 12:405–12.

Nigg EA (1993). Targets of cyclin-dependent protein kinases. *Curr. Opin. Cell Biol.* 5:187–93.

Norbury C and P Nurse (1992). Animal cell cycles and their control. *Ann. Rev. Biochem.* 61:441–70.

Nurse P (1975). Genetic control of cell size at cell division in yeast. *Nature* 256:547–51.

Nurse P (1990). Universal control mechanism regulating onset of M phase. *Nature* 344:503–508.

Nurse P (1994). Ordering S phase and M phase in the cell cycle. *Cell* 79:547–50.

Nurse P and Y Bisset (1981). Gene required for G1 for commitment to cell cycle and in G2 for control of mitosis in fission yeast. *Nature* 292:558–60.

Nyberg KA, RJ Michelson, CW Putnam and TA Weinert (2002). Toward maintaining the genome: DNA damage and replication checkpoints. *Ann. Rev. Genet.* 36:617–58.

Osmani SA and SY Xiang (1997). Targets of checkpoints controlling mitosis: Lessons from lower eukaryotes. *Trends Cell Biol.* 7:283–88.

Peter M and I Herskowttz (1994). Joining the complex: cyclin-dependent kinase inhibitory proteins and the cell cycle. *Cell* 79:181–84.

Reed SI (1992). The role of p34 kinase in the G_1 to S transition. *Ann. Rev. Cell Biol.* 8:529–61.

Rouse J and SP Jackson (2002). Interphases between the detection, signaling and repair of DNA damage. *Science* 297:547–51.

Sherr CJ (1993). Mammalian G_1 cyclins. *Cell* 73:1059–65.

Vaux DL and A Strasser (1996). The molecular biology of apoptosis. *Proc. Natl. Acad. Sci.* USA. 93:2239–44.

Wake RG and J Errington (1995). Chromosome partitioning in bacteria. *Annu. Rev. Genet.* 29:41–67.

Walker DH and JL Maller (1991). Role for cyclin A in the dependence of mitosis on completion of DNA replication. *Nature* 354:314–17.

Weintraub SJ, CA Prater and DC Dean (1992). Retinoblastoma protein switches the E2F site from a positive to negative element. *Nature* 358:259–61.

Zhou BB and SJ Elledge (2000). The DNA damage response: putting checkpoints in perspective. *Nature* 408:433–39.

CHAPTER 8

DNA Replication

LEARNING OBJECTIVES

After reading this chapter, you will be able to:

- List the general features of DNA replication
- Differentiate the mechanism of DNA replication in prokaryotes and eukaryotes
- Explain how DNA synthesis is catalysed by a family of DNA polymerases and other components along with their basic functions
- State the different models of DNA replication
- Explain the activity of telomerase in eukaryotes

8.1 INTRODUCTION

The processes of development, growth and repair of an organism involve cell division. Each daughter cell produced during cell division contains an identical copy of the genetic material. The complete process of cell division requires the chromosomal material inside the cell to double just prior to the actual splitting of the cell. The replication of DNA is one of the fundamental functions that a biological organism must carry out. It is an essential process for the propagation of cellular genomes and also those of 'molecular parasites'—viruses, plasmids and transposable elements. The genome to be duplicated is the parental genome and the copies are daughter genomes.

Cells can have one or many chromosomes, but the entire chromosome must be replicated precisely once in every cell division. The unit of DNA in which an individual act or replication occurs is called the **replicon**. Each 'fires' once, and only once, in each cell cycle. The replicon has an **origin** at which replication is initiated and also a **terminus** to stop. A genome in a prokaryotic cell contains a single replicon. Therefore, the units of replication and segregation coincide. In contrast, eukaryotic chromosomes have many replicons and initiate DNA replication at a number of regions in every chromosome. Watson and Crick intuitively gave the molecular mechanism for DNA replication as strand separation followed by the copying of each strand (**Fig. 8.1**). In the process, each separated strand acts as a template for the synthesis of a new complementary strand whose nucleotide sequence is fixed by the **base pairing rules**. This process entails the recognition of each nucleotide in the DNA template strand by a free complementary nucleotide and aligning it for its enzyme-catalysed polymerisation into a new DNA chain.

8.2 GENERAL FEATURES OF DNA REPLICATION

The primary function of DNA replication is to reproduce the base sequence of the parent molecules accurately. The specificity of base pairing (A with T and G with C) provides the mechanism for all the genetic replication

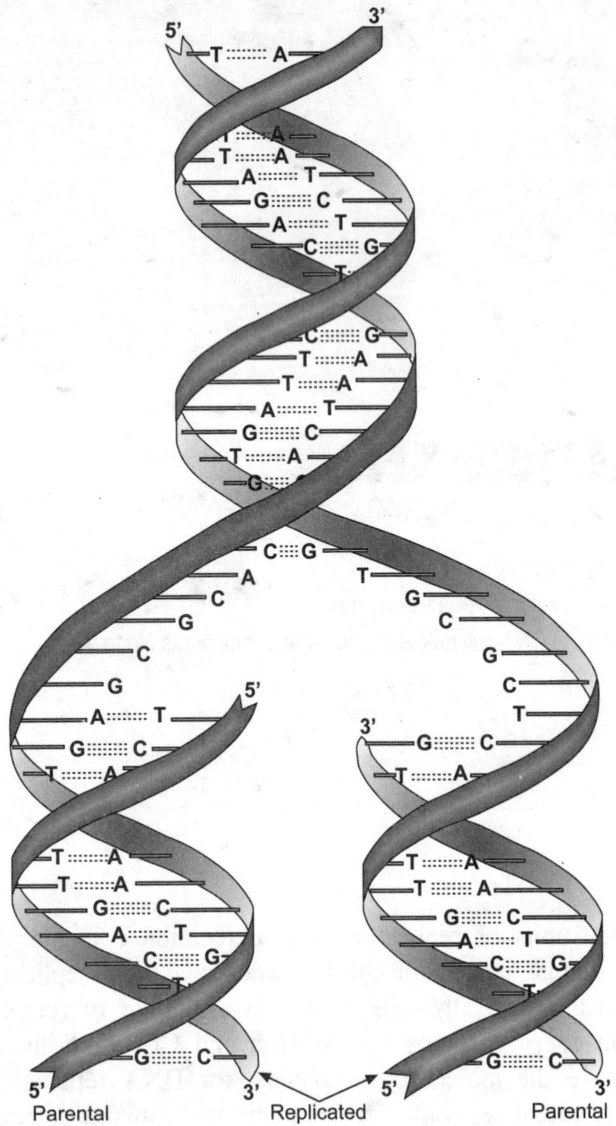

Fig. 8.1 A schematic view of Watson–Crick model of DNA replication

systems. Before studying the mechanism of DNA replication, we shall discuss the general features of DNA replication.

Similarity in replication mechanism

Much of the information about DNA replication available now has been derived from studies of purified bacterial and bacteriophage multi-enzyme systems capable of DNA replication *in vitro*. Mutants in a number of replication genes have facilitated the identification and purification of the corresponding replication proteins. The mammalian replication system that accurately synthesised DNA *in vitro* was described in the yeast, *Saccharomyces cerevisiae* and mutations in genes encoding nearly all of the replication components have now been isolated and analysed. These mutants were exploited to isolate and study many enzymes and protein components involved in the DNA replication in eukaryotes. It is now clear that the fundamental features of DNA replication—including the geometry of the replication fork and use of a multiprotein replication machine—have been conserved during the evolutionary process. Even though there are more protein components in eukaryotic replication machines than there are in bacterial analogues, the basic functions are the same in both.

DNA replication is semiconservative

Before 1950, only two models for replication of cellular DNA were postulated: **conservative** and **dispersive**. In conservative replication, the parental strands remain unchanged and are passed on to one of the daughter cells and the newly synthesised DNA is passed to the other. The two strands do not come apart but become a template for the completely new double-helical molecule. In this case, one daughter molecule would be wholly new and the other totally derived from the parent. In the dispersive mechanism, the parental molecule is partly fragmented and the fragments are incorporated into the two new daughter double helices (**Fig. 8.2**). The third method is the **semiconservative replication model** proposed by Watson and Crick following their discovery of the double-helical structure of DNA. In

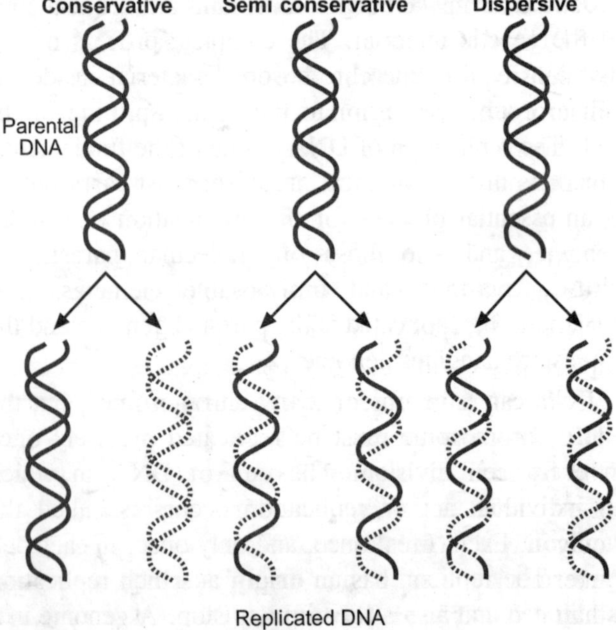

Fig. 8.2 Three probable DNA replication models

this model, one DNA molecule undergoes replication to produce two identical new molecules such that each of the new molecules has one polynucleotide strand from the original molecule bonded to one newly synthesised polynucleotide strand. Each parental strand acts as a template for replication and the daughter duplexes have one each of parental and daughter strands. Cellular DNA replicates in this method but the single-stranded genomes of viruses and some plasmids replicate conservatively. Let us now look at some experiments that prove the semiconservative model.

Meselson–Stahl's heavy isotope labelling

A convincing evidence for semiconservative replication of DNA was obtained from Matthew Meselson and Franklin Stahl's experiment (1958) in *E. coli*. The technique they used is called **heavy isotope labelling**. They grew *E. coli* in a medium containing a heavy isotope of nitrogen ($^{15}NH_4Cl$) of 96.5% isotopic purity for 14 generations so as to label the DNA very heavily with ^{15}N (denser DNA, i.e., dense/dense, D/D). The cells were then transferred to a medium containing $^{14}NH_4Cl$ (the normal form of nitrogen i.e., lighter DNA or L DNA) and allowed to undergo one or more rounds of replication. At a particular interval of time, samples of bacteria were withdrawn for several generations. Each sample was lysed by means of SDS and centrifuged (known as density-gradient centrifugation) in a concentrated solution of cesium chloride (CsCl) at 140,000 g for 20 hours to enable the DNA to attain sedimentation equilibrium. The bands of DNA were found in the CsCl gradient in the region of density 1.71 g/cm^3 and were isolated from all other macromolecular components of the bacterial lysate. Ultraviolet absorption (nucleic acids absorbs strongly at 260 nm) photographs taken during the course of the run revealed the position of the DNA bands.

At the start of the experiment the DNA appeared as one single band corresponding to the heavy ^{15}N-labelled (D/D) nucleic acid (density corresponding to 1.724 g/ml). After each harvest from the ^{14}N-medium, intermediate molecules of DNA ($^{15}N/^{14}N$, H/L) began to appear. Analysis of isolated DNA from the bacterial cells after a single round of replication showed that those cells which had replicated contained only DNA molecules whose molecular weights were intermediate i.e., lighter than ^{15}N-DNA and heavier than ^{14}N-DNA (1.717 g/ml). DNA isolated after a second round of replication contained both molecules of intermediate density (1.717 g/ml) and molecules of lighter density (1.710 g/ml) in equal amounts, but no heavy DNA molecules. During subsequent generations, the molecular weight was 1.710 g/ml (L/L). Moreover, when the hybrid ^{15}N /^{14}N molecules were heated, they separated to give ^{15}N and ^{14}N strands (**Fig. 8.3**).

After the first round of replication, all DNA molecules were made up of one heavy and one light strand, thus yielding molecules of intermediate weight. After a second round of replication, one-half of the DNA strands still contained the intermediate strand while the other half of the remaining molecules contained two light strands. Thus, in DNA replication each strand acts as a template for the synthesis of a new complementary strand, and each new DNA molecule is made up of one newly synthesised strand and one strand passed down from the parent molecule. From their experiment, however, it was not clear whether replication is **sequential** and **ordered**. Replication forks emanate from the fixed origins and each fork would, thus consist of two daughter strands. Daughter strands (new strands) elongate on parental templates (old strands) along with the unwinding of the parental duplex. However, the sequential nature of replication has been shown by genetic experiments and by direct visualisation of replicating molecules.

Autoradiographic evidence of semiconservative DNA replication

Experiments at the chromosomal level also support the view that DNA replication is semiconservative. In 1963, J Cairns grew *E. coli* bacteria in a medium containing radioactive thymine, a component of one of the DNA nucleotides. He then carefully extracted the DNA from the bacteria and placed it on photographic emulsion. He developed the emulsion to produce autoradiographs which he then examined under the electron microscope (**Fig. 8.4**).

The DNA in *E. coli* is circular

The DNA is replicated while maintaining the integrity of the circle and the intermediate structure forms. Replication of the DNA seems to occur at one or two moving Y-junctions in the circle, which further supports the semiconservative mode of replication.

These steps by themselves do not support either a unidirectional or a bidirectional mode of replication. That is, a structure will develop if either one or both Y-junctions are active in replication. When the DNA replication was allowed to begin in the absence of ^3H,

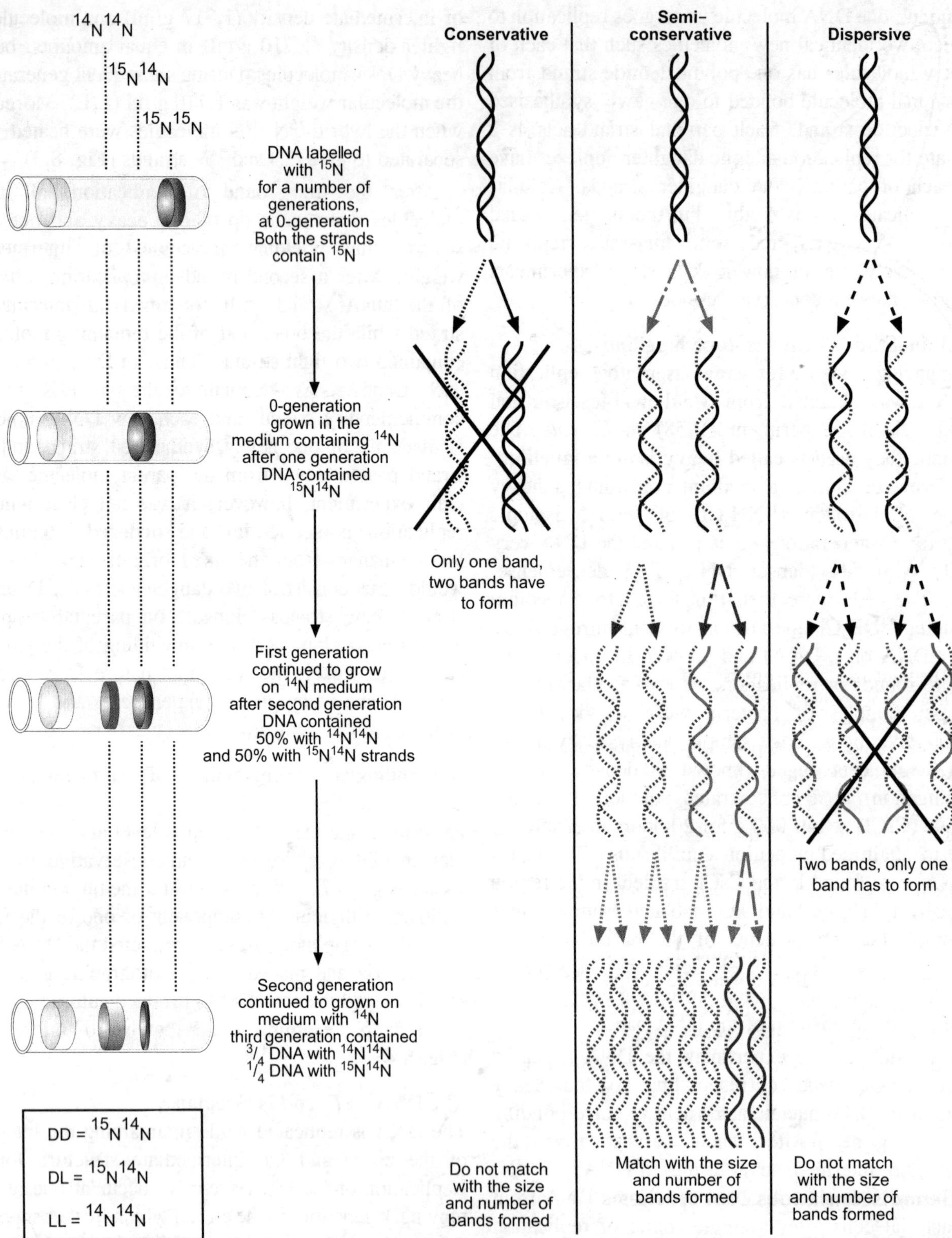

Fig. 8.3 Experimental evidence for semiconservative DNA replication by Meselson and Stahl; the diagrams show the composition of the DNA duplexes after 0, 1, 2 and 3 rounds of replication. DNA strands labelled with ^{15}N are shown in solid line, those labelled with ^{14}N in dotted line. The diagrams at the left show the positions at which bands are expected in the CsCl gradient.

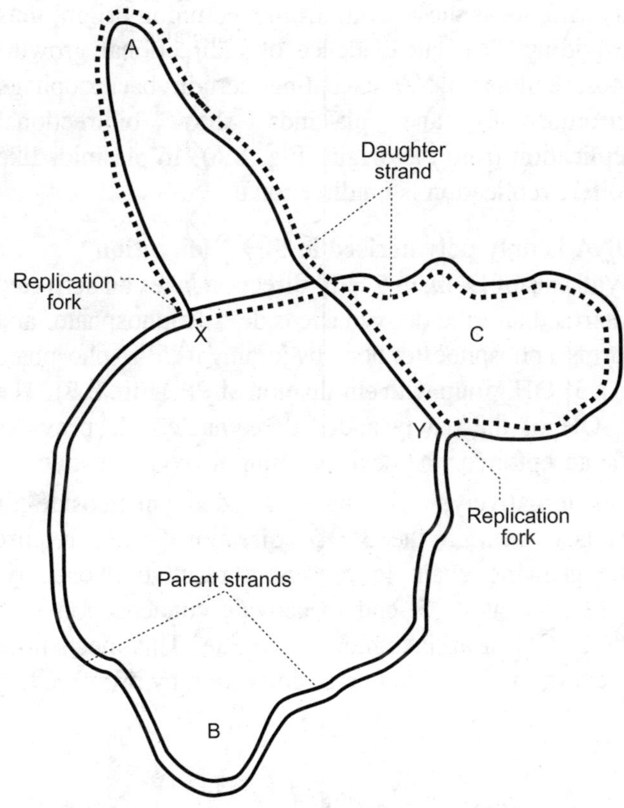

Fig. 8.4 Diagrammatic representation of autoradiograph of *E. coli* DNA during replication; diagram has labels ([^{3}H] thymidine) on the three segments A, B and C, created by the existence of two forks, X and Y, in the DNA. Replication forks are created when the circle opens for replication.

then a radioactive label was added. After a short period of time, the process was stopped and the autoradiographs prepared. In bidirectional replication, the label appears at both Y-junctions.

In plants to which [^{3}H]thymidine has been given as a specific precursor of DNA during the period of DNA synthesis, both the daughter chromatids were found to be labelled at the time of cell division. However, at the next round of duplication after withdrawal of the [^{3}H] thymidine, these two chromosomes each produced one each of labelled and unlabelled chromatid as would be expected.

8.2.1 Semidiscontinuous DNA synthesis

In the early models for replication of DNA double helix, both strands were believed to be replicated as continuous units. However, no known DNA molecule replicates in this manner as DNA polymerase can elongate a newly synthesised DNA strand only at its 3'-OH end group. Thus, if both daughter strands grew in the same overall direction, then each growing strand would need a 3'-OH terminus. However, the two strands of DNA are antiparallel and DNA polymerase can extend DNA only in the 5'→3' direction. One daughter DNA chain grows in the 5'→3' direction by the addition of an incoming deoxyribonucleotide to the 3'-OH at the end of the growing primer chain. However, the other strand would appear to grow by addition to its 5'-end, which is not the same reaction. Therefore, only one of the growing strands can terminate in a free 3'-OH group; the other must terminate in a free 5' end. **Figure 8.5** illustrates the possible three types of models: continuous, semidiscontinuous and discontinuous DNA replication.

This topological problem (**polarity** problem) is overcome within a single replication fork. One strand of the newly-made DNA is synthesised continuously, but the other strand is made in small **precursor fragments** i.e., discontinuously. The two daughter strands synthesise in different ways, but simultaneously, so that replication is **semidiscontinuous**. The physical basis of the terms leading strand (**forward template**) and lagging strand (or **retrograde template**) are used for the continuously and discontinuously replicating strands, respectively.

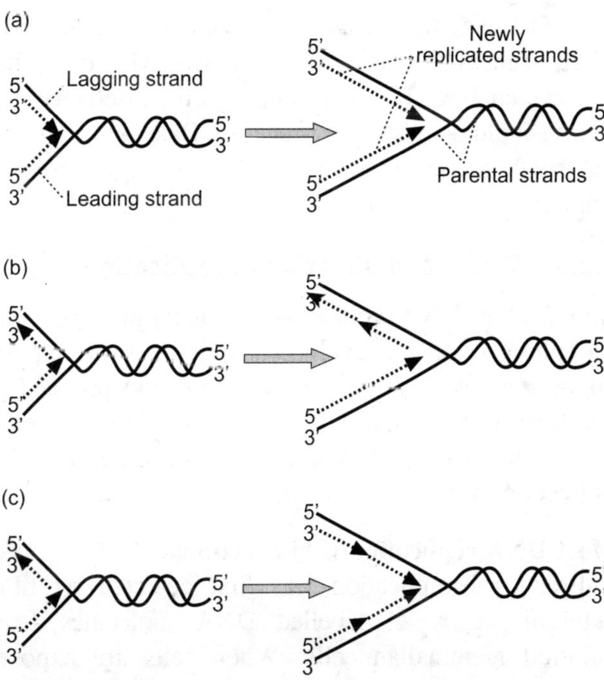

Fig. 8.5 Three replication models (a) Continuous (b) Semi-discontinuous (c) Discontinuous

Okazaki fragments: Tuneko and Reiji Okazaki (1968) provided the biochemical evidence of the semidiscontinuous pattern of DNA replication. Okazaki exposed a rapidly dividing *E. coli* culture to [^{3}H]-thymidine for 30 seconds and quickly collected the cells. He found that one-half of the label incorporated into nucleic acid appeared in short ssDNA chains just 1,000–2,000 nucleotides long. The precursor fragments are also known as *Okazaki fragments* and their size varies from 1,000 bp to 2,000 bp in prokaryotic cells and from 100 bp to 200 bp in eukaryotic cells.

8.2.2 Priming of DNA synthesis

As previously discussed, RNA polymerase can initiate transcription (i.e., *de novo* synthesis) simply by starting a new RNA chain; it puts the first nucleotide in place and then joins the next to it by formation of ester bonds. But DNA polymerases cannot perform the same process with initiation of DNA synthesis. DNA polymerases cannot synthesise DNA *de novo*; they need a primer, either an RNA or DNA. A **primer** is a piece of nucleic acid that the polymerase enzyme can hold onto and extend by adding nucleotides to its 3'– end. Generally, under *in vivo* conditions, the primer is a short piece of RNA. The evidence supporting RNA priming came from the inhibition by rifampicin of the replication of M13 phage DNA by an *E. coli* extract. The reason is that M13 uses the *E. coli* RNA polymerase to make RNA primers for its DNA synthesis. However, the best evidence for RNA priming was the discovery that DNase could not completely destroy Okazaki fragments and small pieces of RNA (10–12 bp long) were left behind.

8.2.3 Origin and direction of replication

How does the DNA replication machinery proceed along a duplex? Available evidence suggests that synthesis might start at a single site or origin and proceed in both directions so that both strands are copied at each of the two growing forks. This phenomenon is called **bidirectionality**.

Most DNA replication is bidirectional

Bidirectional replication was first detected by fibre autoradiography of labelled DNA molecules from cultured mammalian cells. When cells are exposed alternatively to high and low levels of [^{3}H]thymidine, the labelled DNA tracks are first very dark and get lighter progressively. Clusters of active replicons with growing forks move away from a common origin, thus providing clear-cut evidence of bidirectional growth. Most cellular DNA excepting certain bacteriophage chromosomes and plasmids show bidirectional replication from the origin (**Fig. 8.6**). In plasmids like colE1, replication is unidirectional.

DNA is only polymerised in 5'→3' direction

Synthesis of DNA in 5'→3' direction takes an activated intermediate, the deoxynucleoside-5'-triphosphate, and forms a phosphoester bond by joining its 5'-α-phosphate to a 3'-OH group with elimination of PPi (**Fig. 8.7**). The 3'-OH of the newly added deoxynucleoside provides the acceptor for the next incoming deoxynucleotide.

Alternatively, linking of deoxynucleoside-5'-triphosphates in the 3'→5' direction would require the growing chain to present a reactive phosphoric anhydride at its 5'-end to activate condensation with the 3'-OH of an incoming nucleotide. This mechanism is energy feasible, but error correction by proofreading

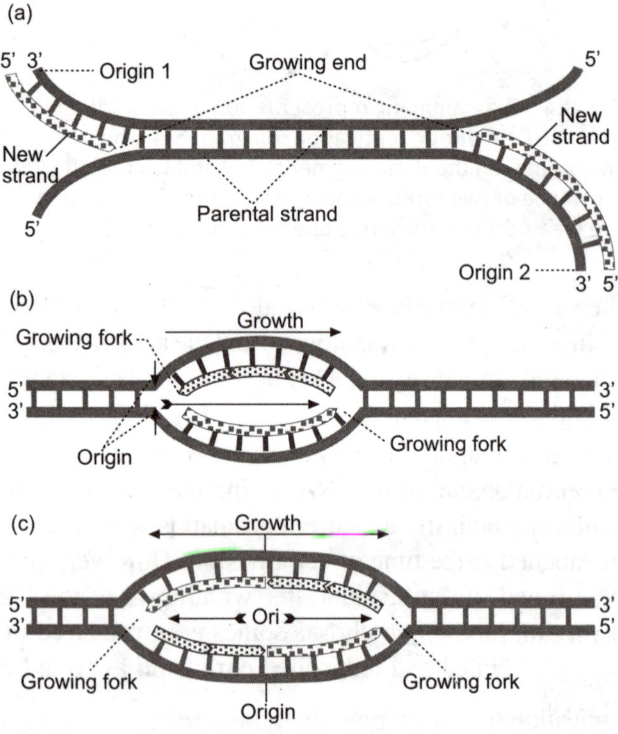

Fig. 8.6 Three mechanisms of growth of DNA strand are consistent with semiconservative replication.
(a) Unidirectional growth of single strands from two origins (2 origins and 2 growing ends) (b) Unidirectional growth of both strands from one origin (1 origin and 1 fork)
(c) Bidirectional growth of both strands from one origin (1 origin and 2 forks)—a most common type of replication in eukaryotes and prokaryotes

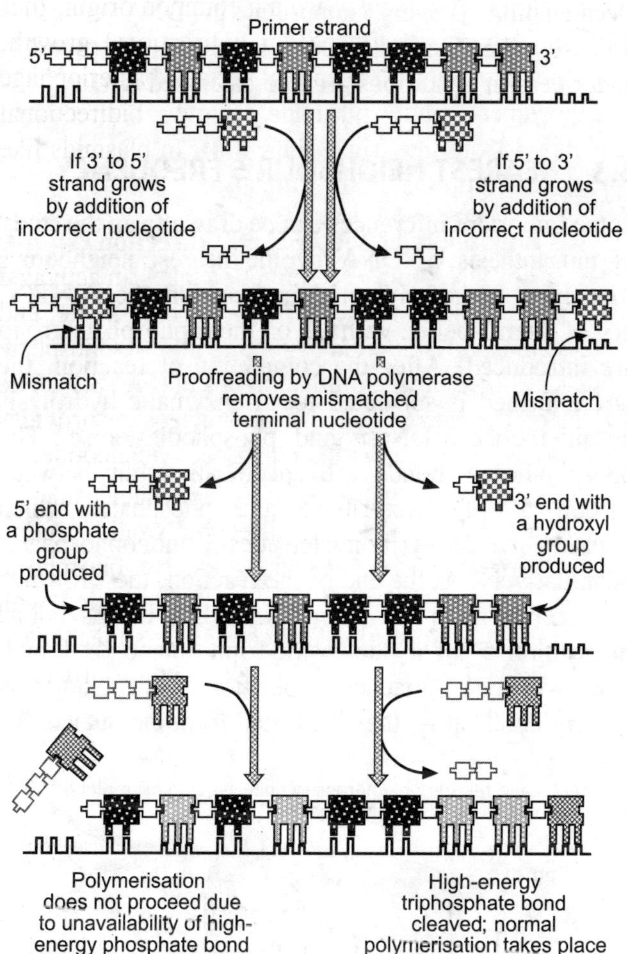

Fig. 8.7 DNA synthesis always favours 5′ to 3′ direction of growth

would create an energy-wise unfavourable situation. The proofreading activity would be by a 5'→3' exonuclease and its action would leave a 5'-OH or a 5'-phosphate ester group at the growing end of the chain, neither of which will form a phosphodiester linkage unless chemically activated. Without activation, chain extension would be aborted.

Unwinding the DNA helix: Semiconservative replication depends on unwinding the DNA double helix to expose single-stranded templates to polymerase action. DNA gyrase, by introducing negative supercoils at the expense of ATP hydrolysis, acts to overcome the torsional stress imposed by unwinding. The unwinding reaction is driven by helicases, a class of proteins that catalyse the ATP-dependent unwinding of DNA double helices.

8.2.4 Removal of RNA primers

In *E. coli*, when the pol III reaches the RNA primer of the preceding Okazaki fragment it disengages from the DNA, leaving a nick at the DNA/RNA junction. The pol I attaches to the nicks, or breaks, between successive Okazaki fragments, and adds nucleotides to the 3'–OH of the preceding fragment, moving in the 5'→3' direction, as with polymerising enzyme (e.g., pol III). As pol I moves, it encounters the RNA of the next Okazaki fragment, and its nucleotides are chopped out. The front end of pol I chops nucleotides out (5'→3' exonuclease activity), and a site further back adds nucleotides to fill the gap with DNA (**Fig. 8.8**). Pol I at the rear end both chops out and synthesises DNA. But there are other vital facts. The 3'→5' exonuclease hydrolyses the terminal nucleotide of the newly formed DNA strand only if it is unpaired with its corresponding base on the template strand and DNA polymerase adds nucleotides only if the preceding one is properly base paired. Thus, pol I checks the preceding base for correctness and cuts it out if unpaired, and replaces it. This is called **proofreading**. Since, pol I does not have the annular clamp to hold the template, it detaches relatively soon after the RNA has been replaced. When it detaches, a nick is left in the chain, which is healed by a separate enzyme, called **DNA ligase.**

Evolutionary significance for use of RNA primers: Many biochemical features provide a plausible

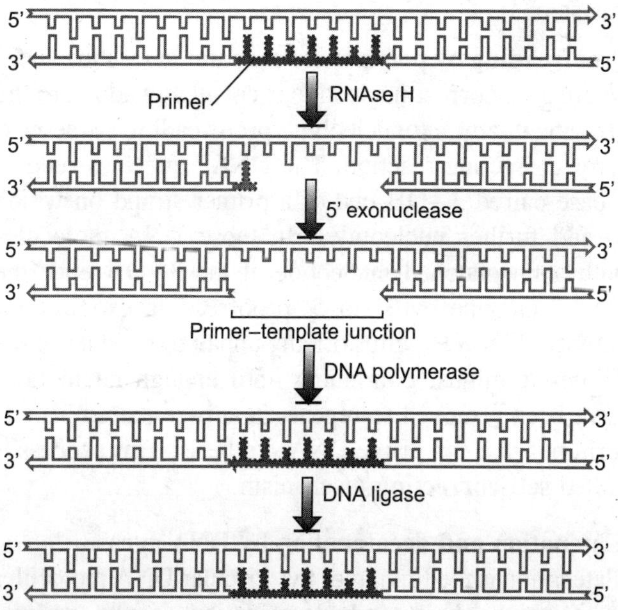

Fig. 8.8 Removal of RNA primers from newly synthesised DNA of the replication fork

explanation for why RNA primers initiate DNA synthesis. Unlike DNA polymerases, RNA polymerases can initiate polynucleotide synthesis *de novo*. Also, RNA polymerases do not check the pairing accuracy of the preceding base pair; RNA polymerases are low-fidelity enzymes in comparison to DNA polymerases; thus, primer is synthesised faster. Further, RNA as the priming oligonucleotide, provides a chemical signal that the primer is temporary and will be replaced.

8.2.5 Several proofreading mechanisms

DNA polymerase can certainly incorporate wrong nucleotides into the growing daughter strands. Mistakes associated with replication are generally very few (error is 10^{-7} nt). The important reasons are the stability afforded by the hydrogen bonding between G/C and A/T, preferential catalysis by DNA polymerase of those nucleotides in correct base pairing and proofreading.

Polymerase's affinity for correctly matched nucleotides

The DNA polymerase enzyme carries out the first proofreading step which occurs just before a new nucleotide is added to the growing chain. The correct nucleotide has a higher affinity for the moving polymerase because only then correct base pairing occurs in the template. Further, before the nucleotide is covalently added to the growing chain, the enzyme must undergo a conformational change and an incorrectly bound nucleotide is more likely to dissociate during this step.

Proofreading by polymerase enzyme: In rare instances where an incorrect nucleotide is covalently added to the growing chain, exonucleolytic proofreading is the next error-correcting reaction. The DNAP enzymes require a base-paired 3'–OH end of a primer strand on which to add further nucleotides. In those DNA molecules with a mismatched nucleotide at the 3'–OH end, the DNAP enzyme by 3'–to–5' proofreading exonuclease activity (**Fig. 8.9**), clips off any unpaired residues at the primer terminus, continuing until enough nucleotides have been removed to regenerate a base-paired 3'–OH terminus that can prime DNA synthesis. This process is called **self-correcting** mechanism.

Catenation and decatenation by DNA gyrase

Catenation is the linking of two circular DNA molecules to form a chain and decatenation is the reverse process. One catenane will produce two DNA circles. Decatenation is very important in DNA synthesis because, when a circular DNA replicates, the products under certain conditions are two catenated circles.

8.3 NEAREST NEIGHBOUR'S FREQUENCY

Some important inferences can be drawn from the study of biosynthesis of DNA by the nearest neighbour's frequency. In the polymerisation reaction, the 4 dNTPs, one of them labelled with ^{32}P on the α phosphate group are introduced. After the completion of reaction, the DNA formed is subjected to an enzymatic hydrolysis by micrococcal DNase and phosphodiesterase. The phosphodiester bonds are specifically split between carbon 5' of deoxyribose and phosphate, which liberates the deoxyribonucleosides-3'-monophosphates quantitatively. At the end of the reaction, the ^{32}P which was incorporated into the DNA when it was bound in position 5' of a nucleoside-triphosphate, is found, after hydrolysis, attached in position 3' of the adjacent nucleotide. That is, the ^{32}P shifted from the nucleotide,

(a) Mistake retards polymerase activity, leaving mismatched nucleotide at the 3' terminus

pol I
3' exonuclease site
3'
5'
3'
5'
Template strand
Mismatched last base pair
Polymerase site

(b) Delay allows spontaeous melting and releases of 3' end to contact exonuclease site, which excises mismatch

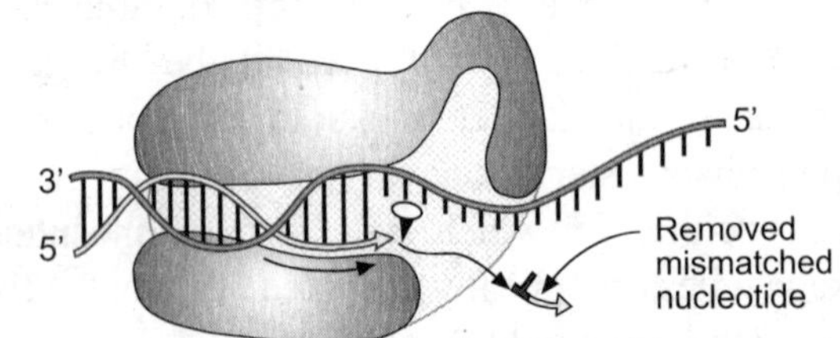

(c) Resume DNA synthesis

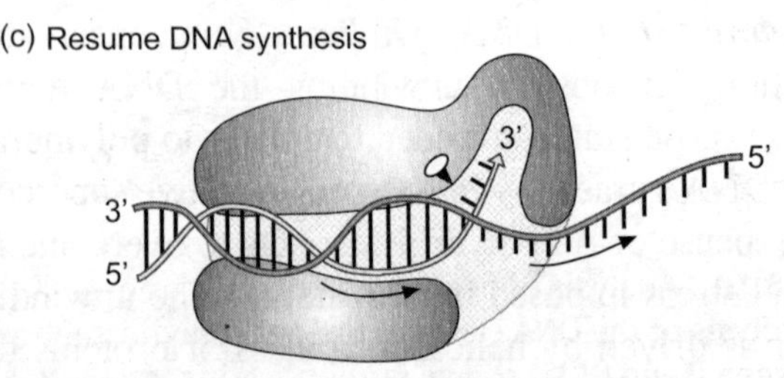

Fig. 8.9 Schematic diagram of proofreading function of DNA polymerase

which carried it, to its nearest neighbour in the chain (**Fig. 8.10**). This study reveals that:

- The frequencies of the nearest neighbours of the template DNA and newly synthesised DNA are identical, which confirms that the DNA added to the reaction does serve as a template.
- Any sequence of the DNA template is copied and there is no preference or specificity.
- The polarity of the newly synthesised chain is the opposite of that of the template chain. The growing chain and the template chain are also said to be 'anti-parallel'. This is also verified by experiments that have shown equal frequencies of dinucleotide sequences.
- The frequency of distribution of CG doublets is useful for assessing the properties of the nucleotide sequences.

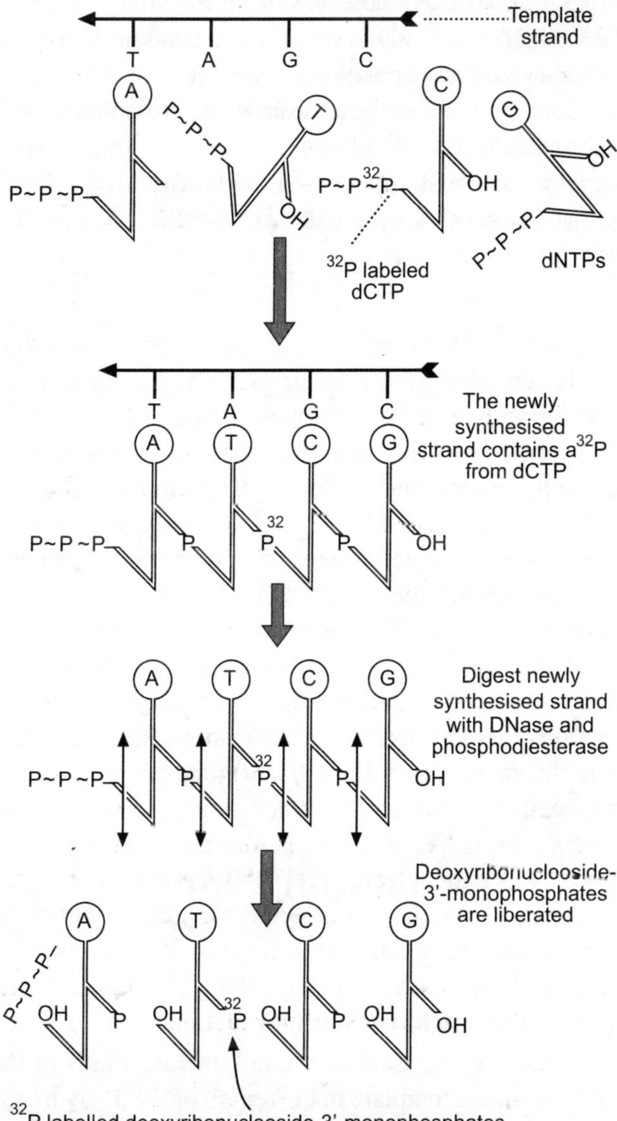

Fig. 8.10 Analysis of nearest neighbour's frequency; hydrolysis of the DNA cleaving the phosphodiester bonds between P and C5' is shown (arrows). It is seen that 32P initially carried by the dCTP is found in the form of dTMP (i.e., in the nucleosides 3'-monophosphates which were adjacent to the adenylic nucleotides in the polymer).

8.4 ENZYMOLOGY OF DNA REPLICATION

DNA synthesis occurs at a replication fork, which progresses along the DNA molecule and a number of proteins are required to bring about this process in an efficient manner. For example in *E. coli*, over 30 different polypeptides have been found to cooperate in replication.

8.4.1 DNA polymerases (DNAP)

DNAP is required to add dNTPs to the growing new strands. It acts alongside other proteins, which increase their fidelity and processivity. There are two types of mechanisms of synthesis of nucleic acids: template-directed and template-independent.

Template-directed nucleic acid synthesis

A pre-existing nucleic acid acts as the template for the assembly of nucleotides to form a new strand, and dictates the order in which the bases are incorporated by complementary base pairing. There are two types of nucleic acids: DNA and RNA, and there are four types of nucleic acid synthesis reactions, each catalysed by specific template-dependent nucleic acid polymerases as shown in **Table 8.1**. These enzymes are unique because the template as well as the enzyme determines the specificity of the substrate.

Template-independent nucleic acid synthesis

Nucleotides are added to the ends of pre-existing nucleic acids without template instructions. Such reactions fall into two classes: specific reactions, where the substrate is chosen by the enzyme itself, and non-specific reactions, where any nucleotide can be used for generating random sequences. Some template-independent enzymes and their cellular functions are also listed in the **Table 8.1**.

8.4.2 Mechanism of action of DNA polymerases

Nucleic acid synthesis in cells is catalysed by different DNA polymerases. The enzyme that forms the sugar–phosphate bond (the phosphodiester bond) between

Table 8.1 Copying process and enzymes synthesising nucleic acid

Copying process	*Enzyme: systematic nomenclature*
Template-directed nucleic acid synthesis	
DNA→DNA (DNA replication)	DNA-dependent DNA polymerase (DNA polymerase)
RNA→RNA (RNA replication)	RNA-dependent RNA polymerase (RNA replicase)
DNA→RNA (DNA transcription)	DNA-dependent RNA polymerase (RNA polymerase)
RNA→DNA (reverse transcription)	RNA-dependent DNA polymerase (reverse transcriptase)
Non-templated nucleic acid synthesis	
3' end extension of DNA	Responsible for adding extra nucleotides into the junctions during V-D-J recombination; adds deoxyribonucleotides non-specifically to DNA with terminal 3'–OH groups: Terminal deoxynucleotidyl transferase, TdT
3' end extension of RNA	Responsible for adding adenosine residues to the 3' end of eukaryotic mRNAs e.g., polyadenylation: Polyadenylate polymerase, PAP.
Addition of G-nucleotide to 5' end of RNA	Responsible for adding 7-methylguanosine cap to the 5' end of eukaryotic mRNAs, e.g., capping (mRNA guanyltransferase)
Nucleic acid synthesis	Bacterial non-specific ribonucleotide polymerase; unique in using nucleoside diphosphates rather than triphosphates as substrates (Polynucleotide phosphorylase)

adjacent nucleotides in a nucleic acid chain is called a DNA polymerase (DNAP). The DNA polymerase I was first identified in the lysate of *E. coli* by Arthur Kornberg (1956). Later, a variety of DNAPs were purified from various sources (bacterial, plant and animal). All DNA polymerases, irrespective of sources, need to satisfy three principal requirements:

- The substrate 5'-triphosphates of the four deoxyribonucleotides (dATP, dCTP, dGTP and dTTP) must be present.
- A pre-existing single strand of the DNA that is to be replicated must be present. Such a strand is called a template strand.
- A short nucleic acid segment must be present and must be hydrogen-bonded to the template strand. This segment is called a **primer**. In living cells, the primer is a short segment of RNA.

Unlike RNA polymerases, DNA polymerases cannot initiate *de novo* strand synthesis and therefore require a primer. The reaction catalysed by the DNA polymerase is the formation of a phosphodiester bond between the free 3'-OH group of primer and the innermost phosphorous atom of the nucleoside triphosphate being incorporated. Thus, DNA synthesis proceeds by the elongation of primer chains, always in the 5'→3' direction and the order in which the deoxyribonucleotides are added is dictated by base pairing to a template DNA chain.

However, in some special cases of replication, 3'–OH terminus of the linear ssDNA may loop back on itself and serve as a priming strand (**Fig. 8.11**) or short lengths of oligonucleotide may act as primers by becoming hydrogen-bonded to the template. DNAP cannot use double-stranded DNA (dsDNA) as a primer or template. However, when a nick or gap is introduced into one strand, the exposed 3'-OH group acts as a primer and the new growing strand will fill the gap and then may continue displacing the complementary strand from the 5'-end. In a strand displacement reaction, this can lead to complications since at some point the enzyme may leave the original template strand and begin to copy the complementary strand so that *in vitro* a branched structure is formed. In the presence of a 5'→3' exonuclease, the 3'–OH group at the nick can act as a primer for polymerase action while the exonuclease digests the DNA from the 5'-phosphate at the nick. The nick thus moves along the duplex DNA in a so-called **nick translation reaction**.

During polymerisation, the polymerase binds to the single-stranded template in the region of the 3'-hydroxyl end of the primer. An incoming deoxyribonucleoside trisphosphate containing a base, which can pair with the corresponding base on the template, becomes attached to the triphosphate binding sites. The DNAP then catalyses a nucleophilic attack by the 3'-phosphate of the deoxyribonucleoside 5'-triphosphate (**Fig. 8.12**).

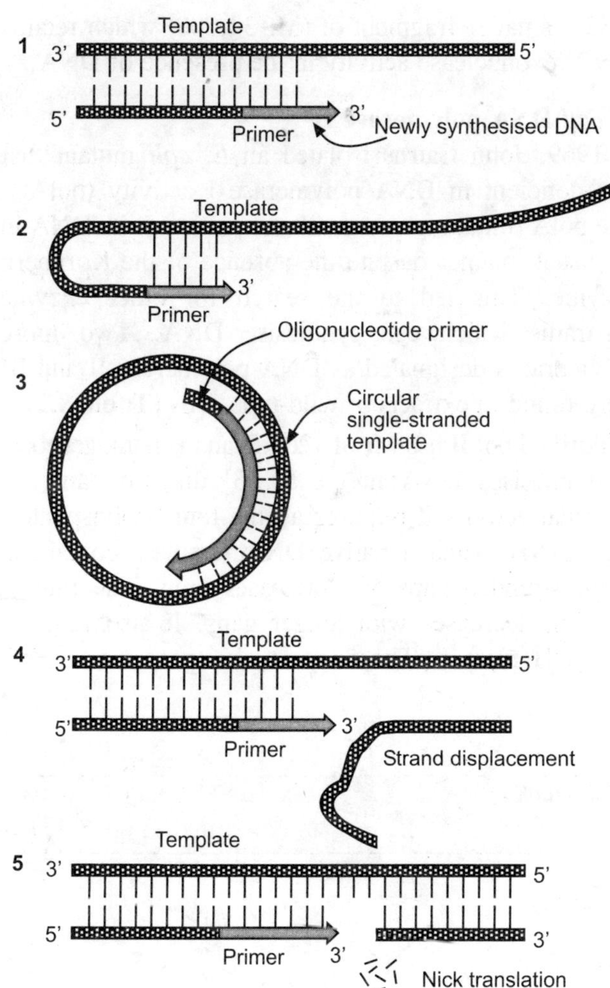

Fig. 8.11 Primer/templates for the replication of DNA

Inorganic pyrophosphate is released, a phosphodiester bond is formed, and the chain lengthens by one unit. The enzyme moves along the template by the distance of one unit and the newly added nucleotide with its 3'-hydroxyl group now occupies the primer terminus site. The process is then repeated until the enzyme reaches the end of the template strand. *In vivo*, the pyrophosphate is hydrolysed by a pyrophosphatase, which drives the equilibrium in favour of DNA synthesis.

DNA polymerases have the ability to catalyse the addition of any of the four deoxynucleoside triphosphates. Only correct base pairing takes place when the 3'-OH of the primer and the α-phosphate of the incoming nucleoside triphosphate are in the optimum position. DNAP shows an incredible ability to distinguish between ribo- and deo-xyribonucleoside triphosphates. This discrimination is mediated by the steric exclusion of rNTPs from the DNAP active site. In DNAP, the nucleotide binding pocket is too small to allow the presence of a 2'-OH on the incoming nucleotide. This space is occupied by two amino acids that make van der Waals contacts with the sugar ring.

8.4.3 DNA polymerases in prokaryotes

E. coli DNA polymerase I

DNA polymerase I catalyses the synthesis of DNA *in vitro* if provided with all the four deoxynucleotide-5'-triphosphates (dNTPs), a template DNA strand to copy and a primer to extend.

Structure and action: The purified enzyme is a protein of Mr 103,000 in the form of a single polypeptide chain (928 amino acid residues). This chain can be unfolded in guanidine-HCl-mercaptoethanol, so as to reversibly denature the protein. Pol I is a single polypeptide containing one sulphydral group and one disulphide group; the residue at the N-terminus is methionine. This enzyme shows 5'→3' polymerase activity and both 3'→5' (degrades single-stranded DNA from the

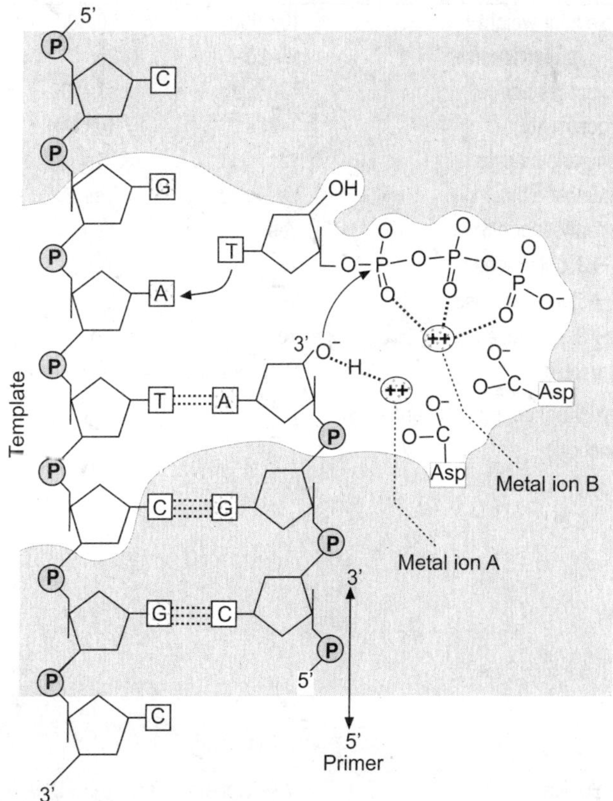

Fig. 8.12 Mechanism of action of DNA polymerase; two metal ions bound to DNA polymerase catalyse addition of nucleotides

3' end) and 5'→3' exonuclease activity (degrades 5' base-paired DNA from the 5' terminuses). This enzyme also cleaves RNA from a duplex containing one strand each of DNA and RNA. The 3' exonuclease serves a proofreading function, to improve the accuracy with which a DNA template is copied. The 5' exonuclease activity plays two known roles. The first of these is excision of RNA primers in lagging strand replication, and the 5' exonulcease activity can cleave deoxyribonucleotides during DNA repair.

The three catalytic activities of DNA polymerase I have been localised to regions of the long polypeptide chain. Treatment of the enzyme with the protease subtilisin or trypsin, breaks it into two fragments (shown by Hans Klenow). A larger fragment (residues 324–928), known as the **Klenow fragment** (Mr 76,000) retains polymerase activity and 3'→5' exonuclease activity and the other is a 323-amino acid long (residues 1–323) smaller fragment of (Mr 34,000) which retains 5'→3' exonuclease activity in the presence of DNA.

E. coli DNA polymerase II

In 1969, John Cairns isolated an *E. coli* mutant that was deficient in DNA polymerase I activity (polA$^-$). The polA$^-$ mutant grows well and replicates its DNA in the usual manner despite the absence of the Kornberg enzyme. This led to the search for other enzyme apparatus which can synthesise DNA. Two more polymerases designated as DNA polymerase II and III were found in extracts of wild-type cells (**Table 8.2**).

Purified pol II has Mr of 120 kD and is homogeneous. It synthesises DNA in the 5'→3' direction and for maximal activity it requires all the four triphosphates, Mg^{2+}, NH_4^+ and a native DNA template containing single-stranded gaps 50–200 bases long. The rate of reaction decreases with longer gaps. It also requires a 3'-OH primer. The enzyme also possesses 3'→5'

Table 8.2 Properties of three DNA polymerases of *E. coli*

Parameters	*Pol I (pol A)*	*Pol II (pol B)*	*Pol III (polC)*
Molecular weight	103,000	90,000	130,000
V_{max}, nucleotides/sec	16–20	25	250–1,000
Molecules per cell	400	17–100	–
Processivity	3–200	10,000	500,000
Template binding	Yes	Yes	Yes
Nucleotide binding	Yes	Yes	Yes
Pyrophosphorylysis	Yes	Yes	Yes
5'→3' polymerase	Yes	Yes	Yes
3'→5' exonuclease	Yes	Yes	No
5'→3' exonuclease	Yes	No	No
Structure	Polypeptide	Polypeptide	Multimeric complex
Structural gene(s), subunits and functions	*pol*A	*pol*B	α = *polC, dnaE*, 5'→3' polymerisation ε = *dnaQ, mutD* 3'→5' exonuclease β = *dnaN*, sliding clamp, initiator complex θ=*holE* core enzyme ($\alpha\varepsilon\theta$) τ = *dnaZX* core dimerisation, holoenzyme γ =*dnaXZ* part of γ-complex/processivity δ =*holA*, part of γ-complex/processivity δ' = *holB*, part of γ-complex/processivity χ = *holC*, part of γ-complex/processivity ψ = *holD*, part of γ-complex/processivity
Biological function (s)	DNA repair, Polymerase, RNA primer excision	SOS DNA repair (error-prone) inducible	Elongation of replication chain

exonuclease activity, but no 5'→3' exonuclease activity. The function of pol II *in vivo* is unknown and may be involved in repair, removal of primer and joining the Okazaki fragments.

E. coli DNA polymerase III

DeLucia and Cairns (1969) isolated an *E. coli* mutant (named polA1) having only 1% of wild-type DNA polymerase I activity, but the polA1 mutant cells grew at essentially normal rates, but DNA repair was defective. Later on, two more (pol I and pol III) polymerases were isolated. DNA pol III exists in a complex holoenzyme form in the cell and it has been difficult to determine the composition of its subunits.

Core DNA polymerase III

The simplest form of DNA polymerase III showing any DNA-synthesising activity *in vitro* is 165 kD in size and consists of three polypeptides: α (130kD), ε (29.5 kD) and θ (10 kD). This 'core' DNA pol III binds at short, single-stranded regions (<100 nucleotides) created by nuclease treatment of dsDNA and fills in the gaps to re-establish the duplex structure. It cannot initiate synthesis on intact duplex DNA. The α subunit carries out the catalytic polymerase function. The ε subunit performs the 3'→5' exonuclease activity and contributes proofreading ability to the 'core' polymerase. The role of subunit θ is unknown. Core enzyme cannot use long ssDNA molecule as template even when provided with a primer. It requires the addition of up to seven auxiliary subunits to increase its efficiency and to endow it with processivity.

DNA polymerase III holoenzyme

Pol III is a multisubunit complex (**pol III holoenzyme**). The holoenzyme functions as a heterodimer of complexes at the replication fork, with each monomer overseeing the synthesis of one daughter strand. In the cell, it exists as a large complex composed of two pol III protein subunits plus at least ten other proteins (**Table 8.2**). *In vitro*, the α, ε and θ subunits associate to form the core enzyme, which contains the essential enzyme activities. Addition of the other subunits promotes dimerisation and increases the processivity of the enzyme. This complex can replicate an entire strand of the *E. coli* genome (about five megabases) without dissociating (processivity of DNA pol I, which is only 20).

8.4.4 DNA polymerases in eukaryotes

Eukaryotic cells contain several DNA polymerases, with a typical cell having more than 15. They resemble the bacterial enzymes in their requirement for a template, a 3'–OH primer and four deoxyribonucleoside triphosphates. In higher eukaryotes, there are at least five DNA polymerases, four nuclear (α, β, δ and ε) and one organellar (γ) genome replication (**Table 8.3**) are the major ones. The nuclear enzymes are DNA **polymerases** α, β, δ and ε, which are responsible for nuclear chromosomal replication. DNA polymerase α has an associated primase activity but no 3'→5' exonuclease activity, whereas DNA polymerase δ has a proofreading capability. DNA polymerase δ binds an accessory factor called **proliferating cell nuclear antigen (PCNA)**, a cyclin analogous to the *E. coli*

Table 8.3 Comparison of properties and functions of eukaryotic DNA polymerases

Name and functions	α	ß	γ	δ	ε
Yeast name	pol1	pol4	polM	pol3	pol2
Yeast gene	Pol1	Pol4	MIP1	Pol3	Pol2
Location	Nuclear	Nuclear	Mitochondrial	Nuclear	Nuclear
Number of subunits	4	1	4 (identical)	2–3	4/?
5'→3' polymerase	Yes	Yes	Yes	Yes	Yes
3'→5' exonulcease	No	No	Yes	Yes	Yes
Primase	Yes	No	No	No	No
Associated factors	None	None	None	PCNA	None
Processivity	Moderate	Low	High	Low, high with PCNA	High
Sensitivity to aphidicolin	High	Low	Low	High	High
Biological function	Lagging strand priming	Repair polymerase	Organelle polymerase	Leading strand replication	Replication

pol III ß subunit which acts as a sliding ring to increase enzyme processivity. DNA polymerase δ synthesises both leading and lagging strands. The function of DNA polymerase α is to extend the RNA primer on the lagging strand and provide the template for an accessory factor, replication factor C (**RF-C**). The role of DNA polymerase ε is unclear. The polymerase ß the smallest of the five enzymes and the one with the lowest fidelity and processivity. Mitochondria have their own DNA polymerase, polymerase γ, which replicates the mitochondrial DNA and a similar enzyme has been isolated from plant chloroplasts.

Processivity of DNA polymerases

The processivity of DNA polymerases is facilitated by their ability to slide along the DNA template. Once bound to a primer–template junction, it interacts tightly with the double-stranded portion of the DNA in a sequence non-specific manner. The incoming nucleotide that matches correctly will have conformation that results in a high affinity for polymerase to catalyse the polymerisation reaction. Proofreading allows the mistakes to be corrected. Most DNA polymerases are capable of exonuclease activities that break phosphodiester bonds in the sugar-phosphate backbones of nucleic acid chains. This exonuclease activity provides a built-in mechanism for correcting rare errors in polymerisation. Occasional addition of incorrect nucleotide to the growing end activates the 3'→5' exonuclease activity, which cleaves the growing chain. Because it cleaves off an incorrect nucleotide and gives the polymerase another chance to get it right, the 3'→5' exonuclease activity of DNA polymerase is also called the proofreading or editing function (**Fig. 8.13**).

8.4.5 Terminal transferase

In addition to replicative DNA polymerase activities, animal cell extracts contain a separable enzyme responsible for the addition of nucleotidyl units to the ends of polynucleotide chains. This activity, which was originally described by Krakow in 1962, does not require a template strand but catalyses the incorporation of nucleotide units from single triphosphates into terminal positions in the DNA primer molecule. This enzyme has been called **terminal deoxynucleotidyl transferase (TdT)**. It shows strong amino acid homology to DNA polymerase ß suggesting it may have arisen from a variant gene. TdT has a M_r of 58,000–62,000 and highly sensitive to EDTA.

8.4.6 Reverse transcriptase

In the 1970s, RNA-dependent DNA polymerases, which use RNA as a template for the synthesis of DNA, were discovered. These enzymes were discovered simultaneously in many viruses like the Rous sarcoma virus (RSV) and Rauscher mouse leukemia virus (R-MLV) and in more than half a dozen other RNA viruses.

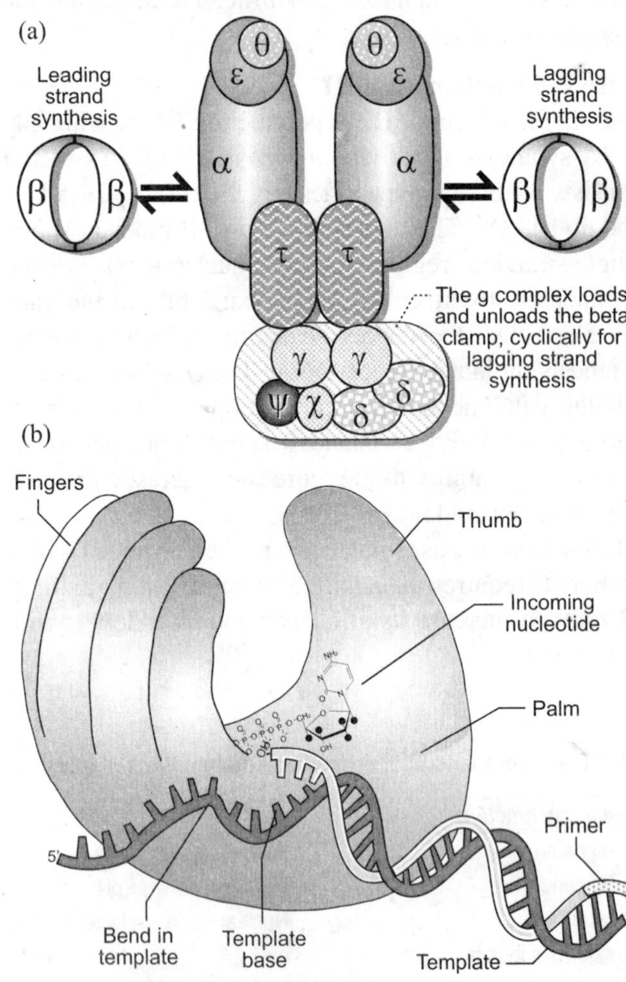

Fig. 8.13 DNA polymerase III (a) Subunits and subassemblies of DNA polymerase III; the a, e and h subunits form the catalytic core, and two t subunits are thought to hold the two cores in a single large complex. One g complex is present in the DNA polymerase III subassembly and in the holoenzyme, forming an asymmetric dimer for the overall structure. Addition of the ring-shaped b2 dimers greatly increases processivity. (b) A schematic illustration of the three-dimensional structure of DNA polymerase resembles the palm of the right hand

The retroviruses contain a 70S RNA genome which, when treated with agents that have hydrogen bonds, dissociates into two genetically identical RNA molecules, several molecules of tRNA and also a unique RNA-directed DNA polymerase. The enzyme uses one of the tRNA molecules as a primer to synthesise DNA on single-stranded RNA template. The immediate product of the reaction is a double-stranded, RNA:DNA hybrid which is the result of the synthesis of a complementary strand of DNA using the single-stranded viral RNA template. The C-terminal region of the enzyme shows ribonuclease H activity which digests the RNA strand of the RNA: DNA hybrid molecule. A DNA-dependent DNA polymerase activity will then convert the hybrid into a duplex DNA. Cellular DNA polymerase can then replicate the duplex DNA so as to provide more copies which may be integrated into the genome of the host cell. The enzyme shows only low fidelity and lacks associated exonuclease activity although it is capable of specifically hydrolysing the ribo strand of an RNA-DNA duplex. The 3'-end of the *pol* gene also encodes an endonuclease important for replication and integration, which altogether makes this a remarkable enzyme.

Reverse transcriptase may be used *in vitro* to make DNA copies of eukaryotic messenger RNA molecules. It is required not only for the replication of RNA retroviruses but also for the replication of DNA viruses such as hepatitis B virus and cauliflower mosaic virus, indicating that these viruses replicate *via* an RNA intermediate. Reverse transcriptase is also involved in the mobility of some transposons and the telomerase is, by definition, also a reverse transcriptase. It is also found in bacteria where it functions in the production of multicopy small, single-stranded DNA molecules.

8.4.7 DNA helicases

DNA helicases are enzymes which translocate along single strands of DNA and use energy derived from ATP hydrolysis to break hydrogen bonds and separate duplex molecules. Helicases are required for many cellular processes (e.g., nucleotide excision repair, homologous recombination, transcriptional termination, conjugation, Ti plasmid) and are essential during replication to provide single-stranded templates. In *E. coli* at least seven helicases have been partially characterised, which implies the multiple functions for these proteins. All known helicases are multimeric proteins and most of them are homodimers, but a few are homohexamers. All these proteins initially bind to ssDNA and then translocate in either a 5'→3' or 3'→5' direction until they reach a duplex region when unwinding starts. Within the helicases, homo-oligomers are structurally asymmetric. The ATP binding and DNA-binding properties of the two subunits are quite different, suggesting a rolling or 'hand-over-hand' mechanism, in which each subunit alternates between tight-binding and loose-binding conformations, depending upon whether ATP or ADP is bound at a given instant.

It is clear that the helicase acting together on both sides of the fork would bring about efficient unwinding of the DNA duplex but it is not clear which pair of enzymes acts in any particular situations. Two molecules of ATP are hydrolysed per nucleotide unwound (**Fig. 8.14**). It has been recently discovered that two inherited diseases in humans, Werner syndrome and Bloom syndrome, result from helicase defects. Both conditions involve elevated frequencies of cancer, premature ageing and death before age 50.

8.4.8 DNA ligase

The discovery of DNA ligase in 1967 provided important information needed for understanding how two DNA strands of opposite polarity extend within the same replication fork. They were first identified in *E. coli* and have since been described in most other organisms. DNA ligases catalyse the formation of a phosphodiester bond between the free 5'-phosphate end of an oligo- or polynucleotide and the 3'-OH group of a second oligo- or polynucleotide are juxtaposed. This enzyme is responsible for joining Okazaki fragments together to make the lagging strand a covalently contiguous polynucleotide chain. In bacteria (e.g., *E. coli* and *B. subtilis*), DNA ligases require NAD cofactor, whereas in eukaryotes, T4 phage infected cells and archaea, ligases require ATP. In both cases, the cofactor supplies an adenylate group, which becomes covalently linked to the active site of the enzyme (**Fig. 8.15**). DNA ligase closes single-strand breaks in dsDNA or in either strand of a RNA:DNA hybrid polymer; double-stranded ribopolymers are not substrates. DNA ligase is an important enzyme involved in many biological processes including DNA synthesis, DNA repair and genetic recombination. In all these cases, its postulated role is in re-establishing continuity by joining a stretch of newly synthesised DNA to pre-existing DNA by the

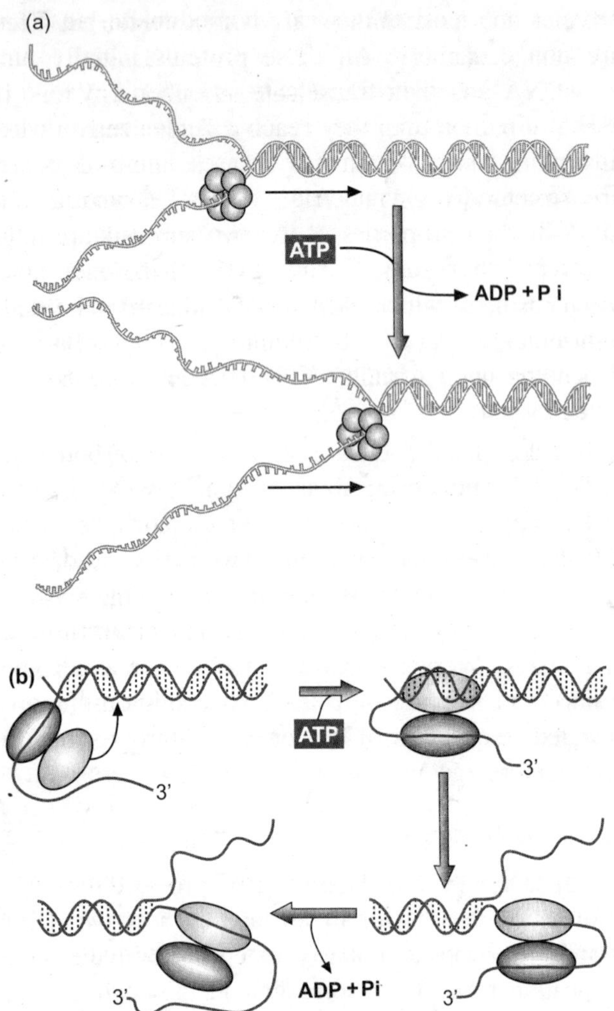

Fig. 8.14 (a) Homohexameric DNA helicase action at the replication fork; helicases separate the two DNA strands of the double helix by using ATP. DNA helicase binds to ssDNA and moves with a defined polarity (5'—>3') on the ssDNA. (b) A model for separation of DNA strand by *Rep* helicase of *E. coli*

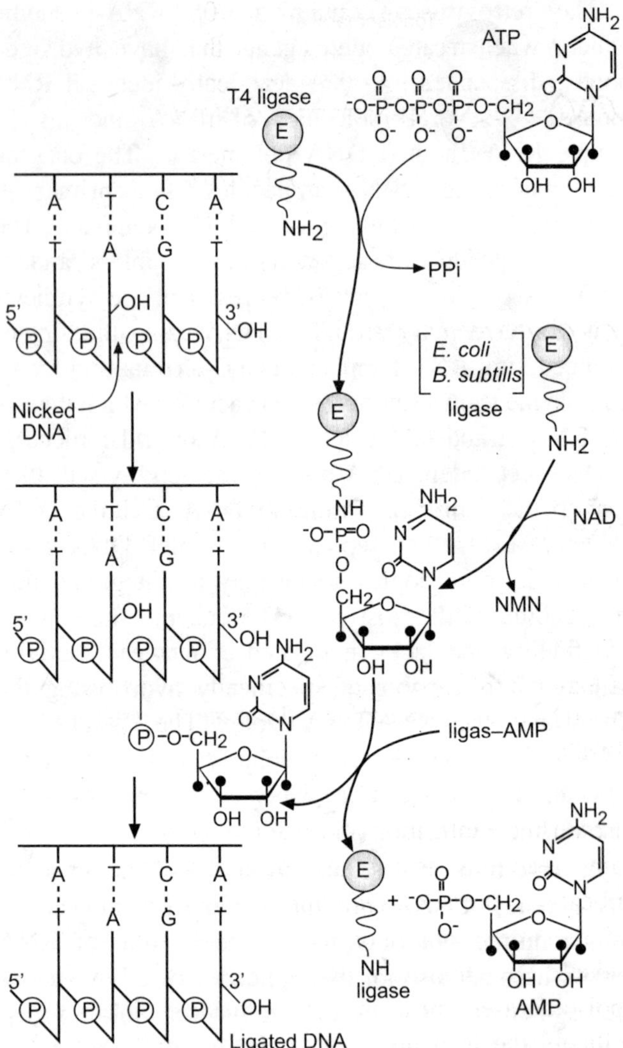

Fig. 8.15 Action of DNA ligases

formation of a phosphodiester bond. There are at least four DNA ligases present in cells of higher eukaryotes but the principal enzyme is the DNA ligase I. The smaller, nuclear DNA ligase II does not change in activity with the growth rate of the cells.

8.4.9 Topoisomerases

DNA topoisomerases are molecular machines that maintain the topological state of the DNA in the cell. Topoisomerases, in addition to managing DNA topology, are also involved in many other important functions. First, a wide variety of topoisomerase-targeted drugs have been identified, many of which generate cytotoxic lesions by trapping the enzymes in covalent complexes on the DNA. Second, the crystal structures of numerous topoisomerase fragments provide key insights into how these machines work.

As the replication fork moves forward, the two DNA strands unwind. Either the whole DNA molecule has to rotate at 10,000 rev/min or some sort of swivel must be present. The two functions are present in a single enzyme, topoisomerase. Topoisomerases catalyse the interconversion of DNA topoisomers (different topological forms of DNA) and are required during replication to relax the torsional strain generated when helicases unwind the duplex (**Fig. 8.16**). They also do similar functions during transcription and recombination and also remove knots and resolve catenanes, which

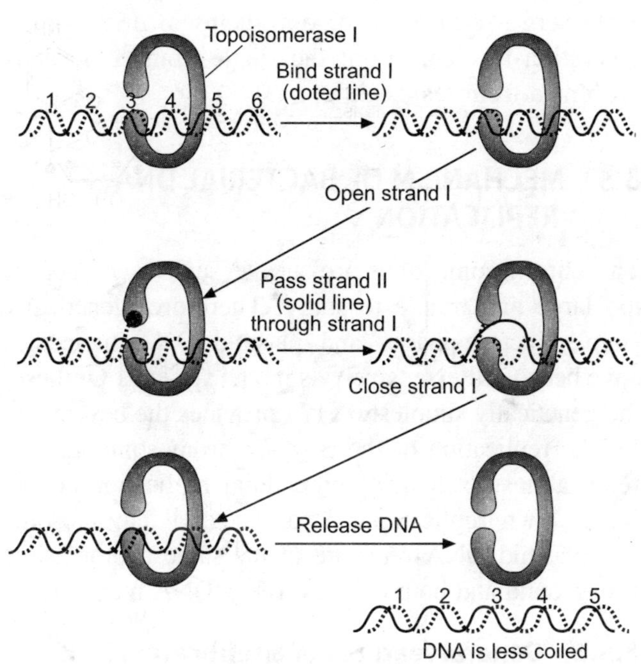

Fig. 8.16 *E. coli* topoisomerase I can reduce DNA coiling in one strand of the double helix and passing the other strand through it.

arise during replication and recombination. In *E. coli*, many reactions involving nucleic acids (including replication, transcription, homologous recombination and the transposition of some mobile elements) are favoured by negative supercoiling. In eukaryotes, topoisomerases are a part of the nuclear scaffold and facilitate the organisation of DNA into functional domains; they also play a major role in the separation of sister chromatids during mitosis.

Classification of topoisomerases

DNA cleavage by all topoisomerases (topo) is accomplished by the formation of a transient phosphodiester bond between a tyrosine residue in the protein and one of the ends of the broken strand. There are two types of topoisomerases, based on reaction mechanism (**Table 8.4**). Topo I cuts only one strand of duplex DNA whereas both strands are cut by topo II and can therefore catenate/decatenate covalently closed circles. Type I topos are further classified as either type IA subfamily member if the protein link is to a 5' phosphate or type IB subfamily members if the protein is attached to a 3' phosphate. Topos that cleave both strands to generate a staggered double-strand break are grouped in the type II subfamily of topos. Further division of the subfamilies (type IIA and type IIB) is based on structural consideration.

8.4.10 DNA primases

DNA primases are enzymes which make the RNA primers required for initiation of DNA synthesis. Primers are short oligonucleotides, containing 6 to 60 nucleotides. They can be made of ribonucleotides or a mixture of deoxyribonucleotides and ribonucleotides. Primases perform two functions. The principal primase in *E. coli* is the 60 kDa protein called **DnaG protein**, the product of the *dnaG* gene. The major primase in eukaryotic cells is the **DNA polymerase** $\tilde{\alpha}$. During DNA replication, primases initiate the synthesis of leading strands at the origin (once at each replication origin), and they facilitate the repetitive initiation of Okazaki pieces during elongation. Most primases can be classified into two main groups:

- Primases from bacteria and their phages and
- Eukaryotic DNA primases.

All primases share common properties, but the two groups differ both in structure and in their relationship with other proteins in the replication complex. The prokaryotic primases are generally found associated with the replicative DNA helicases whereas the eukaryotic primases are found as complexes with DNA polymerase α (pol α).

Table 8.4 **Types of topoisomerases and their action**

Type	*Action on DNA*	*Effect on supercoiling*
Topoisomerase I		
E. coli	Cuts one DNA strand	Relaxes negative supercoils
Eukaryotes	Cuts one DNA strand	Relaxes positive and negative supercoils
Topoisomerase II (gyrase)		
E. coli	Cuts two DNA strands; ATP-dependent	Relaxes positive supercoils; inserts negative supercoils; separates interlinked circles
Eukaryotes	Cuts two DNA strands; ATP-dependent	Relaxes positive supercoils but cannot insert negative supercoils

8.4.11 Nucleases

Nucleases are enzymes which digest nucleic acids by hydrolysing the phosphodiester bonds. Nucleases are of two types, deoxyribonucleases (**DNases**) if their substrate is DNA and ribonucleases (**RNases**) if their substrate is RNA. Further their action can be exonucleolytic or endonucleolytic. Nucleases vary widely in their substrate specificity. Exonucleases may have a specific polarity and may require a particular single- or double-strand specific, end structure to initiate digestion. Endonucleases may also possess single- or double-stranded substrate preference, and may demonstrate various degrees of sequence specificity. During replication, several sources of nuclease activities are required. Generally, the recycling of nucleotides by DNases and RNases provides substrates for both replication and transcription (**salvage pathways**), and mutations in genes controlling these pathways generate quick-stop phenotypes. Three specific nuclease activities are needed during replication.

8.4.12 Single-stranded DNA binding proteins (SSB)

These are a part of a group of replication accessory proteins lacking enzymatic activity, but required for efficient activity of other enzymes in the replisome. All the SSBs have common DNA-binding properties, but perform many functions in the cell related to the stability of single-stranded regions of DNA (e.g., homologous recombination). In replication, this involves stabilising the melted origin, sustaining the activity of helicases, removing secondary structures from the DNA template and the inhibition of nuclease activity. They have been isolated from many sources and have the ability to convert dsDNA into ssDNA form at a temperature 40°C below the normal melting temperature; thus, they are also known as **helix destabilising (HD) proteins**. There are about 300–800 molecules of SSB proteins per uninfected *E. coli* cell, which is sufficient to cover about 1,600 at each replication fork. Purified polymerases are often stimulated in a very specific manner by the presence of their complementary DNA-binding proteins. The human SSB is a heterotrimer that stimulates unwinding of DNA and activity of DNA polymerase α and δ.

The SSBs have a high affinity for ssDNA (but not for dsDNA or for RNA), but with no base sequence specificity—it binds anywhere on the separated strands. The energy released in binding helps to drive strand separation to completion; the single-stranded form is prevented from reannealing.

8.5 MECHANISM OF BACTERIAL DNA REPLICATION

The chromosome of a prokaryote such as *E. coli* is too large and fragile to study. Therefore, genetically simple bacteriophages and plasmid DNA molecules have been used extensively as model systems. Of these, the genetically simplest øX174 provides the best model for the replication of the *E. coli* chromosome since it relies almost exclusively on cellular replication factors for its own replication. Also, bacterial chromosome and plasmid DNA replicate in the same cyotoplasmic background and both exist as free dsDNA molecules.

8.5.1 General features of a replication fork

Figure 8.17 describes the enzymatic apparatus assembled at a replication fork. DNA gyrase and helicase unwind the DNA double helix, and the single-stranded regions of DNA are maintained through interaction with SSB. Primase synthesises an RNA primer on the lagging strand. The leading strand needs priming only once when replication is initiated. The lagging strand template is looped around and the replicative DNA polymerase (DNA polymerase III holoenzyme in *E. coli*), as a dimer, moves 5'→3' relative to each strand, copying the template and synthesising new DNA strands. The replicative polymerase is tethered to the DNA by its ß-subunit-sliding clamps. The DNA pol III γ complex periodically unclamps and then re-clamps the ß subunits on the lagging strand as the primer for each new Okazaki fragment that is encountered. Downstream on the lagging strand, DNA pol I removes the RNA primer and replaces it with DNA. DNA ligase seals the nick.

Origins of replication in bacteria

The specific site on the chromosome where DNA synthesis begins is known as the **origin of replication**. Depending on the organism, there may be as few as one or as many as thousands of origins per chromosome. The origin of the *E. coli* chromosome is designated **oriC** (oriC stands for Col E chromosome) and is essential for normal replication and is bound to the cell membrane. Bacterial chromosomes contain a single origin of replication. Origins of other DNA are also

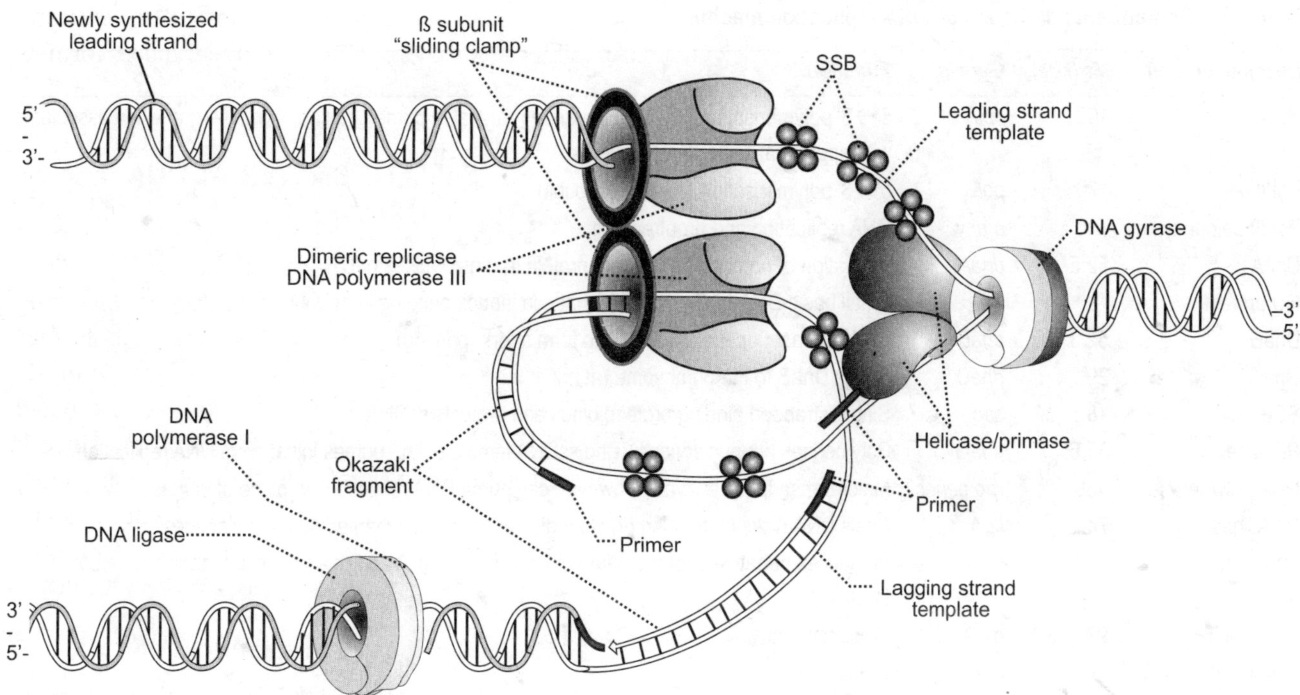

Fig. 8.17 General features of a replication fork; DNA double helix is unwound by DNA gyrase and helicase and the single strands are stabilised by SSB (ssDNA-binding protein)

named similarly, e.g., Ori-λ, Ori-T7, OriCol-E1 and so on. Each of these origins has its own structural features (**Fig. 8.18**).

Francois Jacob, Sydney Brenner, and Jacques Cuzin (1963) proposed a model to explain the events controlling the initiation of replication in bacteria. They stated that DNA is replicated from a particular origin known as a replicon. In this model, there are two components that controlled the initiation of replication: the replicator and the initiator. The **replicator** is defined as the entire set of *cis*-acting DNA sequences that direct the initiation of DNA replication. The second component, the **initiator protein,** specifically recognises a DNA element in the replicator and activates the initiation of replication. OriC is located between gid 'A' (8.5 KB) to the left of the origin and mioc (2.2kb) to the right. The *dnaA* gene is also on its left. Actually this origin is located in a spacer region between the said genes.

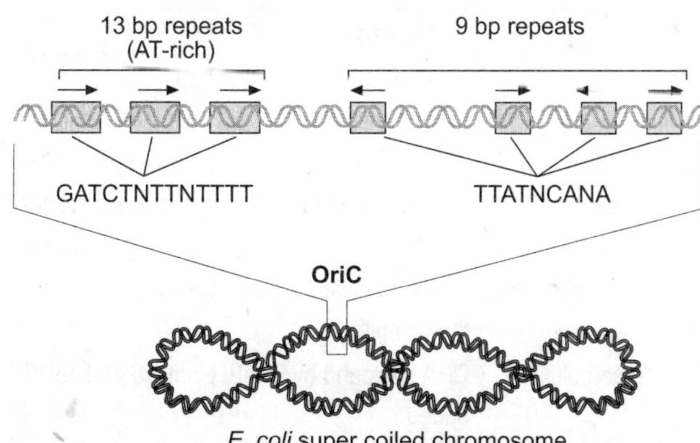

Fig. 8.18 Structural organisation of *E. coli* chromosome having a single origin of replication (oriC)

The minimal sequence of OriC is 245 bp long, and contains a series of tandemly repeated elements, or **iterons** which consist of:

- Four copies of a 9 bp sequence, the *dnaA* box [TT(A/T)T(A/C)CA(A/C)A] arranged in inverted orientations at the other end of the origin and spread over a recognition site. DnaA binds to these boxes with high affinity.
- Three tandem repeats of a 13 bp sequence (5'-GATCTNTNTTNTT) at one end of the origin, which is generally AT-rich. The left most of these three elements is helically unstable. In other words, it can unwind in the absence of DnaA; unwinding is induced by supercoiling alone.
- Six copies of a 6 bp sequence (5'-AGATCT) that is bound by DnaA in complex with ATP. These sites overlap the 13 bp tandem sites.

Table 8.5 Components of the *E. coli* DNA replication machine

Enzyme/subunit	*Mr (kDa)*	*Gene(s)*	*Function*
Pol I	102	*pol*A	5'→3' polymerisation (catalytic subunit); removal of primer, editing, filling the gap when associated with appropriate accessory proteins
Pol II	120	*pol*B	5'→3' polymerisation (catalytic subunit)
Pol III holoenzyme	740	*many*	DNA replication and proofreading
DnaA	52.5	*dna*A	Formation of primosome (initiator protein that recognises oriC)
HU protein			Small basic DNA-binding protein that can induce bending in dsDNA
DnaB	52.3	*dna*B	DNA helicase; opens up dsDNA to form a fork (priming)
DnaC	29	*dna*C	Brings DnaB to the primosome (at oriC)
SSB	18.6	*ssb*	Single-stranded binding protein; binds and protects ssDNA
Primase	65.6	*dnaG*	Polymerises ribonucleotides using ssDNA template, i.e., primes initiation of DNA replication
RNA polymerase	460	*rpo genes*	Major transcription enzyme, however, can prime DNA replication in place of primase
DNA ligase	74	*ligA*	Seals DNA nicks by forming phosphodiester bonds between adjacent nucleotides
DNA gyrase (TopoII)	374		Introduces negative supercoils into closed circular DNA and relaxes (decatenates) supercoiled DNA
α subunit	97	gyrA	Closing the nick
ß subunit	89.8	*gyr*B	ATPase
DnaT	19.3	*dna*T	Termination of replication

- Within oriC there are also 14 or more *dam* methylation sites (GATC sites) which are methylated by the *dam* methylase.

Methylation is important for initiation. There is a lag of about 10 minutes between replication of origin DNA and its methylation and this transient hemimethylation is caused by the sequestration of the origin to membrane sites and is probably essential for chromosome segregation. In addition, this sequestration blocks the initiation of new rounds of replication.

Although there are two promoters for RNA polymerase within the site of origin, there is a strong promoter (*mioC* or p16 kD) just outside the origin region. Transcripts from this promoter enter oriC in a counterclockwise direction and some terminate at transition sites associated with the GATC and DnaA protein binding sites. The principal function of transcription is to alter local DNA structure and allow an initiation complex to be formed at the origin. **Table 8.5** describes the components of the *E. coli* DNA replication machine.

8.5.2 Initiation of replication

Initiator proteins typically perform three different functions:

- They bind a specific DNA sequence within the replicator.
- Once bound to the DNA, they frequently distort or unwind a region of DNA adjacent to their binding site.
- They also interact with additional factors required for initiation.

Table 8.6 and Fig. 8.19 summarises the initiation process in *E. coli*.

Table 8.6 Stages in the initiation of replication in E.coli

Initiation stage	*Characteristics*
Initial complex	DnaA binds to monomer DnaA boxes, no ATP required
Open complex	DNA melted in 13-mer region, requires ATP, and HU protein which controls DNA structure (nucleoid)
Pre-priming complex	DnaBC helicase loaded. DnaC relesed from the complex
Priming complex	DNA unwound by DnaB. Primase and pol III added
Elongation	Topoisomerase and SSB proteins recruited, elongation begins

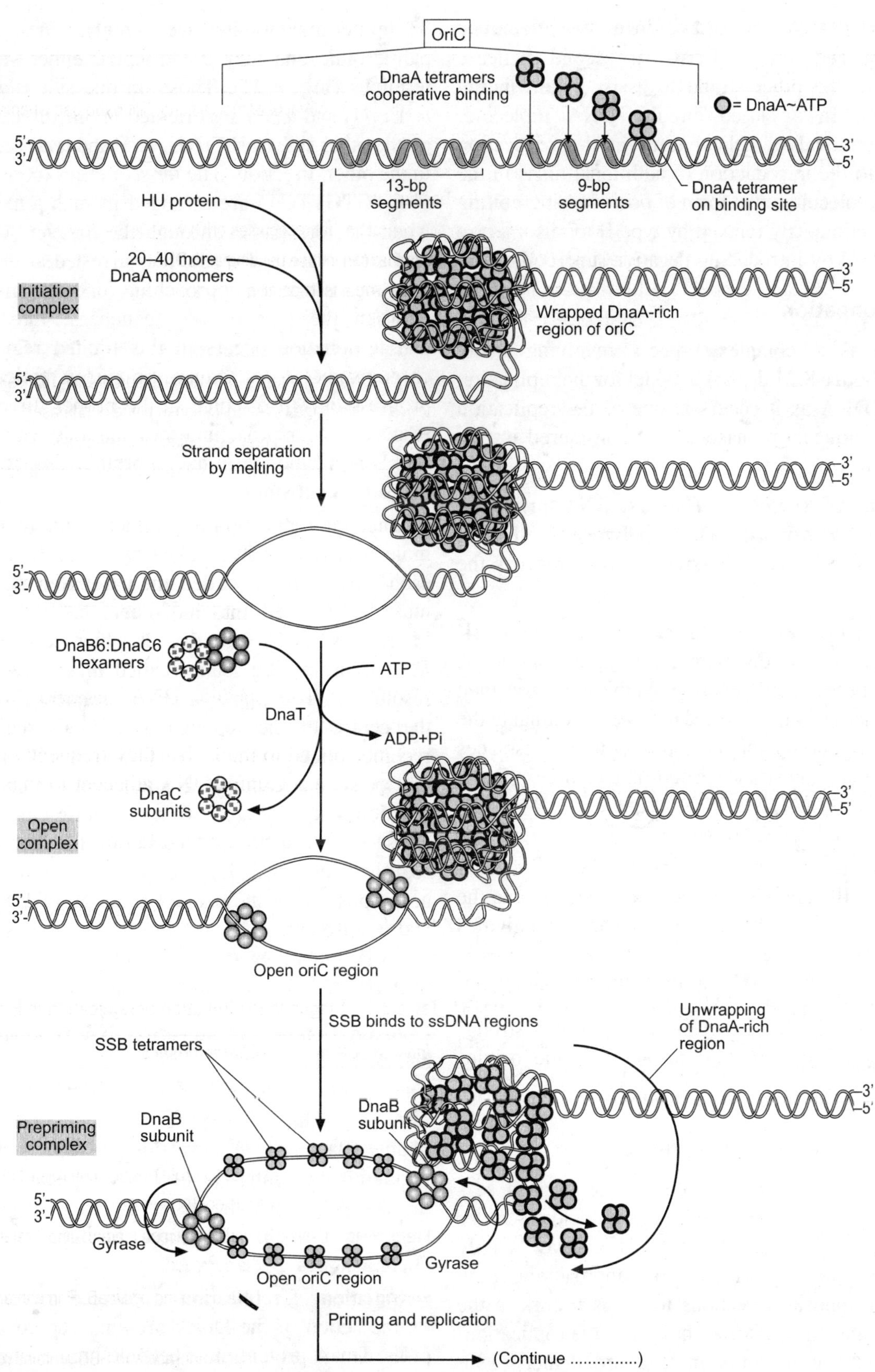

Fig. 8.19 Initiation of DNA replication in bacteria

Unwinding: DNA helicases open the template strands. In addition to DnaB, a second helicase may bind to the other strand to assist in unwinding (**Fig. 8.20**). In a closed circular DNA molecule, however, removal of helical turns at the replication fork leads to the introduction of additional turns in the rest of the molecule in the form of positive supercoiling which is continuously relaxed by type II topoisomerase (DNA gyrase) by introducing negative supercoils.

8.5.3 Elongation

Elongation is a complex process involving many proteins. **Figure 8.21** depicts a model for the replication of *E. coli* DNA as it occurs at one of the replication forks. The elongation phase can be considered in two parts:

Leading strand synthesis: The first RNA primer is synthesised by primase. DNA polymerase III can then synthesise DNA processively until it reaches the terminus.

Lagging strand synthesis: This requires that the pol III complex be partially disassembled from this strand. A new RNA primer must then be synthesised, and then the pol III complex must be reassembled to elongate the primer. As the newly formed replication fork displaces the parental lagging strand, primosome complex (DnaB and primase) synthesises RNA primers every 1,000–2,000 nt on the lagging strand.

Once the lagging strand primers have been elongated by DNA pol III, the RNA primers are removed and the gaps filled by DNA pol I. This enzyme has the activities of all the three enzymes: 5'→3' polymerase, 5'→3' exonuclease and 3'→5' proofreading exonuclease. Pol I removes the last ribonucleotide of the RNA primer along with a few more deoxynucleotides and simultaneously fills the gaps. The final phosphodiester bond between the fragments is made by *DNA ligase*.

8.5.4 Termination and segregation

The replication of circular genomes terminates at a specific sequence, a **terminus region** (defined as *ter* or τ locus). Alternatively, replication may simply cease when two replication forks approaching each other from opposite directions meet, as occurs in the SV40 genome. In *E. coli*, at the end of the replication, the two replication forks meet approximately 180° opposite of oriC. The terminator site is a large region (350 kb) flanked by 10 nearly identical 23 bp terminator sites (***ter*A–*ter*J**). These are non-palindromic and they are oriented either side of the terminus (**Fig. 8.22**). Those on one side (*ter*A, *ter*D, *ter*E, *ter*I and *ter*H) are oriented in one direction; those on the other side (*ter*B, *ter*C, *ter*F and *ter*J) are oriented in the other direction. The *ter* sequences (core sequence 5'-GTGTGTTGT) are arranged in such a manner that when the fork passes through the first *ter* sequence it encounters (from 3'-end), it is arrested at the second. This means that the approaching forks become trapped between the pair of *ter* sequences, thus ensuring that termination occurs in this limited region of the chromosome. A *ter* sequence impedes replication fork progress only if oriented in the proper direction with respect to the approaching fork and only with a specific 36-kD replication termination protein **Tus** (**terminator utilisation substance**).

Once the replication is completed, the two daughter molecules remain interlinked by 20 to 30 coils about each other. This is called the **catenated** state. These must be resolved into monomers before segregation of the new DNA into two daughter cells occurs. In *E. coli*, a specific locus—called ***dif***—is required for resolution. The *dif* is a 28-bp sequence located in the centre of the replication terminus region. Two recombination proteins, **XerC** and **XerD** bind to the *dif* locus and catalyse a site-specific recombination reaction that generates two monomeric chromosomes. These two must be resolved before the new DNA can segregate into two daughter cells. In *E. coli*, this is accomplished by the concerted action of DNA gyrase and topoisomerase IV (a type II DNA topoisomerase). Topo IV mutations are found to be lethal and the dead bacteria exhibit incompletely separated chromosomes.

Coordination of prokaryotic DNA replication with cell division

The rate of bacterial growth is assessed by the **doubling time**, that is, the time required for cell number to become double. Bacteria can solve this problem, generally by regulating the initiation of DNA replication. Shorter the doubling time, the faster the bacteria grow. Fast rates of growth, therefore, produce **multiforked chromosomes**.

Initiation of replication is coupled to cell division by the action of the DnaA protein. The concentration of the DnaA protein must become high enough to bind all the *dnaA* boxes. Since it takes time to accumulate

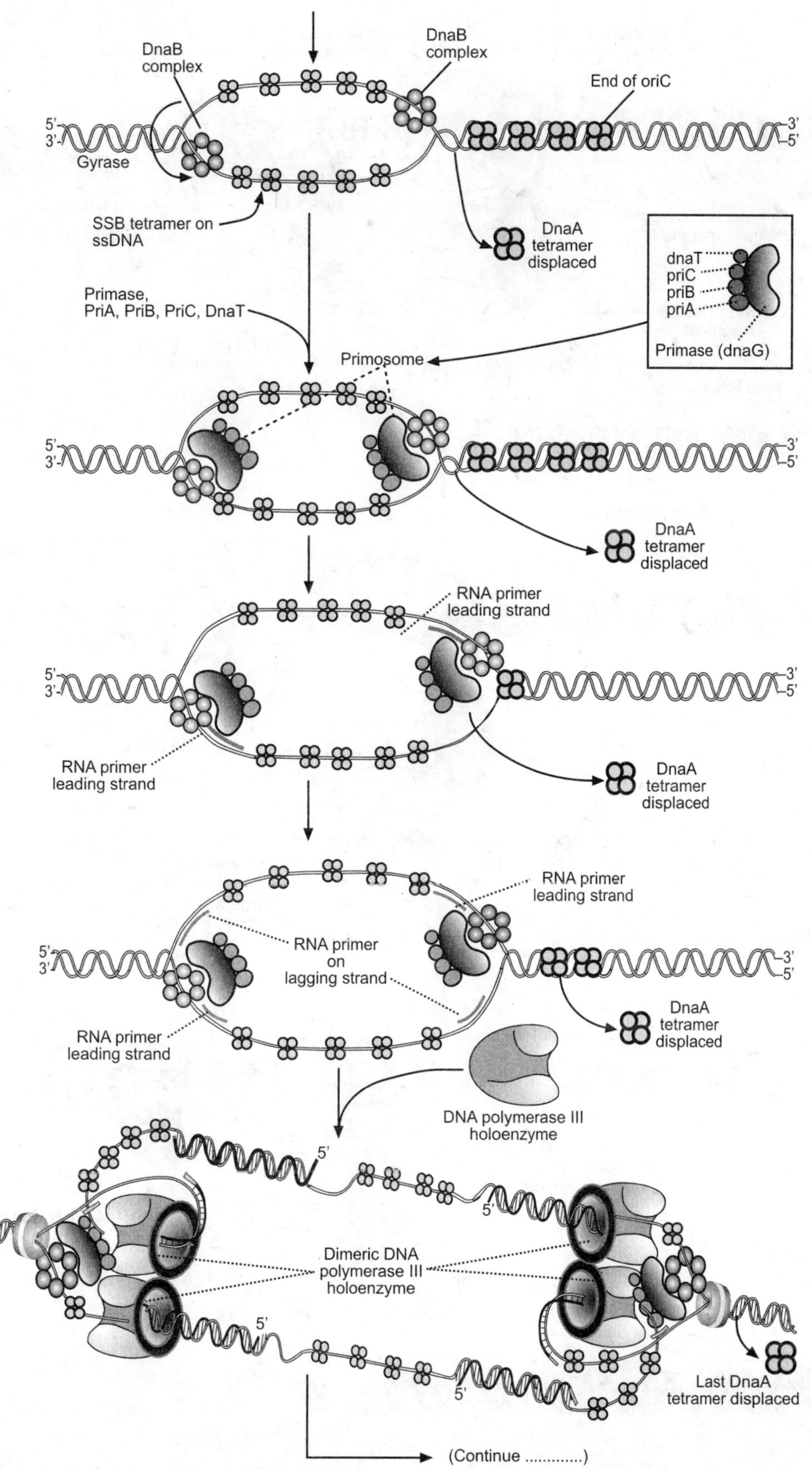

Fig. 8.20 DNA replication: Unwinding

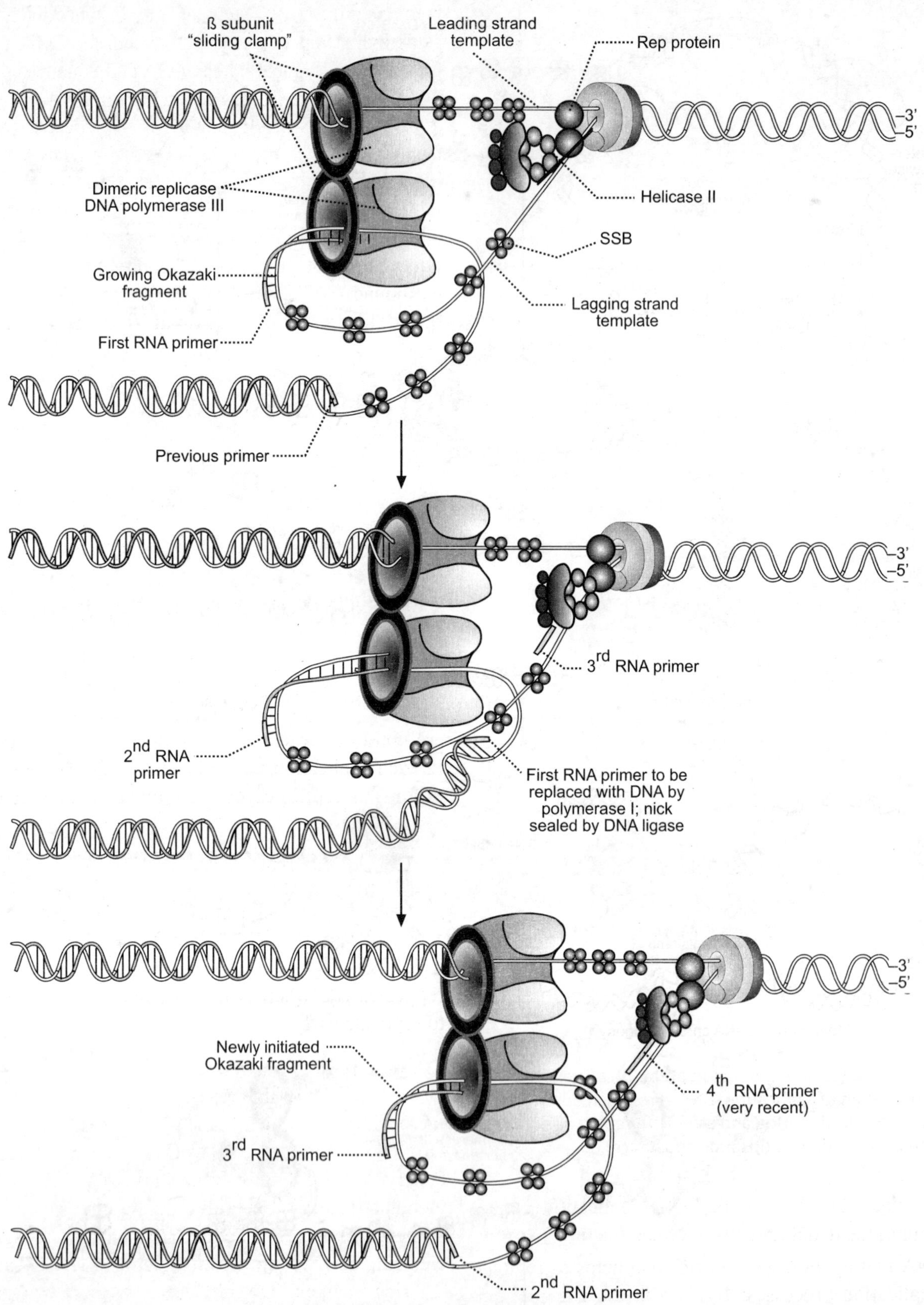

Fig. 8.21 Elongation during DNA replication

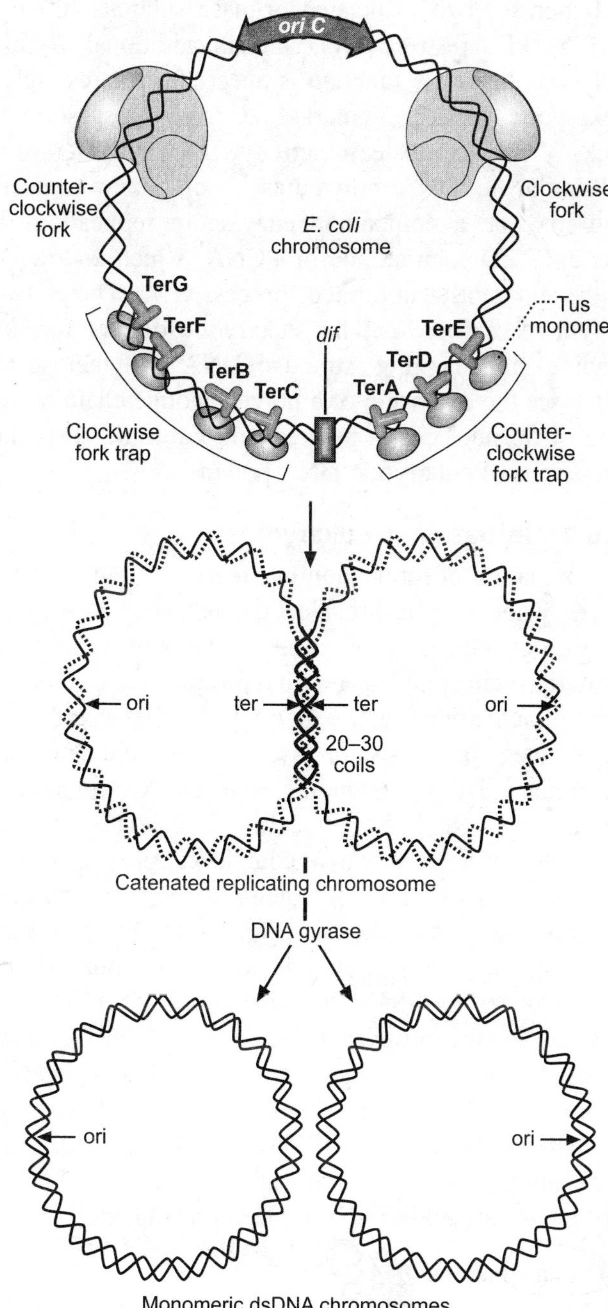

Fig. 8.22 The termination region of the *E. coli* genome and termination of replication; DNA gyrase can untangle chromosomes both during and after DNA replication (ter: terminus tus: terminus utilisation substance)

newly made DnaA protein, DNA replication cannot occur until the daughter cells have had the time to grow.

DNA methylation within oriC also helps to regulate the replication processes. In *E. coli,* the enzyme *dam* methyl transferase adds a methyl group to the A within every GATC sites within oriC. The *dam* methylase recognises the GATC sequence, binds there and attaches a methyl group onto the adenine base. Typically the genome is fully methylated at GATC sequences prior to DNA replication. This is necessary for DNA replication to be initiated at the origin. However, after replication, the A residues in the newly synthesised DNA become hemimethylated. This unmethylated state of the newly replicated oriC is detected by a protein called SeqA. It binds tightly to the GATC sequence with two important effects: first, it dramatically reduces the rate at which the bound GATC sites are methylated. Second, when bound to these oriC proximal sites, SeqA prevents DnaA from associating with oriC and initiating a new round of replication. Together, these methods rapidly and dramatically reduce the ability of *E. coli* to initiate replication from new copies of oriC.

8.6 MECHANISM OF EUKARYOTIC DNA REPLICATION

In eukaryotes, DNA replication occurs only during the S phase of the cell cycle (see **Chapter 7**). In mammalian cells, the replication of DNA in the region between one origin and the next should normally require only about an hour to complete, yet the S phase usually lasts for about 8 hours. This implies that the origins are not all activated simultaneously and that the DNA in each replication unit (clusters of 20–80 replication origins) is replicated during only a small part of the S-phase interval. However, different regions on the same chromosome replicate at distinct points during the S phase.

It seems that the order in which replication origins are activated depends, in part, on the chromatin structure in which the origins reside. Highly condensed chromatin replicates late whereas less condensed chromatin tend to replicate early.

Multiple replication origins

DNA synthesis begins at replication origins which are the positions at which the DNA helix is first opened. Origins in simple eukaryotic systems (unicellular eukaryotes) are similar to prokaryotic origins in their modular nature. Generally, they comprise a core sequence (***ori-core***) containing an origin recognition element, a **DNA unwinding element (DUE)** and an AT-rich element. Outside the core there are one or more auxiliary elements, which act as binding sites for transcription factors. These are thought to increase the efficiency of initiation in the same way that they affect transcriptional initiation, i.e. by interacting with core components of the system, in this case the **orisome** (basal transcriptional apparatus). Simple origins can function as autonomously replicating sequence (ARSs), i.e., sequences which when linked to any other fragment of DNA confer upon it the ability to replicate autonomously in the cell from which the ARS was derived (yeast cloning vectors).

8.6.1 Chromosomal origins in multicellular organisms

In contrast to simple origins, origins in metazoan chromosomes have been difficult to characterise. Functional studies have suggested that the length of the DNA, rather than any specific sequence, is the critical factor for replication. However, the physical analysis of origins by identifying sequences at replication initiation by PCR has shown specific sequences to be involved. The site at which DNA synthesis begins, a primary origin (or **origin of bidirectional replication, OBR**) may be flanked by a number of secondary origins. The primary origin (**Fig. 8.23**) is usually 0.5–2 kbp long, whereas secondary origins may span up to 50 kbp, delineating the so-called **initiation zone**.

The DNA duplex in eukaryotic chromosomes is linear and replicates bidirectionally. Further, replication is initiated at many sites (replication bubbles) in the DNA. Multiple initiations are a means of reducing the total replication time of a large molecule. In a somatic mammalian cell, there are typically about 60,000 replicons. Individual replicons in eukaryotic genomes are relatively small and highly variable, typically ~40 kb in yeast or fly and about 100 kb in animal cells. In eukaryotic cells, movement of each replication fork proceeds at a rate of approximately 10 to 100 nucleotide pairs per second. In a typical eukaryotic cell, origins are spaced about 40,000 nucleotide pairs apart, which allow each chromosome to replicate in 15–30 min. However, the time required for the completion of replication in each chromosome varies.

Eukaryotic DNA polymerases

Eukaryotes use a number of different DNA polymerases named according to the chronology of their discovery. DNA polymerase α, a complex multimeric protein, is found only in the nucleus. It requires a template and a primer and carries out chain extension in the 5'→3' direction. It consists of four subunits: a 180 kD polymerase, two associated primase subunits (60 kD and 50 kD, respectively), and an additional 70 kD polypeptide whose function is uncertain. Interestingly, eukaryotic DNA polymerase α from most sources lacks 3'→5' exonuclease activity. DNA polymerase δ is similar to DNA polymerase α in many respects and may be a companion enzyme in replication. It has a 37 kD subunit, the or PCNA, which endows it with apparently unlimited processivity. These two polymerases represent the eukaryotic dimers for the leading and lagging strands. DNA polymerase α catalyses the extension of a polynucleotide chain at the rate of 50 nucleotides per second. **Table 8.7** lists the properties of eukaryotic DNA polymerases.

8.6.2 Initiation in eukaryotes

The initiation of replication in eukaryotic cells consists of two stages: replicator selection and origin activation. Replicator selection process begins with the assembly of a multiprotein complex at each replicator in the genome. Origin activation occurs only in S phase and also accelerates the replicator-associated protein complex to initiate DNA unwinding and DNA polymerase recruitment. Replicator selection is mediated by the formation of pre-replicative complexes (pre-RCs). The pre-RC consists of four proteins, which assemble in sequence at each replicator (**Fig. 8.24**). The first step in initiation of replication is the recognition of the replicator by the ORC. ORC upon binding recruits two helicase-loading proteins (Cdc6 and Cdt1). ORC and the loading proteins together recruit a protein replication fork helicase (the Mcm 2-7 complex). Phosphorylation of these proteins results in the assembly of additional replication proteins (includes DNA polymerases) at the origin and the initiation of replication (**Fig. 8.25**).

Replication forks

During replication, the DNA must be unwound from the nucleosomes at the replication forks (**Fig. 8.26**). This slows down the movement of the forks to about 50 bp per sec. As in prokaryotes, one or more DNA helicases and a single-stranded binding protein, **replication protein A (RP-A)** are required to separate the strands. Three different DNA polymerases are involved in elongation. The leading strands and each lagging strand fragment are initiated with RNA by the primase that is an integral part of DNA polymerase α. This polymerase continues elongation with DNA but is quickly replaced by DNA polymerase δ on the leading strand and DNA polymerase ε on the lagging strand. Both these enzymes

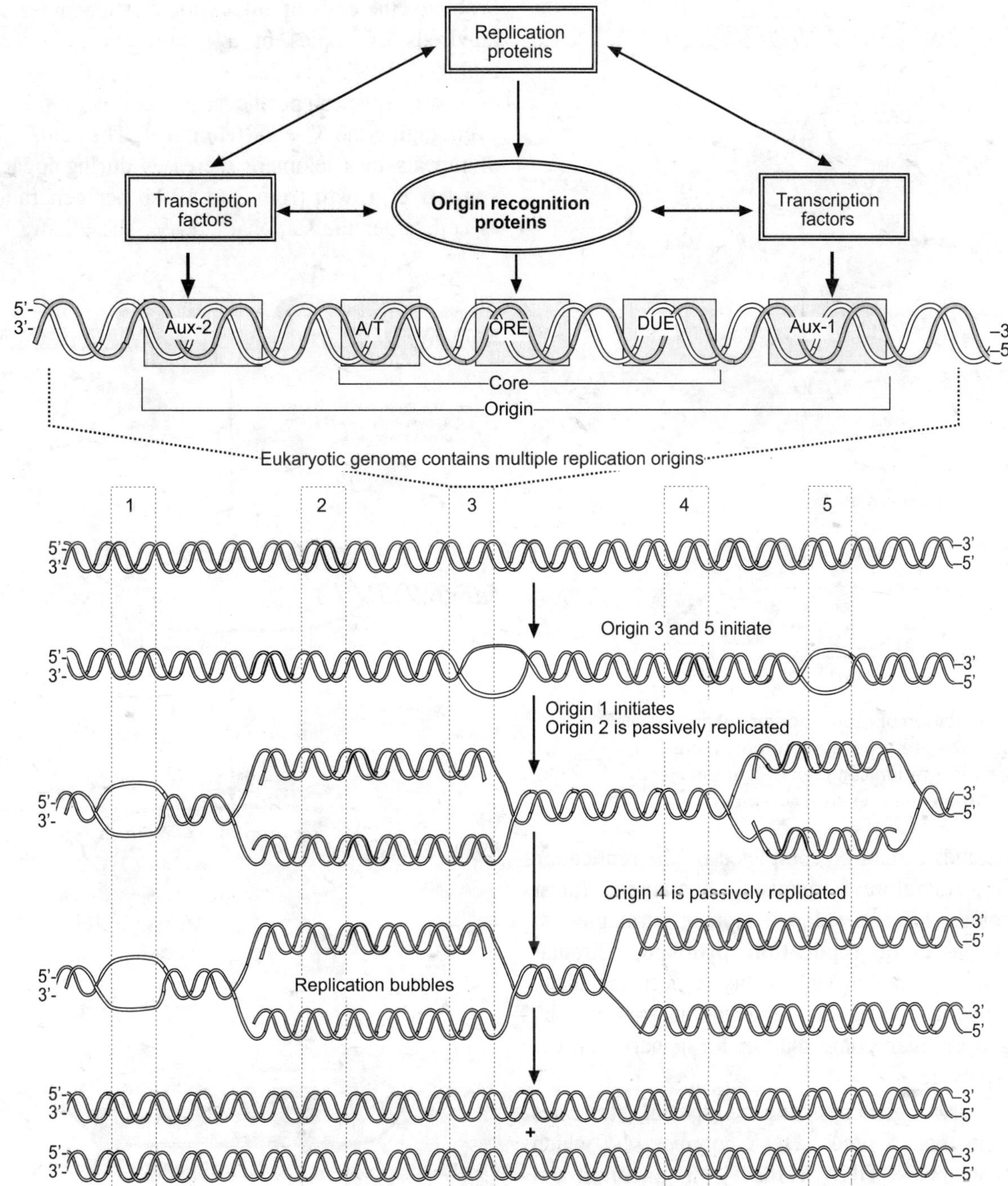

Fig. 8.23 Structure of eukaryotic replicator and multiple origins of initiation bidirectionally in eukaryote genomes

have associated proofreading activity. The ability to synthesise long DNA is conferred on DNA polymerase δ by PCNA, the functional equivalent of the ß subunit of *E. coli* DNA pol III.

The components of the cellular replisome are: DNA helicase, DNA ligase, DNA polymerase, DNA primase, DNA topoisomerase, RNase H and single-stranded binding proteins.

8.6.3 Termination

Completion of DNA replication consists of a sequence of specific events that are different for circular and linear chromosomes. The conventional replication fork machinery can replicate the entire molecule, but the resulting daughter molecules are topologically linked to one another (catenated). In contrast, replication of the ends of linear chromosomes (specifically on

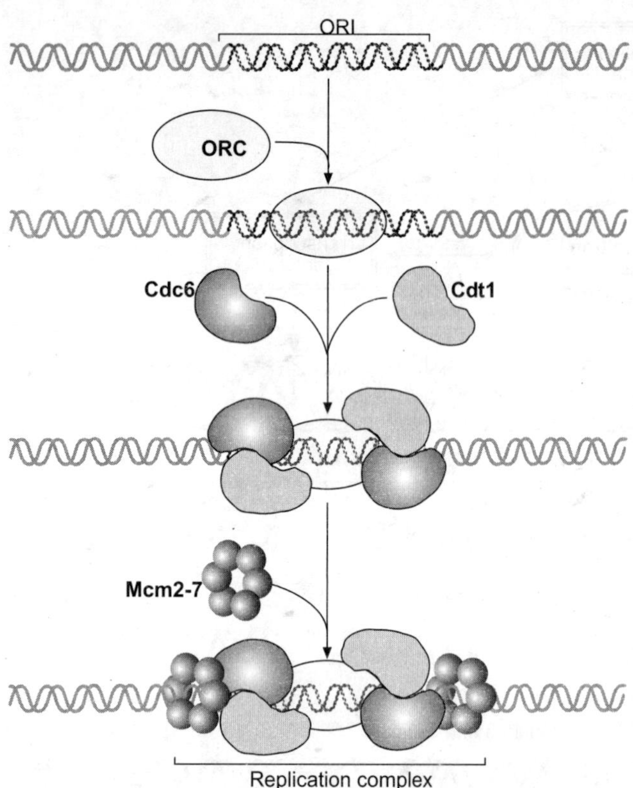

Fig. 8.24 Initiation of replication of eukaryotic DNA: formation of the pre-replicative complex (pre-RC) ORC: origin recognition complex ORI: origin Mcm: minichromosome maintenance complex

lagging strand) cannot be completed by the replication machinery; therefore, organisms containing linear chromosomes have developed unique strategies to overcome this **end replication problem**. Circular genomes have an advantage in this respect and many linear replicons solve the termination problem by adopting a circular conformation for a part of their replication cycle.

Linear DNA genomes require special strategies to complete their 5' ends. DNA polymerases, which can only synthesise DNA in the 5'→3' direction and require a preformed primer. A number of further, elegant strategies have evolved for completing lagging strand synthesis and these are listed in **Table 8.8**.

8.6.4 Telomere maintenance

Eukaryotic DNA molecules are long and linear. The ends of linear chromosomes cannot be fully replicated by semidiscontinuous replication, as there is no DNA to elongate and replace the RNA removed from the 5' end of the lagging strand. Thus, genetic information could be lost from the DNA (**Fig. 8.27**). To overcome this problem, the ends of eukaryotic chromosomes contain hundreds of copies of a simple, non-informational multiple (30–70) repeats or blocks of nucleotides of short, species-specific sequences with the 3'-end overhanging the 5'-end (**telomere**). The actual number of repeats in a telomere increases during logarithmic vegetative growth from 3 to 10 bp per generation but as cells enter the stationary phase, these extra repeats

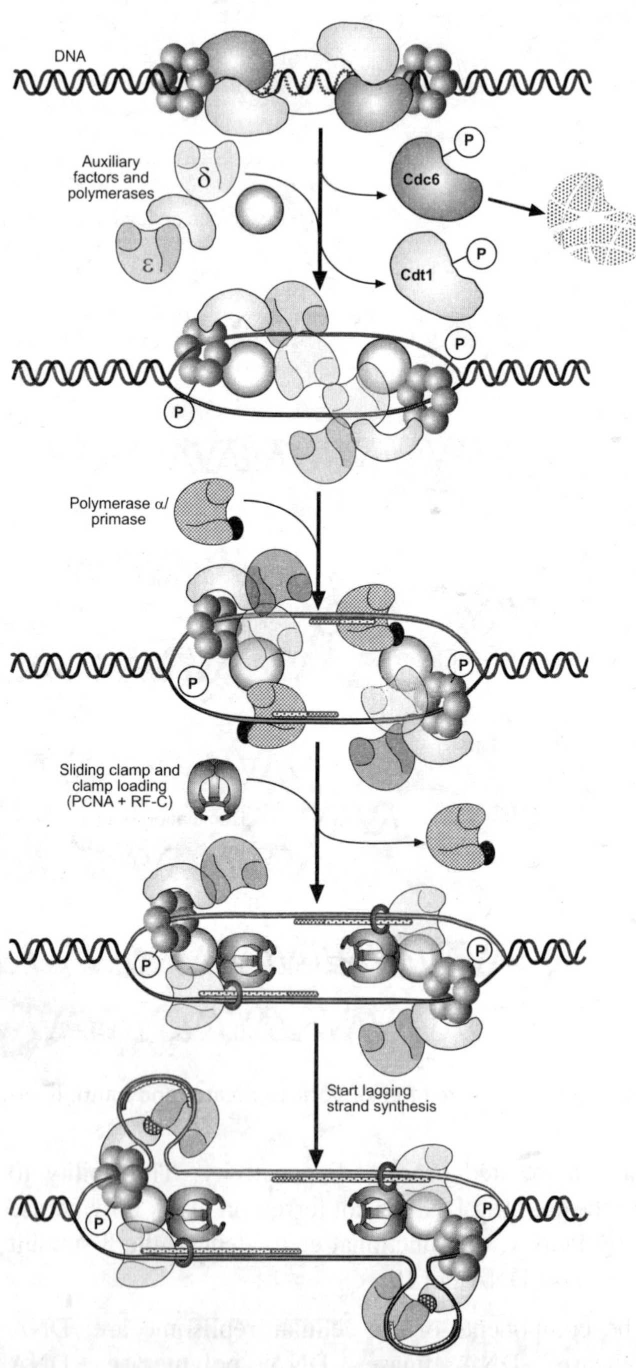

Fig. 8.25 Activation of the pre-RC leads to the assembly of the replication fork

Table 8.7 General properties of eukaryotic DNA polymerases

Properties	α	δ	ε	β	γ
Mass (kD)					
Native	>250	170	256	36–38	160–300
Catalytic core	165–180	125	215	36–38	125
Other subunits	70, 50, 60	48	55	None	35, 47
Location	Nucleus	Nucleus	Nucleus	Nucleus	Mitochondria
Associated functions					
3'→5' exonuclease	No	Yes	Yes	No	Yes
Primase properties	Yes	No	No	No	No
Processivity	Low	High	High	Low	High
Fidelity	High	High	High	Low	High
Replication	Yes	Yes	Yes	No	Yes
Repair subunits	No	?	Yes	Yes	No

are removed. Telomere sequences are extended by a specific telomere terminal transferase (**telomerase**). Telomerase is a ribonucleoprotein, containing a short RNA molecule with the sequence complementary to the repeat. This RNA acts as a template for the addition of these repeats to the 3'-overhang by repeated cycles of elongation and translocation. The complementary strand is then synthesised by normal lagging strand synthesis, leaving a 3'-overhang (**Fig. 8.28**). Interestingly, telomerase activity is repressed in the somatic cells of multicellular organisms, resulting in a gradual shortening of the chromosomes with each cell generation. As this shortening reaches informational DNA, the cells senesce and die. This phenomenon may be important in cell ageing.

8.7 CHROMATIN REMODELLING

Once the DNA has been replicated, new chromosomal proteins should be assembled onto the DNA behind each

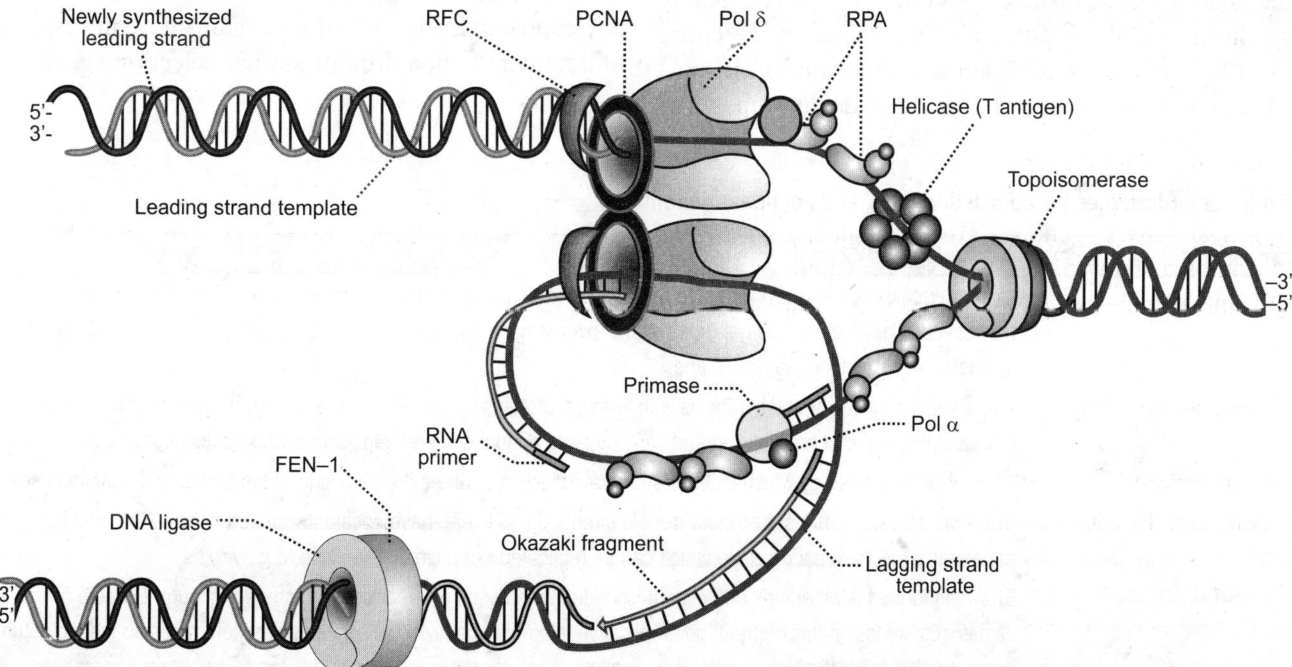

Fig. 8.26 A schematic view of the major components at the eukaryotic replication fork

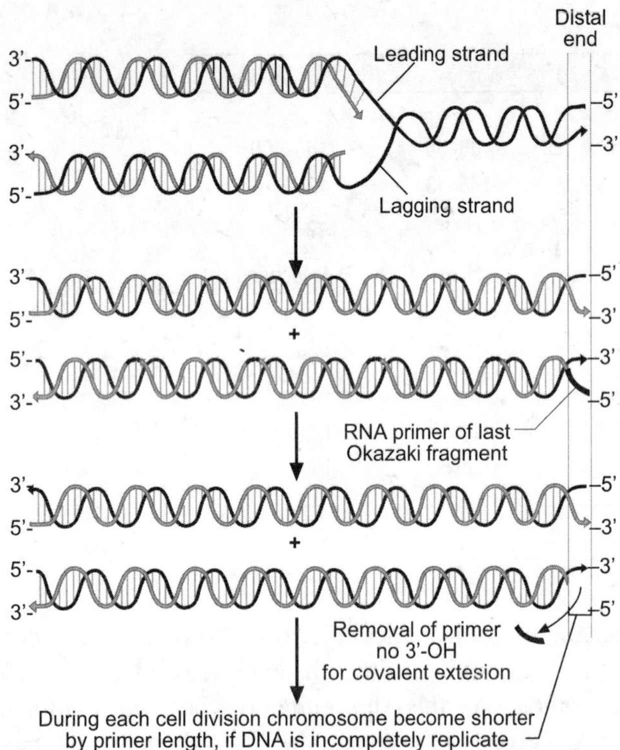

Fig. 8.27 Replication of chromosome ends: telomere lagging strands

replication fork. Since a large amount of new histone proteins is required to make the new nucleosomes, most eukaryotic organisms posses multiple copies of the gene for each histone. Histone proteins are mainly synthesised in the S phase and are stable for the entire life of a cell. As a replication fork advances, it must somehow pass through the parental nucleosomes and the mechanism can be explained by the chromatin-remodelling proteins.

The newly synthesised DNA helices behind a replication fork inherit old histones. Because the amount of DNA has doubled, an equal amount of new histones is needed to complete the packaging of DNA into chromatin. The addition of new histones to the newly synthesised DNA is aided by **chromatin assembly factors (CAFs)**, which are proteins that associate with replication forks and package the newly synthesised DNA as soon as it emerges from the replication machinery. The newly added H3 and H4 histones are rapidly acetylated on N-terminal tails, after they have been incorporated into chromatin, these acetyl groups are removed from the histones.

8.8 OTHER DNA REPLICATION STRATEGIES

There are other methods of DNA replication like amplification and chemical selection. These are discussed in the following sections.

8.8.1 DNA amplification

We have seen so far that complete replication must occur once only within each cell cycle in order to maintain the constancy of the amount of DNA per cell. However, there are some exceptions to this rule. This selective replication of specific sequences of DNA is known as **amplification**. It can occur as part of the developmental program of an organism or in response to treatment with a drug or similar selection agent.

Table 8.8 Strategies for completing the 5' ends of linear genomes

Strategy	*Mechanism and example*
Concatenation	Linear genomes possess redundant termini, which allow circularisation or concatenation. These structures can be cleaved to generate single genomes with 5' overhanging termini which can be filled by conventional DNA synthesis, e.g., bacteriophages T7 and λ.
Terminal protein priming	Viruses which initiate strand synthesis with terminal proteins do not need a specific termination mechanism: they initiate at the 5' end of each strand and can complete up to the 3' end, e.g., adenovirus, bacteriophage ø29.
Hairpin priming	Another priming strategy which initiates strand synthesis from the extreme 5' end of the strand, e.g., parvoviruses
Covalently sealed ends	Some viruses, which are superficially double stranded and linear, have covalently sealed ends so that melting generates a single-stranded circle which can be replicated like a circular replicon, e.g., viroids.
Telomeres	Enzymes termed telomerases add oligonucleotides to the ends of linear chromosomes. Although extreme 5' sequences are lost in this method, postreplicative telomerase activity can replenish the telomeres so that no actual genes are lost in successive rounds of replication.

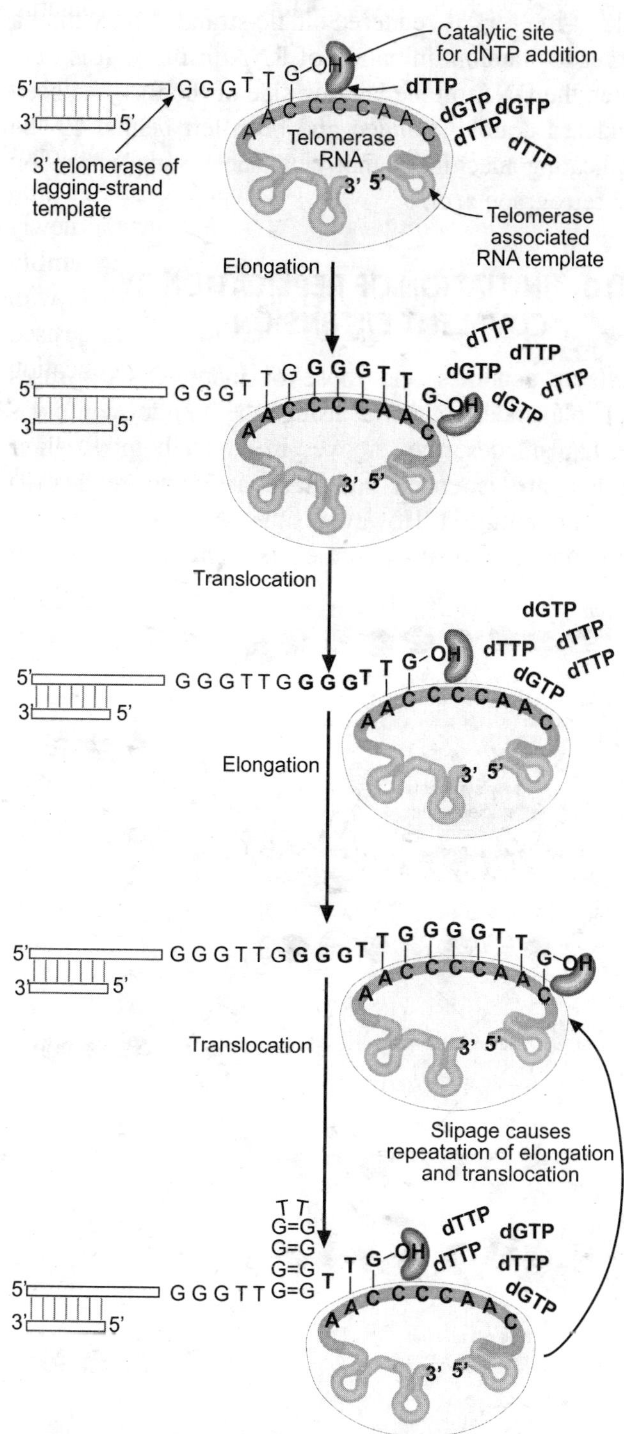

Fig. 8.28 Mechanisms of synthesis of the G-rich strand of telomeric DNA

Developmental amplifications in *Xenopus laevis*

Gene amplification is a cellular process by which a cell can produce vast quantities of a specific gene product. The best-understood example of gene amplification occurs in the development of oocytes in the toad *Xenopus laevis*, in which the genes for rRNA increase many-fold. During oogenesis in *Xenopus,* the ribosomal RNA genes are amplified as much as 2,500–4,000 times the normal haploid value. To accomplish this task, the number of rRNA genes increase through gene amplification.

The amplification process involves an unusual DNA replication event in which the rRNA genes are replicated to form many copies of circular DNA molecules called **minichromosomes** (**Fig. 8.29**). Each circular DNA molecule contains 1–20 copies of the rRNA gene. These minichromosomes accumulate within the nucleoli of the oocyte. Amplification occurs in two phases. The first phase results in the 40-fold amplification of selected ribosomal repeat units in both sexes when only 9–16 primordial germ cells are present. In the male, these amplified sequences are lost at meiosis but in the female oocyte, there is a further amplification. Presumably, to satisfy the special demand of the oocyte, large amounts of ribosomal DNA are produced in the form of small extrachromosomal circles and replicating rolling circles. Each replication rolling circle consists of several copies of the ribosomal DNA repeat and capable of autonomous replication. The precise mechanism of formation of these structures is not known but some information about it is based on the size of the spacer between the 18S and 28S components. The essential feature of these molecules is that each of them contains a spacer having one size, suggesting that a single rDNA unit is excised from the chromosome and circularises, then begins to replicate by the rolling circle mode, generating long branches.

Interestingly, in most of the rolling circles the circular portion does not contain one rDNA unit, but as many as 16. Finally, each replicating unit forms one of the many nucleoli of the mature oocyte. Once the oocyte is mature, the excess rDNA serves no purpose and it is slowly degraded.

Amplification by chemical selection

When animal cells in culture are stressed with drugs, they undergo selection for drug resistance by overproducing a gene product. For example, cells treated with methotrexate, an inhibitor of dihydrofolate reductase (DHFR), may respond by overproducing the inhibited enzyme. This is often a result of selective amplification of the gene coding for dihydrofolate reductase (*dhfr*) gene. Corresponding results are found with other drugs, but it is not always the gene for the target enzyme which is amplified and chromosomal location markedly affects the frequency of amplification.

8.9 CHROMATIN REPLICATION

What happens to the nucleosomal structures at the replication fork and to what extent is the structure of chromatin responsible for the mechanism of DNA synthesis which involves Okazaki fragments and replicons? Immediately after the new strand is synthesised, nucleosomes quickly become associated. However, the rate and extent of cleavage of nascent chromatin is greater than that of mature chromatin indicating a somewhat different structure, which requires 15 minutes to mature.

As the replicating fork progresses along a region of nucleosome-associated DNA, a number of different situations might result, some of which are illustrated in **Fig. 8.30**.

Thus, movement of the DNA fork may be considered to occur as a series of steps caused by relocation of intact, old nucleosomes on the leading side of the fork, which is always made up of duplex DNA. Either the parental nucleosome remains associated with the leading strand or is only transiently displaced and rapidly rebinds to duplex DNA on the leading strand but not to ssDNA on the retrograde arm. At each relocation initiation, a region of about 200 bases of DNA on the lagging side of the fork is rendered single-stranded providing a target for random initiation of RNA primers. It is only when the DNA on the lagging side of the fork has been rendered double-stranded and been left behind by the replicating machinery that new nucleosomes form on the retrograde arm.

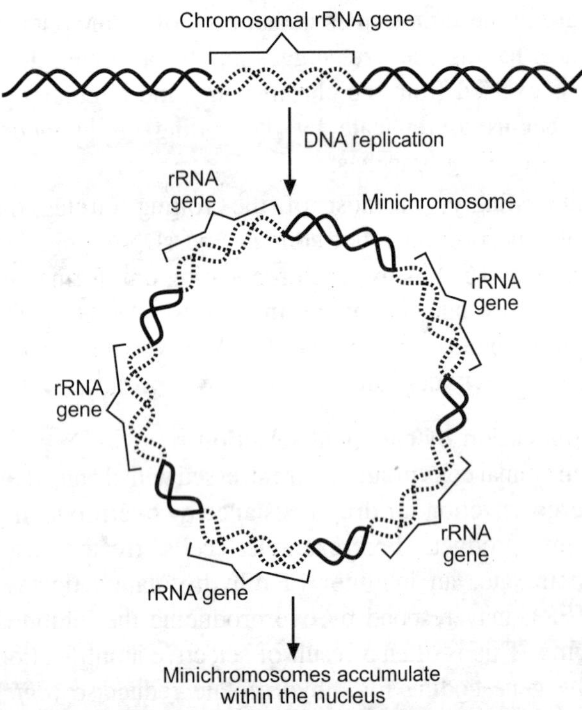

Fig. 8.29 The amplification of the DNA that encodes rRNA genes within the oocytes in *Xenopus laevis*. The rRNA genes are specifically replicated to generate minichromosomes.

8.10 INITIATION OF REPLICATION BY COVALENT EXTENSION

Cellular genomes and those of many DNA viruses and plasmids replicate using the semidiscontinous mechanism described above, in which both daughter strands are extended simultaneously and replication forks are required. However, other dsDNA and ssDNA replicons use distinct strategies. The replication of

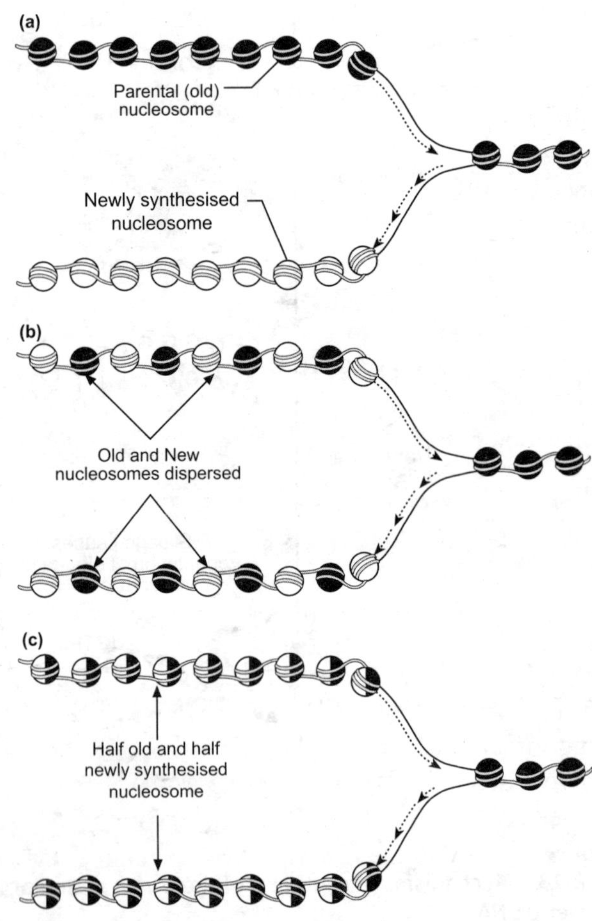

Fig. 8.30 The possible mechanisms for the distribution of nucleosomes at a replication fork (a) All the old nucleosomes remain with one of the parental strands when they separate at the replication fork (corporative segregation). (b) The old nucleosomes are dispersed at random between the two strands of DNA. (c) The old nucleosomes dissociate into half nucleosomes, one half going to each side of the fork

ssDNA and RNA genomes always involves a replicative intermediate of opposite sense to the genome (sometimes called as **antigenome**). The replicative intermediate may act as a distinct template for synthesis of the daughter genome, or the genome and antigenome may stay associated in a double-stranded **replicative form (RF)** from which daughter genomes are produced by various ways such as strand displacement, replication or rolling circle mechanism.

Initiation from a single-stranded template

G4 phage: The mechanism of initiation of the synthesis of precursor fragments in bacterial phages has been well-studied. *E. coli* phages (e.g., phage G4) have single-stranded DNA. The first step in replication is the conversion of circular ssDNA into a double-stranded covalent circle. G4 phage DNA molecule has a small palindromic segment, which forms a double-stranded hairpin. The initial step common to the replication of all single-stranded molecules is the coating of all single-stranded regions with SSB protein. The double-stranded region remains uncoated and is the site of binding of primase, which synthesises a small segment of RNA. Pol III holoenzyme then extends this RNA primer and displacing the SSB proteins, proceeds to replicate around the circle. When the 5'-P terminus is reached, pol I replaces the pol III holoenzyme and removes the RNA, replacing it with DNA. Ligase seals the nick and replication is complete.

Rolling circle model

Initiation of replication in certain circular ssDNAs occurs by a mechanism called rolling circle replication (**Fig. 8.31**). For example, replication of small, ssDNA phages such as M13 and øX174 adopt a distinct way of replication in which the ssDNA is converted into replicative form (RF). These RFs now replicate *via* a rolling circle mechanism, in which a double-stranded RF gives rise to many copies of a ssDNA progeny. This intermediate RF also looks like the lowercase Greek letter σ (sigma), so the mechanism is sometimes called the σ mode.

DNA transfer

Transmission of plasmid DNA from one bacterial cell to another is somewhat different from the rolling circle model. This kind of replication also takes place while transferring plasmid DNA from the soil bacterium *Agrobacterium tumefaciens* to the plant genome.

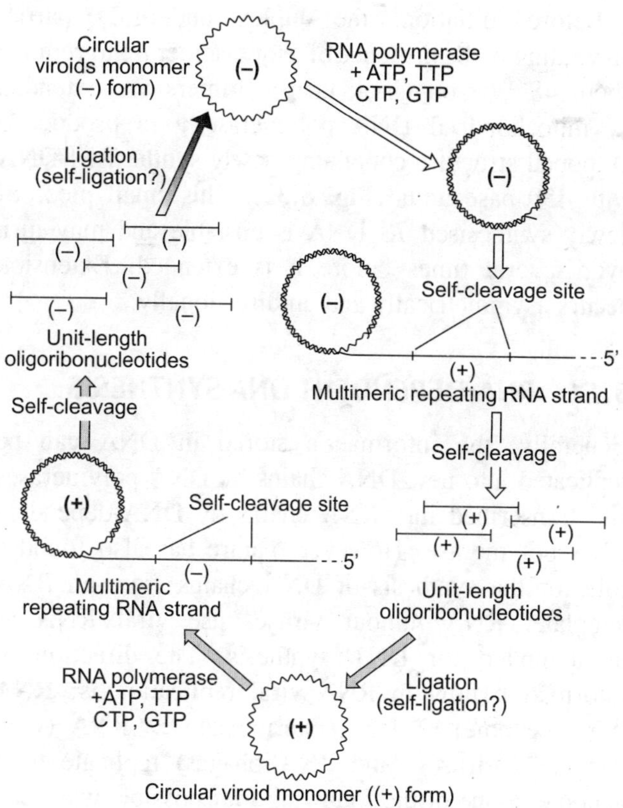

Fig. 8.31 Scheme proposed for the rolling circle replication of circular viroid RNAs making more than one copy of the original circle

Loop model

Eukaryotic mitochondria must be duplicated during the cell cycle and segregated to the daughter cells. Mitochondria contain a closed circular duplex DNA molecule of about 16 kb that replicates unidirectionally from a unique origin. This origin region also contains the starting site for transcription of both heavy and light strands of mitochondrial DNA. Transcription from both heavy and light strand promoters requires mitochondrial RNA polymerase and a mitochondrial transcription factor. The mouse mitochondrial heavy (H) strand transcript has a unique 5' end at nucleotide 16,183. This transcript usually contains all the mitochondrial genome but it can be processed by an RNase H-like activity. This enzyme (RNase MRP) is a ribonucleoprotein containing a nuclear-coded RNA of 265–275 nucleotides and it cleaves some copies of the transcript near a conserved sequence block (CSB) to provide a primer for DNA synthesis. Three such blocks (CSBI, II and III) are found in mouse mitochondrial DNA.

Before initiation, the duplex undergoes partial unwinding in the presence of a topoisomerase to remove about 48 Watson–Crick turns. Primers are extended by mitochondrial DNA polymerase γ to produce a D-looped structure containing newly synthesised DNA with 450 base units (**Fig. 8.32**). This small piece of newly synthesised 7S DNA is unstable and may turn over several times before it is extended. Extension occurs asymmetrically and unidirectionally.

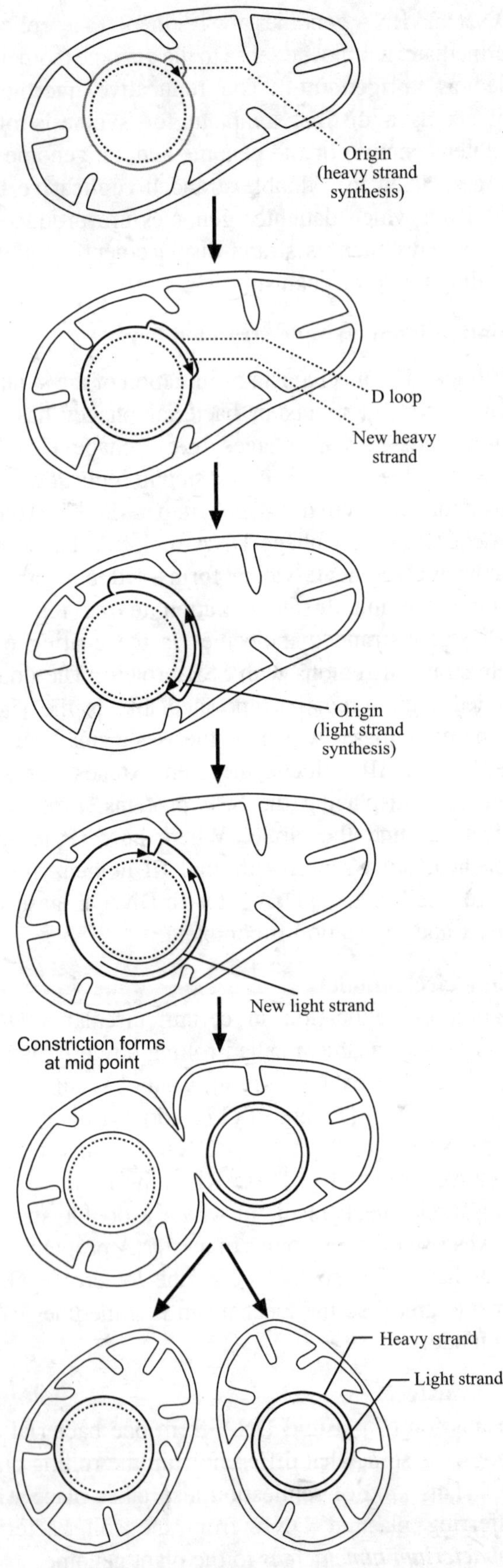

Fig. 8.32 Model for replication of mitochondrial DNA

8.11 RNA-DEPENDENT DNA SYNTHESIS

Generally, the information stored in DNA can be replicated into new DNA chains by DNA polymerases and transcribed into RNA chains by DNA-dependent RNA polymerase. However, nature has also found a role for the synthesis of DNA chains from an RNA template. RNA tumour viruses use viral RNA as the template for DNA synthesis. The direction of information flow in RNA viral replication is, **RNA** (viral genome) → **DNA** intermediate → **RNA** (viral genome). Viruses (and RNA phages) replicate their genome in one of the following four possible ways:

- Viruses of the first group (e.g., poliomyelitis virus or phages of *E. coli*, such as f_2, MS_2, R_{17}, Qß) contain a strand of positive RNA (equivalent of mRNA). During viral replication after infection, a negative strand is synthesised, which serves a template for synthesis of new positive strands.
- Viruses like that of rabies contain a negative strand of RNA which will serve as template for the synthesis of (+) RNA.
- The third group of viruses (e.g., reoviruses) consists of double-stranded RNA viruses, in which the $\pm$ duplex serves as template for the asymmetric synthesis of positive strands.
- The fourth group consists of the retroviruses (e.g., Rouse Sarcoma Virus) in which the positive RNA strand of the virion permits the intermediate synthesis of a DNA, which in its turn serves as template for the synthesis of new positive RNAs.

8.11.1 RNA replicase

The RNA-containing phages use RNA replicase or RNA synthetase (or RNA-dependent RNA polymerase) for the replication of their RNA genome. This enzyme appears only in the cells infected by RNA-containing phage like f_2, MS_2, R_{17}, Qß. A similar type of enzyme is also found in cells infected by some plant viruses (e.g., Tobacco Mosaic Virus, TMV) or animal viruses (e.g., poliomyelitis virus), in which one of the polypeptide chains is coded by the viral genome. RNA replicase is highly specific in the sense that it requires the homologous RNA; for example, the RNA replicase of phase Qß does not accept as template, the RNA of phage MS_2 though the latter is close. This specificity facilitates the virus for preferential replication of the viral RNA.

8.11.2 Replication of retroviral RNA genome

Retroviruses replicate by first converting their RNA genome into DNA using RNA-directed DNA polymerase (reverse transcriptase). Models for reverse transcription of the RNA genome of retroviruses have been proposed to explain certain unexpected observations, i.e., the duplex DNA produced is longer at both ends than the viral RNA and during the reaction short molecules of both plus and minus strand DNA ('strong stop DNAs') are formed.

The enzyme, **reverse transcriptase** (Temin 1970 and Baltimore 1970) synthesises polynucleotides in the 5'→3' direction and it also requires a primer. Interestingly, this enzyme uses specific tRNA molecule captured from the host cell as a primer. The 3'-end of the tRNA is base-paired with the viral RNA template at the site where DNA synthesis initiates and its free 3'-OH accepts the initial dNTP once synthesis commences. Reverse transcriptase then transcribes the RNA template into a complementary DNA (cDNA) strand to form a double-stranded DNA:RNA hybrid.

The reverse transcriptase can perform three essential enzymatic activities necessary for viral replication:

- It is responsible for RNA-directed DNA polymerase activity.
- It can carry out RNaseH activity. It degrades the template genomic RNA and also removes the priming tRNA after DNA synthesis is completed.
- It can also perform DNA-directed DNA polymerase activity. This activity replicates the ssDNA remaining after RNaseH degradation of the viral genome, yielding a DNA duplex. This DNA duplex directs the remainder of the viral infection process or becomes integrated into the host chromosome, where it can lie dormant for many years as a **provirus**. Activation of the provirus restores the infectious state.

8.12 FIDELITY OF DNA REPLICATION

Fidelity of DNA replication is a key determinant of the stability genome and is central to the evolution of species and the basis for many hereditary diseases. Replication errors contribute to the diversity of sequence upon which selection can act for evolution. On the other hand, mutations underlying these effects are modulated by various forms of repair that operate before or after replication. Some estimates suggest that a mistake occurs in the complementary base pairing of DNA replication only about three times during the replication of the whole genome of the DNA in a germ cell. This may not seem like much—since there are approximately 3×10^9 base pairs in the average germ cell. Special repair enzymes are responsible for this remarkable base accuracy and repair. These enzymes 'read' the nucleotides in the newly synthesised DNA strand. Whenever they come to a base that is not complementary to the base on the template strand, they excise or remove the incorrect base. Soon the correct base hydrogen bonds to the template, making it a perfect match for the original parent molecule. If an incorrect base is not removed from the new DNA molecule, it results in a mutation or change in hereditary information.

Fidelity of DNA replication depends on a number of steps:

- Even in the absence of any enzyme, complementary base pairs are about hundred-fold more stable than the non-complementary base pairs.
- The polymerase must selectively bind the complementary deoxynucleoside triphosphate.
- A kinetic proofreading model proposed by Hopfield predicts that a high-energy [enzyme:template:dNMP] complex is produced with the release of pyrophosphate. The polymerase can check the complex to see whether or not the correct base has been incorporated. Following incorporation, an exonucleolytic proofreading mechanism could remove incorrect 3'-terminal nucleotides, which are mismatched with the template strand before elongation. These nucleotides are also released as monophosphates.

Error avoidance mechanisms can increase the fidelity of replication to give an error rate of one mistake in 10 million base pairs replicated.

8.13 REGULATION OF REPLICATION

During growth and division, eukaryotic cells duplicate their genomes with remarkable fidelity. The high accuracy of the process depends largely on stringent regulatory mechanisms that couple DNA replication to cell cycle progression.

DNA replication in all cells coordinates with cell division to prevent gain or loss of DNA, whilst the replication of plasmids is often copy-number-dependent. In contrast, in eukaryotes, we find slower rate of elongation and a much larger genome. So, rapid replication is possible by increasing the number of functional origins per chromosome. Replication in eukaryotes correlates with genetic activity, i.e., housekeeping genes replicate early whereas cell-type-specific genes generally replicate early in the cells where they are expressed and late in cells where they are transcriptionally silent. A relationship between replication timing, chromatin structure, DNA methylation and transcriptional activity has thus been established. However, the complete mechanism of regulation DNA replication is not understood.

KEYPOINTS

- The replication of DNA is one of the most fundamental functions that a biological organism must carry out. Replication is an essential process for the propagation of cellular genomes and 'molecular parasites'—viruses, plasmids and transposable elements.
- Cells can have one or many chromosomes, but the entire chromosome must be replicated precisely once in every cell division.
- DNA replication is mechanistically very complex. The molecular mechanism for DNA replication is, generally the separation of strands followed by the copying of each strand.
- In the process, each separated strand acts as a template for the synthesis of a new complementary strand whose nucleotide sequence is fixed by the base-pairing rules.
- The process of replication, in which each strand of the double helix serves as a template for the synthesis of a new strand, is simple in principle. However, it is a very complex process involving geometric problems and requiring a variety of enzymes and other proteins.
- Though the mechanism of DNA replication is similar in both prokaryotes and eukaryotes, each system has some unique features. The common features are: semiconservative DNA replication, sequential nature, semidiscontinuity, requirement for initiation primers, bidirectional replication, synthesis in 5'→3' direction and proofreading.
- DNA synthesis occurs at a replication fork, which progresses along the DNA molecule and a variety of proteins is required to bring about this process in an efficient manner.
- Nucleic acid synthesis in cells is catalysed by different DNA polymerases. During polymerisation, the polymerase binds to the single-stranded template in the region of the 3'–OH end of the primer. An incoming deoxyribonucleoside trisphosphate containing a base, which can pair with the corresponding base on the template, becomes attached to the triphosphate binding sites. The DNAP then catalyses a nucleophilic attack by the 3'-phosphate of the deoxyribonucleoside 5'-triphosphate.
- Prokaryotes contain three and eukaryotes contain at least five DNA polymerase enzymes in their cells. In addition to their ability to polymerise as many as 1,000 nucleotides per second, DNA polymerases have processive ability.
- In addition to DNA polymerase activities, many other enzymes and proteins are essential for DNA replication, which include terminal deoxynucleotidyl transferase (TdT), DNA helicases, ligases, topoisomerases, primases, single-stranded DNA-binding proteins (SSB), primase and so on.

- DNA replication in eukaryotic cells shows strong parallels with prokaryotic DNA replication, though it is more complex in eukaryotes.
- DNA synthesis begins at replication origins. The DNA replication in eukaryotic chromosome is initiated at many sites. The initiation of replication in eukaryotic cells begins at different stages and times during the cell division cycle.
- In eukaryotes, the ends of linear chromosomes cannot be fully replicated by semidiscontinuous replication, as there is no DNA to elongate and replace the RNA removed from the 5'-end of the lagging strand. To overcome this problem, the ends of eukaryotic chromosomes consist of hundreds of copies of a simple, multiple repeats of short sequences called telomere. Telomere sequences are extended by a specific telomere terminal transferase (telomerase).
- There are some exceptions to the general mechanism of replication. Selective replication of specific sequences of DNA, known as amplification, is found in many organisms. It can occur as a part of the developmental programme of an organism or in response to treatment with a drug or similar selection agent.
- Fidelity of DNA replication is a key determinant of genome stability and is central to the evolution of species and is the basis for many hereditary diseases.
- Replication errors contribute to the sequence diversity upon which selection can act for evolution. The high accuracy of the process depends largely on stringent regulatory mechanisms that couple DNA replication to progression of the cell cycle.

Self-assessment Questions

1. Discuss in detail the basic rules of DNA replication.
2. Explain Meselson and Stahl's experiment that proves that DNA replication is semiconservative.
3. What are Okazaki fragments?
4. Why is DNA replication always in the 5'→3' direction and not otherwise?
5. Discuss the significance of the RNA primer in the initiation of replication.
6. What is semidiscontinuous DNA replication?
7. What factors contribute to the high fidelity of DNA replication?
8. Write a brief note on DNA helicase, topoisomerases and ligase.
9. Give the general features of *E. coli* DNA polymerases.
10. Briefly explain the roles of eukaryotic DNA polymerases.
11. Explain in detail the different stages of DNA replication.
12. Discuss briefly the different DNA replication models.
13. What is the role of the origin of replication?
14. How is the problem associated with initiation of terminal replication in linear strands overcome?
15. Explain briefly the mechanism of regulation of DNA replication.

FURTHER READING

Textbooks

Adams RLP, JT Knowler and DP Leader 1992. *The Biochemistry of the Nucleic Acids.* London: Chapman and Hall.

Alberts B, A Johnson, J Lewis, M Raff, K Roberts and P Walter 2002. *Molecular Biology of the Cell,* 4th ed. Garland Science.

Channarayappa (2006). *Molecular Biotechnology: Principles and Practices.* Hyderabad: Universities Press, CRC Worldwide.

Cooper GM and RE Hausman (2004). *The Cell: A Molecular Approach.* 3rd ed. ASM Press.

De Pamphilis ML (ed.) (1996). *DNA Replication in Eukaryotic Cells.* New York: Cold Spring Harbor Press.

Kornberg A and TA Baker (1992). *DNA replication,* 2nd ed. New York: WH Freeman and Co.

Lodish H, A Berk, SL Zipursky et al. (2000). *Molecular Cell Biology.* 4th ed. New York: WH Freeman.

Watson JD, NH Hopkins, JF Roberts, JA Steitz and AM Weiner (1987). *Molecular Biology of the Gene.* 4th ed. The Benjamin/ Cummings Publishing Co Inc.

References

Baker TA and SH Wickner (1992). Genetics and enzymology of DNA replication in *E. coli. Annu. Rev. Genet.* 26:447–79.

Berger JM, SJ Gamblin, SC Harrison and JC Wang (1996). Structure and mechanism of DNA topoisomerase II. *Nature* 379:225–32.

Blackburn EH (1990). Telomeres: Structure and synthesis. *J. Biol. Chem*. 265:5919–21.

Brautigam CA and TA Steitz (1998). Structural and functional insights provided by crystal structures of DNA polymereases. *Curr. Opin. Struct. Biol.* 8:54–63.

Cairns J (1963). The chromosome of *Escherichia coli.* Cold Spring Harbor Symposium on Quantitative Biology 28:43–46.

Champoux JJ (2001). DNA topoisomerases: Structure, function and mechanism. *Ann. Rev. Biochem.* 70:369–413.

Channarayappa (1990). Cytogenetic analysis of clastogens and aneuploidogens by immunofluorescent and micronucleus methods. PhD thesis submitted to the West Virginia University, Morgantown, West Virginia, USA. **W Microfilm 3365**.

Channarayappa (1993). Resistance mechanism of Lycopersicon species to the vector Bemisia tabaci and Ultrastructural changes in leaf curl virus infected tomato leaves. PhD thesis submitted to the University of Agricultural Sciences, GKVK, Bangalore.

Chong JPJ, P Thommes and JJ Blow (1996). The role of MCM/P1 proteins in the licensing of DNA replication. *Trends Biochem. Sci.* 21:102–6.

Doublie S and T Ellenberger (1998). The mechanism of action of T7 DNA polymerase. *Curr. Opin. Struct. Biol.* 8:704–79.

Frick DN and CC Richardson (2001). DNA primases. *Ann. Rev. Biochem.* 70:39–80.

Jager J and JD Pata (1999). Getting a grip: Polymerases and their substrate complexes. *Curr. Opin. Struct. Biol.* 9:21–28.

Kelly TJ and GW Brown (2000). Regulation of chromosome Replication. *Ann. Rev. Biochem.* 69:829–80.

Kunkel TA and K Bebenek (2000). DNA replication fidelity. *Ann. Rev. Biochem.* 69:497–529.

Marians K (1996). Replication fork propagation in *Escherichia coli* and *Salmonella typhimurium,* In *Cellular and Molecular Biology*, 2nd ed. (eds) FC Neidhardt, R Curtiss III, JL Ingrahm, ECC Lin, KB Low, B Hagasanik, WS Rexnikoff, M Riley, M Schaechter and HE Umbarger. Washington DC: ASM Press.

Meselson M and FW Stahl (1958). The replication of DNA in *E. coli. Pro. Natl. Acad. Sci.* USA 44:671–82.

Ogawa T and T Okazaki (1980). Discontinuous DNA replication. *Annu. Rev. Biochem.* 49:421–59.

Okazaki R, T Okazaki, K Sakabe, K Sugimoto, R Kainuma, A Sugino and N Iwatsuki N (1968). *In vivo* mechanism of DNA chain growth, IV. Possible discontinuity and unusual secondary structure of newly synthesized chains. *Proc. Natl. Acad. Sci.* USA 59:598–605.

Schimmel PR, D Soll D and JN Abelson (1979). Transfer RNA: Structure, properties, and recognition. Cold Spring Harbor Laboratory.

Steitz TA (1998). A mechanism for all polymerases. *Nature* 391:231–232.

Sutton MD and Walker GC (2001). Managing DNA polymerases: Coordinating DNA replication, DNA repair, and DNA recombination. *Proc. Natl. Acad. Sci.* 98:8342–49.

Taylor DJ, NPO Green and GW Stout (1997). In *Biological Science.* (ed) Soper R. Cambridge University Press.

Wang JC (1985). DNA topoisomerases. *Ann. Rev. Biochem.* 54:665–99.

Wells RD (1988). Unusual DNA structures. *J. Biol. Chem.*, 263:1095–98.

Wilson K and J Walker J (eds) (2000). *Practical Biochemistry: Principles and Techniques.* 5th ed. Cambridge University Press.

Wold MS (1997). Replication protein A: A heterotrimeric, single-stranded DNA-binding protein required for eukaryotic DNA metabolism. *Annu. Rev. Biochem.* 66:61–92.

Wynford-Thomas D and D Kipling (1997). Telomerase: Cancer and the knockout mouse. *Nature* 389:551–52.

CHAPTER 9

Transcription

LEARNING OBJECTIVES

After reading this chapter, you will be able to:

- Understand the general features and components of transcription
- Learn the mechanism and different stages of transcription
- List the main differences between transcription in prokaryotes, eukaryotes and viruses

9.1 INTRODUCTION

DNA stores genetic information in a stable form that can be readily replicated. However, the transfer of this genetic information requires information to flow from the DNA to proteins through RNA. This transfer is also known as the central dogma (**Fig. 9.1**). Thus, the order of the nucleotide sequence determines the order of amino acids in a polypeptide chain. In this series of events, the sequence of bases in DNA is first copied into an RNA molecule and then the RNA is used, either directly or after some chemical modifications, to determine the amino acid sequence of a protein molecule. The conversion of DNA information into that of RNA is the first step in gene expression. Thus, the sequential flow of genetic information in normal cells begins with transcription and ends in translation. Indeed, transcription is a vital control point in the expression of many genes.

The process of **transcription** or the **biosynthesis of RNA** is called the polymerisation of ribonucleoside-5'-triphosphates to RNA under the direction of DNA. The mechanism of transcription is chemically and enzymatically very similar to that of DNA replication. In both the processes, polymerase enzymes synthesise a new strand of nucleic acid complementary to a template strand. Like DNA, RNA is also a linear polymer made of four different types of nucleotide subunits linked by phosphodiester bonds. RNA differs from DNA chemically in two aspects:

- The nucleotides in RNA are ribonucleotides (containing ribose sugar).
- Uracil (U) is the base in RNA as opposed to thymine (T) in DNA.

These small differences bring about dramatic differences in the overall structure and functions of both DNA and RNA. In this chapter, we shall first learn how RNA is synthesised in *E. coli*. Transcription in eukaryotes will be discussed later.

9.2 GENERAL FEATURES OF TRANSCRIPTION

Although the basic enzymology of transcription is common in both prokaryotes and eukaryotes, its regulation is more complex in the latter. A large enzyme, the RNA polymerase, catalyses RNA synthesis. Transcription, like nearly all biological polymerisation reactions, takes place in three stages: (i) **initiation** (ii) **elongation** and (iii) **termination**. Signals encoded in DNA tell RNA polymerase where to start and stop so as to accurately transcribe a gene. Initiation of transcription

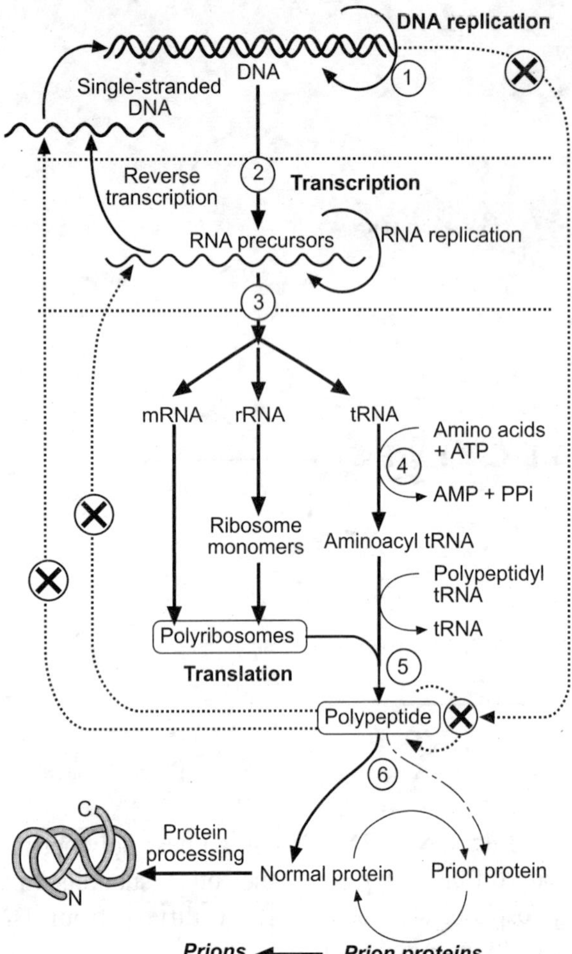

Fig. 9.1 The flow of genetic information: established (bold arrows) and not yet proven (dotted lines). DNA, RNA and proteins are the three key molecules involved in the flow of genetic information in biological systems. There are nine possible information transfers among these three molecules. According to the general transformation: information flows between DNA → DNA (DNA replication), DNA → RNA (transcription) and RNA → proteins (translation). These transfers are proven. According to special transformation: RNA → RNA(RNA replication), RNA → DNA (reverse transcription) and RNA protein (translation) are established. The unknown transfers: Protein → protein (prions?), protein → RNA, protein → DNA and DNA → protein transfers are not yet established

is an important step in gene expression as it is the main point at which the cell regulates which proteins are to be produced and at what rate.

9.2.1 Structure of a typical gene

The gene contains a transcribed sequence of DNA, which is bounded by bases at which transcription is initiated and terminated. By convention, the initiation point is termed the 5' end of the gene and the termination point as the 3' end. This reflects the formation of the transcript rather than the direction in which the DNA strand is transcribed (from 3'→5'). Similarly, all DNA sequences are given in the same direction as the transcript (5'→3') so that the sequences reported are in fact those of the non-transcribed strand of DNA.

9.2.2 Operons in prokaryotes

Bacteria have a simple mechanism for coordinating the regulation of genes that encode products involved in a set of related processes; these genes are clustered on the chromosome and are transcribed together. Most prokaryotic mRNAs are polycistronic—multiple genes on a single transcript—and the single promoter that initiates transcription of the cluster is the site of regulation for expression of all the genes in the cluster. The gene cluster and promoter, plus additional sequences that function together in regulation are called an **operon** (**Fig. 9.2**), often coding for enzymes in a metabolic pathway. Seventy-five different operons controlling 250 structural genes have been identified for *E. coli*. Operons can contain more than 20 genes, but 2–6 genes transcribed as units are very common. Perhaps the best known example of this arrangement is the **lactose** (*lac*) **operon**, which codes for the enzymes responsible for lactose catabolism. A typical operon contains codes for proteins, termed **structural genes** and an upstream control region encompassing the promoter and a regulatory site called the **operator**. In this control region there is also a site, which binds a complex of cyclic AMP (cAMP) and CRP (cAMP receptor protein), which regulates gene expression positively. Lying outside the operon is the repressor gene which codes for a protein that binds to the operator

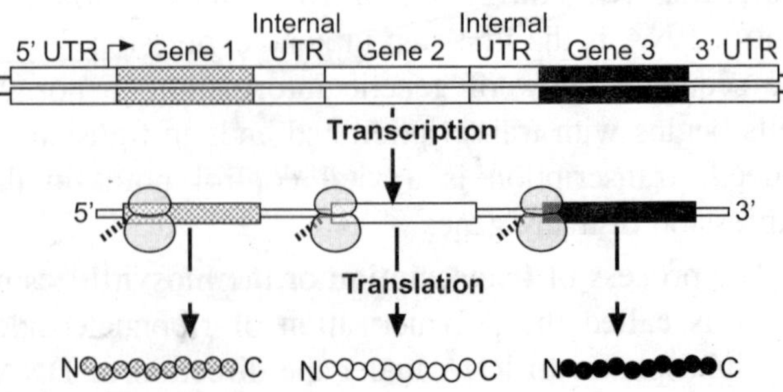

Fig. 9.2 Typical operon structure of protein coding genes in prokaryotes; in bacteria, several genes are present in the same transcript unit and transcribed as a common mRNA which can be divided into coding regions. UTR: Untranslated region

site and is responsible for negative control of the operon by blocking the binding of polymerase.

Eukaryotic genes are monocistronic and express only a single gene product. Structurally, eukaryotic genes are more complex and have many regulatory elements in addition to *cis*-acting sequences (**Fig. 9.3**).

9.2.3 Collinearity in genes

Benzer's (1961) work established that the gene was a linear entity, but he could not demonstrate the collinearity of the gene and its protein product. However, Yanofsky (1967) and his colleagues showed that the order of a polypeptide's amino acids corresponded to the nucleotide sequence in the gene that specified it. Experiments in the tryptophan synthase gene *trp*A in *E. coli*, a gene in which many mutations had been accurately mapped, provided adequate proof. Each mutation was found to result in a single amino acid substituting for the wild-type amino acid in the enzyme. Further, the order of the mutations in the genetic map was the same as the order of the affected amino acids in the polypeptide chain (**Fig. 9.4**). This attribute of genes and polypeptides is called collinearity, which means that the sequence of base pairs in the DNA determines the sequence of amino acids in the polypeptide in a collinear or corresponding manner.

9.2.4 *Cis*- and *trans*- acting elements

Both prokaryotic and eukaryotic genes require promoters for initiating transcription. Each of the three types of polymerase has distinct promoters. Besides the promoter genes, there are also **regulator genes** that are independently transcribed. They are responsible for the formation of proteins that influence the frequency of transcription. Segments of DNA that influence the activities of genes with which they are contiguous are described as ***cis*-acting**. The ***trans*-acting elements**, on the other hand, give rise to diffusible products that

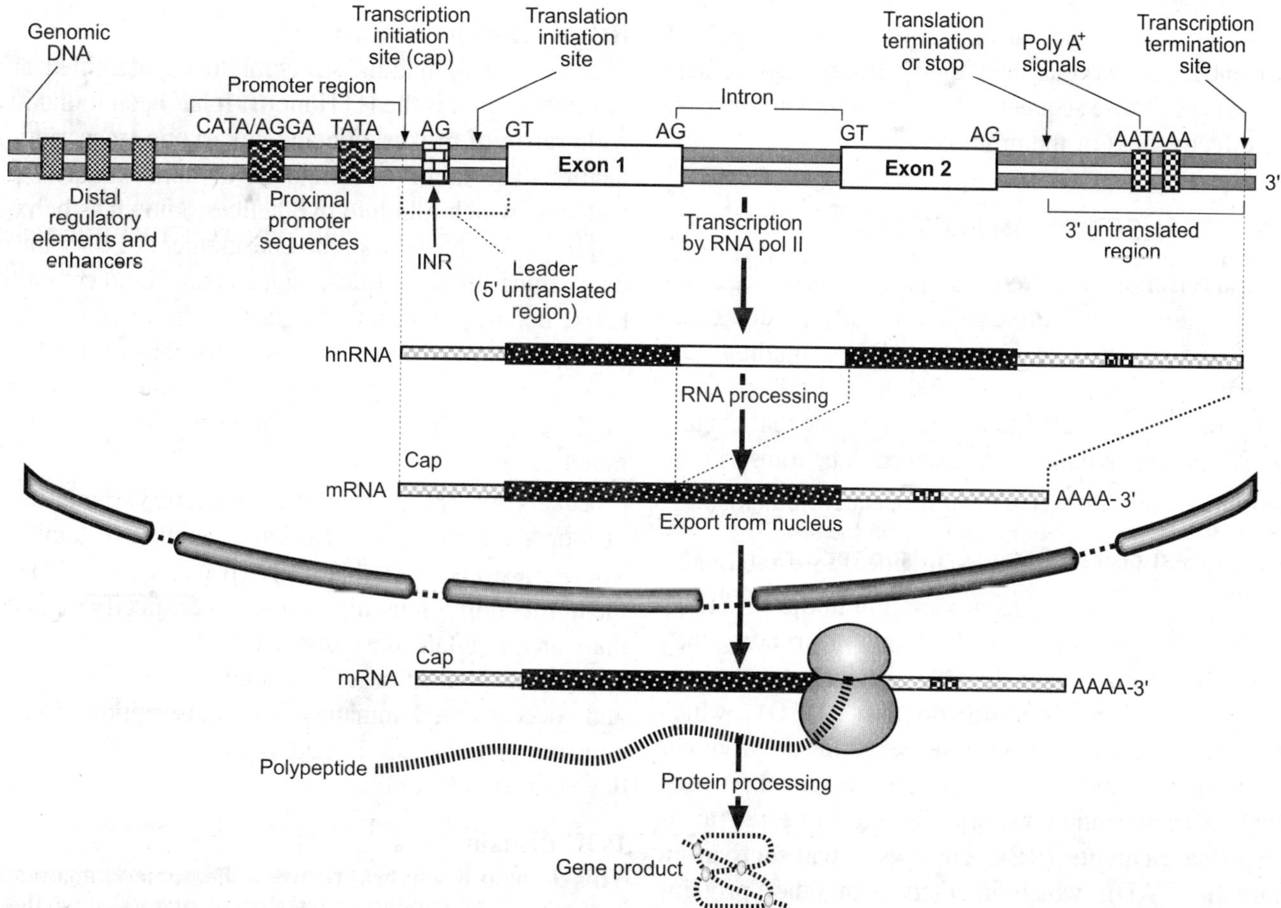

Fig. 9.3 An ideal gene structure of a eukaryotic polypeptide-encoding nuclear gene; in eukaryotes, transcription units comprise a single gene and contain many *cis*- and *trans*-acting elements

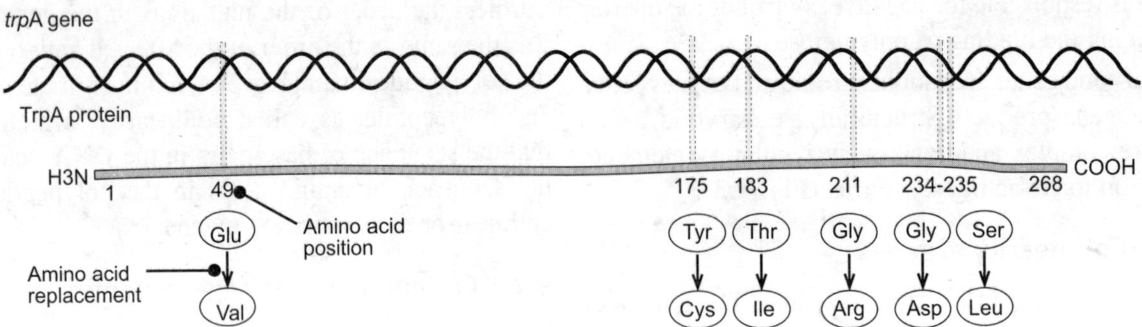

Fig. 9.4 Collinearity of the gene and the amino acid sequence of the α chain of tryptophan synthetase; correlation of the positions of mutations in the genetic map of the *trpA* gene in *E. coli* with positions of amino acid replacements in the TrpA protein

can act at many sites whether on the same or different chromosomes. They are transcribed and translated into proteins called *trans*-acting factors or **transcription factors**. These influence transcription by interacting with promoters, enhancers and other *cis*-acting elements.

Promoters are referred to as *cis*-acting elements. However, they are not the only types of *cis*-acting DNA sequences. Eukaryotes and their viruses also contain **enhancers**. These sequences have the ability to increase the effectiveness of the promoters.

9.3 TRANSCRIPTION FACTORS

A transcription factor is a protein that regulates transcription. These proteins are usually modular and have several effector domains (DNA binding and activation domain). Most of the about 2,600 proteins (10% of the genome) in the human genome that contain DNA-binding domains are presumed to be transcription factors.

9.3.1 Structure of transcription factors

Although there is significant variation in the structures of these factors, they share a common property they have: functional domains or regions (domains). There is a **DNA-binding domain (DBD)**, which recognises the specific DNA sequence of the promoter or enhancer, and binds to it. DNA sequences that bind to transcription factors are often referred to as **response elements** (RE). There is a **transactivation domain** (TAD), which interacts with other proteins (e.g., transcription coregulators) and thereby increases the rate of transcription. An optional signal sensing domain (SSD) transmits external signals to the rest of the transcription complex, resulting in up- or down-regulation of gene expression.

Specialised factors with DNA-binding domain

A partial list of some of the major families of DNA-binding domains/transcription factors are discussed below.

Helix-turn-helix motif (HTH)

The first DNA-binding structure to be identified and studied in detail is the HTH motif. It has been found in a wide range of prokaryotic and eukaryotic transcription factors. The motif is composed of approximately 21 residues, which fold into two helices, known as helix-2 and helix-3, which dock onto one another at 120° with an intervening three- or four-residue turn. The recognition helix, helix-3, is an integral part of the folded protein structure, but the surface-exposed portions of this helix make contact with the edges of the base pairs in the major groove along with neighbouring phosphodiester moieties.

Eukaryotic HTH proteins have more variable turns or loops, and can be divided into a number of families. The largest and best-characterised eukaryotic HTH is the **homeodomain** family. These genes specify the body plan and regulate development in higher organisms. The homeodomain was first identified in *Drosophila* and occurs predominantly in transcription factors controlling homeostasis and regional specification in development (*Hox* genes).

POU domain

This domain is a hybrid between the homeodomain and a POU-specific domain. The term POU was derived from the names of the first three factors shown to have a region of similarity: Pit-1 (a pituitary-specific transcription factor), Oct-1 (an octamer-binding protein first shown to regulate immunoglobulin gene transcription) and *unc-86* (a nematode gene).

9.3.2 Functions of transcription factors

Transcription factors are vital for many cellular processes. Some of the important biological roles they play are listed below.

Regulation of basal transcription: In eukaryotes, transcription factors called general transcription factors (GTFs) like TFIIA, TFIIB, TFIID, TFIIE, TFIIF and TFIIH function as a part of the large transcription pre-initiation complex that interacts with RNA polymerase directly.

Differential enhancement of transcription: Some other transcription factors differentially regulate expression of various genes by binding the enhancer regions of DNA. These factors regulate gene expression spatially and temporally and in the right amount depending on the requirement.

Development: A number of transcription factors in multicellular organisms are involved in development. These switch on or switch off the transcription of the appropriate genes.

Response to intercellular signals: Cells can communicate with each other by releasing molecules that produce the signalling cascade in the recipient cell.

Response to the environment: Many transcription factors function in the downstream of the signalling cascade while responding to environmental stimuli.

Control of cell cycle: Many transcription factors including oncogenes and tumour suppressor genes are involved in the regulation of the cell cycle.

Regulation of gene expression: In biological systems, many important processes have multiple layers of regulation and control. The transcription factors control the rates of transcription and amounts of gene products.

Synthesis: Transcription factors, like other proteins, are transcribed into RNA and translated into protein. Therefore, any of these steps can be regulated to affect the production of transcription factors.

Nuclear localisation: Like other gene products, transcription factors are transcribed in the nucleus and translated in the cytoplasm. Thus, these proteins have the nuclear localisation signals, which direct them to the nucleus with the help of importins. Therefore, this step can become a key point for their regulation.

Activation: Transcription factors may be activated or deactivated through their signal-sensing domain. This process can take place by various mechanisms, which include ligand binding, phosphorylation or interaction with other transcription factors.

Availability of other cofactors/transcription factors: Most transcription factors do not work individually; they always work by combining with other proteins and bind with DNA regulatory sequences.

9.4 MECHANISM OF TRANSCRIPTION

Transcription factors bind to either enhancer or promoter regions of DNA adjacent to the genes that they regulate. Depending on the nature of the transcription factor, the gene is either up- or down-regulated. Transcription factors use a variety of mechanisms for the regulation of gene expression:

- Stabilise or block the binding of RNA polymerase to DNA.
- Catalyse the acetylation or deacetylation of histone proteins; the transcription factor can either do this directly or recruit other proteins with the catalytic activity.
- Recruit coactivator or corepressor proteins to the transcription factor–DNA complex.
- Transcription factors are classified according to the type of the DNA-binding domain they contain. Most of the structural classes of DNA-binding domains have characteristic consensus amino acid sequences.

9.4.1 Transcription unit

A *transcription unit* is a segment of DNA (gene or genes) that is transcribed (copied) into a single RNA molecule by the enzyme, RNA polymerase. Associated with this segment of the genome are various non-transcribed DNA elements such as promoters, enhancers and operators that influence the frequency with which any transcription unit is transcribed. These elements may include nucleotide sequences that are recognised by RNA polymerase or some other DNA-binding protein that modulates RNA synthesis. A given recognition sequence may thus be found associated with all the genes influenced by a particular protein. Such sequences are known as **consensus sequences**.

Direction of RNA synthesis: 5'→3' pathway

RNA is synthesised in the 5'→3' direction on **template strand** (the strand that is transcribed or copied into an RNA molecule) of the DNA. The point at which transcription starts is called the **initiation site** or **start site**. Sequences conventionally written to the left of the transcription unit are described as 5' to the start site (also called as **upstream**), whereas those on the 3' side of the start site are described as **downstream**. Individual nucleotides are numbered away from the start site and are given positive values when downstream and negative values when upstream. Thus, a nucleotide of 10 bp after the start site towards upstream is at position -10.

9.4.2 Promoters and terminator sequences

The 'functional anatomy' of a promoter region can be analysed by determining which of its bases are necessary for efficient transcription. The results of mutagenesis experiments, footprinting studies and comparisons of many higher eukaryotic genes have demonstrated the importance of several upstream regions. The DNA region that RNA polymerase associates with immediately before transcription begins is known as the **promoter**. It is an important part of RNA synthesis in both prokaryotes and eukaryotes. They not only bind to RNA polymerase, but also specify the places of and frequency at which transcription can occur from the gene.

Prokaryotic promoters

Prokaryotic promoters that attract RNA polymerase have some conserved regions. The first is a sequence of 6- or 7-bp centred approximately 10 bp upstream of the start of transcription. This was originally called the **Pribnow box**, but is now usually called the **–10 box**. The second region is another short sequence centred approximately 35 bp upstream of the transcription start site; it is known as the **–35 box**. Although there is substantial variation in the sequence of promoter regions, certain patterns or 'motifs' are quite frequent. In a compilation of 112 well-defined *E. coli* promoter regions, a consensus sequence for the region –50 to +10 was derived. Each base in the consensus sequence is the base most often observed at that position in actual sequences. The consensus sequences in the promoter regions in *E. coli* are TTGACA, centred approximately 35 bp upstream from the transcription start site (+1), and TATAAT, centred approximately 10 bp upstream from the +1 site. The –10 sequence, which is called the **TATA box**, is similar to sequences found at corresponding positions in many eukaryotic promoters. The positions of the promoter sequences determine

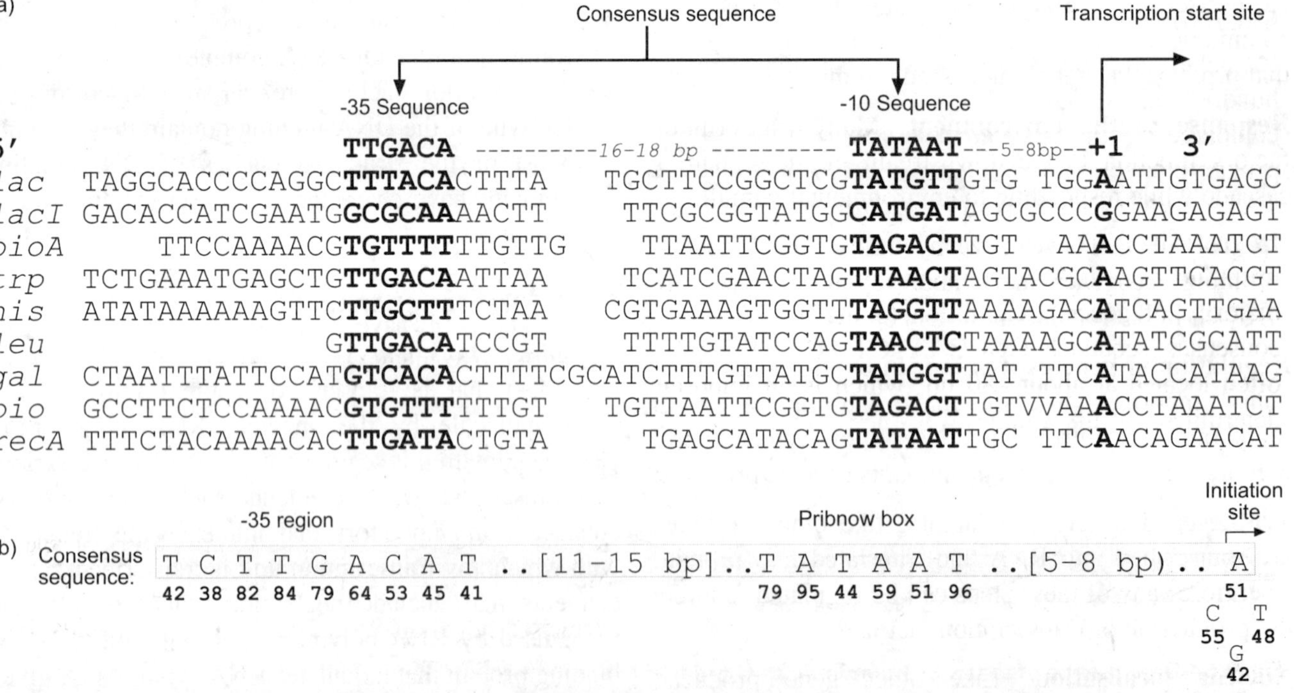

Fig. 9.5 Promoters (a) Consensus sequences in promoter regions of several genes in *E.coli*; the consensus sequences located 10 and 35 upstream from the transcription start site (=1) are indicated (the most conserved sequences are shown in bold) (b) Promoter region of *E.coli* with consensus sequences

where the RNAP begins synthesis: A or G (94% of the time purine) is often the first nucleotide in the transcript (**Fig. 9.5**). The strength of the binding of RNAP to different promoters varies greatly, which causes differences in the extent of expression from one gene to another. These promoters typically require accessory proteins to activate transcription by RNA polymerase.

In addition to the -10 and -35 boxes, which we call the **core promoter elements**, some very strong promoters have an additional element further upstream called an **UP element**. In rapidly dividing *E. coli* cells, the P1 promoters of the seven ribosomal RNA (**rrn**) operons direct more transcription than all of the other promoters in the cell combined. The promoters driving these genes are extraordinarily powerful due to the UP elements. **Figure 9.6** shows the structure of one of these promoters, the rrnB P1 promoter. There are three so-called *Fis* sites located between positions -60 and -150, which are the binding sites for the transcription-activator protein, Fis. These Fis elements are called enhancers. Fis is a site-specific DNA binding protein and a global regulator of gene expression in *E. coli*. In addition to its roles as an activator and repressor of transcription, it also plays a role in site-specific DNA inversion, phage λ excision, Tn5 transposition and DNA replication at oriC.

Eukaryotic promoters

In eukaryotic genes, promoters (spanning several hundred bases) may be important for transcription, compared to the normally fewer than 100 bp for prokaryotic promoters. Particularly critical is the TATAAAA (TATA box) sequence (or Goldberg–Hogness box), located approximately -25 to -35 bp upstream from the transcription initiation point in the majority of genes. In addition to the TATA box, a consensus sequence GGCCAATCT (**CAAT box**) is often located at about -80 bp, which is an important determinant of the efficiency of the promoter. A third cluster involved in the initiation of transcription is the CACCC region, about 95 to 87 base pairs upstream from the cap site. Many upstream promoter elements (UPEs) or upstream activating elements (UAS) have been described that are either general in their action or tissue- or gene-specific. Elements that contain the sequence GGGCGG (**GC box**) may be present at multiple sites and in either orientation or are often associated with housekeeping genes such as those encoding enzymes involved in general metabolism. Some genes contain a consensus sequence of ATTTGCAT (**octamer box**) in different copy numbers. Except the TATA homology, all other consensus sites affect binding efficiency of RNA polymerase or transcription factors. There are many genes whose promoters lack the TATA sequence. They also use RNA polymerase II (mostly those encoding general metabolic proteins and not cell-specific proteins).

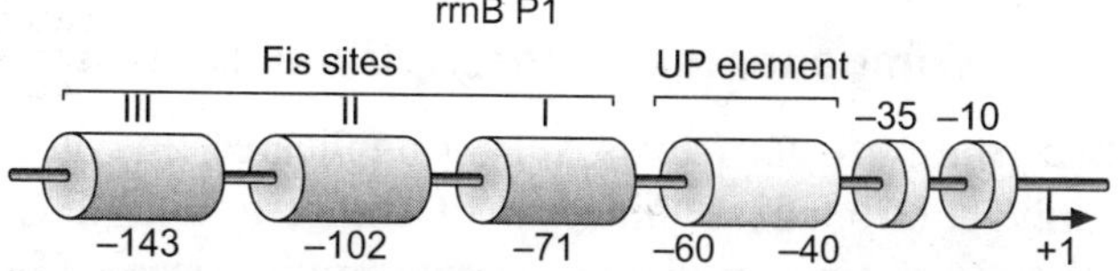

Fig. 9.6 Organisation of the rrnBP1 promoter; the core promoter region containing the -10 and -35 recognition hexamers and the UP element -40 to -60 region are shown by open boxes. The Fis binding sites are centred at -71, -102 and -143 respectively. +1 is the transcription start site

Terminator sequences

Terminator sequences are less well characterised, but are thought to involve nucleotide sequences near the end of mRNA with the capacity to form a hairpin loop, followed by a run of U residues, which may constitute a termination signal for RNA polymerase.

9.5 TYPES OF RNA

Several types of RNA are synthesised in the nucleus of eukaryotic cells. They have been broadly divided into two groups: **coding RNA** and **non-coding RNAs**. Only mRNA encodes polypeptides and all the other classes of RNA are non-coding.

Coding RNA

Heterogenous RNA (hnRNAs): They are well diversified and have a short life; these are later converted into mRNA.

Messenger RNA (mRNA): They are a heterogeneous class of RNA and are later translated into a polypeptide.

Non-coding RNAs

Ribosomal RNA (rRNA): They are abundant and are used in the building of ribosomes.

Transfer RNA (tRNA): They carry amino acids to the growing polypeptide.

Small nuclear RNA (snRNA): Transcription of the genes for mRNA, rRNA and tRNA produces large precursor molecules (primary transcripts) that must be processed within the nucleus to produce the functional molecules for export to the cytosol. Some of these processing steps are mediated by snRNAs.

Small nucleolar RNA (snoRNA): These RNAs within the nucleolus have several functions.

MicroRNA (miRNA): These are tiny (~22 nts) RNA molecules that appear to regulate the expression of messenger RNA (mRNA) molecules.

XIST RNA: This deactivates one of the two X-chromosomes in female vertebrates.

9.5.1 Messenger RNA

The mRNA in particular is a central player (as both structural and functional) in gene expression. In a typical chromosome, there are thousands of different genes. Only one of the two strands of DNA is transcribed into RNA. The mRNA molecule represents a definite number of nucleotide sequences (gene) in a DNA and its concentration depends on the respective gene activity. Therefore, there are diverse mRNA types and of varying concentration in a cell. It carries the DNA sequence information to proteins by the process of translation by ribosomes and other molecules in the cytoplasm. Many mRNAs are common to most cells, encoding 'housekeeping' proteins needed by all cells. Others are specific for only certain types of cells. DNA is immortal in cellular terms, but mRNA is ephemeral with a half-life of perhaps 20 minutes to several hours in eukaryotes and less than two minutes in bacteria. The genes of an *E. coli* are in contact with the cytoplasm since there is no nuclear membrane. In eukaryotes, mRNA always corresponds to single genes, but in prokaryotes the RNA is polycistronic (codes for several proteins).

9.5.2 Transfer RNA

Transfer RNA (tRNA) brings the amino acids to the ribosomes—the site where protein synthesis takes place. There are about 32 different kinds of tRNAs in a typical eukaryotic cell. Each type is the product of a separate gene. They are small (~4S), containing 73–93 nucleotides. Many of the bases in the chain pair with each other forming sections of double helix. The unpaired regions form mainly into three loops. Each kind of tRNA carries (at its 3'end) one of the 20 amino acids. At one loop, three unpaired bases form an **anticodon**. Base pairing between the anticodon and the complementary codon on an mRNA molecule brings the correct amino acid into the growing polypeptide chain.

9.5.3 Ribosomal RNA

Ribosomal RNA (rRNA) is a structural and functional part of the ribosome machinery for synthesising proteins by translating mRNA. There are four kinds of rRNA molecules in eukaryotes: (i) **18S rRNA**: One of these molecules, along with some 30 different protein molecules, is used to make the small subunit of the ribosome. (ii) **28S, 5.8S and 5S rRNA**: One each of these molecules, along with some 45 different proteins, is used to make the large subunit of the ribosome. The 28S, 18S and 5.8S rRNA molecules are produced by the processing of a single primary transcript from a cluster of identical copies of a single gene.

9.5.4 Small RNAs

Small RNAs (e.g., snRNAs, scRNAs, micRNAs, gRNAs, iRNA and others) are involved in various processes of cellular metabolism including RNA processing, regulation of gene expression, building resistance to RNA viruses and so on.

9.6 RNA POLYMERASES

The principal enzyme responsible for RNA synthesis is RNA polymerase, which catalyses the polymerisation of ribonucleoside 5'-triphosphates (NTPs) as directed by a DNA template. RNA polymerases are huge multisubunit protein complexes. Only one RNA polymerase is present in prokaryotes, whereas, there are three kinds in eukaryotes. All RNA polymerases perform essentially the same function in most cells. Therefore, RNA polymerases share many features. Unlike DNA polymerases, RNA polymerases do not require a preformed primer for initiation of transcription (called *de novo* initiation). RNA polymerases carry out helicase and primase activity. RNA polymerases can initiate synthesis on their own.

9.6.1 Components of transcriptional control

Transcriptional activity is very complicated and it is mainly controlled at initiation of transcription; thus, initiation of transcription is a rate-limiting step. The various factors involved in the control of initiation are divided into three components: basal, constitutive and regulated. **Basal components** serve primarily to recruit RNA polymerase for the start of transcription, permitting the initiation of RNA synthesis. **Constitutive components** control the efficiency of initiation in the absence of regulation and, therefore, dictate the default level of transcriptional activity of each gene, allowing different genes to be expressed at different rates to suit the fundamental needs of the cell. **Regulatory components** alter efficiency, allowing the transcriptional activity of individual genes to be modulated in response to changes in the environment. In multicellular organisms, the regulation of transcription also plays a major role in specifying and maintaining differentiated cells. These three components of transcriptional control differ considerably between prokaryotes and eukaryotes.

9.6.2 Prokaryotic RNA polymerases

Prokaryotes have only one RNA polymerase (**Fig. 9.7**), which transcribes all types of gene products (mRNAs, rRNA and tRNA). In addition to the RNA polymerase enzyme, some other polymerases (**Table 9.1**) responsible for various cellular functions are also present.

9.6.3 Eukaryotic RNA polymerases

The nucleus of eukaryotic cells is more complex and contains three types of RNA polymerases (RNAP)

Table 9.1 Prokaryotic RNA polymerases

Enzyme	*Number of subunits*	*Function*
Polynucleotide-phosphorylase	1	Catalyses the formation of polyribonucleotides without the use of DNA template strand
RNA polymerase	5	A single enzyme transcribes DNA template into different RNA molecules (mRNAs, tRNAs and rRNAs)
	α_2	Serves to assemble the RNA polymerase holoenzyme
	β	DNA binding and phosphodiester bond formation between nucleotides (catalysis)
	β'	DNA binding
	ω	Not clear
	σ	Can recognise and bind to a promoter region
Primase		Synthesis of primer during DNA replication

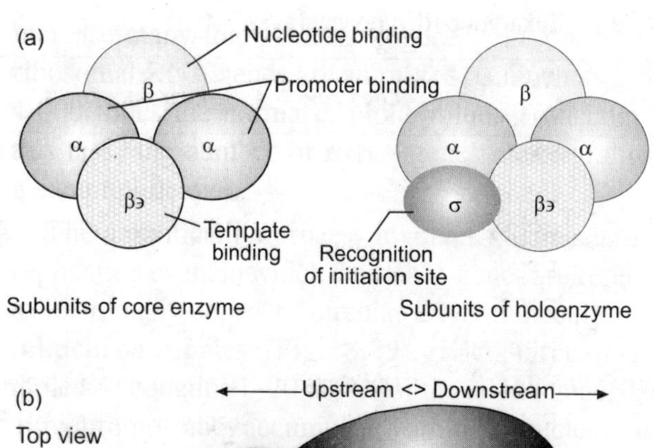

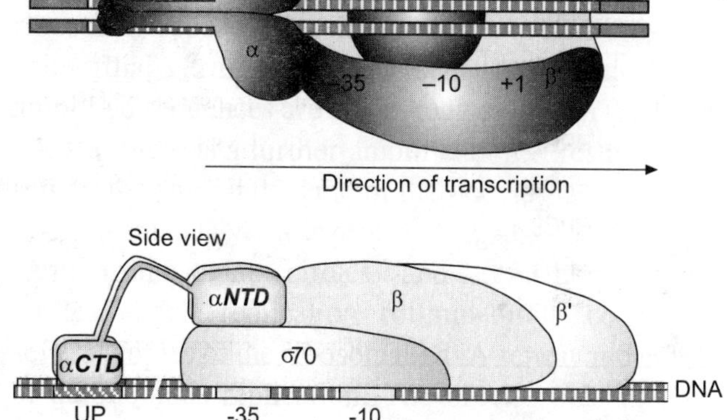

Fig. 9.7 A model for the structure of prokaryotic RNA polymerase (a) Assembly of different subunits of core and holoenzymes (b) Assembly of holoenzyme and the contact points of different subunits in holoenzyme

differing in template specificity, localisation and susceptibility to inhibitors. RNAPs in eukaryotes are divided into three classes. Each class is transcribed by a different RNAP (**Table 9.2**). The accessory factors needed for initiation are similar for all eukaryotic RNAP. The subunit structure of the eukaryotic RNA polymerases is relatively poorly characterised. All three enzymes catalyse RNA synthesis on a DNA template similar to that of the *E. coli* enzyme and they require a divalent metal ion, which can be Mg^{2+} or Mn^{2+}. All three eukaryotic RNAPs contain bound zinc, probably associated with a conserved amino acid sequence in the largest subunits.

Similar to the effects of some antibiotics on RNA synthesis in prokaryotes, α amanitin, a toxin produced by the death cup mushroom, binds to RNA polymerase II, preventing elongation of mRNA.

Perhaps the most studied enzyme is RNAP II, which has been purified to near homogeneity from more than

Table 9.2 Eukaryotic RNA polymerases

Type	*Number of subunits*	*Location*	*Cellular transcripts*	*Effect of α amanitin*
Pol I	RPA1 RPA2 RPC5 RPC9 RPB6 [+9 others]	Nucleolus	Transcription of the 45S rRNA precursors, a single polycistronic unit containing 18S, 5.8S and 28S rRNA (**class I genes**). Responsible for 50%–70% of total RNA transcription	Insensitive, sensitive to actinomycin D
Pol II	RPB1 RPB2 RPB3 RPB11 RPB6 [+7 others]	Nucleoplasm	Transcription of all protein-encoding genes (mRNAs) and most genes for small nuclear RNAs (snRNA) (**class II genes**). It accounts for 20%–40% of total RNA transcription.	Strongly inhibited at 50 ng/ml
Pol III	RPC1 RPC2 RPC5 RPC9 RPB6 [+11 others]	Nucleoplasm	Transcription of tRNA genes, 5S rRNA genes and genes encoding U6 snRNA and the various scRNAs (**class III genes**). It is responsible for the synthesis of 10% of the total RNA transcribed.	Inhibited at high concentrations, inhibited at 5 µg/ml
Primase		Nucleoplasm	Primer synthesis during DNA replication	-

20 different species. In *Saccharomyces cerevisiae,* there are ten subunits (RPB1–RPB10). Each subunit is encoded by a single copy gene but the enzyme appears to be composed of one polypeptide each of RPB1, 2, 6, 8 and 10, two copies of each of RPB3, 5 and 9 and less than stoichiometric amounts of subunits RPB4 and 7. The largest subunit of 220 kDa is the largest of any of the RNAP and experiments involving cross-linking, together with susceptibility to antibodies and proteases, indicate that it is involved in binding to the DNA template. The tandem repeats are common to all known RNAP II enzymes but the number of repeats varies from 17 in the malarial parasite to 52 in mammals.

The second largest subunit of RNAP II, RPB2, appears to participate in substrate binding and formation of phosphodiester bond, thus showing functional homology with the ß subunit of bacterial enzymes. The third subunit is also an essential component of the transcription apparatus, and is probably the functional homologue of prokaryotic sigma factor. At least three other subunits, RPB5, RPB6 and RPB8 are immunologically and biochemically indistinguishable in RNAP I, II and III. It would, thus, appear that many of the enzymatic components are substantially preserved in the three eukaryotic RNAPs.

9.7 TRANSCRIPTION IN PROKARYOTES

Let us learn the general principle of transcription before looking at the process in prokaryotes in detail.

General principle of transcription

Transcription is the enzymatic synthesis of RNA. It requires RNA polymerase enzymes (RNAP) that use single-strand DNA as a template to make complementary copies of information stored in the DNA sequence. This is the first stage in the overall process of gene expression and ultimately leads to synthesis of the protein encoded by a gene. Although RNA polymerase normally uses dsDNA in making RNA, the actual template is ssDNA that the enzyme exposes by unwinding a bit of the helix as it passes. RNA, like DNA, is a long, unbranched chain of four different nucleotides indicating that the genetic information of DNA chains is transferred to a complementary sequence of RNA nucleotides. However, there are two important differences between RNA and DNA synthesis.

Transcription of nuclear genes is carried out by one of the three different types of RNA polymerase, according to the type of gene product. The ribosomal RNA genes, which give rise to the two large ribosomal

RNA molecules (18S and 25S, and a small 5.8S) are transcribed by RNA polymerase I. RNA polymerase III is responsible for transcription of the genes for transfer RNAs (tRNAs) and other small, stable RNAs, such as those involved in intron processing. All mRNAs are transcribed by RNA polymerase II enzyme. Each of the RNA polymerases is a multisubunit enzyme, of complex structure; there are at least 10 subunits. Each polymerase catalyses the addition of nucleotide triphosphates to the 3'-OH of a nucleotide ribose sugar ring, forming a 3'→5' phosphodiester linkage (5'→3' direction), the NTP being determined by complementary nature to the transcribed template strand.

The mechanism of RNA biosynthesis can be broadly described under three steps: (1) initiation, (2) elongation and (3) termination.

9.7.1 Initiation of transcription

First, the holoenzyme recognises and binds to specific promoters (**promoter recognition**). The initial binding of polymerase to a promoter to form DNA enzyme complex is known as a **closed complex**. Second, this binding leads to the formation of the so-called '**open complex**' in which a portion of the double helix is unwound. Finally, once an enzyme gets further than 10 bp, it is said to have escaped the promoter. At this point, it has formed a **stable ternary complex** and RNA synthesis is initiated (**chain initiation**). The sequence of events in the initiation phase of transcription is shown in **Fig. 9.8**.

Recognition of promoters

Experiments carried out *in vitro* with purified enzymes have shown that RNAP can, on its own, transcribe RNA starting from any free 3' ends, but it is not capable of binding to specific DNA sequences, and is thus not sufficient for accurate, efficient initiation of transcription. For this to occur *in vivo*, a specific multiprotein complex (**transcriptional activation complex**) which includes RNAP, must assemble in such a way that the polymerase is aligned with its active site fitting onto the starting point of the transcription. RNAP binds to DNA non-specifically (**loose binding**) within a base sequence from 20 to 200 bases long called a promoter. Von Hippel and coworkers have suggested that non-specific binding is electrostatic and that, in this mode, facilitated transfer may occur by sliding along the DNA or by hopping or a combination of both processes.

Regulation by the addition of sigma (σ) factor: Although the bacterial core RNA polymerase can, in principle, initiate transcription at any point on a DNA molecule, in cells, however, polymerase initiates transcription only at promoters. This specificity is regulated by the addition of an initiation factor called sigma (σ). There are several sigma factors in *E. coli* (**Table 9.3**); the predominant sigma factor is called σ^{70}. Although the vast majority of σ^{70} promoters contain recognisable –35 and –10 regions, the sequences are not always identical. Promoters with sequences closer to the consensus are generally 'stronger' (that is, generate more transcripts per unit of time) than those that match less.

Formation of an open complex

With the tight binding of RNAP to a promoter site, there is a rapid transition to the so-called open complex in which the DNA is partially unwound (**DNA melting**). For the template strand to become accessible for base pairing, the DNA duplex must be unwound by the polymerase. Negative supercoiling enhances the transcription of many genes, since it facilitates unwinding by the RNAP. The initial unwinding of the DNA results in the formation of an open complex with

Table 9.3 Some important sigma factors found in *E. coli*

*Factor**	*Gene*	*Consensus sequences*			*Use*
		–10 sequence	*separation*	*–35 sequence*	
σ^{70}	*rpoD*	TATAAT	16–18 bp	TTGACA	General
σ^{S}	*rpoS*	TATACT	–	GTTAAGC	Stress
σ^{32}	*rpoH*	CCCCATNT	13–15 bp	CNCTTGAA	Heat shock
σ^{E}	*rpoE*	–	–	–	Heat shock
σ^{54}	*rpoN*	TTGCA	6 bp	CTGGNA	Nitrogen starvation
σ^{28} (σ^{F})	*fliA*	GCCGATAA	15 bp	TAAA	Flagellar synthesis

*A number in the name of a factor indicates its mass

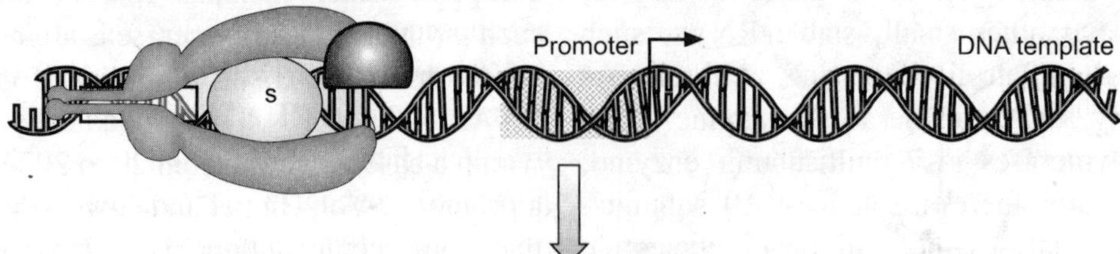

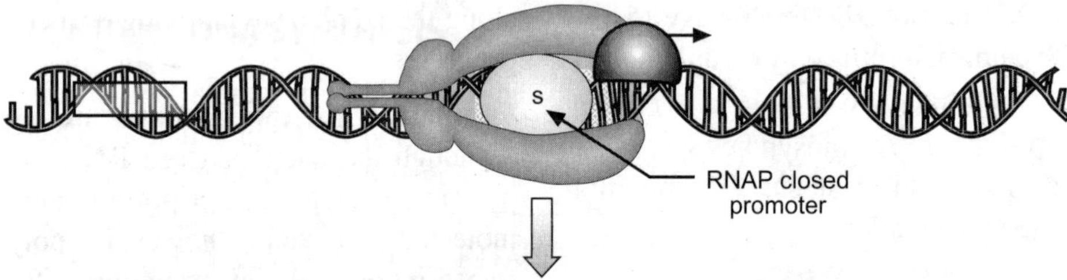

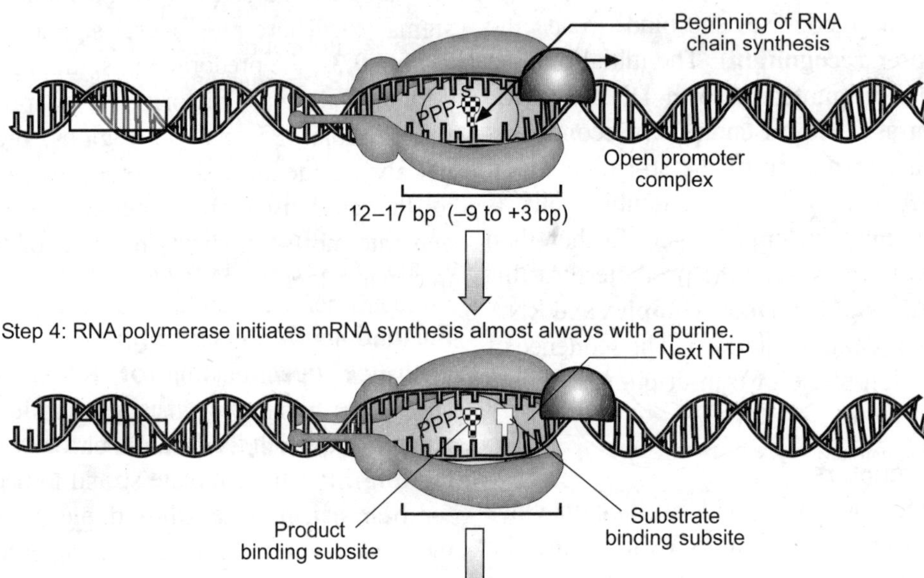

Fig. 9.8 Sequence of events in the initiation phase of transcription

the polymerase and this process is referred to as **tight binding**.

Chain initiation

After the initial binding step, the RNAP 'melts' the DNA double helix, causing the strand that is to be transcribed to separate from its partner strand and become accessible to the polymerase. The RNAP then initiates RNA synthesis at a nearby transcription start site, denoted the +1 site. The RNA polymerase and its cofactors, when assembled on the DNA template, are often referred to as the **transcription complex**. The binding of the initiation nucleotide to RNAP is an order of magnitude stronger than that of the succeeding nucleotides and it occupies a separate initiating site on the enzyme. At an early point in the incorporation of nucleoside phosphates into new transcripts, two further changes occur in the initiation complex. First, the open complex is transformed into a ternary one which is more resistant to dissociation by high salt concentrations, is resistant to inhibition by rifampicin and the RNAP becomes more resistant to denaturation and to proteolytic enzymes. The sigma factor is still attached to the enzyme and the promoter shows increased sensitivity to DNase at position –25, perhaps indicating bending of the DNA in this region. Sigma factor is released after the polymerisation of 10 nucleotides and the release is apparently dependent on and driven by the nucleotide sequence of the promoter. The released sigma subunit will be recycled and continued elongation of the transcript will be catalysed by the core enzyme.

9.7.2 Chain elongation

When initiation succeeds, the RNAP releases the sigma factor (**promoter clearance**) and forms a **ternary complex** containing the three components: RNAP–DNA–nascent RNA. It is also called as **elongation complex**, causing the core polymerase to progress along the DNA (**elongation**). The region of unwound DNA, which is called the **transcription bubble (Fig. 9.9)**, appears to move along the DNA with the polymerase. The size of this region remains to be constant at around 17 bp. The RNAP moves progressively along the transcribed DNA strand, adding nucleotides to the growing RNA chain (**Fig. 9.10**). Only one DNA strand, the **template strand**, is transcribed. The RNA polymerase covalently adds ribonucleotides to the 3'-end of the growing RNA chain in the 5'→3' direction. As the enzyme moves on the template strand, it locally unwinds the DNA, separating the DNA strands, to expose the templatc strand for ribonucleotide base pairing and covalent addition to the 3'-end of the growing chain. The nature of the incoming nucleotide is governed by Watson–Crick base pairing rules and

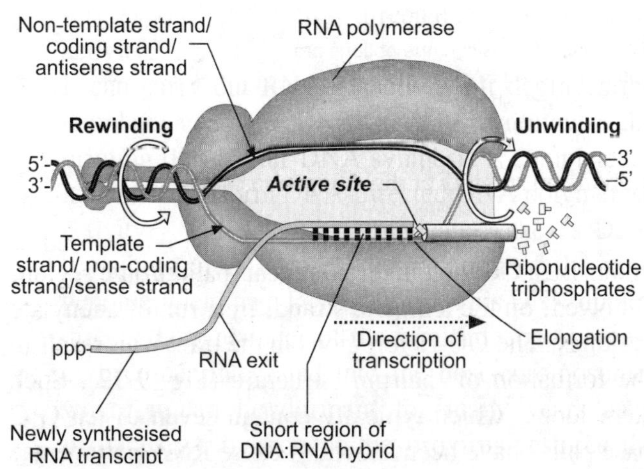

Fig. 9.9 Transcription bubble during elongation in RNA synthesis by RNA polymerase enzyme

bond formation is accomplished by the release of pyrophosphate. The reaction can thus be represented as:

$$\text{pppXpY} + \text{nZTP} \xrightarrow{\textit{RNA polymerase}} \text{pppXpY(pZ)n} + \text{nPPi}$$

where X, Y and Z can be any of the four ribonucleotides.

Elongation produces superhelical tension at the transcription bubble in DNA. In a cell, over-winding must be released for easy elongation. The supercoiling generated by the transcribing polymerase would be released enzymatically. DNA gyrase (topoisomerase II) would remove the positive supercoiling ahead of the RNAP and DNA topoisomerase I would remove negative supercoiling behind it.

The *E. coli* RNAP catalyses this reaction at a rate of around 40 bases per second at 37°C. Many groups have shown that elongation is a discontinuous process and that there are pause sites along the DNA at which RNAP slows or stops. Pausing occurs at GC-rich regions and at points 16–20 nucleotides downstream of regions of dyad symmetry. At these latter regions, the delay is associated with the formation, by the nascent RNA, of stem loop structures either because it base pairs with itself or with the coding strand of the DNA. Pausing is a necessary preliminary to termination and premature termination can occur at internal pausing sites.

9.7.3 Termination of transcription

The RNAP remains bound to the DNA and continues transcription until it reaches a **terminator sequence** (**stop signal**) at the end of the transcription unit (**Fig. 9.11**). When RNAP reaches a **stop signal**, both the newly synthesised RNA molecule and the polymerase are released. Three events are required when the transcription of a gene is terminated. Elongation ceases and the transcript and RNAP are released. The stop signal (terminator) sequences often contain self-complementary regions, which can form a **stem loop** or hairpin secondary structure in the RNA product. These cause the polymerase to pause and subsequently cease transcription. Usually the stem structure is GC-rich favouring stability of its base pairing. The hairpin is often followed by a sequence of four or more U residues. It seems that the polymerase pauses immediately after it has synthesised the hairpin RNA. The subsequent stretch of U residues in the RNA base pairs favours dissociation of the RNA from the transcription complex. Two kinds of termination

Step 5: RNA polymerase holoenzyme-catalysed elongation of mRNA by about four more nucleotides

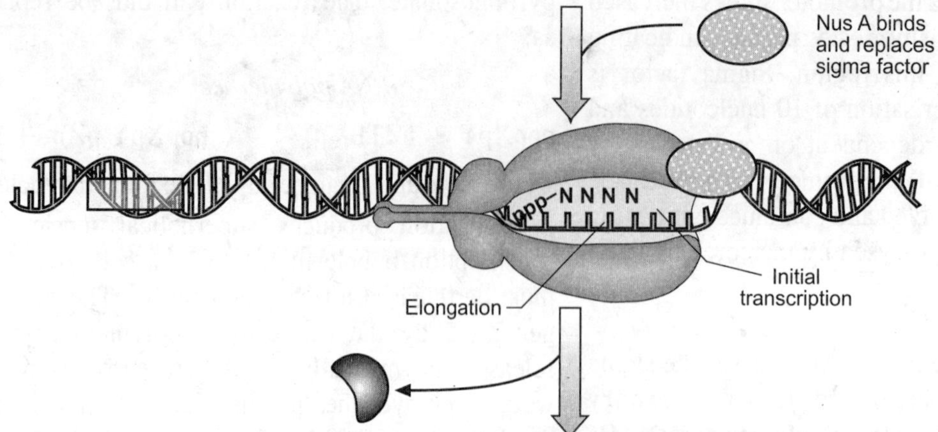

Step 6: Release of sigma subunit as nus A binds and core RNA polymerase proceeds down the template, elongating RNA transcript

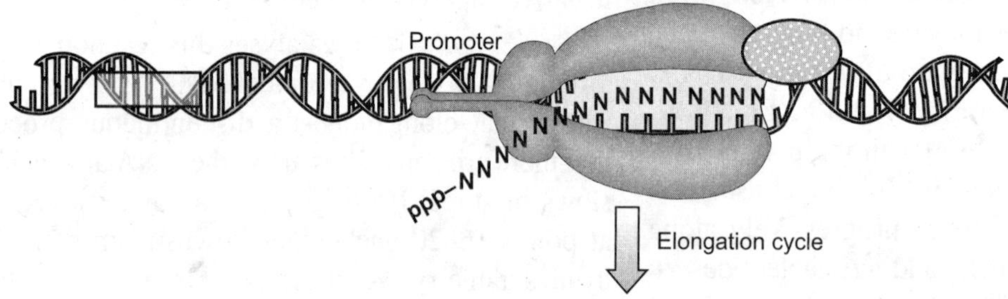

Step 7: Stability of the elongation complex

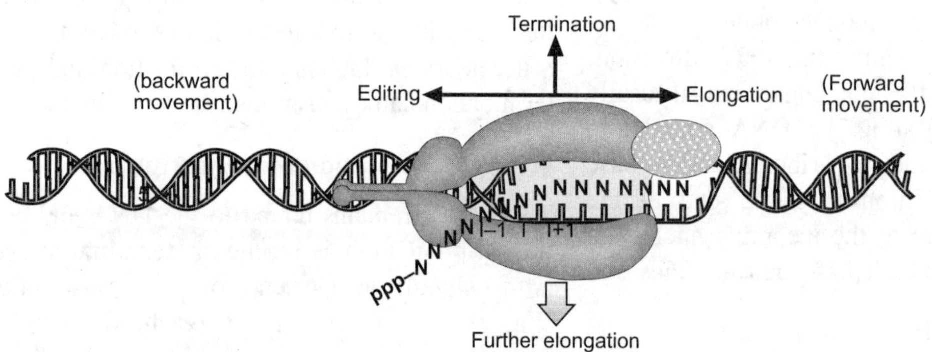

Three competing pathways for the transcription complex at a given template position; the options are: (1) To move forward along the template with the concomitant elongation of the RNA chain by one nucleotide residue (2) To move backward along the template (with or without shortening the transcript by one or more residues) (3) To dissociate from the template, resulting in release of the nascent RNA.

Fig. 9.10 Sequence of events in the elongation phase of transcription and stability of the elongation complex

events are known in prokaryotes: those that are self-terminating and depend only on the base sequence in the DNA template (**rho-independent**) and those that require the presence of a termination protein (**rho-dependent**).

Rho-independent termination (intrinsic termination): Sites on the DNA at which rho (ρ)-independent termination occurs have a characteristic structure that comprise a GC-rich inverted repeat (palindrome region) followed, on the template strand, by a run of adenylate residues. The GC-rich regions in the transcript result in the formation of 'hairpin' structure (**Fig. 9.12**). Such stem loops, which typically contain seven to ten G:C base pairs, have been shown to cause RNA polymerase to pause. Shortly after the GC-rich region, a run of four-to-eight-a-string of uridylate residues are transcribed

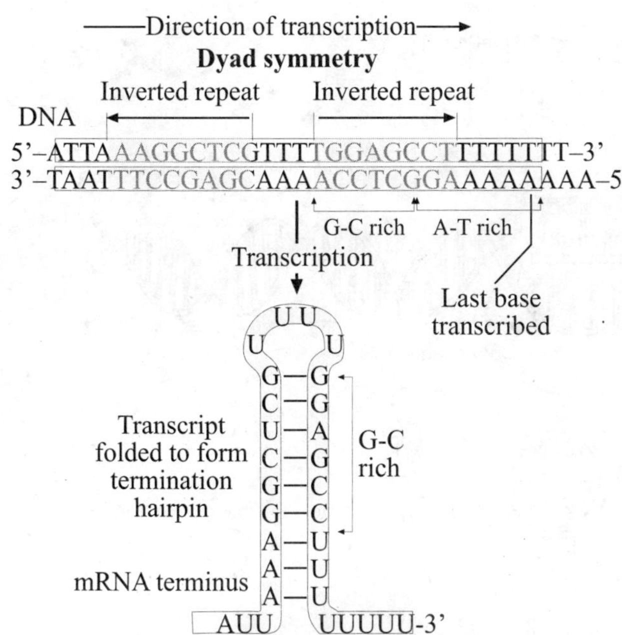

Fig. 9.11 Structure of a *rho*-independent terminator sequence

into the nascent RNA. Decreasing the number of these residues has been shown to reduce the efficiency of termination and it appears that the importance of the sequence lies in the very unstable nature of the rU:dA hybrid.

Rho-dependent termination: Roberts (1969) discovered a protein, which he called rho (ρ), which when added to an *E. coli* transcription system *in vitro* causes the generation of transcripts with discrete 3'-ends. The rho protein is basic in character and in solution consists of six identical, 419 amino acid subunits arranged in a ring structure (**Fig. 9.13**). It recognises and binds to the nascent RNA transcript and the RNA-binding domain is in the first 151 amino acids of the subunit. It binds to 13 ± 1 nucleotides per monomer and 78 ± 6 nucleotide per hexamer such that the polynucleotide appears to be wrapped around or condensed within the protein oligomer. The protein exhibits RNA-dependent ATPase activity that is located in the carboxyl two-thirds of the polypeptide and is required for the helicase activity that brings about the unwinding of RNA:DNA hybrids.

9.7.4 Proofreading by RNA polymerase

During elongation only eight or nine nucleotides of the growing RNA chain remain base paired to the DNA template at any given time; the remainder of the RNA chain is peeled off. In addition to different functions performed by it, the polymerase enzyme also carries out two proofreading functions (**Fig. 9.14**). The first of these is called **pyrophosphorolytic editing**. In this, the enzyme uses its active site, in a simple back-reaction, to catalyse the removal of an incorrectly inserted ribonucleotide, by reincorporation of PPi. The enzyme can then incorporate another ribonucleotide in its place in the growing RNA chain. In the second proofreading mechanism, called **hydrolytic editing**, the polymerase backtracks by one or more nucleotides and cleaves the RNA product, removing the error-containing sequence.

9.7.5 Inhibitors of transcription

The mode of action of some antibiotics is through the inhibition of transcription. For example, rifampicin affects the β subunit of RNA polymerase, blocking the initiation of transcription. This is a useful experimental tool, because the drug does not affect transcription of messages, which have already been initiated. Actinomycin D is effective because it intercalates between GC residues in the DNA helix, making it an ineffective template for RNA synthesis. It can also inhibit rapidly growing eukaryotic cells, allowing it to be an effective agent against some cancers.

9.8 TRANSCRIPTION IN EUKARYOTES

The transcription process in eukaryotes is closely related to prokaryotic transcription. There are however, major differences in the machinery used and complexity of the processes. In eukaryotes, the genetic information carried by DNA is contained within the nucleus. This information is transcribed into RNA, which must then be carried across the membrane boundary into the cytosol prior to translation. This spatial and temporal separation of transcription and translation is just one of many differences between eukaryotic and prokaryotic transcription. Eukaryotes have three different polymerases (pol I, II and III), whereas bacteria have only one. Bacteria require only one additional initiation factor (σ), but eukaryotes need several transcription factors. In eukaryotes, transcription takes place in the membrane-bound nucleus.

9.8.1 Eukaryotic transcription factors

Eukaryotic transcription factors generally belong to the following three groups.

Basal or general transcription factors (GTFs): Eukaryotic RNA polymerases require accessory protein

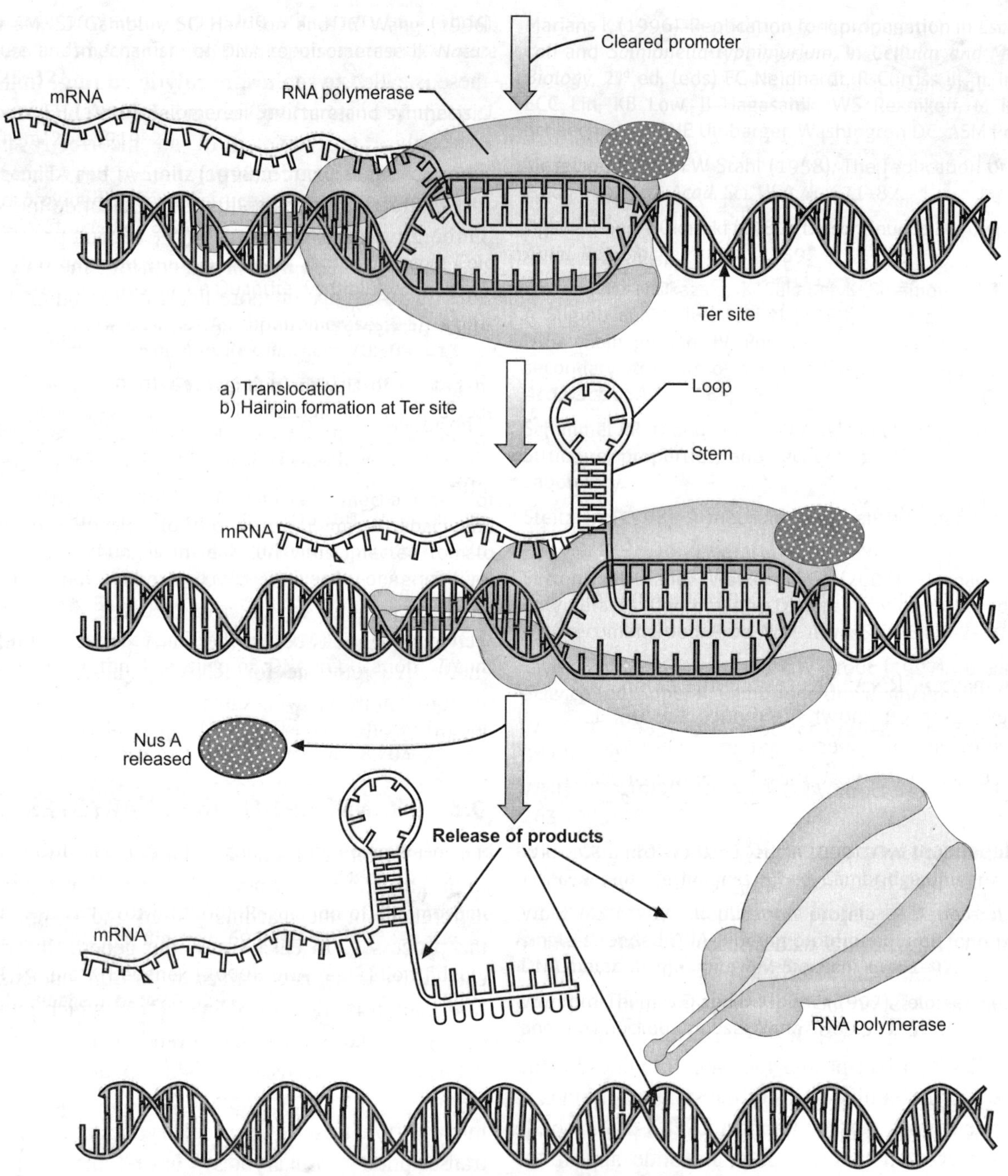

Fig. 9.12 Termination of transcription in prokaryotes (rho factor-independent)

factors—the transcription initiation factors (TIFs) or *basal* or general transcription factors (GTFs), to assist them in the recognition of the promoter. GTFs form a complex with RNA polymerase at the start point, and thus they determine the site of initiation. Recruitment of the enzyme and further basal factors to the DNA forms the pre-initiation complex (PIC). The GTFs together with RNAP constitute the basal transcription apparatus.

Upstream factors: The upstream factors are DNA-binding proteins that recognise specific short consensus elements located upstream of the start point. These are unregulated, ubiquitous and act upon any promoter that contains the appropriate binding site on the DNA. They increase the efficiency of initiation and are required for a promoter to function at an adequate level.

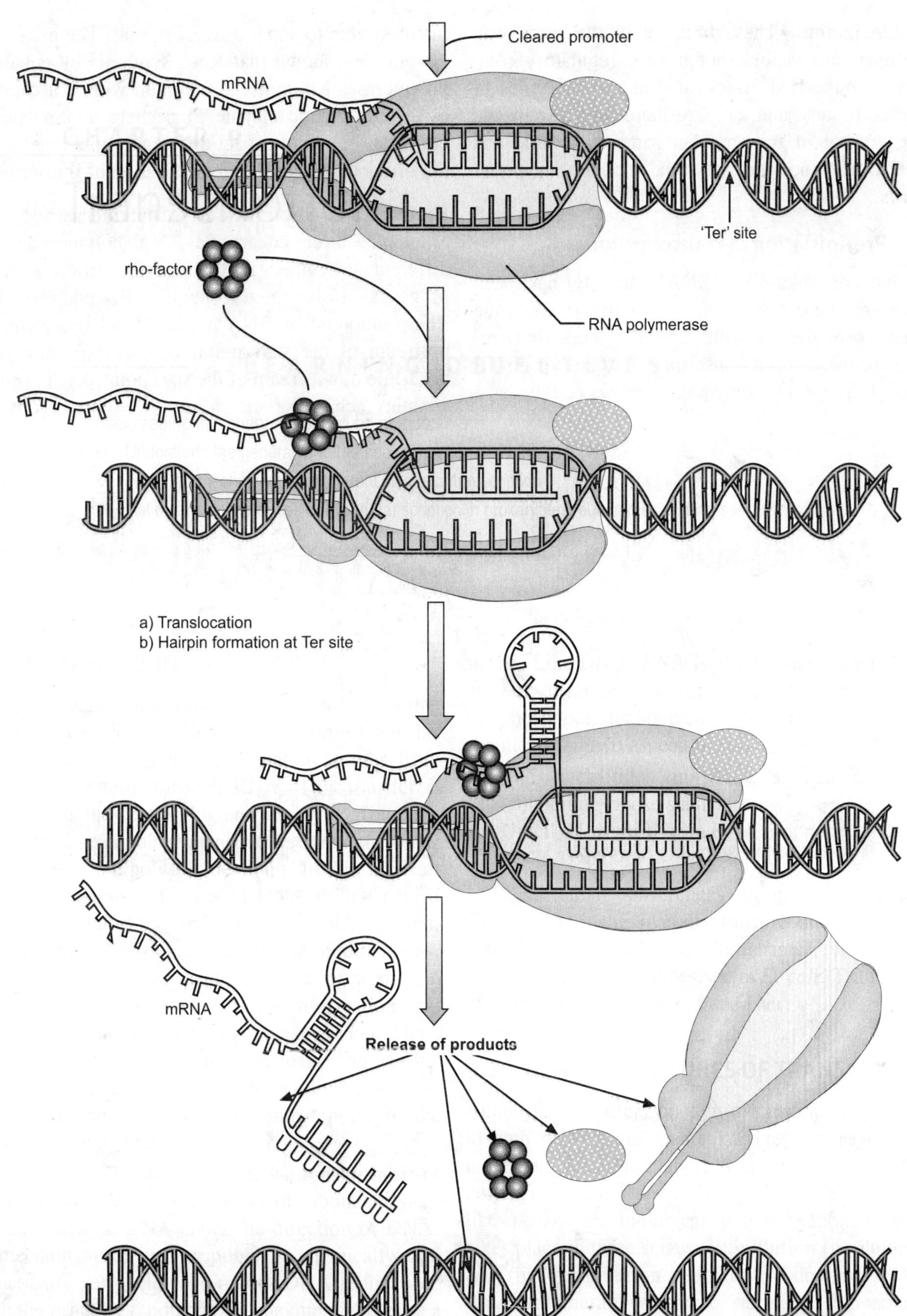

Fig. 9.13 Termination of transcription in prokaryotes (rho factor-dependent)

Inducible factors: These function in the same way as the upstream factors, but have a regulatory role. They are synthesised or activated at specific times or in specific tissues, and they are therefore, responsible for the control of transcription patterns in time and space. The sequences that they bind are called **response elements**.

9.8.2 Pre-initiation of transcription

The study of eukaryotic RNAP activity has been enhanced by the development of systems in which the enzymes correctly transcribe specific genes *in vitro*. The first of these was a transcription system for pol III and led to the use of extracts of *Xenopus* oocytes as a major system to support transcription. The use of such systems has shown that RNA synthesis by eukaryotic polymerases has much in common with that catalysed by the prokaryotic enzyme. Since there are three nuclear enzymes, one may expect differences in recognition of the promoter, control and termination of transcription.

Promoters of eukaryotic RNA polymerases

Since the three eukaryotic RNA polymerases have different structures, they transcribe different classes of genes. Thus, they recognise different promoters. The promoters for RNA pol I and pol II are (mostly) upstream of the start point, but some promoters for pol III lie downstream of the start point. Each promoter contains characteristic sets of short, conserved

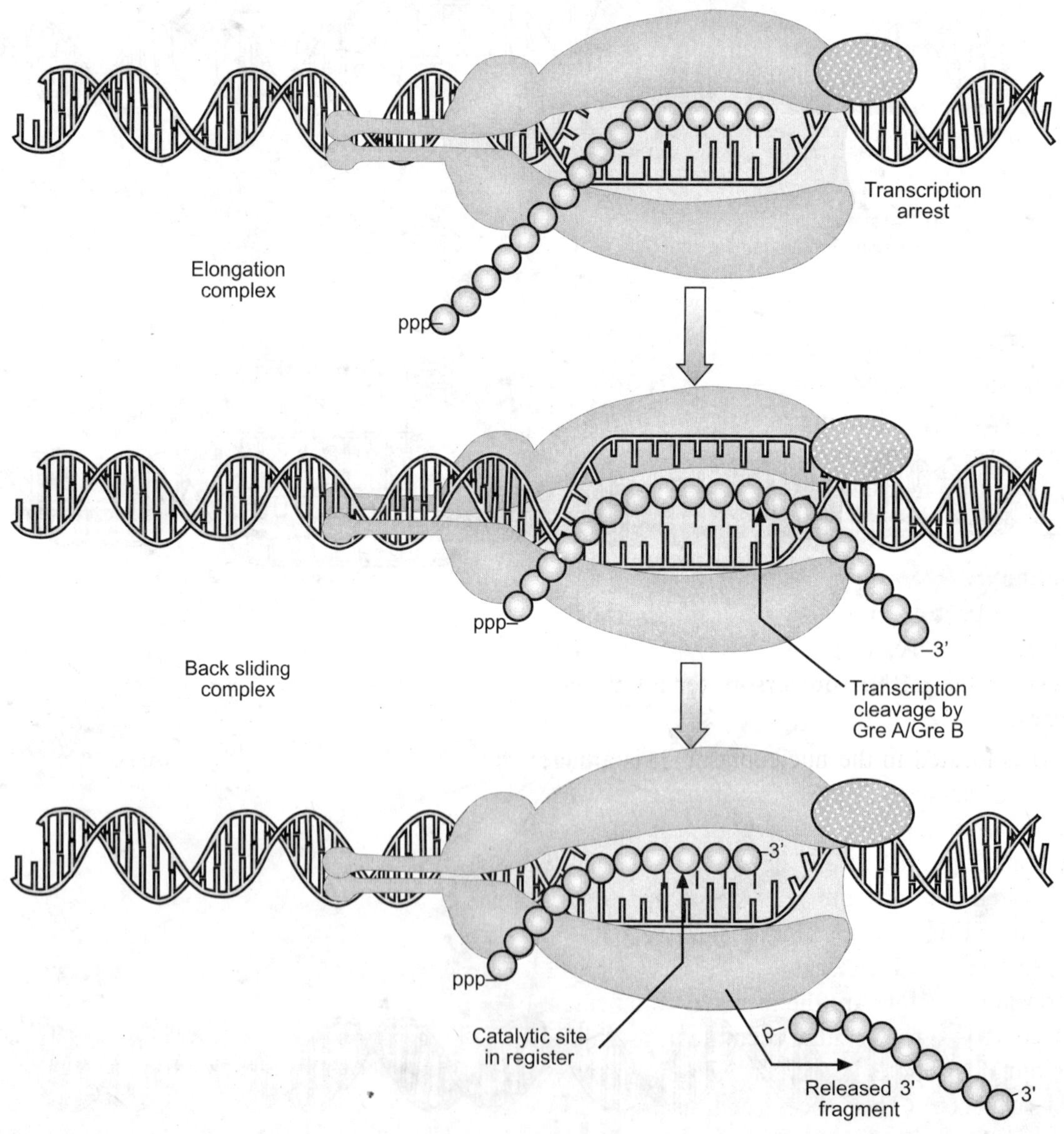

Fig. 9.14 Proposed sliding and editing mechanism due to arrested transcription

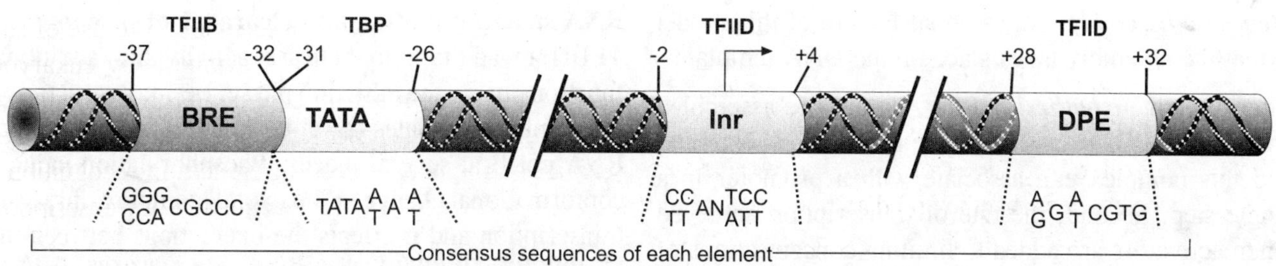

Fig. 9.15 Pol II core promoter: the positions of various DNA elements relative to the transcription start site; BRE: TFIIB recognition element; TATA: TATA box); Inr (initiator element and DPE downstream promoter element

sequences that are recognised by the appropriate class of factors (**Fig. 9.15**). Pol I and pol III each recognises a relatively restricted set of promoters. And they rely upon a small number of accessory factors. The promoter of pol I has two critical regions, in which mutations cause a great reduction in promoter strength.

Initiation of transcription by RNA polymerase I

Initiation by RNA polymerase I is the simplest one. There is a single, class I transcription unit in eukaryotic genomes, encoding several ribosomal RNAs as a polycistronic transcript (**Fig. 9.16**). The vertebrate RNA polymerase I promoter is bipartite, consisting of a core promoter which surrounds the transcriptional start site and an **upstream control element** (**UCE**) about 100 bp 5' to the start site. The core promoter alone is sufficient for basal transcription, but the presence of the UCE strongly increases transcriptional activity. There are at least three controlling elements: proximal promoters, upstream promoter elements and terminators.

Transcription by RNA polymerase II

RNA pol II is central to eukaryotic gene expression and has been studied extensively. RNAP II catalyses the synthesis of the mRNA precursors for all protein-coding genes.

RNAP II is located in the nucleoplasm. It is a huge enzyme with 12 subunits. It requires an array of other proteins called transcription factors in order to become the active transcriptional unit. The basal transcription factors required at all the RNAP II promoters (designated as TFII with an additional identifier) are highly conserved in all eukaryotes (**Table 9.4**). The basal transcription factors are the minimal complement proteins necessary to reconstitute accurate transcription from a minimal promoter (such as a TATA element or initiator sequence). The process of transcription by RNAP II can be described at different steps—assembly, initiation, elongation and termination—each associated with characteristic proteins.

9.8.3 Assembly of RNA polymerase and transcription factors

There are two models that explain the assembly of the pre-initiation complex: **step-wise assembly** and **recruitment hypothesis**.

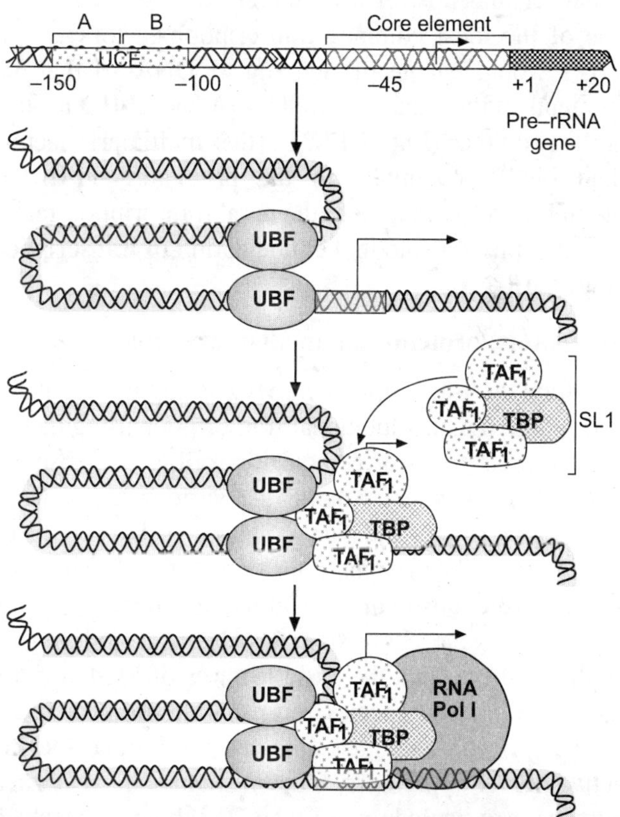

Fig. 9.16 Model for the assembly of the RNA polymerase I initiation complex; UCE: upstream control elements; UBF: upstream binding factors; SL1: selectivity factor; TAF1-, TBP-associated factors

Step-wise assembly: An essential feature of this model is that the assembly takes place on the DNA template.

Recruitment hypothesis: First, large pre-assembled complexes of GTFs and pol II are found in cell extracts, and this complex can associate with a promoter in a single step. Second, the rate of transcription achieved when activators are added to limiting concentrations of pol II holoenzyme can be matched by increasing the concentration of the pol II holoenzyme in the absence of the activators.

9.8.4 Initiation of transcription

The first stage of transcriptional initiation is the recognition of the TATA box by the TFIID factor (also called as **general transcription factor**), which comprises the TATA-binding protein TBP and a number of **TBP-associated factors (TAFs)**. TBP makes sequence-specific contact with DNA at the TATA box through the minor groove. The association of TBP with different TAFs enables the transcription complex to be activated by proteins bound at the enhancer and upstream promoter sites. For example, one of the most common transcription factors is *Spl*. TAFs would enable the looping of DNA so that *Spl* elements in the enhancer would meet the TFIID protein at the promoter (**Fig. 9.17**). Further multimeric factors then bind sequentially to the promoter. TFIID is pivotal in the formation of the basal transcription factor and also plays a role in the modulation of transcription (**Fig. 9.18**).

Regulatory proteins and mediator complex

Initiation of transcription *in vivo* is highly regulated and requires many additional transcriptional regulatory proteins and mediator complex as well as nucleosome-modifying enzymes. Among the various components of the pol II pre-initiation complex (PIC), Mediator appears to be the most important integrator of regulatory signals from upstream activators that define specific patterns of gene expression. Biochemical and genetic studies indicate that the basic features of Mediator are conserved across species, and that it projects many regulatory surfaces that interact with gene-specific activators or repressors. The mediators in yeast and humans each include more than 20 subunits, of which about seven show significant sequence homology. The mediator is organised into modules, which can be dissociated from one another under certain conditions *in vitro*.

RNA strand initiation and clearance of promoters

TFIIH in addition to helicase activity has additional function during the initiation phase, i.e., kinase activity, which phosphorylates the C-terminal domain (CTD) of RNA pol II at several places. Phosphorylation induces conformational change in the overall complex, initiating transcription and it affects the interactions between the transcription complex and other enzymes involved in processing the transcript. During the initial synthesis of 60–70 nt of RNA, first RFIIE and then TFIIH are released, and RNA pol II enters the elongation phase of transcription.

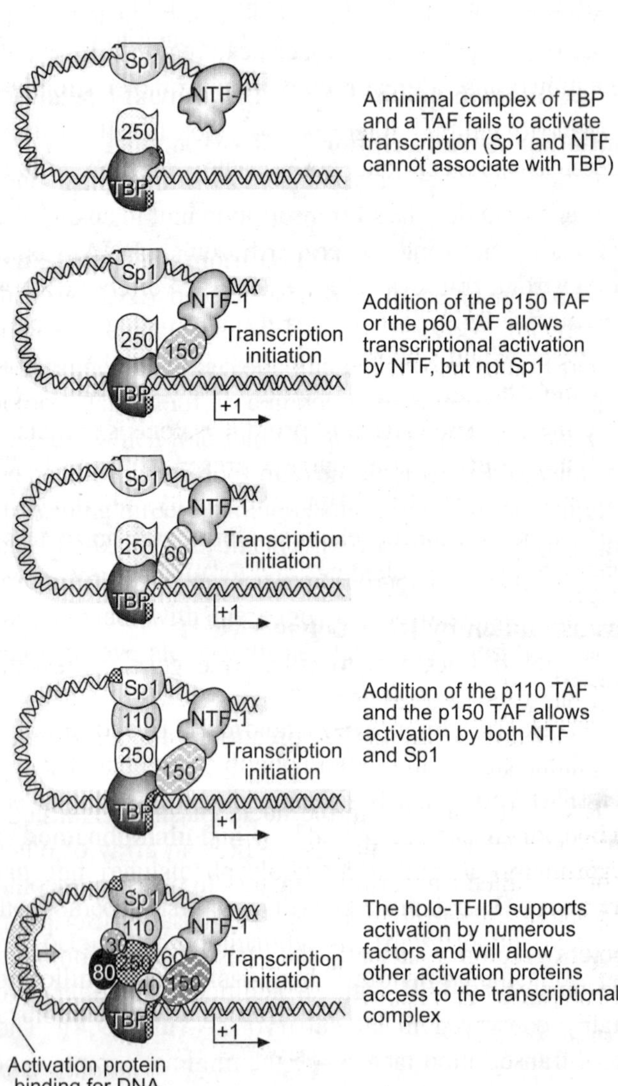

Fig. 9.17 Schematic illustration of experiments suggesting that different TAFs interacted with different transcription factors to activate transcription (***Source:*** Chen et al 1994)

Table 9.4 Proteins required for transcription by RNA polymerase II

Transcription factor	*Number of subunits*	*Subunit Mr*	*Functions*
Initiation			
RNA polymerase II	12	10,000–	Catalyses RNA synthesis
RPB1	1	220,000	PTSPSYS CTD
RPB2	1		NTP binding
RPB3	1		Core assembly
RPB4	1		Promoter recognition
RPB5	1		In pol I, II and III
RPB6	1		In pol I, II and III
RPB7	1		Unique to pol II
RPB8	1		In pol I, II and III
RPB9	1		
RPB10			In pol I, II, and III
TBP (TATA-binding protein)	1	38,000	Specifically recognises the TATA box
TFIIA	3	12,000, 19,000, 35,000	Stabilises binding of TFIIB and TBP to the promoter
TFIIB	1	35,000	Binds to TBP; recruits RNA polymerase–TFIIF complex
TFIID	12	15,000– 250,000	Interacts with positive and negative regulatory proteins
TFIIE	2	34,000, 57,000	Recruits TFIIH; ATPase and helicase activities
TFIIF	2	30,000, 74,000	Binds tightly to RNA pol II; binds to TFIIB and prevents binding of RNA polymerase to non-specific DNA sequences
TFIIH	12	35,000– 89,000	Unwinds DNA at promoter; phosphorylates RNA polymerase; recruits nucleotide-excision repair complex
TAFs	11		Two of the TAFs bind DNA elements at the promoter, e.g., initiator element (Inr) and downstream promoter element (DPE). Several of the TAFs have structural homology to histone proteins, therefore may like histones bind to DNA. Another TAF found to regulate the binding of TBP to DNA
Elongation			
ELL	1	80,000	
P-TEFb	2	43,000, 124,000	
SII (TFIIS)	1	38,000	
Elongin (SIII)	3	15,000, 18,000, 110,000	

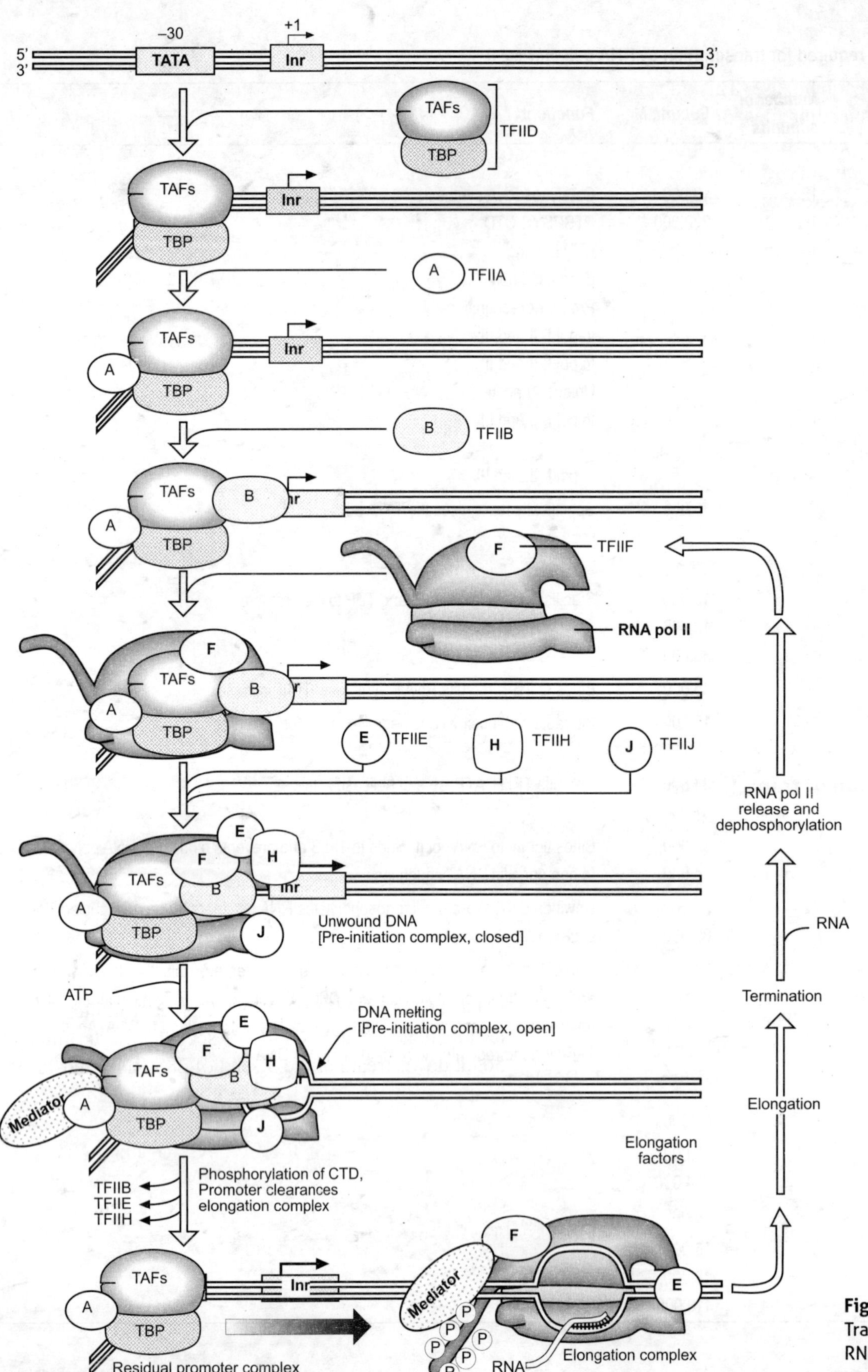

Fig. 9.18 Transcription at RNA polymerase II promoters

RNA proofreading

Soon after polymerase has initiated transcription, it shifts into the elongation phase. This transition involves the pol II enzyme shedding most of its initiation factors, for example, the general transcription factors and mediator proteins. In their place another set of factors elongation factors is recruited. Others are required for RNA processing. Various proteins are thought to stimulate elongation by pol II. Another factor that directly stimulates elongation is TFIIS which works by limiting the duration of the polymerase pauses when it encounters sequences that would otherwise tend to slow the enzyme's progress. TFIIS also contributes to proofreading by polymerase.

9.8.5 Elongation

Throughout elongation, TFIIF remains associated with RNA pol II and the activity of the polymerase is greatly enhanced by proteins called elongation factors (**Table 9.4**). Others required for RNA processing are generally recruited to the C-terminal tail of the large subunit of pol II, the **CTD**. The factors, however, favour the phosphorylated form of the CTD. Thus, phosphorylation of the CTD leads to an exchange of initiation factors required for elongation and RNA processing.

9.8.6 Termination of transcription

Once the RNA transcript is completed, transcription is terminated by mechanisms not yet well understood. Often it is difficult to even be sure of the relationship between observed 3'-termini of the newly made RNA molecules and the real point at which transcription terminates. This is especially in transcripts of RNAP II, the majority of which is post-transcriptionally modified by 3'-polyadenylation. The *in vitro* transcription with the incorporation of radioactive nucleotides and hybridisation with corresponding 3'-end DNA fragments, demonstrated that many genes are transcribed well past their apparent 3'-termini and are then subject to processing to generate their 3'-termini.

Termination by RNAP II

This refers primarily to the termination of polyadenylated mRNA. Use of the run-off transcription assay has demonstrated that RNAP II characteristically terminates at heterogeneous sites up to several thousand base pairs beyond the polyadenylation site (downstream). It has become clear that the factors involved in recognition of AAUAAA and GC-rich sequences, signals for the 3'-processing of pre-mRNA, are also involved in termination. This was first identified in a patient with α thalassaemia in whom a base change in the α_2 globin gene converted the AATAAA motif into AATAAG. This not only deactivated the processing of the α_2 globin pre-mRNA but also caused transcription to continue well past the normal termination sites.

Termination by RNA polymerase I

The ribosomal genes of *Xenopus* have three major sites that have been associated with RNA 3'-end formation and are designated t_1, t_2 and t_3. Their relationship with termination showed that t_1 was at the 3'-end of 28S RNA but was not a terminus for transcription. Transcription continued from t_2 to across the intergenic spacer to t_3 just 215 bp upstream of the initiation site for the next tandemly arranged rRNA gene and about 60 bp from its promoter. The transcript between t_2 and t_3 was rapidly degraded and was only detected by run-off transcription or electron microscopy. Thus, t_2 is a processing site and it defines the 3'-end of the largest relatively stable precursor, 40S pre-RNA. The true terminator t_3 is a 12 bp element that is conserved between different amphibians.

Termination elements have also been identified in mammalian rDNA particularly that of the mouse where a repetitive 18 bp motif, known as the Sal box (AGGTCGACCAG$^{A}/_{T}{}^{T}/_{A}$NTCCG) constitutes the termination signal. RNAP I specifically terminates 565nt downstream of the 3'-end of 28S RNA and 11 bp upstream of the first termination signal. The 3'-end of the transcript is then trimmed by 10 nt to produce the 3'-terminus of the first relatively stable precursor, 45S RNA. Termination again requires the binding of a specific factor (TTF-1) to the Sal box sequence.

KEY POINTS

- The flow of genetic information from DNA to RNA is accomplished by a series of events. The sequence of bases in DNA is first copied into an RNA molecule and then the RNA is used to determine the amino acid sequence of a protein molecule. It is the first step in gene expression.
- The basic enzymology of transcription is common to prokaryotes and eukaryotes, although its regulation is more complex in the latter. Transcription, like nearly all biological polymerisation reactions takes place in three stages: initiation, elongation and termination.
- Bacteria have a simple general mechanism for coordinating the regulation of genes that encode products involved in a set of related processes.
- The gene cluster and promoter, plus additional sequences that function together in regulation are called an operon. In contrast, eukaryotic genes are monocistronic, express only a single gene product.
- Structurally, eukaryotic genes are more complex and have many regulatory elements in addition to *cis*-acting sequences.
- A transcription factor is a protein that regulates transcription. These factors are the *trans* regulators of promoters and enhancers and regulate the binding of RNA polymerase.
- RNA is synthesised in a 5'→3' direction on template strand of the DNA. Promoter sequences for RNA polymerases are generally located on the 5′ side of the start site. Terminator sequences are near the end of coding region of mRNA.
- Several types of RNA are synthesised in the nucleus of eukaryotic cells. RNA molecules can be broadly grouped into: mRNA, tRNA, rRNA and small RNAs.
- Transcription is an enzymatic process requires RNA polymerase enzymes (RNAP) that use single-strand DNA as a template to make complementary copies of information stored in the DNA sequence.
- Transcription of nuclear genes is carried out by one of the three different types of RNA polymerase, according to the type of gene product. However, prokaryotes have only one RNA polymerase, which transcribes all types of gene products.

Self-assessment Questions

1. Explain the general features of transcription.
2. How is the operon of the bacteria transcribed?
3. Describe with the help of a diagram the *cis*- and *trans*-acting elements of a gene.
4. Explain briefly the role of different DNA-binding proteins in transcription.
5. Distinguish between the structure and function of prokaryotic and eukaryotic promoters.
6. List the different types of RNAs and their functions.
7. Describe briefly the different steps involved in RNA synthesis in prokaryotes.
8. What are the mechanisms involved in termination of transcription in prokaryotes?
9. Discuss the functions of different types of eukaryotic RNA polymerases.

FURTHER READING

Textbooks

Adams RLP, JT Knowler and AD Leader (1992). *The Biochemistry of the Nucleic Acids*. Chapman and Hall.

Cooper GM and RE Hausman (2004). *The Cell: A Molecular Approach*, 3rd ed. ASM Press.

Hartl DL and EW Jones (1998). *Genetics: Principles and Analysis*, Jones and Bartlett Publishers.

Krebs JE, ES Goldstein and ST Kilpatrick (2011). Lewin's *Genes X*. Jones and Bartlett Publishers, LLC.

Lodish H, D Baltimore, A Berk, L Zipursky and P Matsudaira (1995). *Molecular Cell Biology*. Scientific American Books.

White RJ (2001). *Gene Transcription: Mechanisms and Control*. Connecticut: Blackwell Science.

References

Borukhov S and E Nudler (2003). RNA polymerase holoenzyme: Structure, function and biological implications. *Curr. Opin. Microbiol.* 6:93–100.

Busby S and RH Ebright (1994). Promoter structure, promoter recognition and transcription activation in prokaryotes. *Cell* 79:743–46.

Butler JE and JT Kadonaga (2002). The RNA polymerase II core promoter: A key component in the regulation of gene expression. *Genes Dev.* 16:2583–92.

Campbell EA, N Korzheva, A Mustaev, K Murakam, S Nair, A Goldfarb and SA Darst (2001). Structural mechanism for rifampicin inhibition of bacterial RNA polymerase. *Cell* 104:901–12.

Darst SA (2001). Bacterial RNA polymerase. *Curr. Opin. Struct. Biol.* 11:155–62.

Fickett JW and AG Hatzigeorgiou (1997). Eukaryotic promoter recognition. *Genome Res.* 7:861–71.

Henkin TM (2000). Transcription termination control in bacteria. *Curr. Opin. Microbiol.* 3:149–53.

Keaveney M and K Struhl (1998). Activator-mediated recruitment of the RNA polymerase machinery is the predominant mechanism for transcriptional activation in yeast. *Mol Cell* 1:917.

Korzheva N, A Mustaev, M Kozlov, A Malhotra, V Nikiforov, A Goldfarb and SA Darst (2000). A structural model of transcription elongation. *Science* 289:619–25.

Lee TI and RA Young (2000). Transcription of eukaryotic protein-coding genes. *Ann. Rev. Genet.* 34:77–137.

Muller CW (2001). Transcription factors: Global and detailed views. *Curr. Opin. Struct. Biol.* 11:26–32.

Murakami KS and SA Darst (2003). Bacterial RNA polymerases: The whole story. *Curr. Opin. Struct. Biol.* 13:31–39.

Paget MS and JD Helmann (2003). The sigma 70 family of sigma factors. *Genome Biol.* 4:203.

Reines D, JW Conaway and RC Conaway (1996). The RNA polymerase II general elongation factors. *Trends Biochem. Sci.* 21:351–55.

Richardson JP (2002). Rho-dependent termination and ATPases in transcript termination. *Biochem. Biophys. Acta.* 1577:251–60.

Roeder RG (1996). The role of general initiation factors in transcription by RNA polymerase II. *Trends Biochem. Sci.* 21:327–35.

Schramm L and N Hernandez (2002). Recruitment of RNA polymerase III to its target promoters. *Genes Dev.* 16: 2593–2620.

Uptain SM, CM Kane and MJ Chamberlin (1997). Basic mechanisms of transcript elongation and its regulation. *Annu. Rev. Biochem.* 66:117–72.

Verrijzer CP and R Tjian (1996). TAFs mediate transcriptional activation and promoter selectivity. *Trends Biochem. Sci.* 21:338–42.

Woychik NA and M Hampsey (2002). The RNA polymerase II machinery: Structure illuminates function. *Cell* 108:453–63.

Yanofsky D, BC Carlton , JR Guest , DR Helinski and U Henning (1964). On the collinearity of gene structure and protein structure. *Proc. Nat. Acad. Sci.* 51:266–72.

Young BA, TM Gruber and CA Gross (2002). View of transcription initiation. *Cell* 109:417–20.

CHAPTER 10

RNA Processing

LEARNING OBJECTIVES

After reading this chapter, you will able to:

- List the different types of RNA processing
- Understand the mechanism of RNA processing
- State the biological significance of RNA processing

10.1 INTRODUCTION

Transcription and translation are simultaneous in prokaryotes, but in eukaryotes, the former takes place in the nucleus and the latter on ribosomes in the cytoplasm. Consequently, primary or heterogeneous transcripts (hnRNAs) must move from the nucleus to the cytosol to be translated. Additionally, these hnRNAs undergo processing i.e., alterations that convert the nascent RNAs into mature RNAs. Thus, RNA processing is the collective term used to describe these alterations to the primary transcript. The commonest types of alterations include: removal of nucleotides by nucleases; addition of nucleotides to the 5'- or 3'- ends of the primary transcripts or their cleavage products; modification of certain nucleotides on either the base or sugar moiety; and an additional process referred to as RNA editing. These events take place in both prokaryotes and eukaryotes on the major classes of RNA (**Table 10.1**). The multiple steps in the synthesis and processing of RNAs provide opportunities for additional levels of gene control beyond the regulation of transcription.

Most eukaryotic genes are split ones. Interruptions in genes are found in all organisms to varied degrees. Protein-coding genes in eukaryotes are interrupted with sequences that are faithfully transcribed by RNA polymerase II, but are later spliced out of the primary transcript and so do not appear in the mature messenger RNA (mRNA). These **intervening sequences (introns** or **IVS)** can comprise up to 99% of the primary transcript (pre-mRNA). An average eukaryotic exon is only 140 nts long, but one human intron stretches for 480,000 nucleotides. Even the genes for rRNA and tRNA are split by introns. Pre-mRNA splicing entails identification of the precise junctions between an intron and its flanking exons, namely the 5' and 3' splice sites and removal of the intervening sequence via a two-step chemical reaction. This process is catalysed by the spliceosome, a megaDalton ribonucleoprotein complex assembled from five subcomplexes called snRNPs.

Recently, several types of catalytic reactions have been discovered to reside in RNA (called **catalytic RNAs**). Ribozyme is a good example for RNA with catalytic activity. Introns of the group I and II classes splice themselves out of pre-mRNA (**autosplicing**) that contains them, like protein enzymes. The enzyme ribonuclease P is a ribonucleoprotein that contains a single RNA molecule bound to a protein. The RNA component of ribonuclease P possesses the ability to catalyse the cleavage in a tRNA substrate. Some RNAs

Table 10.1 Different categories of RNA processing reactions

Processing reaction	*Examples*
Cleavage	Processing of pre-rRNA and pre-tRNA from polycistronic transcripts; termination of eukaryotic mRNA transcription
Exonucleolytic degradation	Processing rRNA and tRNA to generate mature ends
Nucleotidyl transfer	Transfer of CCA trinucleotide to 3' end of some tRNAs
Chemical modification of bases	Occasional methylation in mRNA and rRNA; extensive base modification in tRNA and snRNA
Nucleotide excision and replacement	Hypermodification of guanosine residues in tRNA to produce queuosone and wyosine
Capping	Addition of 7-methylguanosine to 5' end of eukaryotic mRNA
Polyadenylation	Addition of polyadenylate tail to 3' end of most eukaryotic mRNAs and a few bacterial mRNAs
Splicing (*trans*-esterification)	Removal of most introns (usually in *cis*, but occasionally in *trans*)
Splicing (ligation)	Removal of tRNA introns
RNA editing	Changing the information carried in mRNA by:
Base modification	Base modification of human apolipoprotein RNA
Insertion and deletion	Insertion and deletion of residues from the coding region

also perform RNA editing. Changes are introduced into mRNA at particular positions by the modification of an individual base or insertion or deletion of bases (most commonly uridine residues).

10.2 NATURE OF GENE TRANSCRIPTS

In this section, we shall see some of the important features of gene transcripts in prokaryotcs and eukaryotes.

10.2.1 Prokaryotic transcripts

In prokaryotes, ribosomes attach to the 5'-end of the mRNA and commence translation before its 3'-synthesis is complete. Since the rates of the two processes are similar, the unit can be regarded as a **transcription–translation complex** in which RNA polymerase transcribing the gene is followed by a string of ribosomes translating the message (**Fig. 10.1**). Many transcripts are considerably long because prokaryotic genes are often grouped into functional units known as operons in which, for instance, the genes encoding the enzymes of a particular metabolic pathway might occur consecutively on the DNA. Typically, an operon has a single 5'-promoter and the genes within it are transcribed by RNA polymerase into a single polycistronic mRNA. Such messages may encode up to 20 proteins.

Another feature of prokaryotic mRNA is a very short half-life, which averages 0.7 min although it varies by a factor of at least 50 and can depend on the growth rate.

10.2.2 Eukaryotic transcripts

Eukaryotic transcription follows a series of RNA processing events (**Fig. 10.2**), which includes the covalent modification of both ends of the RNA and the removal of intron sequences that are discarded from the middle of the RNA transcript by the process of RNA splicing. Three characteristics of eukaryotic mRNAs make them easier to study than those of prokaryotes:

- They are more stable and have half-lives ranging from 1 hour to 24 hours.
- They have 3'-polyadenylate tails.

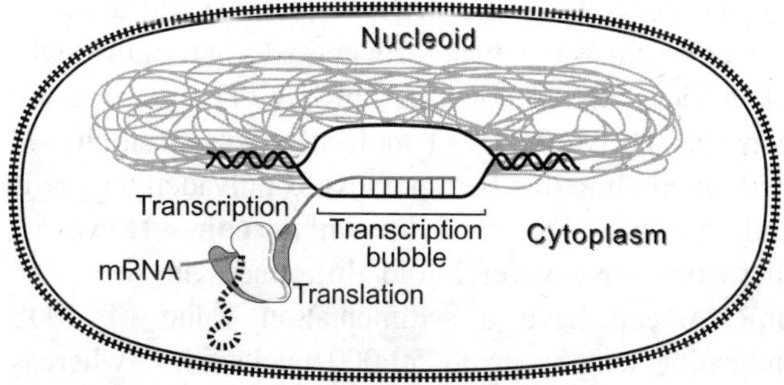

Fig. 10.1 Transcription–translation complex

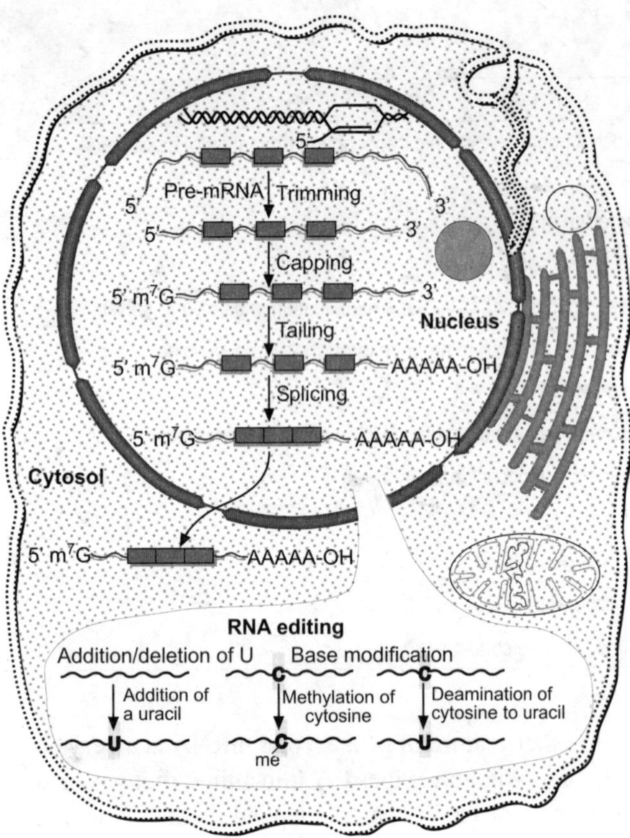

Fig. 10.2 Major types of RNA modifications in eukaryotes

In some specialised cells, a high percentage of the total mRNA codes for one or a few abundant proteins. For example, 50% of the mRNA produced by the tubular gland cells of the magnum portion of the hen oviduct encodes the egg white protein, ovalbumin. Similar cell- or tissue-specific abundance led to the early isolation of globin mRNA from reticulocytes, and silk fibroin mRNA from the silk glands of the larvae of the silk moth, *Bombyx mori*.

It is now established that cytoplasmic mRNA of eukaryotic cells is the product of the maturation of longer precursors, which form at least a part of the high-M_r nuclear RNA known as hnRNA. The relationship between the two types of molecules was strengthened with the finding that both possess 3'-polyadenylate tails and modified 5'-termini known as **caps**. However, these two have several fold differences in size. The hnRNA can have a sedimentation value of 100S indicating lengths up to 50,000 nucleotides, whereas even a large mRNA, such as that of the avian egg yolk precursor protein, vitellogenin (6700 nucleotides) is much smaller.

Association with abundant nuclear proteins

Nascent RNA and mRNA intermediates in the nuclei of eukaryotic cells do not exist as free RNA molecules. From the time the newly synthesised transcripts emerge from RNA polymerase II until they are transported into the cytoplasm, they are associated with an abundant set of nuclear proteins. These proteins were first characterised as being the major protein components of hnRNA particles. High-dose UV irradiation causes covalent crosslinks to form between RNA bases and closely associated proteins. This technique has been used to identify a large number of hnRNP proteins. The hnRNP proteins identified are designated as A1–U and ranging in size 34–120 kDa. The hnRNP proteins bind to RNA with extremely high affinity but differing specificity.

Like transcription factors, most hnRNP proteins have a modular structure. They contain one or more RNA-binding domains and at least one other domain that is thought to interact with other proteins. The association of mRNA precursors with hnRNP proteins could prevent the formation of short secondary structures dependent on base pairing of complementary regions, thereby making the pre-RNA accessible for interaction with other macromolecules. This function would be analogous to that of single-strand-binding protein at a DNA replication fork; however, the diversity of hnRNP proteins suggests that they have other functions as well. For example, various hnRNP proteins may interact with different portions of a pre-mRNA and contribute to the structure recognised by the RNA-processing factors.

Role of snRNAs in processing events

The members of a group of small nuclear RNA molecules, collectively known as snRNA or U snRNA, play a considerable role in the processing of gene transcripts. In vertebrate cells, six of these, U1–U6, are present in nuclei at 10^5–10^6 copies per cell whereas others, of which a further six have been characterised, are less abundant. Until now, about 30 species of snRNAs have been described from yeast and plant cells. The snRNAs are very stable, have highly conserved primary and secondary structures, unique modified nucleotides at their 5'-caps and multiple base modifications clustered at their 5'-ends. They occur as ribonucleoprotein particles (snRNP or Snurps) in which one (U1, U2, U5) or two (U4/U6) molecules of RNA are associated with at least seven polypeptides that are common to all snRNPs.

Maturation of untranslated RNAs: Approximately 80% of the total RNA in rapidly growing mammalian cells contain rRNAs and the rest are tRNAs. The primary transcripts produced from most rRNA and tRNA genes, like pre-mRNAs, are extensively processed to yield the mature, functional forms of these RNAs. Both bacteria and eukaryotes synthesise polycistronic rRNA transcripts. In addition to this, eukaryotic cells contain a class of small RNA molecules (100–300 nt). These are called small nuclear RNA (**snRNA**) and the cytoplasmic species are small cytoplasmic RNA (**scRNA**).

10.3 PROCESSING OF RIBOSOMAL RNA

Ribosomes are macromolecular structures composed of ribosomal RNA (rRNA) bound to protein. They occur in the cell cytoplasm where they bind to mRNA and translate it to produce proteins. Each ribosome is composed of two parts called the large and small subunits. In prokaryotes, the ribosome is 70S and is made up of 50S and 30S subunits and the 50S subunit contains two rRNAs (23S and 5S) complexed with 31 polypeptides. In eukaryotes, the ribosome is 80S and is composed of 60S and 40S subunits. The 60S subunit contains three RNAs (28S, 5.8S and 5S) and about 49 polypeptides; the 40S subunit contains one rRNA (18S) and about 33 proteins. The rRNAs are transcribed together from a single gene present in the genome as multiple copies.

10.3.1 Processing of prokaryotic pre-rRNA

In *E. coli*, there are seven different operons (rrn) for rRNA that are dispersed throughout the genome and which are called rrnH, rrnE and so on. Each operon contains a copy of each of the 5S, 16S and 23S rRNA sequences. One and four coding sequences for tRNA molecules are also present in these rRNA operons, and these primary transcripts are processed to give both rRNA and tRNA molecules. Some of the major steps involved are described below.

Processing of 16S and 23S transcripts

The post-translational processing of *E. coli* rRNA takes place in a well-defined series of steps (**Fig. 10.3**). After the 6,000-nt primary transcript has been transcribed, the RNA folds up into a number of stem loop structures (secondary structures) by base paring between complementary sequences in the transcript. These secondary structures of stems and loops allow some proteins to bind to form a **ribonucleoprotein (RNP) complex**. Many of these attached proteins become a part of the ribosome. After the binding of proteins, modifications such as 24-specific methylation take place. The sequences flanking both the 16S and 23S gene transcripts form inverted repeats, the two halves of which are, respectively, 1,700 and 2,900 nucleotides apart. These sequences interact to form base-paired stems at the base of giant loops that contain the complete sequences of 16S and 23S and which, in the nascent transcript, are already in the form of a ribonucleoprotein.

Processing of 5S transcript

At least three enzymatic steps are required for the generation of 5S rRNA from the transcripts of the *E. coli* rrn operons. The first of these is the removal of the precursors to 23S rRNA from the growing nascent rRNA precursor chain, which is catalysed by RNase III. The second enzyme RNase E has been purified as a 66,000 M_r protein but perhaps requires additional factors. It generates a precursor of 5S RNA known as p5S which as compared to the mature molecule has three extra nucleotides at both the 5'- and 3'-termini. Processing apparently occurs on the ribosome after the binding of the three 5S-associated ribosomal proteins. The third processing step is catalysed by uncharacterised enzyme activities that complete the processing of 5S RNA by trimming the residual 5'- and 3'-tails.

10.3.2 Processing of eukaryotic pre-rRNA

The pre-rRNA in eukaryotes are processed in the following steps.

Processing of 18S, 5.8S and 28S transcripts

Eukaryotic cells contain multiple pre-RNA genes that are arranged in tandem. The transcription of rRNA, together with processing and assembly into ribosomal particles, occurs in the nucleolus. Electron micrographs of rRNA in oocytes during transcription reveal that the rDNA fibre forms a Christmas–tree-like structure with the 'trunk' as the non-transcribed spacer which separates the rRNA gene units. The 'branches' on the transcription unit are represented by about 100 nascent transcripts, along the length of the rDNA fibre axis and have at their base a transcribing molecule of RNA polymerase (**Fig. 10.4**). The longest branches are complete transcripts of the transcription unit and are partially wrapped into ribonucleoproteins.

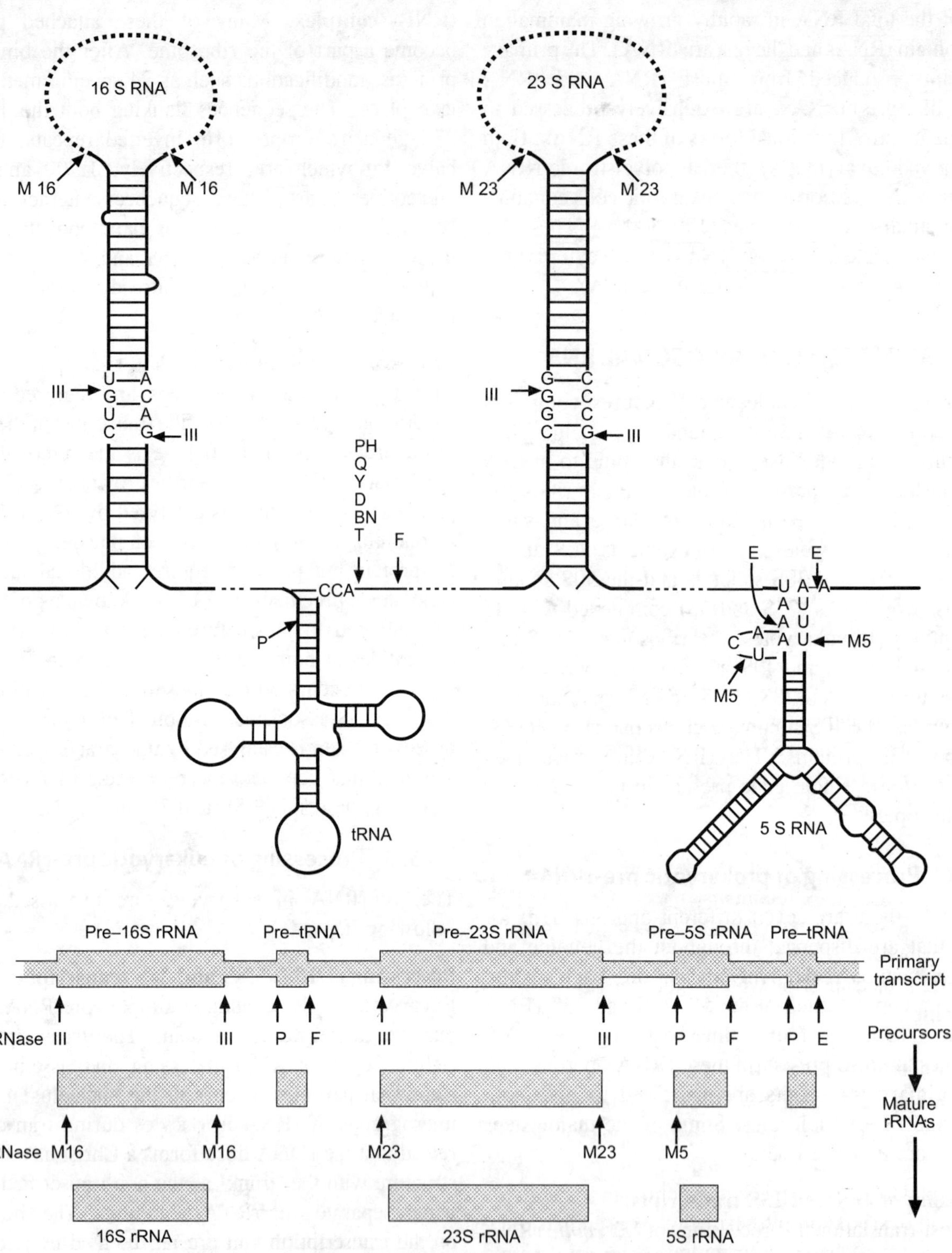

Fig. 10.3 The processing of prokayotic rRNA, tRNA and 5S rRNA precursors; arrows and associated symbols indicate sites cleaved by the following ribonucleases: III indicates sites cleaved by RNase III in the double-stranded regions flanking 16S and 23S RNA. P indicates the site of cleavage by RNase P to generate the 5′ end of mature RNA. F indicates RNase F and M16 and M23 are ill-characterised nucleases. PH, Q, Y, D, BN, T and E are endonucleases involved in cleavage

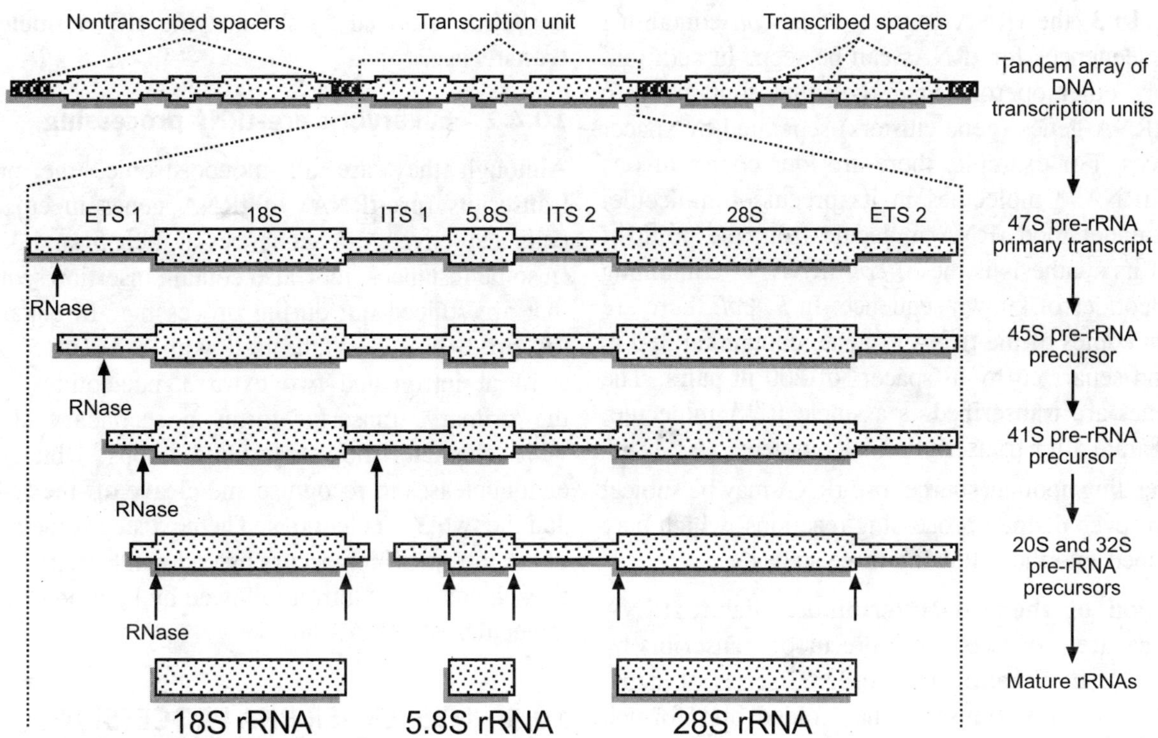

Fig. 10.4 Processing of eukaryotic pre-ribosomal RNA (primary transcript)

In eukaryotes, rRNA is also generated from a single, long precursor molecule by specific modification and cleavage steps. The rRNA genes are present in a tandemly repeated cluster containing 100 or more copies of the transcription unit and they are transcribed in the nucleus by RNA pol I. The precursor size varies from about 7,000 nt in yeast and 135,000 nt in mammals (**Fig. 10.4**). It contains one copy each of 18S, 5.8S and 28S coding regions. The eukaryotic 5S rRNA is transcribed by RNA pol III from unlinked genes to give a 121 nt transcript which undergoes little or no processing.

Post-trancriptional base modification of rRNA

Early studics established that rRNA contains methylated nucleotides and it has since become apparent that it also contains pseudouridine and in eukaryotes, both N4-acetylcytosine and the hypermethylated nucleotide 3-(3-amino-carboxypropyl)-1-methyl pseudouridine. Some modified bases are strongly conserved and presumably have important functions. A conserved sequence, $m^6{}_2Am^6{}_2A$, found at the 3'-ends of both prokaryotic and eukaryotic small subunit rRNAs seems to have a role in the decoding phase of protein synthesis. Of 70 methyl groups on HeLa cell 28S rRNA, 65 are 2'-O-riboses on 18S rRNA and six base methylations. Methylation takes place at over 100 sites to give 2'-O-methylribose and this is now known to be carried out by a subset of small nuclear RNP particles which are abundant in the nucleolus.

10.4 PROCESSING OF tRNA

The tRNAs are synthesised by transcription of tRNA genes by the same RNA polymerase that synthesise mRNA and rRNA in case of prokaryotes, but in eukaryotes it is exclusively by RNA polymerase III enzyme (see **Chapter 9**). The tRNA genes exist as multiple copies, especially in eukaryotic cells, reflecting the large requirement of cells for tRNA. The tRNAs are produced as precursor RNA molecules called **pre-tRNAs** which are processed to give the mature tRNA sequence. Several tRNA genes may be transcribed together as a single pre-tRNA, which is then processed by ribonucleases that cleave at the 5' and 3' ends of each tRNA sequence.

10.4.1 Prokaryotic pre-tRNA processing

Some tRNA genes may be transcribed singly or as part of a polycistronic unit. In *E. coli*, some tRNA genes are co-transcribed with rRNA genes in a common operon.

In **Fig. 10.3**, the rRNA operons of *E. coli* containing coding sequences for tRNAs can be seen. In addition, there are other operons in *E. coli* that contain up to seven tRNA genes (gene clusters) separated by spacer sequences. For example, there are four copies of one of the $tRNA^{Leu}$ molecules in its precursor molecule. A well-understood tRNA molecule from the point of view of its synthesis is the *E. coli* $tRNA_I^{Tyr}$, containing 85 nucleotides of known sequence. In *E. coli* there are adjacent copies of the $tRNA_I^{Tyr}$ gene, containing 350 nt each and separated by a 'spacer' of 200 nt pairs. The two genes are transcribed as a single RNA molecule, and separate after transcription is complete.

Depending upon its source, pre-tRNA may be subject to up to seven distinct processing reactions, which may be grouped into the following three stages.

Formation of the 3'-OH terminus: Mature tRNA molecules are processed from precursor transcripts by RNases: D, E, F and P in an ordered series of steps. Once the primary transcript has folded and formed characteristic stems and loops, an endonuclease (RNase E or F) cuts off a flanking sequence at the 3'-end, at the base of a stem, to leave a precursor with nine extra nucleotides. The exonuclease RNase D then removes seven of these 3'-nucleotides one at a time.

Formation of the 5'-P terminus: RNase P can then make an endonucleolytic cut to produce the mature 5'-end of the tRNA. RNase P is probably responsible for generation of the 5'-P terminus of all *E. coli* tRNA molecules. RNase P does not recognise a specific base sequence at the cleavage site or anywhere else, and responds to the overall 3D conformation of the tRNA molecule with its several hairpin loops and then makes a cut at the right place.

Production of modified bases: Finally, the tRNA undergoes a series of base modifications. Different pre-tRNAs are processed in a similar manner, but the base modifications are unique to each type of tRNA. In the case of $tRNA_I^{Tyr}$, two uridines are converted to pseudouridine (Ψ), two uridines to two 4-thiouridines (4tU), one guanosine to 2'-*o*-methylguanosine (2mG), and one adenosine to isopentenyladenosine (2ipA). These and other modified bases are formed in most tRNA molecules also.

All tRNA molecules end with CCA-3'-OH by appropriate cut. However, the precursors of some tRNA molecules lack a terminal CCA. In these molecules, the CCA is added by the enzyme **tRNA nucleotidyl transferase**.

10.4.2 Eukaryotic pre-tRNA processing

Although they are all monocistronic, the primary transcripts (pre-tRNA) of tRNA genes in eukaryotic cells also have extra sequences at their 5' and 3' ends. In some instances, they also contain insertion sequences that are spliced out during processing. In eukaryotes, the pre-tRNA is synthesised with a 16 nt 5'-leader, a 14 nt intron and two extra 3'-nucleotides. Again, the primary transcript forms a secondary structure with characteristic stems and loops which allow endonucleases to recognise and cleave off the 5'-leader and the two 3'-nucleotides. The next step is the removal of the intron, which occurs by endonucleolytic cleavage at each end of the intron followed by ligation of the half molecules of tRNA.

10.5 MESSENGER RNA PROCESSING

In prokaryotes, the mRNA molecule is used directly in polypeptide synthesis. In eukaryotes, mRNA is generally stable, encodes only one polypeptide (strictly **monocistronic**) and undergoes extensive processing in the nucleus before it is exported. However, not all the base sequences in mRNA molecule are translated into the amino acid sequences of polypeptides. Initiation of protein synthesis may begin many nucleotides downstream from the 5' end of RNA called a **leader** and in some cases contains regulatory sequences that affect the rate of protein synthesis. The leader is followed by a **coding sequence**, also called an **open reading frame** (**ORF**), which specifies the order in which the amino acids are present in the polypeptide chain. Like the leader sequence, the 3' end of an mRNA molecule following the coding sequence is not translated.

In prokaryotes, the primary product of transcription is mRNA; in contrast, the primary transcript in eukaryotes must be converted into mRNA. Because eukaryotic RNA pol II transcribes a wide variety of different genes, the collection of products is referred to as **heterogeneous nuclear RNA** (**hnRNA**). Immediately after transcription, hnRNA is rapidly covered by the proteins to form heterogeneous nuclear ribonucleoprotein (**hnRNP**). The proteins involved have been classified as hnRNP proteins A–U. The modification of hnRNA into mRNA is called **RNA processing**.

10.5.1 Capping or 5'-end modification

Capping is the first process of modification of eukaryotic pre-mRNAs (or hnRNA). As soon as transcription commences, nascent eukaryotic pre-mRNA (about 25 nt) are capped by the addition of an inverted guanosine triphosphate residue at the 5'-end. During the end processing (immediately after 20–30 nt long), the 5' end is chemically altered by the addition of a modified guanosine, 7-methyl guanosine (m^7G), linked 5'→5' direction. This 5' terminal group is called a cap and added to the new RNA in the reverse orientation (5'→5' bridge). This capping reaction is catalysed by three enzymes, which act in succession. (**Fig. 10.5**)

Capping has many biological functions: The 5'-methyl cap signals the 5' end of eukaryotic mRNAs, and helps the cell to distinguish mRNAs from the other types of RNA molecules. For example, RNA polymerases I and III produce uncapped RNAs during transcription, in part because these polymerases lack tails.

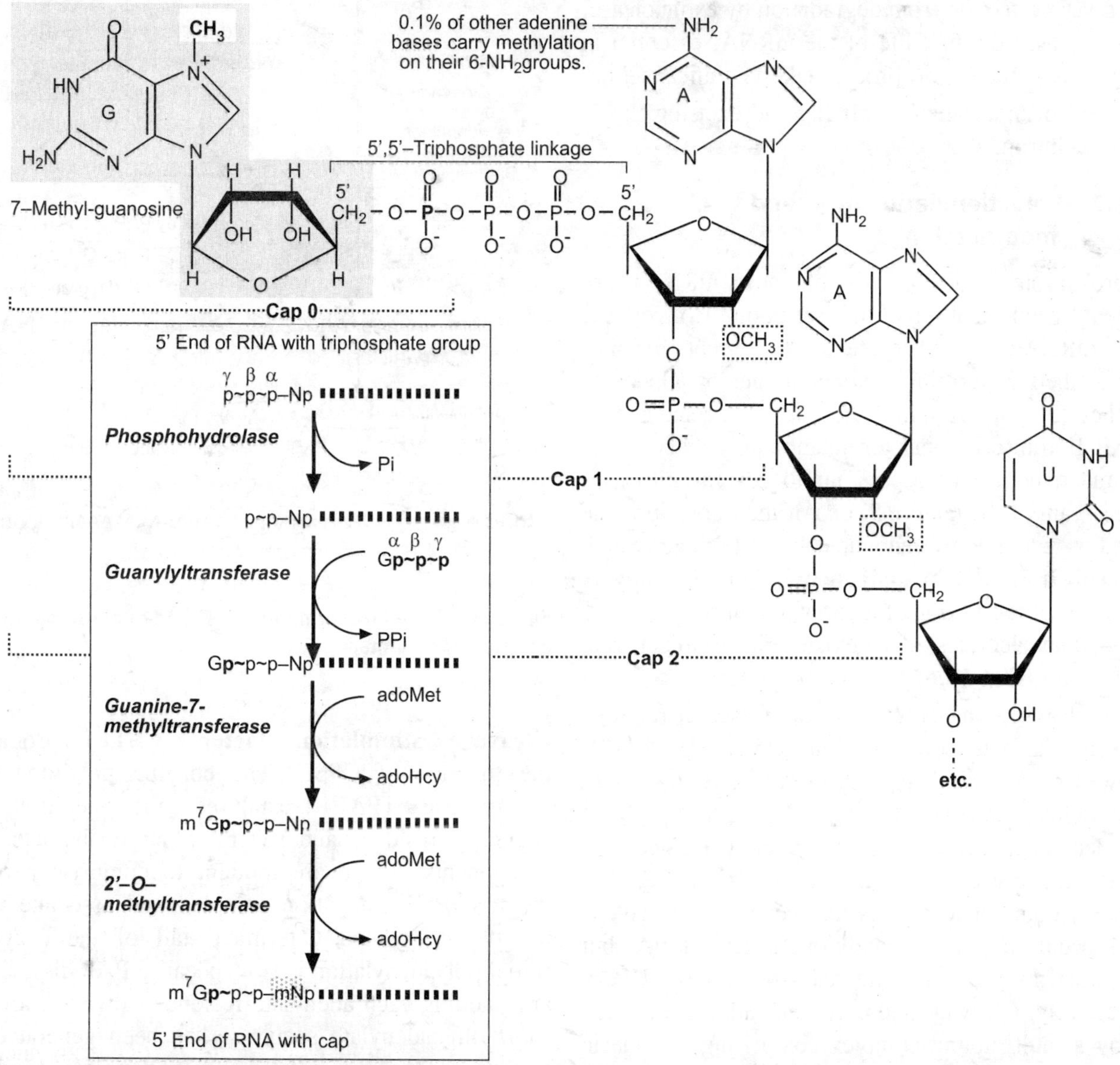

Fig. 10.5 The 5' cap of mRNA; methylation of several specific sites located at the 5'-end of eukaryotic pre-mRNAs is an essential step in mRNA maturation. Inside the box: Generation of the 5' cap involves four to five separate steps. adoHcy: S-adenosylhomocysteine

In the nucleus, the cap binds a protein complex called **CBC (cap-binding complex)**, which helps the RNA to be properly processed and exported. Although it has long been suggested that caps may function in the transport of mRNA from the nucleus, it is only recently suggested that the 5'-cap structure is essential for several RNA functions in eukaryotes. It is required for export through nuclear pores.

It is essential for efficient translation (ribosome binding and initiation of translation). The presence of the cap stimulates transformation of mRNA about 300-fold.

It gives protection from degradation by exonuclease. This increases the half-life of the mRNA, essential in eukaryotes as the export process takes significant time.

It is also important for splicing and excising the 5' proximal intron.

10.5.2 Polyadenylation or 3'-end modification

In prokaryotes, the 3'-ends of most mRNAs are formed by termination of transcription. Conversely, most mRNAs of eukaryotic cells except histones have at their 3'-termini a homopolymer of adenosine residues (3' poly A tail). The correct mechanism of RNAP II transcriptional termination in eukaryotes is not understood, and occurs up to several kilobases downstream of the mature 3' end of the transcript. The 3'-end is generated by endonucleolytic cleavage, which is usually followed by **polyadenylation** (the **poly A tail)*n***, i.e., the addition of a variable number (as many as 20–300 nucleotides) of adenylate residues to generate a polyA tail (**Fig. 10.6**).

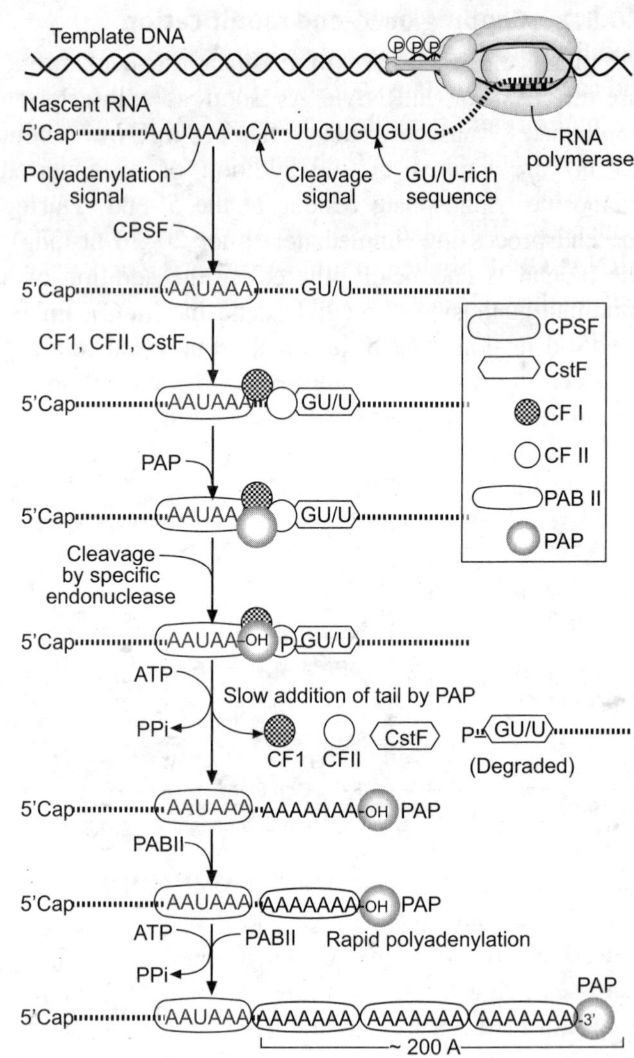

Fig. 10.6 Model for formation of 3'-end of mRNA by the addition of polyA tail

The cleavage and polyadenylation reaction requires a specific hexanucleotide sequence (5'-AAUAAA-3', followed by 5'-YA-3') present in pre-RNA transcript. It is highly conserved (5'-$A_{98}A_{86}U_{98}A_{98}A_{95}A_{96}$-3') in polyadenylated mRNAs and is typically located 15–25 residues upstream of the polyA addition site. One relatively common variant is the sequence AAUAAA, which occurs in approximately one in ten mRNAs but other variants are very rare and are processed very ineffectively. Cleavage and polyadenylation are carried out by a multisubunit complex comprising a trimeric **cleavage polyadenylation specificity factor** (**CPSF**) recognising the polyadenylation site—an endonuclease comprising two cleavage factors which carries out the cleavage reaction. Another multiprotein complex, **Cleavage stimulation factor** (**CSTF**) promotes the function of CPSF. The enzyme **polyadenylate polymerase** (**PAP**) catalyses the addition of adenylate residues and several other uncharacterised components. These are thought to comprise part of the RNAP II elongation complex and associate with the phosphorylated C-terminal tail of the enzyme. Initial polyadenylation is slow because PAP dissociates after adding each adenylate residue. However, after a short oligoadenylate sequence has been generated, a further component, **nuclear polyadenylate binding protein** (**PABPN**), attaches to the tail and increases the processivity of PAP.

The polyadenylation process (i.e., 3' end modification) takes place sequentially by many enzymes and accessory proteins: CPSF and CstF, both of which are multiprotein complexes, start binding to the rear of the advancing RNA polymerase II. As the RNA polymerase II advances over the adenylation signal sequences CPSF and CstF transfer to the new pre-mRNA, CPSF binding to the AAUAAA sequence, and CstF to the GU- or U-rich sequence following it.

CPSF and CstF promote cleavage (endonucleolytic cleavage) of the transcript approximately 35 nucleotides after the end of the AAUAAA sequence. This occurs while it is still being synthesised and probably with advancing RNAP at a considerable distance downstream. The cleavage generates a 3'–hydroxyl and a 5'-phosphate. Immediately, PAP starts writing the polyadenosine tail, adding adenosine residues one at a time to the former. Cleavage will not occur unless PAP is bound to the complex, preventing the chances of premature cleavage.

The exact role of polyadenylation is not clear. But it may influence stability of transcript, and in certain cases has been shown to play a critical role in translation. The *Drosophila bicoid* mRNA, for instance, is not translated until after fertilisation when three proteins facilitate the extension of the polyadenylate tail. Some eukaryotic transcripts are not polyadenylated, the most notable examples being the histone mRNAs and the genomes of certain plant viruses. While most bacterial mRNAs lack polyadenylate tails, short and relatively unstable oligoadenylate tails have been exploited for *in vitro* RNA manipulation or purification by using synthetic oligo-dT sequences which selectively hybridise to them. The cellular RNA purified in this manner consisting mostly of mRNA, is termed the **polyA$^+$ RNA fraction**; the remainder is **polyA$^-$ fraction**.

10.6 SPLICING

Another kind of RNA modification in eukaryotic primary transcripts is the removal of some internal segments of RNA sequences in primary transcripts. It is now conclusively established that many eukaryotic genes are mosaics of nucleotide sequences (split genes) that code for protein structure along with others that do not. For example, the gene for the ß chain of haemoglobin is interrupted within its amino acid coding sequence by a long intervening sequence of 550 base pairs and a short one of 120 base pairs. Thus, the ß-globin gene is split into three coding sequences.

Many chromosomal genes contain interruptions, which are called intervening sequences (IVS) or introns, and the sequences which code for proteins are called exons. There was a great deal of speculation about how the corresponding RNAs without introns could be generated. Many lines of evidence quickly converged to establish that RNA splicing must be the mechanism for removal of introns in both protein coding and RNA-specifying genes. There are four types of introns according to the splicing mechanism (**Table 10.2**). With the exception of the pre-tRNA introns, all splicing mechanisms involve a pair of sequential transesterification reactions. In the first reaction, a nucleotide carrying a free hydroxyl group attacks the phosphodiester bond joining the intron and the upstream exon. This releases the 3' end of the exon, which carries its own hydroxyl group. In the second reaction, the free OH–group attacks the phosphodiester bond joining the intron to the downstream exon. The second reaction joins the exons together and ejects the free intron, which is usually degraded. The three types of transestrification introns differ in their structure, the source of the initial hydroxyl group donor, the nature of the intermediate formed and whether the reaction is autocatalytic or requires a splicing apparatus to be supplied in *trans*.

10.6.1 Splicing of nuclear pre-mRNA introns

RNA splicing was discovered in mRNA synthesis in adenovirus. During these studies, the abundant viral mRNA molecules encoding the major virion capsid protein, called hexon, was isolated by gel electrophoresis of cytoplasmic polyadenylated RNA. While mapping the region of the viral DNA coding for hexon mRNA, researchers hybridised the isolated mRNA to the coding strand and the RNA–DNA hybrid was visualised in the electron microscope (**Fig.10.7**). Three loops of single-stranded DNA (A, B and C) were observed; these correspond to the three introns in the hexon gene. Since these intron sequences in the viral genomic DNA are absent in mature hexon mRNA, they loop out between the exon sequences that hybridise to their complementary sequences in the mRNA.

Similar analyses of hybrids between RNA isolated from the nuclei of infected cells and viral DNA revealed RNAs that were collinear with the viral DNA (primary transcripts) and RNAs with one or two of the introns removed. These results, together with the findings that the 5'-cap and 3'-poly A tail of mRNA

Table 10.2 Properties of different classes of introns and splicing mechanisms

Introns	*Characteristics of introns and splicing mechanism*
Transesterification introns: Splice recognition sites within intron	
Nuclear pre-mRNA introns	Structurally diverse, initial donation of free hydroxyl group made by adenosine residue at internal branch site; lariat intermediate formed; Large *trans*-acting splicing assembly, spliceosome complex required
Self-splicing introns class I	Conserved secondary structure, initial donation of free hydroxyl group made by guanosine nucleotide cofactor (GNP, GDP or GTP), no lariat formed; autocatalytic splicing
Self-splicing introns class II	Conserved secondary structure, initial donation of free hydroxyl group made by adenosine residue at internal site, lariat formed; autocatalytic splicing
Self-splicing introns class III	Similar to group II introns but smaller (100–120 bp); containing a restricted number of domains
Twintrons	Multiple embedded self-splicing introns, often group II or mixed group II and group III; usually spliced in a particular order
Non-transesterification introns: Splice recognition involves exon structure	
Nuclear tRNA introns	Splicing mechanism similar to tRNA maturation—involves cleavage followed by ligation; no intermediate formed—intron excised as linear fragment; several processing enzymes required in *trans*.

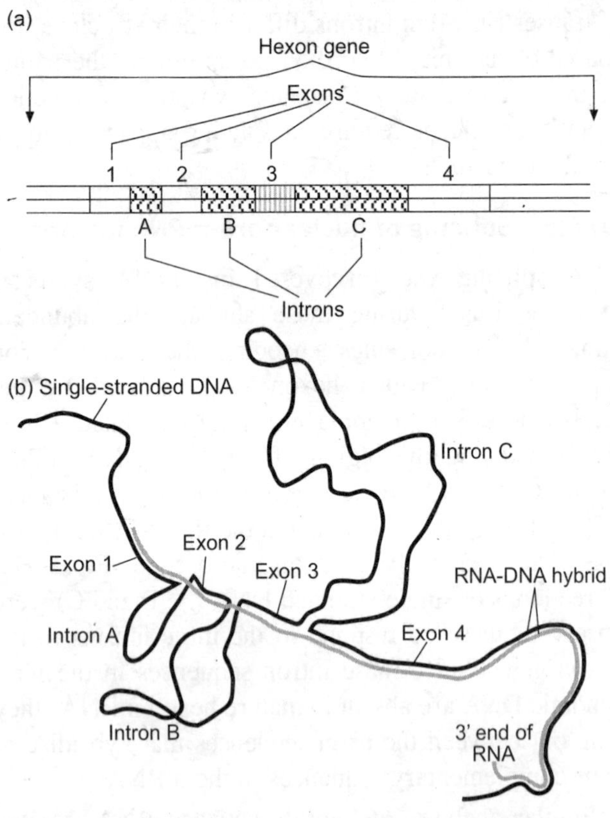

Fig. 10.7 RNA–DNA hybridisation reveals spliced-out introns.

precursors are retained in mature cytoplasmic mRNAs, led to the realisation that introns are removed from primary transcripts as exons are spliced together. For short transcription units, RNA splicing usually follows cleavage and polyadenylation of the 3' end of the primary transcript. But, for long transcription units containing multiple exons, splicing of exons in the nascent RNA sometimes begins before transcription of the gene is complete.

10.6.2 Splicing of pre-mRNA

One remarkable aspect of nuclear pre-mRNA splicing is the precise manner in which most genes are spliced. All splice sites are essentially the same, and therefore, any 5' site can in principle join any 3' site, but in most genes with multiple introns there is usually no **exon skipping** and only the principal splice sites, not cryptic sites, are used (**alternative splicing**). Investigation of splicing intermediates has revealed a preferred splicing pathway for many genes. This suggests that choice of the splice site may be an intrinsic property of the transcript, perhaps reflecting changes in the secondary structure as splicing occurs. That is, the order and specificity of splicing may be controlled by making illegitimate splice sites unavailable by sequestration into the secondary structure, but once a splice has been successfully completed, a change in conformation reveals further splice sites. The precise basis of recognition remains unknown, but commitment may be mediated by splicing factors known as **SR proteins** (because they are serine-/arginine-rich), which interact with both the U1 snRNP and U2AF.

10.6.3 Signals for removal of intron transcripts

Figure 10.8 shows the consensus sequence found at the 5' (**donor**) end, at the 3' (**acceptor**) end and the branch site (a run of pyrimidines, Y) in most introns. The donor and acceptor sites possess the highly conserved consensus sequences GU and AG respectively (the **GU–AG** rule or **GT–AG** rule when applied to the coding stand of the genomic DNA) embedded within a weaker consensus. The most striking observation, first made by Chambon, is that almost all introns in nuclear mRNA precursors begin and end in the same way: exon/GU–intron–AG/exon. However, a typical intron will contain several GUs and AGs within it. Why are these not used as splice sites? After analysis of many exon and intron boundaries, a consensus sequence has emerged: 5'-AG/GUAAGU-intron-YNCURAC-Y_nNAG/G-3' where the slashes denote the exon–intron borders, Y is either pyrimidine (U or C), Y_n denotes a string of about nine pyrimidines, R is either purine (A or G), <u>A</u> is a special A that participates in forming a branched splicing intermediate and N is any base.

The mutations in intron boundaries alter or delete them, which disrupt splicing. This has been well exemplified in ß-thalassaemias; a group of hereditary anaemias in which the production of ß-globin is either diminished ($ß^+$-thalassaemia) or absent (ß-thalassaemia). In one form of ß-thalassaemia, a mutation in the GGT splice point at the exon–intron boundary of the large intron of the gene caused no globin to be produced. As a result, processing of ß-globin transcripts was impaired. In $ß^+$-thalassaemia, defective globin synthesis was shown to be associated with a single G→A mutation which created a sequence CTATTAG within the large intron. The sequence closely resembled the normal intron–exon sequence CCGTTAG and competed with it as a splice point to the extent that 90% of transcripts were incorrectly spliced. It is clear that mutations in the GT and AG splice site sequences severely affect splicing and mutations near the GT may also have an effect. Further, a mutation in, for instance, the 5'-splice site may result in abnormal splicing such that the 3'-exon is ligated to a new 5'-site which may show strong homologies with the normal splice site but would not normally be used.

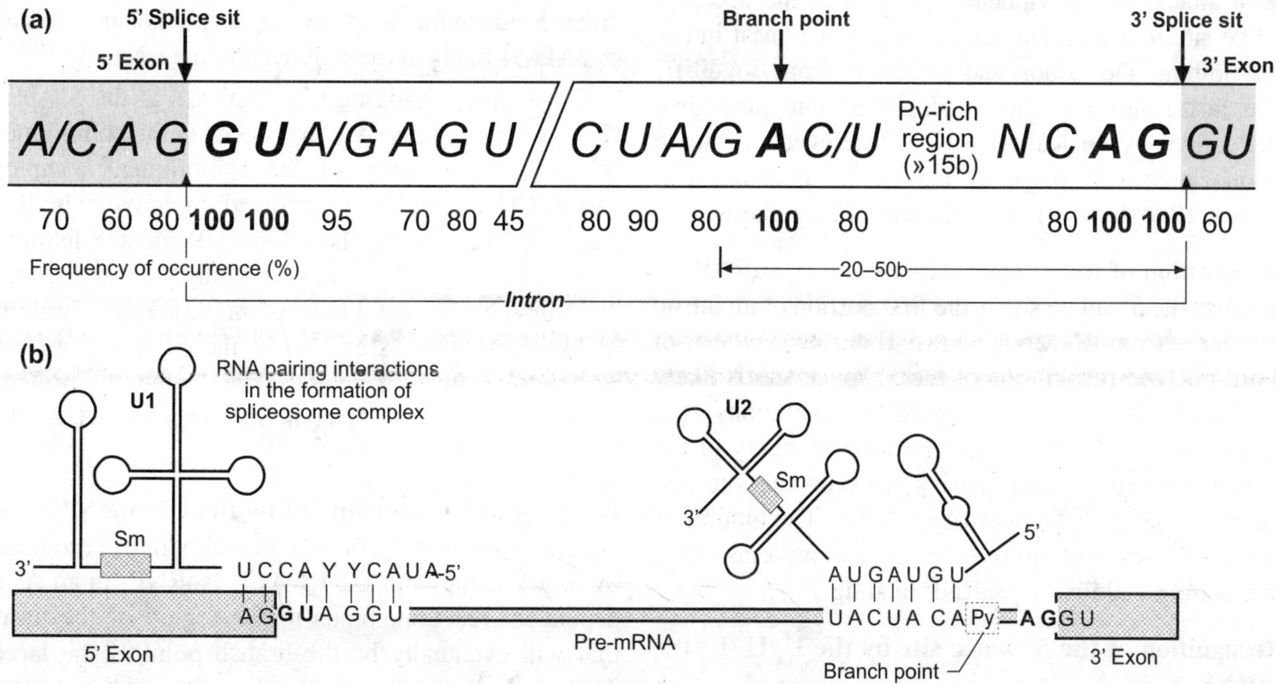

Fig. 10.8 (a) Consensus sequences around 5' and 3' splice sites (intron–exon boundary) in vertebrate pre-mRNAs, Py = pyrimidines (C or U) (b) U1 and U2 recognise specific, conserved splice sites (E1 and E3) on mRNA primary transcripts

10.6.4 Spliceosome assembly: Synthesis of holospliceosome

Incubation of synthetic pre-mRNA with yeast or human cell extract in the presence of ATP results in the production of splicing intermediates and products. When splicing reactions containing radioactively labelled pre-mRNA are subjected to non-denaturing gel electrophoresis, at least four distinct complexes (called the **commitment complex**, **pre-spliceosome**, **complete spliceosome** and **active spliceosome**) can be resolved. These complexes differ in their snRNP composition and order of appearance. All contain pre-mRNA, but only active spliceosomes contain products of the first and second catalytic steps of splicing. Based on these and other findings, a sequential model of spliceosome assembly was developed.

In the first step of splicing, the 2'-OH of the adenosine (A) at the branch site, which is located proximal to the acceptor site, attacks the phosphodiester bond at the donor splice site junction (i.e., upstream exon). The attack results in cleavage at the donor splice site and formation of a branched molecule known as a **lariat** because it has a covalently closed loop and a tail. The A–G linkage at the 'knot' of the lariat is usual in being 2'→5' (instead of the usual 3'→5'). In the final step of splicing, the 3'-OH of the guanosine of the donor exon attacks the phosphodiester bond at the acceptor splice site (downstream exon), freeing the lariat intron and joining the donor and acceptor exons together. The lariat intron is rapidly degraded into individual nucleotides by nucleases. All of these events occur in ribonucleoprotein particles known as **spliceosomes**. **Figure 10.9** shows the steps in the splicing pathway.

Recognition of the 5' splice site by the U_1 snRNP

Because the 5' splice site is the first portion of an intron to emerge from RNA polymerase II during synthesis of a pre-mRNA, recognition of the 5' splice site is likely to precede other snRNP–pre-mRNA interactions on a given intron. U_1 snRNP recognises and binds to the 5'-splicing site by base pairing between the site and the 5'-terminal nucleotides of U_1 RNA. The binding is rapid and does not require ATP. Several proteins–RNA interactions stabilise U_1 snRNP binding.

Recognition of the 5' splice site by the $U_4/U_5/U_6$ tri-snRNP

RNA cross-linking studies reveal an early interaction of U_5 RNA with the 5' splice site region while the U_1 snRNP is bound there.

Recognition of the branchpoint/3' splice site by BBP/SF1 and Mud2/U2AF

Immediately after a 5' splice site emerges from a transcribing RNA polymerase, the branchpoint and 3' splice site appears. In human pre-mRNAs, which tend to have multiple, long introns and short and short exons, branchpoint/3' splice site recognition is often coupled to U1 snRNP binding to the next downstream 5' splice site. This mechanism is known as **exon definition** and involves accessory splicing factors called SR proteins. The consensus branch point sequence is UACUA<u>A</u>C, where the underlined 'A' is the branch nucleotide. In yeast, this sequence is often preceded by two U residues and followed by an A residue. The 3' splice site consensus is much more limited, consisting of YAG in both yeast and mammals, where Y is pyrimidine base and the slash indicates the 3' splice site. In addition, many yeast introns have a pyrimidine-rich (more U) tract of 8–12 nt that ends one nucleotide upstream of the 3' splice site. The pyramidine-rich tracts are more ubiquitous and generally longer in human introns. The KH domain of BBP/SF1 specifically binds the branch residue and flanking nucleotides. Mud2 is the yeast homologue of the large subunit of U2AF, a heterodimeric human splicing factor that binds to the polypyrimidine tract and 3' splice site. The small subunit of U2AF (U2AF35) directly contacts the AG dinucleotide at the 3' splice site, while the large subunit (U2AF65) binds to the polypyrimidine tract.

Cross-intron bridging of complexes at the 5' splice site and branchpoint may first occur in the commitment complex. Two forms of the commitment complex, named CC1 and CC2 are resolved on native gels. Both the complexes contain the U1 snRNP and are disrupted by mutations in the 5' splice site; only CC2 is disrupted by mutations in the branchpoint sequence. Bridging of the complexes assembled at the 5' splice site and branchpoint in CC2 may be facilitated by protein–protein interactions between BBP and the U1 snRNP protein, Prp40.

Recognition of branchpoint by the U2 snRNP

After it is bound by BBP, the branchpoint is recognised by base pairing with U2 RNA. U_2 snRNP, in an ATP-dependent reaction, binds to the region of the intron that will eventually be the branch point of the lariat. Thus, at both the 5' and 3' splice sites, the conserved elements are initially recognised by ATP-independent factors and this is followed by the ATP-dependent and

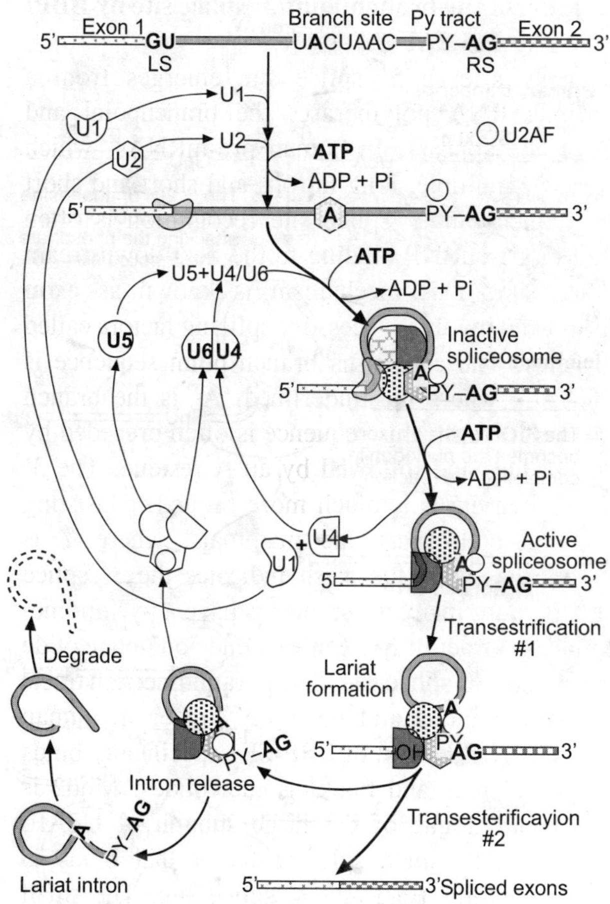

Fig. 10.9 Splicing mechanism in mRNA primary transcript

highly stable binding of U_2 snRNP in what may be the rate-limiting step in the spliceosome assembly. To bind productively to the branchpoint region, the U2 RNA must be in the correct conformation and be associated with the proper proteins.

Unwinding of U4/U6 and the switch of U6 for U1 at the 5' splice site

In the next step, a larger complex is formed, by the association of the above aggregates with a multi-snRNP particle containing U4, U5 and U6 snRNAs (**Fig. 10.9**).

Genetic evidence suggests that U6 RNA may transiently exist as a free RNA, forming a 3' **intermolecular stem loop** (ISL) that is thought to persist in the U2/U6 pairing, and a potential pseudoknot structure called the **telestem**. Prp19 is an essential splicing factor that associates with the spliceosome around the time U4 is unwound from U6 RNA. Prp19 is a complex with at least seven other 'Ntc' (Nineteen complexes) proteins.

The splicing reaction described above takes place in association with and as a result of the formation of the spliceosome complex and leads to the efficient and accurate removal of the intron transcript. Splicing results in the ligation of the two exons and the release of the intron as a lariat-shaped RNA in a complex with U_2, U_5, U_4 and U_6 snRNPs (**Fig. 10.9**).

The above splicing mechanism was verified by Sharp and his colleagues.

The phosphorus atom (in phosphodiester bond) between the two exons in the spliced product comes from the 3'-(downstream) splice site. Sharp and colleagues approached this problem by using a technique called **nearest neighbour analysis**. **Figure 10.10a** shows that the labelled phosphate that came into the RNA on a cytosine nucleotide has been passed to its nearest neighbour on the left, a uracil nucleotide in this case. **Figure 10.10b** shows how Sharp and colleagues modified the nearest neighbour method to show where the phosphate joining the two exons comes from.

(a) Nearest neighbour analysis:

1. Incorporation of (α–^{32}P) nucleotides:

pppApGpU–OH + pp****p***C-OH ⟶ pppApGpU****p***C–OH

2. Base hydrolysis:

pppApGpU****p***C-OH $\xrightarrow{OH^-}$ pppAp + Gp + U****p*** + C–OH

↓ Analyse labelled products

(b) Identifying origin of phosphate between spliced exons:

1. Incorporation of (α–^{32}P) nucleotides:

exon 1 — GpGpG****p*** ⋮ GpU — intron — ApG–OH + pp****p***C

exon 1 — GpGpG****p*** ⋮ GpU — intron — ApG****p*** ⋮ Cp — exon 2

2. Splice: exon 1 — GpGpG****p*** ⋮ Cp — exon 2

3. RNase A: GpGpG****p***Cp

4. Base hydrolysis: Gp + Gp + G****p*** + Cp

Fig. 10.10 To find the source of the phosphate between the two exons in a spliced transcript (a) Nearest neighbour analysis (b) Between spliced exons

The intermediate (exon 2 plus intron) and the spliced intron contain a branched nucleotide that has both its 2' and 3' hydroxyl groups bonded to other nucleotides.

Minor spliceosomal introns: The minor spliceosome is very similar to the major spliceosome. However, it splices rare introns with different splice site sequences. Here, the 3' and 5' splice sites are AU and AC respectively. While the minor and major spliceosomes contain the same U5 snRNP, the minor spliceosome has different, but functionally analogous snRNPs for U1, U2, U4 and U6, which are respectively called U11, U12, U4atac' and U6atac.

Trans-splicing: Generally, all functional mRNAs of vertebrates and insect cells are derived from a single pre-mRNA by removal of introns and splicing of exons. In contrast, in trypanosomes and euglenoid protozoans, mRNAs are constructed by splicing together separate RNA molecules. This process, called trans-splicing, is also used in the synthesis of 10%-15% of the mRNAs in the roundworm, *C. elegans*. In both cases, a single exon is spliced onto the 5' end of many different RNA transcripts produced by the cell. Thus, all the products of trans-splicing have the same 5' but different 3' exons.

10.6.5 Self-splicing introns

Self-splicing occurs for rare introns that form a ribozyme, performing the functions of the spliceosome by RNA alone without the need of proteins. There are two types of self-splicing introns: Group I and Group II.

Group I self-splicing introns

Group I introns are found in diverse locations. They are found in mitochondrial genomes and, more rarely, in the nuclear genomes of unicellular eukaryotes. They are present in three genes of phage T4 and also found in bacteria (e.g., in the bacteriophage T4 thymidylate synthase gene). Group I introns have an intrinsic ability to splice themselves (called **self-splicing** or **autosplicing**). Splicing activity is intrinsic to the precursor rRNA molecule itself. In this fascinating mechanism, introns of the 413-nucleotide are excised and circularised by a series of phosphoester transfer reactions (**Fig. 10.11**). The self-splicing reaction requires only a monovalent cation, a divalent cation, and a guanine nucleotide cofactor (GTP, GDP, GMP and guanosine itself). No other base can be substituted for G.

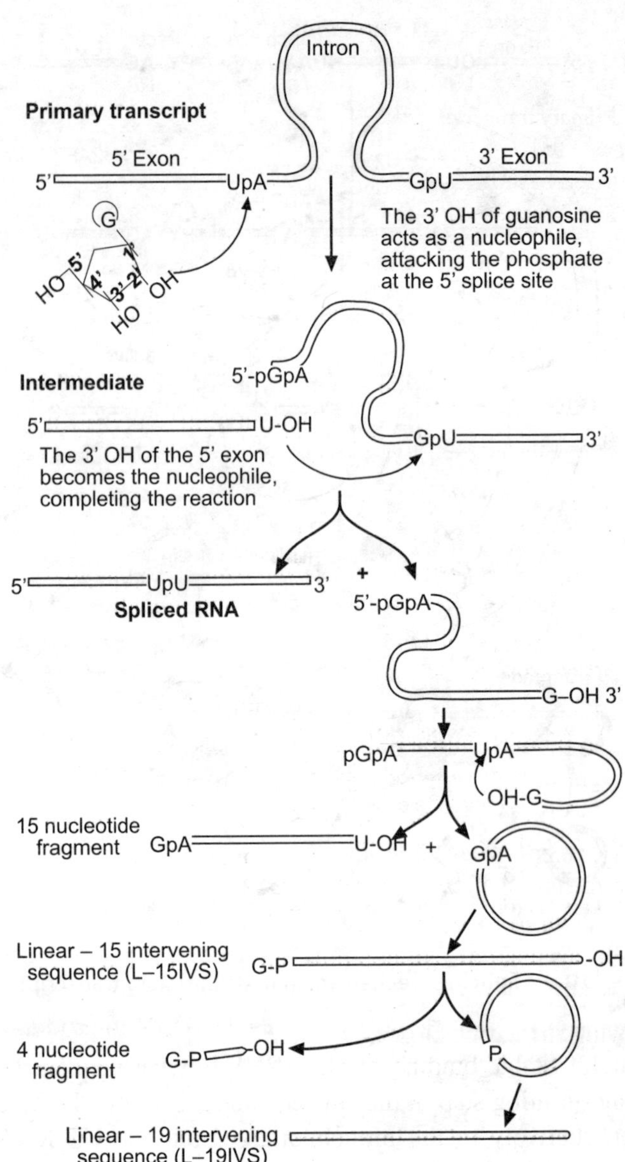

Fig. 10.11 Self-splicing mechanism of group I introns

Group-II self-splicing introns

Group II introns are found in genomes of plant and lower eukaryotes. Like group I introns, they splice autocatalytically, but the splicing mechanism is closely related to that of pre-mRNA introns. Splicing is initiated by the 2'-OH group of an internal adenosine residue which attacks the phosphodiester bond linking the intron and upstream exon, generating a lariat and a free exon with a terminal 3'-OH group (**Fig. 10.12**). This group then attacks the phosphodiester bond linking the intron to the downstream exon in a second transestrification reaction which joins the exons together and releases a lariat intermediate. The autocatalytic activity of

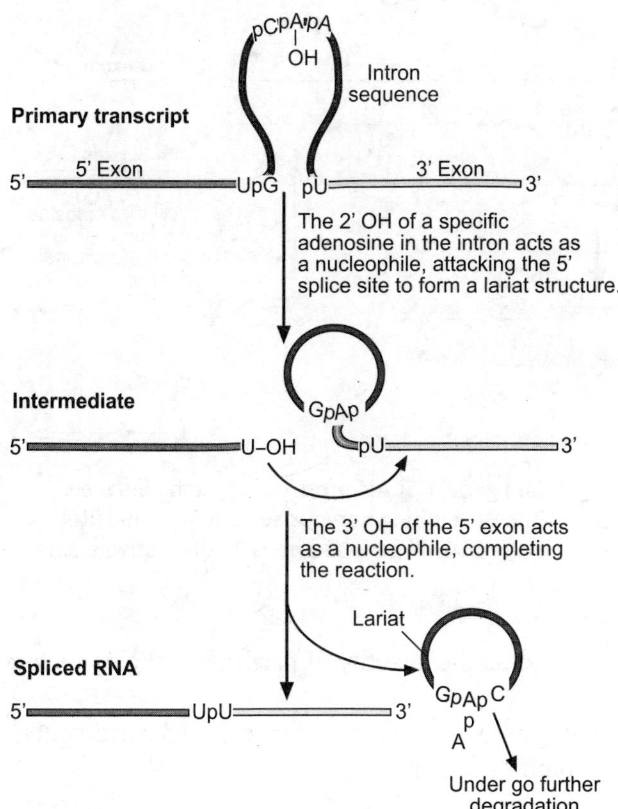

Fig. 10.12 Autocatalytic mechanism of the self-splicing intron (group II) from *Tetrahymena* rRNA

group II introns arises from a characteristic secondary structure comprising six stem loop domains which bring the exons closer.

10.7 ALTERNATIVE OR DIFFERENTIAL SPLICING

Humans may have only 25,000 to 30,000 genes, but probably make at least 10 times that number of different proteins. More than 50% of genes produce pre-mRNAs that are alternatively spliced. Splicing can also generate diversity in RNA molecules, and consequently, in proteins. The processing of pre-mRNA for many proteins proceeds along various paths in different cells or under different conditions. For example, early in the differentiation of a B cell, the cell first uses an exon that encodes a transmembrane domain that causes the molecule to be retained at the cell surface. Later, the B cell switches to using a different exon whose domain enables the protein to be secreted from the cell as a circulating antibody molecule. Thus, a single gene can produce multiple products by alternative splicing.

In some cases, the number of potential alternatives that can be generated from a single gene is hundreds (e.g., in the rat *Slo* gene) or even thousands (e.g., the *Drosophila* DSCAM gene).

Alternative splicing not only provides different proteins from a single gene but also different 3' UTRs and 5' UTRs. Although not translated into protein, these untranslated regions contain signals that, for example, dictate the point inside the cell where proteins can accumulate. The family of related mature transcripts generated produces different gene products, which may be termed **splice variants** or **splice isoforms**. Alternative splicing occurs in several ways (**Fig. 10.13**).

- Omission of 5' exon(s) by using alternative promoters
- Omission of 3' exon(s) by alternative polyadenylation at the site
- Choosing between 5' or 3' splice sites while defining exons
- Choosing between 5' or 3' splice sites to skip an exon

(a) Alternative selection of promoters (e.g., myosin primary transcript)

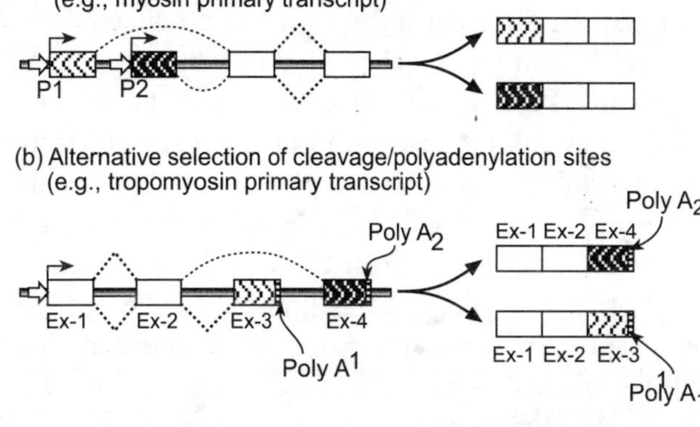

(c) Intron retaining mode (e.g., transposase primary transcript)

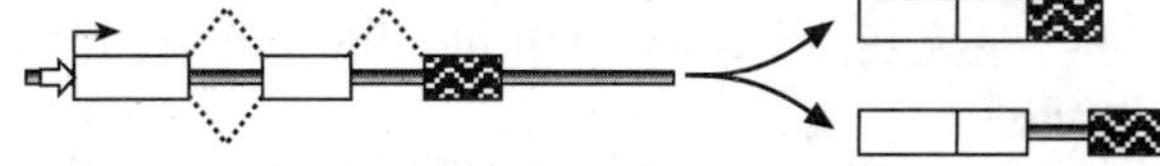

(d) Exon cassette mode (e.g., troponin primary transcript)

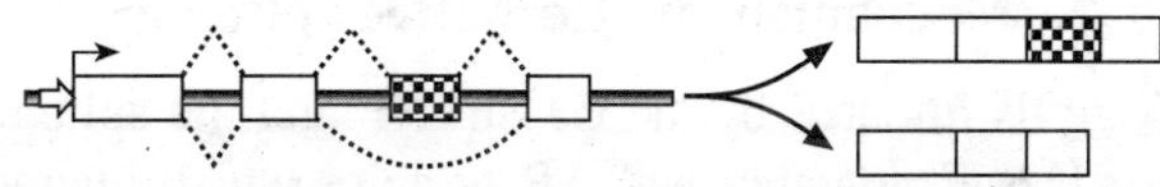

Fig. 10.13 Different modes of alternative splicing of primary transcripts (a) Alternative selection of promoters (b) Alternative selection of cleavage sites (c) Intron retention of exon cassette mode

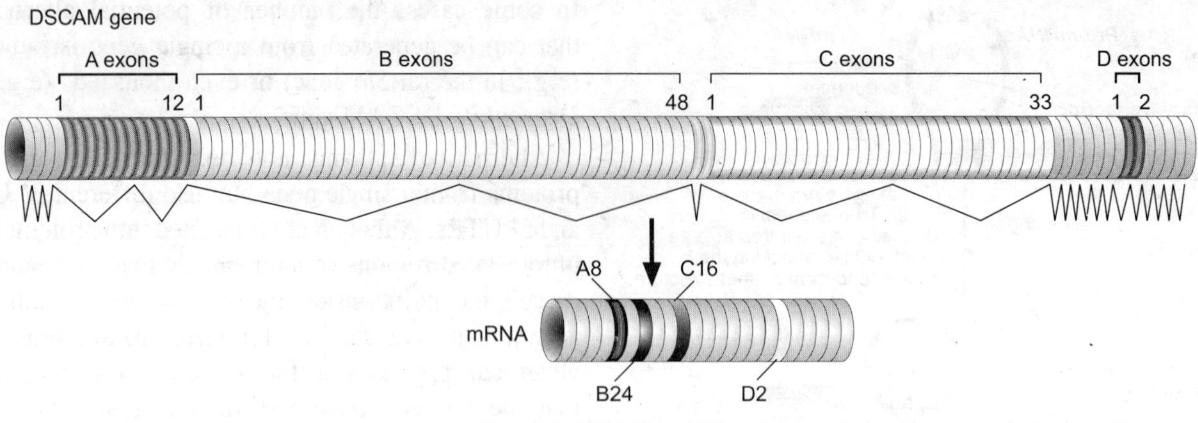

Fig. 10.14 Alternative splicing of RNA transcripts of the *Drosophila* DSCAM gene; DSCAM proteins are axon guidance receptors that help to direct growth cones to their appropriate targets in the developing nervous system. The final RNA contains 24 exons, four of which (denoted by A, B, C, and D) are present in the DSCAM gene as arrays of alternative exons (***Source***: Alberts et al 2002)

Mutual exclusion of exons in DSCAM gene of *Drosophila*

The single gene contains some 116 exons of which 17 are retained in the final mRNA. Some exons are always included; others are selected from an array. The vast number of different alternative exons found in the DSCAM gene of *Drosophila* could result in over 38,016 different mRNAs being produced, which is greater than the number of different genes in this organism (**Fig. 10.14**). The DSCAM proteins are involved in guiding neurons to their proper destination and probably recognition and phagocytosis of invading bacteria. Perhaps the incredible diversity of synaptic junctions in the mammalian CNS ($\sim 10^{14}$) is mediated by alternative splicing of a limited number of gene transcripts. So whether a particular segment of RNA will be retained as an exon of excised as an intron can vary under different circumstances, such as:

- The type of cell the gene is in
- The stage of differentiation that the cell is passing through
- The extracellular signals that the cell is receiving.

10.7.1 Mechanism of alternative splicing

Splicing is initiated by the U1 snRNP and the splicing factor U2AF, together with SR proteins which interact with them. The alternative splicing mechanism directs these splicing components to specific sites of the pre-mRNA, either by stimulating or inhibiting the use of particular splice sites. There are two methods of controlling the splicing process:

- Based on the quantity of particular splicing factors, and
- In a tissue-specific manner with dedicated splicing control proteins.

Quantitative control often occurs where alternative 5' or 3' sites are used for differential splicing. For example, analysis of differential splicing in the SV40 T/t antigen gene identified an alternative splice factor (ASF), and SR protein, which was shown to be identical to the splice factor SF2. Subsequently, it was shown that where alternative 5' splice sites are available, higher concentrations of ASF2/SF2 favour the use of the most proximal site, whereas lower concentrations favour the use of the most distal site. A second factor, hnRNPA1, has the opposite effect. Differences in the relative levels of the two factors in different cells can explain cell type-specific differences in the predominance of the two splice isoforms. However, the quantitative model cannot adequately explain situations where splice isoforms are produced in a strict cell type-specific manner. Many situations where this occurs have been described in eukaryotes. In *Drosophila*, the most common example of alternative splicing is the hierarchy of splice control signals controlling sex determination (**Fig. 10.15**).

Exon shuffling

All eukaryotes have introns but these elements are rare in bacteria. The two likely reasons are as follows.

Introns early model: According to this model, introns existed in all organisms but the bacteria lost them

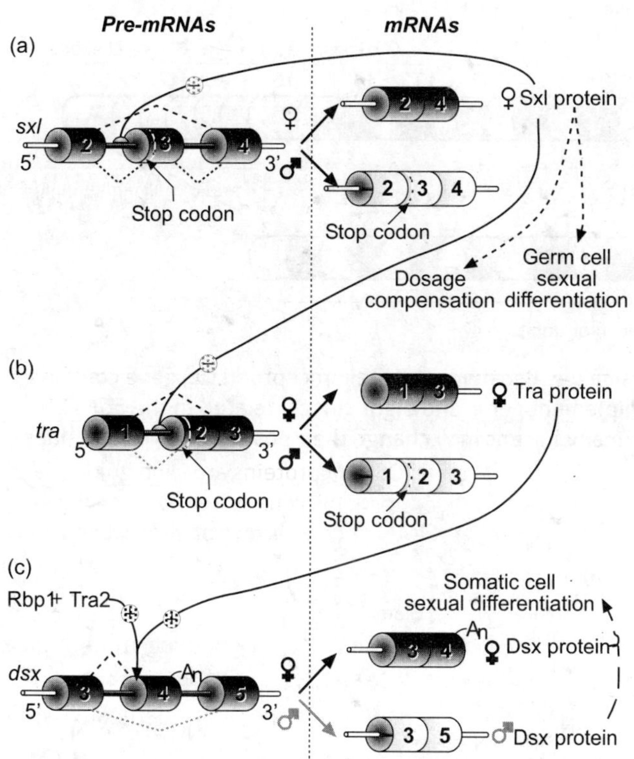

Fig. 10.15 Series of regulated splicing that controls sex determination via expression of sex-lethal (*sxl*), transformer (*tra*) and double sex (*dsx*) genes in *Drosophila*. The cylinders represent the exons and the region between the introns are shown by solid lines. Splicing is indicated by dashed lines above for females and dashed lines below for males in the pre-mRNAs. Dotted lines in the exons indicate in-frame stop codons. Only female embryos produce functional Sxl protein, which represses splicing between (a) exons 2 and 3 in *sxl* pre-mRNA (b) between exons 1 and 2 in *tra* pre-mRNA. In contrast, the cooperative binding of Tra protein and two SR proteins (Rbp1 and Tra2) activates splicing between exons 3 and 4 and cleavage/polyadenylation (An) at the 3′ end of exon 4 in *dsx* pre-mRNA in female embryos. In male embryos, which lack functional Tra, the SR proteins do not bind to exon 4, and consequently exon 3 is spliced to exon 5. The distinct Dsx proteins produced in females and male embryos as the result of this cascade of regulated splicing represses transcription of genes required for sexual differentiation of the opposite sex. (***Source:*** Lodish et al 2004)

because these 'gene-rich' organisms have streamlined their genomes in response to selective pressure to increase the rate of chromosome replication and cell division.

Introns late model: This model states that introns never existed in bacteria but arose later in the evolution of eukaryotes. Introns might have been inserted into genes, probably by a transposon-like mechanism.

So, the important question is why have the introns been retained in eukaryotes? Retaining introns and being able to remove them, facilitates alternative splicing which can generate multiple protein products from a single gene. Further, having the coding sequences of genes divided into several exons allows new genes to be created by **reshuffling** the exons (**Fig. 10.16**). Three observations strongly support this process:

- The borders between exons and introns within a given gene often coincide with the boundaries between domains, i.e., each exon very often encodes an independently folding unit of protein.
- Many genes and the proteins they encode have apparently arisen during evolution in part via exon duplication and divergence. For example, proteins with repeating units like the immunoglobulins, have probably arisen this way.
- Related exons are sometimes found in otherwise unrelated genes. That is, there is evidence that exons really have been reused in genes encoding different proteins. For example, LDL receptor gene contains some exons that are evolutionarily related to exons found in the gene encoding the EGF precursor.

Significance of RNA splicing

RNA splicing plays multiple roles in biological systems.

Evolutionary implications: Since exons often coincide with protein 'domains', they are parts of the protein with a specific function. Exons can be readily 'exchanged' between different genes by recombination. This means that new types of proteins can be formed relatively easily.

Swapping exons during gene expression: For example, during development, some genes are spliced one way, and then spliced a different way later on. Changing the way an mRNA is spliced changes the amino acid sequence in the protein made from it, so cells can 'modify' the sequence and function of a protein. Splicing thus offers yet another opportunity for regulation of gene expression in eukaryotic cells. It also regulates the making of a specific version of a protein.

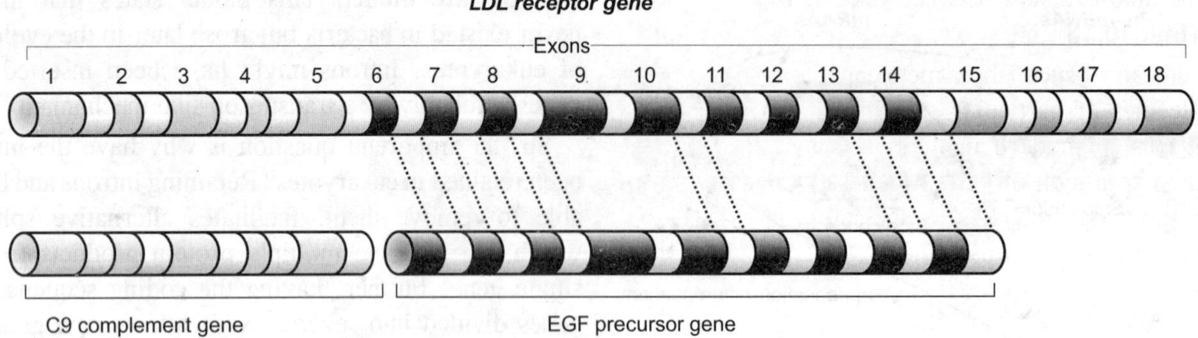

Fig. 10.16 Genes reconstituted by parts of other genes; the plasma low density lipoprotein receptor (LDL) gene contains a stretch of six exons closely related to six exons from the *C9* complement gene, and eight closely related to the *EGF* (epidermal growth factor) precursor gene. In *EGF* precursor gene, many introns may change their position. The dotted lines in the diagram indicate the introns are in the same position.

10.8 SELF-CLEAVING RNA SEQUENCES

Splicing is catalysed largely by RNA molecules with the associated proteins playing only a secondary role. The self-cleaving RNA sequence first discovered in plant viruses has recently been found in many higher organisms including vertebrates. Introns are also present in some genes in organelles. Since they do not contain spliceosomes, the splicing mechanism is very different. For example, in some organelles, the intron contains a sequence coding for a protein that participates in removing it.

The splicing mechanism in the ciliate *Tetrahymena* is intrinsic to the folding of the precursor; that is, the RNA precursor is self-splicing (**Fig. 10.17**). The self-splicing *Tetrahymena* RNA was the first example found of an RNA molecule that could function as an enzyme in catalysing a chemical reaction; such enzymatic RNA molecules are called **ribozymes** (**catalytic RNA**s). In this case, the elimination of the intron requires, in addition to the RNA itself, a guanosine nucleotide (G, denoted GMP, GDP and GTP any one) as the only factor. The 3'-OH group of GMP attacks the phosphodiester bond corresponding to the 5'-end of the intron (5' splice site) and due to a transestrification reaction displaces the 3' end of the exon. The free OH–group of the latter then attacks the phosphodiester bond of the 3' end of the intron, causing its cleavage. This second transestrification reaction joins the two exons and leads to the release of the 414 nt intron.

In the Lucerne Transient Virus (LTSV) the auto-excision of the RNA (+ strand) of a virusoid takes place. This RNA of 324 nucleotides can take a **hammerhead shape** by pairing of complementary bases, which

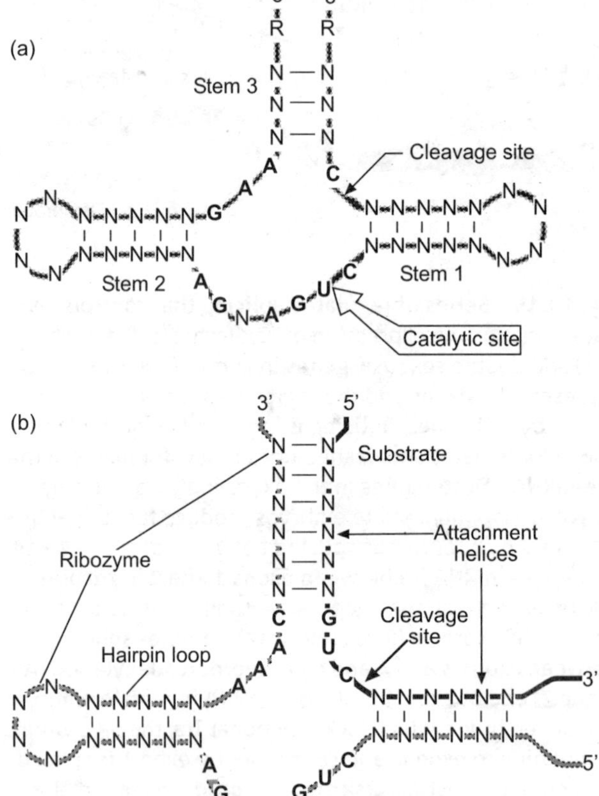

Fig. 10.17 (a) A consensus sequence of hammerhead structure; the bases in bold are conserved among various hammerheads and are necessary for activity. The R groups indicate a small section of a larger RNA molecule which has folded up into the hammerhead structures at the cleavage site. (b) Simplified diagram of an artificial ribozyme (gene shears); after cleavage, the two halves of the substrate can separate, enabling the ribozyme to attack another substrate molecule. The ribozyme acts in *trans* in a truly catalytic manner. The presence of GUC sequence of the substrate is important and the ribozyme's base sequence is designed to be complementary to the bases of the substrate which form the attachment helices.

permits the auto-excision between the nucleotides 167 and 168 (**Fig. 10.18**).

The synthesis of such ribozymes, capable of cleaving RNA molecules at particular sequences, can play an important role in modern medicine because they can prevent the expression of certain genes (by destroying the corresponding mRNAs) especially in viruses pathogenic for plants, animals and humans.

Introns: Functions and evolution

Since the discovery of introns, scientists have been trying to understand their functions. Walter Gilbert suggested that introns separate exons into functional domains; that is, they separate different exons that presumably have specific tasks. In a given protein, one exon might code for a membrane-binding region, another for the active site of the enzyme, and a third one might code for ATPase activity. By recombinant mechanisms, or by excluding an exon during intron removal, exon shuffling would allow the rapid evolution of new proteins whose structures would be conglomerates of various functional domains (**Fig. 10.19**). Darnell and Doolittle have expanded the early view of exon shuffling, and suggested that introns arose before the first cells evolved.

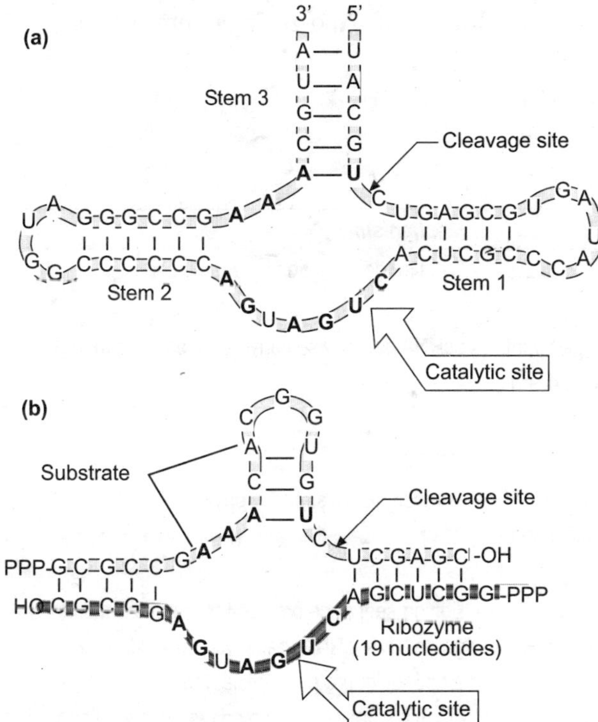

Fig. 10.18 Ribozyme structures (a) Auto-excision (at the place indicated by arrow) of the RNA (324 nt) of a Lucerne Transient Streak Virusoid (LTSV) having a hammerhead structure; this permits the auto-excision between the nucleotides 167 and 168. (b) Hammerhead structure (synthesised from Avocado Sunblotch Viroid, ASBV) resulting from the pairing of a ribozyme (the lower RNA strand comprising 19 nucleotides) with an RNA cleaved at the position indicated by the arrow

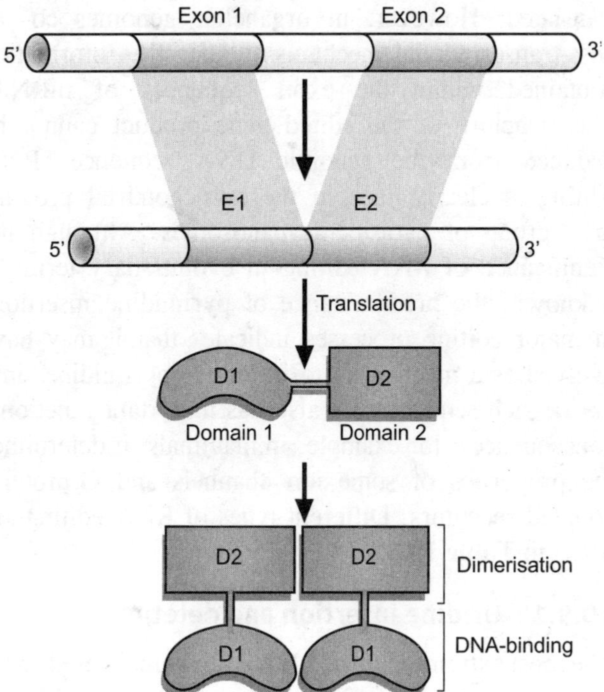

Fig. 10.19 Exons encode protein domains. In this case, the DNA-binding domain of protein is encoded by one exon, while the dimerisation domain of that same protein is encoded by a separate exon. This mechanism suggests that exons can be exchanged between proteins productively. (*Source*: Watson et al 2004)

10.9 RNA PROCESSING BY EDITING

Events other than splicing during RNA processing that can alter the protein coding sequences of some mRNAs are referred to as editing. It alters the mRNA sequence post-transcriptionally. The mechanisms of RNA editing include nucleoside modifications such as C to U and A and I deaminations, as well as non-templated nucleotide additions and insertions.

Normally the information in a protein-coding gene, except removal of introns by RNA splicing without affecting the exons, is unaltered before it has been translated. However, in organellar genomes co- or post-transcriptional mechanisms alter the information contained within the exon sequences of mRNA. The structure of the edited gene product cannot be deduced from the genomic DNA sequence. RNA editing is clearly seen in the mitochondrial proteins of a group of parasitic trypanosomes. Although the significance of RNA editing in evolutionary terms is unknown, the predominance of pyrimidine insertions in major editing processes indicates that it may have evolved as a mechanism to introduce pyrimidines into purine-rich sequences. It also has important functional consequences; for example, in mammals, it determines the properties of some ion channels and G-protein-coupled receptors. Different types of RNA editing are listed in **Table 10.3**.

10.9.1 Uridine insertion and deletion

The most extreme form of RNA editing uncovered so far occurs in the mitochondria of trypanosomes, which was discovered by Benne et al (1986) in the mitochondria of the kinetoplastid flagellate, *Trypanosoma brucei*. The discovery of RNA editing came from the realisation that the nucleotide sequences of a number of mRNAs from trypanosomes species differed from that of the maxicircle genes from which they transcribed. These genes, now known as cryptogenes, were transcribed into unedited pre-mRNA and that editing gave rise to mature message (**Fig. 10.20**). Thus, a total of four uridines were inserted into one region of the mRNA encoding cytochrome oxidase subunit II and it appeared that the process might involve a relatively simple sequence recognition event by an editing enzyme.

10.9.2 Base modification during editing

RNA editing regulates protein function in mammals. A classic example of site-specific conversion of C to U residues has been described in mammalian apolipoprotein B (**Apo-B**) mRNA (**Fig. 10.21**) and various plant mRNAs. The *apo-B* gene encodes two alternative forms of Apo-B, a serum protein. The

Table 10.3 Different mechanisms of RNA editing

Editing process	*Properties*	*Mechanism*
Simple editing (post-transcriptional)	Single residue conversions: C→U transition in the mammalian apolipoprotein mRNA	Cytidine deamination by deaminase
	A→G transition in mRNA for mammalian glutamate receptor subunits and serotonin receptors; editing also occurs in the introns.	dsRNA deaminase converts adenosine to inosine.
	C→U transition in plant organelles	Not clear
Insertional editing (co-transcriptional)	Insertion of single nucleotides or small runs of nucleotides, e.g., G insertions during transcription of the paramyxovirus P gene	Transcriptional strand slipping
Pan-editing (post-transcriptional)	Insertion/deletion of multiple uridine residues e.g., U insertions/deletions in the kinetoplast mRNA of trypanosomes	Editing sequence provided by external antisense guide RNA (gRNA) pairs with pre-edited mRNA in a ribonucleoprotein particle, the editosome, and identifies positions to be edited as mismatches. The gRNA has polyuridylate tail, which supplies uridine residues for insertion. Editing has 3'→5' polarity
	Insertion of multiple cytidine residues, e.g., C insertions in at least four mitochondrial mRNAs of *Phisarum polycephalum*	Unknown
Polyadenylation editing (post-transcriptional)	Addition of adenosine residues at the end of the transcript to complete stop codons, e.g., polyadenylation of several vertebrate mitochondrial mRNAs	Since the pre-mRNA lacks a stop codon, it is polyadenylated, with the first one or two adenylate residues providing the missing information.

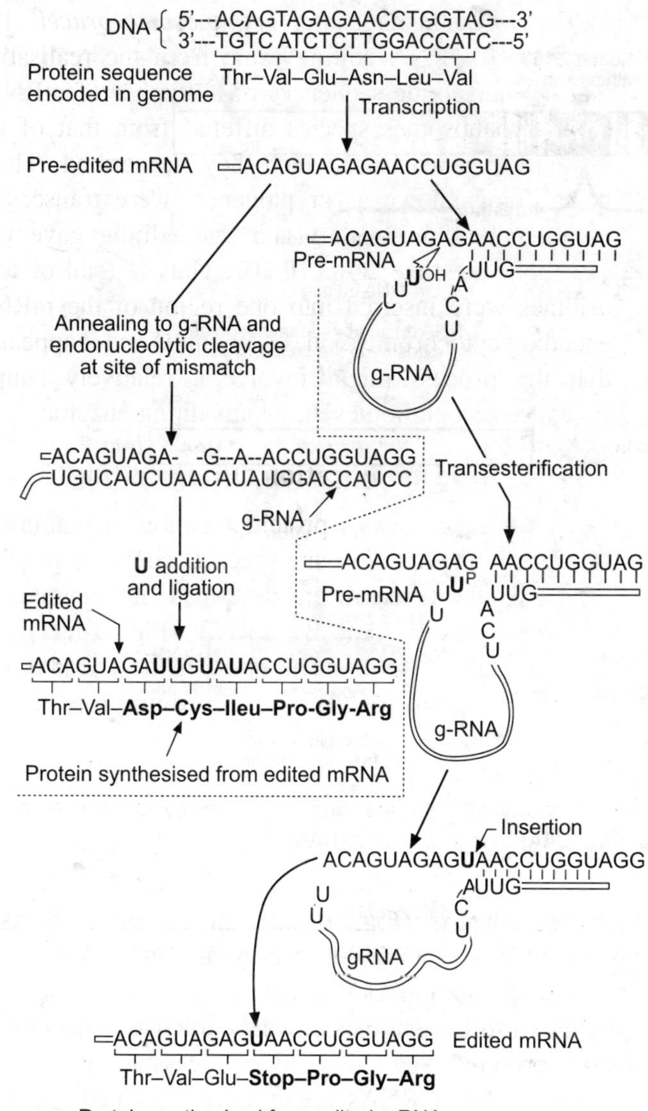

Fig. 10.20 Two models for the role of guide RNAs (g-RNA) in the insertion of a single uridine into pre-mRNA

cell type-specific expression of these two forms in hepatocytes and intestinal epithelial cells is regulated by mRNA editing. In hepatocytes, a 512-kDa protein called Apo-B100 is expressed whereas Apo-B48 is expressed in intestinal epithelial cells. Both Apo-B proteins transport lipids in the serum in the form of large lipoprotein complexes. However, only the large protein made in the liver delivers cholesterol to body tissues by binding to the low-density lipoprotein (LDL) receptor present on all cells. Comparison of Apo-B mRNA from hepatocytes and intestinal cells showed that nucleotide 6,666 of the mRNA was changed from the C encoded in the DNA of both cell types to a U in the mRNA from intestinal cell. This change, which converts a CAA codon for glutamine to a UAA stop codon, results in synthesis of the shorter Apo-B48 in the intestinal cells.

10.9.3 Other mechanisms

A number of other modifications that alter the coding potential of mRNA have been described in various organisms. The addition of guanosine residues occurs in specific internal sites in the mRNA of paramyxoviruses. The mechanism of this is uncertain but some evidence suggests that it involves RNA polymerase 'stuttering' when transcribing a stretch of template homopolymer. Thus, the modification may be transcriptional rather than post-transcriptional.

10.9.4 Significance of RNA editing

RNA editing like alternative splicing, probably, is a mechanism to increase the number of different proteins available without the need to increase the number of genes in the genome. So, it can create proteins with slightly different functions to use in specialised circumstances. It has accomplished the same results as alternative splicing. RNA as well as DNA editing may help protect the genome against some viruses (which they often evolve as countermeasures) and damage by retrotransposons. There is also evidence that RNA editing of precursor mRNA is a signal to retain them within the nucleus ready to be quickly exported, if needed by the cell. Defects in RNA editing are associated with some human cancers as well as with amyotrophic lateral sclerosis (ALS).

10.10 DEGRADATION OF RNA

The processing steps so far discussed in this chapter result in the formation of mature functional RNAs. These RNAs are then transported to the cytoplasm and execute their functions. Over 90% of hnRNA sequences are introns, which are degraded within the nucleus following their excision by splicing. This degradation process is catalysed by an enzyme that can recognise the unique 2'→5' bond formed at the branching point, as well as either the 5' or 3' unprotected ends of RNA molecules in either direction. In addition to degrading introns, cells possess a quality control system (called **nonsense-mediated mRNA decay**) that leads to the degradation of mRNAs that lack complete open reading frames. This eliminates defective mRNA molecules and prevents the synthesis of abnormal truncated proteins.

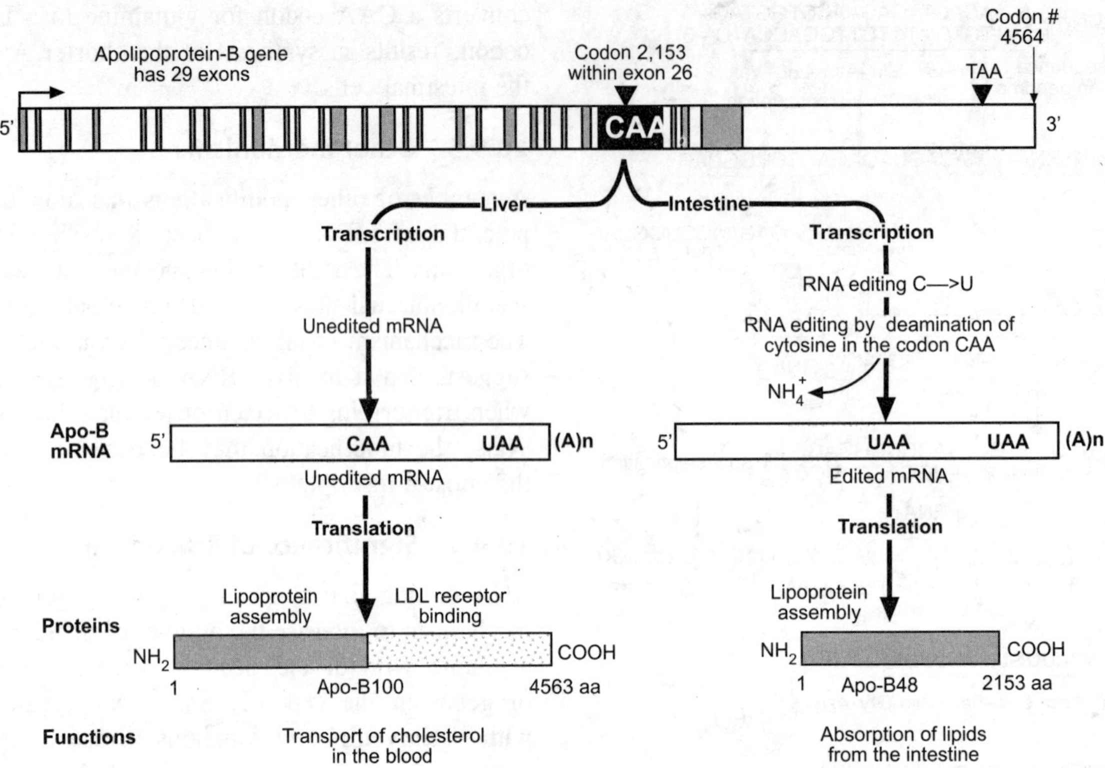

Fig. 10.21 RNA editing of Apo-B mRNA by base modification. Enzyme-catalysed deamination of a specific cytidine residue in the mRNA for apolipoprotein B-100 changes a codon for glutamine (CAA) to a stop codon (UAA)

The nonsense-mediated mRNA decay takes place in the cytoplasm when a premature termination codon is encountered by a ribosome during protein synthesis.

Since, the intracellular level of any RNA is determined by a balance between synthesis and degradation, the rate at which individual RNAs are degraded is another level at which gene expression is regulated. The kinetics of mRNA turnover is very rapid, which facilitate cells to respond quickly to alterations in its environment. In eukaryotic cells, different mRNAs are degraded at different rates. The half-lives of mRNAs in mammalian cells vary from less than 30 minutes to about 20 hours.

10.10.1 Regulation of RNA processing

As described in previous chapters, conversion of an RNA transcript at the 5' end into a functional mRNA involves two primary steps:

- Cleavage and polyadenylation at the 3' end and
- Ligation of exons with the concomitant excision of introns, or RNA splicing.

The expression of both simple and complex transcription units could be controlled by on-off regulation of the cleavage and polyadenylation of their pre-mRNAs (**Fig. 10.22**). In the pre-mRNAs produced from complex transcription units, cleavage and polyadenylation at different poly A sites and/or splicing of different exons yields mRNAs encoding distinct proteins. Such alternative processing pathways of complex pre-mRNAs are usually regulated, often in a cell type-specific manner. Since about 5% of all transcription units in higher eukaryotes are complex units, post-transcriptional regulation is a significant gene control mechanism in such organisms (see **Chapters 14** and **15**).

10.10.2 Transport of mRNA

RNA is prepared for export while processing goes on. Once all the processes (capping, intron splicing and polyadenylation) are completed, mRNA is transported out of the nucleus through nuclear pore complex into the cytoplasm (**Fig. 10.23**) where it is translated to produce its protein product. Movement of mRNA from the nucleus to the cytoplasm is an active process. Transport of mRNAs has to be very selective, since nucleus contains, for example, damaged or misprocessed RNAs, a large number of liberated introns and many other RNA molecules.

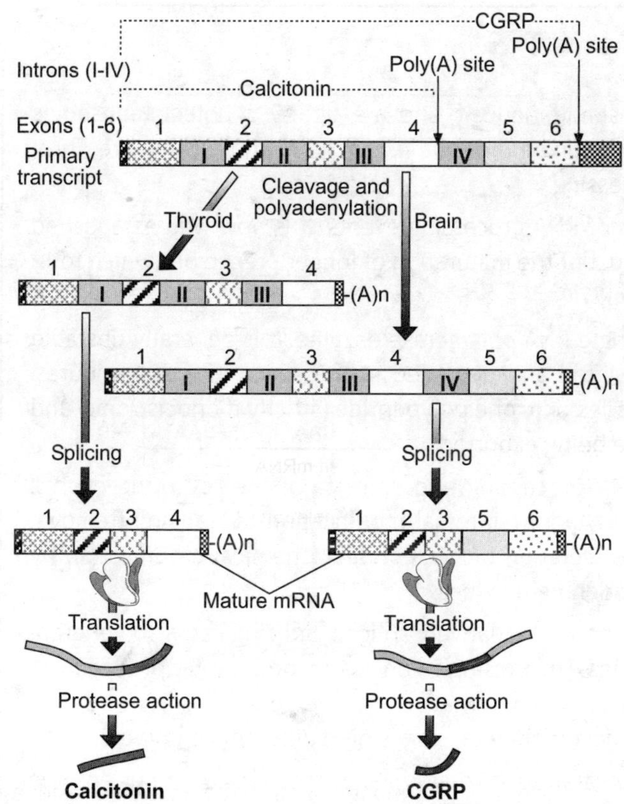

Fig. 10.22 Alternative processing of the calcitonin gene transcription in rats

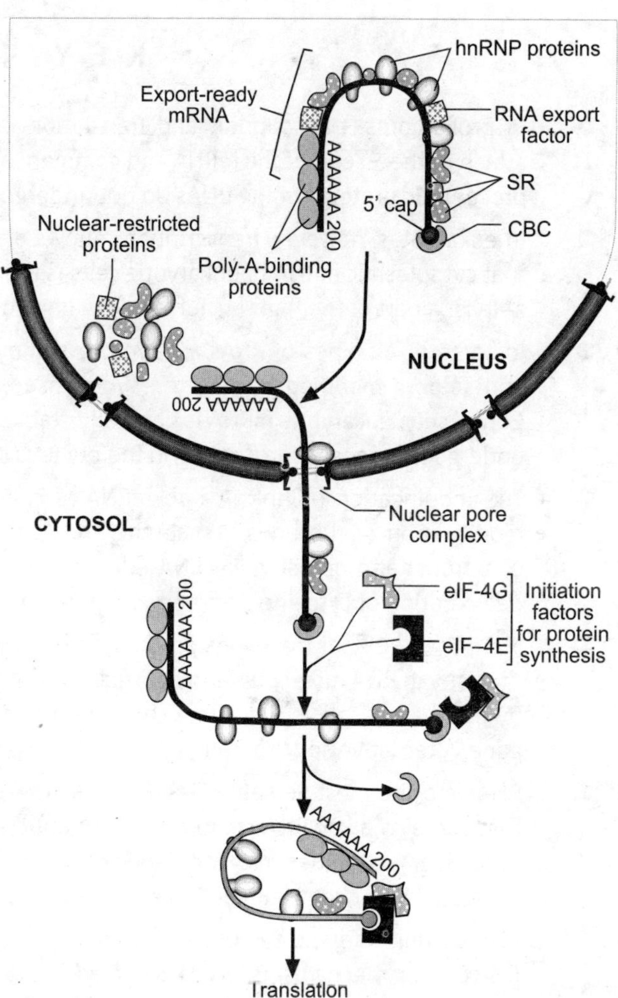

Fig.10.23 Schematic illustration of an 'export-ready' mRNA molecule and its transportation through the nuclear pore

Cells have evolved a very efficient mechanism for transportation of mRNAs selectively into the cytoplasm. From the moment an RNA molecule starts to be transcribed, it becomes associated with a variety of proteins. Although some of these proteins are replaced at various steps along the processing path, others like SR proteins remain associated with mRNA and additional proteins also join in. As a result, a typical mature mRNA carries a collection of proteins that identifies it as being mRNA destined for transport. Other RNAs not only lack these associated signature proteins, but have their own set of proteins that actively blocks export. RNA export takes place through a special nuclear structure in the nuclear membrane called the **nuclear pore complex**. Molecules smaller than 50 kd can diffuse through these pores unaided; but larger molecules and complexes, including mRNAs and their associated proteins, require active transport. Once in the cytoplasm, the proteins are discarded, and will be recycled. But the mRNAs will be immediately recognised by the translation machinery and will enter into translation. RNA export requires energy and is supplied by the hydrolysis of GTP by a GTPase protein called Ran.

KEY POINTS

- In prokaryotes, transcription and translation occur simultaneously and are tightly coupled. Ribosomes attach to the 5′-end of the mRNA and commence translation before synthesis of its 3′-end is complete. Thus, prokaryotic protein coding RNAs do not undergo processing.
- In eukaryotes, however, transcription follows a series of RNA processing events. It is now well established that cytoplasmic mRNA of eukaryotic cells is the product of the maturation of longer precursors, which form at least a part of the high-M_r nuclear RNA known as hnRNA.
- In bacteria, all types of RNAs are synthesised by a single RNA polymerase enzyme. It is generally unstable and seldom modified—it is often synthesised, translated and degraded in the space of a few minutes. Conversely, eukaryotic mRNA is generally stable, encodes only one polypeptide (strictly monocistronic) and undergoes extensive processing in the nucleus before being exported.
- The modification of hnRNA into mRNA is called RNA processing and consists of the following events: modification of the ends (5′-capping and 3′-polyadenylation), internal modification by splicing (removal of introns) and occasionally RNA editing (addition or deletion of nucleotides), chemical modification by deamination of cytosine and methylation of internal adenine residues.
- More than 50% of the genes produce pre-mRNAs that are alternatively spliced. Splicing can also generate diversity in RNA molecules and consequently in proteins. Thus, a single gene can produce multiple products by alternative splicing. In some cases, hundreds of potential alternatives can be generated from a single gene. Alternative splicing can occur either at specific developmental stages or in different cell types.
- RNAs are the most versatile class of molecules. They catalyse several types of reactions. Another enzyme that contains a key RNA component is ribonuclease P (RNaseP), which catalyses the maturation of tRNA by removing nucleotides from the 5′-end of the precursor molecule. RNA molecules that function as an enzyme in catalysing a chemical reaction are called ribozymes or catalytic RNAs.
- RNA editing refers to the events that alter the protein coding sequences of some mRNAs. The diversity of RNA editing mechanisms includes nucleoside modifications such as C to U and A and I deaminations, as well as non-templated nucleotide additions and insertions. RNA editing in mRNAs effectively alters the amino acid sequence of the encoded protein so that it differs from that predicted by the genomic DNA sequence.
- RNA prepares for export while processing goes on. Transport of mRNA has to be very selective and highly regulated, since the nucleus could contain, for example, damaged or misprocessed RNAs, a large number of freed introns and many other RNA molecules.
- RNA degradation not only removes unwanted RNAs, but also regulates gene expression by altering the turnover kinetics.

Self-assessment Questions

1. Discuss the main differences in transcription between prokaryotes and eukaryotes.
2. Explain briefly the mechanism of ribosomal RNA and tRNA processing in prokaryotes.
3. Discuss the steps involved in the processing of rRNA in eukaryotes.
4. How are tRNAs processed?
5. Explain with the help of a diagram the modification of the 5′- and 3′- ends of mRNA in eukaryotes.
6. How are group I and group II introns processed?
7. What are the different mechanisms of RNA editing?

FURTHER READING

Textbooks

Alberts B, A Johnson , J Lewis , M Raff, K Roberts and P Walter (2002). *Molecular Biology of The Cell*, 4th ed. Garland Science, Taylor and Francis Group.

Hartl DL and EW Jones (1998). *Genetics: Principles and Analysis*. Jones and Bartlett Publishers.

Lewin B (2000). *Genes VII*. Oxford University Press.

Lodish H, A Berk , L Zipursky , P Matsudaira, D Baltimore and J Darnell (2000). *Molecular Cell Biology*. 4th ed. New York: WH Freeman.

Turner PC, AG McLennan, AD Bates and MRH White (1998). *Instant Notes in Molecular Biology*. Viva Books Private Limited.

Twyman RM (1998). *Advanced Molecular Biology*. Oxford: Bios Scientific Publishers.

References

Aphasizhev R, S Sbicego , M Peris , SH Jang and L Aphasizheva (2002). Trypanosome mitochondrial 3′ terminal uridylyl transferase (TUTase): The key enzyme in U-insertion/deletion RNA editing. *Cell* 108:637–48.

Bachellerie JP and J Cavaille (1997). Guiding ribose methylation of rRNA. *Trends Biochem. Sci.* 22:257–61.

Barass JD and JD Beggs (2003). Splicing goes global. *Trends Genet.* 19:295–98.

Bass BL (1997). RNA editing and hypermethylation by adenosine deamination. *Trends Biochem. Sci.* 22:157–62.

Blane V and NO Davidson (2003). C-to-U RNA editing: Mechanisms leading to genetic diversity. *J. Biol. Chem.* 278:1395–98.

Brow DA (2002). Allosteric cascade of spliceosome activation. *Annu. Rev. Genet.* 36:333–60.

Cech TR (1990). Self-splicing of group I introns. *Ann. Rev. Biochem.* 59:543–68.

Cech TR (1990). Self-splicing and enzymatic activity of an intervening sequence RNA from *Tetrahymena*. *Biosci. Rep.* 10:230–61.

Chabot B (1996). Directing alternative splicing—cast and scenarios. *Trends Genet.* 12:472–78.

Chen CYA and AB Shyu (1995). AU-rich elements characterization and importance in messenger RNA degradation. *Trends Biochem. Sci.* 20:465–70.

Chester A, J Scott, S Anant and N Navaratnam (2000). RNA editing: Cytidine to uridine conversion in apolipoprotein B mRNA. *Biochem. Biophys. Acta* 1494:1–3.

Copertino DW and RB Hallick (1995). Group II and group III introns of twintrons-potential relationships with nuclear premessenger RNA introns. *Trends Biochem. Sci.* 18:467–71.

Decatur WA and MJ Fournier (2003). RNA-guided nucleotide modification of ribosomal and other RNAs. *J. Biol. Chem.* 278:695–98.

Doherty EA and JA Doudna (2000). Ribozyme structures and mechanisms. *Annu. Rev. Biochem.* 69:597–615.

Dreyfuss G, VN Kim and N Kataoka (2002). Messenger RNA binding proteins and the messages they carry. *Nat. Rev. Mol. Cell Biol.* 3:195–205.

Fedor MJ (2000). Structure and function of the hairpin ribozyme. *J. Mol. Biol.* 297:269–91.

Gott JM and JL Emeson (2000). Functions and mechanisms of RNA editing. *Ann. Rev. Genet.* 34: 4911.

Graveley BR (2001). Alternative splicing: Increasing diversity in the proteomic world. *Trends Genet.* 17:100–107.

Jurica MS and MJ Moore (2003). Pre-mRNA splicing: Awash in a sea of proteins. *Mol. Cell* 12:5–11.

Kable ML, S Heidmann and KD Stuart (1997). RNA editing: Getting U into RNA. *Trends Biochem. Sci.* 22:162–66.

Keegan LP, A Gallo and MA O'Connell (2001). The many roles of an RNA editor. *Nat. Rev. Genet.* 2:869–78.

Maniatis T and B Tasic (2002). Alternative pre-mRNA splicing and proteome expansion in metazoans. *Nature* 418:236–43.

Mass S, A Rich and K Nishikura (2003). A to I RNA editing: recent news and residual mysteries. *J. Biol. Chem.* 278: 1391–94.

Minvielle-Sebastia L and W Keller (1999). mRNA polyadenylation and its coupling to other RNA processing reactions and to transcription. *Curr. Opin. Cell Biol.* 11: 352–57.

Moore MJ (2002). Nuclear RNA turnover. *Cell* 108:431–34.

Newman A (1998). RNA splicing. *Curr. Biol.* 8: R903–R905.

Reed R (2000). Mechanisms of fidelity in pre–mRNA splicing. *Curr. Opin. Cell Biol.* 12:340-54.

Reed R and E Hurt (2002). A conserved mRNA export machinery coupled to pre-mRNA splicing. *Cell* 108:623–531.

Sarker N (1997). Polyadenylation of mRNA prokaryotes. *Annu. Rev. Biochem.* 66:173–97.

Simpson L, S Sbicego and R Aphasizhev (2003). Uridine insertion/deletion RNA editing in trypanosome mitochondria: A complex business. *RNA* 9:265–76.

Simpson L, OH Thiemann, NJ Savill, JD Alfonzo and DA Maslov (2000). Evolution of RNA editing in trypanosome mitochondria. *Proc. Natl. Acad. Sci. USA* 97:6986–93.

Stark H, P Dube, R Luhrmann and B Kastner (2001). Arrangement of RNA and proteins in the spliceosomal U1 small nucleic ribonucleoprotein particle. *Nature* 409:539–42.

Strehler EE and DA Zacharias (2001). Role of alternative splicing in generating Isoform diversity among plasma membrane calcium pumps. *Physiol. Rev.* 81:21–50.

Tarn WY and JA Steitz (1997). Pre-mRNA splicing: The discovery of a new spliceosome doubles the challenge. *Trends Biochem. Sci.* 22:132–37.

van Hoof A and R Parker (2002). Messenger RNA degradation: beginning at the end. *Curr. Biol.* 12:R285–R287.

Wahle E and U Kuhn (1997). The mechanism of 3′ cleavage and polyadenylation of eukaryotic pre-mRNA. *Prog. Nucliec Acid Res. Mol. Biol.* 57:41–71.

CHAPTER 11
Genetic Code

LEARNING OBJECTIVES

After reading this chapter, you will be able to:

- Understand the general nature of the genetic code
- Discuss the techniques of deciphering the genetic code
- Appreciate the mechanism of transfer of genetic information from DNA to proteins *via* RNA

11.1 INTRODUCTION

DNA has been well established as the genetic material in all the cellular organisms as also in many viruses. It is also responsible for the phenotypic expression through transcription of DNA templates to produce mRNA and specific nucleotide sequences on tRNA that interact with complementary base groups of mRNA to translate mRNA base sequences into proteins on ribosomal surface.

Studies of amino acid (phenylalanine) metabolism on nutritionally deficient strains of the fungus *Neurospora* and a host of others make it clear that occurrence of a given biochemical reaction depends on the presence of a specific enzyme (protein) that, in turn, is due to the action of a particular genetic locus. From these results it emerges that the classical particulate gene of Mendel is a series of deoxyribonucleotides with the relationship of gene→RNA→polypeptide→phenotype. This transfer of information raises the question: What is the specific relationship between sequences of the four bases in DNA and sequences of the 20 amino acids in proteins? Since a one-to-one relationship cannot exist between nucleotides in a DNA and amino acids in the protein, this implies the existence of biological language, a **genetic code**, in which some number of nucleotides specify the incorporation of a single amino acid into a protein. This genetic code relates the information found in the sequence of nucleotides in DNA to the information found in the sequence of amino acids in proteins. In this chapter, the nature and the underlying logic of the genetic code, the experiments conducted to 'crack' the code and the use of mutations and the mutants to elucidate the genetic code have been discussed.

11.2 COLLINEARITY BETWEEN GENES AND PROTEINS

The gene can be defined as the collection of adjacent nucleotides that specify the amino acid sequences of the cellular polypeptide chains. Direct evidence for the specific relationship between genes and proteins came from studies that compared the genetic location of a mutation within a gene to the location of the change in the corresponding amino acid in the protein coded for it. The classic example is from the work of Charles Yanofsky (1967) on the tryptophan synthetase gene in *Eschericia coli*. Yanofsky and his colleagues were able to map various mutations and identify their specific site within the 'A' gene (an enzyme is a complex of two different subunits, A and B). By sequencing the 'A' polypeptide, they could determine where each amino acid change had occurred in the protein. Their finding was that the location of each mutant site within the 'A' corresponded exactly to the mutations with the corresponding amino acid in the 'A' polypeptide. Thus, there is a direct or collinear relationship between the nucleotide and the amino acid sequences.

However, the molecular processes underlying collinearity were not understood completely, because the 20 different amino acids far exceeded the number of different nucleotides in DNA. In fact, groups of nucleotides must somehow specify or code for each amino acid.

From DNA to the sites of protein synthesis

Since almost all DNA is located on the chromosomes in the nucleus and almost all cell protein synthesis takes place in the cytoplasm, the genetic information of DNA must be transferred through an intermediate molecule. Now it is confirmed that RNA serves as the template that orders the amino acids within the polypeptide chains of proteins during the process of translation.

11.3 NATURE OF THE GENETIC CODE

Once it was established that the sequence of bases in a gene specified the sequence of amino acids in protein, various possibilities for the genetic code came under consideration. How many bases are required to specify each amino acid? Is the codon degenerate and unambiguous? Is the code overlapping or non-overlapping? Is the code punctuated or continuous? Is the code universal?

The first model of the genetic code presented by the physicist George Gamow was that each amino acid in a polypeptide was encoded by three sequential nucleotides. He reasoned that if the four bases in DNA (A, T, G and C) specify the 20 amino acids in proteins and if all codons were assumed to have the same number of bases, then each codon would have to contain at least three bases. Codons consisting of pairs of bases would be insufficient, because four bases can form only $4^2 = 16$ pairs; triplets of bases would suffice, because four bases can form $4^3 = 64$ triplets (**Table 11.1**).

In fact, the genetic code is a **triplet code**, and all 64 possible **codons** (three nucleotide code words in mRNA) carry information of some sort. Out of the 64 codons, 61 represent amino acids and 3 codons cause the termination of protein synthesis.

11.3.1 Genetic code

Most amino acids are encoded by more than one codon. Because there are more codons (61) than there are amino acids (20), almost all amino acids are represented by more than one codon. The only exceptions are methionine and tryptophan. The codons that have the same meaning are called **synonyms**. Because the genetic code is actually read on the mRNA, usually it is described in terms of the four bases present in RNA: U, C, A and G (**Table 11.1**). Distribution of codons in the genetic code suggests that the code evolved in such a way that the deleterious effects of point mutations were minimised. For example, mutations in the first position of a codon will often code for changes in the corresponding amino acid, indicating the stringency of a first nucleotide in a codon. Codons with pyrimidines in the second position specify mostly hydrophobic amino acids and those with purines in the second position correspond mostly to polar amino acids.

The genetic code can be expressed as either RNA codons or DNA codons. The former occur in messenger RNA (**mRNA**) and are the codons that are actually 'read' during the synthesis of polypeptides (the process called **translation**). But each mRNA molecule acquires its sequence of nucleotides by transcription from the corresponding gene. Each triplet nucleotide sequence or codon (**Table 11.1**) refers to the nucleotide sequence in mRNA that specifies the incorporation of the indicated amino acid or termination of the polypeptide chain. Ochre, Opal and Amber are stop (**terminator**) codons.

11.3.2 Frameshifts

Sydney Brenner and Francis Crick (1961) studied a number of acridine-induced mutations in the rapid lysis (rII) region of the DNA of phage T4. The rII mutant types arose through additions or deletions at any of a large number of sites within the rII region, but these may revert to the wild type or a very similar one through deletions or additions elsewhere in the rII region. That is, in many cases, one mutation may be suppressed by another, particularly if the second change is located relatively close to the first. These results led them to the major statement that **each codon contains three bases**. They made genetic crosses to show that the addition or subtraction of either one or two base pairs (bp) results in highly abnormal non-functional proteins. In contrast,

Table 11.1 Standard genetic code

FIRST POSITION		SECOND POSITION 5'→3' U	C	A	G	THIRD POSITION
	U	UUU Phe	UCU Ser	UAU Tyr	UGU Cys	U
		UUC Phe	UCC Ser	UAC Tyr	UGC Cys	C
		UUA Leu	UCA Ser	UAA* Ochre	UGA* Opal/Sec	A
		UUG Leu	UCG Ser	UAG* Amber	UGG Trp	G
	C	CUU Leu	CCU Pro	CAU His	CGU Arg	U
		CUC Leu	CCC Pro	CAC His	CGC Arg	C
		CUA Leu	CCA Pro	CAA Gln	CGA Arg	A
		CUG Leu	CCG Pro	CAG Gln	CGG Arg	G
	A	AUU Ile	ACU Thr	AAU Asn	AGU Ser	U
		AUC Ile	ACC Thr	AAC Asn	AGC Ser	C
		AUA Ile	ACA Thr	AAA Lys	AGA Arg	A
		AUG† Met	ACG Thr	AAG Lys	AGG Arg	G
	G	GUU Val	GCU Ala	GAU Asp	GGU Gly	U
		GUC Val	GCC Ala	GAC Asp	GGC Gly	C
		GUA Val	GCA Ala	GAA Glu	GGA Gly	A
		GUG Val	GCG Ala	GAG Glu	GGG Gly	G

*Chain-terminating or nonsense or stop codons

†Used as an initiation codon; in bacteria, it specifies the initiator formyl-Met-$tRNA^{fMet}$

if three base pairs were either added or subtracted, the resulting proteins frequently were totally active. They concluded that the genetic code is read in step-wise groups of three base pairs.

To demonstrate, assume the code is triplet and non-overlapping and that a repeated deoxyribonucleotide sequence is involved. The order of reading, or **reading frame**, is indicated by:

..CAT..CAT..CAT..CAT..CAT..CAT..CAT... (1)

...His......His......His....His....His.....His...His....

Now if the second T (the sixth base) in this hypothetical series is deleted, the reading frame is shifted by one base starting right after the deletion and becomes

..CAT..CAC..ATC...ATC...ATC...ATC...AT... (2)

...His....His....Ile...Ile....Ile....Ile....

A deletion thus alters the reading of all the subsequent groups. This alteration is termed as a **frameshift**. Transcription from the original DNA sequence (1) produces mRNA of repeating GUA codons and determines a peptide consisting only of the amino acid sequence coded by GUA. Transcription from (2) produces the codon sequence GUA...GUG... UAG...UAG...UAG...UAG...UA..... This could well be expected to result in a change in the composition of amino acids of the peptide chain beyond the deletion, if it is assumed that codons GUA and UAG code for different amino acids.

Now, still carrying this deletion, assume a thymidylic acid (T) should be inserted at a different position in sequence (2), between the sixth and seventh nucleotides. The DNA sequence then becomes

...CAT...CAC...TAT...CAT...CAT...CAT... CAT..... (3)

...His....His....Tyr....His...His....His....His...

and the wild-type reading frame is restored with the fourth triplet. Only the second (triplet altered but codes for the same amino acid) and third (changed both triplet and amino acid) produces a misreading. Insertion of any of the other three nucleotides at the same point produces only a slightly different faulty segment. Suppose deoxycytidylic acid (C) is inserted instead of T at the same point, the sequence then becomes

...CAT...CAC...CAT....CAT...CAT...CAT... CAT.... (4)

...His...His....His...His....His...His....His..,

Wild-type sequence in (4) is restored after one 'altered' triplet instead of two. Thus, an insertion farther down the reading sequence corrects for an earlier deletion, regardless of whether the inserted base is the same as the deleted one or not. A single frameshift, therefore, may be expected to result in a protein so altered in amino acid sequence that it is non-functional. On the other hand, if the altered reading between two opposing events (a deletion followed by insertion) is of a small enough magnitude, and if the mis-sense mutation is such that it does not alter or only slightly alters the function of the protein, then the second event suppresses the first.

11.3.3 Nature of the code: Triplet and non-overlapping

Whatever the may be the codon size, the form of the code may be overlapping or non-overlapping. If, for example, the codons were triplets, this would mean that a sequence of 18 nucleotides codes either six amino acids, if it were non-overlapping, or more amino acids, if it were overlapping. A decision between these alternatives can be made by determining whether a single mutation produces a change in only a single amino acid in a protein product (non-overlapping code) or in more than one amino acid (overlapping code). All evidence to date indicates that single nucleotide mutations affect only a single amino acid in a protein chain, meaning that the code is non-overlapping. For non-overlapping codes further two possibilities exist: (i) the code has 'commas', consisting of one or more nucleotides placed between the codons; (ii) the code is 'commaless', without being interspersed with nucleotides among the codons. A commaless code would have to be 'read' from a particular starting point on the mRNA chain in a 'reading frame' consisting of codons placed side by side (**Fig. 11.1**). This allows any nucleotide sequence to be divided into codons in three ways i.e., there are three reading frames (**Fig. 11.2**). In addition, the restoration of sense sequence clearly suggests that there is no punctuation between the codons; that is, each codon is immediately adjacent to the next with no intervening 'spacer' bases or gaps with non-coding nucleotides.

The reading frame used for protein synthesis is determined by the position of the initiation codon, although the frame may occasionally be changed during translation (frameshift mutations). In genome sequence data, potential genes are often identified as **open reading frames (ORFs)**, i.e., long sequences of uninterrupted sense codons, because incorrect reading frames usually contain one or more stop codons. Where an open reading frame has been identified but the product has not been characterised, it may be termed an **unassigned reading frame (URF)**.

11.4 GENETIC CODE FROM *IN VITRO* PROTEIN SYNTHESIS

11.4.1 Translation

It is during translation in protein synthesis that the codon in mRNA is matched to its amino acid. This process is facilitated by transfer RNA (tRNA). The tRNA is charged when attached to an amino acid and is referred to as an aminoacyl–tRNA. This complex enters the ribosomal A site and the anticodon pairs with the codon. This places the amino acid adjacent to the ribosomal peptidyltransferase site which catalyses the transfer of the amino acid from the tRNA to the existing polypeptide chain.

Deciphering the code was done in two stages. In the first stage, specific triplet RNA sequences were synthesised. The researchers then determined the amino acid–tRNA complex that was bound by each sequence. In the second stage, long artificial mRNAs (synthetic mRNAs) determined the amino acids that were incorporated into protein.

11.4.2 Binding tRNA to trinucleotide codons

This approach is a direct way of ordering nucleotides within some of the codons. This method does not involve any enzymatic reactions; specific aminoacyl–tRNA molecules can bind to ribosome–mRNA complexes. For example, when poly U is mixed with ribosomes, only phenylalanyl tRNA will attach (**Fig. 11.3**). Correspondingly, poly C promotes the binding of prolyl–tRNA. The addition of the trinucleotide UUU results in phenylalanyl–tRNA attachment, whereas if AAA is added, lysyl–tRNA specifically binds to ribosomes. The method is a straightforward and simple way of determining the order of nucleotides within many codons (**Table 11.2**). By this method, even though all 64 possible nucleotides were synthesised and assayed, not all codons were

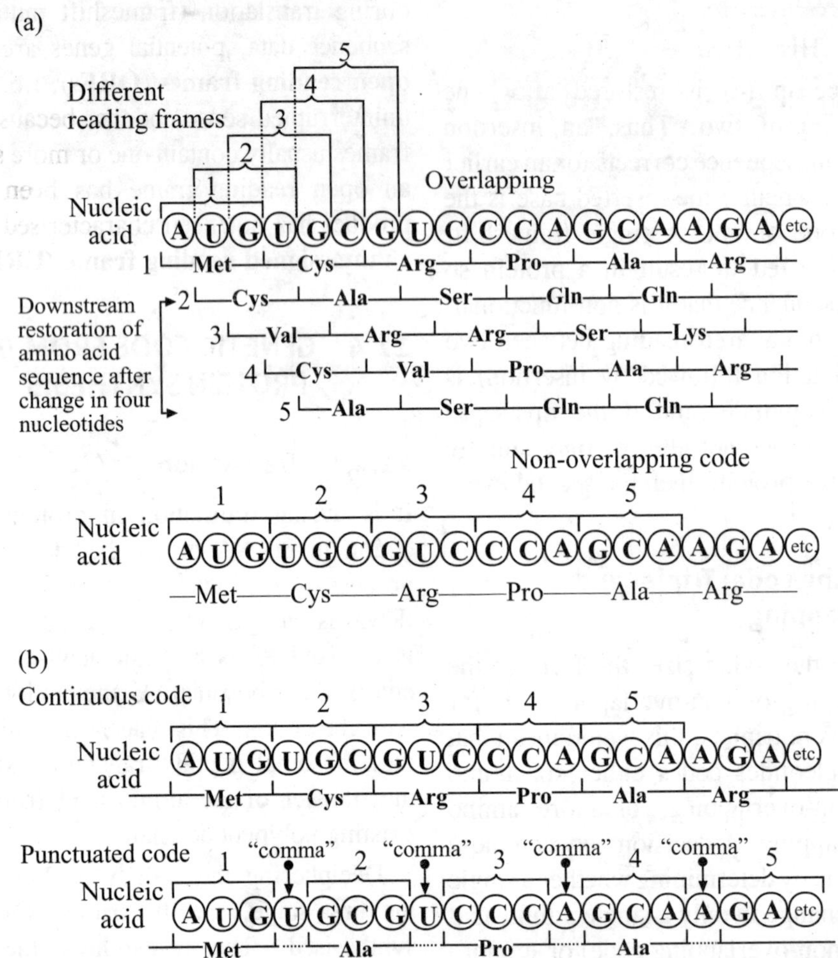

Fig. 11.1 Genetic code (a) Overlapping versus non-overlapping code (b) Continuous versus punctuated code

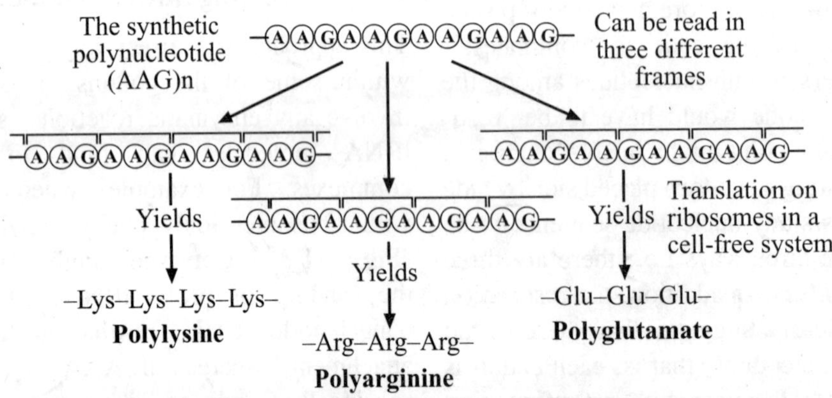

Fig. 11.2 Use of synthetic polynucleotides with repeating sequences to decipher the code

Table 11.2 Assay of binding of aminoacyl-tRNA molecules to trinucleotide-ribosome complexes

Trinucleotide (5'à3')						*AA–tRNA bound*
UUU	UUC					Phenylalanine
UUA	UUG	CUU	CUC	CUA	CUG	Leucine
AAU	AUC	AUA				Isoleucine
AUG						Methionine
GUU	GUC	GUA	GUG	UCU*		Valine
UCU	UCC	UCA	UCG			Serine
CCU	CCC	CCA	CCG			Proline
AAA	AAG					Lysine
UGU	UGC					Cysteine
GAA	GSG					Glutamic acid

*Mis-assigned codon by this method

determined, since some trinucleotides (UUU, GUU) have low affinity to ribosomes, making it impossible to know whether or not they code for specific amino acids.

The coding ability of the synthetic mRNA had produced reliable results in experiments designed to decipher the code. But, true proteins were programmed by mRNAs only when natural mRNAs were added to the bacterial extract. The first successful synthesis of a specific protein occurred when the mRNA of bacteriophage F2 was added to bacterial extracts and the coat protein was formed.

11.4.3 Poly U codes for polyphenylalanine

Nirenberg and Matthaei (1961) used a cell-free system extracted from *E. coli* containing ribosomes, radioactively labelled amino acids, tRNA and the necessary enzymes and ATP for protein synthesis. When TMV virus RNA was added, there was a marked incorporation of amino acids into proteins and they were TMV-like proteins. However, their most remarkable discovery occurred when they added the synthetic polyribonucleotide, poly rU, to this system and obtained a polypeptide whose amino acids were all phenylalanine (Phe-Phe-Phe) in spite of the presence of other amino acids in the original mix. From this simple result, and knowledge that the code is a triplet code, it was concluded that UUU must be a codon for the amino acid, phenylalanine. Variations on this basic experiment identified other codons. Likewise, poly C (polypeptides with only proline) and poly A (strings of lysine) coded for single amino acids. Poly G did not work because it assumes usable stacked structures.

11.4.4 Codon assignments from repeating copolymers

Most of the remaining unidentified codons were established when Hargobind Khorana found ways to make codons by using repeating copolymers. He used synthetic mRNAs with altering bases (e.g., GUGUGU...., AAGAAG... and GUUGUU...). For example, when a long sequence of guanines was added at the terminus of the poly U, the polyphenylalanine was terminated by a sequence of glycines, indicating that GGG is a gylcine codon (**Fig. 11.4**). A trace of leucine or tryptophan was also present in the glycine-terminated polyphenylalanine. Incorporation of these amino acids was directed by the codons UUG and UGG at the transition point between U and G.

11.5 GENERAL NATURE OF THE GENETIC CODE

The genetic code consists of the following:

- A group of three base codes for one amino acid
- Non-overlapping
- The base sequence read from a fixed starting point without punctuation

11.5.1 Triplet nature of the code

In experiments by Francis Crick and his colleagues, a chemical mutagen, the acridine dye, proflavin, caused deactivation of the rapid lysis (rIIB) gene of the bacteriophage T4. This chemical deactivates the gene by either adding or deleting a nucleotide from the DNA. The rII gene controls the morphology of this

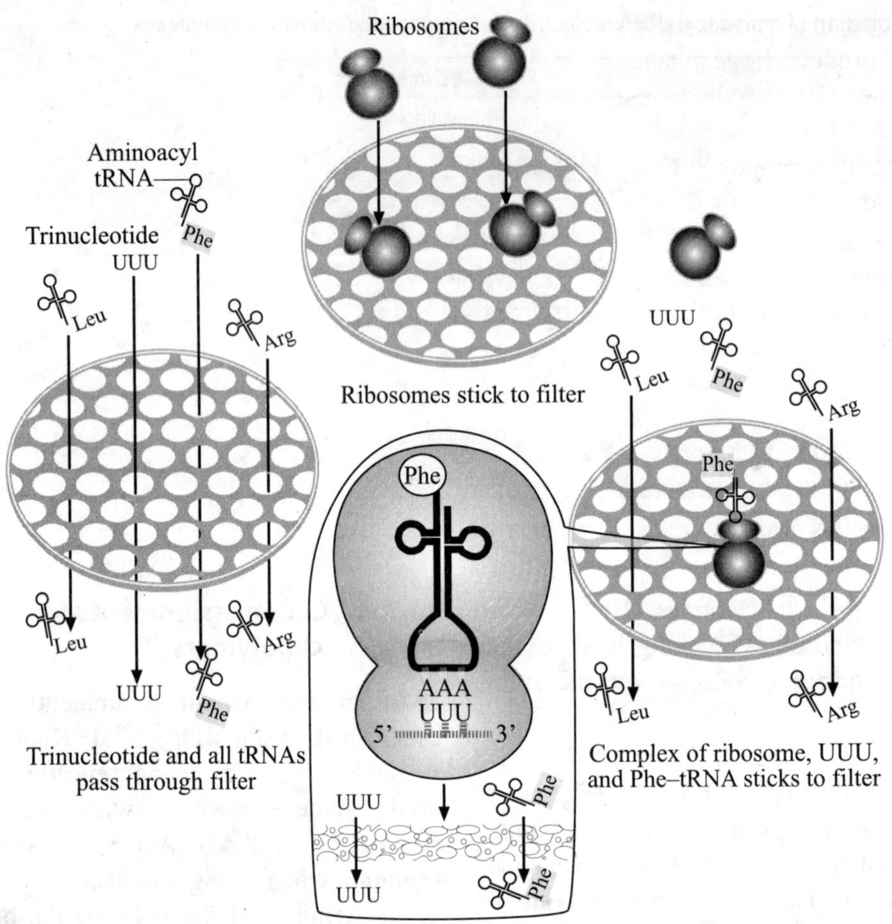

Fig. 11.3 Extracts of *E.coli* to which chemically synthesised trinucleotides were added to decipher the entire genetic code (***Source***: Nirenberg et al 1964)

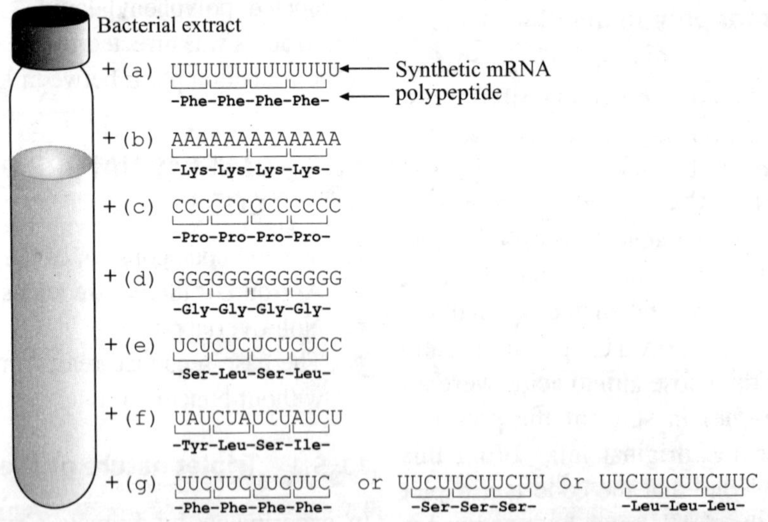

Fig. 11.4 Coding properties of several synthetic mRNAs (a) Poly(U) contains one alternative codon UUU, which code for Phenaylalanine (Phe) (b) Poly(A) contains one codon AAA, which codes for Lysine (Lys) (c) Poly(C) contains one codon CCC, which codes for Proline (Pro) (d) Poly(G) contains one codon GGG, which codes for Glycine (Gly) (e) Poly(UC) contains two alternating codons, UCU and CUC, which code for Ser and Leu respectively (f) Poly(UAUC) contains four codons, UAU, CUA, UCU and AUC, which codes for Tyrosine (Tyr), Leu, Ser and Isoleucine (Ile) respectively (g) Poly(UUC) contains three codons, UUC, UCU, and CUU, which code for Phe, Ser and Leu respectively

bacteriophage's plaque that grows on *E. coli* cells. Rapid-lysis mutants produce large plaques; the wild-type form of the gene, rII^+, results in normal plaque morphology. **Figure 11.5** shows the consequences of adding or deleting a nucleotide. From the point of addition or deletion onward, a frameshift causes codons to be read in different groups of three. If a deletion is combined with an addition to produce a double-mutant gene, the frameshift occurs only in the region between the two mutants. If this region is small enough or does not contain coding for vital amino acids, the function of the gene may be restored. However, two deletions or two additions combined will not restore the reading frame. But the combinations of three additions or three deletions would.

The complete translation of the genetic code is presented in **Table 11.1**. Codons, like other nucleotide sequences, are read 5'→3'. Codons represent triplets of bases in mRNA or, replacing U with T, triplets along the non-transcribed (sense) strand of DNA. Following are the important features of the genetic code:

- All the codons have meaning: 61 out of 64 codons specify particular amino acids. The remaining three, UAA (**ochre**), UAG (**amber**) and UGA (**opal** or **umber**) specify no amino acid. So, they are called nonsense codons or termination codons or stop signals.
- The genetic code is unambiguous. Each of the 61 'sense' codons encodes only one amino acid.

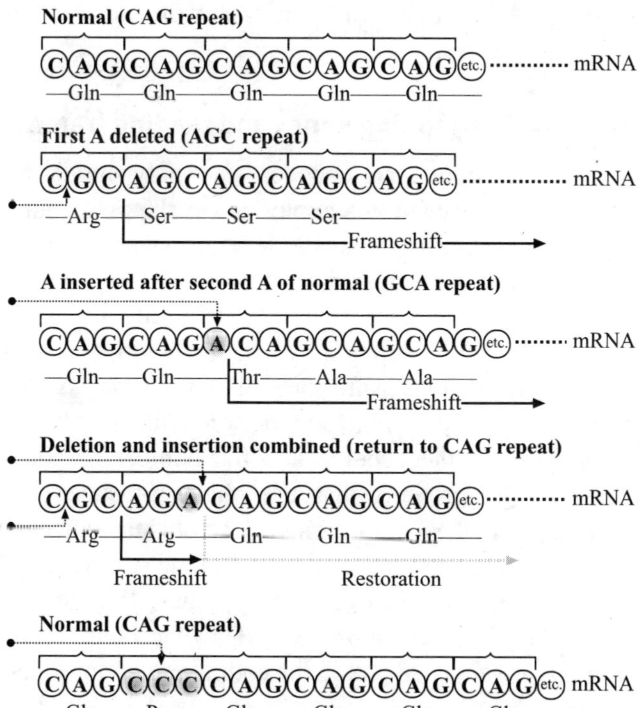

Fig. 11.5 Frameshift mutations in a gene result from the addition or deletion of one or several nucleotides (any number other than multiple of three) in the DNA. The mRNA shown here normally has a CAG repeat. A single-base deletion shifts the three-base reading frame to a series of AGC repeats. However, insertion restores the reading frame. The addition of CCC frame to CAG repeats adds extra code but no change in frames downstream to the addition.

Codons representing the same amino acid or chemically similar acids tend to be similar in sequence. Often the third base in a codon is irrelevant, so that, for example, all the four codons in the GGX family specify Gly and the UGX family specifies Ser (**Table 11.1**). This feature is known as third-base degeneracy or **wobble hypothesis**.

The genetic code is 'universal'. While certain minor exceptions are seen in the codons of the genomes of mitochondria, some prokaryotes and certain lower eukaryotes, the most striking feature of the code is its universality: codon assignments are virtually the same throughout life forms.

It can be seen from **Table 11.1** that 20 different amino acids are specified by the genetic code. Other amino acids are found in proteins, but almost all of these are generated by post-translational enzymatic modification of these 20 genetically-defined amino acids. The one known exception is the rare amino acid, **selenocysteine**, which is encoded by UGA, otherwise a termination codon. It is also apparent that three of the codons (UAA, UAG and UGA) designated as 'stop' codons) do not normally specify an amino acid, but are all signals for the termination of the polypeptide chain. Although the termination process does not involve tRNA, certain mutant tRNAs have been found to recognise individual termination codons. These tRNAs occur in the 'suppressor' strains of *E. coli*, so called for their ability to suppress particular classes of 'nonsense' mutants. Even though the main role of AUG is to specify the start of a polypeptide chain, it also encodes for methionine residues in the body of the polypeptide chain. In bacteria, although AUG is the main initiation codon, it is not the sole one. GUG and certain other much rarer codons are involved (minor initiation codons).

11.5.2 *In vivo* evidence for the code

In vivo evidence supporting the proposed genetic code from studies of proteins have been carefully determined by biochemical means and then correlated with genetic data or exact nucleotide sequences. In *E. coli*, for example, Yanofsky and coworkers have shown that mutations at the tryptophan synthetase locus correlate closely with specific amino acid changes in the A protein of the tryptophan synthetase enzyme. In two instances, mutations A23 and A46, only the same glycine amino acid is changed to arginine (inA23) and to glutamic acid (inA46). These mutants are generally stable, except rare revertants occasionally occur either back to glycine (complete revertants) or to other amino acids (partial revertants) caused by single nucleotide changes (**Fig. 11.6**). Direct sequencing studies of a number of nucleic acids and the polypeptide chains derived from them have confirmed all of the codon designations in **Table 11.1**. The first such study, comparing genes and proteins in the RNA bacteriophage MS2, can be considered illustrative of many others. MS2 is a small *E. coli* phage containing about 3,000 nucleotides, which are known to possess only three cistrons (coding for coat protein replicase and maturation proteins).

There is an exact correspondence between each amino acid in the protein and one or more of its designated codons as given in the **Table 11.1**. Furthermore, among the 49 different codons used in this gene, there is, as expected, an obvious degeneracy.

The initiation codon for translation, AUG, is exactly in its expected place at the beginning of the coat protein gene and also starts the codon sequence for the replicase gene.

The entire nucleotide sequence of the coat protein gene can be arranged in a unique secondary structure based on the maximum amount of base pairing and the most molecularly stable configuration.

Questions such as the specificity of mutagens and the existence of mutational hot spots may well be resolved by these multidimensional considerations. A further consequence of the molecular structure of MS2 is on the question of polarity mutations, in which a mutation in one cistron affects the expression of one or more adjacent cistrons.

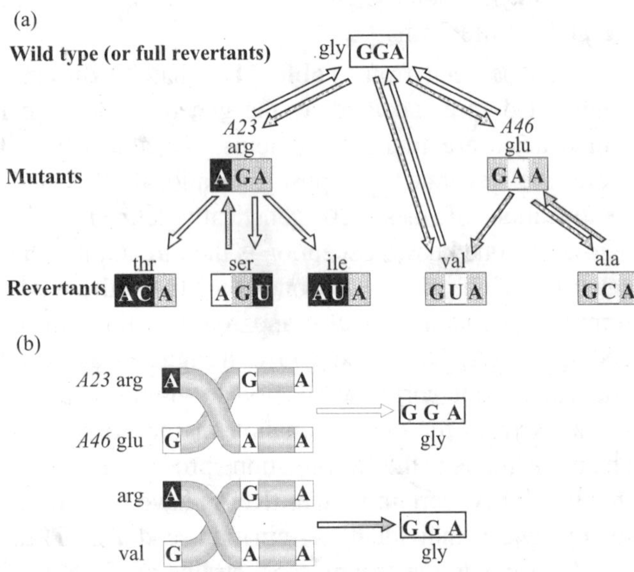

Fig. 11.6 *In vivo* evidence for the code (a) Some of the observed amino acid replacements at the number 210 amino acid position on the protein A chain of *E. coli* tryptophan synthetase and an interpretation of these changes in terms of single nucleotide changes in the codons (b) Production of the wild type in crosses between A23 and A46 and between arginine and valine mutants results from a single cross over event that actually takes place between DNA nucleotides, but shown here at the level of RNA codons. (*Source*: Strickberger MW 1999)

11.5.3 Overlapping genes and reading frames

In general, one gene encodes one polypeptide but now we know several examples of overlapping coding regions (**overlapping genes**). Generally, these genome-economising mechanisms occur where the genome is small and there is higher density of storage of information. For example, when quantitative nucleic acid determinations were carried out for the ssDNA phage øX174, disproportion between the number of nucleotides in the genome and the number of amino acids in the proteins that the phage produced was noticed. It makes 11 proteins of combined molecular mass 262 kDa (or 2,327 amino acids long) from a 5,386 bp genome. Without overlapping genes, this genome could encode at the most 200 kDa of protein. Three proteins are encoded within the coding region for longer proteins, which is equivalent of at least 2,327 × 3, or about 7,000 nt long mRNA. The answer to this problem came from an exact nucleotide sequence determined by Sanger and coworkers which helped construct a detailed genetic map of øX174. In prokaryotes, the ribosomes simply have to find the second start codon to be able to translate the overlapping gene and they may achieve this without detaching from the template (**Fig. 11.7**).

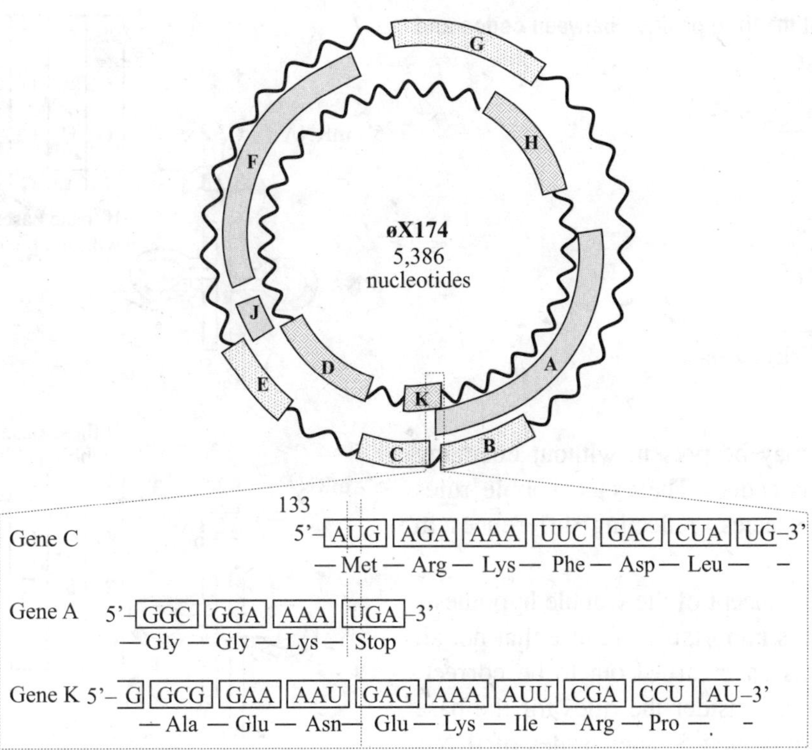

Fig. 11.7 Overlapping genes; the circular DNA of øX174 contains ten (A to K), gene B lies within the sequence of gene A but uses a different reading frame. Gene E lies within gene D but also overlaps with gene J, each gene using a different reading frame. Gene K overlaps with genes A, B and C.

Eukaryotes have a different way of economising their genetic material by initiating protein synthesis and tend to make use of alternative RNA processing to generate variant proteins from one gene (see **Chapter 10**).

11.6 DEGENERACY OF THE CODE

The genetic code displays two types of degeneracy: **First and second base degeneracy** (where codons with different bases in the first two positions may encode the same amino acid) and **third base degeneracy** (where codons with different bases in the third position may encode the same amino acid). Triplets coding for the same amino acids are not distributed at random but are grouped together so that they generally share the same 5' and middle base. The maximum size of a codon family is six and the minimum size is one. The first and second base degeneracy reflects the existence of **isoaccepting tRNAs**. Both prokaryotic and eukaryotic cells encode about 30 distinct species of tRNAs, but because only 21 amino acids are specified by the genetic code, some of the tRNA species must bind the same amino acid. Such duplicate tRNAs may have the same anticodon sequence, in which case they are functionally interchangeable.

In 1966, Crick proposed the wobble hypothesis to account for the third base degeneracy (i.e., lack of specificity in the 5' base of the anticodon). The hypothesis predicts a relaxation in the normal base pairing rules (**Table 11.3**) between the third base of the codon and the first base of the anticodon (the wobble position), allowing a single tRNA species to recognise several different codons (**Fig. 11.8**). It suggests that normal bases become less discriminating in the wobble position, and in some cases can only recognise the type of base (purine or pyrimidine) rather than the specific base in the opposite strand. This explains the existence of degenerate codon families of two, in which the third position can be either purine or pyrimidine (e.g., AAR encodes lysine, AGY encodes serine). There are eight codon families in which all four codons sharing the same first two bases have the same meaning, so that the third base has no role at all in specifying the amino acid. There are seven codon pairs in which the meaning is the same whichever pyrimidine is present at the third position and there are five codon pairs in

Table 11.3 Wobbling at the third position between codon and anticodon

Anticodon	*Codon*
A	U
C	G
G	U, C
I (inosine)	U, C, A
U	A, G

Based on the work of Crick, 1966

which either purine may be present without changing the amino acid that is coded. The exact wobble rules may differ between species, reflecting differences in tRNA modification.

Although the basic concept of the wobble hypothesis has been validated, it is important to realise that not all its detailed predictions have turned out to be correct. Furthermore, one must consider the rules for the base pairing of important modified 5'-nucleosides of tRNA which were unknown at the time of formulation of the hypothesis. Although the first and second bases of a codon are paired with the third and second bases of the anticodon, respectively, in standard Watson–Crick fashion, a number of non-standard interactions can occur between bases in the wobble position (**Fig. 11.9**). The tRNA modifications in and around the anticodon exhibits altered properties to pair with the mRNA codon. This alteration also facilitates docking of the tRNA into the A site of the ribosome during translation and these modifications have an influence on codon pairing. Thus, they function directly to determine how the cell assigns the meaning of the tRNA. Particularly important is the G:U base pair, which structurally fits almost as well as the standard G:C pair. One of the most important unusual wobble-position bases in plants and animals is **inosine**, a deaminated product of adenine, which can base-pair with A, C and U (**Fig. 11.9**). A tRNA with inosine in the wobble position, can, thus recognise the corresponding mRNA triplets with A, C, or U in the third codon position. For this reason, inosine-containing tRNAs are heavily employed in translation of the synonymous codons that specify a single amino acid (e.g., the six codons for leucine; **Table 11.4**).

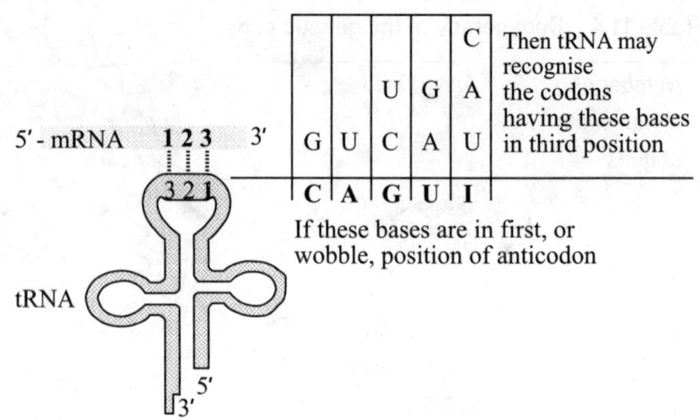

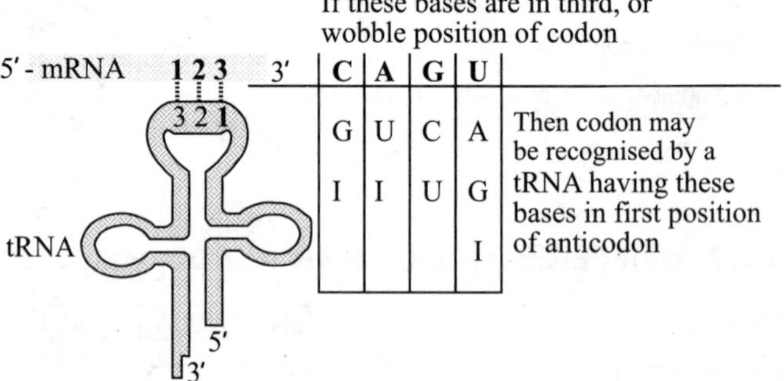

Fig. 11.8 Pairing relationship of codon and anticodon; the first and second bases in an mRNA codon form Watson–Crick base pairs with the third and second bases, respectively, of a tRNA anticodon

Fig. 11.9 The non-standard, wobble base pairs U–G, C–I, A–I and U–I

Table 11.4 Degeneracy of the genetic code

Number of synonymous codons	*Amino acid*	*Total number of codons*
6	Leu, Ser, Arg	18
4	Gly, Pro, Ala, Val, Thr	20
3	Ile	3
2	Phe, Tyr, Cys, His, Gln, Glu, Asn, Asp, Lys	18
1	Met, Trp	2
Total number of codons for amino acids		61
Number of codons for termination		3
Total number of codons in the genetic code		64

Table 11.5 Codon choice for arginine between humans, *Drosophila* and *E. coli*

Codon	*Human*	*Drosophila*	*E. coli*
Arginine:			
AGA	22%	10%	1%
AGG	23%	6%	1%
CGA	10%	8%	4%
CGC	22%	49%	39%
CGG	14%	9%	4%
CGU	9%	18%	49%
Total number of arginine codons	2403	506	149
Total number of genes	195	46	149

***Source*:** Grantham et al (1986)

11.7 DIFFERENTIAL CODON USAGE

Because the genetic code is degenerate, there exist alternative choices between 'synonymous' codons for most of the amino acids, for the termination signal and even for the initiation signal. **Codon usage** (choice, bias or preference) refers to the phenomenon where particular codons in a family are used preferentially in a particular organism. The usage of such synonymous amino acid codons is definitely non-random, and differs between prokaryotes and eukaryotes, viruses and their hosts. In some cases, e.g., certain eukaryotic viruses and mycoplasma, codon choice seems to be determined by the pressure to a particular extreme base composition. However, in bacteria and also in yeast, there is evidence that it can influence the rate of translation. In *E. coli*, it was observed that certain codons (**Fig. 11.10**) that were rarely used in the genes for highly expressed proteins (e.g., ribosomal proteins) were found at a significantly greater frequency in the genes for weakly expressed proteins (e.g., the *lac* and other repressors).

Codon choice differs between degenerate amino acids encoded by only a proportion of their representative codons, but different codons are predominant in different organisms. First and second base preference may reflect complex wobble rules where particular pairing conformations are more stable (**Table 11.5**). In either case, codon usage can be used as a mechanism of gene regulation. For example, rare codon can delay protein synthesis. Codon bias may also arise due to global effects, for example, thermophiles favour GC-rich codons.

11.8 AMBIGUITY IN THE GENETIC CODE

The genetic code is for the most part unambiguous *in vivo* under natural conditions. This means that each codon specifies the same amino acid all the time. This property is essential for the faithful translation of genetic information. However, in two special circumstances ambiguity is tolerated, but because of the uniqueness of each situation, there is no loss of fidelity.

11.8.1 At initiation

The universal initiation codon AUG encodes methionine both at internal sites and at the initiator position (initiator methionine is modified to form N-formyl methionine in prokaryotes and organelles). In bacteria, alternative initiation codons may be used: GUG is common in *Micrococcus luteus* and GUG and UUG are used occasionally in *E. coli*. These codons specify valine and leucine at internal sites but always N-formyl methionine at the initiator position. Alternative initiator codons are very rare in eukaryote nuclear genomes, although CUG is used occasionally. This type of ambiguity reflects the distinct molecular environments of the initiation and elongation stages of protein synthesis—methionine residues are not inserted at internal CUG and UUG sites in *E. coli*. This type of variation of codon reading is also called **site-specific variation**, in which the interpretation of a codon depends on its specific location.

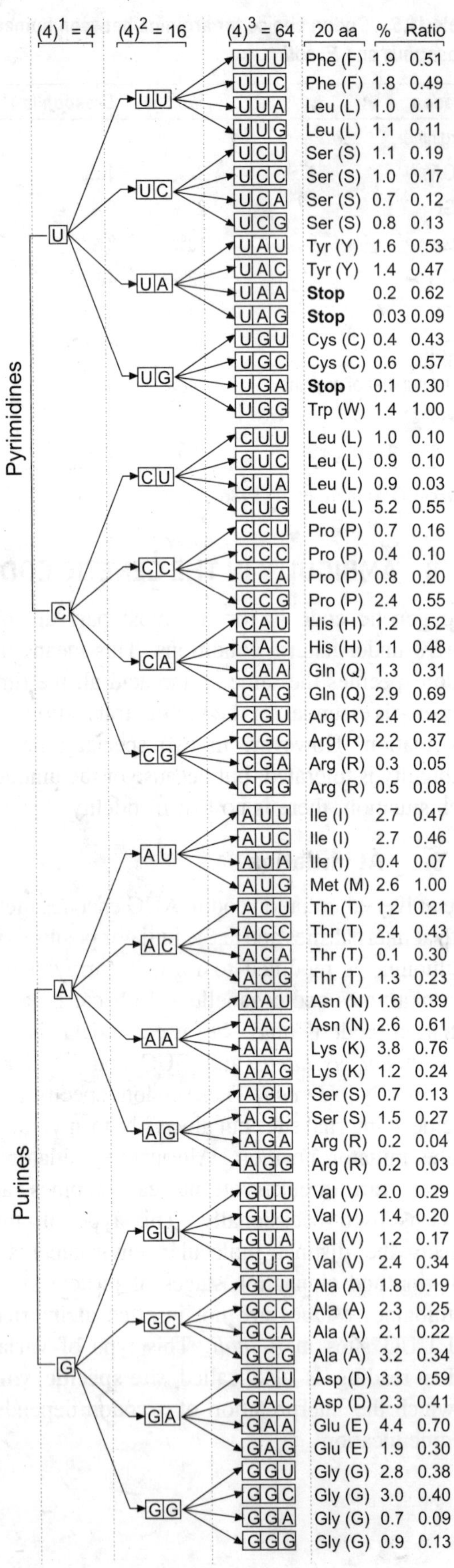

11.8.2 Insertion of rare amino acid

A striking example is the ambiguity involved during insertion of the rare amino acid selenocysteine. This amino acid is similar to cysteine but contains selenium instead of sulphur, and is required for the efficient function of several gene products termed **selenoproteins** or **selenoenzymes**. Most of the unusual amino acids found in proteins are post-translational modification products, but selenocysteine is generated by modification of serine before incorporation, and the amino acid, therefore, has its own cognate tRNA. Selenocysteine is specified by the codon UGA, which is usually recognised as a termination codon. In selenocysteine-encoding mRNAs, however, secondary structures known as **selenocysteine insertion sequences (SECIS)** cause the protein synthesis machinery to translate the codon; the exact mechanism is unclear. In bacteria, SECIS are found in the coding region of the mRNA, whereas in eukaryotes, they are found in the 3' UTR. Mutation in 4*sel* genes causes a deficiency in selenoprotein synthesis. *selC* codes fro tRNA that is charged with serine. *selA* and *selD* are required to modify the serine to selenocysteine. SelB is an alternative elongation factor. It is a guanine nucleotide-binding protein that acts as a specific translation factor for entry of seleno-Cys-tRNA into the A site. It thus provides a replacement for factor EF–Tu.

11.8.3 Deviation from the standard genetic code

One of the first indications that mRNA codons of one species could be translated into the correct amino acid sequence by the tRNAs of another species was shown by the *in vitro* experiments performed on haemoglobin synthesis by von Ehrenstein and Lipmann (1961). They found that *E. coli* tRNA to which labelled amino acids were added would form haemoglobin when incubated with the mRNA and ribosomes of rabbit reticulocytes.

Fig. 11.10 All the triplet codons convey genetic message: 61 codes represent amino acids and three are stop codons (do not represent any amino acids). The letter in parenthesis represents the one-letter code for the amino acid. Codon usage in *E. coli*: the % represents the average frequency this codon is used per 100 codons and the ratio represents the abundance of that codon relative to all of the codons for the particular amino acid.

(***Source:*** www.weeds/mgh.harvard.edu and others)

The evidence for universality of the code has since been reinforced by many *in vitro* and *in vivo* experiments. With the exception of some codons in mitochondria, the genetic code of all prokaryotes and eukaryotic nuclei may be considered to be universal.

The universality of genetic code is remarkable, but as more organisms have been studied, subtle variations in codon assignment have been discovered (**Table 11.6**). Most deviation occurs in mitochondrial genomes, probably reflecting the small number of proteins synthesised. However, in plant organelles, RNA editing is prevalent and it is not clear whether all instances of deviation from the genetic code in plants are true variations or consequences of RNA editing prior to translation. Occasional changes also occur in bacterial genomes and eukaryote nuclear genomes, but usually involve the termination codons. The phylogenetic distribution of these changes indicates that the code is still evolving.

11.8.4 Misreading of the genetic code

Mutations arise as a change in the coding sequence of a gene. Mutations have two important features.

Table 11.6 Unusual codon usage in nuclear and mitochondrial genes

Occurrence/Source	*Codon*	*Universal code*	*Unusual code*
Some animal mitochondria	AAA	Lys	Asp
Drosophila mitochondria			Ser
Most animal mitochondria	AGR	Arg	Ser
Vertebrate mitochondria			STOP
Some animal and yeast mitochondria	AUA	Iso	Met
Mitochondria of higher plants	UGA CGG	Stop Arg	Trp Trp
Candida cylindracea nuclear genome	CUG	Leu	Ser
Yeasts mitochondria	CUN	Leu	Thr
Mycoplasma, Spiroplasma, mitochondria of many species	UGA	Stop	Trp
Acetabularia, Tetrahymena, Paramoecium	UAR	Stop	Gln
Euplotes	UGA	Stop	Cys

R = purines and N = both purines and pyridines

They are responsible for inherited disorders and cancers. At the same time, mutations are the source of phenotypic variation on which natural selection acts. In particular, mutations in tRNAs can have a drastic effect on misreading of the genetic code. The nonsense mutations can be suppressed by mutant tRNA suppressor genes. The tRNAs coded for by these genes can be charged normally with an amino acid, but because of substitutions in their bases, the anticodons of these tRNAs are now complementary to one of the stop codons. For example, there are mutant genes for tyrosine tRNA that code for the anticodon (3') AUC(5') instead of the correct anticodon for tyrosine (3')AUG(5'). The mutant anticodon recognises the stop codon (5') UAG(3') instead of the appropriate tyrosine codon (5')UAC(3'). When a mutant tyrosine tRNA comes across an internal, mutant UAG stop codon, it recognises the stop codon and inserts a tyrosine into the polypeptide chain, allowing translation to continue. The end result is a full-length, functionally normal protein (**Fig. 11.11**). Cells containing such a mutant tRNA suppressor gene can survive only if there are multiple copies of the gene for that tRNA. UGA-suppressor tRNA occasionally inserts amino acids at the normal sites of chain termination, leading to many oversized polypeptide chains. Therefore, the cells carrying such mutations may grow much less than their normal equivalents.

11.9 SECOND GENETIC CODE

Interaction between an aminoacyl-tRNA synthetase and tRNA is termed as a second genetic code. Although many other biological systems also rely on the recognition of information in nucleic acids, primarily the DNA-binding proteins, which control transcription and other DNA functions, the recognition of nucleotide sequence information by tRNA is the cornerstone of the genetic code (see **Chapter 9**). Efforts to identify an 'amino acid code' governing sequence-specific, protein–DNA interaction have shown that no universal sequence-to-sequence correlation exists. However, for certain protein families, recognition codes are beginning to be characterised.

An individual aminoacyl-tRNA synthetase must be specific not only for a single amino acid but also for certain tRNAs as well. Discriminating among dozens of tRNAs is just as important for the overall fidelity of protein biosynthesis as distinguishing among

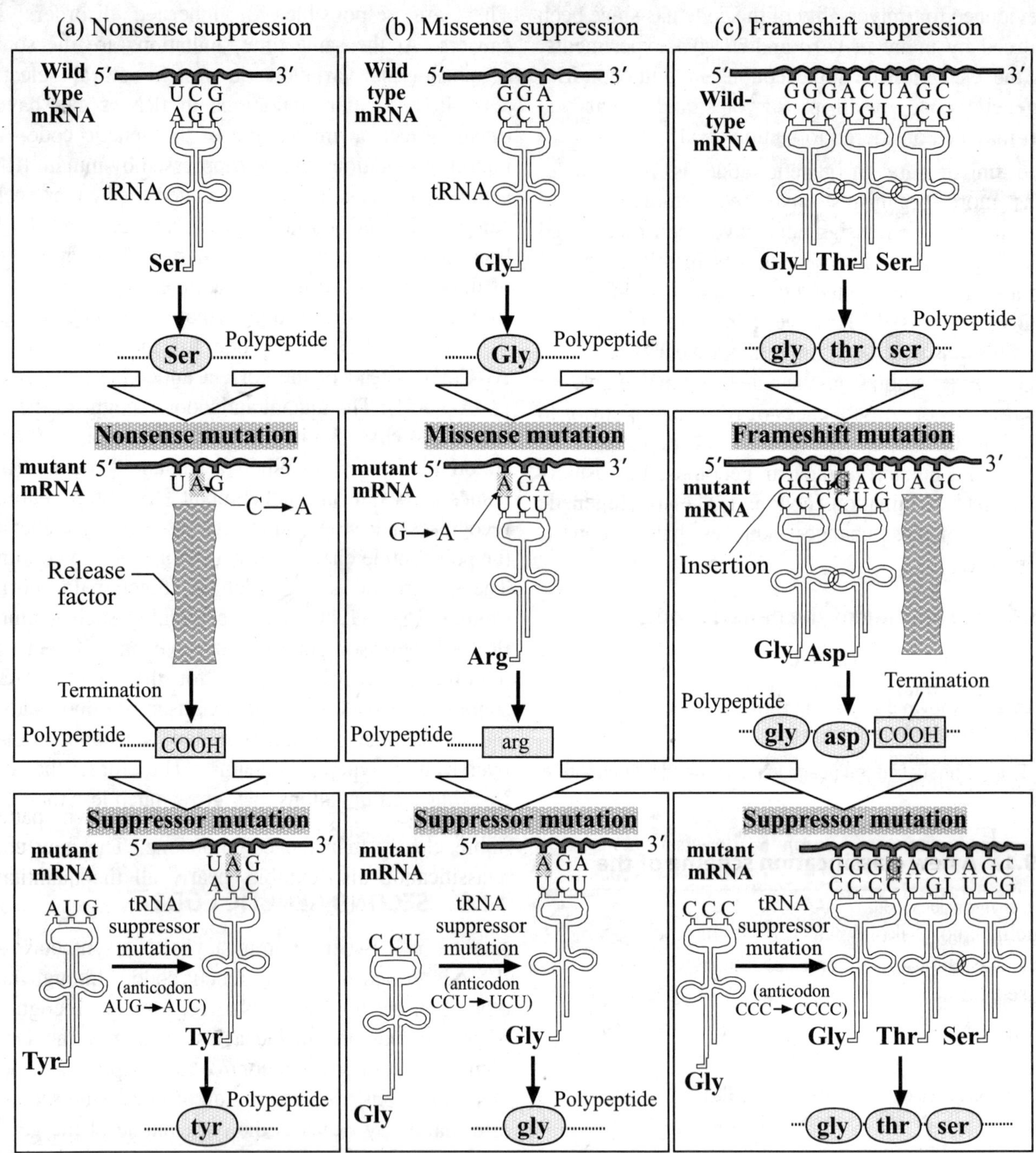

Fig. 11.11 Scheme by which suppressor tRNA mutations can act to correct or modify the effects of some nucleotide changes on polypeptide synthesis (***Source:*** Strickberger MW 1999)

amino acids. The 'coding' rules appear to be more complex than those in the 'first' code. As described in **Fig. 11.12**, some nucleotides are conserved in all tRNAs and therefore cannot be used for discrimination. Nucleotide positions that are involved in discrimination by the aminoacyl-tRNA synthetases have been identified by observing that changes at these nucleotides alter the enzyme's substrate specificity (**Fig. 11.13**). These interactions seem to be concentrated in the amino acid arm and the anticodon arm, including the nucleotides of the anticodon itself, but are also located in other parts of the molecule.

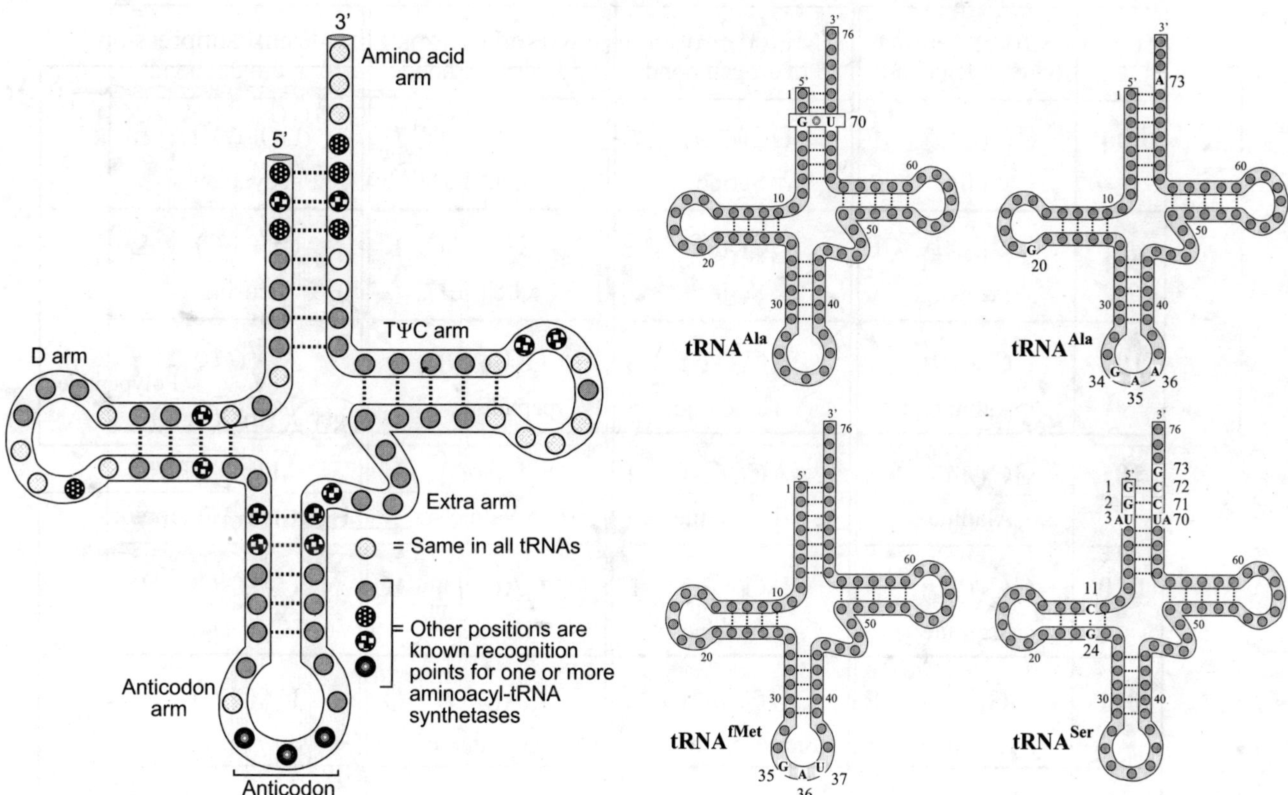

Fig. 11.12 Known positions in tRNAs recognised by aminoacyl-tRNA synthetases; structural features other than nucleotide sequence are also important for recognition by some of the synthetases

Fig. 11.13 The tRNA structural elements recognised by the tRNA synthetase

11.9.1 A new classification scheme of the genetic code

Wilhelm and Nikolajewa (2004) have developed a novel classification scheme of the genetic code that is based on a binary representation of the purines and pyrimidines. This scheme reveals known patterns more clearly than the common one. The new codon classification also exhibits nearly all the quantitative amino acid properties. It also shows that evolution of codons started with a binary doublet code and developed via a quaternary doublet code into the contemporary triplet code (**Fig. 11.14**).

Codon	Strong codons 6 hydrogen bonds	Mixed codons 5 hydrogen bonds	Mixed codons 5 hydrogen bonds	Weak codons 4 hydrogen bonds
000	CC(*C/U*) **P** Proline	UC(*C/U*) **S** Serine	CU(*C/U*) **L** Leusine	UU(*C/U*) **F** Phenylalanine
001	CC(*A/G*) **P** Proline	UC(*A/G*) **S** Serine	CU(*A/G*) **L** Leusie	UU(*A/G*) **L** Leusine
100	CC(*C/U*) **A** Alanine	AC(*C/U*) **T** Threonine	GU(*C/U*) **V** Valine	AU(*C/U*) **I** Isoleusine
101	GC(*A/G*) **A** Alanine	AC(*A/G*) **T** Threonine	GU(*A/G*) **V** Valine	AU(*A/G*) **I/M** Isoleusine/Methionine
010	GC(*C/U*) **R** Arginine	UG(*C/U*) **C** Cysteine	CA(*C/U*) **H** Histidine	UA(*C/U*) **Y** Tyrosine
011	CG(*A/G*) **R** Arginine	UG(*A/G*) **W** Stop/Tryptophan	CA(*A/G*) **Q** Glutamine	UA(*A/G*) *
110	CG(*C/U*) **G** Glycine	AG(*C/U*) **S** Serine	GA(*C/U*) **D** Aspartic acid	AA(*C/U*) **N** Aspargine
111	GG(*A/G*) **G** Glycine	AG(*A/G*) **R** Arginine	GA(*A/G*) **E** Glutamic acie	AA(*A/G*) **K** Lysine

Wilhelm T and Nikolajewa S (2004). A new classification scheme of the genetic code. *J. Mol. Evol.* 59:598–605.

✸ is Halitsky's family-nonfamily symmetry operation; ⟷ is the codon–anticodon symmetry axis

Fig. 11.14 The new classification scheme of the genetic code is based on a binary representation of purines (G,A) = 1 and pyrimidines (C, U) = 0. Therefore, the binary representation $2^3 = 8$ is represented by 000, 001, 100, 101, 010, 011, 110, 111.

KEY POINTS

- It is now well established that DNA is the genetic material in all the cellular organisms and also in many viruses. It is also responsible for the phenotypic expression through transcription of DNA templates to produce mRNA and specific nucleotide sequences on tRNA that interact with complementary base groups of mRNA to translate mRNA base sequences into proteins on ribosomal surface.
- The genetic information in DNA has to be conveyed solely by the linear sequences of the four bases (A, T, G and C) in the DNA alphabet.
- The evidence for the specific relationship between genes and proteins came from studies that compared the genetic location of a mutation within a gene to the location of the corresponding amino acid change in the protein coded for by the gene. It also became evident that a one-to-one correspondence between nucleotides and amino acids could not exist. Instead, groups of nucleotides must somehow specify (code for) each amino acid.

- The first model of the genetic code presented by the physicist George Gamow was that each amino acid in a polypeptide was encoded by three sequential nucleotides. In fact, the genetic code is a triplet code, and all the 64 possible codons (three nucleotide code words in mRNA) carry information of some sort.
- Most amino acids are encoded by more than one codon. Because there are more codons (61) than there are amino acids (20), almost all amino acids are represented by more than one codon. The only exceptions are methionine and tryptophan. Each triplet nucleotide sequence or codon refers to the nucleotide sequence in the mRNA that specifies the incorporation of the indicated amino acid or termination of the polypeptide chain.
- Sydney Brenner and Francis Crick (1961) studied a number of acridine-induced mutations in the rII region of the DNA of phage T4. Their results led them to the major statement that each codon contains three bases.
- The code is a triplet, commaless and non-overlapping. Deciphering the code was done in two stages: specific triplet RNA sequences were synthesised and then determined the amino acid–tRNA complex that was bound by each sequence. In the second stage, MW Nirenberg and his colleagues made long artificial mRNAs (synthetic mRNAs) and determined the amino acids these mRNAs incorporated into protein.
- A group of three base codes for one amino acid, the code is not overlapping and the base sequence is read from a fixed starting point without punctuation. Except AUG for methionine and UGG for tryptophan, many of the 20 amino acids are encoded by more than one triplet; thus, the code is said to be degenerate.
- The genetic code is for the most part unambiguous *in vivo* under natural conditions. This means that each codon specifies the same amino acid all the time. However, two special circumstances exist where certain ambiguity is tolerated, but there is no loss of fidelity.
- The origin and evolution of the genetic code is still an unsolved problem. However, the discovery of variations in the code of mitochondrial DNA and the wobble nature of genetic code aided the considerations regarding the origin and development of the present code.

Self-assessment questions

1. Explain briefly the collinearity of a gene and a protein.
2. Explain the nature of the genetic code.
3. How did scientists decipher the genetic code?
4. Discuss the special properties of a codon.
5. What are the exceptions to the general genetic codes?

FURTHER READING

Textbooks

Cold Spring Harbor Symposia on Quantitative Biology (1966). *The Genetic Code.* Volume 31. New York: Cold Spring Laboratory.

Hartl DL and EW Jones (1998). *Genetics; Principles and Analysis.* Jones and Bartlett Publishers.

Lodish H, D Baltimore, A Berk, L Zipursky and P Matsudaira (1995). *Molecular Cell Biology.* Scientific American Books.

Strickberger MW (1999). *Genetics*, 3rd ed. New Delhi:Prentice-Hall India.

Woese CR (1967). The *Genetic Code:The Molecular Basis for Genetic Expression.* New York: Harper and Row.

References

Bonitz SG *et al* (1980). Codon recognition rules in yeast mitochondria. *Proc. Natl. Acad. Sci.* USA. 77:3167–70.

Brenner S, AOW Stretton and S Kaplan (1965). Genetic code: The 'nonsense' triplets for chain termination and their suppression. *Nature* 206:994–98.

Capecchi MR and G Gussin (1965). Suppression *in vitro*: identification of serine-sRNA as a 'nonsense' suppressor. *Science* 149:417–22.

Crick FHC (1963). On the genetic code. *Science* 139:461–64.

Crick FHC (1963). The recent excitement in the coding problem, *Prog. Nucleic Acid Res.* 1:164.

Crick FHC (1966). Codon–anticodon pairing: The wobble hypothesis. *J. Mol. Biol.* 19:548–55.

Crick FHC (1966). The genetic code III. *Scientific American* 215(4):55–62.

Crick FHC, L Barnett, S Brenner and RJ Watts–Tobin (1961). General nature of the genetic code for proteins. *Nature,* 192:1227–32.

Fox TD (1987). Natural variation in the genetic code. *Annu. Rev. Genet.* 21:67–91.

Frisch L, (ed) (1967). The genetic code Cold Spring Harbor Symposia, *Quant. Biol.* 31.

Garen A (1968). Sense and nonsense in the genetic code. *Science* 160:149–59.

Jukes TH (1983). Evolution of the amino acid code, In *Evolution of Genes and Proteins,* (eds) Nei M and RK Koehn. Sunderland: Sinauer Associates Inc.

Khorana HG (1966). Polynucleotide synthesis and the genetic code. *Harvey Lects.* 62:79–106.

Khorana HG (1968). Nucleic acid synthesis in the study of the genetic code. *In Nobel Lectures: Physiology of Medicine* (1963–1970). American Elsevier (1973).

Khorana HG (1968). Polynucleotide synthesis and the genetic code. *Harvey Lecture Series,* 62: 1966–67.

Knight RD and LF Landweber (2000). The early evolution of the genetic code. *Cell* 101:569–72.

Lagerkvist U (1981). Unorthodox codon reading and the evolution of the genetic code. *Cell* 23:305–6.

Leder P and MW Nirenberg (1964). RNA code words and protein synthesis II. Nucleotide sequence of a valine RNA code word. *Proc. Natl. Acad. Sci.* (USA) 52:420–27.

Leder P and MW Nirenberg (1964). RNA code words and protein synthesis III. On the nucleotide sequence of a cysteine and a leucine RNA code word. *Proc. Natl. Acad. Sci.* (USA) 52:1521–29.

Low SC and JM Berry (1996). Knowing when not to stop: Selenocysteine incorporation in eukaryotes. *Trends Biochem. Sci.* 21:203–208.

Nirenberg M and P Leder (1964). The effect of trinucleotides upon the binding of sRNA to ribosomes. *Science* 145: 1399–1407.

Nirenberg M and P Leder (1964). RNA code words and protein synthesis. *Science* 145:1319–1407.

Nirenberg MW (1963). The genetic code II. *Sci. Am.* 208: 80–94.

Nishimura S, DS Jones and HG Khorana (1965). The *in vitro* synthesis of a copolypeptide containing two amino acids in alternating sequence dependent upon a DNA-like polymer containing two nucleotides in alternating sequence. *J. Mol. Biol.* 13:302–24.

Sarabhai AS, AOW Stretton and S Brenner (1964). Collinearity of the gene with the polypeptide chain. *Nature* 201:13–17.

Speyer JF, P Lengyel, C Basilio, AJ Wahba, RS Gardner and S Ochoa (1963). Synthetic polynucleotides and the amino acid code. Cold Spring Harbor Symp. *Quant. Biol.* 28:559–68.

Szathmary E (1999). The origin of the genetic code: Amino acids as cofactors in an RNA world. *Trends Genet.* 15:223–29.

Vogel G (1998). Tracing the history of the genetic code. *Science* 281:329–31.

Wilhelm T and S Nikolajewa (2004). A new classification scheme of the genetic code. *J. Mol. Evol.* 59:598–605.

Yanofsky C (1967). Structural relationships between gene and protein. *Ann. Rev. genet.* 1:117–38.

Yanofsky D, BC Carlton, JR Guest, DR Helinski and U Henning (1964). On the collinearity of gene structure and protein structure. *Proc. Nat. Acad. Sci.* 51:266–72.

Zinder ND, DL Englehardt and RE Webster (1966). Punctuation in the genetic code. Cold Spring Harbor Symposia *Quant. Biol.* 31:251–56.

CHAPTER 12

Translation: Decoding the Genetic Information

LEARNING OBJECTIVES

After reading this chapter, you will be able to:

- List the components of the machinery for protein synthesis
- Discuss the mechanisms of protein synthesis in both prokaryotes and eukaryotes
- Appreciate the regulation of protein synthesis

12.1 INTRODUCTION

Translation or biosynthesis of proteins means the flow of information whereby the genetic information carried as a messenger RNA (mRNA) molecule is translated into a polypeptide. This is the final step in the central dogma, i.e., information flow from RNA to protein. The synthesis of every protein molecule in a cell is directed by an mRNA, which is originally copied from DNA. Although mRNA forms the template for correct addition of individual amino acids, both transfer RNA (tRNA) and ribosomal RNA (rRNAs) are also involved in the process. All proteins are synthesised from only 20 naturally occurring amino acids (there is one exception, selenocysteine). These are called α-amino acids because the α-carbon has four specific groups attached to it, imparting specific properties to the amino acid. Based on the optical properties, the two forms of each amino acid are referred to as D and L and interestingly, all biologically active amino acids are of the L form.

12.2 PROTEIN SYNTHESIS: A HIGH-ENERGY PROCESS

Among the many biological pathways, translation is the most conserved across all organisms and also the most energy-intensive for the cell. For example, in actively growing bacterial cells, protein synthesis requires up to 80% of the cell's energy and 50% of its dry weight. Therefore, synthesis of a single protein requires the coordinated action of well over 100 proteins and RNAs to be completed without errors. In Chapter 11, we discussed the nature of the genetic code and how tRNAs recognise the elements of the code. In this chapter, events that allow decoding of the mRNA will be discussed. The primary cellular machine involved in protein synthesis is the ribosome. This very abundant macromolecular particle contains a well-characterised arrangement of large ribosomal RNA molecules and numerous ribosomal proteins.

Rate of protein synthesis

The rate of protein synthesis is usually higher in prokaryotes (up to 20 amino acids per second). But, in some specialised eukaryotic cells, the rate of protein synthesis may be remarkably high. For example, silkworm (*Bombyx mori*) has two silk glands that constitute approximately one-quarter of the worm's mass and produce liquid silk (**Fig. 12.1**). This polymer is composed of a 350-kDa fibroin heavy chain, a 25-kDa fibroin light chain and a family of proteins called sericins that bind the two threads together as they emerge from the glands and harden on exposure to air. The genome of these cells is amplified several thousand-fold, resulting in a single nucleus that contains the equivalent of 400,000 haploid genomes. The specific protein synthesis activity in the posterior gland cells is remarkable: a total of approximately 1,000 cells produce up to 300 mg of protein per gland in four days. This corresponds to a specific productivity of about 80 μg protein/cell/day. By comparison, the best recombinant high-yield mammalian cell culture systems only produce about 50 pg protein/cell/day.

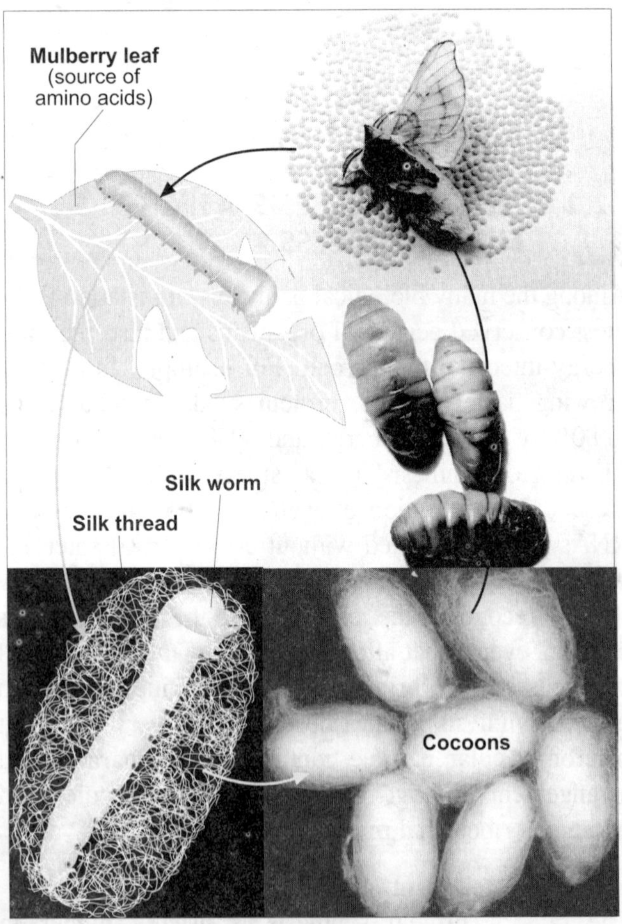

Fig. 12.1 Protein synthesis; photomicrograph of silkworm (*Bombyx mori*) cocoon woven with silk fibres (***Source:*** Channarayappa 2008)

It is obvious that the protein production capacity of silkworms exceeds that of any other industrial system in use today. Very interestingly, the glands of these caterpillars produce an almost pure product.

12.3 MAJOR COMPONENTS OF PROTEIN SYNTHESIS

Protein synthesis is a multistep process and involves all the three types of RNA and about 200 distinct accessory molecules. We shall now discuss in detail about mRNA which provides the template on which the code would be translated.

12.3.1 Messenger RNA

Messenger RNA (mRNA) is produced by transcription and forms the template for translation (see Chapter **9**). It brings the ribosomal subunits together and provides the coding sequence of bases (template) that determines the amino acid sequence in the resulting polypeptide chain. It has two essential features: open reading frame (ORF, a sequence of translatable codons) and a ribosome binding site (where the small ribosome subunit binds and the rest of the ribosome assembles). Translation starts at the 5' end of the ORF and proceeds one codon at a time to the 3' end. The first and last codons of an ORF are known as the start and stop codons. In bacteria, the start codon is usually 5'-AUG-3' but 5'-GUG-3' and sometimes even 5'-UUG-3' are also used. However, in eukaryotes, the start codon is always 5'-AUG-3'. The start codon provides two important functions: (i) it specifies the first amino acid to be incorporated into the growing polypeptide chain and (ii) it defines the reading frame for all subsequent codons. This is very important since, misreading of one or two nucleotides in any direction may result in drastic changes in the amino acid sequence. The initiation codon and the stop codons may form at new positions or they may be lost from the sequence itself (**Fig. 12.2**).

The differences between the mRNAs of prokaryotes and eukaryotes are as follows.

Site of translation: In bacteria, transcription and translation occur simultaneously in the same cellular compartment, whereas in eukaryotes, transcription is restricted to the nucleus and RNA must be exported into the cytoplasm for translation (**Fig. 12.3**).

Half-life of RNA: Bacterial mRNA has a limited half-life (several minutes for the most stable transcripts).

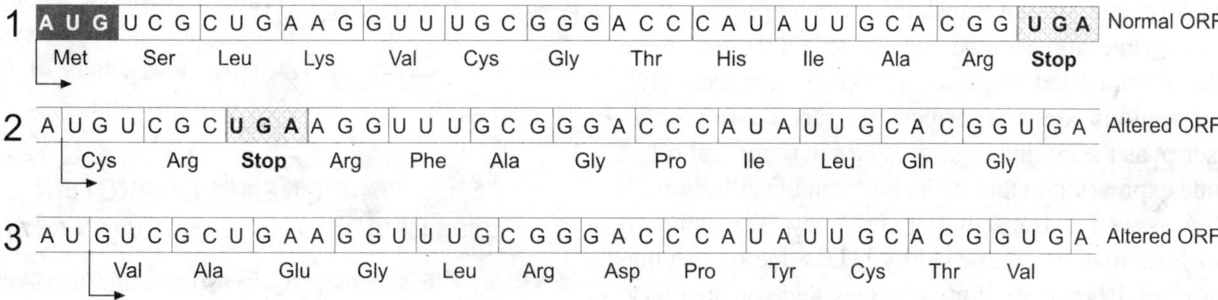

Fig. 12.2 Three possible reading frames of the *E. coli* tryptophan leader sequence; start codons are shown with shaded boxes and stop codons with hatched boxes.

Some eukaryotic mRNAs are also unstable, but most are stable for hours or even days (e.g., in eggs).

RNA processing: Bacterial transcripts are used directly for translation, whereas eukaryotic transcripts are extensively processed and modified. Processing reactions include the splicing of introns, 5' capping and 3' end polyadenylation; these changes may regulate the efficiency of translation, either directly or by modulating mRNA stability. Some transcripts are also edited.

Binding site: Ribosome binding in bacteria is also referred to as a **Shine–Delgarno sequence (SD)**, typically located on 3–9 bp on the 5' side of the start codon, which complements a part of the ribosomal 16S rRNA (**Table 12.1**). Binding occurs wherever this sequence is found, including internally, allowing bacterial transcripts to be **polycistronic**. On the other hand, eukaryotic ribosomes are docked onto the mRNA by a protein which recognises the modified 5' cap. Therefore, binding cannot occur internally and eukaryotic mRNAs are almost universally **monocistronic**. Some eukaryotic mRNAs contain a purine at three bases upstream of the start codon and a guanine immediately downstream (5'-G/ANNAUGG-3'). Presence of this sequence increases the efficiency of translation and was originally identified by Marilyn Kozak and is referred to as the **Kozak sequence**.

Table 12.1 Representative Shine–Delgarno (SD) sequences

Message for	*SD sequences and start codon*
Ribosomal protein L10	5'-AGGAGCAAAGCUA**AUG**-3'
E. coli lac z	5'-AGGAAACAGCU**AUG**-3'
λ phage Cro	5'-UAAGGAGGUUGU**AUG**-3'

Coding and non-coding (untranslated) regions of mRNA

Coding regions are composed of codons, which are decoded and translated into polypeptide by the translation machinery (i.e., on ribosomes). As previously mentioned, coding regions begin with the start codon and end with one of the three possible stop codons. In addition to their role in protein synthesis, portions of coding regions may also serve as regulatory sequences (**exonic splicing enhancers** or **exonic splicing silencers**).

Untranslated regions (UTRs) are sections of the RNA transcript before the start codon and after the

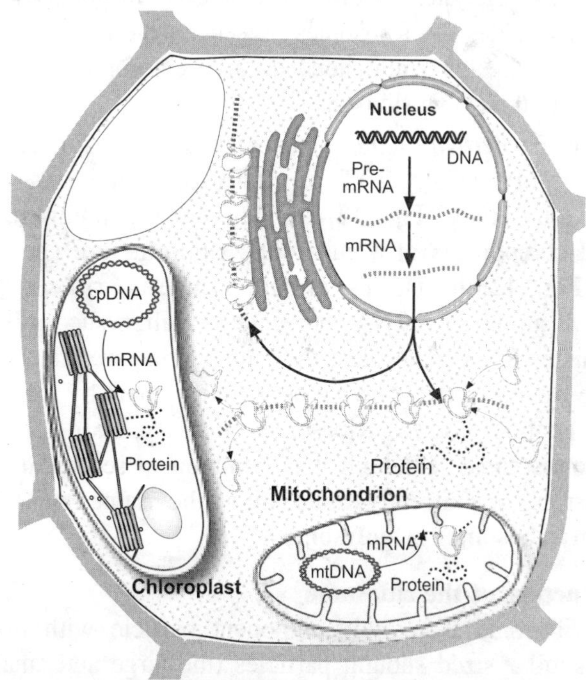

Fig. 12.3 Sites of protein synthesis in a plant cell; a typical plant cells synthesises proteins in three distinct compartments—the cytosol, plastids and mitochondria

stop codon that are not translated, which are termed as the five prime untranslated regions (5'UTR) and three prime untranslated regions (3'UTR), respectively. These regions are transcribed as part of the same transcript as the coding region. UTRs play several roles in gene expression including stability and localisation of mRNA as well as translational efficiency. The stability of mRNAs may be controlled by UTRs due to varying affinity for RNA degrading enzymes and for ancillary proteins that can promote or inhibit degradation. Some of the elements contained in the untranslated regions form a characteristic secondary structure when transcribed into RNA. One class of mRNA element, the **riboswitches**, directly binds small molecules, changing their fold to modify levels of transcription or translation. Thus, mRNA regulates itself.

12.3.2 Ribosomes

The highly specific chemical interactions involved in translating mRNA into protein occur on specialised macromolecular complexes called ribosomes. The mRNA with its encoded information and the individual tRNAs loaded with their correct amino acids are brought together in the ribosome. It directs the elongation of a polypeptide at a rate of three to five amino acids added per second. Thus, the primary role of ribosomes is that of being the sites of protein synthesis. In eukaryotic cells, besides the ribosomes present in the cytoplasm, cell organelles (mitochondria and chloroplast) also contain their own ribosomes.

In *Escherichia coli,* ribosomes account for 25% of the dry weight, 10% of total protein and 80% of the total RNA (**Fig. 12.4a)**. The corresponding size and components of a eukaryotic ribosome are shown in **Fig. 12.4b**. During protein synthesis, ribosomes move along an mRNA molecule and align successive transfer RNA molecules; the amino acids are attached one by one to the growing polypeptide chain by means of peptide bonds. Ribosomes may function alone (**monosomes**), although it is common to see them in clusters concurrently translating the same mRNA (**polyribosomes** or **polysomes**).

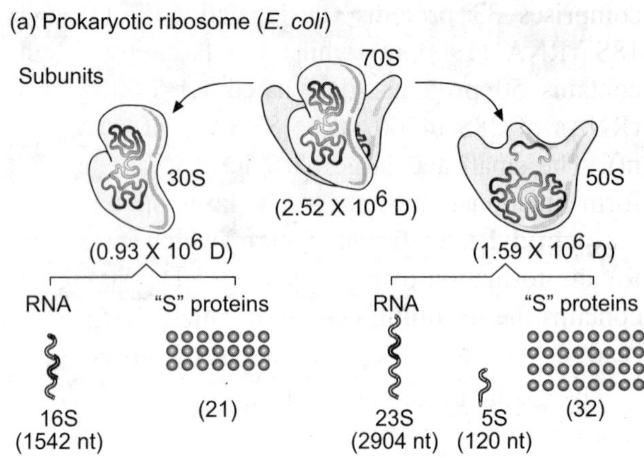

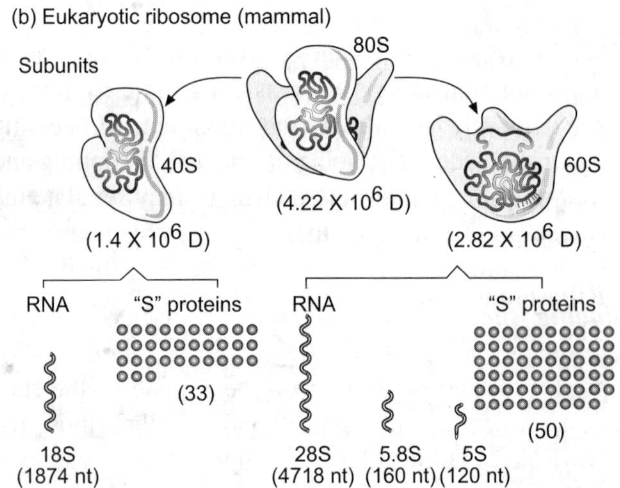

Fig. 12.4 The organisation and composition of ribosomes (a) Prokaryotic (b) Eukaryotic

Structure of the ribosome

A ribosome is a multicomponent particle with two dissimilar sized subunit particles (the large and small subunits), which are held by the magnesium ion. Each subunit consists of several ribosomal RNAs (rRNAs) and numerous ribosomal proteins (r-proteins). Size of the intact ribosome and its subunits is often expressed in Svedberg units.

An *E. coli* ribosome has a mass of about 27 kd, diameter of approximately 200 Å and sedimentation coefficient of 70S. A 70S ribosome can be dissociated into a large subunit 50S and a small subunit 30S (**Fig. 12.4a**). These subunits can be further split into their constituent proteins and RNAs. The 30S subunit contains 21 different proteins (designated S1–S21) and a 16S RNA molecule (1,542 nt). The 50S subunit contains 34 different proteins (designated L1–L34, L12 is present in four copies) and two RNA molecules, a 23S (2904 nt) and a 5S (120 nt) species. Both the 30S and the 50S subunits can be reconstituted *in vitro* from their constituent proteins and RNA. Some proteins are common to both subunits (e.g., L6+S20). The counterparts in eukaryotes are 40S and 60S and both together make 80S ribosomes. The small (40S) subunit comprises 33 proteins (designated S1–S33) and the 18S rRNA (1900 nt) whilst the large (60S) subunit contains 50 proteins (designated L1–L50) and three rRNAs of 28S (4700 nt), 5.8S (156 nt) and 5S (120 nt). The small and large particles combining together form functional ribosomes. The large subunit contains the **peptidyl transferase centre**, which is responsible for the formation of peptide bonds. The small subunit contains the **decoding centre** in which charged tRNAs read or 'decode' the codon units of the mRNA.

The spatial organisation of the ribosome is complex. In a ribosome, rRNA makes up 60%–65% and protein is about 40% of the total mass and these are essential for structural integrity and function. The rRNA adopts complex tertiary and quaternary conformations through base pairing within and between the molecules. The manner in which proteins interact with each other and RNA within this network has been studied in great detail for the *E. coli* ribosome using cross-linking and footprinting studies and structural analysis by neutron scattering and diffraction.

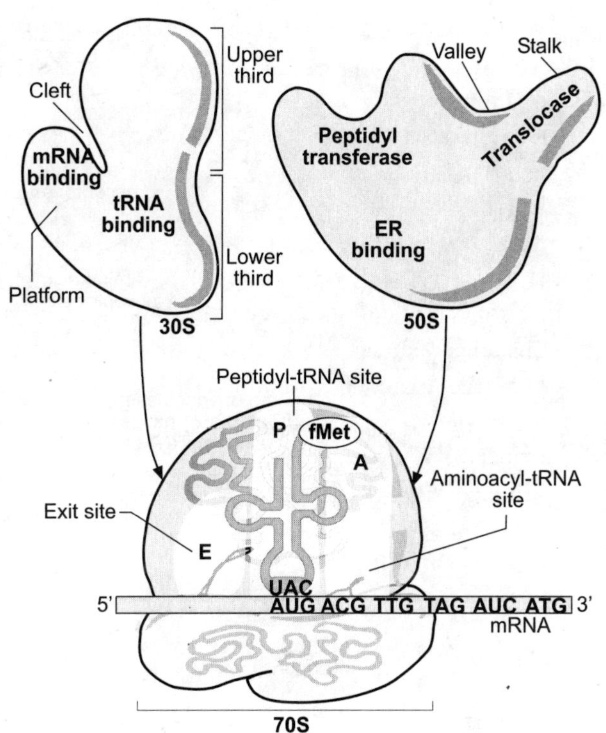

Fig. 12.5 Functional view of the *E. coli* large and small ribosomal subunits; A, P and E are tRNA binding sites

Ribosomal subunits as functional sites for translation

The ribosomes in a typical eukaryotic cell can collectively make about one million peptide bonds per second. Several domains of the ribosome have particularly important functions during protein synthesis (**Fig. 12.5**).

The ribosomes have three binding sites for tRNA. During prokaryotic translation, mRNA lies between the 30S and 50S subunits. The small subunit contains an mRNA binding site and two major binding sites for tRNA (**Fig. 12.5**). These sites, or cavities in the ribosome, are referred to as **A-sites** (amino acyl-tRNA site) which bind incoming charged tRNAs during elongation, whilst the **P-site** (peptidyl-tRNA site) binds the tRNA carrying the nascent polypeptide chain. Bacterial ribosomes further contain an **E-site** (exit site) to which spent tRNAs are despatched prior to ejection. The tRNAs in the A and P sites are angled at 26° relative to each other at their anticodons. The peptide chain is transferred from the peptidyl-tRNA in the P site to the aminoacyl-tRNA in the A site. The large subunits possess a peptidyl transferase domain, which provides the catalytic activity for the formation of peptide bond and a GTPase domain, which is required for translocation of the ribosome along the mRNA.

The structure of tRNA

All transfer RNA molecules have a common design.

Primary structure: The tRNAs are the adaptor molecules that deliver amino acids to the ribosome and decode the information in mRNA. Their primary structure is 60–95 nt long (18–28 kDa), but most commonly 76, which are extensively processed during their production. They have many modified and hypermodified bases. About 50 different types of modified bases in any tRNA molecule have been identified. About 20% of the bases in tRNA is conserved and play an essential role in the adoption of tertiary structure.

Secondary structure: All tRNAs exhibit cloverleaf-like secondary structure (**Fig. 12.6**).

Acceptor stem: The stem is called the amino acid acceptor stem, working 5' to 3' (anti-clockwise). It consists of a helix of seven pairs of bases generally making Watson–Crick base pairs (i.e., A:U or G:C) but occasionally with a G:U base pair, together with an unpaired sequence of four bases at the end of the 3'-strand of the stem. The last three bases are always 5'-CCA-3' and it is to the 3'-terminal adenosine residue that the amino acid is attached.

Dihydrouridine (D) arm: The next secondary structural feature is called the dihydrouridine arm or the D or DHU arm. It is composed of a 3- or 4-bp stem and a

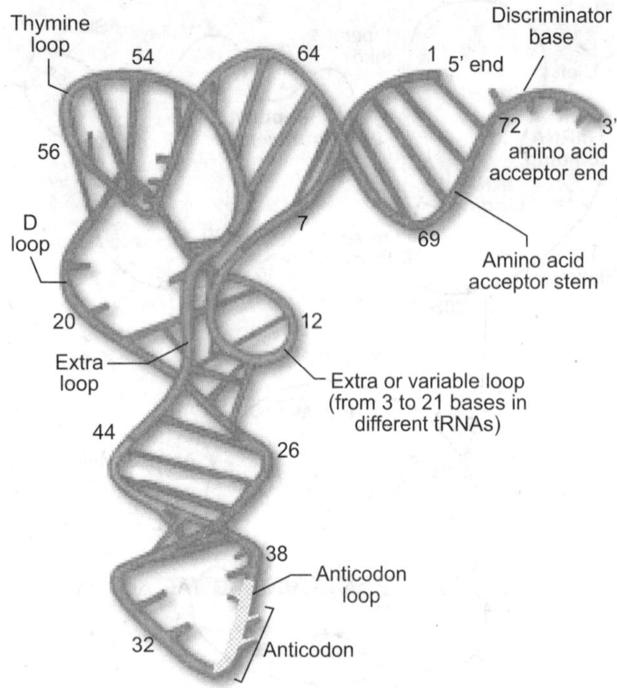

Fig. 12.6 A three-dimensional representation of $tRNA^{Phe}$ showing the L form of the molecule according to x-ray diffraction studies

loop called the D-loop (DHU-loop I) usually containing eight to eleven nucleotides, some of which are invariant. There are several species without this arm.

The anticodon arm: The next structural feature consists of a stem with 5 bp and loop with seven residues in which there are three adjacent nucleotides called anticodon which are complementary to the codon sequence that the tRNA recognises. The 5'-base of the anticodon, the base in the 'wobble' position, is frequently a modified or non-standard base.

The extra arm (III): This is an extremely variable arm and can have between three and 21 residues and may form a stem of up to 7 bp. It may be small containing only 2–3 nt (class I tRNAs) or larger with 13–21 nt with up to five base pairs in a stem (class II tRNAs).

The TψC arm (loop IV): It consists of a 5-bp stem ending in a loop containing the invariant residues GYTC, where ψ is a modified nucleotide called pseudouracil.

There are nine hydrogen bonds (tertiary hydrogen bonds) that help to form the cloverleaf structure of tRNA molecules. Base pairing between residues in the D- and T-arms fold the tRNA molecule over into an L-shape, with the anticodon at one end and the amino acid acceptor site at the other. All tRNAs have a 5'-terminal phosphate. They contain 13 invariant positions and 8 semi-variant positions.

Tertiary structure: The tertiary structure of several tRNAs has been studied by X-ray crystallography. The molecules adopt a compact, L-shaped conformation with *TψC* and acceptor stems forming one continuous double helix and the D loop and anticodon loop forming another. These helices are held perpendicular to each other, with the *TψC* loop and D-loop forming a rigid core and the acceptor stem and anticodon loop forming flexible arms at opposite ends of the molecule. The amino acid is attached to the 3'-end terminal nucleotide (always an adenine). The tertiary structure is stabilised by base pairing and base stacking.

Aminoacylation of tRNA

The main function of tRNA is to ensure that each amino acid incorporated into a protein corresponds to a particular codon in the mRNA. Cells usually contain between 31 and 40 individual species of tRNA, each of which binds specifically to one of the 20 amino acids. Consequently, there may be more than one tRNA for an amino acid. The tRNAs that bind to the same amino acid are called **isoacceptors**. How does a tRNA recognise the codon(s) specifying a given amino acid and attach to it? This crucial process is catalysed by the different **aminoacyl-tRNA synthetases (ARSs)**. In bacteria, there are 21 different ARSs one for each amino acid, except methionine, which has two ARSs. In eukaryotes, there are separate sets of 21 each of cytoplasmic and mitochondrial synthetases, all coded in the nucleus. Each of these recognises one amino acid and all its compatible or **cognate tRNAs**. This mapping is achieved through the **recognition domains** on the tRNA. Each enzyme catalyses the ATP-dependent esterification of its specific amino acid to the 3'-end of its cognate tRNA molecules. A tRNA attached to its amino acid is called an **aminoacylated tRNA** or a **charged tRNA (Fig. 12.7)**.

12.3.3 Factors required for translation

In prokaryotes, the factors are abbreviated as IF or EF for **initiation** and **elongation** factors respectively, whereas in eukaryotes they are called eIF and eEF (**Table 12.2**).

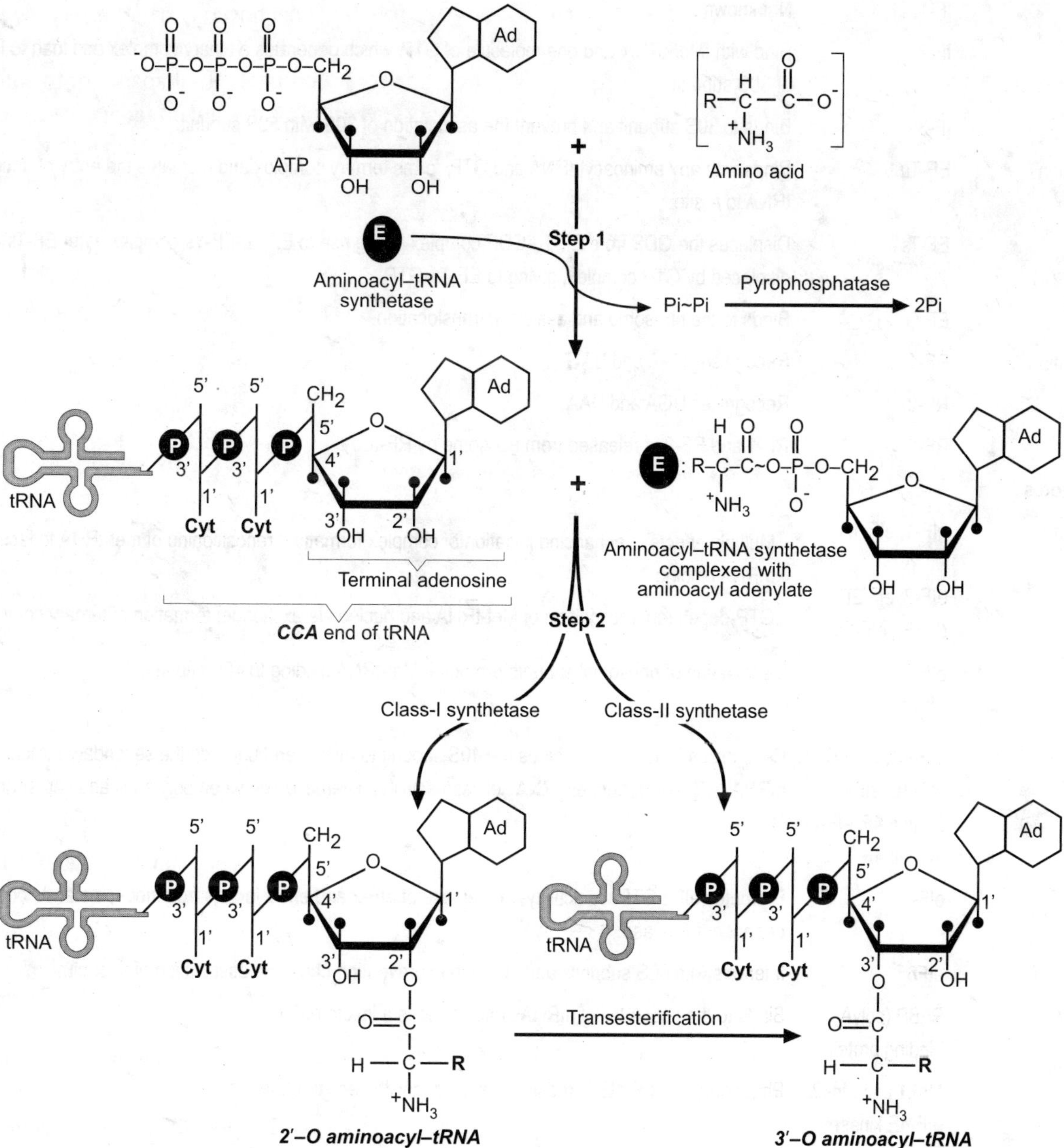

Fig. 12.7 Aminoacylation of tRNA; amino acids are covalently linked to cognate tRNAs by aminoacyl–tRNA synthetase

Table 12.2 Translation factors and their functions

Role	Name of class	General role
Prokaryotes		
	IF-1	Not known
Initiation	IF-2	Bind with fMet-tRNA$_f$ and one molecule of GTP, which generates a binary complex and load to P site of 30S subunit.
	IF-3	Binds to 30S subunit and prevent the association of 30S with 50S subunit.
Elongation	EF-Tu	Binds with any aminoacyl-tRNA and GTP, forms ternary complex and mediates the entry of aminoacyl-tRNA to A-site.
	EF-Ts	Displaces the GDP from EF-Tu-GDP complex giving rise to EF-Tu-EF-Ys complex, later EF-Ts itself displaced by GTP complex, giving to EF-Tu-GTP.
	EF-G	Binds to the ribosome and assists in translocation.
Termination	RF-1	Recognises UAA and UAG.
	RF-2	Recognises UGA and UAA.
	RF-3	RF-1 and RF-2 is released from ribosome by RF-3.
Eukaryotes		
Initiation	eIF-1, eIF-1A	Multiple effects in enhancing initiation of complex formation; repositioning of met-tRNA to facilitate mRNA binding.
	eIF-2, eIF-2B (GEF), eIF-2c	GTP-dependent recognition of Met-tRNA and nucleotide exchange; formation of ternary complex.
	eIF-3	Dissociation of ribosomal subunit; promotes Met-tRNA binding to 40S subunit.
	eIF-4A, eIF-4B, eIF-4E, eIF4F, eIF(iso)4F, eIF-4G, eIF4H	Recognises 5' cap mRNA; binds the 40S subunit to mRNA and unwinds the secondary structure of mRNA (ATPase-dependent RNA helicase activity); interaction between poly A tail and cap structure.
	eIF-5	Promotes eIF2 GTPase activity and release of eIF-2 and eIF-3 factors from ribosome and ribosome-dependent GTPase.
	eIF6	Interacts with 60S subunits and involved in preventing ribosome association of subunit.
	PABP (polyA-binding protein)	Binds to the poly A tail of mRNAs and provides a link to eIF-4G.
	Mnk1 and Mnk2 eIF-4E kinases	Phosphorylate eIF-4E increasing association with cap structure.
	4E-BP (also called PHAS) 3 known forms	When dephosphorylated 4E-BP binds eIF-4E and represses its activity, phosphorylation of 4E-BP occurs in response to many growth stimuli leading to release of eIF-4E and increased translational initiation.
Elongation	eEF-1α, eEF-1$\beta\gamma$,	Forms ternary complex with GTP and the charged tRNA and promotes binding of the charged tRNA to the A-site of the ribosome.
	eEF-2	Formation of ternary complex.
Termination	eRF-1,	Recognises the stop codon at A-site.
	eRF-3	Binds with GTP.

12.3.4 Start signal and initiator region

The start signal is AUG (or GUG) preceded by several bases that pair with 16S rRNA. Binding of ribosome to AUG codon automatically puts the ribosome in the proper reading frame so that it correctly reads the entire message from that point on. All the known mRNA molecules contain signals that define the beginning and end of each encoded polypeptide chain (**Fig. 12.8**). For example, in -CAUTGC**AUG**CAGTUGAACGCU- the ribosome moves from the initiation codon, AUG, to the next three nucleotides, CAG and so on along the entire line. Thus, two kinds of interactions determine where protein synthesis starts: (i) the pairing of mRNA base with the 3'-end of 16S rRNA and (ii) the pairing of the initiator codon on mRNA with the anticodon of an initiator rRNA molecule.

12.4 TRANSLATION ON PROKARYOTIC RIBOSOME

In bacteria, the ribosomes bind to an mRNA chain and move along mRNA chain in the direction of 5'→3'. In this section, bacterial protein synthesis will be described using the three-site model of the ribosome.

12.4.1 Chain initiation

For successful initiation, the following three steps are necessary:

- The ribosome must be recruited to the mRNA.
- A charged tRNA must be placed into the P-site of the ribosome.
- The ribosome must be precisely positioned over the start codon. This step is very critical.

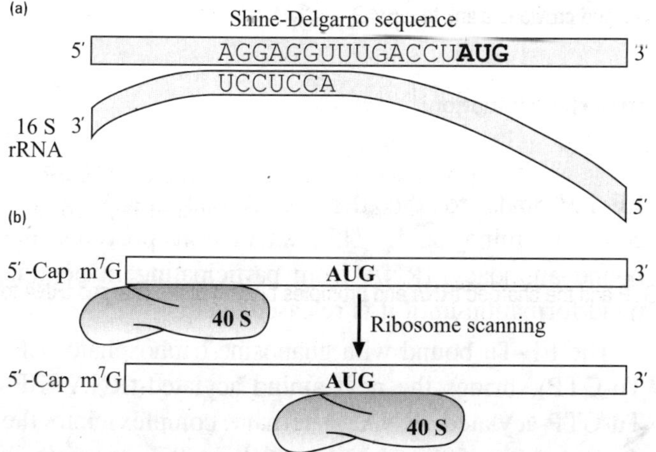

Fig. 12.8 Signal sequence for translation initiation in mRNA (a) Prokaryotic mRNA (b) Eukaryotic mRNA

The components involved are the large and small ribosomal subunits, the mRNA, the initiator tRNA in its charged form and initiation factors. Initiation of protein synthesis in *E. coli* requires three initiation factors IF1, IF2 and IF3 and GTP.

Formation of the 30S initiation complex

The initiation of prokaryotic translation begins with the small subunit and is catalysed by three translation initiation factors, **IF1**, **IF2** and **IF3**.

Recruitment of small ribosomal subunit to the initiation codon: Once ribosomes are dissociated into 50S and 30S subunits, the cell builds a complex on the 30S ribosomal subunit, including mRNA, charged tRNA and initiation factors. The initiation factors IF1 and IF3 interact with the 30S subunit; IF2 binds with a special initiator tRNA charged with methionine. These components come together (known as **30S initiation complex**) and combine with the AUG codon on the mRNA. The 30S initiation complex binding to the mRNA typically involves base pairing between the ribosome binding site and the 16S rRNA in the small subunit. IF3 either directly or indirectly facilitates the recognition of the untranslated 'leader' sequences that precede the initiating AUGs of natural mRNAs (**Fig. 12.9**).

Formation of the 70S initiation complex: For elongation to occur, the large ribosomal subunit (50S) must join the 30S initiation complex to form the **70S initiation complex**. The 30S + $tRNA_f^{Met}$ + mRNA complex recruits a 50S subunit, which results in the release of IF3. The $tRNA^{Met}$ is positioned in the P-site (the only charged $tRNA_f^{Met}$ that can be placed directly in P-site) and aligned with the AUG initiation codon, forming the 70S initiation complex. Binding with tRNA is accomplished by hydrogen bonding between the AUG codon in the mRNA and the three-base anticodon in the tRNA.

12.4.2 Elongation

Once the ribosome is assembled with the charged initiator tRNA in the P-site, polypeptide synthesis can begin. Elongation requires participation of elongation factors **EF-Tu** (also called EF1A), **EF-Ts** (EF1B) and **EF-G** (EF2). The elongation (**Fig. 12.10**) stage of translation consists of four steps:

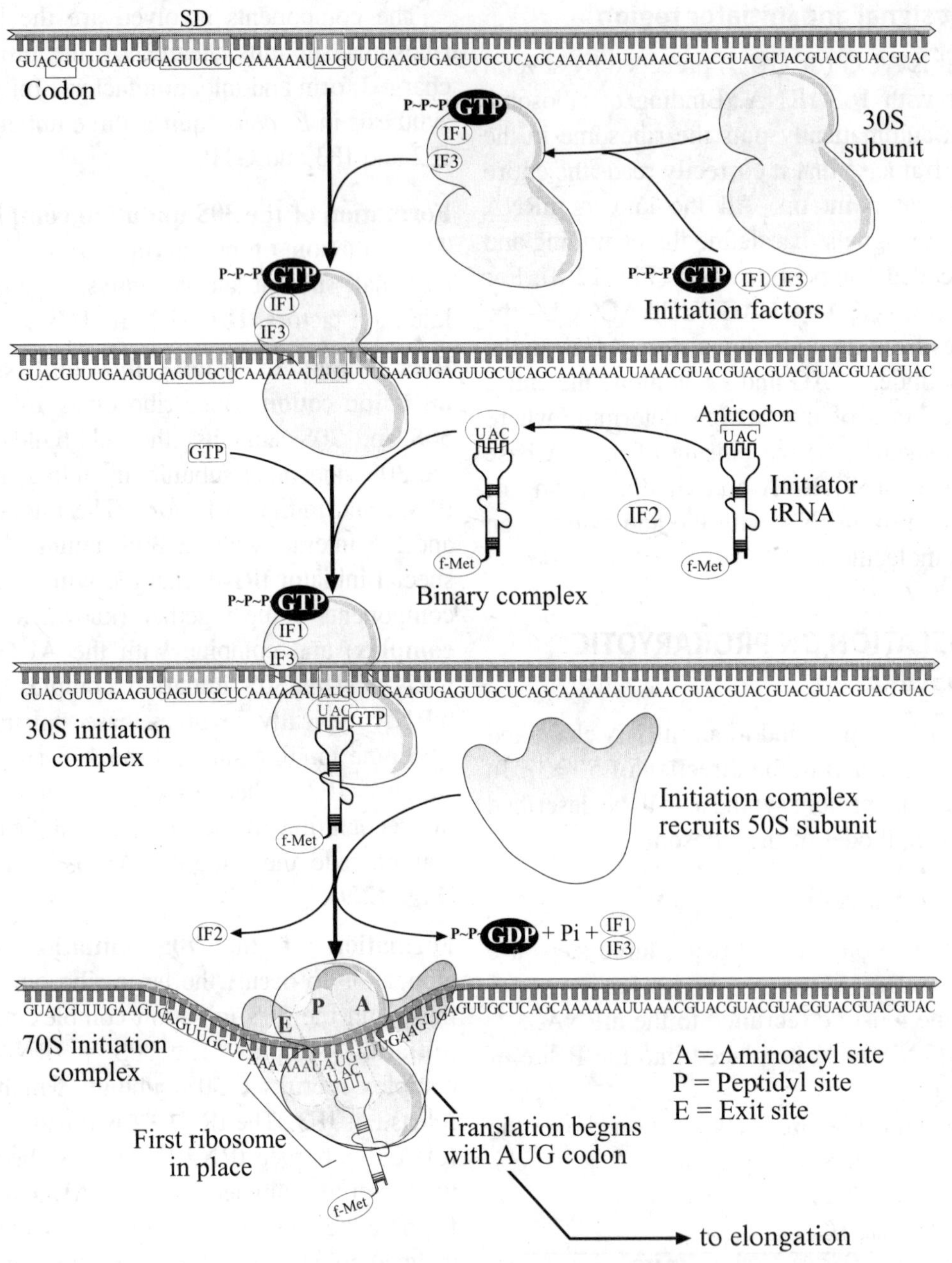

Fig. 12.9 Translation in bacteria: The sequence of events in peptide chain initiation

Selection of aminoacyl-tRNA

With the charged initiator tRNA in place within the P-site, the ribosome is available for the second aminoacyl-tRNA to enter into the vacant A-site. This is the first step in elongation. The positioning of the second tRNA is specified by the codon at the A-site. Aminoacyl-tRNAs do not bind to the ribosome on their own. Instead, they are 'escorted' to the ribosome by the elongation factor. Once a tRNA is aminoacylated, EF-Tu binds to the tRNA's 3'-end, masking the coupled amino acid. This interaction prevents the bound aminoacyl-tRNA from participating in peptide bond formation until it is released again.

The EF-Tu bound with guanosine triphosphate (EF-Tu-GTP), brings the next amino acylated-tRNA (EF-Tu-GTP-acylated tRNA, a ternary complex) into the A-site on the 50S subunit, which in this example is $tRNA^{Phe}$. When EF-Tu delivers the aminoacyl-tRNA

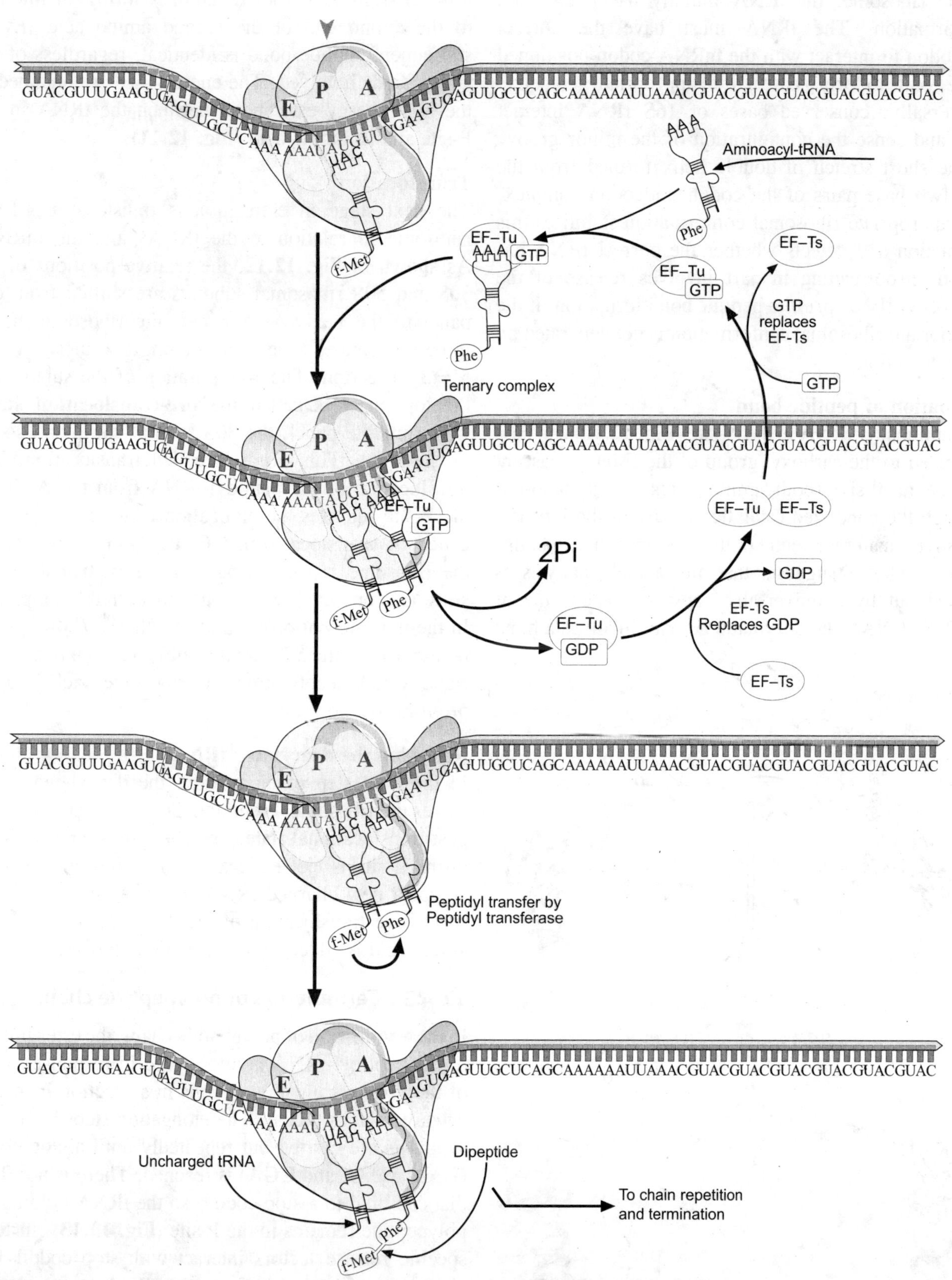

Fig. 12.10 Translation in bacteria: The cycle of events in peptide chain elongation

to the ribosome, the tRNA initially has a distorted conformation. The tRNA must have the correct anticodon to interact with the mRNA codon positioned at the A-site to form a base pair of appropriate geometry. Universally, conserved bases of 16S rRNA interact with and sense the configuration of the minor groove of the short stretch of double helix formed from the first two base pairs of the codon–anticodon complex. The appropriate ribosomal conformation stabilises this interaction and detects whether the correct tRNA has bound. Proofreading in part involves release of the aminoacyl-tRNA prior to peptide bond formation, if the appropriate ribosomal conformation is not generated by this interaction.

Formation of peptide bond

The aminoacyl-tRNA in the A-site (**Fig. 12.10**) can now be linked to the carboxyl group of the fMet or nascent peptide at P-site (both amino acids are juxtaposed) through the concerted catalytic activity of the intrinsic peptidyl transferase centre of the 50S ribosomal subunit. Some evidence suggests that the actual catalysis is carried out by a universally conserved adenosine of the 23S rRNA. The enzymatic activity involves a bond transfer from the carboxyl end of N-formyl methionine to the amino end of the second amino acid. Every subsequent peptide bond is identical, regardless of the amino acids involved. The energy used is contained in the high-energy ester bond between the tRNA in the P-site and its amino acid (**Fig. 12.11**).

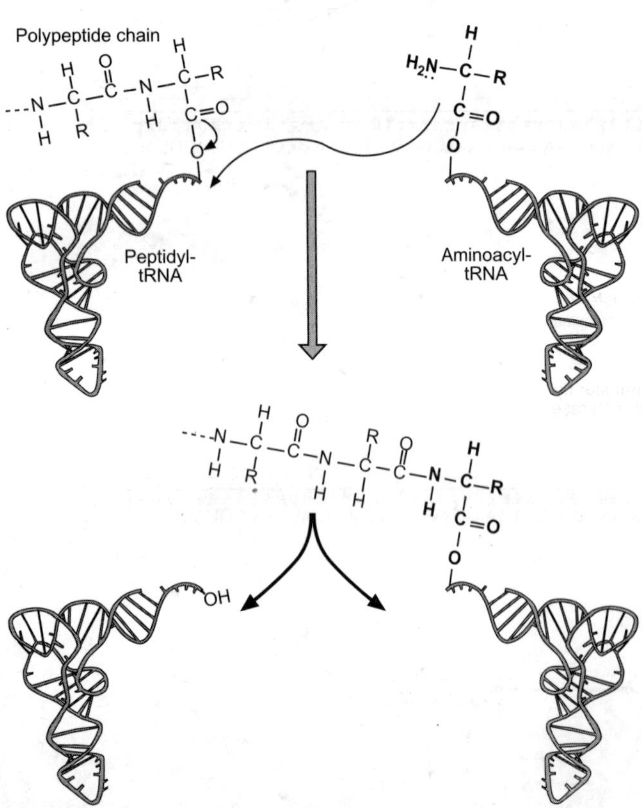

Fig. 12.11 The peptidyl transferase reaction

Translocation

The next stage in elongation is translocation of the ribosome in relation to the tRNAs and the mRNA. As shown in **Fig. 12.12**, the relative positions of the 30S and 50S ribosomal subunits are shifted from one panel to the next. As a result, the ribosome moves three nucleotides (one codon) along the mRNA in the 5'→3' direction. The configuration of the subunits in the top panel is called the **pre-translocation state**. In the middle panel, the 30S subunit shifts one codon to the right. This event is called translocation. The translocation of the peptidyl-tRNA from the A-site to the P-site requires the elongation factor EF-G (earlier called as **translocase**) and GTP. After translocation, the ribosome is said to be in the **post-translocation state** and one cycle of elongation is thus, completed. In the next step of polypeptide synthesis, shown in the bottom panel, the 50S subunit shifts one step over to the right, which reconfigures the ribosome back into the pre-translocation status.

Releasing the deacylated tRNA

In the final step of elongation, the deacylated tRNA leaves the ribosome, emptying the E-site. After the first post-translocational state is reached, the depleted tRNA in the E-site is ejected, leaving the ribosome ready to accept a new charged tRNA in the A-site. With each successive translocation of the 30S subunit, one more amino acid is added to the growing polypeptide chain.

12.4.3 Termination of polypeptide chain

In the termination of polypeptide chain, the ester linkage of the peptidyl-tRNA is hydrolysed in response to one of the three termination codons in a reaction involving **release factors** (**RFs**). The elongation steps of protein synthesis are carried out repeatedly until a stop codon (**UAA**, **UAG**, and **UGA**) is reached. There is no tRNA that can bind to a stop codon, so the tRNA holding the polypeptide remains in the P-site (**Fig. 12.13**). Instead, specific release factors interact with stop codons and cleave the polypeptide from the tRNA to which it is attached as well as to disassociate the 70S ribosome from the mRNA, after which the individual 30S and

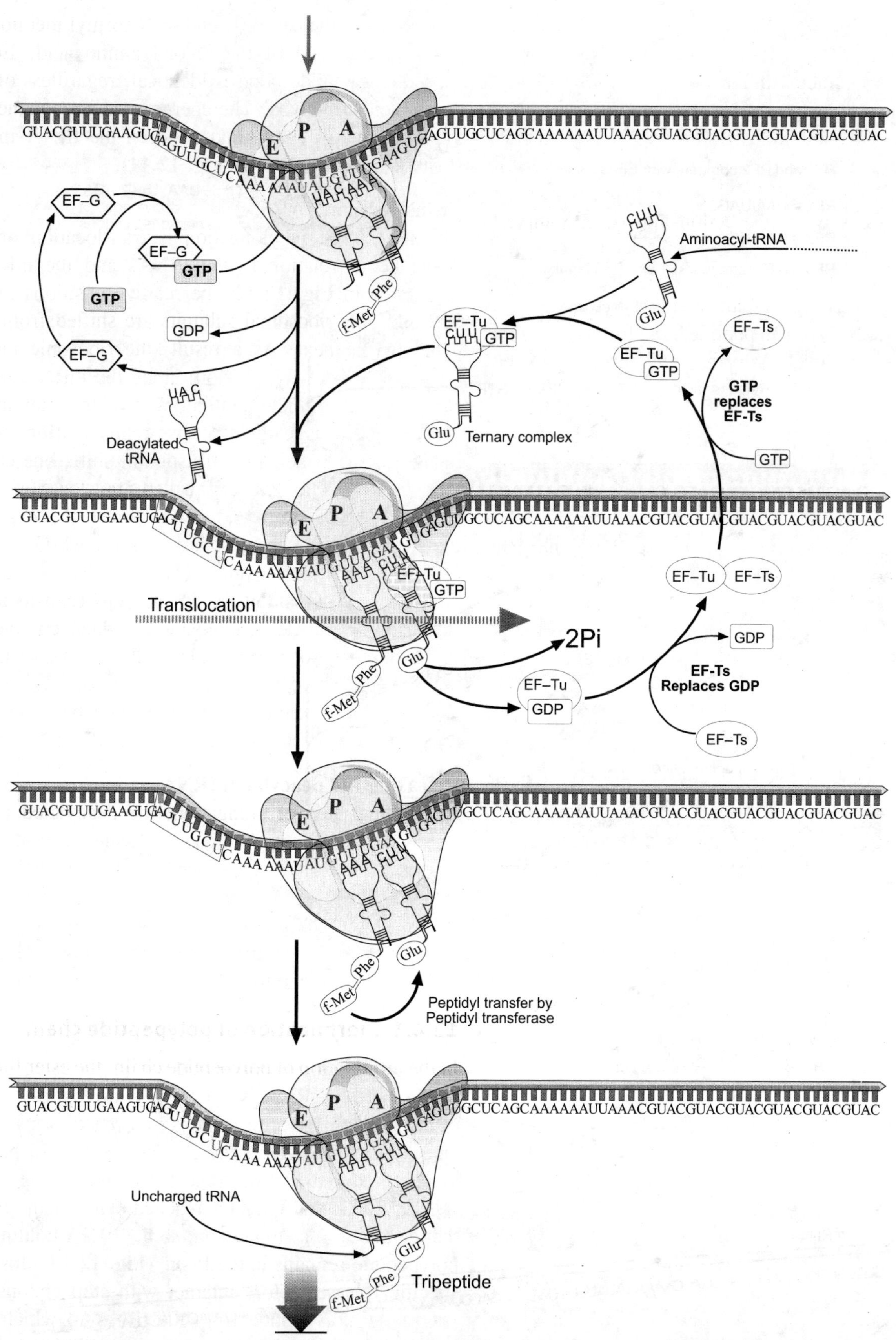

Fig.12.12 Translation in bacteria: Continues elongation cycle until 'n' amino acids added

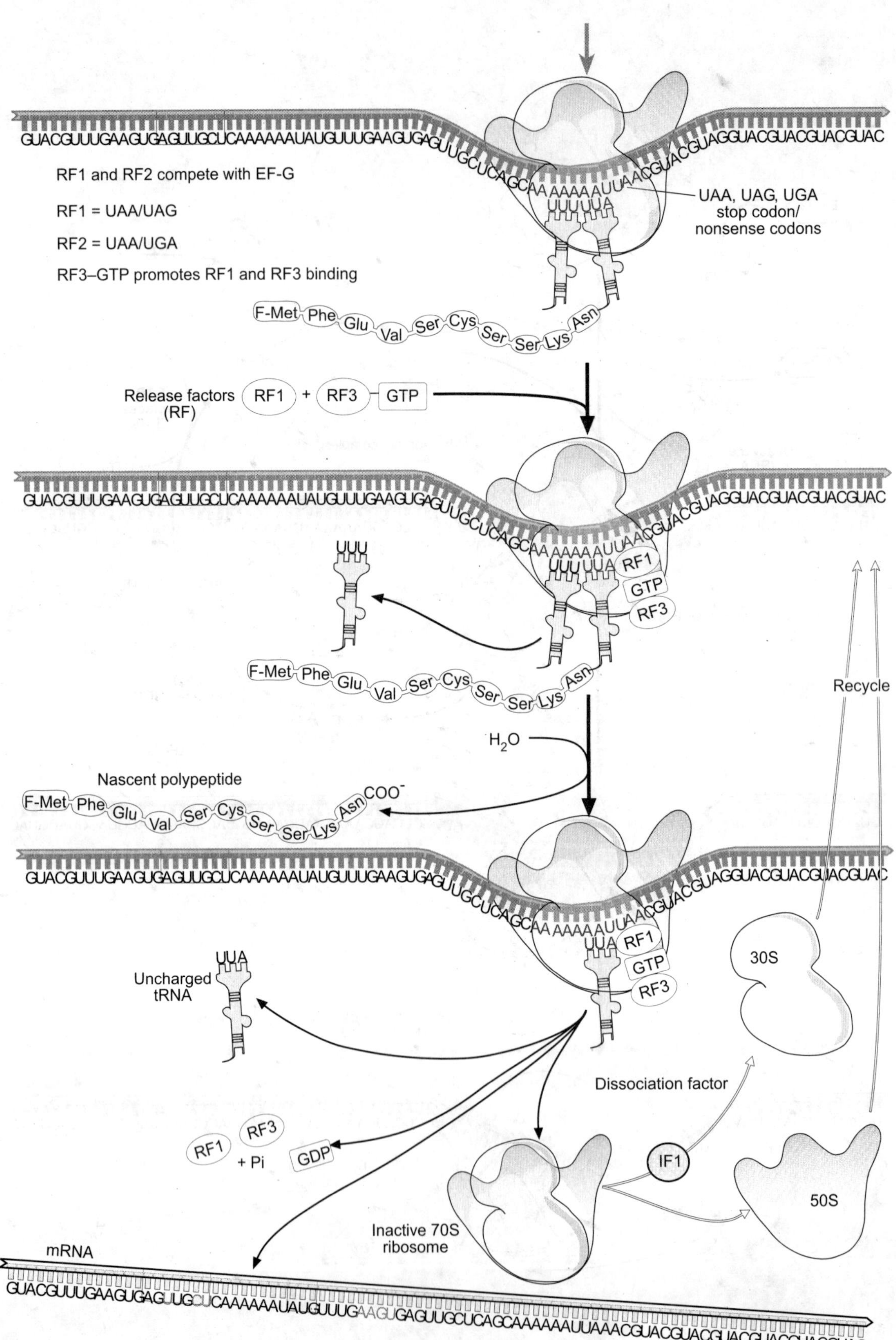

Fig. 12.13 Translation in bacteria: The events in peptide chain termination

50S subunits are recycled to initiate translation of another mRNA.

How do the release factors recognise stop codons? Experiments in RF1 and RF2 proteins have pinpointed that just three amino acids are responsible for the specificity of recognition of stop codons. The region defined by these amino acids represents a peptide anticodon that interacts with and recognises stop codons.

After the class I release factor has triggered the hydrolysis of the peptidyl-tRNA linkage, it will be removed from the ribosome by RF3 factor. RF3 is a GTP-binding protein but this factor has a higher affinity for GDP than GTP. The binding of GTP to RF3 leads to the formation of a high-affinity interaction with the ribosome that displaces the class I factor from the ribosome. This change also allows RF3 to associate with the factor-binding centre of the large subunit and also stimulates the hydrolysis of GTP.

12.5 TRANSLATION IN EUKARYOTES

The mechanism of protein synthesis differs from prokaryotic translation. Most of these differences are in the initiation phase where more eIFs are involved (**Fig. 12.14**). The eukaryotic initiation begins with methionine, not N-formyl-methionine. However, the initiator $tRNA_i^{Met}$ is different from the one that adds methionines to the interiors of polypeptides ($tRNA_m^{Met}$). Another major difference is that eukaryotic mRNAs contain no Shine–Dalgarno sequence; instead most eukaryotic mRNAs have caps at their 5'-ends.

The translation process in eukaryotic cells has the following phases: scanning, initiation, elongation and termination.

12.5.1 Scanning

The method of finding the correct start codon is through the **Kozak scanning mechanism**. Marilyn Kozak developed this model in 1978, based on four considerations:

- Eukaryotic translation was not initiated at an internal AUG.
- Initiation did not occur at a fixed distance from the 5'-end of the mRNA.
- In all the mRNA molecules analysed, the first AUG downstream of the cap was used for initiation.
- The cap at the 5'-end of the mRNA facilitates initiation.

Since there is no ribosome-binding sequence (similar to the Shine–Delgarno sequence), the mechanism of selecting the start codon must be different. The 40S subunit already containing the initiator tRNA attaches to the 5'-end of the mRNA and scans along the mRNA until it finds an appropriate AUG initiation codon. In most cases, the ribosome recognises the first AUG it encounters and initiates translation there. However, a survey of 699 eukaryotic mRNAs revealed that the first AUG is not the initiation site in 5%–10% of the cases. In these mRNAs, ribosome skips one or more AUGs before encountering the right one and initiating translation. In these cases, the consensus sequence surrounding the initiation codon is 5'-CCRCC**AUG**G-3', where R means purine (A/G). Scanning begins from the 5'-end of mRNA but it bypasses a region of the mRNA upstream of the initiation codon, called leader or 5' untranslated region (**5' UTR**), in favour of an AUG further down the mRNA.

The cap-independent initiation: Under some circumstances, eukaryotic ribosomes can initiate protein synthesis within the mRNA if that mRNA contains a sequence called an **internal ribosome entry site (IRES)**. These sequences were discovered in the poliovirus RNA and in several cellular mRNAs. They are about 400 nt long. The ribosome can be recruited directly to the start site by IRES *trans*-acting factors ITAFs bypassing the need to scan from the 5' end of the untranslated region of the mRNA.

12.5.2 Initiation

Ribosomal dissociation: In eukaryotes, two initiation factors, eIF-3 and eIF-1A, bind to the newly dissociated 40S ribosomal subunit. This delays its reassociation with the 60S subunit and allows other translation initiation factors to associate with the 40S subunit.

The eukaryotic initiator tRNA is unique as it functions only during initiation. Like the prokaryotic initiator tRNA, the eukaryotic version carries only Met. However, unlike prokaryotic fMet-$rRNA_i^{Met}$, the Met on this tRNA is not N-formylated and designated as $tRNA_i^{Met}$, with the 'i' indicating 'initiation'. In eukaryotes, although there are at least 12 (nine well-defined) initiation factors (which are preceded by an 'e' to denote they are eukaryotic, eIF1, eIF2 and so on) involved in protein synthesis, some have analogous functions to the three prokaryotic IFs.

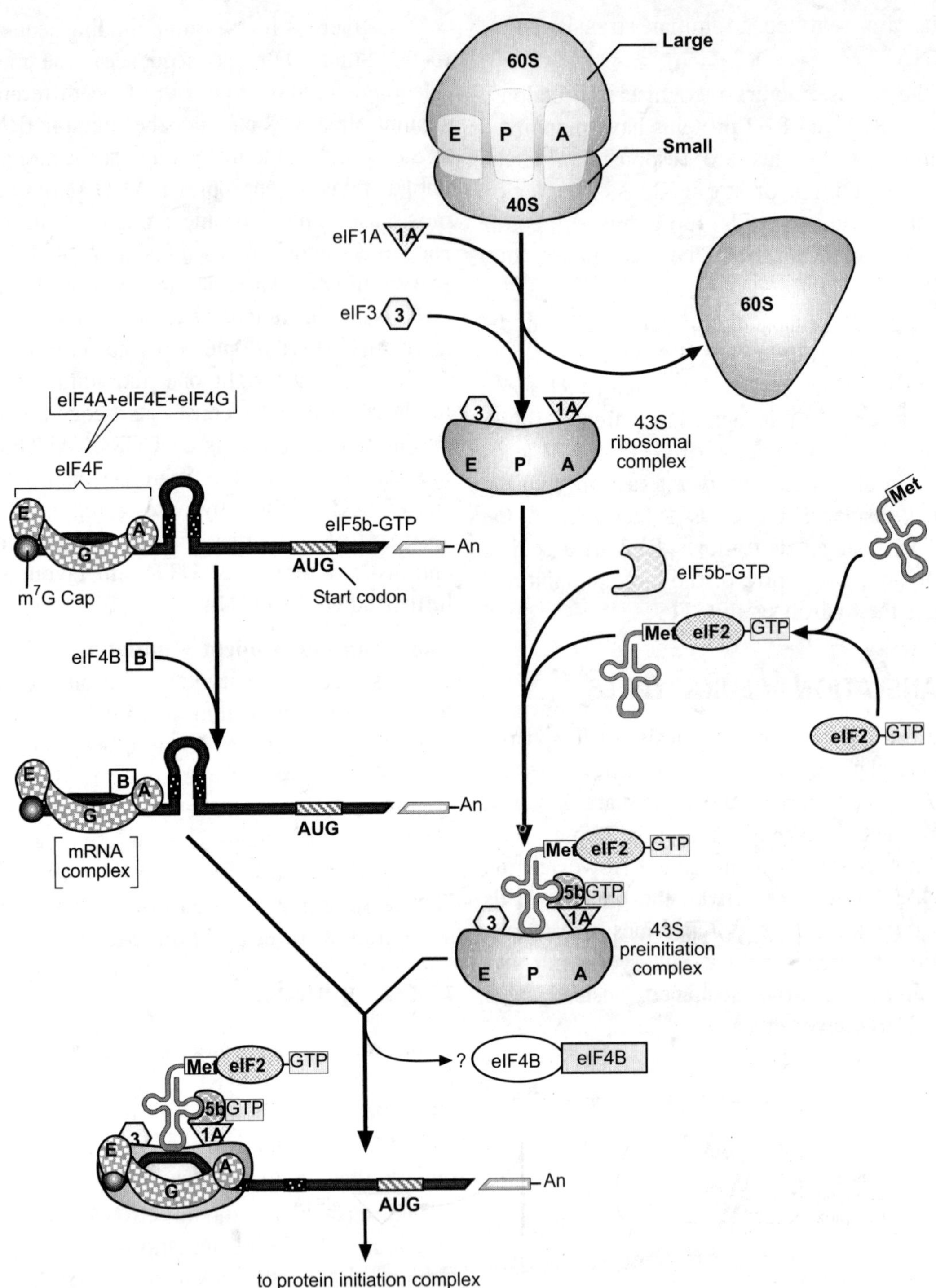

Fig. 12.14 Overview of the mechanism of polypeptide chain initiation in eukaryotes: Assembly of the eukaryotic small ribosomal subunit and initiator tRNA onto the mRNA

There are three basic steps during initiation in eukaryotes.

Formation of the 43S pre-initiation complex: Met-tRNA$_i^{Met}$ and the initiation factors eIF2 and GTP associate to form a stable ternary complex (eIF2:GTP:Met-tRNA$_i^{Met}$). This complex then binds to a 40S ribosomal subunit bearing eIF3 and eIF4C, giving a 43S **pre-initiation complex** (**PIC**).

Formation of the 48S initiation complex: Here the 43S PIC binds to the 5'-end of the mRNA (**Fig. 12.15**)

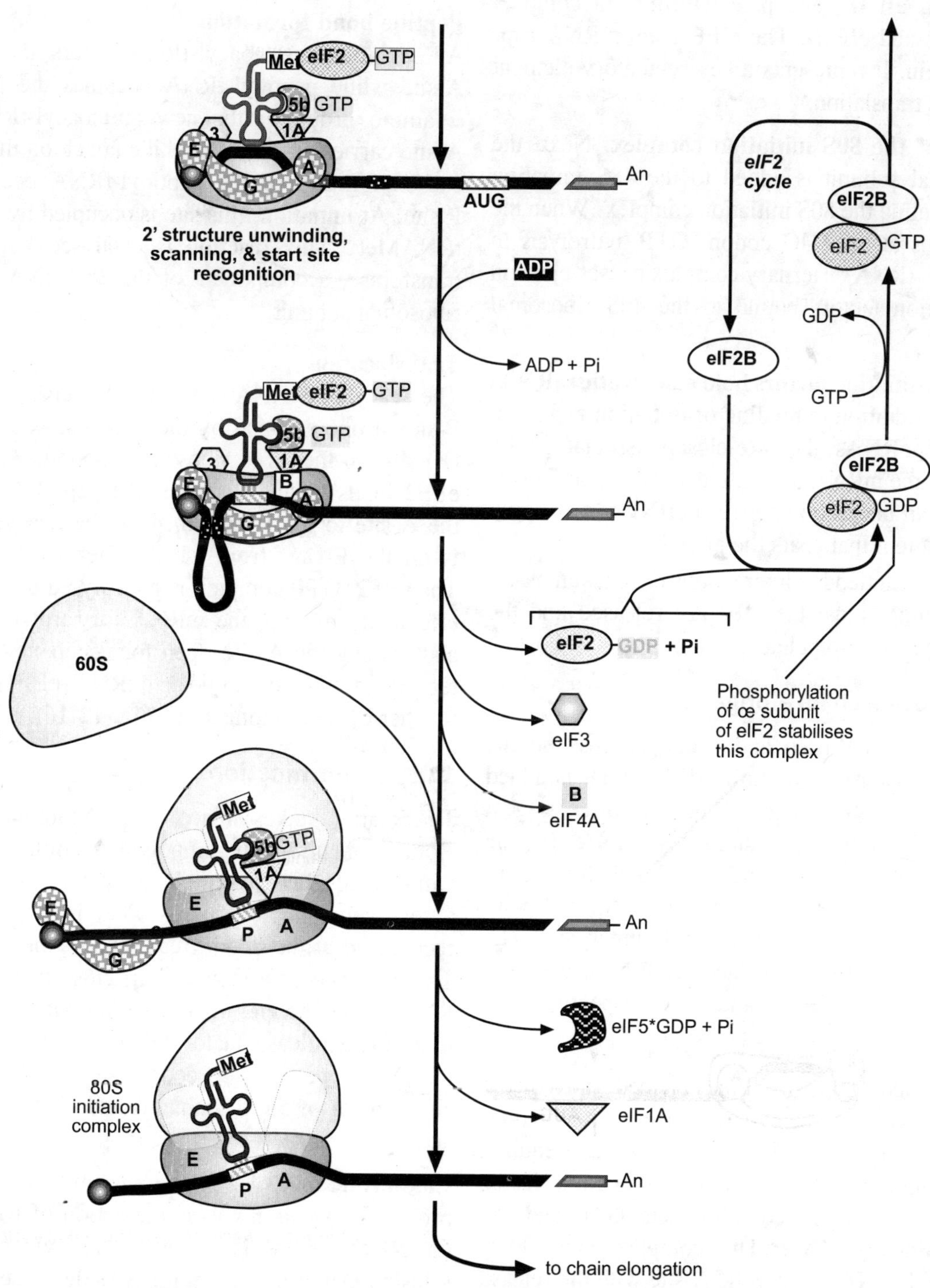

Fig. 12.15 The initiation of protein synthesis in eucaryotes: Identification of the initiating AUG by the small ribosomal subunit and formation of 80S initiation complex

and the 40S ribosomal subunit migrates to the correct AUG initiation codon. Four factors are involved at this stage: eIF4A, eIF4E, and p330 (forming a complex called eIF4F) and eIF4B. The eIFE is an mRNA cap-binding protein. It represents a key regulatory element in eukaryotic translation.

Formation of the 80S initiation complex: Next, the 60S ribosomal subunit is added to the 48S initiation complex, forming the 80S initiation complex. When the 43S PIC stops at an AUG codon, GTP hydrolysis in the eIF2: Met-tRNA$_i^{Met}$ ternary complex causes ejection of the initiation factors bound to the 40S ribosomal subunit.

Translation initiation factors hold eukaryotic mRNAs in circles: In addition to binding of initiation factors to the 5'-end of mRNAs, they are closely associated with the 3'-end of the mRNA through its poly A tail. This is mediated by an interaction between eIF4F and the poly A binding protein that coats the poly A tail.

Once codon–anticodon interactions have taken place, the 60S subunit binds, the eIFs are released and the initiation process is complete.

12.5.3 Peptide chain elongation

Elongation is a multistep and cyclic process on the ribosome in which one amino acid at a time is added to the nascent peptide chain. There are three stages: **recruitment** (binding of aminoacyl-tRNA to the A-site), **transpeptidation** (peptide bond formation) and **translocation**. Three factors are required with properties similar to their prokaryotic counterparts. The factor eEF1α, eEF1ß, eEF1γ and eEF2 have the roles described for EF-Tu, EF-Ts and EF-G respectively.

Recruitment

In the complete 80S ribosome formed during the process of initiation, the A-site is free. The binding of the proper aminoacyl-tRNA in the A-site requires correct recognition of the codon. Elongation factor eEF1α forms a ternary complex with GTP and the entering aminoacyl-tRNA. This complex then allows the aminoacyl-tRNA to enter the A-site with the release of eEF1A–GDP and phosphate. GTP hydrolysis is catalysed by an active site on the ribosome. The eEF1A–GDP then recycles to eEF1A–GTP with the aid of other soluble protein factors and GTP. The EF1ßγ (or perhaps EFß$\gamma\delta$), functionally analogous to the monomeric EF-Ts, promotes release from EF1α of GDP and exchange for GTP.

Peptide bond formation

An incoming aminoacyl-tRNA enters the ribosomal A-site while peptidyl-tRNA occupies the P-site. The α amino group of the new aminoacyl-tRNA in the A-site carries out a nucleophilic attack on the esterified carboxyl group of the peptidyl-tRNA occupying the P-site. At initiation, this site is occupied by aminoacyl-tRNAMet^i. This reaction is catalysed by a peptidyl transferase, a component of the 28S RNA of the 60S ribosomal subunit.

Translocation

The deacylated tRNA is attached by its anticodon to the P-site at one end and by the open CCA tail to an **exit (E) site** on the large ribosomal subunit. At this point, eEF2 binds to and displaces the peptidyl tRNA from the A-site to the P-site. In turn, the deacylated tRNA is on the E-site, from which it leaves the ribosome. The eEF2–GTP complex is hydrolysed to eEF2–GDP, effectively moving the mRNA forward by one codon and leaving the A-site open for occupancy by another ternary complex of amino acid tRNA–eEF1A–GTP and another cycle of elongation (**Fig. 12.16**).

12.5.4 Termination

Termination takes place after multiple cycles of elongation culminating in polymerisation of the specific amino acids into a protein molecule. Termination is a relatively simple process wherein a stop codon is recognised at the downstream end of the coding gene. The presence of a stop codon, close to C-terminal of RNA marks the end of translation (**Fig. 12.17**). There is a single release factor (**eRF**) which recognises all the three types of stop codons. The eRF (110 kD) is an α_2 dimer of 55 kD subunits and it requires GTP for activity.

Eukaryotic mRNAs: Although initiation of eukaryotic protein synthesis involves the action of RNA helicase on eIF4A, there is experimental evidence that the translational efficiency of mRNAs decreases if artificial stem-loop structures are inserted in the 5'-untranslated region. It has recently been suggested that the secondary structures could be a reason for the translational repression of maternal mRNAs of *Xenopus* oocytes. After fertilisation, when translation of maternal mRNAs occurs, the oocyte can become capable of translating

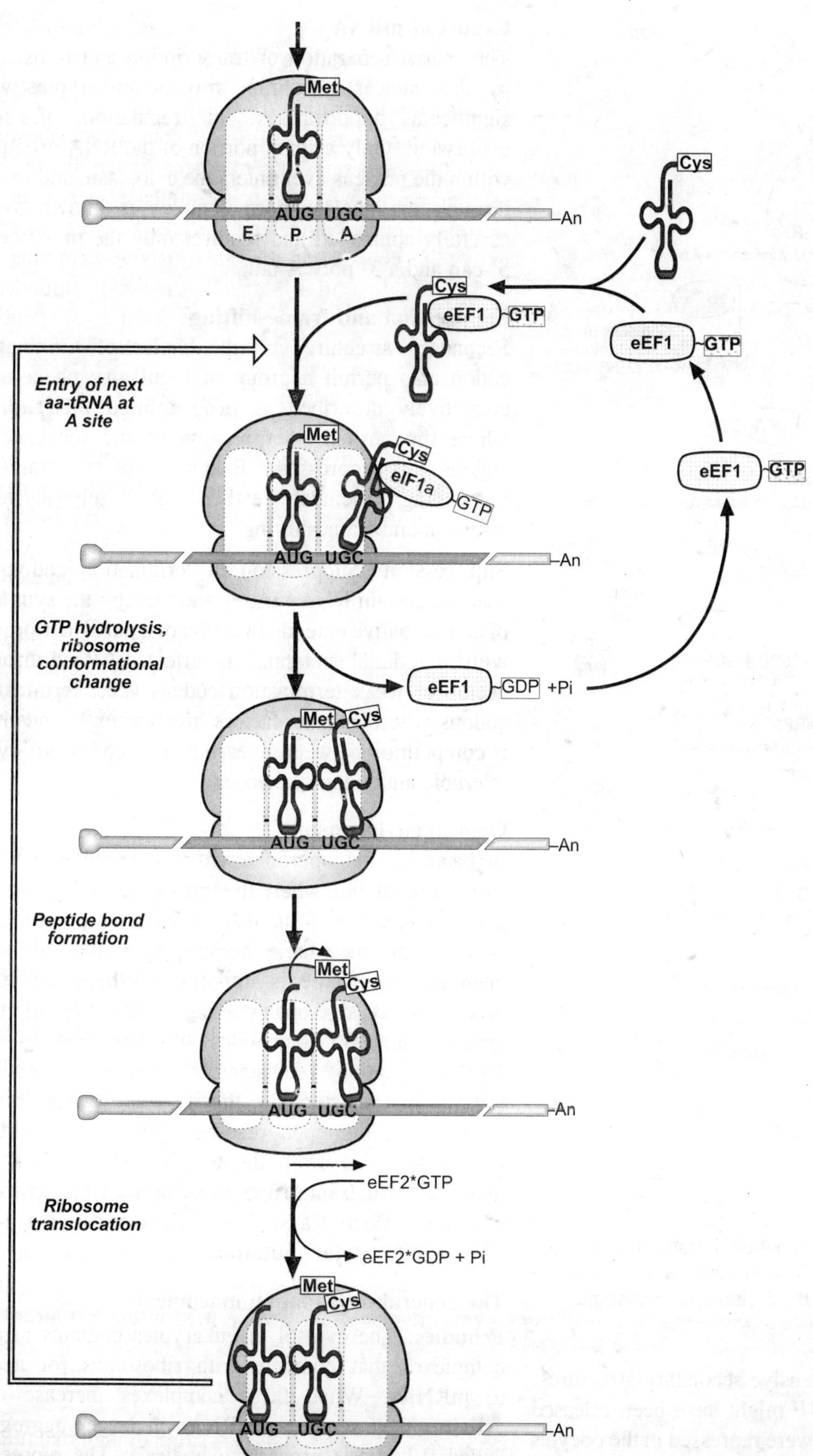

Fig. 12.16 Cycle of peptidyl chain elongation during translation in eukaryotes

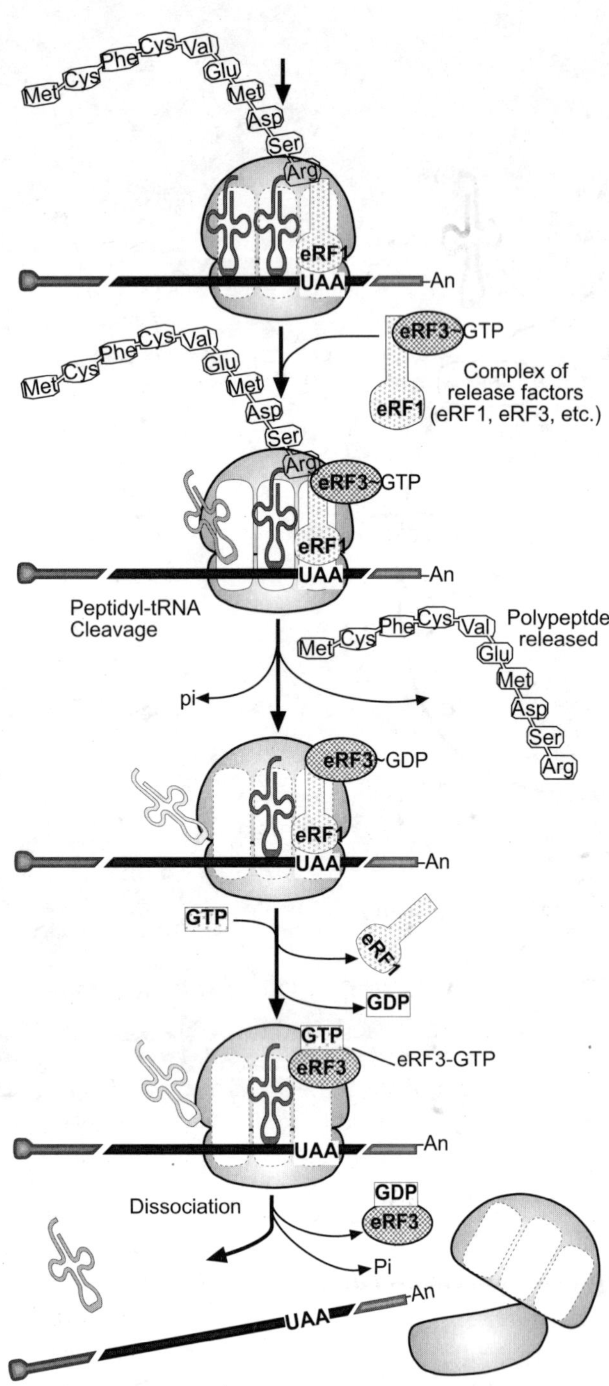

Fig. 12.17 Events during termination of translation in eukaryotes; polypeptide termination occurs when one of three stop codons appears in the A-site of the ribosome

artificial mRNAs with extensive secondary structures, probably because the eIF4F might have been released from the inhibitors, which were repressed in the oocytes by translation.

Export of mRNA

The spatial separation of transcription and translation by the nuclear membrane provide eukaryotes with significant opportunities for regulation of gene expression. Only a small portion of the RNA produced within the nucleus ever enters the cytoplasm and export through the nuclear pore complex is known to be carefully controlled and requires only the presence of 5'-cap and a 3' poly A tail.

Suppression and frameshifting

Secondary structure and the choice of termination codon also permit a group of regulatory phenomena collectively described as **programmed misreading**, where the normal interpretation of the sequence of codons is suppressed. Examples of programmed misreading include read-through, selenocysteine insertion and frameshifting.

Suppression: Suppression of termination codons is used by certain RNA viruses and thereby the synthesis of an alternative extended version of a particular protein with functional potential is affected. Read-through occurs at weak termination codons (i.e., termination codons where release factors are limiting), and there is competition between release factors and tRNAs with tolerable anticodon sequences.

Operon environment

In bacteria, the operon environment allows a novel form of regulation where the translation of downstream genes is dependent on that of upstream genes in a polycistronic message. The spacing of the individual open reading frame is important. Where the space between successive open reading frames is greater than approximately 30 nucleotides, translation begins with discrete initiation events and each locus is independent. If there is a shorter gap, the ribosome can re-initiate by bridging the open reading frames, i.e., without first dissociating from the template (**Fig. 12.18**). A mutation, which interrupts translation of the upstream gene also affects translation of the downstream gene. This is called **polar mutation** (e.g., *lac* operon).

The general degradation machinery

P bodies: The cytosol of eukaryotes contains protein complexes that compete with ribosomes for access to mRNAs. When these complexes increase their activity, they sequester mRNAs in larger aggregates called P bodies ('processing bodies'). The repression proteins breakdown the mRNA by removing its 'cap'

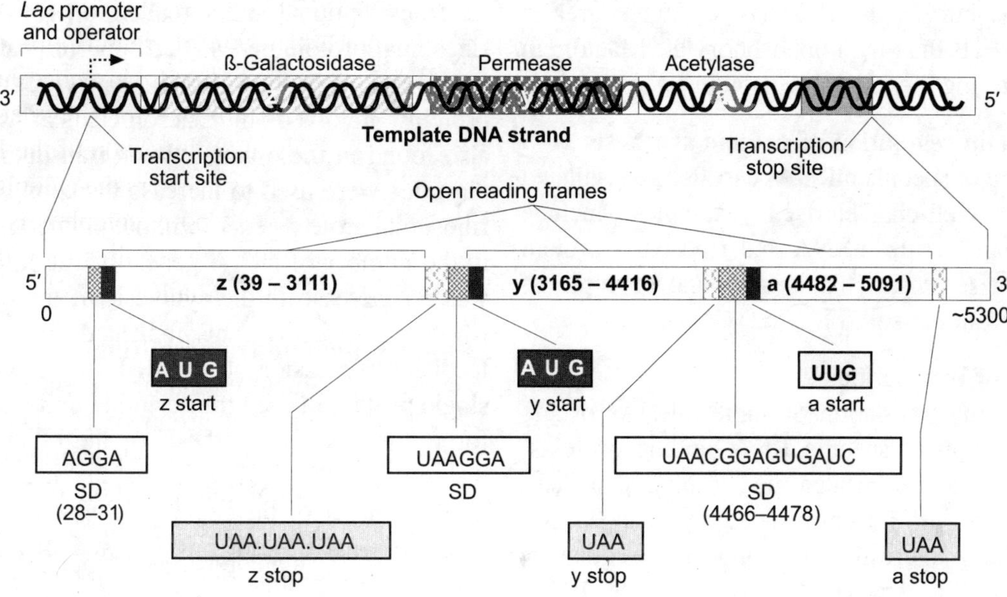

Fig. 12.18 The structure of *lac* operon mRNA

and removing its poly A tail, degrading the remaining message in the 5'→3' direction. This mechanism is very useful for destruction of 'bad' mRNAs (e.g., those with premature stop codons) and increasing the turnover of mRNAs and therefore the flexibility of gene expression in the cell.

Exosomes: These are hollow macromolecular complexes with two openings. They work on unfolded RNA molecules by taking in and degrading them in the 3'→5' direction.

MicroRNAs: These are small RNA molecules which bind to a complementary portion in the 3'-UTR of the mRNA and prevent it from being translated by ribosomes or trigger its destruction. Both these activities take place in the P bodies.

Riboswitches

Regulation of the level of certain metabolites is controlled by riboswitches. A riboswitch is a part of a molecule of mRNA with a specific binding site for the metabolite. For example, if thiamine pyrophosphate is available in the culture medium of *E. coli*, it binds to a messenger RNA whose protein product is an enzyme needed to synthesise thiamine from the ingredients in minimal medium. Binding induces an allosteric shift in the structure of the mRNA so that it can no longer bind to a ribosome and thus cannot be translated into the enzyme. *E. coli* no longer wastes resources on synthesising a vitamin that is available preformed. A thiamine pyrophosphate riboswitch has also been found in plants, *Archaea* and *Neurospora*. In *Neurospora,* riboswitches regulate the genes involved in the metabolism of vitamin B by alternative splicing of their transcripts. Other riboswitches also act on transcription rather than translation.

12.6 GLOBAL REGULATION OF PROTEIN SYNTHESIS

The components of the 'basal apparatus' of protein synthesis (the initiation, elongation and termination factors and the components of the ribosome) may be regulated and may exercise a global control over protein synthesis.

Protein phosphorylation

Initial studies of the phosphorylation of eIF2 was performed in reticulocyte lysates as they provide a very active protein-synthesising system, but if haem is not added there is a rapid cessation of protein synthesis. It was demonstrated that in the absence of haem, the binding of Met-tRNA$_f$ to the 40S ribosomal subunit was affected. It is a reaction that requires eIF2. A haem-controlled repressor/inhibitor (HCR or HRI) is a cytoplasmic protein which is inactive in the normal cells. It is a specific protein kinase that phosphorylates the α subunit of eIF2. Thus, it prevents the liberation of GDP from the eIF2 released from the ribosome, although eIF2B still forms a complex with the released species. Phosphorylated eIF2 has a much greater affinity for eIF2B than the unphosphorylated factor, in effect sequestering the exchange protein.

Narrow domain regulation of protein synthesis

The translation of specific mRNAs can also be regulated individually. Well-characterised examples include the mammalian ferritin mRNA and ribosomal protein L5 mRNA in *E. coli*. This is a typical example of negative translational control.

Iron control of translation

Regulation of the translation of some mRNAs occurs through the action of specific RNA-binding proteins. These proteins have been identified to bind sequences in either the 5' or 3' untranslated regions (UTR). Two particularly interesting and important regulatory schemes related to iron metabolism include RNA-binding proteins that bind to either the 5'-UTR of one mRNA or the 3'-UTR of another.

The transferrin receptor is a protein located in the plasma membrane that binds the protein transferrin, a major iron transport protein of the membrane. When iron levels are low, the rate of synthesis of the transferrin receptor mRNA increases so that cells can take up more iron. This regulation occurs through the action of an iron response element binding protein (**IRBP**) that binds to specific iron response elements (IREs) in the 3'-UTR of the transferrin receptor mRNA. These IREs form hairpin loop structures that are recognised by IRBP. When iron levels are low, IRBP is free of iron and can therefore, interact with the IREs in the 3'-UTR of the transferrin receptor mRNA. Conversely, when iron levels are high, IRBP binds iron which then cannot interact with the IREs in the transferrin receptor mRNA. The effect is an increase in degradation of the transferrin receptor mRNA.

RNA masking

Most mRNAs are bound to a variety of proteins that prevent association with ribosomes. When the appropriate signal is received, the proteins dissociate from mRNA, leaving the transcript free to associate with the ribosome. The signal is usually in the form of phosphorylation/dephosphorylation. Masking is a major form of regulation in early embryonic development.

Autogenous regulation of bacterial ribosomal protein synthesis

The synthesis of ribosomal proteins in *E. coli* is closely related to that of rRNA. There are about 25 transcriptional units (called operons) for the 54 ribosomal protein genes. Each operon may contain one or several ribosomal protein genes. Genes for other proteins involved in macromolecular synthesis are also found in these units. When transducing phages or plasmids were used to increase the copy number of the ribosomal protein genes, without a concomitant increase in the number of rRNA genes, it was found that there was no increase in the synthesis of the corresponding ribosomal proteins. This indicated that some form of feedback repression occurred. It was discovered that a single protein in each transcription unit was responsible for repressing the synthesis of the other members of the unit *in vivo*. Experiments *in vitro* demonstrated that the target of the regulation was the translation of the ribosomal protein mRNA, and that the repressor proteins also inhibited their own translation.

Gene-specific proteins

Translation of at least one mRNA in humans is repressed by a proteinaminoacyl-tRNA synthetase. In response to the inflammatory cytokine interferon-gamma (IFN-γ), the synthetase abandons its normal function (adding Glu and Pro to their respective tRNAs) and instead binds to the mRNA blocking its translation. In virus-infected cells, translation is regulated by turning off protein synthesis to prevent propagation of the viruses in a cell. The complete switching off of the protein synthesis is achieved by the induced synthesis of interferons (IFs). At least three classes of IFs are known: the leukocyte or α-IFs, the fibroblast or β-IFs and the lymphocyte or γ-IFs. IFs are induced by dsRNA and they further induce a specific kinase termed RNA-dependent protein kinase (PRK) that phosphorylates eIF2 thereby shutting off translation. Additionally, IFs induce the synthesis of 2'-5'-oligoadenylate, pppA(2'p5'A)n, that activates RNAase L, a pre-existing ribonuclease. RNAse L degrades all classes of mRNAs thereby shutting off translation.

Antisense RNA

A recent discovery is that antisense RNA fragments synthesised at different loci control translation. These fragments correspond to the non-coding strand of DNA and are complementary to the 5' end of an mRNA. An antisense sequence blocks the initiation site by formation of a double-strand structure. Antisense RNA technique is very specific and has been mainly found in prokaryotic systems (e.g., in the control of Q protein synthesis in bacteriophage λ, control of plasmid replication genes and control of transposase synthesis in various transposons) as well as in a few eukaryotes (e.g., in the control of Fgf-2 synthesis in the chick limb bud).

The selective blocking of the translation of specific messages by antisense RNA has potential application in the treatment of diseases including cancer. More recently, antisense suppression of translation has been used as a strategy to block gene function without *in vitro* disruption. However, a major difficulty in using RNAs as gene-inactivating drugs is the entry of small RNA molecules into cells. Negatively charged RNAs do not pass through membranes easily. Even if they get inside the cells, they are subject to enzymatic degradation. This limitation has led to attempts to modify the RNA so that it can reach the cell and resist degradation by the nucleases.

12.7 OTHER FUNCTIONS OF THE PROTEIN SYNTHESIS APPARATUS

In addition to involvement in protein synthesis, the ribosomes of bacteria synthesise guanine nucleotides which act as '**alarmones**' to integrate the regulation of RNA and protein synthesis. This phenomenon (absent in eukaryotes) occurs when amino acids are limiting, and is known as the 'stringent response'. It results in the selective inhibition of the synthesis of rRNA and tRNA, but not bulk mRNA. Hence, wasteful synthesis of more ribosomes and tRNA is avoided and the amino acids provided by protein turnover are conserved for the synthesis of essential proteins.

In addition to their role in the ribosomes, specific tRNAs are important for several other cellular processes. Thus, certain aminoacyl-tRNAs act as donors of amino acids in the synthesis of the peptide components of the peptidylglycans of bacterial cell walls. A more recently described related function appears to be in the post-translational tagging of the N-terminus of proteins with chemically basic amino acids to prepare them for proteolysis by the ubiquitin system. Further, $tRNA^{Glu}$ is involved in the synthesis of δ-aminolevulinate from glutamate in plants, where it is an important precursor to chlorophyll. For its NADPH-linked reduction to glutamate 1-semialdehyde, glutamate must be in the form of Glu-tRNA.

Retroviruses use tRNAs to prime reverse transcription, where a specific sequence on the tRNA hybridises to the viral 5'-LTR. However, in certain plant viruses, tRNA-like structures are found at the 3'-ends of the genomic RNAs (e.g., TMV). These can be aminoacylated by specific amino acids.

12.8 MONOCISTRONIC AND POLYCISTRONIC mRNA

The process of selecting the correct AUG initiation codon is significant in understanding many features of gene expression. In prokaryotes, mRNA molecules commonly contain information for the amino acid sequences of several different polypeptide chains. Such a molecule is called a **polycistronic mRNA.** A cistron refers to a base sequence that encodes a single polypeptide chain. In a polycistronic mRNA, each region coding for a polypeptide is preceded by its own ribosome-binding site and AUG initiation codon. The genes contained in a polycistronic mRNA molecule often encode the different proteins of a metabolic pathway. For example, in *E. coli*, three proteins needed to metabolise lactose are encoded by one polycistronic mRNA molecule (**Fig. 12.19**). The use of polycistronic mRNA is an economical way for a cell to regulate the synthesis of related proteins in a coordinated manner. With a polycistronic mRNA molecule, the synthesis of several related proteins can be regulated by a single signal, so that appropriate quantities of each protein are made at the same time. This is termed **coordinated regulation**.

In eukaryotes, there is no mechanism for initiating polypeptide synthesis at any AUG other than the first one encountered. Eukaryotic mRNA is always **monocistronic**. An mRNA is said to be monocistronic when it contains the genetic information to translate only a single protein. A single mammalian ribosome is capable of synthesising about 400 peptide bonds each minute. Polyribosomes actively synthesising proteins can exist as free particles in the cellular cytoplasm or may be attached to the ER.

Polarity of protein synthesis: The definitive feature of translation is that it proceeds in a particular direction along the mRNA and the polypeptide. The mRNA is always translated from an initiation codon to a stop codon in the 5'→3' direction and the polypeptide is synthesised from the amino end towards the carboxyl end by the addition of amino acids, one by one, to the carboxyl end.

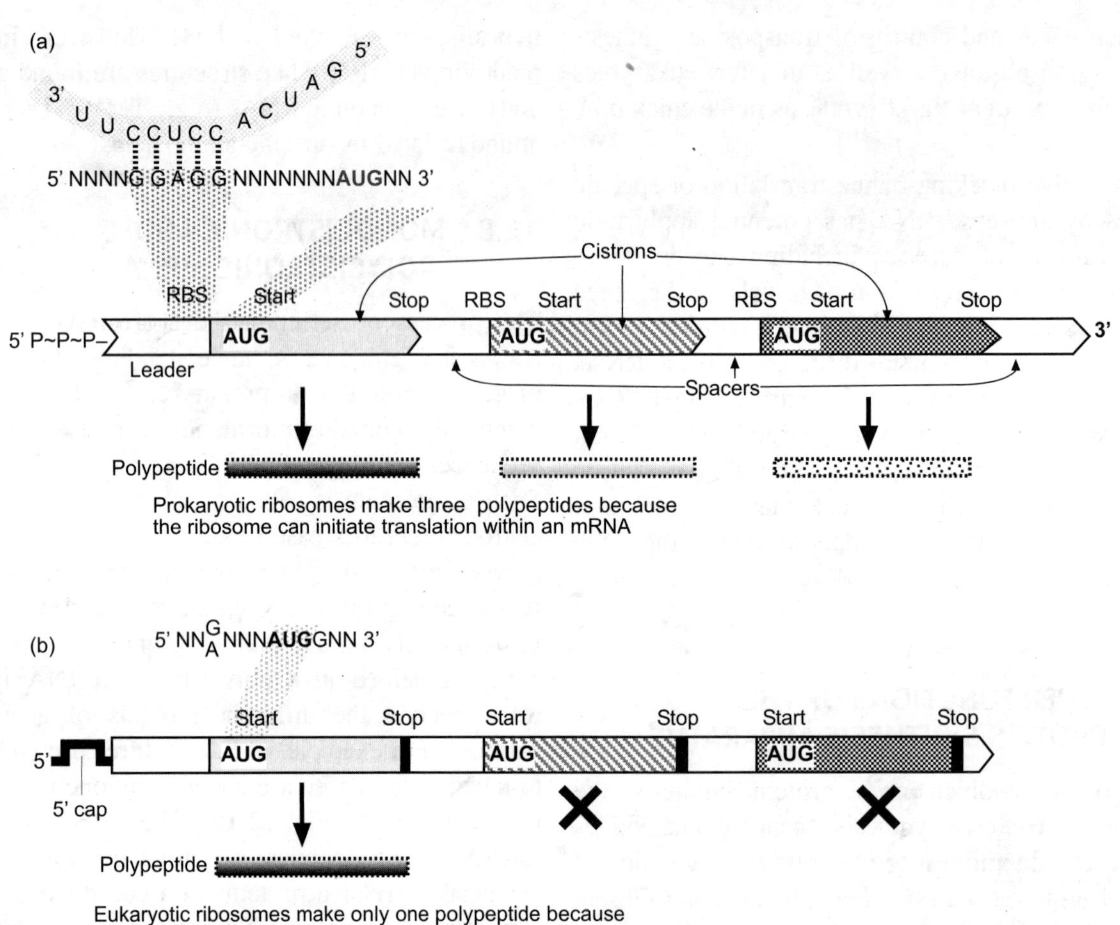

Fig. 12.19 Different products are translated from a three–cistron mRNA molecule by the ribosomes of (a) Prokaryotes and (b) Eukaryotes. The prokaryotic ribosome translates all of the genes, but the eukaryotic ribosome translates only the gene nearest the 5'-terminus of the mRNA

12.9 INHIBITORS OF PROTEIN SYNTHESIS

Many inhibitors of protein synthesis are utilised in the study of translation. Some of the important inhibitors are discussed below.

12.9.1 Antibiotics

Hundreds of antibiotics have been isolated from bacteria and fungi and are used to inhibit protein synthesis in bacteria. Antibiotics have different modes of action (**Fig. 12.20**). Some of these are explained below:

Cell wall synthesis inhibitors: Bacteria contain murein or peptidoglycan, which are essential for maintaining the structure of the cell wall. Inhibitors such as β-lactams, cephalosporins and glycopeptides block the microorganisms from synthesising their cell wall by inhibiting the synthesis of peptidoglycan.

Interfering with protein synthesis: Antibiotics such as tetracyclins, chloramphenicol, aminoglycosides and macrolides inhibit protein synthesis.

Effect on nucleic acids: Nucleic acids (DNA and RNA) present in every living cell, are involved in the storage and transfer of genetic information. Antibiotics such as quinolones and rifamycins bind to the proteins that are required for the processing of DNA and RNA. By blocking protein synthesis they affect the growth of the cells.

Cell membrane inhibitors: They disrupt the integrity and structure of cell membranes. Such antibiotics (such as polymyxins) are mostly effective on Gram-negative bacteria because only those contain a definite cell membrane.

Competitive inhibitors: Sulfonamides such as gantrism and trimethoprim inhibit the important metabolic

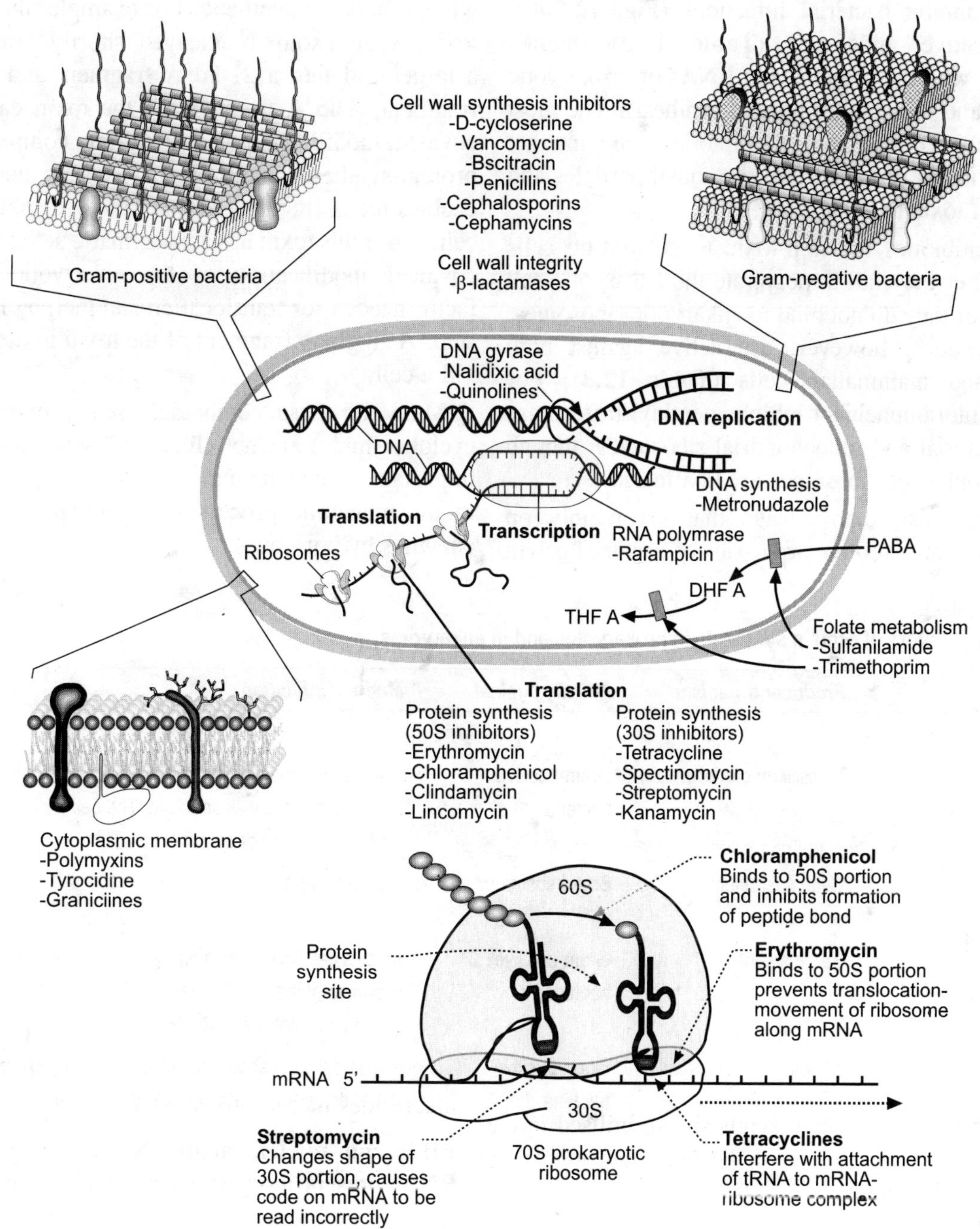

Fig. 12.20 Inhibitors of different antimicrobial agents: Protein synthesis, nucleic acid synthesis, cytoplasmic membrane and metabolism

pathways inside the bacterial cell competitively. They are also called antimetabolites or growth factor analogues.

Other antibiotics: Macrolides (erythromycin and azithromycin) are products of actinomycetes (soil bacteria). Erythromycin and other macrolide antibiotics inhibit protein synthesis by binding to the 23S rRNA molecule (in the 50S subunit) of the bacterial ribosome blocking the exit of the growing peptide chain of sensitive microorganisms. Gram-positive bacteria accumulate about 100 times more erythromycin than do Gram-negative microorganisms.

Ribosomes in bacteria and in the mitochondria of higher eukaryotic cells differ from the mammalian ribosome in many respects. These differences are very important for the development of drugs, which selectively inhibit bacterial infections (**Fig. 12.20b**). Many effective antibiotics (**Table 12.3**) interact specifically with the proteins and RNAs of prokaryotic ribosomes and thus inhibit protein synthesis. The most useful members of this class of antibiotics do not interact with components of eukaryotic ribosomal particles and thus are not toxic to eukaryotes.

Many antibiotics are non-toxic to eukaryotic cells because either they fail to penetrate the eukaryotic cell membrane or they do not bind to eukaryotic ribosomes. Some antibiotics, however, are active against both bacterial and mammalian cells (**Table 12.3**). For example, chloramphenicol inhibits peptidyl transferase in both bacterial and mitochondrial ribosomes, though normal cytoplasmic ribosomes are unaffected.

There are also many drugs that act mainly on eukaryotes. Some of these are toxins of bacterial origin while others are synthetic. For example, the diphtheria toxin is an exotoxin, cleaved shortly after entering a target cell into a 21-kd A fragment and a 40-kd B fragment. The A fragment of the toxin catalyses the covalent modification of an important component of the protein-synthesising machinery, whereas the B fragment enables the A fragment to enter the cytosol of its target cell. Since this toxin has an enzymatic activity, it causes covalent modification of the eukaryotic elongation factor needed for translocation and thereby inhibits that step. A single A fragment of the toxin in the cytsol can kill a cell.

Many of these compounds (e.g., puromycin and cycloheximide) are not clinically useful but have been important in elucidating the role of protein synthesis in the metabolic processes, particularly induction of enzymes by hormones.

Table 12.3 Inhibitors of protein synthesis in prokaryotes and in eukaryotes

Inhibitor	*Producer organism*	*Activity against*	*Mode of inhibition*
Cell wall inhibitors			
Penicillin	*Penicillium chrysogenum*	Gram-positive bacteria	Inhibits formation of the bacterial cell wall by blocking cross-linking of the cell wall structure. The cell wall is needed for a protective casing of bacterial cells.
Cephalosporin	*Cephalosporium acremonium*	Broad spectrum	Wall synthesis
Bacitracin	*Bacillus subtilis*	Gram-positive bacteria	Inhibits cell wall production by blocking step in the process: recycling of the membrane lipid carrier, which is needed to add on new cell wall subuits.
Vancomycin	*Streptomyces orientalis*	Gram-positive bacteria	Interferes with cell wall development by blocking the attachment of new cell wall subunits (muramyl pentapeptides).
Griseofulvin	*Penicillium griseofulvum*	Dermatophytic fungi	Microtubules
Cell membrane inhibitors			
Polymyxin B	*Bacillus polymyxa*	Gram-negative bacteria	Cell membrane
Amphotericin B	*Streptomyces nodosus*	Fungi	Cell membrane
Protein synthesis inhibitors			
Erythromycin	*Streptomyces erythreus*	Gram-positive bacteria	Binds to the 50S subunit and prevents formation of the 70S ribosome; consequently no ribosomal translocation. Has no effect on an active 70S ribosome.

Streptomycin	*Streptomyces griseus*	Gram-negative bacteria	Inhibits initiation by binding to the S12 protein of the 30S ribosomal subunit and thereby inhibits binding of $tRNA_f^{Met}$ to the P-site. Also causes misreading in a system during synthesis.
Kanamycin		Broad spectrum	Binds to 30S subunit of ribosome irreversibly.
Gentamicin	*Micromonospora purpurea*	Broad spectrum	Binds to 30S subunit.
Neomycin	*Streptomyces fradiae*	Broad spectrum	Binds to 30s subunit.
Tetracyclin	*Streptomyces rimosus*	Broad spectrum	Binds to the 30S subunit and inhibits binding of charged tRNA to the acceptor site on the 70S ribosome.
Borrelidine			Specifically inhibits the aminoacylation of $tRNA^{thr}$.
Spectinomycin			Reversibly interferes with mRNA interaction with the 30S ribosome. It is structurally similar to aminoglycosides.
Hygromycin B			Prevents translation of A-site tRNA to P-site and inhibits both prokaryotic and eukaryotic cells.
Edeine A			Miscoding in prokaryotes.
Paromycin			Increases error rate during translation by decreasing selectivity of codon: anticodon pairing adjacent to A-site in prokaryotes.
Kasugamycin			Inhibits binding of tRNA.
Lincomycin			Inhibits peptidyl transferase complex.
Kirromycin			Inhibits the function of EF-Tu. The binding of kirromycin to EF-Tu stimulates formation of EF-Tu-GTP and binding of ternary complex to ribosome, but inhibits release of EF-Tu-GDP complex from the ribosome. Its continued presence prevents formation of the peptide bond between the peptidyl-tRNA and the aminoacyl-tRNA.
Thiostrepton			Prevents translocation by inhibiting EF-G.
Abrin, ricin			Inhibits binding of aminoacyl tRNA.
Chloramphenicol		Broad spectrum	Protein synthesis: inhibits peptidyl transferase activity of 50S ribosomal subunit (prokaryotic and mitochondrial ribosomes). Inactive against cytoplasmic ribosomes.
Fusidic acid		Gram-positive bacteria	In prokaryotes, it binds to EF-G;GTP is still hydrolysed and translocation occurs. However, EF-G and GDP are not released from ribosome, so the ribosome cannot bind another aminoacyl tRNA. In eukaryotes, inhibits translocation by altering an elongation factor. It 'jams'.
Viomycin			Inhibits translocation in prokaryotes.
Pactamycin			Inhibits positioning of $tRNA_f^{Met}$ on the 40S ribosome.
Showdomycin			Inhibits formation of the eIF2-$tRNA_f^{Met}$-GTP complex.
Sparsomycin			Inhibits translocation in both prokaryotes and eukaryotes.

Ricin and α-Sarcin (protein toxins)			Ricin has been isolated from the castor bean; inactivates eukaryotic 28S ribosomal RNA by providing the N-glycolytic cleavage or removal of a single adenine. Prevents activation of translation factor GTPases, affects in both prokaryotic and eukaryotic large ribosomal subunit.
Puromycin			Puromycin resembles an amino acid attached to the terminal adenosine of tRNA. The polypeptide attached to peptidyl tRNA is transferred to the NH_2 group of the puromycin, then the polypeptidyl puromycin adduct is released from the ribosome in the form of polypeptidyl puromycin. This premature chain termination, by acting as an analogue of charged tRNA is responsible for the lethal action.
Cycloheximide	*Ricinus cummini*	Eukaryotes	Inhibits the peptidyl transferase activity of the 60S ribosomal subunit of eukaryotes; hence inhibits formation of the peptide bond.
Diphtheria toxin			Enzymatic catalysis of a reaction between NAD^+ and eEF2 to yield an inactive factor. Inhibits translation in eukaryotic cells.
Nucleic acid inhibitors			
Nalidixic acid (Quinolines)		Gram-positive cocci	Binds to the A subunit of DNA gyrase (topoisomerase) and prevent supercoiling of DNA, thereby inhibiting DNA synthesis.
Rifamycin	*Streptomyces mediterranei*	Broad spectrum	Binds to DNA-dependent RNA polymerase and inhibits initiation of RNA synthesis.
Antimetabolite antimicrobials			
Sulfonamides, sulfones		Broad spectrum	Analogue of para-aminobenzoic acid and competitively inhibit formation of dihydropteric acid.
methatrexate		Broad spectrum	Binds to dihydrofolate reductase and inhibits formation of tetrahydrofolic acid.
Isoniazid (INH)		Tuberculosis	Inhibits synthesis of mycolic acids.

KEY POINTS

- In translation or biosynthesis of proteins, the messenger RNA (mRNA) molecule is translated into a polypeptide. This is the final step in the information flow from RNA to protein. The synthesis of every protein molecule in a cell is directed by an mRNA, which is originally copied from DNA.
- Among the many biological pathways, translation is the most highly conserved across all organisms and the most energetically costly for the cell. In actively growing bacterial cells, up to 80% of the cell's energy and 50% of the cell's dry weight are dedicated to protein synthesis.
- Protein synthesis is a multistep process and involves many components. The main components of the translation machinery are mRNA, ribosomes, tRNA and a number of accessory proteins. The mRNA is the template for translation and is produced by transcription.
- Protein synthesis occurs on specialised macromolecular complexes called ribosomes. The ribosomes in a typical eukaryotic cell can collectively make about a million peptide bonds per second.
- The tRNA plays a direct role in the recognition of the codons within the mRNA. Transfer RNAs are small molecules that act as adaptors during protein synthesis; they link the nucleotide sequence of mRNA to the amino acid sequence of the polypeptide. The main function of tRNA is to ensure that each amino acid incorporated into a protein corresponds to a particular codon in the mRNA.
- Activation of amino acids requires energy in the form of ATP and occurs in a two-step reaction catalysed by the aminoacyl-tRNA synthetases. Each aminoacyl-tRNA synthetase is highly specific for a given amino acid. Most aminoacyl-tRNA synthetases contain editing sites in addition to acylation sites.
- Initiation, elongation and termination factors are essential for translation. In prokaryotes, the factors are abbreviated as IF or EF for initiation and elongation factors respectively, whereas in eukaryotes they are called eIF and eEF.
- Polypeptide synthesis takes place in the direction parallel to the 5′ to 3′ orientation of the mRNA. During each cycle of elongation, a peptide bond is formed between the carboxyl group in the last amino acid of the polypeptide chain and the amino group in the amino acid being added by sequentially.
- Eukaryotic translation control mechanisms are exceptionally complex. In general, the translation rate of any mRNA species is related to its abundance. The mRNA half-lives range from about 20 minutes to over 24 hours. Several features of mRNA secondary structure influence translation and it differs in prokaryotes and eukaryotes because of their different modes of initiation.
- The spatial separation of transcription and translation by the nuclear membrane provide eukaryotes with significant opportunities for regulation of gene expression.
- In bacteria, the operon environment allows a novel form of translational regulation where the translation of downstream genes is dependent on the translation of upstream genes in a polycistronic message. The genes contained in a polycistronic mRNA molecule often encode the different proteins of a metabolic pathway.
- In eukaryotes, the 5′-terminus of mRNA molecule binds to the ribosome, after which the mRNA molecule slides along the ribosome until the AUG codon nearest the 5′-terminus is in contact with the ribosome, then translation proceeds until it reaches the stop codon.
- Many inhibitors of protein synthesis are well known and utilised in the study of translation mechanism.

Self-assessment Questions

1. Explain with examples how protein synthesis consumes a high amount of energy.
2. Explain briefly the different components of protein synthesis.
3. List the major differences between prokaryotic and eukaryotic mRNAs.
4. Describe the composition and structure of a ribosome.
5. How are different tRNAs aminoacylated?
6. List the various factors required for the process of protein synthesis.
7. Briefly explain the stages of protein synthesis.
8. What are the factors responsible for the termination of protein synthesis?
9. Explain the different mechanisms for the regulation of protein synthesis.
10. List the different types of protein synthesis inhibitors with examples.

FURTHER READING

Textbooks

Adams RLP, JT Knowler and DP Leader (1992). *The Biochemistry of the Nucleic Acids*. London: Chapman and Hall.

Brown TA (2002). *Genomes*, 2nd ed. Oxford BIOS Scientific Publishers Ltd. UK.

Lodish H, D Baltimore, A Berk, L Zipursky and P Matsudaira (1995). *Molecular Cell Biology*. Scientific American Books.

Lewin B (2000). *Genes VII*. Oxford University Press.

References

Arnez JG and D Moras (1997). Structural and functional considerations of the aminoacylation reaction. *Trends Bichem. Sci.* 22:211–16.

Beuning PJ and K Musoer–Forsyth (2000). Hydrolytic editing by a class II aminoacyl-tRNA synthetase. *Proc. Natl. Acad. Sci.* USA. 97:8916–20.

Brodersen DE, WM Clemons Jr, AP Carter, RJ Morgan-Warren, BT Wimberly and V Ramakrishnan (2000). The structural basis for the action of the antibiotics tetracycline, pactamycin and hygromycin B on the 30S ribosomal subunit. *Cell* 103: 1143–54.

Carter CW Jr. (1993). Cognition, mechanism and evolutionary relationships in aminoacyl-tRNA synthetase. *Annu. Rev. Biochem.* 62:715–48.

Cusack S (1997). Aminoacyl-tRNA synthetases. *Curr. Op. Struct. Biol.* 7:881–89.

Dahlberg AE (2001). Ribosome structure: The ribosome in action. *Science* 292:868–69.

Dever TE (2002). Gene-specific regulation by general translation factors. *Cell* 108:545–56.

Farabaugh PJ (1996). Programmed translational frame-shifting. *Annu. Rev. Genet.* 30:507–28.

Frank J (2000). The ribosome-A macromolecular machine par excellence. *Chem. Biol.* 7:R133–R141.

Green R (2000). Ribosomal translocation: EF-G turns the crank. *Curr Biol.* 10:R369–R373.

Green R and HF Noller (1997). Ribosomes and translation. *Annu. Rev. Biochem.* 66:679–716.

Hershey JWB (1991). Translational control in mammalian cells. *Annu. Rev. Biochem.* 60:191–227.

Ibba M and D Soll (2000). Aminoacyl-tRNA synthesis. *Annu. Rev. Biochem.* 69:617–650.

Jacobson A and SW Peltz (1996). Interrelationships of the pathways of messenger RNA decay and translation in eukaryotic cells. *Annu. Rev. Biochem.* 65:693–739.

Kisselev LL and RH Buckingham (2000). Translational termination comes of age. *Trends Biochem. Sci.* 25:561–66.

Kozak M (1999). Initiation of translation in prokaryotes and eukaryotes. *Gene* 234:187–208.

Merrick WC (1992). Mechanism and regulation of eukaryotic protein synthesis. *Microbiol. Rev.* 56:291–312.

Moor PB and TA Steitz (2002). The involvement of RNA in ribosome function. *Nature* 418:229–35.

Nissen P, J Hansen, N Ban, PB Moore and TA Steitz (2000). The structural basis of ribosome activity in peptide bond synthesis. *Science* 289:920–30.

Noller HF (1984). Structure of ribosomal RNA. *Annu. Rev. Biochem.* 53:119–62.

Pestova TV and CU Hellen (2000). The structure and function of initiation factors in eukaryotic protein synthesis. *Cell Mol. Life Sci.* 57:651–74.

Preiss T and MW Hentze (1999). From factors to mechanisms: Translation and translational control in eukaryotes. *Curr. Opin. Genet. Dev.* 9:515–21.

Porse BT and RA Garrett (1999). Ribosomal mechanics, antibiotics and GTP hydrolysis. *Cell* 97:423–26.

Ramakrishnan V (2002). Ribosome structure and the mechanism of translation. *Cell* 108:557–72.

Rodnina MV and W Wintermeyer (2001). Ribosome fidelity: tRNA discrimination, proofreading and induced fit. *Trends Biochem. Sci.* 26:124–30.

Sachs AB, Sarnow P and MW Hentze (1997). Starting at the beginning, middle and end: translation initiation in eukaryotes. *Cell* 98:831.

Wittman HG (1983). Architecture of prokaryotic ribosomes. *Annu. Rev. Biochem.* 52:35–65.

CHAPTER 13

Protein Processing and Transportation

LEARNING OBJECTIVES

After reading this chapter, you will be able to:

- Understand protein sorting and targeting
- List the different mechanisms of protein processing
- List the routes of protein transportation

13.1 INTRODUCTION

Recent studies suggest that newly synthesised proteins undergo processing and approximately one-half of them are transported into or across the cell membrane. Structurally, prokaryotes consist of a single intracellular compartment surrounded by a plasma membrane. However, the eukaryotic cell is elaborately subdivided into functionally distinct compartments (organelles) enclosed within membranes. Each organelle contains a set of enzymes and other specialised molecules and complex distribution systems that transport a specific product from one compartment to another. A typical mammalian cell can have 10,000–20,000 different kinds of proteins (e.g., yeast cells can have about 5,000 proteins). Each of these proteins must be incorporated into the correct aqueous space (e.g., lumen and cytosol), membrane or organelle in order to function properly. In addition, each of the membranes and aqueous spaces in a cell must contain a discrete set of proteins in order to carry out its unique functions. Thus, a critical factor in the building and functioning of a cell from its constituent proteins, lipids and nucleic acids is **protein sorting** and **targeting**. A nascent protein in *E. coli*, for example, can stay in the cytosol or be sent to the plasma membrane, the outer membrane, the space between them (periplasmic space) or to the extracellular medium.

The polypeptide produced by mRNA translation is often not the final biologically functional form. They undergo many modifications (**post-translational modifications**) in the lumen of the endoplasmic reticulum (ER) and transported to the Golgi complex, where they are modified further. Post-translational modification (PTM) is the chemical modification of a protein after its translation. It is one of the later steps in biosynthesis for many proteins. Some of the important protein modifications are the removal of N-terminal methionine, oxidation of cysteine residues to give a disulfide bond, attachment of carbohydrate moieties to the side chains of asparagine, serine and threonine to yield glycoproteins and so on. Other types of chemical modifications include hydroxylation, phosphorylation and acetylation. The above processes allow the cell to compartmentalise protein synthesis and the activation of different specific proteins at locations within or outside the cell.

Proteins synthesised on the cytoplasmic ribosomes of eukaryotic cells can be retained in the cytoplasm (released directly from ribosomes in the cytoplasm) or transferred to subcellular organelles (**post-translational transport**). Proteins synthesised on ribosomes of ER are inserted into the plasma membrane (through transport vesicles) or transported into the lumen of the ER (known as **cotranslational transport**) or secreted out (exocytosis). The molecular mechanisms governing these results involve recognition of features in amino acid sequences—**targeting signals**. There are two main classes of such signals: those that direct the translocation of proteins across a membrane separating two compartments and those that cause the retention of a protein in a particular cellular compartment.

The opposite of protein targeting is their import by **receptor-mediated endocytosis** and **pinocytosis**. Proteins in the extracellular space bind to specific receptors in specialised regions of the plasma membrane called **coated pits**. Clathrin, a protein that forms polyhedral lattices, converts coated pit regions into membrane-bound vesicles inside the cell. When these vesicles become acidified, the imported proteins dissociate from their receptors.

Besides protein targeting, sorting, processing and import, another important mechanism is **programmed destruction** of proteins. The half-life of a protein is influenced by the nature of its amino terminal residues. Ubiquitin, a protein found in all eukaryotes becomes conjugated to proteins marked for degradation.

13.2 INTRACELLULAR COMPARTMENTS

Membranes not only create compartments but are also the site of a number of vital biochemical processes that occur within or on their surface. For example, lipid metabolism is catalysed mostly by membrane-bound enzymes; oxidative phosphorylation and photosynthesis both require a membrane to couple the transport of H^+ to the synthesis of ATP. Intracellular membranes create enclosed compartments that are separate from the cytosol, thus providing the cell with functionally specialised aqueous spaces. Since the organelle's membrane is impermeable to most hydrophobic molecules, the membrane of each organelle thus contains membrane transport proteins that are responsible for the import and export of specific metabolites.

A unique and complex membrane compartment of a eukaryotic cell is the nucleus. It contains the main genome and is the principal site of DNA and RNA synthesis. It is surrounded by cytoplasm. About one-half the total area of membrane in a eukaryotic cell is taken up by the endoplasmic reticulum (ER), where protein processing and lipid synthesis occurs. ER sends many of its proteins and lipids to the Golgi apparatus. It consists of organised stacks of disc-like compartments called **Golgi cisternae**. Cell organelles, mitochondria and chloroplasts generate most of the ATP used by cells. Lysosomes and peroxysomes are small vesicular compartments that contain enzymes utilised in a variety of digestive and oxidative reactions.

13.2.1 Different methods of movement of proteins

Most of the cellular proteins are synthesised on ribosomes in the cytosol, except for a few that are synthesised on the ribosomes of mitochondria and plastids. Their subsequent fate, however, depends on their amino acid sequence, which can contain sorting signals that direct their delivery to locations outside the cytosol. Some proteins, which have specific sorting signals can direct their transport from cytosol into the nucleus, the endoplasmic reticulum (ER), mitochondria, plastids, peroxisomes or secrete out of the cell through the ER and Golgi apparatus. Those proteins, which do not have sorting signals, remain in the cytosol as permanent residues.

Intracellular transportation of proteins, in general, takes place in three fundamentally different ways by which proteins move from one compartment to another (**Fig. 13.1**). Protein transportation can also be grouped as: post-translational (gated transport and transmembrane transport) and cotranslational (vesicular transportation).

Gated transport: In this type of transportation, the nuclear pore complexes function as selective gates that actively transport specific macromolecules and macromolecular assemblies, in addition to the free diffusion of smaller molecules.

Transmembrane transport: Here the membrane-bound **protein translocators** directly transport specific proteins across a membrane from the cytosol into a space that is topologically distinct. The transported protein molecule usually must unfold to snake through the translocator. The initial transportation of

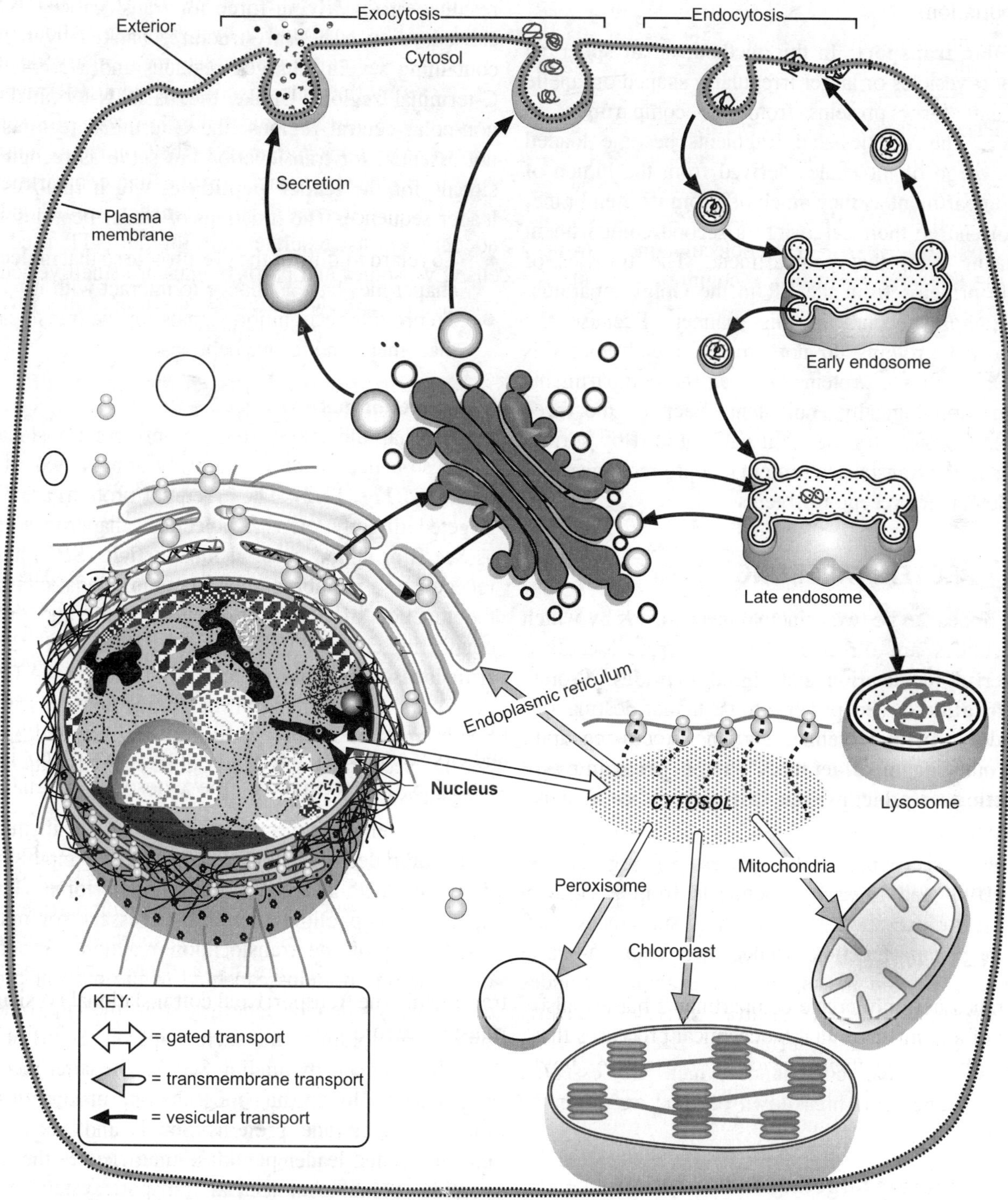

Fig. 13.1 Major intracellular protein trafficking routes: Post-translational transport (gated transport and transmembrane transport) and cotranslational transport (vesicular transport) to various target sites

selected proteins from the cytosol into the ER lumen or into mitochondria are examples of transmembrane transportation.

Vesicular transport: In this method, small spherical transport vesicles or larger irregularly shaped organelle fragments ferry proteins from one compartment to another. The vesicles and fragments become loaded with a cargo of molecules derived from the lumen of one compartment as they pinch off from its membrane; they discharge their cargo into a second compartment by fusing with that compartment. The transfer of soluble proteins from the ER to the Golgi apparatus, for example, occurs in this manner. Because the transported proteins do not cross a membrane, this method can move proteins between the compartments that are topologically equivalent. Each of the three modes of protein transfer is usually guided by sorting signals in the transported protein that are recognised by complementary receptor proteins.

13.3 PROTEIN TARGETING

There appear to be two principal mechanisms by which polypeptides are directed to their correct locations: **transcript localisation** and **signal peptides**. Protein **translocation** directs proteins to particular destinations in the cell. These proteins are synthesised in precursor form carrying an N-terminal stretch of amino acid residues or **leader peptide** that serves as a signal sequence.

Protein secretion is the process of segregating, elaborating and releasing chemicals from a cell. In contrast to excretion, the secreted substances may have a certain function, rather than being a waste product. Secretion in humans, for example, includes gastrointestinal tract, digestive enzymes, gastric acid, pulmonary and surfactant. Bacteria secrete many extracellular compounds which are essential for communication, breakdown of food and immune response.

13.3.1 Protein targeting and secretion in prokaryotes

Protein transfer in bacteria may be cotranslational or post-translational translocation or both (**Fig. 13.2**). Most proteins destined for any location other than the cytoplasm are synthesised with amino terminal leader sequences with 15–26 amino acid residues. These leader sequences (or signal sequences) consist of a basic N-terminal region, a central domain of 7–13 hydrophobic residues (main driving force for translocation), which can adopt an α helical structure, flanked by regions containing several charged residues and a non-helical C-terminal region. Unlike the basic N-terminal and non-polar central regions, the C-terminal features are not essential for translocation but serve as recognition signals for the **leader peptidase**, which removes the leader sequence. The functions of leader peptides are:

- To retard folding of the pre-protein so that molecular chaperones have a chance to interact with it,
- To provide recognition signals for the translocation machinery and leader peptidase.

The secretion pathway

In bacteria, the major route of protein translocation is the so-called secretion pathway abbreviated as **Sec pathway (Fig. 13.2)**. The precursor proteins are either targeted directly or via molecular chaperones, such as Sec B or **signal recognition particle (SRP)** and its receptor to a membrane-bound complex that combines the function of both a molecular motor and protein conducting channel. This system is collectively termed **translocase.** The pre-protein stimulates the ATPase, Sec A and a large integral membrane domain with the Sec Y and Sec E polypeptides at the core and Sec G, Sec D, Sec F and Yaj C as additional subunits of the translocase.

Pre-proteins are translocated across the membrane in an unfolded state and require energy input in the form of ATP and the proton-motive force (PMF). Specific phospholipids are also necessary for optimal functioning of the translocation reaction. A number of chaperone proteins can bind to the nascent protein as the leader peptide emerges from the ribosome, to prevent misfolding. One such chaperon, Sec B plays a predominant role by binding Sec A. The latter mediates translocation by feeding the substrate protein through the transmembrane proteins Sec E and Sec Y. An enzyme called leader peptidase then cleaves the leader peptide. Some other special transport systems are as follows.

Twin-arginine translocation: The twin-arginine translocation (Tat) pathway transports fully folded proteins across the membrane. The system requires two consecutive arginines in the signal sequence.

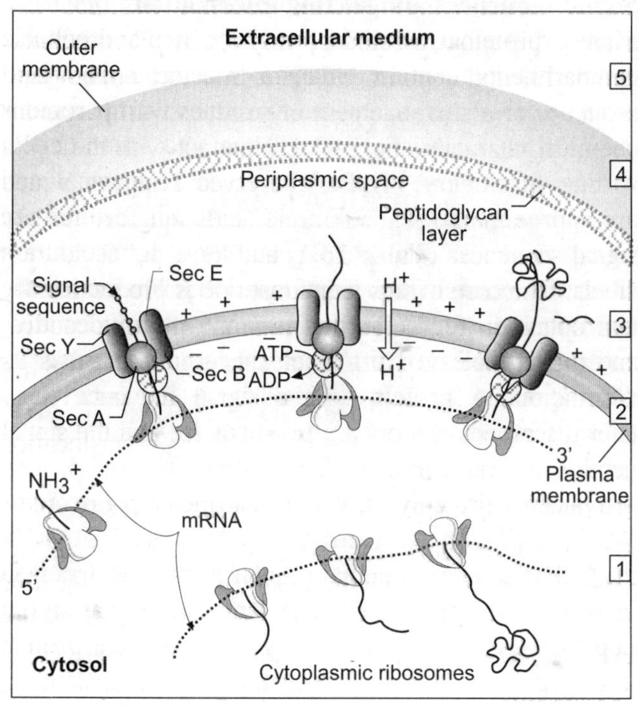

Fig. 13.2 Protein secretion in prokaryotes; the numbers in a box indicate the five different compartments in Gram-negative bacteria

Release of outer membrane vesicles: In addition to multiprotein complexes, Gram-negative bacteria form vesicles on the outer membrane. Portions of the membrane pinch off, forming spherical structures made of a lipid bilayer enclosing periplasmic materials. Vesicles from a number of bacterial species have been found to contain virulence factors; some have immunomodulatory effects while others directly adhere to and intoxicate host cells. The release of vesicles has been demonstrated as a general response to stress conditions.

13.3.2 Eukaryotic protein sorting and translocation

In eukaryotes, proteins are mostly synthesised in the cytoplasm of the cell, except for a small number encoded in the mitochondrial and chloroplast genomes. Eukaryotic cells have a well-developed ER and Golgi apparatus to fulfill these functions. However, there are many proteins like FGF1 (aFGF), FGF2 (bFGF), interleukin1 (IL1) and so on which do not have a signal sequence. They do not use the classical ER–Golgi pathway. These are secreted through various non-classical pathways.

Classical ER–Golgi secretion pathways

Proteins are imported from the cytoplasm into the ER, mitochondria, peroxisomes and chloroplasts by mechanisms that use a targeting sequence and translocation machinery. In higher eukaryotic cells, signal sequences targeting proteins to their appropriate compartments are located at the N-terminus as cleavable pre-sequences, although many proteins have internal, non-cleaved targeting sequences. Thus, the targeting information resides in more generalised features of the leader sequences such as charge distribution, relative polarity and secondary structure.

Protein sorting in eukaryotic cells occurs at several levels and requires multiple sorting signals and events. During the first level of sorting, the proteins synthesised on cytosolic ribosomes are divided into two broad categories: those on membrane-bound ribosomes of the ER and those synthesised on cytosolic ribosomes not attached to membranes. The latter proteins are released into the cytosol; some remain there, while others are specifically incorporated into the nucleus, mitochondrion, chloroplast or peroxisome and are subsequently sorted further to reach their correct destinations within these organelles. Large classes of proteins synthesised on ribosomes bound to the ER enter the secretory pathway (**Fig. 13.3**). This class includes proteins that cross the membrane of the ER cotranslationally (i.e., during synthesis) and become soluble in the ER lumen. Some of these soluble proteins are secreted from the cell and small transport vesicles carry them sequentially through the Golgi complex on to secretory vesicles.

During exocytosis, the secretory vesicle membranes fuse with the plasma membrane, releasing the protein to the cell's exterior. Each protein bears one or more targeting signals that direct it to its destination. Another important class of proteins that is synthesised on the rough ER and thus enters the secretory pathway is that of integral membrane proteins. Each such protein has a unique orientation with respect to the membrane's phospholipid bilayer. Sequence of up to 26 amino acid residues within the protein, called **topogenic sequences** ensure that each protein acquires its proper orientation during its insertion into the ER membrane. These proteins move via transport vesicles from the ER to their final destinations (e.g., plasma membrane or membranes of lysosome and Golgi) on vesicles. In eukaryotic cells, most ribosomes are bound to endoplasmic reticulum (ER). Proteins can be translocated into the

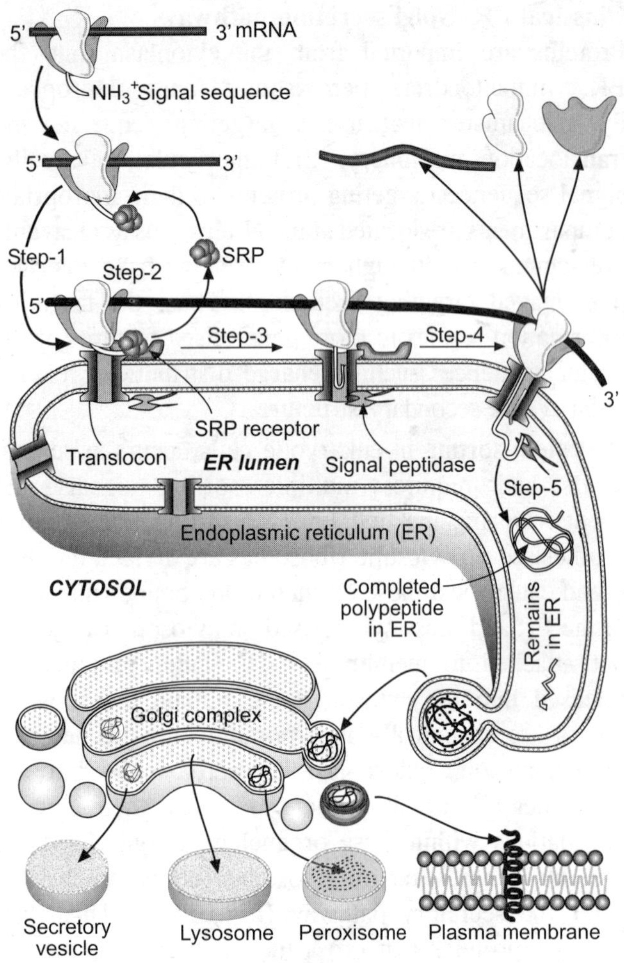

Fig. 13.3 Cotranslational targeting of secretory proteins to the ER lumen, followed by transport to final destination

ER, either during their synthesis on membrane-bound ribosomes (cotranslational translocation) or after their translocation has been completed on free ribosomes in the cytosol (post-translational translocation). The role of the ER in protein processing, sorting and secretion includes: rough ER → Golgi → secretory vesicles → cell exterior pathway. ER-bound ribosomes synthesise three major classes of proteins: secretory proteins, lysosomal proteins and plasma membrane proteins. All integral membrane proteins of the cell except those located in the membranes of mitochondria and chloroplasts, are formed by ribosomes bound to the ER. Nearly all proteins that ultimately reside in different compartments of the secretory pathways or in the plasma membrane are first imported into the ER. To fulfil its role as the primary gateway to the secretory pathway, the ER houses enzymes that process polypeptides and catalyse protein folding.

Signal sequences and sorting information

Many proteins destined for particular cellular compartments contain either a specific amino acid sequence or an arrangement of residues with particular chemical characteristics, which interact with the cell's sorting machinery. These conserved residues which are approximately 15–60 amino acids long are termed signal sequences (**Table 13.1**) and these act as address labels. The core of this signal peptide is predominantly hydrophobic, found at the termini of polypeptides, and may be cleaved off when the protein reaches its destination. A protein with a signal sequence (i.e., later discarded) is termed a **pre-protein**, and the signal peptide a **pre-sequence**. This differs from a protein activated by proteolysis, which is termed a **proprotein**. Where a protein must cross several interfaces within the cell, a number of signal sequences may be arranged in series and act sequentially. Amino terminal **signal peptides** are the first structures of the nascent protein to emerge from the ribosome, allowing cotranslational sorting to occur.

13.4 MECHANISM OF CO-TRANSLATIONAL TRANSPORT OF PROTEINS

The methods of co-translational transport through the ER membrane and from the Golgi apparatus are described below.

13.4.1 Through the ER membrane

Gunter Blobel and Peter Walter (1975) proposed the **signal hypothesis**, which explains the translocation of polypeptides across the rough ER membrane. The two key components in this targeting route are the signal recognition particle (SRP) and SRP receptor. The adaptor coupling the protein synthesising machinery (an amino terminal hydrophobic signal peptide) in the cytosol to the translocating machinery in the ER membrane is an 11S ribonucleoprotein SRP complex. This 325-kDa protein assembly consists of a small cytoplasmic RNA molecule about 260–300 nucleotide long called 7SL and six different polypeptide chains (SRP72, 68, 54. 19. 14 and 9). The SRP binds tightly only to ribosomes containing the nascent chain with a signal sequence (ribosome–nascent chain complex) as seen in **Fig. 13.4**. Elongation of the polypeptide chain is arrested or slowed down while the SRP is binding. The SRP–ribosome complex then diffuses to the ER membrane, where SRP binds to the SRP receptor, a

Table 13.1 Sorting signals in eukaryotic proteins

Type of signal	*Destination*	*Example of signal sequence*
No signal required	Cytosol	None
Nuclear Export (NES)	Cytosol	-Leu-Ala-Leu-Lys-Leu-Ala-Gly-Leu-Asp-Ile-
Nuclear Import (NIS)	Nucleus	-Pro-Pro-Lys-Lys-Lys-Arg-Lys-Val-
Mitochondrial targeting (pre-sequence)	Mitochondrion	$^{+}H_3N$-Met-Leu-Ser-Leu-Arg-Gln-Ser-Ile-Arg-Phe-Phe-Lys-Pro-Ala-Thr-Arg-Thr-Leu-Cys-Ser-Ser-Arg-Tyr-Leu-Leu-
Plastid import (transit peptide)	Plastid	$^{+}H_3N$-Met-Val-Ala-Met-Ala-Met-Ala-Ser-Leu-Gln-Ser-Ser-Met-Ser-Ser-Leu-Ser-Leu-Ser-Ser-Asn-Ser-Phe-Leu-Gly-Gln-Pro-Leu-Ser-Pro-Ile-Thr-Leu-Ser-Pro-Phe-Leu-Gln-Gly-
Peroxisomal import (PTS)	Peroxisome	-Ser-Lys-Leu-Coo^-
ER import	ER	$^{+}H_3N$-Met-Met-Ser-Phe-Val-Ser-Leu-Leu-Lue-Val-Gly-Ile-Leu-Phe-Trp-Ala-Thr-Glu-Ala-Glu-Gln-Leu-Thr-Lys-Cys-Glu-Val-Phe-Gln-
ER return	ER	-Lys-Asp-Glu-Leu-COO^-
SRP and a Golgi retention	Golgi	-
SRP and lysosomal targeting	Lysosome	-
SRP signal and the protein contains a hydrophobic transmembrane domain	Plasma membrane	-
SRP signal only	Secretion	-
Vacuolar sorting signal (VSS)	vacuole	-

heterodimeric **docking protein** (69 and 30 kDa subunits to which GDP is bound) located on the ER membrane. The ribosome then associates with its own receptor, a trimeric transmembrane protein called Sec 61 and protein synthesis continues. In the ER membrane, there are protein assemblies known as translocons spanning the membrane. They are channels formed by several subunits of a protein complex and are normally closed in the absence of ribosomes. Thus, they prevent leakage of small molecules from the cytosol into the ER. As soon as the ribosomes attach to the ER, an exchange of GDP on the SRP receptor for GTP is triggered off. This process is catalysed by an exchange enzyme present in the cytoplasm. This conformational change makes the docking protein bind more tightly to the SRP so that the latter can release the signal peptide. The translocon gets the signal peptide and resembles a gated channel that opens only when the latter is attached. The peptide emerges from the interior of the ribosome so that there is a completely enclosed channel from ribosome to the ER lumen.

The polypeptide is fed into the ER lumen, a process termed **vectorial discharge** or **cotranslational import** through the Sec 61 complex. Soon after it enters the ER lumen, the cleavage site of signal peptide is exposed at the internal face of the membrane. The signal peptide is clipped off by membrane-bound signal peptidase. On completion of the polypeptide synthesis, the protein is released into the lumen of the ER and the ribosome dissociates into its subunits to rejoin the cytoplasmic pool for further use. The signal peptide is destroyed by oligopeptidases and the translocon also closes down.

Retention as membrane proteins

Elongation and translocation are distinct processes. The translocation process for integral membrane is more complex than for secretory and lysosomal proteins, which are threaded through in entirety. Transmembrane proteins contain an internal 20-residue hydrophobic **stop transfer sequence** (also called a **membrane anchor sequence**) which lodges the protein in the ER membrane (**Fig. 13.5**). It should be stressed that there are other orientations of proteins in the plasma

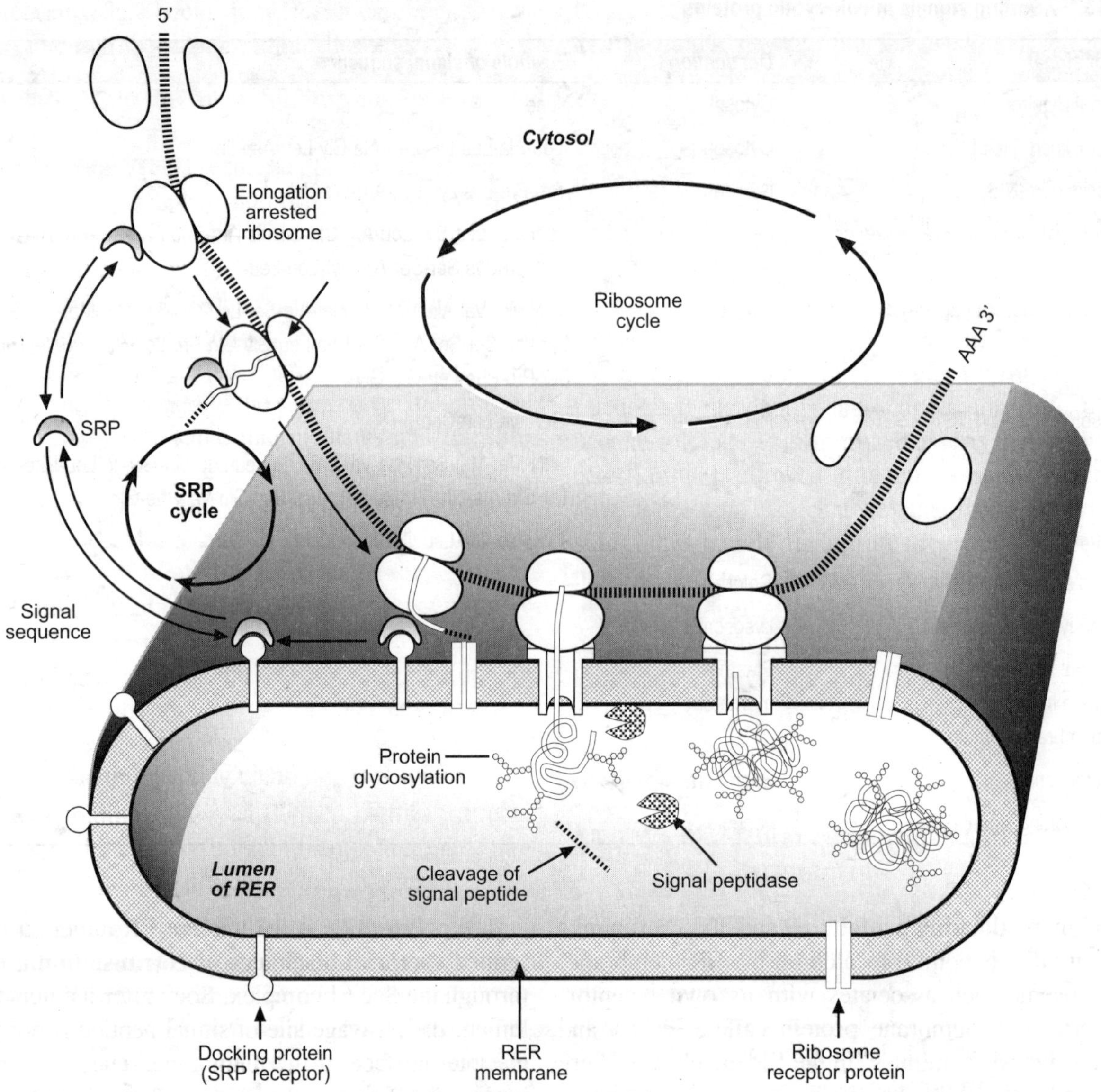

Fig. 13.4 The signal hypothesis; the functioning of SRP and the docking protein in the translocation of proteins across the endoplasmic reticulum. SRP: signal recognition particle RER: rough endoplasmic reticulum

membrane: external orientation of the C-terminus or multiple loops back and forth through the membrane. However, proteins with such orientations are inserted into the membrane through internal signal sequences.

Targeting proteins to membrane

The stop transfer sequence allows a protein to remain in the ER membrane itself, but by combining various similar signals (**membrane anchor signals**) with particular retention signals, a protein can also be targeted to the Golgi or lysosomal membrane. Any membrane-associated protein lacking retention signals will eventually arrive at the cytoplasmic membrane. Proteins in the Golgi which carry a **GPI sequence** become conjugated to a GPI (glycosyl-phosphatidylinositol) anchor which attaches the protein to the inner membrane surface. Proteins in the cytoplasm which have been modified by fatty acylation (e.g., myristoylation) are also targeted to the membrane.

Retention in the lumen of ER

The lumen of the endoplasmic reticulum contains a number of resident soluble proteins (e.g., protein disulfide isomerase). These are prevented from being secreted by a specific C-terminal sequence, which in animal cells is KDEL (Lys-Asp-Glu-Leu), and in

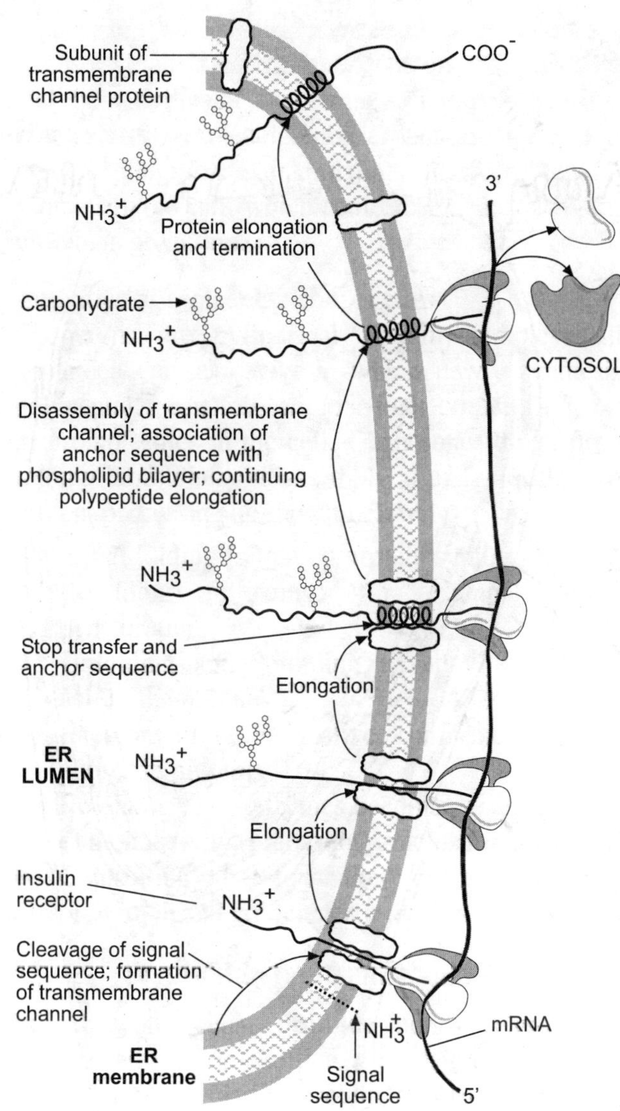

Fig. 13.5 Synthesis and membrane insertion of the insulin receptor

yeast is generally HDEL (called **retention signal**). See **Fig. 13.6**. If this sequence is deleted from a protein that normally functions in the ER, the mutated protein is instead transported to the Golgi and secreted from the cell. Conversely, addition of the KDEL sequence to the carboxy terminus of proteins that are normally secreted blocks their secretion. The KDEL and KKXX signals cause these ER proteins to be selectively retrieved from the ER–Golgi intermediate compartment or the Golgi complex and returned to the ER via a recycling pathway. Thus, the Golgi to the ER is the first branch point encountered by proteins being sorted to their correct destinations in the secretory pathway. Similar branch points arise at each subsequent stage of transport, such as retention in the Golgi versus export to lysosomes or the plasma membrane. In each case, specific localisation signals target proteins to their correct intracellular destinations.

Proteins secreted from the cell

In the Golgi apparatus proteins are packaged into vesicles for secretion. These vesicles then move towards the plasma membrane. There are two types of secretion: continuous release or those requiring a neuronal or hormonal stimulus. Serum proteins, for example, are continuously released from the liver

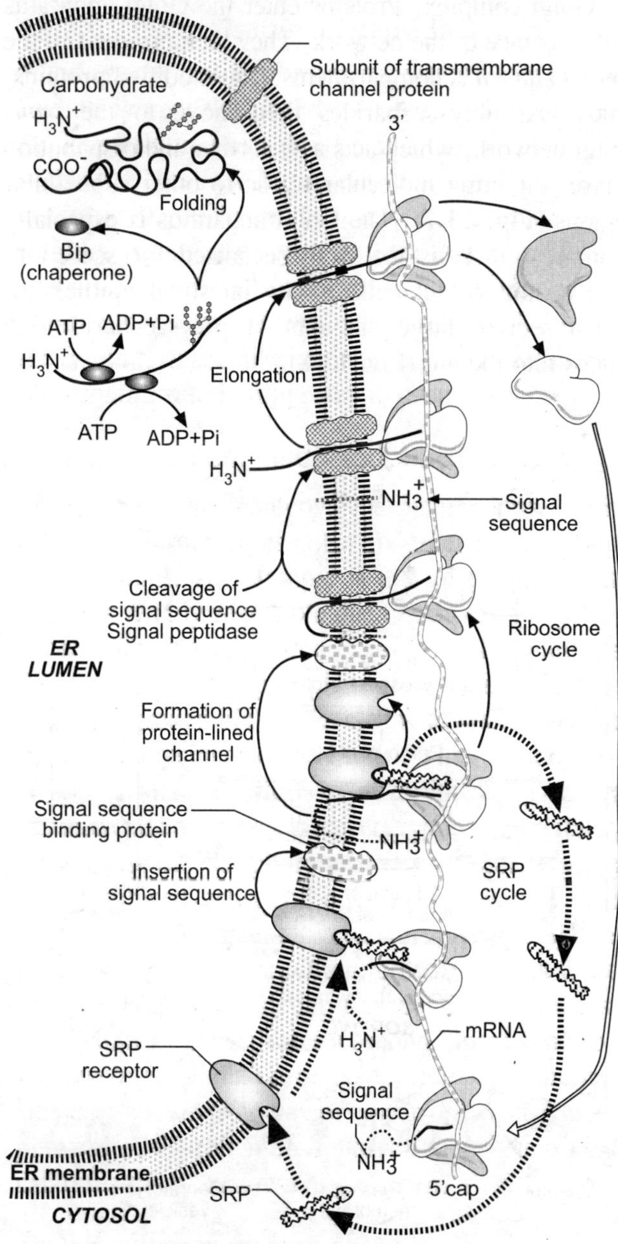

Fig. 13.6 Synthesis of secretory proteins on the endoplasmic reticulum (ER); signal recognition particle (SRP) participates in delivering ribosomes to the ER.

without requiring any signal. The vesicles fuse with the plasma membrane and release their contents by exocytosis. However, in the case of digestive enzymes from the pancreas, release is required only when food enters the gut. In this case, large secretory vesicles (granules) from the Golgi store these enzymes until they are released in response to a stimulus.

13.4.2 Protein processing within the Golgi

Distinct processing and sorting events appear to take place in an ordered sequence within different regions of the Golgi complex. Proteins enter the Golgi apparatus at the *cis* face of the network. They then progress to the medial and *trans* compartments. The modified proteins, lipids and polysaccharides then move to the *trans* Golgi network, which acts as a sorting and distribution centre, directing molecular traffic to other subcellular regions (**Fig. 13.7**). The Golgi apparatus is especially prominent in cells that are specialised for secretion, such as the goblet cells of the intestinal epithelium, which secrete large amounts of polysaccharide-rich mucus into the gut (**Fig. 13.8**).

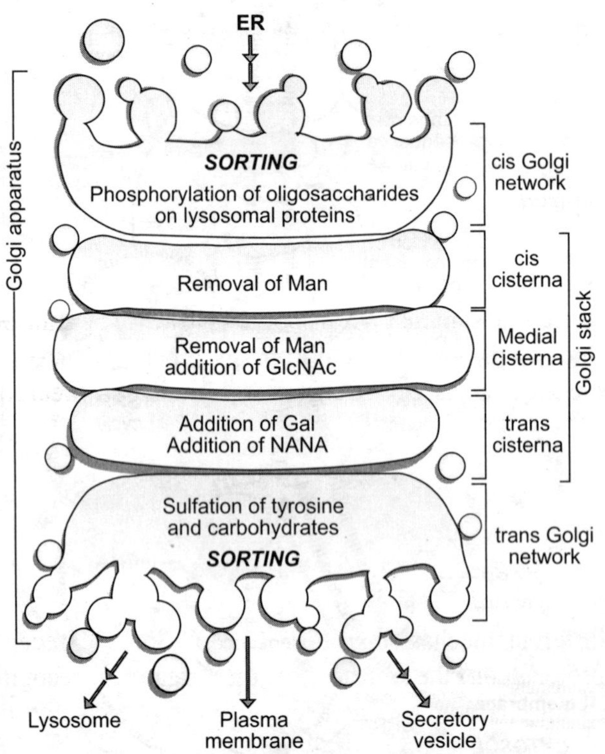

Fig. 13.7 The functional compartmentalisation of the Golgi apparatus; the localisation of each processing step is shown

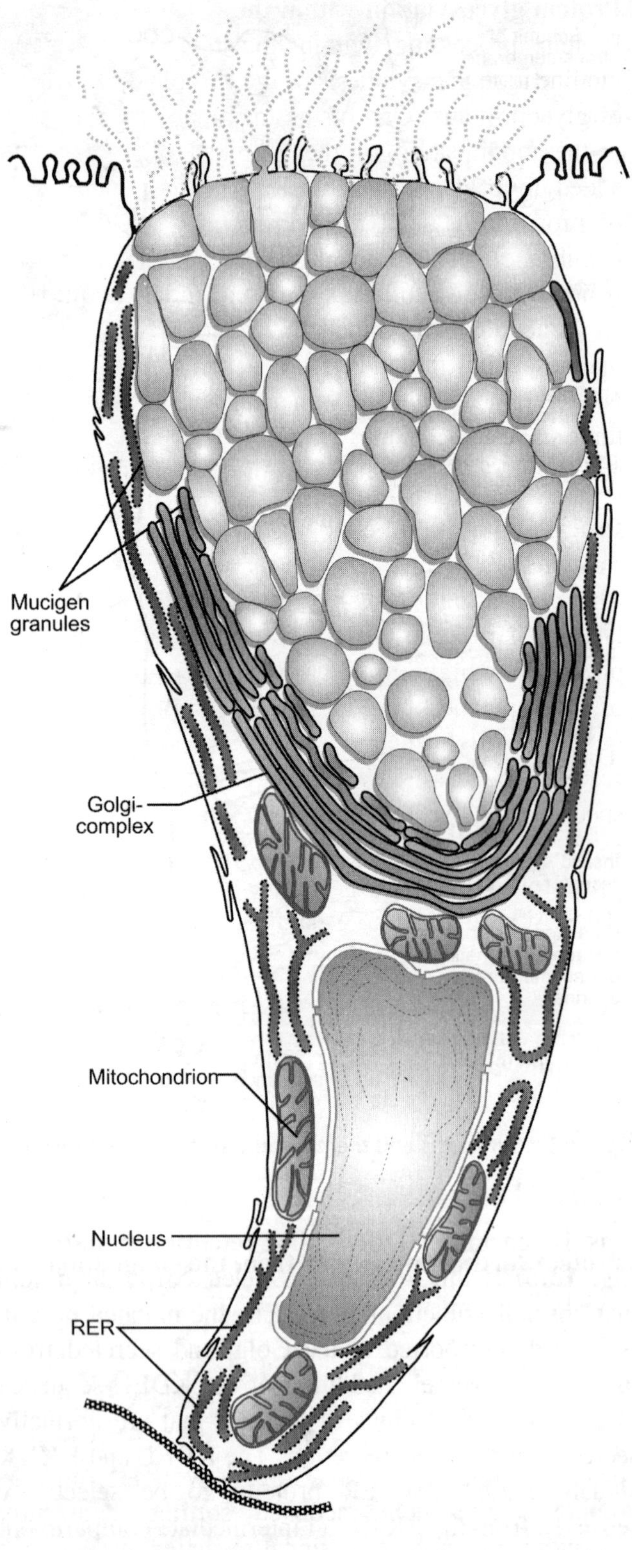

Fig. 13.8 The polarised structure of a secretory cell: A mucus-secreting goblet cell from the rat colon. The cell is specialised for secreting mucus, a mixture of glycoproteins and proteoglycans synthesised in the ER and the Golgi apparatus

Protein glycosylation within the Golgi

Protein processing within the Golgi involves the modification and synthesis of the carbohydrate portions of glycoproteins. One of the major processing aspects is the modification of the N-linked oligosaccharides added to proteins in the ER. The first modification of proteins detained for secretion or for the plasma membrane is the removal of four additional mannose residues. This follows the sequential addition of an N-acetylglucosamine, the removal of two more mannoses and the addition of a fucose and two more N-acetylglucosamines. At the end, three galactoses and three sialic acid residues are added. Glycosyltransferases add sugar residues and glycosidases remove them.

Processing of N-linked oligosaccharides of lysosomal proteins

The processing of the N-linked oligosaccharides of lysosomal proteins differs from that of secreted and plasma membrane proteins. Instead of initial removal of three mannose residues, the proteins are modified by mannose phosphorylation. In the first step of this reaction, N-acetylglucosamine phosphates are added to specific mannose residues, when it is still the *cis* Golgi, followed by removal of the N-linked oligosaccharide. This follows removal of the N-acetylglucosamine (GlcNAc) group, leaving mannose-6-phosphate residues on the N-linked oligosaccharide. The phosphorylated mannose residues are specifically recognised by a mannose-6-phosphate receptor in the *trans* Golgi network, which directs the transport of these proteins to endosomes and on to lysosomes. The phosphorylation of mannose residues is, thus, a critical step in sorting lysosomal proteins to their correct intracellular destination.

Protein sorting and export from the Golgi apparatus

In addition to protein modification, sorting of proteins into different kinds of transport vesicles, which bud from the *trans* Golgi network and delivering their contents to the appropriate cellular locations are the other major activities of the Golgi complex. Three pathways of protein sorting in the *trans* Golgi network are: mannose 6-phosphate receptor-mediated sorting, constitutive secretory pathway and signal-mediated diversion to secretory vesicles for regulated secretion (**Fig. 13.9**). Proteins that function within the Golgi apparatus are retained. The signals responsible for retention of some proteins within the Golgi have been localised to their transmembrane domains, which prevent the protein from being packaged into the transport vesicles. Proteins sorted into the regulated secretory pathway in the *trans* Golgi network are packaged into specialised secretory vesicles. These immature vesicles, further process their protein contents until specific signals direct their fusion with the plasma membrane. Selective transport is the well-characterised pathway from the Golgi apparatus to the lysosomes. In yeast and plant cells, which lack lysosomes, proteins are transported from the Golgi to the vacuole. Vacuoles assume the functions of lysosomes in these cells besides their other roles such as storage of nutrients, maintenance of turgor pressure and osmotic balance and so on.

13.4.3 Intracellular cargo transport

Cargo transport of organelles, secretory vesicles and protein complexes by small molecular motors is an essential intracellular process. Molecular motors that drive cargo transport along the cytoskeletal highway include myosins travelling along actin filaments and kinesin and cytoplasmic dynein motors travelling on microtubules (**Fig. 13.10**). These motors share cargo transport duties and face the challenge of maneuvering through a complex cytoskeleton with numerous microtubule and actin filament intersections. Both the secretory and endocytic pathways require that vesicular cargo is transferred between actin and microtubule tracks.

13.5 POST-TRANSLATIONAL TRANSPORT OF PROTEINS INTO ORGANELLES

All the processes described so far have been related to cotranslational transport. However, transport into mitochondria, chloroplasts, peroxisomes and the nucleus is post-translational transport. Secretion from prokaryotic cells is also post-translational. Most chloroplast and mitochondrial proteins and all peroxisomal and nuclear proteins are synthesised outside the organelle on 'free' cytosolic ribosomes. The newly formed proteins released into the cytosol are taken up specifically into the proper organelle by different mechanisms (**Fig. 13.11**). Specific receptor proteins on the surface of the organelle recognise specific targeting sequences on the new proteins that direct the proteins to the correct organelles.

Proteins enter the nucleus through nuclear pores. As the pores are aqueous-filled channels, proteins and particles do not cross phospholipid bilayers.

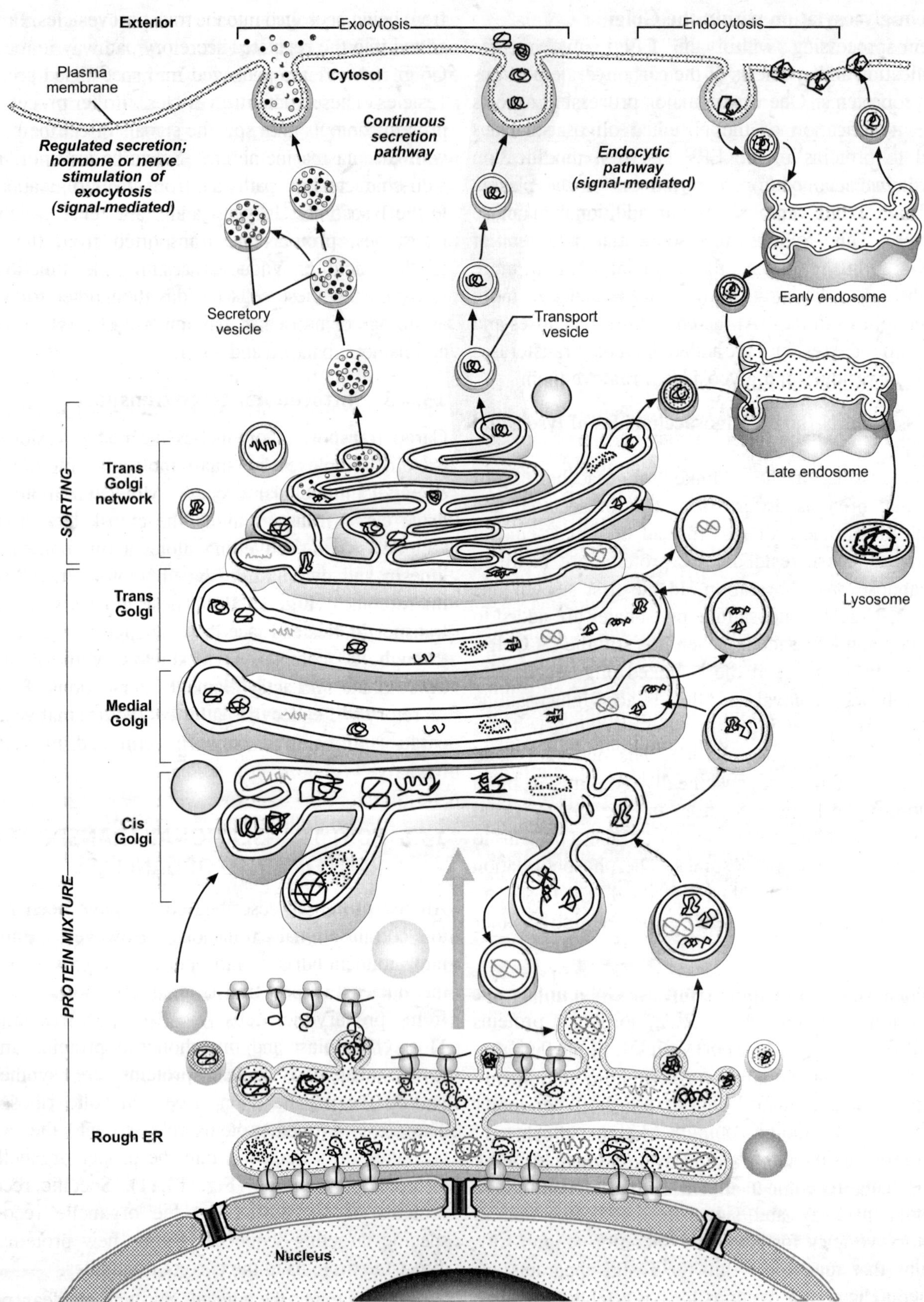

Fig. 13.9 An overview of the biosynthetic/secretory and endocytic pathways that unite endomembranes into an interconnected network; maturation of secretory proteins can also be seen

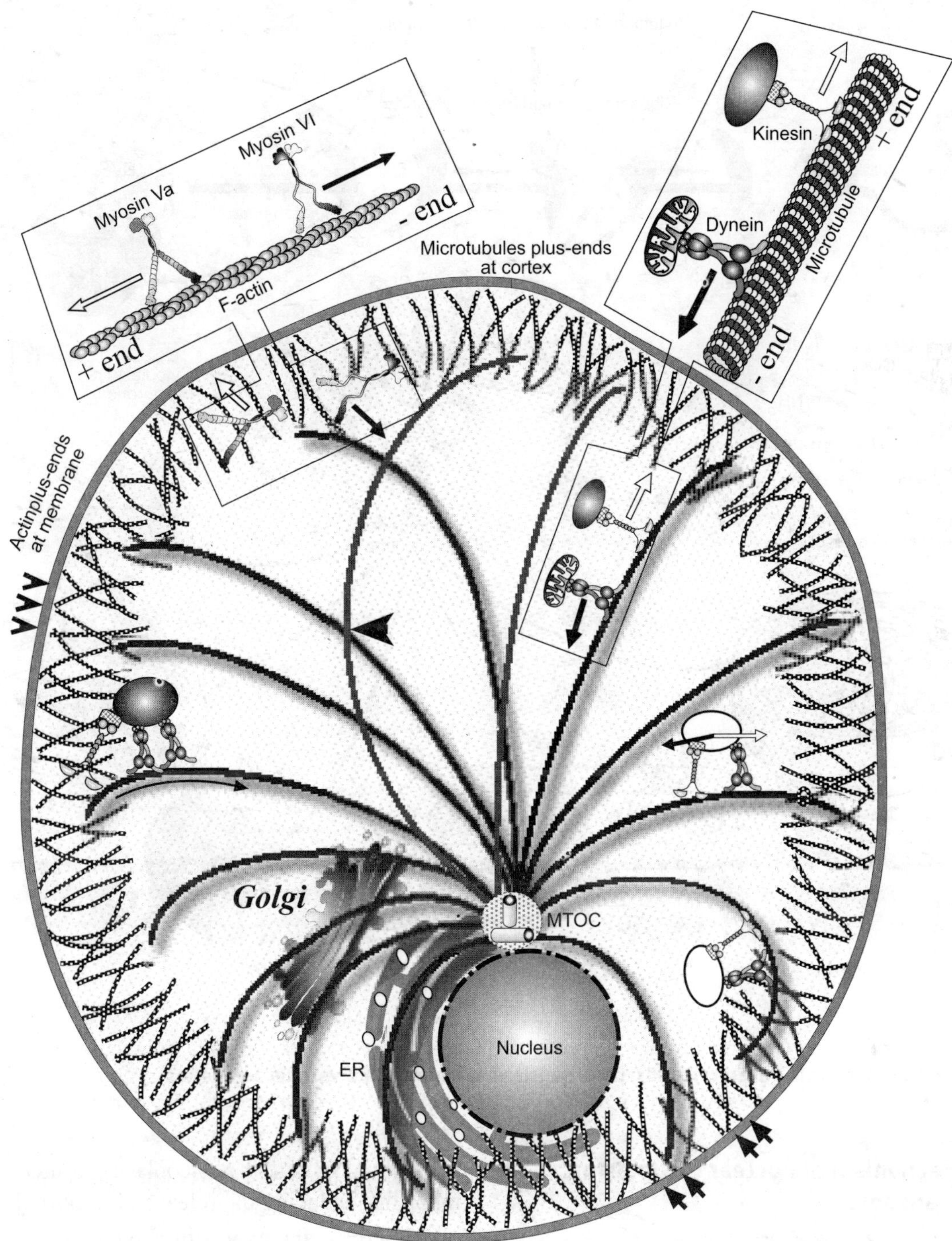

Fig. 13.10 An illustration describing motor proteins and vesicular cargo transport in the cell (***Source:*** Jennifer LR et al 2008)

Consequently, large protein or ribonucleoprotein particles can be formed in the cytosol and then imported into the nucleus and ribosomal subunits that are formed in the nucleus can be transported into the cytosol. In contrast, proteins moving from the cytosol into mitochondria, peroxisomes or chloroplasts must cross one or more phospholipid bilayers. The mitochondrion and chloroplast contain multiple membranes and membrane-limited spaces. The transport of proteins into the organelles occurs through protein-lined channels localised to 'contact sites' where the inner and outer membranes are close together.

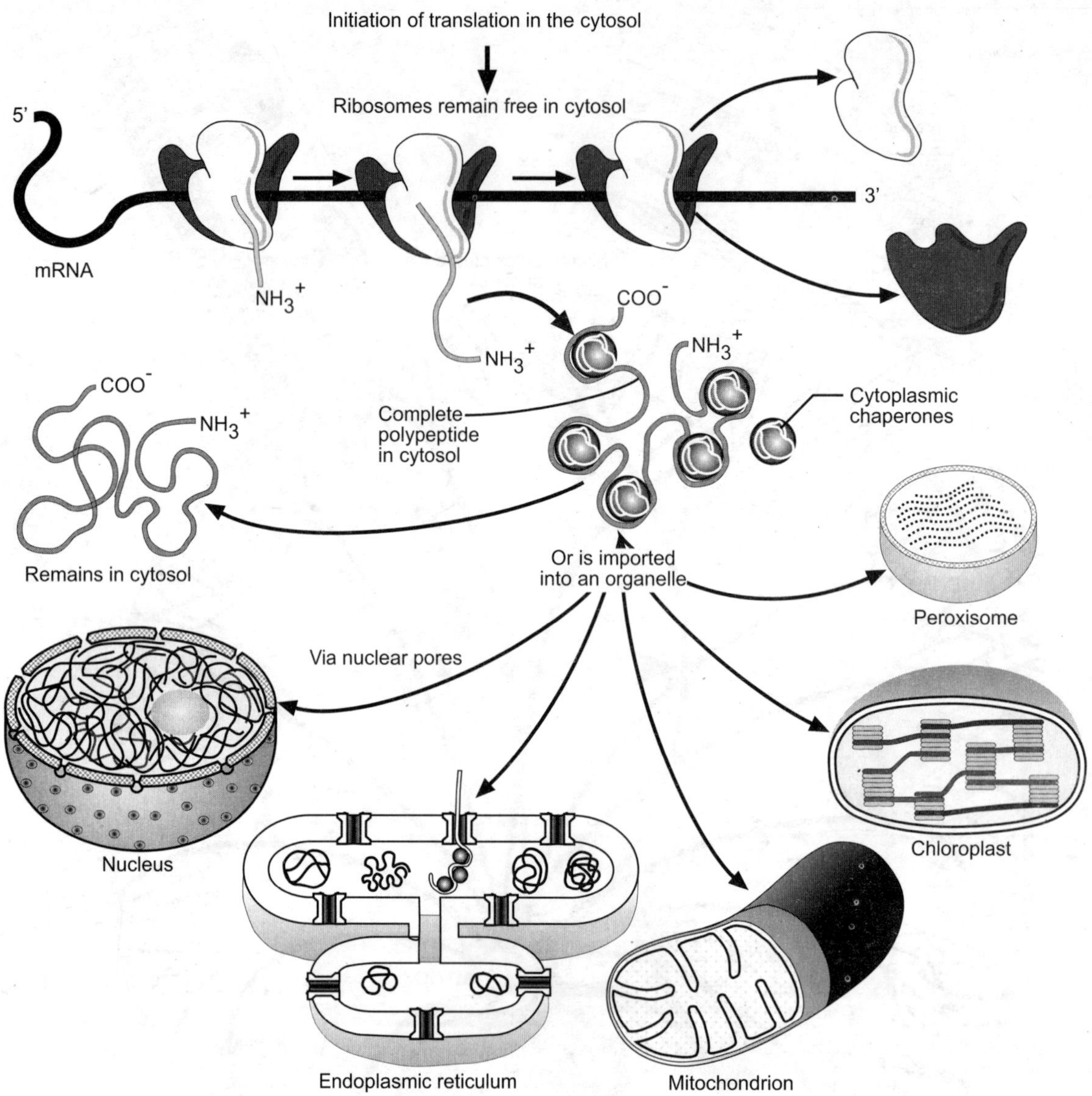

Fig. 13.11 Protein sorting in eukaryotic cells; post-translational import into various organelles

13.5.1 Mechanism of nuclear cytoplasmic transport

Apart from the free diffusion of smaller molecules through the pores, all nuclear cytoplasmic transport depends on a combination of proteins, namely importins and exportins. A dimeric cytoplasmic protein, **importin**, is required for the docking stage of nuclear import. Importins and exportins are regulated by a small GTPase Ran, which is thought to be highly enriched in the nucleus in its GTP-bound form. Each importin is specific to molecules to which they bind *via* the NLS. Importins recognise their substrate in the cytoplasm and transport them through nuclear pores into the nucleus (**Fig. 13.12**). In the nucleoplasm, RanGTP binds to importins, inducing the release of imported molecules.

In contrast, the export of proteins from the nucleus to the cytoplasm carrying a nuclear export signal (NES) occurs in a reverse cycle (**Fig. 13.12**). In a group of **exportin** proteins, each is specific for proteins carrying the appropriate NLS. Attachment of the exportin to the NLS of its target molecule occurs only when Ran-GTP is attached to the exportin. The Ran-GTP-exportin cargo moves into the cytoplasm where the GTP activates the Ran-GTPase causing hydrolysis of the GTP and dissociation of the complex into Ran-GDP and exportin

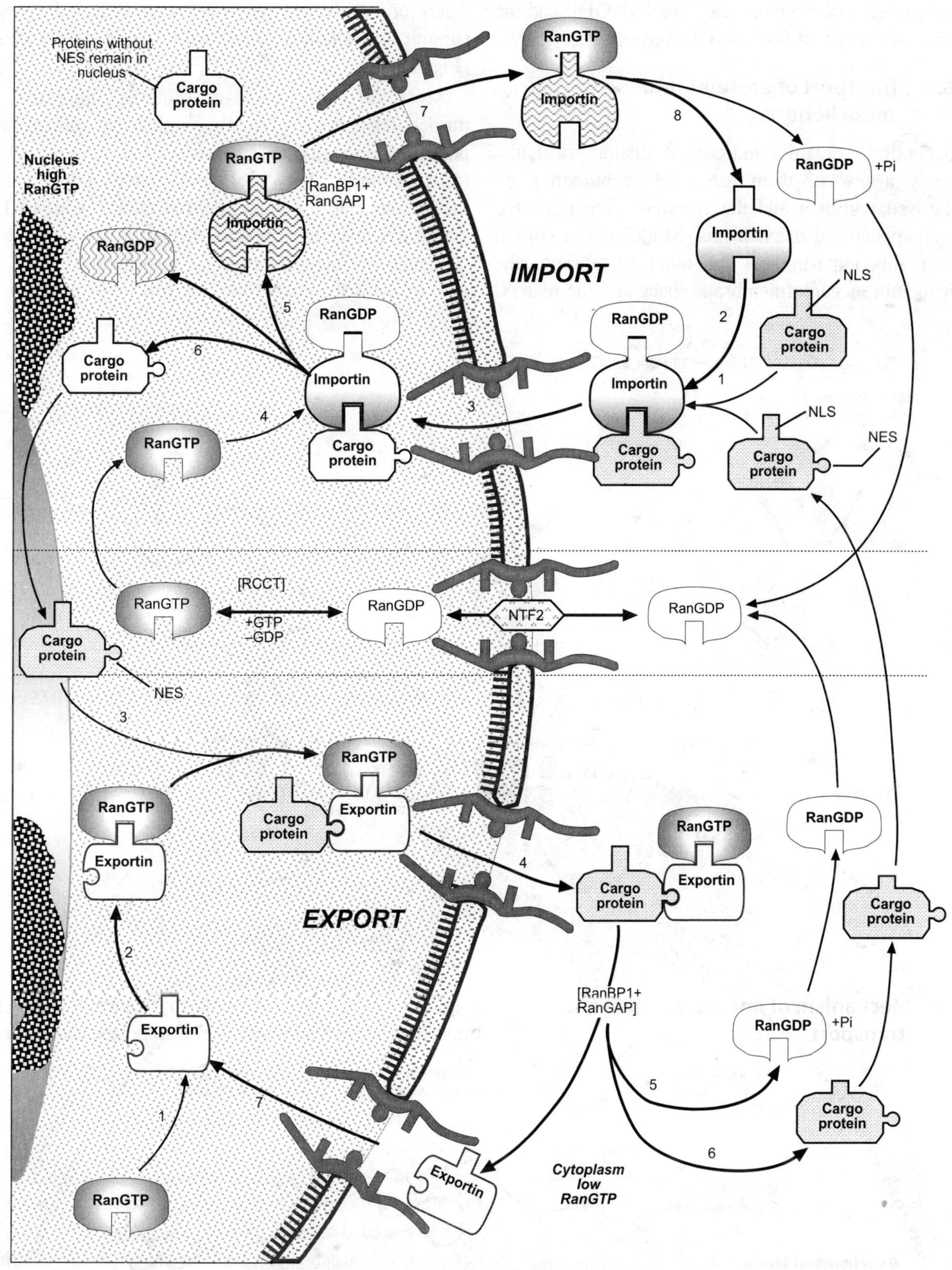

Fig. 13.12 The mechanism of transport cycles of importins and exportins and their coordination by the RanGTP and RanGDP systems. NTF2: nuclear transport factor 2 NLS: nuclear localisation signal NES: nuclear export signal cargo protein: the protein that is to be imported or exported.

with the release of the molecule. The Ran-GDP and the exportin are recycled back into the nucleus.

13.5.2 Transport of proteins into mitochondria

Mitochondria contain hundreds of distinct proteins, but only a few of them (e.g., 13 in humans) are synthesised within it and the remainder are imported through specialised mechanisms. Mitochondria consist of four subcompartments: the outer membrane, the inner membrane, intermembrane space and the matrix. Therefore, mitochondrial proteins must reach the proper subunit and attain a functionally active conformation (**Fig. 13.13**).

Most of the nuclear proteins imported to mitochondria bear a targeting signal that directs them post-translationally to a translocation system, which leads through the outer and inner mitochondrial membranes into the mitochondrial matrix (**Fig. 13.14**). These pre-proteins contain either an N-terminal signal sequence or internal targeting information. With the help of cytosolic chaperone, the precursor proteins are

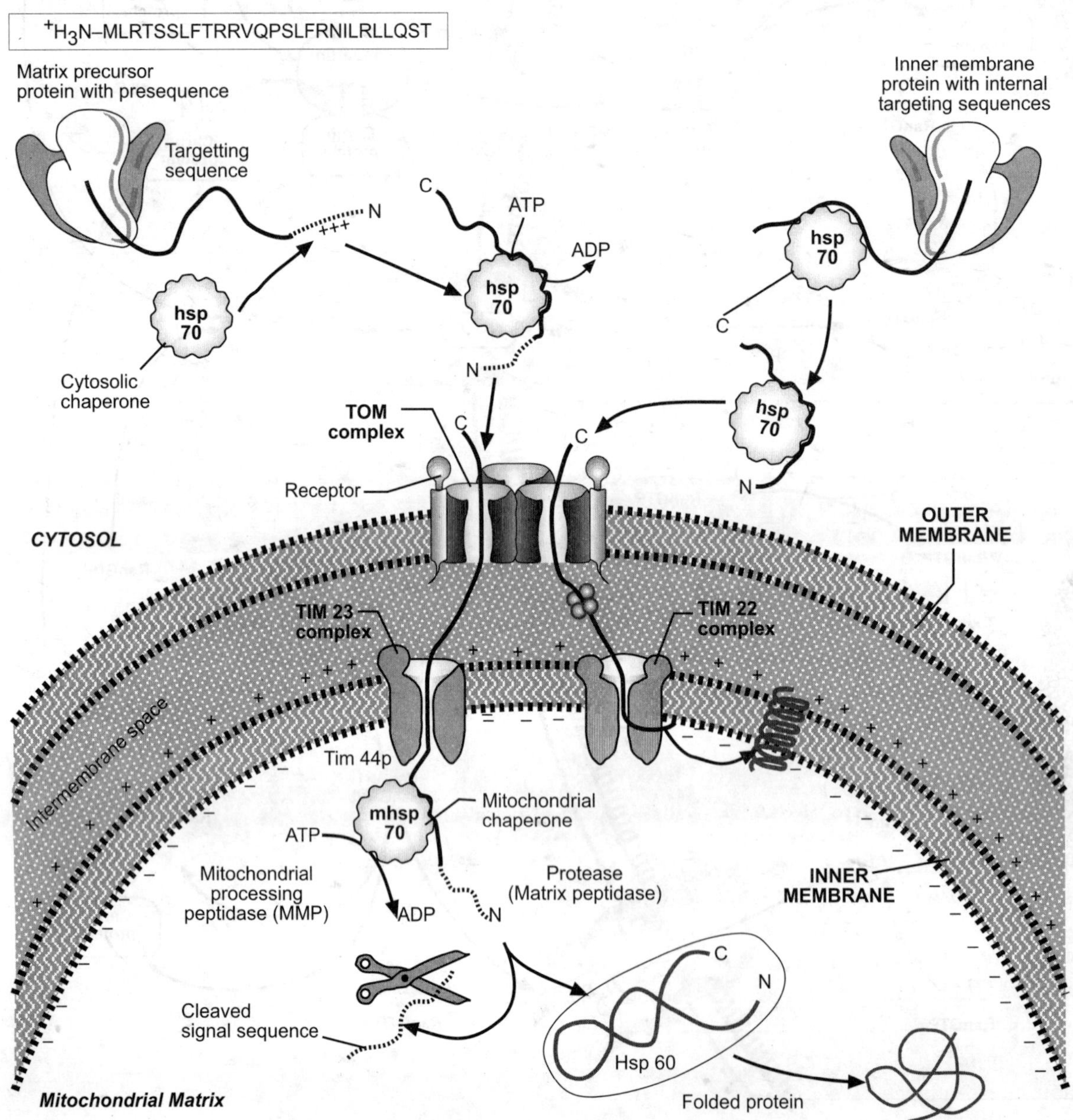

Fig. 13.13 A mitochondrial matrix-targeting sequence TOM: translocase of outer membrane TIM: translocase of inner membrane mhsp: mitochondrial heat shock protein hsp: heat shock protein or chaperone

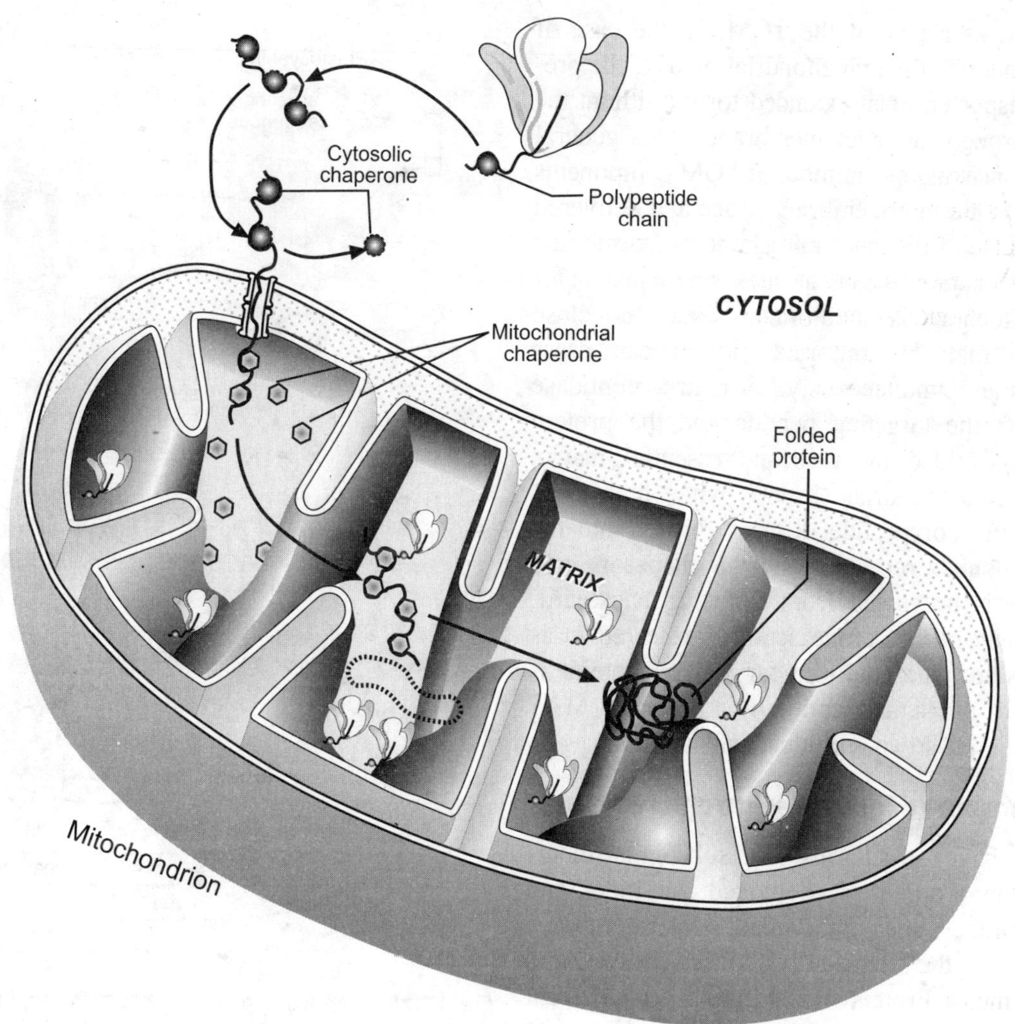

Fig. 13.14 Action of chaperones during protein transport

guided to the mictochondrial surface where they engage the action of specialised translocation components. Mitochondria contain three distinct protein import systems, one in the outer membrane and two in the inner membrane. A multisubunit protein complex of the mitochondrial outer membrane, called **translocase of the outer membrane** (**TOM**), forms the central entry gate for virtually all mitochondrial pre-proteins. The two multisubunit protein complexes of the inner membrane are termed the **translocases of the inner membrane** (TIM).

The mitochondrial targeting sequence takes the form of an N-terminal extension called **transit sequence** (cleavable pre-sequence 10 to 70 residues long) of the mature protein. These mitochondrial pre-sequences lack contiguous hydrophobic regions. Instead, they have positively charged and hydroxyl-amino acid residues spread along their entire length. An important property of the targeting sequence is a potential for forming a positively charged amphiphilic α helix with basic residues on the one side of the helix and the uncharged hydrophobic residues on the other, i.e., mitochondrial pre-sequences are positively charged amphiphiles. In general, mitochondrial targeting sequences share no sequence homology.

The mitochondrial pre-proteins are retained in an unfolded state with their target sequences exposed, through association with heat shock proteins of the hsp70 family (**chaperones**). Soluble cytoplasmic chaperonins prevent the mitochondrial protein from folding completely after translation. The protein is then transferred from the cytoplasmic hsp70 to a receptor on the outer mitochondrial membrane, with the concomitant hydrolysis of ATP. Receptor proteins have been identified as a part of the TOM. In the case of proteins destined for the mitochondrial matrix, the pre-protein is transported in an extended form (without the chaperone) through the outer membrane, via a general import pore consisting of a number of TOM components and then across the intermembrane space and delivered to the translocase of the inner mitochondrial membrane (**TIM**). Translocation occurs at sites where the outer and inner mitochondrial membranes come into close proximity so that the imported protein can cross both membranes simultaneously. A matrix peptidase hydrolyses off the targeting peptide and the protein becomes fully folded by a second essential matrix chaperonin, mitochondrial hsp60 with hydrolysis of ATP. As the polypeptide diffuses further into the matrix, it is bounds repeatedly by the chaperone and at each stage is prevented from diffusing backward. This mechanism of chaperone action is referred to as **biased diffusion.** The final step for matrix proteins is the cleavage of the signal peptide by a protease (MAS protease), which has two non-identical subunits.

13.5.3 Transport of proteins into chloroplast

Chloroplasts have structurally six subcompartments into which the proteins are delivered: an inner and outer membrane and intervening intermembrane space, as well as the stroma, thylakoid membrane and thylakoid lumen. Protein translocation to different compartments of the chloroplast appears similar to the mitochondrion. In fact, the N-terminal leader sequences for the two events are so similar that a chloroplast leader from a green alga can transport an attached protein into yeast mitochondria. How then do the receptors in mitochondria and chloroplasts differentiate between the leader sequence of proteins to be transported respectively into them? Initial targeting to the chloroplast is conveyed by an N-terminal extension somewhat resembling the mitochondrial one. This signal delivers the protein through the double membrane of the chloroplast envelope into the stroma. All proteins translocated through the chloroplast envelope contain a **stroma targeting domain** as a part of their transit peptide, which guarantees that the polypeptide will enter the stroma. From the stroma, certain proteins need to be transported further into the lumen or thylakoid membrane, which bear an additional segment in their transit peptide, called the **thylakoid transfer domain**. Several distinct pathways have been identified by which proteins are either inserted into the thylakoid membrane or translocated into the thylakoid lumen. These transport pathways resemble the transport systems in bacterial cells.

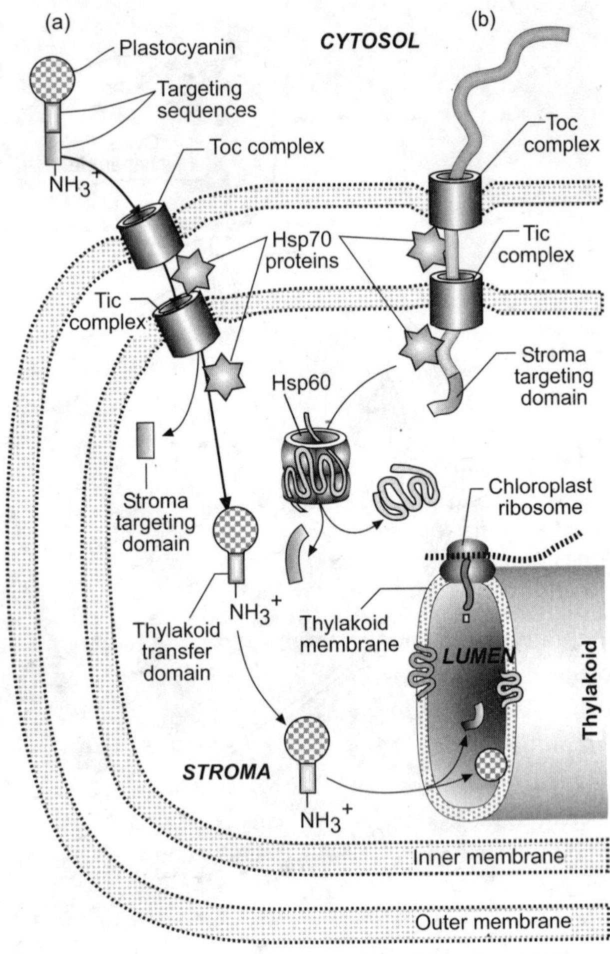

Fig. 13.15 Importing proteins into a chloroplast. (a) Proteins destined for the thylakoid (e.g. plastocyanin) contain both a stroma-targeting domain and a thylakoid transfer domain at their N-terminus (b) Proteins destined for the stroma contain a stroma-targeting domain at their N-terminus.

Proteins destined for chloroplast subcompartments other than the stroma have an additional pathway to travel (**Fig. 13.15**). Proteins are also transported into microbodies such as peroxisomes (**Fig. 13.16**).

13.6 POST-ENTRY SORTING OF MEMBRANE PROTEINS

The plasma membrane is a dynamic structure that regulates the movement of molecules. Small molecules, such as amino acids, sugars and ions can traverse the plasma membrane through the action of integral

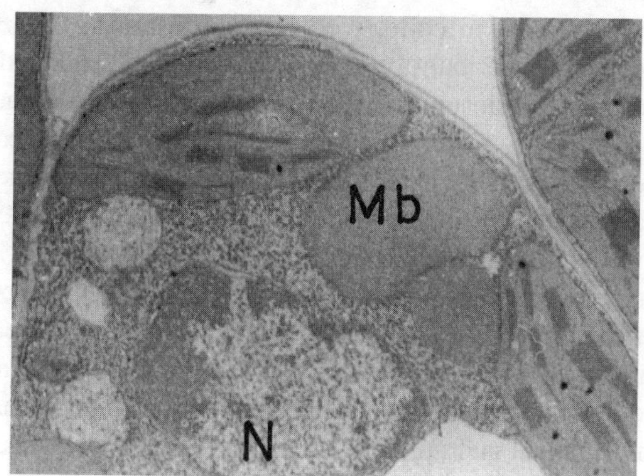

Fig. 13.16 TEM micrograph of peroxisomes found in tomato plant cells; its close association with chloroplasts and mitochodria appears to facilitate exchange of materials during photorespiration (***Source:*** Channarayappa 1992)

membrane protein pumps or channels. But how do large molecules penetrate the membrane? Or how do proteins on the plasma membrane move further into the cell? Both of these processes occur by the endocytic pathway in which macromolecules are carried into the cell by vesicles. Endocytosis is primarily a process by which the cell internalises the cell surface receptors and bound extracellular ligands. There are two broad methods of endocytosis: (1) phagocytosis and pinocytosis and (2) receptor-mediated endocytosis.

This section focuses on the better-understood molecular mechanisms governing receptor-mediated endocytosis, i.e., **clathrin-mediated endocytosis (CME)**.

13.6.1 Receptor-mediated endocytosis

Endocytosis is the process in which a small region of the plasma membrane invaginates to form a new intracellular membrane-limited vesicle about 0.05–0.1 μm in diameter. During this transportation, a specific receptor on the surface of the membrane binds correctly to the extracellular macromolecule that it recognises, called the ligand; the plasma membrane region containing the receptor–ligand complex then undergoes endocytosis, becoming a transport vesicle. In this process, the rate of endocytosis of a ligand is limited by the amount of its corresponding cell surface receptor, and receptor–ligand complexes are selectively incorporated into the intracellular transport vesicles; most other plasma membrane proteins are excluded. Vertebrate cells bear many types of receptors on the cell surface that binds specific ligands tightly and with a high degree of specificity.

Receptor-mediated endocytosis occurs via clathrin-coated pits and vesicles. Barbara Pearse (1975) named the protein isolated from the coated vesicles as clathrin. It is visible in electron micrographs as a proteinaceous layer coating the cytosolic side of pits that come from the *trans* Golgi or from the plasma membrane. Clathrin-coated pits make up about 2% of the surface of cells such as hepatocytes and fibroblasts. Some receptors cluster over clathrin-coated pits even in the absence of the plasma membrane. But they undergo a conformational change when binding to a ligand, so that the receptor–ligand complex can be retained when it diffuses into a clathrin-coated pit. The coated pit appears bristly, consisting of a network of hexagons resembling a honeycomb. Each clathrin molecule consists of three heavy chains and three light chains; joined together to form a three-legged assembly called a **triskelion** (**Fig. 13.17**). It provides a remarkably adaptable module with which a cell can construct different types of polygonal lattices.

The overlapping arrangement of the three-legged clathrin triskelions within the wall of the vesicle coat, and the interaction between the clathrin lattice and the adaptors, are shown in the **Fig. 13.18**. There are more than 24 different accessory proteins that form a dynamic network of interacting molecules. The best studied of these accessory proteins is **dynamin**. It is a GTP-binding protein that is required for the release of a clathrin-coated vesicle. Dynamin self-assembles into a helical collar around the neck of an invaginated and coated pit, just before it pinches off from the membrane. Hydrolysis of the bound GTP by the polymerised dynamin molecules induces a conformational change in the dynamin helix that severs the coated vesicle from the plasma membrane.

13.6.2 The endocytic pathway

Two different types of receptors are involved in endocytosis:

- The **housekeeping receptors** are responsible for the uptake of materials that will be used by the cell. The best-studied examples are the low-density lipoprotein (LDL) and iron-laden transferrin (Tfn).
- The **signalling receptors** are responsible for binding extracellular ligands that include hormones such as insulin and growth factors such as EGF.

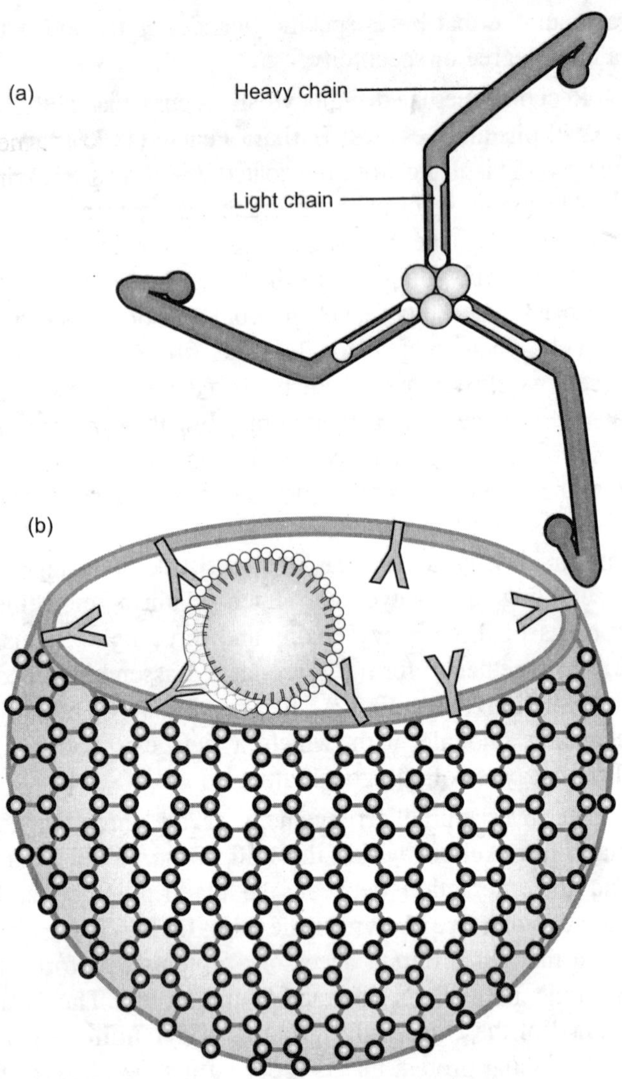

Fig. 13.17 **(a)** Clathrin triskelion; triskelion is composed of three heavy chains. The inner portion of each heavy chain is linked to a smaller light chain. **(b)** Clathrin-coated vesicle carrying LDL by an receptor-ligand-mediated transport.

These bind to the surface receptor and signal a physiological response inside the cell.

Endocytosis of housekeeping receptors leads typically to the delivery of molecules such as iron and cholesterol to the cell and the return of the receptor to the cell surface for additional rounds of uptake. On the other hand, signalling receptors are often destroyed by a process called **receptor downregulation** so that sensitivity of the cell to further stimulation by the hormone or growth factor is reduced. Receptor down-regulation is a mechanism by which cells regulate their ability to respond to extracellular messengers.

Soon after internalisation, vesicle-bound materials are transported to a dynamic network of tubules and vesicles known collectively as **endosomes**, which represent distribution centres along the endocytic pathway. Endosomes are grouped into two distinct classes: **early endosomes** that are typically located near the peripheral region of the cell and **late endosomes** that are located closer to the nucleus. Receptors taken up by endocytosis are transported in vesicles to an early endosome, which serves as a sorting station that directs different types of receptors and ligands along different pathways. Housekeeping receptors typically dissociate from their bound ligands in the acidified endosomal environment and are then concentrated into specialised tubular compartments of the early endosome. Vesicles that bud from these tubules carry receptors back to the plasma membrane for additional rounds of endocytosis.

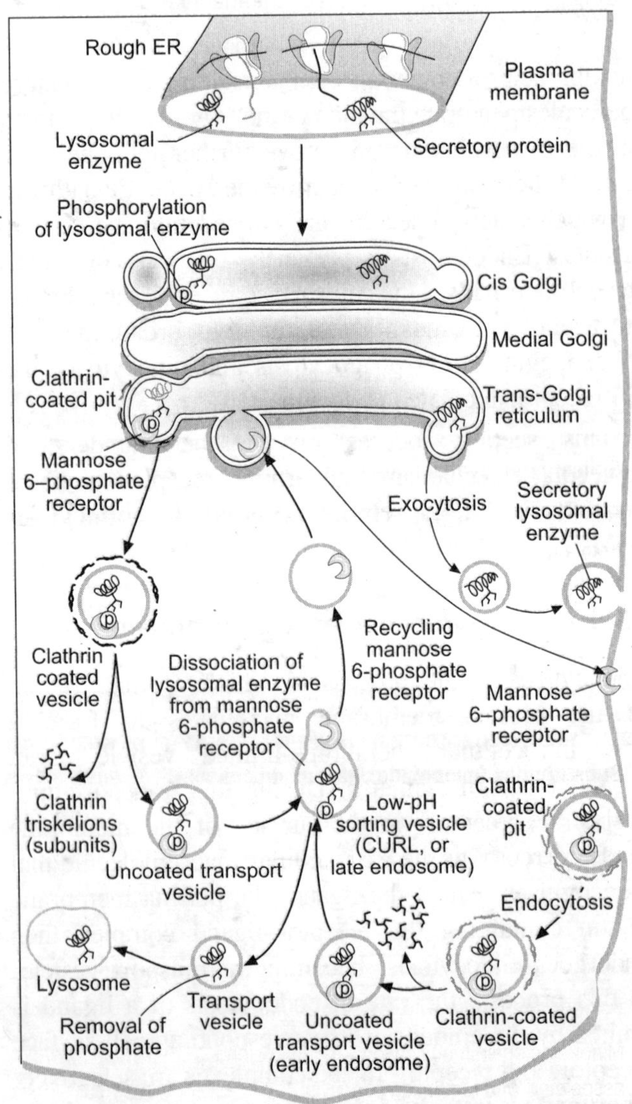

Fig. 13.18 The targeting of lysosomal enzymes to lysosomes

In contrast, released ligands (e.g., LDL) become concentrated into a sorting compartment before being moved to a late endosome and ultimately to a lysosome where final processing occurs.

13.7 POST-TRANSLATIONAL MODIFICATION OF PROTEINS

Release of the completed polypeptide from the ribosome is not necessarily the final step in the covalent construction of a protein. During or following protein synthesis, all polypeptides undergo some form of covalent modification before they form functional proteins. Generally, post-translational processing prepares each polypeptide for its functional role and for folding into its biologically active conformation. In the course of these modifications, the primary structure of a protein may be altered and novel derivations may be introduced into its amino acid side chains (**Table 13.2**). Over 200 different types of post-translational modification reactions have been identified and hundreds of amino acid variations have been described in proteins, virtually all arising post-translationally. Structurally, such modifications can be divided into the following groups.

Table 13.2 Enzymatic protein modifications with different roles

Covalent modification	*Examples*
Cleavage (removal of residues)	
Cleavage of peptide bonds	Cotranslational or post-translational cleavage of initiator methionine occurs in most cytoplasmic proteins. Cotranslational cleavage of signal peptide occurs during translocation across the endoplasmic reticulum membrane for secreted proteins. Maturation of immature proteins (pro-proteins) by cleavage; e.g., activation of zymogens (inactive enzyme precursors) by proteolysis, removal of internal C-peptide of proinsulin, cleavage of hedgehog proteins into N-terminal and C-terminal fragments. Processing of genetic information: Cleavage of polyproteins synthesised from polio virus genome and mammalian tachykinin genes, splicing out of intervening polypeptides (inteins).
Splicing (removal of intervening sequences)	
Removal of intervening amino acid sequence from a polypeptide	Protein splicing is a chemical reaction in which a spliced intein is excised from a precursor protein and the flanking N-terminal (N-extein) and C-terminal (C-extein) regions are ligated with the peptide bond to produce mature proteins. For example, a 120-kDa yeast vacuolar membrane ATPase (VMA1) protein catalyses a self-protein splicing post-translationally to yield the 70-kDa catalytic subunit of the vacuolar H^+-ATPase and the 50-kDa DNA endonuclease (intein).
Substitutions (minor side chain modifications)	
Minor side chain modification—permanent and associated with protein function	Hydroxylation of proline residues in collagen stabilises triple helical coiled tertiary structure. Sulfation of tyrosine residues in certain hormones. Iodination of thyroglobulin. γ-carboxylation of glutamine residues in prothrombin.
Formation of intra- and intermolecular bonds	Formation of disulfide bond in many extracellular proteins, e.g., insulin, immunoglobulins.
Minor side chain modification—reversible and associated with regulation of activity	Phosphorylation of tyrosine, serine and threonine residues regulates enzyme activity, e.g., receptor tyrosine kinases, cyclin-dependent kinases.
Augmentations (major side or main chain modifications)	

Addition of chemical groups to side chains—associated with protein function	Addition of nucleotides required for enzyme activity (e.g., adenyl groups added to glutamine synthase in *E. coli*). Addition of N-acetylglucosamine to serine or threonine residues of some eukaryotic cytoplasmic proteins. Addition of cholesterol to Hedgehog family; signalling proteins control their diffusion. Addition of prosthetic groups to conjugated proteins, e.g., haeme group to cytochrome C or globins.
Addition of chemical groups to side chains—associated with protein targeting or trafficking	Acylation of cysteine residue targets protein to cell membrane. Addition of GPI membrane anchor targets protein to cell membrane. N-glycosylation of asparagine residues in the sequence Asn-Xaa-Se/Thr is a common modification in proteins entering the secretory pathway. O-glycosylation of Ser/Thr occurs in Golgi.
End group modification	Acetylation of N-terminal amino acid of many cytoplasmic proteins appears to relate to the rate of protein turnover. Acylation of N-terminal residue targets proteins to cell membrane, e.g., myristoylation of *Ras*.
Protein degradation	Ubiquitination of proteins targeted for degradation.

Cleavage: In this process, residues are removed from the nascent polypeptide.

Splicing: Proteins are cleaved and joined together resulting in the formation of a new functional protein.

Minor substitutions: These cause only minor changes to amino acid side chains.

Augmentations: Bulky chemical groups are added to particular amino acid residues.

Protein degradation: Some proteins have very high turnover and may be synthesised and degraded at a high rate.

Functionally, modifications may be grouped as neutral or essential for protein functioning or required for protein folding, trafficking and processing. Most augmentations are required for protein sorting. Several types of modifications play an important role in the rate of protein turnover and degradation. Modifications can be permanent or reversible. Cleavages and most augmentations are permanent. Substitutions may be permanent with an essential structural role (e.g., formation of disulfide bonds between cysteine residues), or reversible, fulfilling a regulatory function (e.g., phosphorylation of tyrosine residues). While these are enzymatic modifications programmed by the cell, other modifications occur in a non-enzymatic manner and are usually associated with protein ageing. Such modifications include oxidation, deamination and, for blood proteins, reaction with glucose.

13.7.1 Protein cleavage

Proteolytic cleavage (**proteolysis**) is the most prevalent form of post-translational protein modification. The simplest form of this is removal of the **initiator methionine** from the amino terminus of many polypeptides. Initially all polypeptides begin with formyl-methionine (in bacteria, plastids and mitochondria) or methionine (in the eukaryotic cytoplasm). The formyl group is almost always removed by a ribosome-associated deformylase. In about one-half of all the proteins, the initiating methionine is removed from the nascent chain by a ribosome-associated Met-aminopeptidase. Interestingly, the fate of the N-terminal methionine is determined largely by which amino acid occupies the second position. It has been identified that small neighbouring residues favour removal of the Met. Occasionally, the second amino acid is also removed. In addition, residues at the C terminus may be removed post-translationally on rare occasions.

13.7.2 Protein splicing

Protein splicing is a unique autocatalytic reaction discovered in the yeast vacuolar membrane ATPase (VMA1) protein, encoded by the VMA1 gene of *S. cerevisiae*. It is a chemical reaction in which a spliced intein is excised from a precursor protein and the flanking N-terminal (N-extein) and C-terminal (C-extein) regions are ligated with the peptide bond to produce mature proteins.

Essential residues for the splicing reaction

Inteins are protein insertion sequences that are embedded in host protein sequences, which are removed during post-translational maturation of the host protein. The intein-mediated protein maturation is called **protein splicing**, in which the intein sequence is precisely excised, and the flanking host protein sequences (N- and C-exteins) are joined by a normal peptide bond. Intein and protein splicing may be viewed as the protein equivalent of intron and RNA splicing, respectively. The names 'intein' (internal protein) and 'exeins' (external proteins) were adopted in 1994. The intein database lists about 100 inteins, and this number is bound to increase when more genes and genomes are sequenced.

Mechanism of splicing reaction

The mechanisms of protein splicing have been characterised as a four-step pathway (**Fig. 13.19**) through theoretical modelling and experimental studies, with key early evidence provided by Xu and coworkers (1996).

An N-S (or N-O) acyl rearrangement moves the N-extein to the side chain of the Ser (or Cys) at the upstream splice junction, forming the linear thioester (or ester) intermediate. A *trans*-esterification reaction results in cleavage at the upstream splice junction and transfer of the N-extein to the side chain of Ser (or Thr, Cys) at the downstream splice junction, forming the branched intermediate. Cyclisation of the intein C-terminal Asn results in cleavage at the downstream splice junction and release of the excised intein. A spontaneous O-N (or S-N) acyl rearrangement results in the formation of a peptide bond between the exteins.

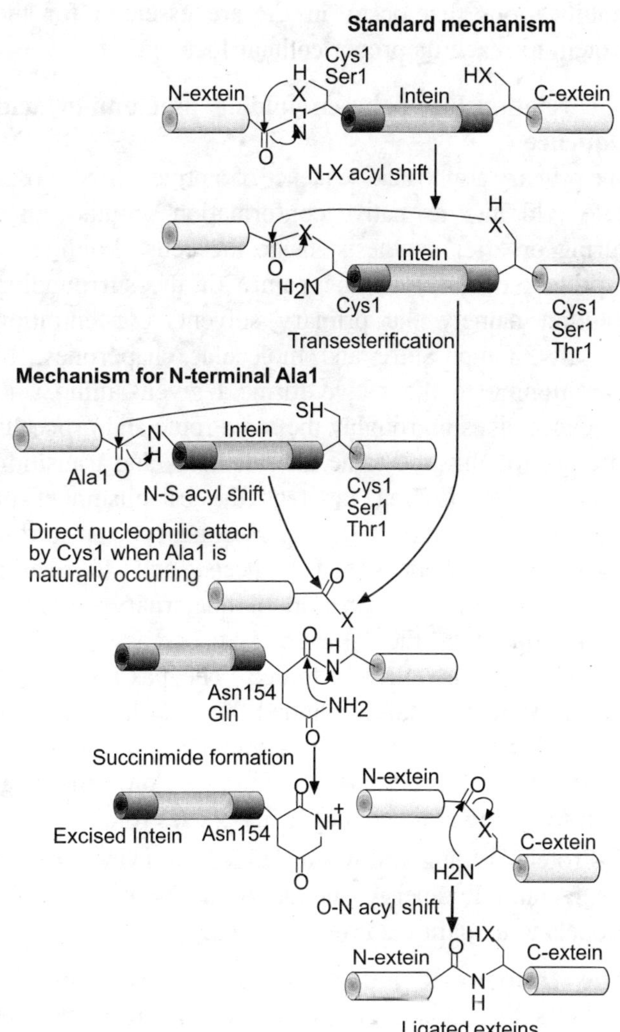

Fig. 13.19 Mechanism of protein splicing and cleavage (***Source:*** Liu XQ, 2000, *Ann. Rev. Genet.* 34:61–76)

13.7.3 Protein folding and processing in the ER

Protein folding is the process by which a protein assumes its characteristic functional shape, also known as the native state. All newly synthesised proteins are linear heteropolymers of amino acids and this sequence is known as the primary structure. Most proteins can carry out their biological functions only when folding has been completed, because the three-dimensional shape of the proteins in the native state is critical to their biological function. Most folded proteins have a hydrophobic core with stabilised side-chain packing and charged or polar side chains on the exposed surface where they interact with surrounding water molecules or solvents.

The process of folding *in vivo* often begins co-translationally, so that the N-terminus of the protein begins to fold while the C-terminal portion of the protein is still being synthesised by the ribosome. Once protein synthesis is complete, the newly introduced polypeptides in the membrane and lumen of the ER are folded, sorted and transported. Secretory and membrane proteins undergo many modifications during their transport to the cell surface. Only properly folded and assembled proteins can be transported from the ER to the Golgi complex and to the cell surface. The modifications that occur in ER are essential for the protein to reach its proper cellular location.

The relationship between folding and amino acid sequence

The primary amino acid sequence of a protein predisposes it to fold into its native conformation spontaneously during or after synthesis inside the cells. Folding of a protein depends on the nature of the surrounding solution namely the primary solvent, concentration of salts, temperature and molecular chaperones. In transitioning to the native form, a given amino acid sequence takes on roughly the same route and proceeds through roughly the same intermediates and transition states. Often folding involves the establishment of regular secondary and supersecondary structures, particularly α-helices and β-sheets, and afterwards tertiary structure. Formation of quaternary structure usually involves the 'assembly' or 'co-assembly' of subunits that are already folded. The passage of the folded state is mainly guided by the hydrophobic interactions, formation of intramolecular hydrogen bonds and van der Waals forces. Conformational entropy of the polypeptide chain opposes folding.

Protein folding follows a defined pathway directed by favourable energy interactions. There are three models to explain the protein folding process:

The framework model: In this model, adjacent residues throughout the polypeptide interact to form secondary structures without forming a defined tertiary structure. The number of possible arrangements of the secondary structures is limited, and random diffusion of these preformed structures is sufficient to identify and adopt the native conformation.

The hydrophobic collapse model: This model predicts that the process of excluding water molecules from non-polar side chains would drive an initial folding reaction to form the hydrophobic protein core due to its favourable energy parameters. The remainder of the protein would then reorganise in the more limited spectrum of conformational arrangements available. The consolidation stage of protein folding would involve a reorganisation of an intermediate to generate the native state.

The nucleation model: This model suggests that specific residues in the polypeptide initiate the protein folding, which forms a nucleus around which other structures build. This global folding needs no intermediates and could occur using a well-defined nucleus or a weakly defined nucleus which condenses as a structure forms around it (nucleation-condensation).

Most studies of protein folding have identified intermediate folding states, which support the framework or collapse models. More recently, however, several small proteins have shown to fold in a simple two-state manner with no intermediates and a single transition state.

Energy landscape theory of protein folding

Joseph Bryngelson and Peter Wolvnes (1990) formulated the energy landscape theory to explain the protein folding. This approach asserts that evolutionary selection has allowed the amino acid sequences of natural proteins in such a way that interactions between side chains largely favour the molecule's acquisition of the folded state. Therefore, proteins generally have a globally 'funnelled energy landscapes' that are largely directed towards the native state. This 'folding funnel' landscape allows the protein to fold to the native state through any of a large number of pathways and intermediates, rather than being restricted to a single mechanism.

13.7.4 Molecular chaperones and folding of nascent proteins

Under *in vivo* conditions, many newly synthesised proteins may fold into their native conformations spontaneously (**self-assembly**). This is either because the folding process requires the transition to an intermediate state with unfavourable energy or several equally stable alternative folding pathways, only one of which is native. In most cases, self-assembly of proteins requires hours to reach completion, yet new secretory proteins generally fold in the ER lumen within minutes after their synthesis. In these cases, correct folding of proteins is directed by a family of helper proteins called **molecular chaperones.** These proteins assist other proteins in achieving proper folding. Many chaperones are heat shock proteins, that is, proteins expressed in response to elevated temperatures or other cellular stresses. The reason for this behaviour is that protein folding is affected by heat and therefore, some chaperones act to repair the potential damage caused by misfolding. Chaperone binding stabilises the amino-terminal protein in an unfolded conformation until the rest of the polypeptide chain is synthesised and the completed protein can fold correctly. Correct folding reflects a cycle of chaperones–substrate binding and release which is often dependent on ATP hydrolysis and the activities of chaperone accessory proteins (**co-chaperones**). Molecular chaperones are necessary to prevent illegitimate interactions between denatured proteins in the cell, which could result, for example, in protein aggregation.

Activities of molecular chaperones

Chaperons are cylindrical protein complexes that contain a central cavity in which newly synthesized polypeptides can fold without interference from other macromolecules in the cell. Chaperones direct protein folding, unfolding and assembly at many levels during protein processing as shown below:

- Initial folding following protein synthesis
- Refolding following, for example, thermal denaturation
- Unfolding and refolding to allow translocation across membranes
- Interconverting alternative conformations of all allosteric proteins
- Refolding e.g., enzymes become denatured as part of their activity (preparing proteins for degradation)
- Controlling the formation of multimeric and protein–ligand complexes.

13.7.5 Disulphide bonds formed in the ER

In eukaryotic cells, disulfide bonds (Cys-S-S-Cys) are formed only in the lumen of the rough ER. This is also called oxidative protein folding in the ER. A disuphide bond is a covalent bond between the thiol groups of two cysteine residues. Such bonds usually stabilise the fold of a protein and are essential for proper structure. Christian Anfinsen (1963) discovered that **protein disulfide isomerase** (PDI) catalyses the breakage and re-formation of these bonds. Disulfide bonds are confined to secretory proteins (e.g., insulin) and to the exoplasmic domains of membrane proteins and they are important stabilising forces in the tertiary structure of these proteins. Cytoplasmic proteins generally do not possess disulfide bonds because of the presence of various reducing agents in the cytoplasm. Recently, functioning of specific catalytic systems in the periplasm and the ER that catalyse the formation of correct disulfides has been revealed.

In some proteins, disulfide bonds are formed while the polypeptide is still growing on the ribosome, as in the case of the immunoglobulin (Ig) light chain, whose disuphide bonds stabilise the two separate domains and form sequentially. However, disulfide bonds do not always form sequentially. For example, proinsulin has three disulfide bonds that are not sequential—pairing is between cysteines 1 and 4, 2 and 6 and 3 and 5. In the ER, PDI accelerates the rearrangement process which is why it is found in abundance in the ER of secretory tissues in the liver and pancreas. PDI acts by catalysing the rearrangement of disulfide bonds in a broad range of protein substrates, allowing them to generate thermodynamically most stable configurations.

13.7.6 Isomerisation of peptide bonds

Another enzyme that plays an important a role in protein folding is **peptidyl prolyl isomerase**, which catalyses the isomerisation of peptide bonds that involve proline residues. The equilibrium between the *cis* and *trans* conformations of peptide bonds that precede proline residues is only slightly in favour of the *trans* form. In contrast, peptide bonds between other amino acids are almost always in the *trans* form. Isomerisation between the *cis* and *trans* configurations of prolyl peptide bonds represents a rate-limiting step in protein folding.

13.7.7 Covalent modifications of protein

Other than the non-covalent interactions, activity of many proteins is also regulated by covalent modifications. An example is that of phosphorylation which is readily reversible due to allosteric regulation. Protein kinases catalyse protein phosphorylation and they mostly transfer phosphate groups from ATP to the hydroxyl groups of the side chains of serine, threonine or tyrosine residues (**Fig. 13.20**). They are one of the largest protein families in eukaryotes. In contrast, **protein phosphorylases** reverse phosphorylation by hydrolysis of phosphorylated amino acid residues. The combined action of protein kinases and protein phosphatases mediates the reversible phosphorylation of many cellular proteins. Often, protein kinases function as components of signal transduction pathways in which one kinase activates a second kinase, and so on. In muscle cells, the hormone epinephrine (adrenaline) signals the breakdown of glycogen to glucose-1-phosphate. Glycogen phosphorylase, a protein kinase, catalyses the breakdown of glycogen.

13.7.8 Hydration

Hydroxylation of the amino acids proline and lysine is required for the structural integrity of the proteins

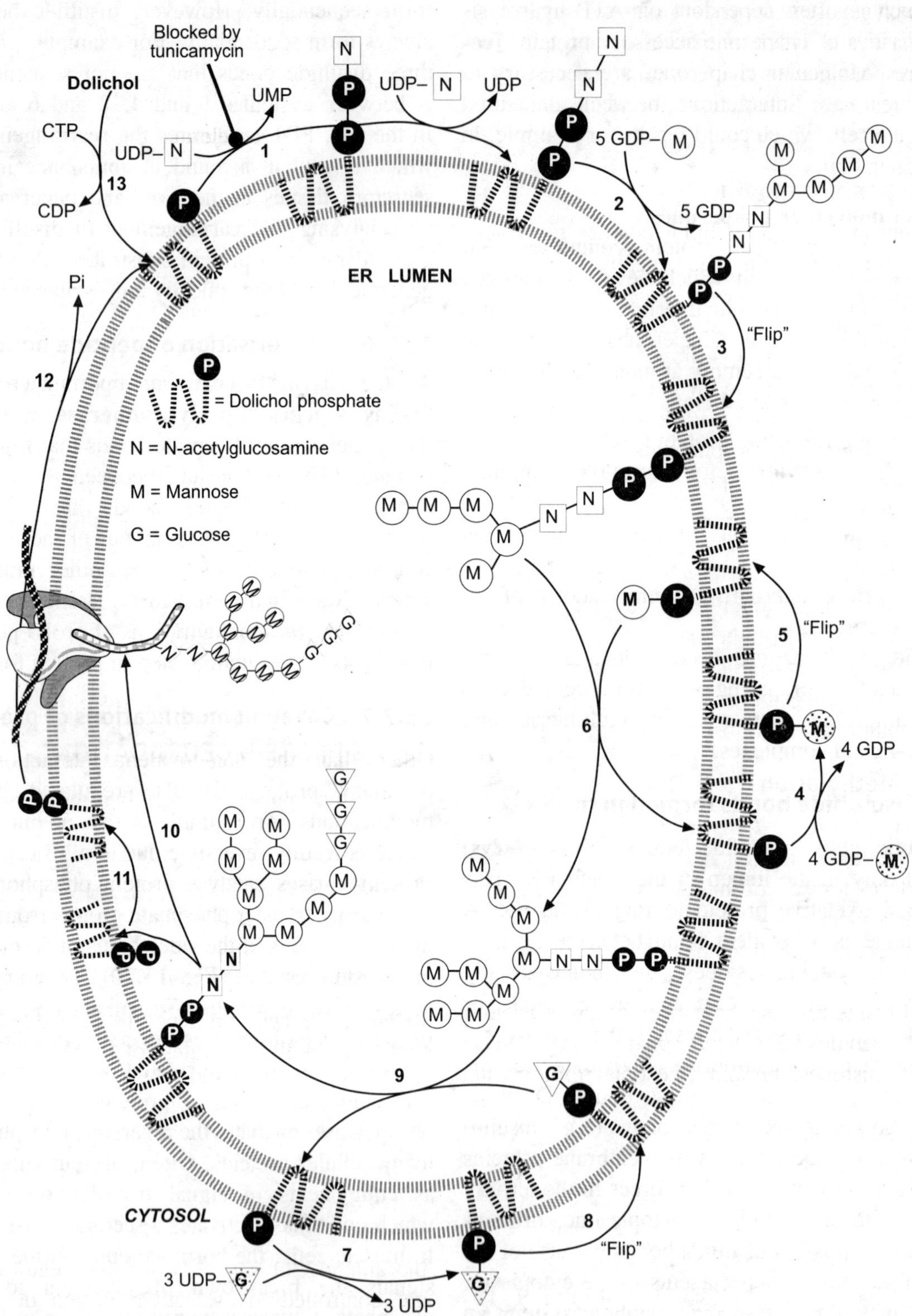

Fig. 13.20 Biosynthesis of the dolichol pyrophosphoryl oligosaccharide precursor of N–linked olgiosaccharides M: mannose; G: Glucose; N: N-Acetylglucosamine (NAG); P: phosphate

of the connective tissue and elastin. Three mixed-function oxygenases (prolyl-4-hydroxylase, prolyl-3-hydroxylase and lysyl hydroxylase) located in the RER are responsible for hydroxylating certain proline and lysine residues. Substrate requirements are highly specific.

13.7.9 Lipophilic modifications

The covalent attachment of lipid moieties to proteins improves membrane binding capacity and certain protein–protein interactions. Three general types of lipid additions—**N-myristoylation (Fig. 13.21)**, **prenylation** and **palmitoylation**—are common in eukaryotic proteins associated with the cytosolic face of the plasma membrane. A fourth type of modification, the addition of **glycolipids**, plays an important role in anchoring some cell surface proteins to the extracellular face of the plasma membrane. Among the most common lipophilic modifications are acylation (the attachment of fatty acids) and prenylation although the fatty acid, myristate (14:0), is relatively rare in eukaryotic cells. Myristoylation is one of the most common forms of acylation. N-myristoylation has been shown to increase the affinity of the α-subunit of certain G-proteins for ß- and γ-subunits attached to the membranes.

13.7.10 Methylation

Protein methylation plays diverse roles in eukaryotes. Aspartate residues methylated by a specific type of methyltransferase promotes either the repair or the degradation of damaged proteins. Other methyltransferases catalyse reactions that alter the cellular roles of certain proteins. For example, methylated lysine residues have been found in proteins such as ribulose-2,3-bisphosphate carboxylase, calmodulin, histones, certain ribosomal proteins and cytochrome c. Other amino acid residues that may be methylated include histidine and arginine.

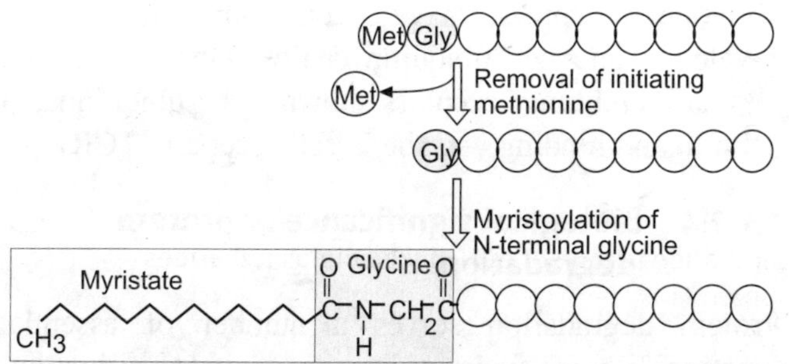

Fig. 13.21 Addition of a fatty acid by N-myristoylation

13.8 SELENOPROTEINS

Selenium (Se) is a trace element found as a component of several prokaryotic and eukaryotic enzymes that are involved in redox reactions. The selenium in these selenoproteins is incorporated as a unique amino acid, selenocysteine, during translation. An important eukaryotic selenoenzyme is glutathione peroxidase, which is required during the oxidation of glutathione by hydrogen peroxide (H_2O_2) and organic hydroperoxides. Selenocysteine is incorporated by the translation machinery through an interesting mechanism. The tRNA for selenocysteine is charged with serine and then enzymatically selenylated to produce the selenocysteinyl-tRNA. The anticodon of selenocysteinyl-tRNA interacts with a stop codon in the mRNA (UGA) instead of a serine codon. The selenocysteinyl-tRNA has a unique structure that is not recognised by the termination machinery and is brought into the ribosome by a dedicated specific elongation factor. An element in the 3' non-translated region (UTR) of selenoprotein mRNAs determines whether UGA is read as a stop codon or as a selenocysteine codon.

13.9 PROTEIN DEGRADATION

Cellular proteins vary widely in stability, with half-lives ranging from a few minutes (e.g., many regulatory proteins) to several weeks or longer (e.g., structural and storage proteins). Proteins are destroyed by inclusion into the **proteosome complex**.

13.9.1 Protein turnover

In an active cell, proteins are in a dynamic state of turnover, with the rates of protein synthesis and degradation ultimately determining the amount of protein present at any point in time. In many instances, transcriptional regulation determines the concentrations of specific proteins expressed within cells, with protein degradation playing a minor role. In other instances, protein synthesis is constitutive, and the amounts of key enzymes and regulatory proteins are controlled via selective protein degradation. In addition, abnormal proteins arising from biosynthetic errors or post-synthetic damage must be destroyed to prevent the deleterious consequences of their build-up. The elimination of proteins typically follows first-order kinetics, with half-lives ($t_{1/2}$) of different proteins ranging from several minutes to many days.

There are several distinct degradation pathways in the cell depending on the nutritional status and signalling from other cells, and may involve protein modification. The two proteolytic systems within eukaryotic cells are lysosomes and proteasomes. Lysosomes are membrane-bound organelles that contain an assortment of acidic hydrolases, including many proteolytic enzymes. They degrade molecules taken up from the extracellular medium and plasma membrane receptors downregulated in response to ligand binding. They are also responsible for turnover of most organellar proteins of the endolytic and late secretory pathways. Lysosome-mediated protein breakdown are largely non-selective. A single, random proteolytic break in the polypeptide backbone of a protein is believed sufficient to cause its rapid disappearance, since no partially degraded proteins are normally observed in cells. Many characteristically unstable proteins contain **PEST sites** (i.e., rich in proline, glutamine, serine and threonine) and target them for rapid proteolysis. The lysosomal pathway is ubiquitin-independent which generally involves relatively stable proteins. Proteins are brought into the cell by the budding off of vesicles by a process termed **microautophagy**. During starvation, some cytoplasmic proteins may be taken into the lysosome directly by the recognition of a signal peptide.

On the other hand, proteosomes contain a selective, ATP-dependent mechanism for protein degradation in eukaryotic cells. Proteosomes are multiprotein complexes (40–80 kDa) found in the cytosol and the nucleus. Each one consists of a tunnel-like core particle with one or two smaller, regulatory particles positioned at either or both ends like caps. The core particle is formed by four stacked rings—each composed of seven subunits—surrounding a central channel that constitutes a proteasome's digestive tract. The two outer rings appear to act as gates to keep stray proteins from accidentally getting into the degradation chamber. The proteins to be degraded are targeted by post-translational modification, either involving chemical modification of specific residues (e.g., phosphorylation) or conjugation to the small protein ubiquitin.

13.9.2 Ubiquitin pathway for protein degradation in eukaryotes

The selective degradation of many short-lived proteins in eukaryotic cells is carried out by the ubiquitin system. Ubiquitin is a highly conserved, 76-residue (8.5 kDa) polypeptide widespread in all eukaryotic cells. The amino acid sequences of yeast and human ubiquitins are 53% identical. Proteins are committed to degradation through covalent ligation to ubiquitin, which is transferred to its substrates by a carrier protein and targets them for degradation in the proteosome. There are now at least 60 known substrates that include transcription factors, cell cycle regulators, kinases, phosphatases, tumour suppressors and a number of different transmembrane proteins.

Ubiquitination is an essential cellular process that is affected by a multi-enzyme cascade involving classes of enzymes known as $E_{1\text{-}3}$ (**ubiquitin-activating enzymes**). Three proteins (E_1, E_2 and E_3) in addition to ubiquitin are involved in the ligation process. E_1 is the ubiquitin-activating enzyme (105 kDa dimer). It becomes attached via a thioester bond to the C-terminal Gly-residue of ubiquitin through ATP-driven formation of an activated ubiquitin-adenylate intermediate. It is then transferred to protein E_2, the **ubiquitin carrier protein**. E_2 is actually a family of at least seven different small proteins, several of which are heat shock proteins. Alternatively, E_2-S ~ ubiquitin transfers ubiquitin to free amino groups on proteins selected by E_3 (180 kDa), the **ubiquitin-protein ligase**. Upon binding a protein substrate, E_3 catalyses the transfer of ubiquitin from E_2-S ~ ubiquitin to free amino groups on the protein. More than one ubiquitin may be attached to a protein substrate and tandemly linked chains of ubiquitin occur via isopeptide bonds between the C-terminus of one ubiquitin and the ε-amino of Lys^{48} of another, resulting in a polyubiquitinated substrate, which is the signal for proteosome-mediated degradation (**Fig. 13.22**).

13.9.3 Ubiquitination of plasma membrane proteins

In addition to the degradation of proteins retained in the ER, ubiquitination has recently been shown to play a role in the transport and degradation of plasma membrane proteins. Cell surface receptors and growth hormone receptors, have been found to contain covalently attached ubiquitin chains. One of the first plasma membrane receptors shown to be ubiquitinated upon ligand binding was the T-cell receptor (TCR).

13.9.4 Biological significance of protein degradation

Protein degradation serves a number of essential functions like cell cycle progression, signal transduction, transcriptional regulation, receptor downregulation and

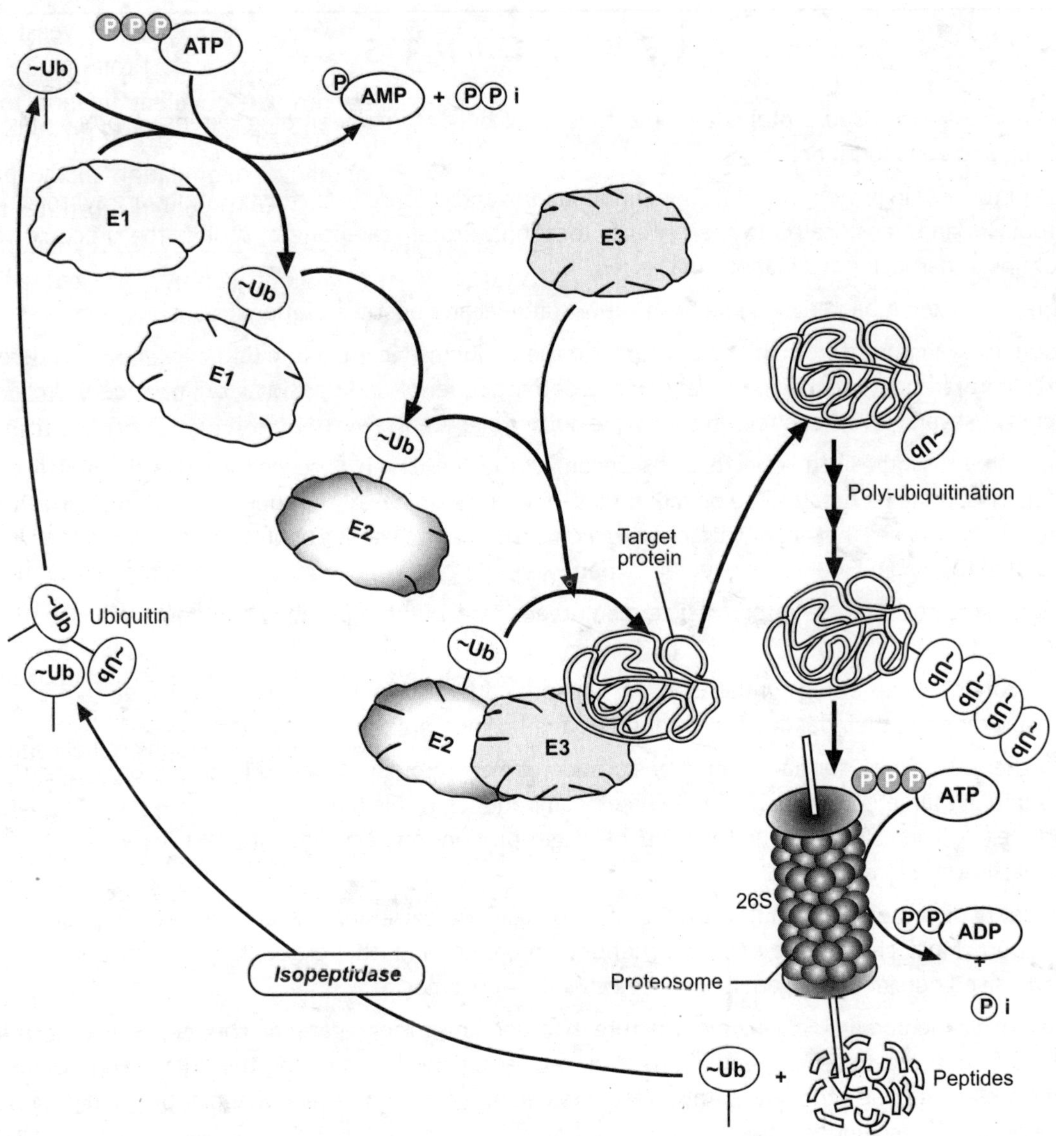

Fig. 13.22 The schematic view of ubiquitin–proteosome pathway; the components of the ubiquitin–conjugating system, thiol–ester bond formation (~) between Ub and E3. The term isopeptidase is used to refer to any protein that cleaves linkages between the carboxy terminus of ubiquitin and the ε–amino group of a lysine residue.

endocytosis. The ubiquitin system has been implicated in immune response, cell development and programmed cell death. Degrading a crucial enzyme or regulatory protein, for example, is the common mechanism that cells use to slow or stop a biochemical reaction. In contrast, many cellular processes are activated by the degradation of a critical inhibitory protein. Protein degradation also plays a special role in the overall regulation of body metabolism. In times of need, such as malnourishment or disease, the proteosome pathway becomes more active in our muscles, providing amino acids that can be converted into glucose to provide energy. During immune response, the antigens present on the cells depend on proteosomes to generate the flags that distinguish such dangerous cells (e.g., virus infected). Protein breakdown by proteosomes also serves as a kind of cellular quality control system that prevents the accumulation of aberrant proteins. The degradation of abnormal proteins is important in a number of human genetic diseases.

Abnormalities in ubiquitin-mediated processes have been shown to cause pathological conditions, including malignant transformation.

KEY POINTS

- The newly synthesised proteins become functional only after undergoing extensive processing, folding, sorting and transportation.
- Except for certain proteins produced in mitochondria and chloroplasts, most proteins are synthesised in the cytoplasm and translocated to their proper locations. Protein targeting describes the processes directing proteins to particular destinations in the cell.
- Protein transfer in bacteria may be both cotranslational and post-translational translocation.
- Proteins are imported from the cytoplasm into the endoplasmic reticulum (ER), mitochondria, peroxisomes and chloroplasts by mechanisms that use a targeting sequence and translocation machinery. Protein sorting occurs at several levels and requires multiple sorting signals and events.
- For proteins synthesised on cytosolic ribosomes, the first level of sorting divides them into two broad categories: proteins synthesised on membrane-bound ribosomes of the rough ER and those synthesised on cytosolic ribosomes that are not attached to membranes. Large classes of proteins synthesised on ribosomes bound to the rough ER enter the secretory pathway.
- During exocytosis, the secretory vesicle membranes fuse with the plasma membrane, releasing the protein to the cell exterior.
- The proteins while being synthesised on ribosome attached to endoplasmic reticulum are prepared for transportation to their destination, through a complex membrane system called cotranslational transport.
- In higher eukaryotes, translation and translocation of many proteins destined for secretion are tightly coupled and initially targeted to the ER. The two key components in this targeting route are the signal recognition particle (SRP) and SRP receptor. The Golgi packages proteins for secretion into vesicles, which move towards the plasma membrane.
- Transport of proteins into mitochondria, chloroplasts, peroxisomes and the nucleus is post-translational transport. Most chloroplast and mitochondrial proteins, and all peroxisomal and nuclear proteins are synthesised outside the organelle on free cytosolic ribosomes.
- The plasma membrane is a dynamic structure that functions to segregate the chemically distinct intracellular milieu from the extracellular environment by regulating and coordinating the entry and exit of small and large molecules. The endocytic pathway fall into two broad categories i.e., phagocytosis and pinocytosis and receptor-mediated endocytosis.
- Over 200 different types of post-translational modification reactions have been identified and hundreds of amino acid variations have been described in proteins, virtually all arising post-translationally. In an active cell, proteins will be in a dynamic state of turnover, with the release rates of protein synthesis and protein degradation ultimately determining the amount of protein present at any point in time.

Self-assessment Questions

1. Explain the different paths of protein movement in eukaryotic cells.
2. Describe the mechanism of protein secretion in bacterial cells.
3. How are proteins sorted into different target sites in a cell?
4. Explain with the help of a diagram the mechanism of cotranslational transport of proteins.
5. Discus the steps involved in post-translational targeting of proteins.
6. List the different types of protein processing.
7. Explain the role of mitochondria in protein processing and transportation.
8. Describe protein trafficking between the nucleus and cytoplasm with the help of a diagram.
9. How is protein from the cytoplasm transported into the mitochondria and chloroplast?
10. Explain receptor-mediated endocytosis in detail.
11. Discuss in detail the steps involved in exocytosis.
12. Explain with the help of a diagram the process of protein folding.

13. Discuss the different mechanisms of post-transcriptional modification of proteins.
14. How does protein degradation play a role in the regulation of protein turnover?

FURTHER READING

Textbooks

Alberts B, A Johnson, J Lewis, M Raff, K Roberts and P Walter (2002). *Molecular Biology of The Cell.* 4th ed. Garland Science, Taylor and Francis Group.

Hartl D L and EW Jones (1998). *Genetics: Principles and Analysis.* Jones and Bartlett Publishers.

Karp G (2005). *Cell and Molecular Biology, Concepts and Experiments.* 4th ed. Wiley International Edition.

Krebs JE, ES Goldstein, and ST Kilpatrick (2011). Lewin's *Genes X.* Jones and Bartlett Publishers Inc.

Lodish H, A Berk, P Matsudaira, CA Kaiser, MP Scott, SL Zipursky and J Darnell (2004). *Molecular Cell Biology,* 5th ed. New York: W. H. Freeman and Company.

Turner PC, AG McLennan, AD Bates and MRH White (1998). *Instant Notes in Molecular Biology.* Viva Books Private Limited.

References

Abeijon C and CB Hirschberg (1992). Topography of glycosylation reactions in the endoplasmic reticulum. *Trends Biochem. Sci.* 17:32–36.

Anraku Y (1997). Protein splicing: Its chemistry and biology. *Genes to Cells* 2:359–67.

Barr FA (2002). The Golgi apparatus: Going round in circles? *Trends Cell Biol.* 12:101–4.

Blobel G and RW Wozniak (2000). Proteomics for the pore. *Nature* 403:835–36.

Bonifacino JS and AM Weissman (1998). Ubiquitin and the control of protein fate in the secretory and endocytic pathways. *Annu. Rev. Cell Dev. Biol.* 14:19–57.

Conner SD and SL Schmid (2003). Regulated portals of entry into the cell. *Nature* 422:37–44.

Craig EA (1993). Chaperones: Helpers along the pathways to protein folding. *Science* 260:1902–3.

Driouich A, L Faye and LA Staehelin (1993). The plant Golgi apparatus: A factory for complex polysaccharides and glycoproteins. *Trends Biochem. Sci.* 18:210–14.

Fersht AR (1997). Nucleation mechanisms in protein folding. *Curr. Opin. Struct. Biol.* 7:3–9.

Fewell SW, KJ Travers, JS Weissman and JL Brodsky (2001). The action of molecular chaperones in the early secretory pathway. *Annu. Rev. Genetics* 35:149–6.

Glick BS (2001). ER export: More than one way out. *Curr. Biol.* 11:R361–R363.

Goldberg AL (2000). Probing the proteasome pathway. *Nature Biotech.* 18:494–96.

Goldberg AL, SJ Elledge and JW Harper (2001). The cellular chamber of doom. *Scientific American,* January 2001, 56–61.

Gorlich D and IW Mattaj (1996). Nucleocytoplasmic transport. *Science* 271:1513–18.

Haucke V (2003). Vesicle budding: A coat for the COPs. *Trends Cell Biol.* 13:59–60.

Hochstrasser M (1996). Ubiquitin-dependent protein degradation. *Annu. Rev. Genet.* 30:405–39.

Jennifer LR, MY Ali and DW Warshaw (2008). Cargo transport: molecular motors navigate a complex cytoskeleton. *Curr. Opin. in Cell Bio.* 20:41–47.

Levitt M, M Gerstein, E Huang, S Subbiah and J Ysai (1997). Protein folding–the endgame. *Annu. Rev. Biochem.* 66: 549–79.

Liu XQ (2000). Protein–splicing intein: Genetic mobility, origin and evolution. *Ann. Rev. Genet.* 34:61–76.

Martin J and FU Hartl (1997). Chaperone-assisted protein folding. *Curr. Opin. Struct. Biol.* 7:41–52.

Mattaj IW and L Englmeir (1998). Nucleocytoplasmic transport: the soluble phase. *Ann Rev. Biochem.,* 67:265–06.

Neypert W (1997). Protein import into mitochondrial. *Annu. Rev. Biochem.* 66:863–914.

Pfanner N and M Meijer (1997). Mitochondrial biogenesis: The TOM and TIM machine. *Current Biology.* 7:R100–3.

Pfeffer S (2003). Membrane domains in the secretory and endocytic pathways. *Cell* 112:507–14.

Priola SA, B Chesebro and B Caughey (2003). A view from the top: Prion diseases from 10,000 feet. *Science* 3000:917–20.

Rapoport TA, B Jungnickel and U Kutay (1996). Protein transport across the eukaryotic endoplasmic reticulum and bacterial inner membranes. *Annu. Rev. Biochem.* 65:801–48.

Rechsteiner M and SW Rogers (1996). PEST sequences and regulation by proteolysis. *Trends Biochem. Sci.* 21:267–71.

Rechsteiner M, L Haffman and W Dubiel (1993). The multicatalytic and 26S proteases. *J. Biol. Chem.* 268:6065–68.

Regeimbal J and JCA Bardwell (2002). Disulfide bond formation in prokaryotes and eukaryotes. In *Protein targeting, transport and translocation,* edited by Dalbey RE and G von Heijne. Academic Press.

Roder H and W Colon (1997). Kinetic role of early intermediates in protein folding. *Curr. Opin. Struct. Biol.* 7:15–28.

Rothman JE (1994). Mechanisms of intracellular transport. *Nature* 372:55–68.

Schatz, G and B Dobberstein (1996). Common principles of protein translocation across membranes. *Science* 271: 1519–26.

Schekman R and L Orci (1996). Coat proteins and vesicle budding. *Science* 271:1526–33.

Schmid SL and H Damke (1995). Coated vesicles: A diversity of form and function. FASEB J, 9:1445–53.

Singer SJ (1990). The structure and insertion of integral proteins in membranes. *Ann. Rev. Cell Biol.* 6:247–96.

Walter P and AE Johnson (1994). Signal sequence recognition and protein targeting to the endoplasmic reticulum membrane. *Annu. Rev. Cell Biol.* 10:87–119.

Wickner W, AJ Driessen and FU Hartl (1991). The enzymology of protein translocation across the *Escheichia coli* plasma membrane. *Annu. Rev. Biochem.* 60:101–24.

Xu MQ and FB Perler (1996). The mechanism of protein splicing and its modulation by mutation. EMBO J. 15: 5146–53.

CHAPTER 14

Regulation of Gene Expression in Prokaryotes

LEARNING OBJECTIVES

After reading this chapter, you will be able to:

- Discuss the principle and regulatory sequences of gene expression
- Understand various mechanisms of regulation of gene expression in prokaryotes
- Appreciate the different roles that regulation plays in prokaryotes

14.1 INTRODUCTION

A gene is a DNA segment (sequence) that codes for a functional product, whereas, the transfer of genetic information from DNA into protein constitutes **gene expression**. We have so far seen how the information stored as a sequence of bases in a region of DNA (a gene) can be transcribed into RNA and then translated into a sequence of amino acids to yield a specific protein. In this chapter, we shall discuss how the major processes of cellular control orchestrate the numerous chemical reactions in an efficient manner. Gene expression can be defined as the ability of a gene to produce biologically active molecules such as protein or RNA. The regulation of gene expression has been studied at the phenotypic level for many years, but only recently the responses of the cells to external factors (e.g., stress, pests and pathogens or applied chemicals) have been understood at the molecular level.

The regulatory controls that act on gene expression are very complex in eukaryotes than in prokaryotes. In fact, in prokaryotes, control of initiation of transcription is a major point of regulation. On the other hand, gene expression is regulated through flow of information in eukaryotes. Generally, regulation of the various steps in this process could lead to differential gene expression (**Fig. 14.1**). Of the 4,000 or so genes in the typical bacterial genome or the estimated 100,000 genes in the human genome, only a fraction is expressed at any given time. Some gene products are present in very large amounts as compared to others. Requirements for some gene products change over time and environment.

14.2 GENERAL MECHANISMS OF REGULATION

Gene expression is controlled by a number of regulatory proteins. In prokaryotes, extracellular signals (mainly molecules present in growth medium) control genes which receive the information through regulatory proteins. These proteins are mainly of two types: activators and repressors. They are DNA-binding proteins that recognise specific sites at or near the genes they regulate. When the expression of genetic information is quantitatively increased by the presence

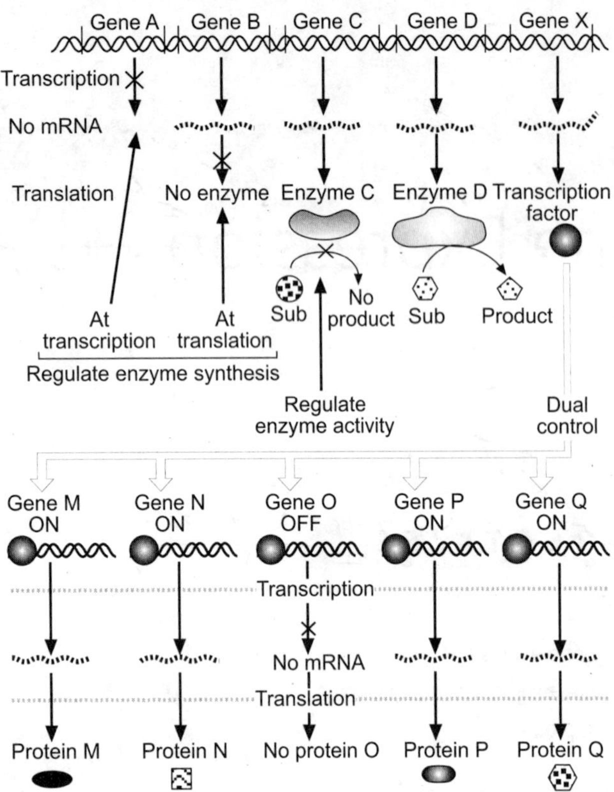

Fig. 14.1 An overview of the mechanisms that can be used in regulation; a gene can regulate the activity of other genes when it codes for a transcription factor

of a specific regulatory element, regulation is said to be positive and the mediator is called an **activator**. In contrast, when the expression of genetic information is diminished by the presence of a specific regulatory element, regulation is said to be negative and the mediator is called a **repressor**.

Prokaryotes provide good and simple models for the study of gene expression. Analysis of the regulation of gene expression in prokaryotic cells helped establish the principle that information flows from the gene to the mRNA to the specific protein molecule. The principles established in early studies of bacterial systems, has led to an understanding of the regulation of gene expression in higher eukaryotes.

14.2.1 Scope of regulation

Regulation of gene expression is essential for an organism to survive and sustain in a given environment. In any cell, a function is turned on when it is needed and switched off when it is not needed. This type of on-off activity is primarily accomplished by regulation at the transcription level and also post-transcriptionally (**Fig. 14.2**). Often an organism needs to regulate many different genes simultaneously in response to a change in its environment. For example, when the bacterium *E. coli* is starved for phosphate, over 80 different genes are transcribed in response. These proteins play roles in adapting the bacterium to a phosphate-deficient environment. There are several such sets of genes in *E. coli* whose products are required to respond to particular conditions. Because these control mechanisms operate on a wide cellular basis, they are referred to as global control systems and may include one or more regulons. In addition to allowing an organism to respond to a signal by activating a network of genes, global regulation can be used to prevent some genes from responding unnecessarily.

Wide and narrow domain regulatory mechanisms refer to the control of specific genes on a group basis or an individual basis respectively. Example of wide-domain regulation in bacteria includes the SOS response in *E. coli*. Narrow-domain control includes examples of individual gene regulation and coordinated regulation of small groups of genes, such as the regulation of the three genes of *E. coli lac* operon by lactose. Enzyme

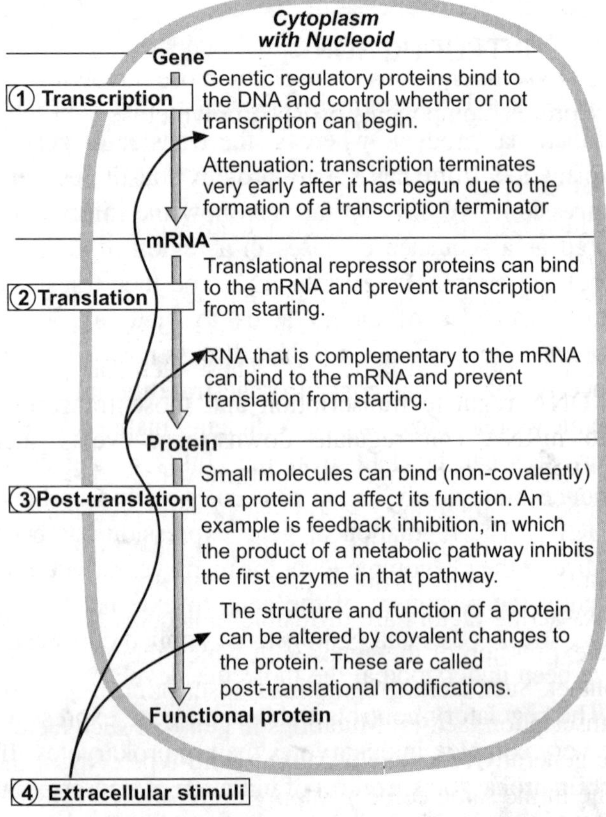

Fig. 14.2 Common points of genetic regulation in bacteria

activity is commonly regulated by the concentration of the reaction product. This type of regulation is called **feedback inhibition** or end-product inhibition.

14.2.2 Principle of gene regulation

Genes for products that are required continuously are expressed at more or less constant levels in virtually every cell of a species or organism. Genes such as those for the cytoskeleton, microtubules, citric acid cycle and so on are called **housekeeping genes**. Genes that express at a low level and are constitutive are called **constitutive** or **basal genes**. However, where cellular levels of gene products rise and fall in response to molecular signals, it is known as regulated gene expression. These genes are in turn of two types:

- Products that increase in concentration under particular molecular circumstances are referred to as inducible and an increase in their expression is called **induction**.
- Gene products that decrease in concentration in response to a molecular signal are referred to as repressible and the process is called **repression**. **Attenuation** is a further mechanism for the control of gene expression, which is involved in the biosynthesis of amino acids in bacteria. Transcription is mediated and regulated by protein–DNA interactions, especially those involving the protein components of RNA polymerase.

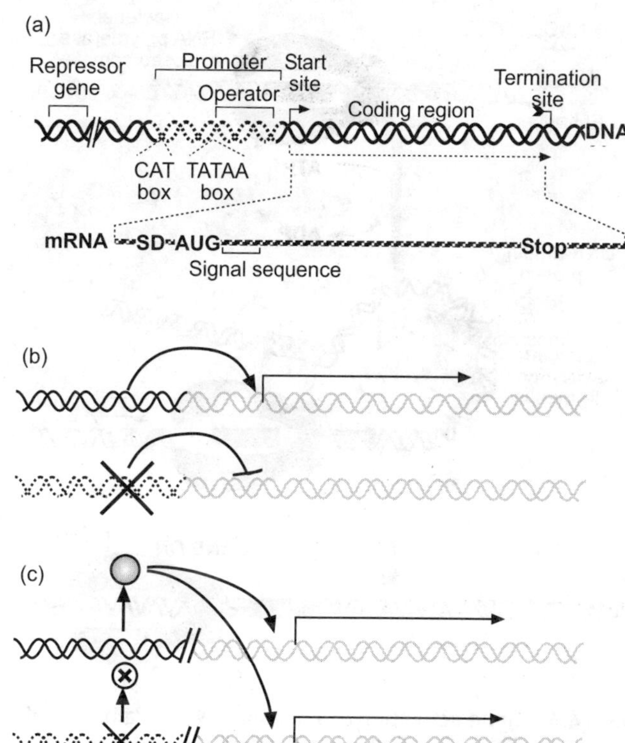

Fig. 14.3 Some major regulatory sequences in prokaryotic gene (a) Detection of *cis*-acting elements and *trans*-acting factors in merodiploids. Mutated elements are shown as dotted lines with crosses. (b) Mutations in *cis*-acting elements are dominant. (c) Mutations affecting (diffusible) *trans*-acting factors are usually recessive

14.2.3 Regulatory sequences

Regulatory processes of nucleic acids work either in *cis* or *trans* effect (**Fig. 14.3**). In DNA, *cis*-acting elements are nucleotide sequences, which regulate genes to which they are attached (adjacent), while having no effect on other chromosomes. These elements present in DNA regulate transcription and those transcribed into mRNA can regulate downstream events such as RNA processing, RNA turnover and protein synthesis. Mutations affecting such elements cannot be complemented by a wild-type copy of the same element supplied in *trans* in the same cell. On the other hand, *trans*-acting factors are diffusible molecules regulating the expression of the genes with which they come into contact. Such factors regulating transcription are termed transcription factors. Mutations in genes of such factors are generally recessive because a wild-type copy of the gene in the same cell can supply more of the diffusible factor.

DNA-binding proteins generally interact and bind to adjacent sites. To accommodate this interaction, the DNA between the sites loops out, bringing the sites into proximity with one another. One way to help bring distant DNA sites together is the binding of other proteins to sequences between those sites. In bacteria, there are a number of instances in which a protein binds between an activator binding site and the promoter and helps the activator interact with polymerase by bending the DNA (**Fig. 14.4**).

14.3 TYPES OF REGULATION OF TRANSCRIPTION

There are a number of common patterns of regulation of transcription depending on the type of metabolic activity of the system being regulated. For example, in a catabolic (degradative) system, the concentration of the substrate of an enzyme early in the pathway often determines whether the enzymes in the pathway would

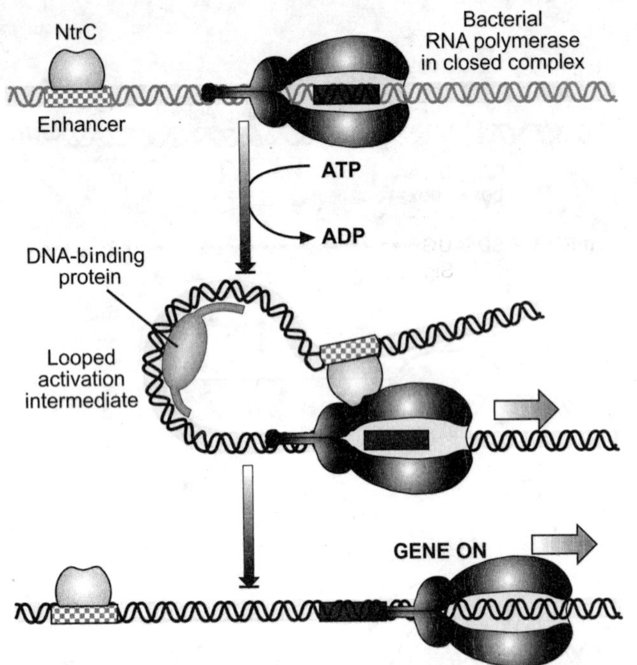

Fig. 14.4 Gene activation at a distance; NtrC is a bacterial gene regulatory protein that activates transcription by facilitating the transition between the initial binding of RNA polymerase to the promoter and the formation of an initiating complex.

be synthesised. In contrast, in an anabolic (biosynthetic) pathway, the final product is the regulatory substance. Even in a system in which a single type of protein molecule is translated from a monocistronic mRNA, the transcriptional activity of the promoter is determined directly by the concentration of the protein. In prokaryotes at least three groups of proteins regulate transcription initiation by RNAP.

(i) **specificity factors** (e.g., sigma factors) alter the specificity of RNA polymerase for a given promoter or set of promoters,

(ii) **repressors** impede access of RNA polymerase to the promoter and

(iii) **activators** enhance the RNA-promoter interaction (**Fig. 14.5**).

Repressors bind to specific sites on the DNA. Regulation by means of a repressor protein that inhibits transcription is referred to as **negative regulation**. The repressor binding to DNA is regulated by a molecular signal (effector), which binds to the repressor and causes a conformational change. The interaction between repressor and signal molecule either increases

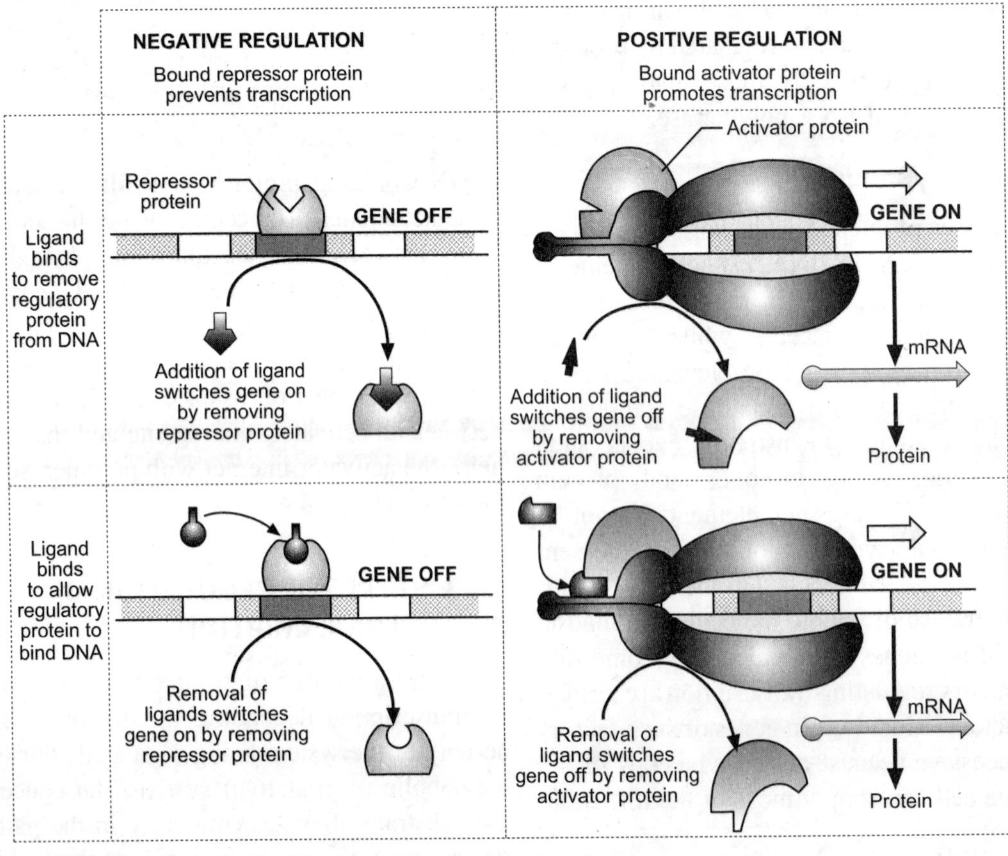

Fig. 14.5 The general mechanism of specific gene transcription in prokaryotes

or decreases transcription. On the other hand, activators bind to DNA and enhance the activity of RNA polymerase (**positive regulation**). An anti-inhibitor, generally called an **inducer**, is needed to turn the system on. Negative and positive regulation, are not mutually exclusive, and some systems are both positively and negatively regulated and need two regulators. Some eukaryotic activators bind to DNA sites called **enhancers** that are quite distant from the promoter, influencing the rate of transcription at a promoter that may be located thousands of base pairs away. Signal molecules can increase or decrease transcription depending on how they affect the activator.

14.3.1 Regulation of enzymatic activity

There are a number of mechanisms of post-transcriptional regulation. In some cases, an enzyme is synthesised as part of a larger inactive precursor protein and activated by removing a portion of it. Another mechanism is to reduce protein degradation.

Product inhibition

A simple mechanism by which an enzymatic reaction may be regulated is a process called product inhibition (**Fig. 14.6**). Enzymatic reactions are generally equilibrium reactions in which the feedback inhibition controls the activity. Feedback inhibition occurs because of the allosteric property of the inhibited enzyme. Such an enzyme has two important binding sites: the active site where the substrate binds, and the allosteric site where the inhibitor binds reversibly. When an inhibitor binds, generally non-covalently, at the allosteric site, the conformation of the enzyme molecule changes so that the substrate no longer binds efficiently at the active site. When the concentration of the inhibitor decreases, the enzyme returns to its catalytic shape. For example, the amino acids proline and arginine are both synthesised from glutamic acid. But the synthesis of one does not affect that of the other.

Covalent modification

There are several examples in bacteria in which enzymes are regulated by covalent modification, usually by addition or deletion of a small organic molecule. The best characterised are enzymes whose activity is affected by attachment of adenosine monophosphate (AMP) or adenosine diphosphate (ADP), by attachment of inorganic phosphate or by methylation.

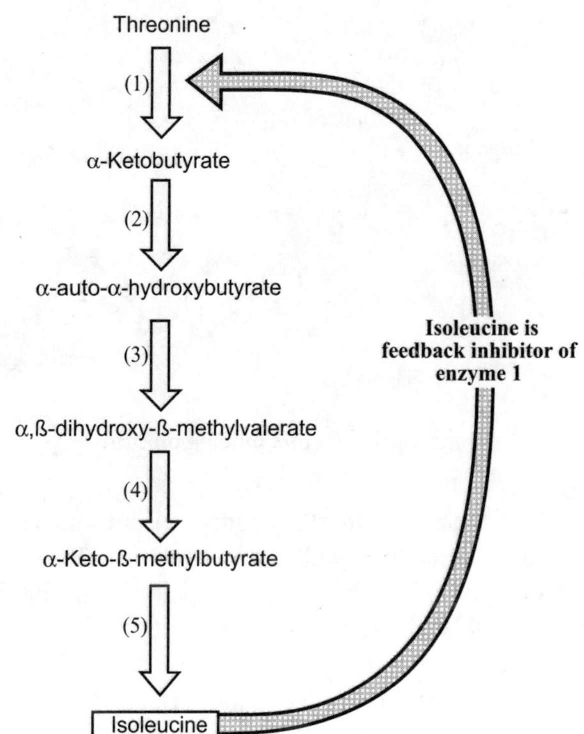

Fig. 14.6 Feedback inhibition of the pathway for synthesising isoleucine

14.3.2 DNA-binding proteins

Although small molecules are often involved directly in regulating transcription, they rarely do so directly. Instead they typically influence the binding of certain proteins, called regulatory proteins, to specific sites on the DNA, and it is these proteins that actually regulate transcription (**Fig. 14.7**). These gene-switching devices contain two types of fundamental components: short stretches of DNA of defined sequence and gene regulatory proteins that recognise and bind to them. One of these regulators, the **lambda repressor** is encoded by a bacterial phage lambda. The repressor shuts off the viral genes that code for the protein components of new virus particles and thereby enables the viral genome to remain a silent passenger (multiply along with bacterial chromosome) in the bacterial chromosome. Other bacterial regulators are the ***lac* repressor**, which turns off the production of the proteins responsible for lactose metabolism when this sugar is absent from the medium.

Interaction of proteins with nucleic acids

Protein–nucleic acid interactions are central to replication, transcription and translation, as well as to the regulation of these processes. Gene regulatory

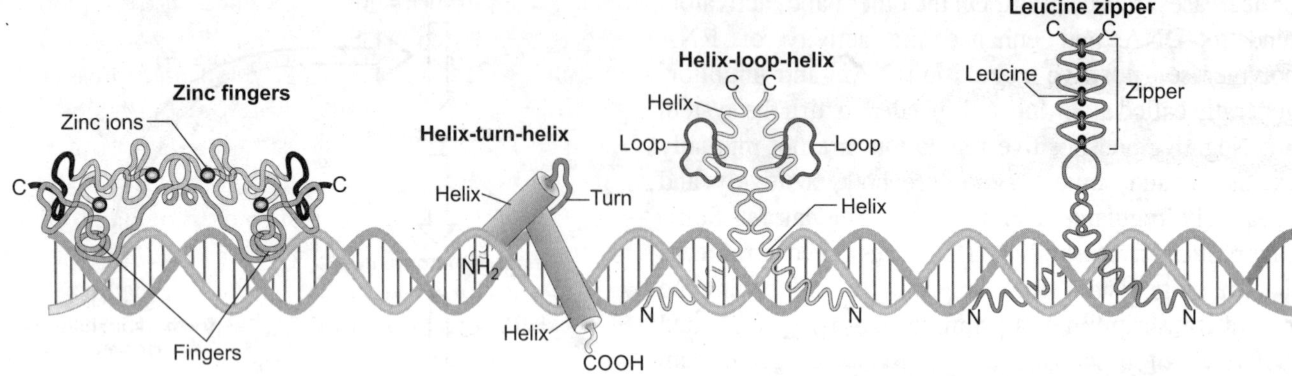

Fig. 14.7 Transcription factors binding on DNA

proteins, therefore, must recognise specific nucleotide sequences embedded within this structure. They need not open the bonds between base pairs in the interior of the double helix. Instead they recognise the DNA sequence on the outside of the double helix. The edge of each base pair is exposed at the surface of the double helix, presenting a distinctive pattern of hydrogen bond donors, hydrogen bond acceptors and hydrophobic patches for proteins to recognise in both the major and minor groove. Proteins that interact specifically with DNA are frequently dimers, composed of two identical polypeptide chains. On each polypeptide chain is a region, called a domain, that interacts specifically with a region of DNA in the major groove. Because the protein recognises contact points associated with specific base pairs, its binding is sequence-specific. The mode of interaction may involve binding to end groups, or enclosing the nucleic acid in a cleft or ring. Most nucleic acid-binding proteins, however, interact in a localised manner at an internal site.

Nucleic acid-binding proteins are classified as follows according to substrate specificity:

- Non-specific binding proteins,
- Sequence-specific binding proteins,
- Proteins that bind unorthodox structures.

14.3.3 Coordinated control of expression of many genes

In bacteria, the most common way to regulate gene expression is by influencing the rate at which transcription is initiated. Synthesis of a particular mRNA takes place only when the gene product is needed. The genes for functionally related proteins, such as enzymes of a metabolic pathway are linked into clusters and controlled by operons. (see **Chapter 10**). Generally the operon is under the control of a single promoter from which RNA polymerase transcribes all of the constituent genes as a single polycistronic mRNA. The promoter is in turn controlled by a series of regulatory elements, which respond to the bacterial environment and modulate the activity of RNA polymerase.

14.3.4 Gene rearrangement

A gene could be turned on, by moving it from a location far from its promoter to a site near its promoter. This change in DNA can affect gene expression, which is termed as the rearrangement of the DNA structure. Regulation by DNA rearrangement occurs in many systems ranging from bacteria to plants and humans. Gene rearrangement in prokaryotes includes activation of nitrogen fixation genes in *Rhizobium* and in heterocyst of cyanobacterium, flagellar antigen or phase variation in *Salmonella* and so on.

In *Anabaena*, the nitrogen-fixation (*nif*) genes were originally mapped using DNA of vegetative cell. A surprise finding was the fact that the genes were not tightly clustered and the *nifK* gene was located 11 kb away from *nifD* (**Fig. 14.8a**). Further research proved that the heterocyst map showed no intervening DNA between *nifK* and *nifD*. The organisation of *nif* genes has been studied most extensively in *Klebsiella pneumoniae* (**Fig. 14.8b**). The *nif* genes of *K. pneumoniae* are all clustered together in a 20-kb segment of the chromosome and they are all expressed as seven or eight operons.

14.4 REGULATION OF INITIATION

There are two main modes of gene regulation in prokaryotes: inducible systems and repressible systems.

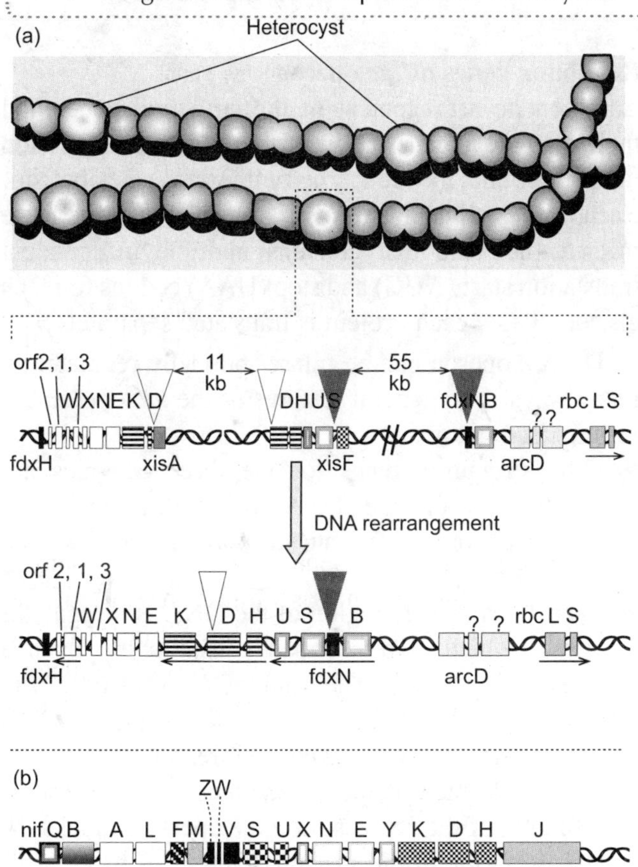

Fig. 14.8 Gene expression regulation by gene rearrangement (a) Developmental DNA rearrangements in cyanobacteria and (b) Organisation of nitrogen-fixing genes in *Klebsiella pneumoniae*

Inducible systems: An inducible system is normally off unless there is the presence of some molecule (called an inducer) that allows for gene expression. In order to survive, bacterial cells must synthesise metabolic enzymes in a dynamic environment. At the same time, economy of activity and energy expenditure dictate shutting off production at the appropriate time. Therefore, a broad category of enzymes, called inducible enzymes are ordinarily absent or only present in very small quantities unless a specific compound, called an inducer is present. These inducers, usually small molecules that serve as metabolic substrates, trigger the production of the enzymes necessary to metabolise them.

Repressible systems: A repressible system is on except in the presence of some molecule (called a **corepressor**) that suppresses gene expression. In this system, generally, an enzyme catalysing the synthesis of a specific product is not synthesised as long as the product is present in the medium. For example, the enzymes involved in the formation of the amino acid tryptophan are synthesised only when the latter is not present in the culture medium; external tryptophan represses the synthesis of these enzymes. Enzyme repression is a very widespread phenomenon in bacteria—it occurs as a means of controlling the synthesis of a wide variety of enzymes involved in the biosynthesis of amino acids, purines and pyrimidines. In almost all cases, it is the final product of a particular biosynthetic pathway that represses the enzymes of the pathway and repression is quite specific.

14.4.1 The operon model

The operon model of prokaryotic gene regulation was first proposed by François Jacob and Jacques Monod in 1961. Their model was based on observations of the regulation of lactose metabolism by the intestinal bacterium *E. coli*. An operon consists of an operator, regulator and structural genes. The regulator gene codes for a repressor protein that binds to the operator, obstructing the promoter of the structural genes. If the repressor molecule is removed, transcription may occur. Operons are either inducible or repressible according to the control mechanism. Seventy-five different operons controlling 250 structural genes have been identified for *E. coli*. Both repression and induction are examples of negative control since the repressor proteins turn off transcription.

The lactose (*lac*) operon

The *lac* operon is the classic example of negative regulation of a bacterial operon by a repressor protein. It consists of one regulatory gene (the *i* gene) and three structural genes (*z*, *y* and *a*). The *i* gene codes for the repressor of the *lac* operon. The *z* gene codes for ß-galactosidase (*lacZ*), which is primarily responsible for the hydrolysis of the disaccharide, lactose into its monomeric units: galactose and glucose (**Fig. 14.9**). The enzyme can additionally convert lactose to allolactose, which is also important. The *y* gene codes for permease (*lacY*) which increases permeability of the cell to ß-galactosides. The *a* gene encodes a thiogalactoside transacetylase (*lacA*).

During normal growth on a glucose-based medium, the *lac* repressor is bound to the operator region of the *lac* operon, preventing transcription. However, in the presence of an inducer (e.g., lactose) in the medium, the repressor protein binds to the inducer and is rendered incapable of interacting with the operator. RNA

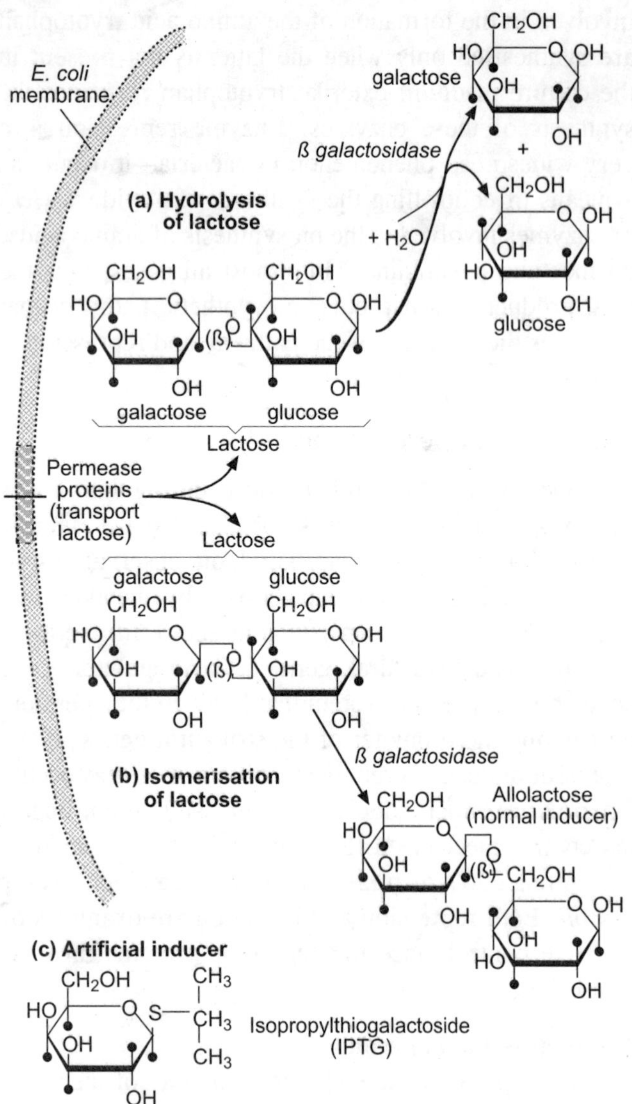

Fig. 14.9 Lactose metabolism of *E. coli* (a) Hydrolysis of the sugar lactose into galactose and glucose by the enzyme ß galactosidase (b) Some of the galactose and glucose molecules hydrolysed by the enzyme are converted into allolactose, the inducer for ß-galactosidase synthesis. (c) Structure of the artificial inducer, IPTG, which can act as an inducer for ß-galactosidase synthesis without being metabolised by the enzyme (gratuitous inducer)

polymerase is, thus able to bind at the promoter region and transcription of the operon ensues. In the presence of the inducer, the concentration of ß-galactosidase rapidly increases from approximately five to several thousand molecules per cell. The structural genes for these three enzymes, along with the *lac* promoter and *lac* operator (a regulatory region), constitute the ***lac* operon (Fig. 14.10).**

Regulator genes of *lac* operon

This genetic arrangement of the structural genes and their regulatory genes allows for coordinate expression of the three enzymes concerned with lactose metabolism. Each of these linked genes is transcribed into one large mRNA molecule that contains multiple independent translation start (AUG) and stop (UAA) codons for each cistron. Thus, each protein is translated separately.

The *lac* operon can be transcriptionally regulated in many ways. The general features of the *lac* operon are summarised below.

- The structural genes for the three enzymes are contiguous on the DNA and transcribed from a single promoter into a polycistronic mRNA transcript.
- The *I* (repressor) region of the DNA is a regulator gene, which is transcribed constitutively from a separate promoter. The resulting *I* mRNA transcript is translated into a monomeric polypeptide, which aggregates into a tetrameric repressor protein and, in the absence of inducers, binds to the operator. Binding of the repressor on the operator blocks the transcription of the structural genes.
- In the presence of lactose, a complex is formed between the inducer (lactose) and the repressor protein. The complex cannot bind to the operator and transcription of the structural genes can thus proceed (**Fig. 14.11**).

The natural inducer is not lactose but allolactose formed from lactose by transglycosylation, which acts as an inducer for ß-galactosidase. The ability to act as an inducer can also be assumed by some molecules such as isopropylthiogalactoside (**IPTG**) and thiomethylgalactoside (**TMG**), called **gratuitous inducers (Fig. 14.12**), which cause the production of ß-galactosidase but are not metabolised by the enzyme. Along with the induction of ß-galactosidase, a second protein called permease is produced, which facilitates entry of lactose into the cell and hydrolysis by ß-galactosidase into its component sugars. Another enzyme, **transacetylase**, also appears upon lactose induction in small quantities. All of these proteins are synthesised from a single polycistronic mRNA transcribed from a cluster of genes that map within the *lac* locus, in the sequence Z (ß-galactosidase), Y (permease) and A (acetylase).

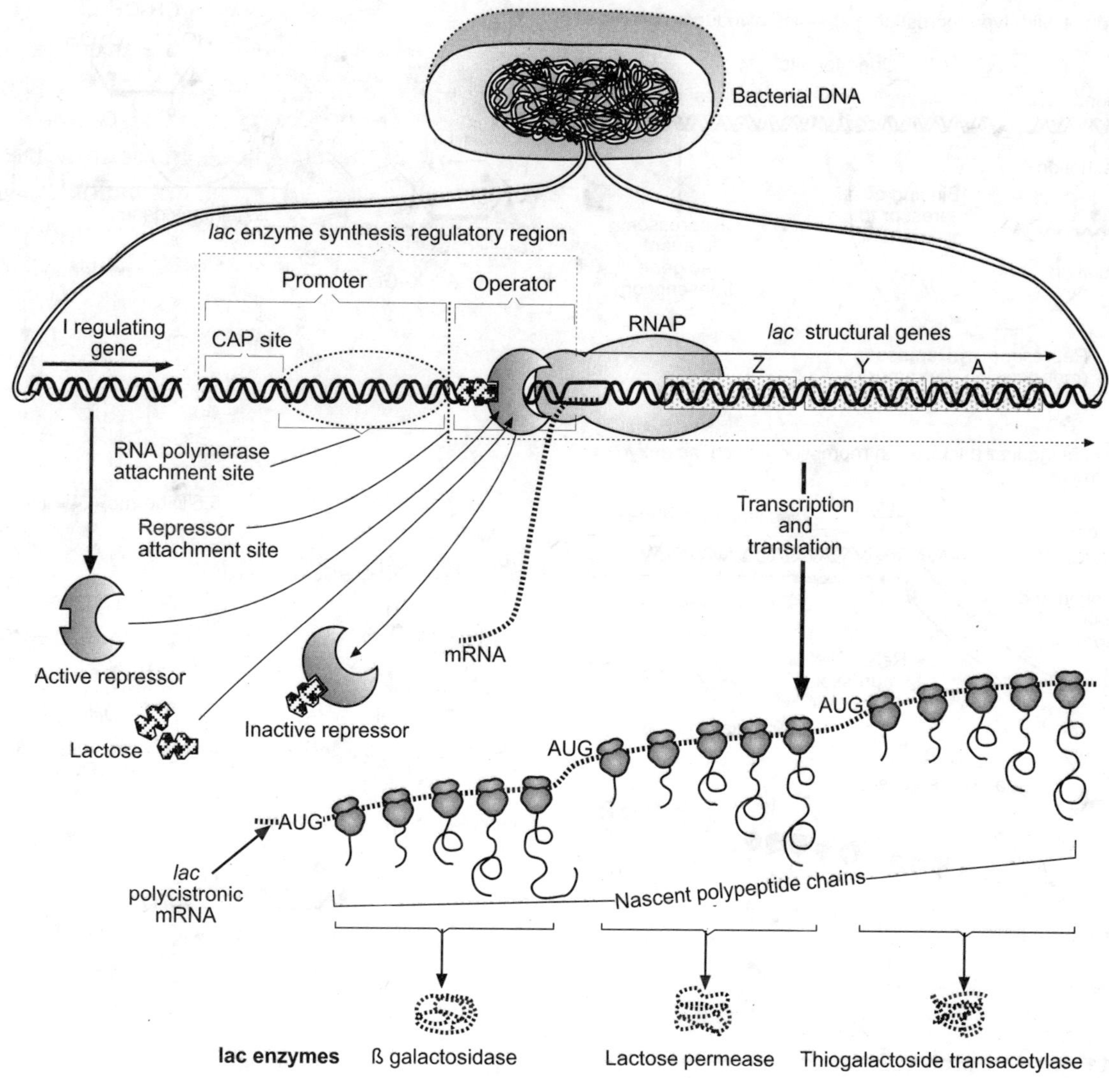

Fig. 14.10 Organisation of a bacterial operon: A model of transcriptional regulation of the *lac* operon by the *lac* repressor

Mechanism of induction of *lac* operon

The first bacterial repressor protein to be isolated was that of the *lac* operon. In wild-type *E. coli,* it represents around 0.001% of the total protein of the cell. The *lac* repressor is an acidic protein of 360 amino acids, with four identical subunits. Each subunit binds one molecule of inducer but the protein interacts with the *lac* DNA as a tetramer. The repressor is an allosteric protein that assumes slightly different shapes. There are two forms for the repressor: one in which it can bind to the operator DNA and the other not so. When the inducer (lactose, allolactose or IPTG or ONPG) is bound to the repressor, it causes an allosteric change in its configuration that greatly reduces its affinity for the operator by a factor of 10^3. Since no covalent bonds are involved, the repressor simply dissociates from the operator. After the repressor releases from the operator, RNA polymerase can begin transcription. These three *lac* operon genes are then transcribed and subsequently translated into their respective proteins. This system of control is very efficient. After all the lactose is metabolised, the repressor returns to its original shape and can again bind to the operator. The system is turned off.

Catabolite repression: A positive control system

Another method of regulation of the *lac* operon is catabolite repression. An interesting property of the *lac* operon and others that code for enzymes that catabolise certain sugars (e.g., arabinose, galactose) is that they are all repressed in the presence of glucose. When grown in the presence of both lactose and

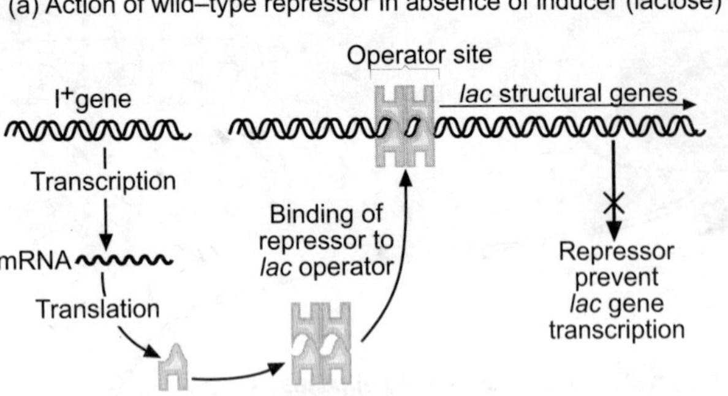

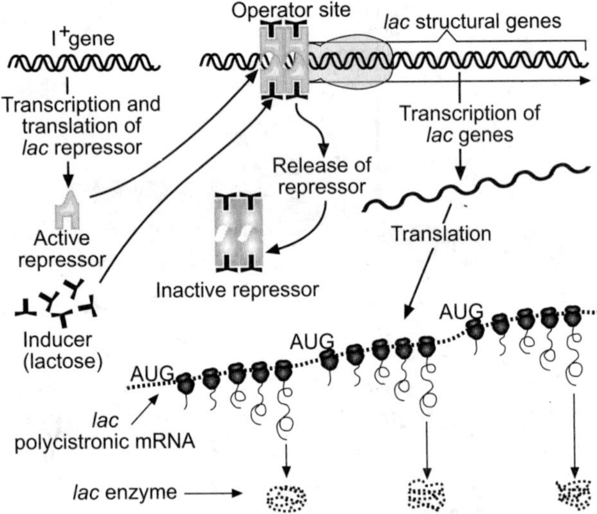

Fig. 14.11 General scheme of *lac* enzyme synthesis and the effect of repressor on this inducible system (a) Absence of inducer and (b) Presence of inducer

Fig. 14.12 Lactose analogues (a) Catalysation of colourless X-Gal by ß-galactosidase into a coloured product 5,5′ dibromo–4,4′–dichloro–indigo (b) Effects of IPTG on *lac* repressor structure (c) Orthonitrophenolgalactoside (ONPG) (d) Allolactose

glucose, *E. coli* selectively metabolises glucose above all other sugars, probably because glucose is a central molecule in carbohydrate metabolism. When glucose enters an *E. coli* cell, it is utilised directly without the induction of any new enzymes. If glucose is present in a medium to which lactose, arabinose, or any of a number of other sugars is added, the enzymes needed to metabolise the other sugars are not induced to high levels until the glucose is used up. For example, in lactose metabolism, this is achieved by control of the *lac* operon that is distinct from the negative control of the repressor protein. No ß-galactosidase is formed until all of the exogenous glucose is consumed.

Catabolite repression depends on the ability of glucose to reduce the cellular levels of cyclic AMP (cAMP). In the absence of glucose, the levels of the nucleotide rise and result in the formation of CAP–cAMP, which in turn activates transcription (**Fig. 14.13**). If the glucose is used up, catabolite repression will be alleviated and the bacterium will then express the *lac* operon.

14.4.2 The galactose (*gal*) operon

Galactose, a monosaccharide, is another sugar that *E. coli* can utilise as a carbon source. It is metabolised by three enzymes: **galactokinase**, **galactose transferase** and **galactose epimerase**, which act in a sequence of steps to yield the overall reaction:

$$\text{Galactose} + \text{ATP} \rightarrow \text{Glucose-1-phosphate} + \text{ADP} + \text{H}^+$$

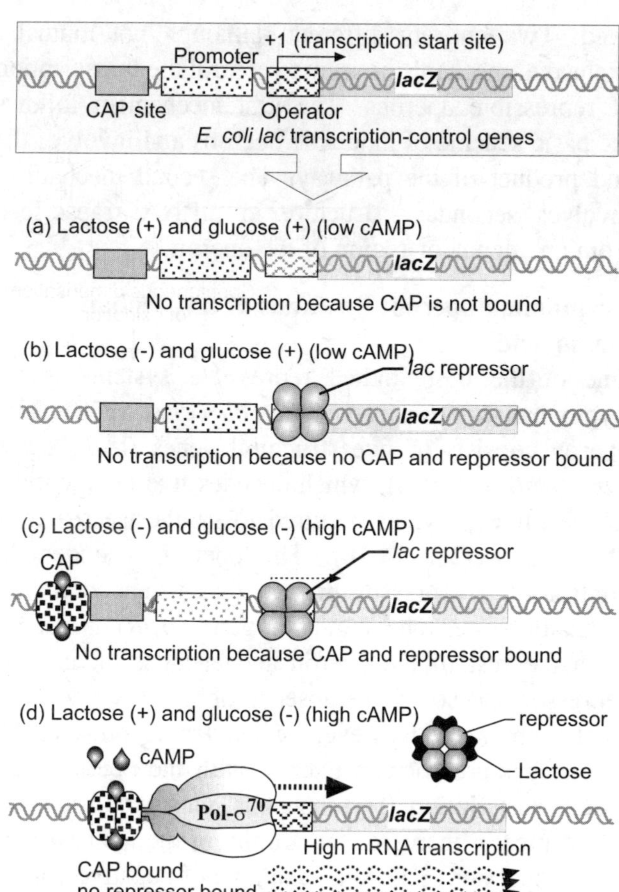

Fig. 14.13 Regulation of transcription from the *lac* operon of *E. coli*. (a–d) the *lac* operon response to the presence or absence of glucose and/or lactose in the medium

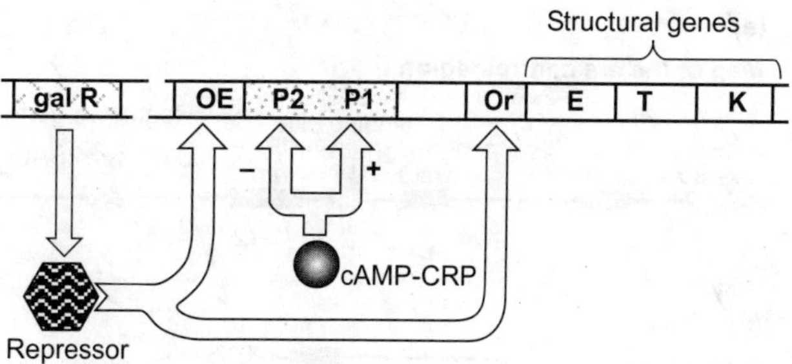

Fig. 14.14 Regulation of the *gal* operon of *E. coli*

The *gal* operon (**Fig. 14.14**) is regulated like the *lac* operon. There are, however, several differences that can be observed. The genes corresponding to the three gal enzymes (I, II, III) are *galK*, *galT* and *galE* respectively. Together with an operator (*gal*O), a promoter region (*galP*) and a repressor gene (*galR*), these genes form an operon. The *gal* operon has two operators: one termed $galO_e$ is upstream and adjacent to *galP* but the other, termed $galO_i$, is 95 bp downstream from $galO_e$ and within the *galE* gene. The action of *gal* repressor is that the *galR* and $galO_c$ mutations confer a constitutive phenotype and purified repressor binds to both of the operator sequences.

Since galactose is used lesser than glucose, the *gal* operon is, therefore, less inducible in the presence of glucose. The mRNA molecules synthesised from the two *gal* promoters differ in length by only five nucleotides and both the transcripts contain the *galK*, *galT* and *galE* genes. The examination of the base sequence of the promoter–operator region shows that it has two start points, S1 and S2 for the two *gal* mRNA molecules. These S1 and S2 differ functionally: the transcription from S1 occurs only when glucose is absent and *in vitro* this transcript is made only if cAMP–CRP is present. Transcription from S2 occurs primarily when glucose is present; cAMP–CRP is not required for initiation at S2 and in fact inhibits the transcription from S2. Glucose does not inhibit induction of the *gal* operon because one of the start sites does not require cAMP–CRP and thus remains active even though glucose is present.

14.4.3 The arabinose (*ara*) operon

Arabinose is another monosaccharide (a pentose) that can serve as a carbon source for metabolism. It is found in the cell walls of many plants and is released in the human intestine after digestion of vegetables. It is not absorbed by the intestine and hence provides a source of carbon for bacterial inhabitants of the intestine such as *E. coli*. The arabinose (*araBAD*) operon allows *E. coli* to use arabinose as an energy source. It is induced by arabinose and subject to catabolite repression when glucose is also present. Its three structural genes (*araB*, *araA* and *araD*) encode the enzymes required to convert arabinose into xylulose 5-phosphate, an intermediate in the pentose phosphate pathway. They are contiguous and are transcribed into a polycistronic mRNA. Two independent operons contain the *araE* and *araF* genes that encode a permease system involved in arabinose uptake into the bacterial cell.

The dual control of the *araBAD* operon is complex. The gene *araC* encodes a regulator protein araC that is involved in both positive and negative control of the *araBAD* genes. It functions by binding to three different control regions of the operon; ***araI***, $araO_1$ and $araO_2$ (**Fig. 14.15**). In fact, *araI* has two regions: $araI_1$ and $araI_2$. In the absence of arabinose, the dimeric araC protein binds to each of the control regions $araI_1$ and $araO_2$ and

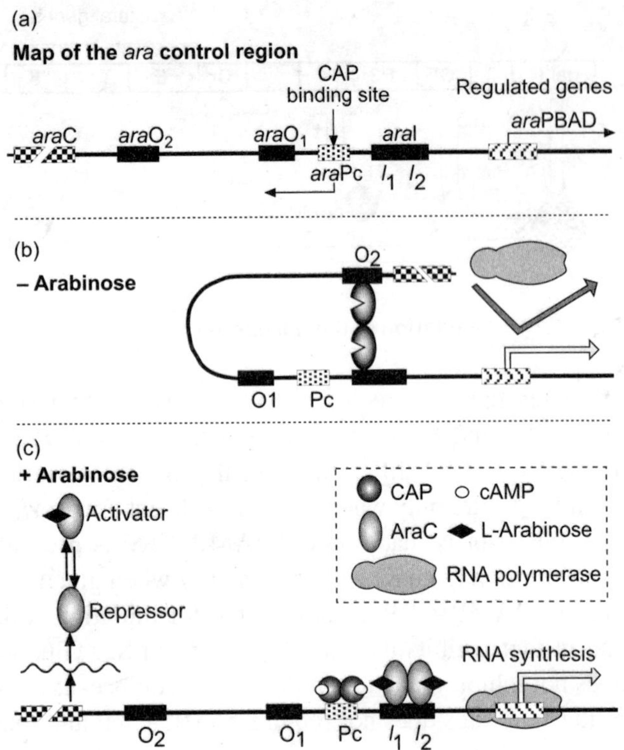

Fig. 14.15 Control of the *ara* operon. There are four araC binding sites (araO1, araO2, araI1 and araI2), all of which lie upstream of the *ara* promoter, araPBAD

also to each other. Two dimers, thus, interact and the intervening 211 bp of DNA forms the **repression loop** that leaves the *araBAD* genes uninducible (**Fig. 14.15**). AraC protein normally has a low affinity for $araO_2$ but the ligand-free protein binds with high affinity to $araI_I$ and also has potential to bind another dimer. This increases the effective concentration of dimer in the region of $araO_2$ leading to cooperative binding and full repression. Such a cooperative interaction is called the **chelate effect**. Analysis of gene expression shows that the control region of the operon has an 11.1 helical repeat and that maximal repression requires that $araO_2$ and $araI_I$ regions side up on the same face of the DNA. Thus, the insertion of five extra base pairs rotates one site halfway around the DNA double helix and greatly diminishes repression.

14.4.4 Regulation of tryptophan (*trp*) operon: A repressible system

The inducible operons are activated when the substrate that is to be catabolised enters the cell. Anabolic operons function in the reverse manner. They are turned off when their end products accumulate in excess of their need. Two entirely different, although not mutually exclusive, mechanisms seem to control the transcription of repressible operons. The first mechanism follows the basic scheme of inducible operons and involves the end product of the pathway. The second mechanism involves secondary structure in mRNA transcribed from an attenuator region of the operon.

Tryptophan operon: Synthesis of tryptophan amino acid

One of the best studied repressible systems is the tryptophan (*trp*) operon in *E. coli* (Platt 1978). This operon consists of five structural genes (*trpE*, *trpD*, *trpC*, *trpB* and *trpA*), which encodes five polypeptides of the three enzymes required to synthesise *trp* from chorismate (**Fig. 14.16**). The operon is under the control of a repressor protein that is encoded in a regulator gene. The regulator gene (*trpR*) is remote from the rest of the operon and encodes an inactive repressor and so, in the absence of *trp*, does not bind to the operator. However, when *trp* is present, the activated repressor can interact with the operator and block transcription. Tryptophan is, thus, a co-repressor as *trp* greatly increases the affinity of the *trp* repressor for its operator. This is an example of negative control because the active repressor switches off transcription.

The *trp* repressor controls more than one (multiple loci) dispersed structural genes by inhibiting the operator. It controls three unrelated sets of genes.

- An operator at the cluster of structural genes *trpEDBCA* controls coordinated synthesis of the enzymes that synthesise tryptophan from chorismic acid,
- An operator at another locus controls the *aroH* gene, which codes for one of the three enzymes that catalyse the initial reaction in the common pathway of biosynthesis of aromatic amino acid,
- The *trpR* regulator gene is repressed by its own product, the *trp* repressor.

Thus, the repressor protein acts to reduce its own synthesis and is an example of **autogenous control**. This type of circuit is common in regulatory genes and may be either negative or positive. As expected, a related 21-bp operator sequence is present at each of the three loci at which the *trp* repressor acts.

Regulation by attenuator

The *trp* operon is also controlled by attenuation. When tryptophan levels are high, RNA polymerase that has initiated transcription pauses at a specific site and

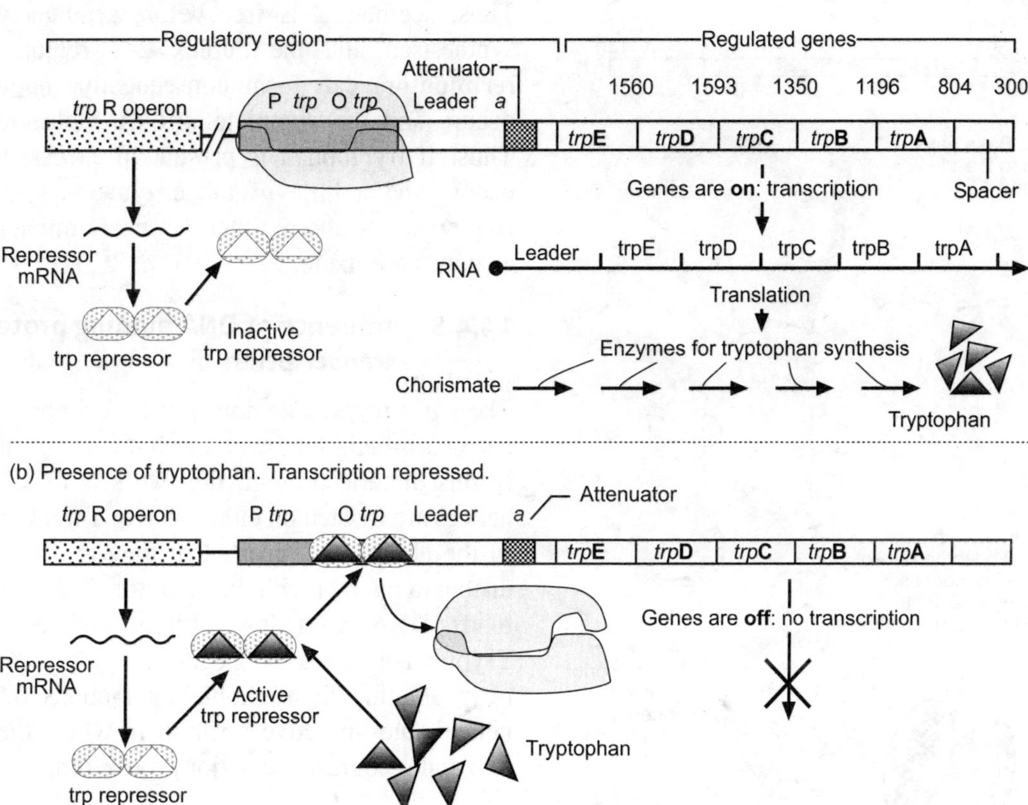

Fig. 14.16 The tryptophan (trp) operon (a) transcription occurs in the absence of tryptophan (b) transcription is repressed in the presence of tryptophan

terminates before getting to *trpE*. But when tryptophan is limiting, polymerase does not terminate, and instead reads through the *trp* genes. During attenuation, transcription actually begins, but is terminated before the entire mRNA is made. A segment of DNA, termed the attenuator, is important in facilitating this premature termination of transcription. When attenuation occurs, the mRNA encoding the *trp* operon is made as only a short piece that terminates shortly past the *trpL* region. Since this short mRNA does not encode the genes required for *trp* biosynthesis, this mechanism can prevent the expression of these genes.

The attenuation mechanism

Complete sequencing of *trp* leader polypeptide has revealed that it has four segments denoted 1, 2, 3 and 4, which are capable of base pairing in two different ways: either the base-paired regions 1–2 and 3–4 or just the region 2–3. If sequences 2 and 3 are paired in the duplex segments 1–2 and 3–4, then the region 2–3 cannot be present simultaneously with 1–2 and 3–4 (**Fig. 14.17**).

The attenuation model proposes that termination of mRNA is mediated through translation of the leader peptide. Because there are two tryptophan codons in this sequence, the ability to translate the sequence should be sensitive to the concentration of charged $tRNA^{Trp}$; that is, if the amount of tryptophan is limiting, there will be insufficient charged $tRNA^{Trp}$ and hence translation across the tryptophan codons would be very slow. In bacteria, transcription and translation are coupled and all base pairing is eliminated in the segment of the mRNA that is in contact with the ribosome. If the ribosome is in contact with segment 2 when synthesis of segment 4 is being completed, then segments 3 and 4 are free to form the duplex region 3–4 without segment 2 competing for segment 3. The presence of the 3–4 stem-and-loop configurations (**terminator loop**) allows termination to occur when the terminating sequence of seven Us is reached. If there is no added tryptophan, the concentration of charged $tRNA^{Trp}$ becomes limiting and occasionally a translating ribosome is stalled for an instant at the tryptophan codons. These codons are

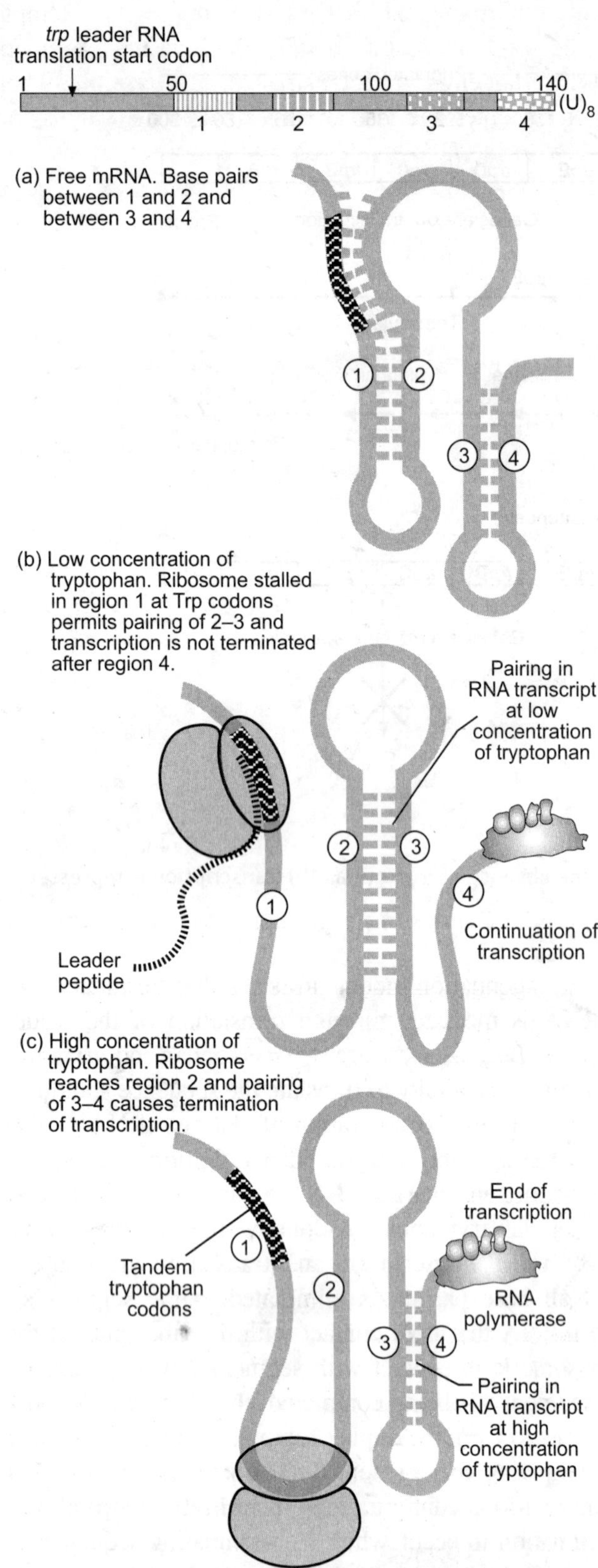

Fig. 14.17 Attenuation in the *E. coli trp* operon; the tryptophan codons are shown in hatched box

located 16 bases before the beginning of segment 2. Thus, segment 2 is free before segment 4 has been synthesised and the duplex 2–3 region (the **anti-terminator**) can form consequently, no termination occurs and the complete mRNA molecule is made. Thus, if tryptophan is present in excess, termination occurs and a little of the enzyme is synthesised; if tryptophan is absent, there is no termination and the enzymes are made.

14.4.5 Influence of DNA binding proteins on transcription

The *trp* repressor demonstrates the importance of the helix-turn-helix interaction in influencing transcription. In this dimeric protein, the two sets of DNA-binding helices are located on either side of a rigid protein core. In the absence of tryptophan they are folded inwards unable to interact with the operator. Thus, the repressor binds DNA with low affinity and no specificity. Tryptophan increases specific binding a thousand-fold by promoting hydrophobic interactions that lock the helices into an active position in which they protrude into and recognise the major groove of the *trp* operator.

14.5 TERMINATION OF TRANSCRIPTION

Attenuation has been recognised in several operons. In transcription, it occurs after initiation of RNA synthesis but before its completion. We shall now discuss some other modes of regulation of termination.

Anti-terminators of transcription

During lytic infection by bacteriophage λ and related phages, the viral genes are expressed sequentially in what is known as the **lytic cascade**. The primary control of this process is actually the termination of transcription, which is inhibited by viral anti-terminator proteins. When phage λ first infects a cell, two genes, *cro* and *N*, expressed by host RNAP are transcribed opposite to that of the promoters P_R and P_L (rightward and leftward promoters). These genes are known as the immediate early genes and they encode the anti-repressor protein cro and an anti-terminator, the N protein. This protein permits the lytic cycle to continue with the expression of the next block of genes: the delayed early genes. It achieves this by allowing RNAP to transcribe past ρ-dependent terminators (tL and t_R1) located at the 3'-ends of the *N* gene and *cro* gene respectively. The delayed early genes encode the cII and cIII proteins. They also encode two proteins involved in the replication of the viral genome, a series of proteins that function in recombination and another anti-termination factor, Q.

The protein Q unlocks the final block of bacteriophage genes, the so called **late genes** that encode the viral head and tail proteins as well as the two proteins that bring about lysis of the host cell. They are transcribed from a separate promoter $P_{R'}$ from which RNAP normally initiates RNA synthesis constitutively. However, in the absence of Q, the transcripts terminate at an adjacent ρ-dependent terminator ($t_{R'}$) with the release of a 194-nt RNA known as 6S RNA. The Q protein suppresses termination at $t_{R'}$ and allows RNAP to transcribe through it and into the late genes. The *N* gene encodes a protein that is a highly specific anti-terminator for λ genes. Related bacteriophages have equivalent proteins but they are not interchangeable. This specificity suggests that the protein recognises unique sites on λ DNA and such sites have been identified from mutations that abolish their function. They are known as N protein utilisation sites (*nutL* and *nutR*) and were identified as 17-bp sequences located between P_L and t_L on the rightward strand of the λ DNA. They are identical for 16 out of 17 bp and contain a 5-bp region of inverted symmetry.

14.6 MODIFICATION OF RNA POLYMERASE

Diversity in sigma (σ) factor

The specificity of transcription in bacteria is dependent on the association of σ factor with RNAP core enzyme, a subunit that allows the enzyme to recognise gene promoters while leaving untouched the ternary complex with transcripts with a given number of nucleotides. A *σ* factor is a prokaryotic transcription initiation factor that enables specific binding of RNAP to gene promoters. Different *σ* factors are activated in response to environmental conditions. Every molecule of RNAP contains exactly one *σ* factor subunit. *E. coli* has at least eight types of *σ* factors; the number of *σ* factors varies between bacterial species. They are classified according to their characteristic molecular weights.

Environmental regulons

Genes can be regulated in a group even though they are located at different parts of the bacterial chromosome. Such a group of genes regulated by the same environmental signal is a regulon. An example is the rpoS starvation regulon. Bacteria have to adapt to different environmental stresses. For example, when bacteria enter the human digestive system it has to adjust to the host's defence mechanisms such as acid and antioxidant stress and cope with starvation after leaving the system. It can cope with starvation in the following manner:

- When bacteria use up their carbon sources, they express rpoS, the starvation σ factor.
- RpoS joins RNA polymerase to initiate transcription of different environmental stress genes. This phenomenon is known as **cross-protection**.
- The stress genes can be used for conditions as unrelated as acid or base resistance.
- The stress genes activated may or may not be part of multigene operons. They may face in opposite directions or arise from different promoters or be located at different loci around the genomic map (**Fig. 14.18**).

14.6.1 Modulation of host RNA polymerase

Similar to other bacteriophages, the development of the *E. coli* phage, T4, is also characterised by the successive expression of several gene classes. All transcriptions are catalysed by the host RNAP but the enzyme undergoes a series of phage-induced modifications that

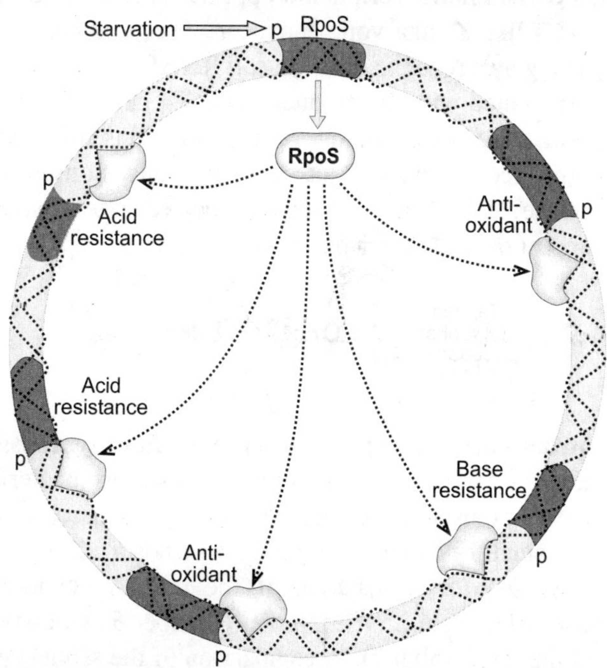

Fig. 14.18 Starvation regulon: A group of genes regulated by the same environmental signal

forces the host enzyme to initiate different classes of T4 promoters. At least three classes of viral genes, early, middle and late are known and each has a distinct promoter type.

The early genes are at first transcribed by unmodified *E. coli* RNAP from promoters identical to the host. However, immediately after infection, the polymerase undergoes its first modification carried out by the product of the T4 *alt* gene. This protein is a component of intact phage particles and is injected with the DNA. It brings about the addition of ADP-ribose to an arginine residue of one of the host enzyme's α subunits. A few minutes later, a second ADP-ribosylation on the same residue (arginine 265) of the other α subunit is catalyzed by the product of one of the early genes (*mod*). This results in a more efficient ρ-dependent termination. The promoters of at least some of the T4 middle genes include the so-called mot box (TA$^{T}/_{A}$GCTT), which occur 13 bp upstream of the –10 element. Viral *motA* and two more, *motB* and *motC*, may also be involved in initiation.

The most dramatic changes in the transcriptional specificity of host RNAP occur with the switch to the transcription of the late genes. Expression of these genes requires concurrent DNA synthesis and is stimulated by three T4-encoded DNA polymerase accessory proteins (gp44, gp45 and gp62). During T4 DNA replication, these proteins bind to the DNA at the primer–template junction and in transcription it appears that this binding site acts like a eukaryotic enhancer. The promoters of the late genes recognised by *E. coli* RNAP core enzyme supplemented with the product of the T4 gene 55. This protein, gp55, is an alternative σ factor and is sufficient for the recognition of the late promoter consensus sequence, TATAAAATA that is located a few bases upstream of the transcription start site.

14.7 REGULATION OF PROTEIN SYNTHESIS

Translational control of ribosomal protein synthesis

Control of ribosomal protein synthesis in bacteria involves ribosomal protein S8 (a protein associated with a small ribosomal subunit). This ribosomal protein inhibits *in vitro* translation of the mRNA encoding ribosomal protein L5 (i.e., protein number 5 associated with the large subunit). A comparison of the sequences of a stem structure in 16S rRNA and of the 5' end of the mRNA encoding L5 protein suggest a physical basis for the arrest caused by excess S8 protein. During the assembly of bacterial ribosomes, S8 protein binds tightly to the stem portion of one particular stem-loop in 16S rRNA. The sequence of L5 mRNA in the region of the start codon can form a stem-loop whose sequence is similar to that of the S8-binding site in 16S rRNA. S8 protein can bind to this sequence in L5 mRNA, but has a higher affinity for its binding site in 16S rRNA. Consequently, when 16S rRNA is available, S8 protein binds to it preferentially and is assembled into ribosomes. But when more S8 protein than 16S rRNA is present, the excess S8 protein binds to L5 mRNA, inhibiting the initiation of translation by sterically blocking the interaction of the 30S ribosomal subunit with the initiation codon and SD sequence in L5 mRNA. When subsequent transcription of rRNA genes produces more rRNA, S8 protein binds to its higher-affinity binding site in 16S rRNA, releasing L5 mRNA and allowing further synthesis of L5 protein.

Antisense RNA regulation of translation

This type of regulation is mediated by antisense RNA, which contains sequences complementary to the region of an mRNA containing an initiation codon. Hybridisation of the complementary antisense RNA blocks recognition of the initiation codon and binding of the 30S ribosomal subunit to the SD sequence, thereby preventing initiation of translation.

14.8 POST-TRANSLATIONAL CONTROL

Feedback inhibition

A common mechanism to regulate the activity of metabolic enzymes is feedback inhibition or end-product inhibition. Even after a gene has been transcribed and mRNA has been translated, a cell can still exert some control over the functioning of the enzymes produced if the enzymes are allosteric proteins. An example of post-translational control occurs with the enzyme aspartate transcarbamylase, which catalyses the first step in the pathway of pyrimidine biosynthesis in *E. coli*. An excess of one of the end products of the pathway, cytidine triphosphate (CTP) inhibits the functioning of aspartate transcarbamylase. This method of control is called feedback inhibition because a product of the pathway is the agent that turns the pathway off. The synthesis of many cellular molecules such as amino acids, vitamins and nucleotides occurs via the action of a series of enzymes that convert precursor molecules to a particular product.

Protein degradation

Another mechanism that controls the amount of gene product in a cell is the rate at which proteins degrade. The normal lifespans of proteins vary greatly. For example, some proteins last longer than a cell cycle, whereas others may be broken down in minutes. Several models have been suggested for control of protein degradation, including the **N-end rule** and the **PEST hypothesis**. According to the N-end rule, the amino acid at the amino- or N-terminal end of a protein is a signal to protease that control the average lifespan of a protein. According to the PEST hypothesis, protein degradation is determined by regions rich in one of the four amino acids: proline (P), glutamic acid (E), serine (S) and threonine (T). Proteins that have these regions tend to degrade in less than two hours.

14.9 REGULATION OF GENE EXPRESSION IN BACTERIOPHAGES

Like many eukaryotic viruses like HSV or HIV, some bacterial viruses can either reside in a dormant state within the host chromosomes or can replicate within the bacterium and eventually lead to lysis and killing of the host cell. *E. coli* harbours such a 'temperate' virus, bacteriophage λ.

Phage λ exhibits a complex system of controls of both early and late operons, as well as controls for the choice of lytic infection versus lysogenic integration. The λ genes are grouped into four operons: left, right, late and repressor. The left and right operons contain the genes for DNA replication and recombination and phage integration. The late operon contains the genes that determine phage head and tail proteins and lysis of the host cell. After infection of the *E. coli* cell by a λ phage, the phage DNA circularises, using the complementarity of the *cos* sites. Transcription begins, and within a very short time, the phage is guided towards either entering the lytic cycle and producing virus progeny or entering the lysogenic cycle and integrating into the host chromosome.

14.9.1 Repressors of the bacteriophage lambda

The repressor-operator interactions of the *E. coli* bacteriophage λ have been studied in considerable detail. In the lysogenic state, the viral DNA, known as prophage, is replicated. Conversely, in the lytic cycle, the genes of the bacteriophages transcribed by the host RNAP lead ultimately to the production of progeny bacteriophages, which are released with the lysis of the bacterial cell. The differential expression of a repressor gene (*cI*) and a regulator gene (*cro*) play an important role in the lifecycle of the phage.

Maintenance of the lysogenic state

Maintenance of the lysogenic state of infection depends on a single protein, the λ repressor. The way in which this protein suppresses the alternative lytic cycle of the phage is illustrated in **Fig. 14.19**. The viral DNA has two **left operator** (O_L) and one **right operator** (O_R) each of which has three 17-bp repressor binding sites designated O_L1, O_L2, O_L3 and O_R1, O_R2 and O_R3, respectively. The binding sites conform to a consensus sequence and show partial dyad symmetry. Each binding site will bind the N-terminal domains of a repressor dimer. The binding sites O_L1 and O_R1 exhibit the greatest affinity for the repressor and they overlap with the leftward and rightward promoters P_L and P_R. The preferential binding of the repressor to these sites inhibits transcription from the promoters and in so

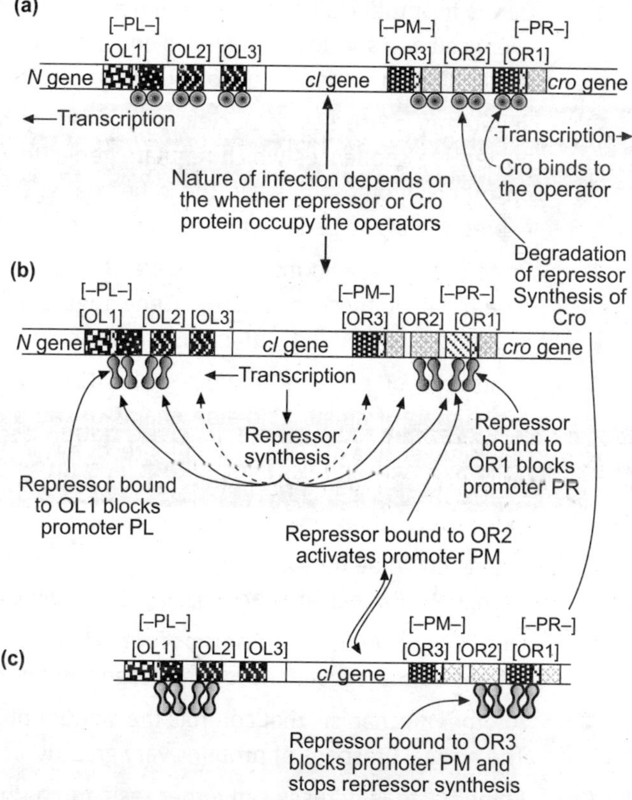

Fig. 14.19 The mechanism of action of lambda repressor protein (a) Lysogeny (b) Entry into the lytic cycle (c) Autogenic control of repressor protein synthesis

doing stops the expression of the proteins required in the lytic cycle of viral infection.

Switching from lysogenic to lytic infection

The λ phage can switch back and forth from lysogenic to lytic state and vice versa. The switching event in λ is centred around an 80-bp region in its dsDNA genome referred to as the right operator (O_R). This region is flanked on its left by the structural gene for the cI protein and on its right by the cro. When in the prophage state, the *cI* repressor gene is the only λ gene expressed. When the bacteriophage is undergoing lytic growth, the *cI* repressor gene is not expressed, but the *cro* gene as well as many others are expressed. That is when one gene (e.g., c*I*) is expressed, the other gene (e.g., *cro*) is turned off and vice versa. This alternative switching between repressor gene and *cro* gene transcription is a paradigmatic example of a molecular switch.

KEY POINTS

- The transfer of genetic information from DNA into protein constitutes gene expression.
- Some of the principal stages at which gene expression is regulated are transcription, RNA processing and translation.
- In prokaryotes, genes are very often controlled by extracellular signals and are communicated to genes by regulatory proteins. These regulatory proteins are generally DNA-binding proteins that recognise specific sites at or near the genes they regulate.
- A system is turned on when it is needed and off when it is not needed. Often, an organism needs to regulate many different genes simultaneously in response to a change in its environment.
- The term gene regulation, however, refers to the many mechanisms used by cells to control the amount of each gene product they produce.
- Genes for products that are required at all times are expressed at more or less constant levels in virtually every cell of a species or organism and are often called housekeeping genes. However, some gene products rise and fall in response to molecular signals. This phenomenon is called regulated gene expression.
- Regulatory processes of nucleic acids work either in *cis* or *trans* effect. In DNA, cis-acting elements are nucleotide sequences which regulate genes to which they are attached (adjacent), while having no effect on other chromosomes. On the other hand, *trans*-acting factors are diffusible molecules regulating the expression of the genes with which they come into contact.
- Protein–nucleic acid interactions are central to replication, transcription and translation, as well as to the regulation of these processes. Gene regulatory proteins, therefore, must recognise specific nucleotide sequences embedded within this structure.
- A number of proteins interact with DNA in a sequence-specific manner. These interactions occur by association of amino acid side chains of the proteins with the bases as well as with the phosphate and sugar molecules of DNA. Molecular recognition generally relies on an exact fit between the surfaces of two molecules. A regulatory protein recognises a specific DNA sequence because the surface of the protein is extensively complementary to the special surface features of double helix in that region.
- A gene could be turned on by moving it from a location far from its promoter to a site near its promoter. This change in DNA can affect gene expression, which is termed as the rearrangement of the DNA structure.
- One of the common mechanisms for translational control is the binding of a regulatory protein to the mRNA. A translational regulator protein recognises sequences within the mRNA and inhibits translation.
- Another mechanism that controls the amount of gene product in a cell is the rate at which proteins degrade. The normal lifespans of proteins vary greatly.
- Some bacterial viruses can either reside in a dormant state within the host chromosomes or can replicate within the bacterium and eventually lead to lysis and killing of the host cell. Phage lambda exhibits a complex system of controls of both early and late operons, as well as controls for the decision of lytic infection versus lysogenic integration.

Self-assessment Questions

1. State briefly the role of *cis*- and *trans*-acting factors in the regulation of gene expression.
2. What is the role of DNA-binding proteins in the regulation of gene expression in bacteria?
3. What is the role of sigma factors?
4. Describe the regulation of the *lac* operon.
5. Explain how the tryptophan operon is doubly regulated?
6. Explain the mechanism of regulation by antisense RNA mediation.
7. List the different methods of post-translational regulation in prokaryotes.
8. Describe briefly the mechanism of switching from lysogenic to lytic phase.

FURTHER READING

Textbooks

Adams RLP, JT Knowler and DP Leader (1993). *The Biochemistry of the Nucleic Acids*, London: Chapman and Hall.

Alberts B, A Johnson, J Lewis , M Raff, K Roberts and P Waler (2002). *Molecular Biology of the Cell*. 4th ed. Garland Science.

Baumberg S (ed) (1999). *Prokaryotic Gene Expression*. Oxford: Oxford University Press.

Hartl DL and EW Jones (1998). *Genetics: Principles and Analysis*. Jones and Bartlett Publishers.

Latchman D (1995). *Gene Regulation: A Eukaryotic Perspective*. London: Chapman and Hall.

Lewin B (2000). *Genes VII*. London: Oxford University Press.

Lin ECC and AS Lynch (eds) (1996). *Regulation of Gene Expression in Escherichia coli*. London: Chapman and Hall.

Malacinski GM (2003). *Essentials of Molecular Biology*, 4th ed. Jones and Bartlett Publishers.

Twyman RM (1998). *Advanced Molecular Biology*, New Delhi: Viva Books.

References

Bell CE and M Lewis (2001). The lac repressor: A second generation of structural and functional studies. *Curr. Opin. Struct. Biol.* 11:19–25.

Busby S and Rh Ebright (1999). Transcription activation by catabolite activator protein (CAP). *J. Mol. Biol.* 292:199–215.

Harrison SC (1991). A structural taxonomy of DNA-binding domains. *Nature* 353:715–19.

Hochschild A (2002). The switch closes the gap in auto-regulation. *Curr. Biol.* 12:R87–R89.

Huffman JL and RG Brennan (2002). Prokaryotic transcription regulators: More than just the helix-turn-helix motif. *Curr. Opin. Struct. Biol.* 12:98–106.

Jacob F and J Monod (1961). Genetic regulatory mechanisms in the synthesis of proteins. *J. Mol. Biol.* 3:318–56.

Laity JH, BM Lee and PE Weight (2001). Zinc finger proteins: new insights into structural and functional diversity. *Curr. Opin. Struct. Biol.* 11:39–46.

Lloyd G, P Landini and S Busby (2001). Activation and repression of transcription initiation in bacteria. *Essays Biochem.* 37:17–31.

Magasanik B (2000). Global regulation of gene expression. *Proc. Natl. Acad. Sci.* 97:14044–45.

Muller CW (2001). Transcription factors: Global and detailed views. *Curr. Opin. Struct. Biol.* 11:26–32.

Muller-Hill B (1998). Some repressors of bacterial transcription. *Curr. Opin. Microbiol.* 1:145–51.

Nomura M, R Gourse and G Baughman (1984). Regulation of the synthesis of ribosomes and ribosomal components. *Ann. Rev. Biochem.* 53:75–117.

Roy S, S Garges and S Adhya (1998). Activation and repression of transcription by differential contact: Two sides of a coin. *J. Biol. Chem.* 273:14059–62.

Schleif R (2003). AraC protein: A love-hate relationship. *Bioessays* 25:274–82.

Wesiberg RA and ME Gottesman (1999). Processive antitermination. *J. Bacteriol.* 181:359–67.

Xu H and HR Hoover (2001). Transcriptional regulation at a distance in bacteria. *Curr. Opin. Microbiol.* 4:138–44.

Yanofsky C (2000). Transcription attenuation: Once viewed as a novel regulatory strategy. *J. Bacteriol.* 182:1–8.

CHAPTER 15

Regulation of Gene Expression in Eukaryotes

LEARNING OBJECTIVES

After reading this chapter, you will be able to:

- Understand the different stages of regulation of eukaryotic gene expression
- List the different mechanisms of regulation of gene expression in eukaryotes
- Understand the role and applications of regulation of gene expression in development

15.1 INTRODUCTION

Regulation of gene expression is the cellular control of the amount and timing of appearance of the functional product of a gene. Any step of gene expression may be modulated, from DNA to RNA transcription, to post-translational modification of a protein. The mechanism of regulation of gene expression facilitates control over cell structure and function, and is the basis for cellular differentiation, morphogenesis, growth, and the versatility and adaptation of an organism.

The control of gene expression in eukaryotic cells is less understood compared to that of prokaryotes. Eukaryotes bear a number of distinctive features of gene regulation. Multicellularity (except unicellular yeast, algae and protozoa) and more complex cell structure demand greater level of gene regulation. The lifecycle involves the progression through several developmental stages to achieve a mature organism. For example, in an embryo, a cell must not only grow and produce many progeny cells but also undergo considerable changes in morphology and biochemistry and maintain the changed state (i.e., differentiated). The phenotypic differences that characterise the multitudes of cells are largely due to differential gene expression within those cells. Some mechanisms control the constitutive expression of 'housekeeping genes', while others determine a rapid increase in expression as a consequence of external stimuli. Another set is responsible for the complete deactivation of gene expression in specific tissues. Thus, a liver cell will not express the genes encoding proteins found only in a brain cell or proteins made only during the differentiation of germ cells.

Another important feature of eukaryotic DNA is that it is entirely coiled and complexed with nucleosome structures. Most or all of eukaryotic RNA transcripts must be processed before translation, and mRNA transcription and translation occur in separate cellular compartments, which add considerable complexity to the regulatory process. The concentrations of individual species of mRNA molecules vary significantly. Thus, different eukaryotic mRNAs have been classified into three groups: **low-copy-number mRNAs** (one-to-several copies), **moderately prevalent mRNAs** (several hundred copies per cell) and **superprevalent mRNAs** (above 10^4 copies per cell). The regulation of synthesis of the third type has been well understood.

Many eukaryotic genes have several binding sites and are controlled by many regulatory proteins. Regulators bind short sequences and often bind at positions further from the start site. The presence of a number of regulatory binding sites in eukaryotes suggests that more signals are required to switch a given gene on at the right time and place. In higher eukaryotes, regulatory sequences can spread over thousands of nucleotides from the promoter—both upstream and downstream. These sites are called enhancers and typically regulate only a specific gene. Other regulatory sequences, called **insulators** or **boundary elements**, are found between enhancers and some promoters. Insulators ensure that activators do not work recklessly.

15.2 REGULATORY STRATEGIES IN GENE EXPRESSION

In prokaryotes, almost all the regulatory mechanisms for gene expression are centred around the initiation or termination of transcription. However, in eukaryotic cells, this mechanism is complex and there are several key points of control (**Fig. 15.1**). These include activation of gene preparation, initiation, termination of elongation, nuclear RNA processing, mRNA transport, mRNA translation and RNA stability. The most striking and unique among the additional steps is **splicing**. Nucleosomes and their modifiers influence access to genes. The genome of eukaryotes is concealed in a cluster of proteins called histones to form nucleosomes and so the DNA is not easily accessible to transcriptional machinery. **Table 15.1** lists the main control points and the corresponding regulatory processes for gene expression in eukaryotes.

15.3 GENOMIC CONTROL

15.3.1 Gene alteration

The RNA and protein content of different cell types show enormous diversity in both quality and quantity. This range of variation is due to differences in the DNA present in each tissue. Thus, the gene products required in one particular tissue may be discarded in all other tissues. Or it would be activated by a selective increase in its copy number in such tissue where the given gene product is required through amplification or by some rearrangement of the DNA necessary for its activation. Also, the production of the most abundant

Table 15.1 Principal levels of gene expression and mechanisms of regulation

Control points	*Regulatory processes involved*
Genomic control	
Gene alteration	Gene loss
	Gene amplification
	Genomic rearrangements
Gene preparation	Genome uncoating and synthesis of template strand (e.g., viruses)
	Modification of chromatin structure
	Changes in topological and conformational properties of DNA
	Changes in DNA methylation states
	Histone acetylation and de-acetylation
Transcriptional regulation	
Transcription	RNA polymerases
	Promoter usage and escapes
	Transcription factors
	Regulation of regulatory sequence activities
	Formation of active transcription complexes
	Elongation versus pausing
Post-transcriptional regulation	
RNA processing (eukaryotes)	Capping
	Polyadenylation
	RNA splicing
	RNA editing (modification, addition and deletions)
RNA export	Export of mRNA from nucleus to cytoplasm
	RNA targeting
RNA turnover	RNA stability
	Retro-regulation (prokaryotes)
Translational regulation	
Protein synthesis	Initiation of protein synthesis
	Codon usage
	Frame shifting
	Read-through/incorporation of selenocysteine
Protein processing	Chemical modification of residues
	Cleavage of polyproteins
	Splicing of introns
	Adoption of quaternary structure
	Interaction with regulatory proteins
Post-translational regulation	
Protein targeting	Targeting to cellular compartments
	Secretion
Protein turnover	Protein loss and degradation
Extracellular signals	
Biotic stimuli	Signals received from other cells
	Signals received by various infectious agents and insects
Abiotic stimuli	Environmental stimuli, such as availability of nutrients, temperature, water, pH, light, altitude, gravity and so on

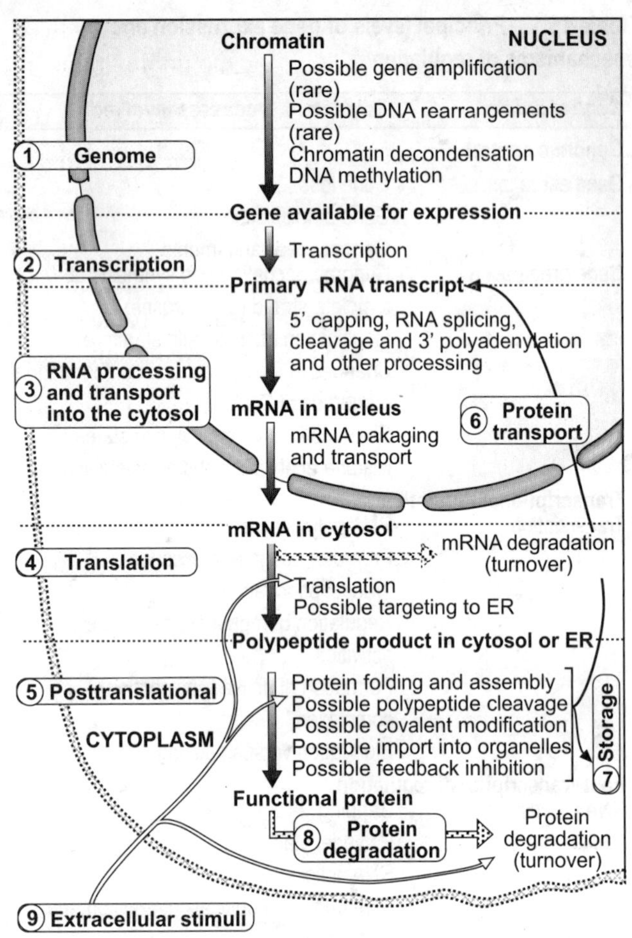

Fig. 15.1 Points of regulation of expression in eukaryotes

mRNA molecules, is usually associated with a change in the stage of differentiation. For example, a mature erythroblast produces a huge amount of mRNA from which adult globins may be translated, whereas little or no globin is produced in the precursor cells. This could be due to a permanent change in the gene(s) or in the number of their copies. Such changes are of three major types—**gene loss**, **amplification** and **rearrangement of genes (Fig. 15.2)**.

15.3.2 Gene loss

Gene loss is the mechanism that removes the activity of certain genes during differentiation so as to eliminate them from the cell. For example, some protozoa, insects, and crustaceans have adapted this mechanism. In this case, entire chromosomes are lost. In these organisms, only those cells destined to produce germ cells maintain the complete DNA complement. Many protozoa have two types of nuclei—the **micronucleus** (small) and the **macronucleus** (large) nuclear bodies (**Fig. 15.3**). The germ cells contain only a micronucleus, but as the cell differentiates, the micronucleus undergoes a variety of changes, resulting in a cell possessing both a micronucleus and a macronucleus. The micronucleus DNA is inactive and is simply a storehouse of genetic information for the production of future germ cells. The DNA molecules of the macromolecules serve as

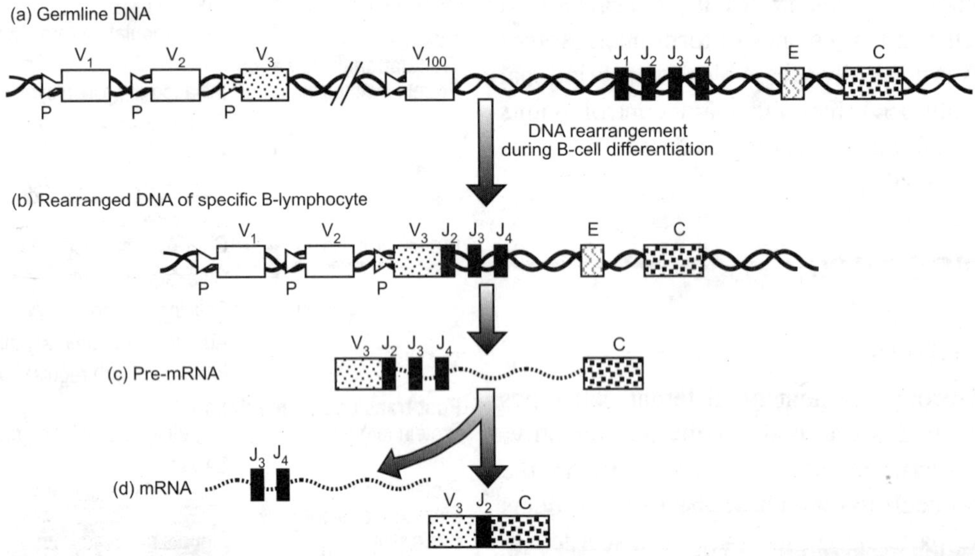

Fig. 15.2 The production of the mature k-type L chain mRNA of a particular IgG molecules involves processing at the DNA and RNA levels. During B-cell difeeerentiation, the DNA between a V and J segment is excised and genomic sequences are rearranged; compare (a) and (b). Further, the enhancer (E) is moved closer to the promoter (P) of V3 and the given B lymphocyte is able to produce a specific k-type L chain pre-mRNA (c). This pre-mRNA undergoes splicing, whereby the C segment is joined to V3J2. This results in the mRNA 9d), which is ready for translation to produce the specific k-type L chain protein.

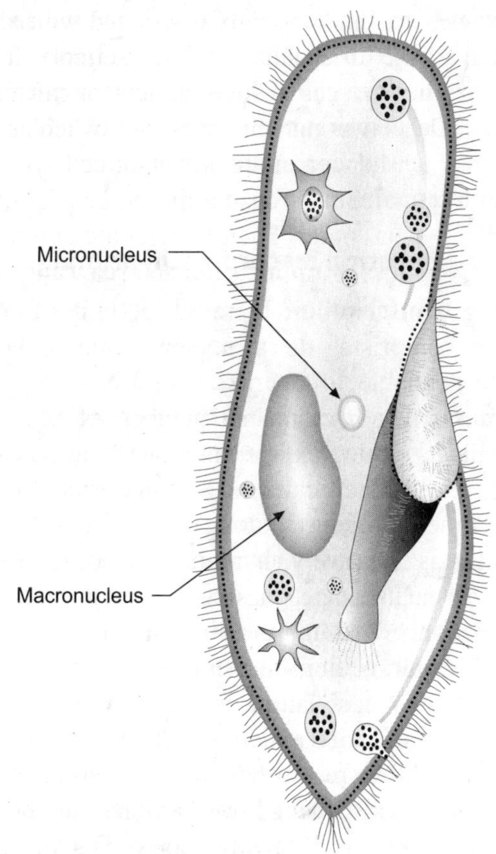

Fig. 15.3 The ciliated protozoa *Paramoecium*; the micronuclear DNA does not code for proteins or RNA but contributes to the production of germ cells. The macronucleus serves as the template for all transcripts

the templates for all transcripts. The formation of the macronucleus of the protozoan *Oxytrichia* has been fairly well understood. The DNA of the micronucleus in the germ cell cleaved during differentiation into linear fragments ranging in sizes of 0.5–20 kbp. Most of the DNA of the micronucleus is completely degraded during the fragmentation. These fragments replicate repeatedly until the cell contains about 1000 copies of 17,000 different fragments that constitute the macronucleus, but in which more than one-half of the DNA sequences are absent. Although, in higher organisms, gene loss is not a general mechanism, expression of a good number of genes is regulated by gene loss in some differentiated cell types.

Loss of the nucleus in mammalian red blood cells: The best known case occurs in the development of the mammalian red blood cell. In the cell, the whole nucleus is pinched off. Although this process is a clear-cut example for a selective DNA loss during differentiation, it is clearly a highly specialised case in which the loss of the nucleus is primarily intended to allow the cell to fill up with haemoglobin, rather than as a means of gene control.

15.3.3 Modification of chromatin structure

The three-dimensional packing of chromatin is an important parameter that largely affects gene expression. Transcription of DNA is dictated by its structure. If the chromatin is very tightly packed (closed conformation) transcription is not easily initiated. In contrast, chromatin that is in an open conformation is more easily accessible to transcription factors and RNA polymerase, so transcription can take place. Most eukaryotic genes are inactive in most tissues, and become active only in a specific tissue or in response

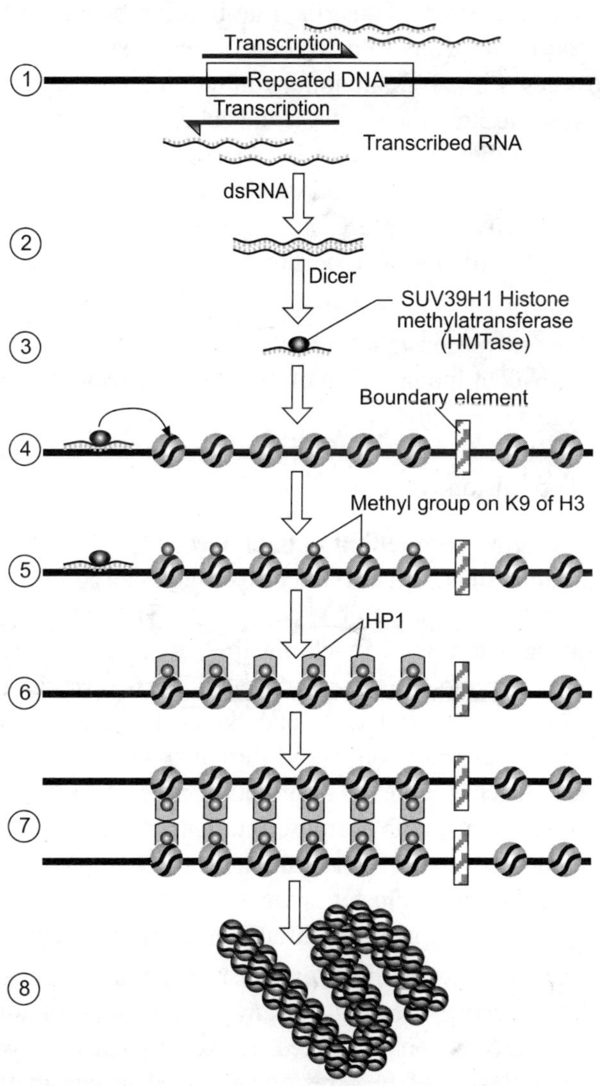

Fig. 15.4 Model of the possible events during the formation of heterochromatin

to a particular signal which suggests that it may be more economical to have a system in which the genes are constitutively inactive in most tissues, without any repressor being required.

In higher organisms, such as yeast and humans, nuclear DNA is coated with at least an equal mass of protein, forming a complex called chromatin, which controls gene activity and the inheritance of traits. DNA in eukaryotic cells is packed inside the nucleus and highly organised into condensed regions (heterochromatin) and more open 'euchromatin' (**Fig. 15.4**). The principal protein components of chromatin are proteins called histones, which organise to form a chromatin subunit called **nucleosome**. Core histones are among the most highly conserved eukaryotic proteins known. Regulatory signals entering the nucleus encounter chromatin first and not the DNA. Thus, the rate-limiting biochemical response that leads to activation of gene expression in most cases involves alterations in chromatin structure. Clues as to how chromatin is unpacked came from the discovery that components of chromatin are subject to a wide range of modifications that are correlated with gene activity. There are three general ways in which chromatin structure can be altered.

- Nucleosome remodelling, which can be induced by complexes designed specifically for the task.
- Covalent modification of histones can occur within the nucleosome.
- Histone variants may replace one or more of the core histones.

Chromatin remodelling to gain access

The nature of chromatin structure provides an additional level of regulation of gene expression. The process of inducing alterations in chromatin structure is called chromatin remodelling. The development of specialised organs, tissues, and cells and their functions in the organism depends on the differential expression of genes, which is achieved by providing access to different regions of chromatin to transcription machinery in cells from various tissues. For example, the DNA containing the ß-globin gene cluster is the active chromatin in the reticulocytes but inactive chromatin in muscle cells.

The mechanism of chromatin remodelling requires input of energy, because many protein–protein and protein–DNA contacts need to be disrupted. Two general classes of proteins are involved in chromatin remodelling. Some proteins acetylate the N-terminal tails of the histones, a process that may cause the nucleosomes to bind the DNA less tightly and thus make it available for attachment of transcription factors. These enzymes are called **histone acetyl transferases (HATs)**. Deacetylating enzymes have the reverse effect; they act to repress transcription and also affect chromatin remodelling (**Fig. 15.5**).

15.3.4 Activation by chromatin priming

Enabling transcription: If transcription has to occur, the DNA comprising the promoter region of the gene must be available to interact with RNA polymerase and *trans*-acting factors. A number of chromatin-remodelling complexes mobilise nucleosomes, causing the histone octamers to move short distances (nucleosome phasing) along the DNA. Active chromatin is relatively hypomethylated. It exhibits changes in nucleosome phasing and may differ in the conformation, winding, bending and looping of the DNA. The unravelling or priming of chromatin may be regulated by acetylation or methylation of the free amino groups of lysine residues at the N-terminal end of histone molecules, resulting in a decreased net positive charge and a subsequent lower affinity for the DNA phosphate backbone. Demethylation of DNA contained in the promoter region of the target gene can also be involved in chromatin priming. Several studies have indicated that histone H1 is non-randomly deposited in the nucleus and depleted in active chromatin.

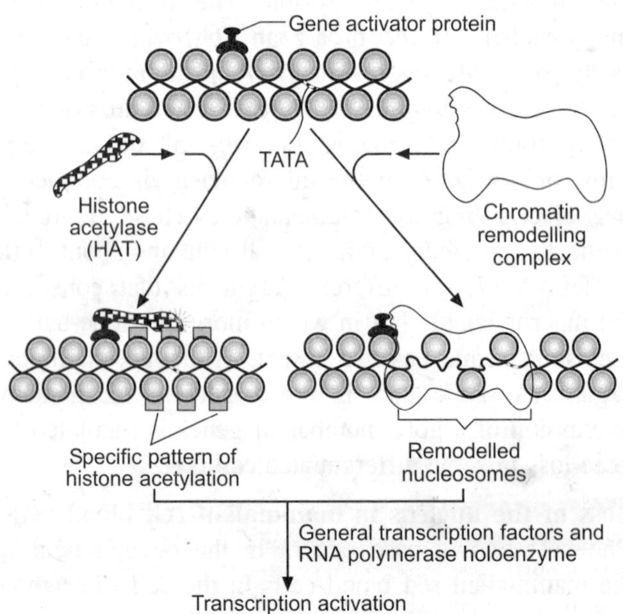

Fig. 15.5 Regulation of gene expression by chromatin remodelling

Ubiquitination of proteins: Ubiquitin is a highly conserved 76-amino acid protein that is required for ATP-dependent non-lysosomal intracellular protein degradation. It is also found conjugated to histones, principally histone H2A. Ubiquitinated histones are present at higher concentrations in actively transcribed chromatin regions and it has been speculated that the ubiquitination of histones by a protein called RAD6 induces chromatin remodelling.

15.3.5 Changes in topological and conformational properties of DNA

Modification of histones: Accessibility of eukaryotic DNA to transcription involves, at least temporarily, modifying the histone components of nucleosomes to enable a nucleosome-free section of DNA to be exposed to RNA polymerase activity. Histone phosphorylation and methylation are generally associated with a decrease in gene activity and, acetylation of histones with increased transcription rates. Thus, for a hormone or any other molecule to act as an inducer or a repressor of transcription, there are some intermediary steps which change the degree or configuration of DNA coiling and chromatin structure.

Z-DNA and gene transcription: The DNA double helix can undergo a conformational change associated with the control of transcription, by switching to the left-handed helical Z-DNA. Most work with Z-DNA has concerned the sequence requirements and conditions required for its formation *in vitro*. Recent investigations have shown the presence of segments as few as 12 bp of alternating CG residues, which can adopt the Z-DNA conformation *in vivo* and there is also the possibility that specific proteins could induce the conformational switch in less CG-rich sequences.

15.3.6 Active genes associate with the nuclear matrix

It is postulated that the attachment of the supercoiled DNA loops to the matrix plays a role in the control of gene expression. This arose from the finding that when the supercoiled loops of chromatin were digested with nuclease, the sequences that remained attached to the matrix were enriched in transcriptionally active genes. More recently, regions of DNA that specifically attach to the matrix have been mapped to the boundaries of genes or gene domains and these elements have been called **matrix attachment regions (MARs)**. In some cases, MARs have been shown to be close to but separate from enhancers. Their importance was further supported by experiments that demonstrated that if MARs were placed on either side of a reporter gene it enhanced position-independent expression of the gene in transfected cells. This suggests that MARs ensured that the reporter gene was inserted into the host genome in a conformation in which it was accessible for transcription.

15.4 CHANGES IN DNA METHYLATION STATES

Another more common way to alter DNA structure is by DNA methylation. In most higher plants and animals, the DNAs are often modified after synthesis by the enzymatic conversion of many cytosine bases (60%–90%) to 5-methylcytosine bases. The most common of these in eukaryotic DNA is 5-methyl cytosine (2%–7% of cytosine methylated). Approximately 90% of this methylated C occurs in the dinucleotide, CG, where the C is followed on its 3' side by a G residue (also called 'CpG islands'). Although some CG sites are always unmethylated and others are always methylated, a number of sites exhibit a tissue-specific methylation pattern, being methylated in some tissues but not in others. For example, a particular site within the chicken globin gene is methylated in a wide variety of tissues, but is unmethylated in erythrocytes. Similarly, the tyrosine amino-transferase gene, which is expressed only in the liver, is under-methylated in this tissue when compared with other tissues where it is not expressed.

The involvement of methylation in the control of gene expression in eukaryotes is based primarily on the indirect evidences such as:

- A correlation between the level of gene expression and the degree of methylation, such that low methylation leads to higher gene expression and high methylation to lower gene expression.
- Methylation patterns are tissue-specific, at least in some cases.
- The drug 5-azacytidine, which cannot be methylated after it is incorporated into the DNA, has been shown to result in the expression of genes in tissues where they are normally not expressed.
- The housekeeping genes, which provide general cell function and are continuously transcribed, are rarely found methylated in or near their initiation regions.

- *In vitro* methylation of the upstream site of a cloned γ-globin gene, followed by transformation of the affected DNA into susceptible cells, prevents transcription, but not, if methylation is outside of the upstream sequence.

What then is the role of methylation in the control of gene expression? It is experimentally proved that alteration in DNA methylation can affect chromatin structure and gene expression. Evidence is accumulating that tissue-specific expression may require demethylation in order that transcription factors can gain access to their recognition sequences. A second possible control mechanism is that methylated CpGs could be a target for specific nuclear proteins. The resulting complex might access the transcription factors. Histone H1 and methylated DNA-binding protein H1 have also been suggested as candidates for generalised methyl-CpG-binding proteins. The discovery that segments of DNA that have sequences in which purines and pyrimidines are alternating along each strand can form left-handed double helices. The novel left-handed double helical form of DNA has been named Z-DNA for the zigzagged paths of the sugar-phosphate backbones of the molecule. There are many evidences to show the possible involvement of Z-DNA in regulation of gene expression.

15.5 MODIFICATION OF HISTONES

Given the essential role of histones in chromatin structure, it is possible that potentially active chromatin might be marked in some way by alteration of the histones within it. Several forms of specific histones encoded by distinct genes have been described. Two of these, the histone H2 variant, H2A.Z and the histone H3 variant, H3.3, have shown to be preferentially localised in active genes. Similarly, the histone H1 variant, H1b, cooperates with the Msx 1 transcription factor to inhibit transcription of muscle-specific genes. This suggests that the variant forms of histones are involved in the regulation of chromatin structure and gene expression. Further, recent studies have revealed a critical role for modified forms of the standard histones in determining whether specific regions of chromatin are active or inactive. Some examples of such modifications are acetylation and deacetylation, ubiquitination, phosphorylation and methylation.

15.5.1 Acetylation of histones

Acetylation of the arms of histones is an important process in transcription and has been correlated with transcriptionally active chromatin in a wide variety of cell types (**Fig. 15.6**). All the histone proteins in a nucleosome can be acetylated, but variations do occur. In the process of acetylation, the free amino groups on specific lysine residues at the N-terminus of the histone molecules are modified by one of the hydrogen atoms in the amino group being substituted by an acetyl group ($-COCH_3$). Hyperacetylated forms of these histones, containing several such acetyl groups, have been shown to be localised preferentially in active genes, while hypoacetylation of the histones is characteristic of transcriptionally inactive regions. Histone acetyl transferases (HATs) are capable of adding acetyl groups to histone molecules.

Acetylation is reversible and the enzymes that can remove the acetyl groups from histones are called **histone deacetylases (HDACs)**. Trichostatin and butyric acid inhibit deacetylases and cause acetylated nucleosomes to accumulate.

15.5.2 Deacetylation of histones

Different proteins with specific acetylase and deacetylase activities are associated with various components of the transcription apparatus. In yeast, mutations in SUN3 and Rpd3 behave as though these loci repress a variety

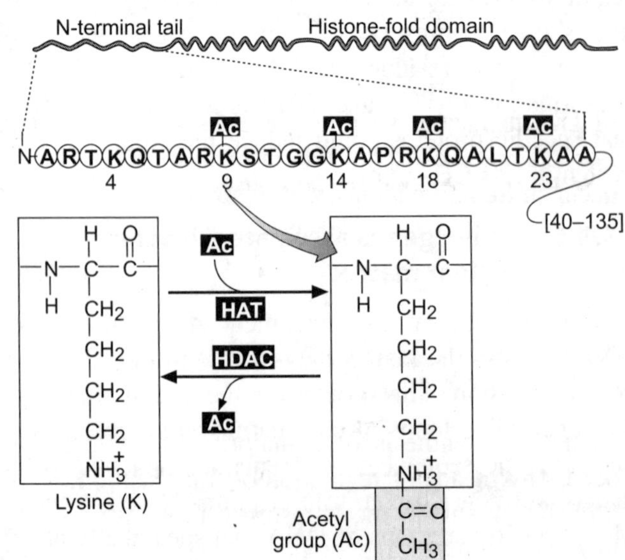

Fig. 15.6 Regulation of gene expression by histone acetylation

of genes. Absence of histone acetylation is also a feature of both constitutive and facultative heterochromatin. Typically the N-terminal tails of histones H3 and H4 are not acetylated in heterochromatic regions.

15.5.3 Ubiquitination and sumoylation of histones

Ubiquitin is a small protein of only 76 amino acids, which forms a conjugate with histone H2A or H2B in which its C-terminal carboxyl group is joined to the free amino group on an internal lysine residue in the histone, to form a branched molecule. This process, called ubiquitination, is associated with the repression of gene expression. For example, H2A histone ubiquitination is involved in gene repression by the polycomb proteins, which promote an inactive chromatin structure. But ubiquitination of H2B has the opposite effect as it stimulates gene expression.

Lysine residues on histones are modified by the addition of small unbiquitin-related modifiers (SUMOs). When a gene is about to be repressed, histones are sumoylated and enzymes that deacetylate histones are recruited. This results in a more closed chromatin structure.

15.5.4 Phosphorylation of histones

Phosphorylation takes place on serine amino acids in the histone molecules. It has been shown that when cells are stimulated with growth factors, histone H3 is phosphorylated on the serine residue at position 10 in the amino terminus of the protein. These phosphorylated histones are localised to genes, which become transcriptionally active following stimulation of growth factors such as the c-*fos* and c-*myc* genes. For example, phosphorylation of histone H3 does not occur in human patients with Coffin Lowry syndrome who lack the phosphorylation enzyme, known as Rsk-2.

15.6 REGULATION AND GENE ORGANISATION

Control of synthesis of a majority of gene products occurs by means of regulation of transcription. This is probably the major mechanism for proteins encoded in low-copy number and moderately prevalent mRNAs. Several different ways that transcription can be switched on and off are described.

Structure of the eukaryotic gene: There are two main differences between the structure of eukaryotic and prokaryotic genes. The former are complexed with chromatin and are not collinear with their peptide products. All functional genes include a transcribed sequence bound by sequences at which transcription starts and stops (**Fig. 15.6**). The transcribed region is flanked by elements necessary for initiation and regulation of transcription and is read by RNA polymerase. Within the transcribed region of structural genes, one or more exons containing the protein coding sequence (ORF) and non-translated 5' and 3' flanking sequences are involved in different aspects of gene expression.

Organisation of proximal and distal regulatory sequences: As in prokaryotes, accurate transcription of eukaryotic mRNA-encoding genes requires a series of discrete DNA sequence elements that function as targets for regulatory proteins. However, these elements may be dispersed over a large region of DNA and may extend 60 kbp or more on either side of the start site of transcription. Operationally, the controlling elements are divided into proximal (promoter) sequences responsible for the correct initiation of transcription, and the distal (regulatory) sequences, which regulate and modulate gene expression (**Fig. 15.7**).

Role of proximal promoter regulatory sequences: Promoter sequences in eukaryotic DNA are used for recognition of RNA polymerase, binding and initiation of transcription. Most of the mRNAs of higher eukaryotes characterised to date are monogenic. The proximal promoter consists of up to three known regions: (a) the start of transcription (TCS) (b) a conserved sequence referred to as the TATA box and (c) a further conserved sequence referred to as'CAAT or AGGA box within 75 bp of the TCS. In addition, a number of promoters show a 9-nucleotides sequence, GGC/TCAATCT (the CAAT box), that lies about 40 or 50 nucleotides farther upstream from the TATA box.

Role of distal regulatory sequences: In addition to nearby promoters, many eukaryotic genes are also regulated by more distant *cis*-acting regulatory elements called enhancers (or **upstream activator sequences, UAS**) and silencers. These distal regulatory sequences have been found to regulate a wide variety of environmentally and developmentally controlled gene systems. As the names suggest, enhancers increase transcription and silencers decrease transcription of

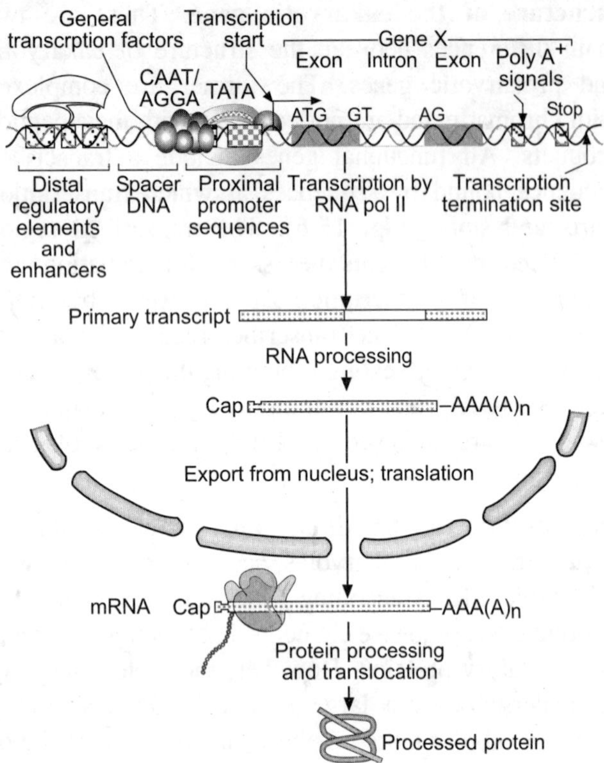

Fig. 15.7 Structure of an ideal plant gene and representation of its expression

the regulated genes. Enhancers, may be located either upstream or downstream, at distances as far as 1,000 bases or more, from the genes they regulate. They are independent of orientation, normal or inverted. They are composed of many different transcriptional elements each of about 10–20 bp and forming the binding site for one or more proteins regulating transcription. The distinction between promoters and enhancers is operational and often they share the same sequence elements.

15.7 REGULATORY MECHANISMS

15.7.1 Signal transduction pathways

The response of higher eukaryotic cells to environmental stimuli and the signals passing between tissues or cell types (hormones, metabolites and ions) are often rapid. There are various ways that signals are detected by a cell and communicated to a gene. Often the effect is direct: a small sugar molecule enters the cell and binds the transcriptional regulator directly. But in most cases, the signal is indirect: the signal induces a kinase that phosphorylates the activator. This type of indirect but very rapid signalling is an example of a signal transduction pathway. In this pathway, the initiating ligand is typically detected by a specific cell surface receptor. The ligand binds to an extracellular domain and the signal is relayed to the relevant transcriptional regulator, often through a cascade of kinases. The transmission of information to the genetic material occurs in stages. The detection of the external stimulus starts with transmission of information to regulatory proteins, which interact with specific DNA sequences. These regulatory proteins are often referred to as transcription factors. In some cases, secondary messengers convey the information between a primary sensor and DNA-binding transcription factors. In response to an external stimulus, genes may either be upregulated or downregulated. These processes are important in gene regulation, for example, via hormones such as insulin (upregulation) and glucagon (downregulation).

15.7.2 Signal control regulator

Once a signal has been communicated, directly or indirectly, to a transcriptional regulator, it controls the activity of regulators in many ways. In eukaryotes, the transcriptional regulators are usually controlled in one of the following ways.

Unmasking an activating region: This is accomplished either by a conformational change in the DNA-bound activator or by the release of a masking protein from an activating region. The conformational changes required in each case can be triggered either by binding the ligand directly or through a ligand-dependent phosphorylation. In many cases, the masking protein is a deacetylase which not only blocks the activating region but also actively represses the gene.

Transport into and out of the nucleus: When a gene is not expressed, many activators and repressors are held in the cytoplasm. The signalling ligand causes them to move to the nucleus where they act. The regulator can be held in the cytoplasm through interaction with an inhibitory protein or with the cell membrane, or it can be in a conformation in which a signal sequence required for its nuclear import is concealed. The release and transport into the nucleus in response to a signal can be mediated through proteolysis of an inhibitor or tethering region, or by allosteric changes.

15.8 ACTIVATION OF TRANSCRIPTION

In eukaryotes, an RNA polymerase holoenzyme with TFIID to recognise promoters, does not generally transcribe on its own. It needs access to the promoter, which is usually wrapped around nucleosomes, and it needs a specific transcription factor to become active (**Fig. 15.8**). Activation of transcription is a two-stage process. In the first stage, chromatin must be decondensed (i.e., chromatin remodelling), since genes exist in an inert state, tightly packaged with histones into chromatin. In the second stage, transcriptional factors interact with specific DNA sequences. The DNA-binding factors are often the final element in the multicomponent signal transduction pathway.

15.8.1 Regulation of transcription

It is generally accepted that the majority of regulatory events that control gene expression occur at initiation. This consists of processes which orchestrate the assembly of the transcription apparatus and more specifically, the positioning of RNA polymerases (RNA polymerases I, II, and III) at the promoter. The presence of a core promoter sequence and assembly of the basal transcription apparatus is all that is required to initiate transcription of any given gene. It now seems clear that genes will be constitutively expressed at a minimal rate determined by its core promoter unless the rate of transcription is increased by an enhancer or is 'switched off' by the actions of a silencer. Some important mechanisms of gene regulation at transcriptional level are discussed in this chapter.

15.8.2 RNA polymerases

Enzymes, which are capable of copying the sequence information in DNA as a complementary RNA copy is produced by the polymerisation of ribonucleotides, are referred to as RNA polymerases. In eukaryotes, there are three RNA polymerases (see **Chapter 16**). All three RNA polymerases are large multisubunit enzymes and several subunits are shared among the three enzymes. In considering the transcriptional regulatory processes that produce tissue-specific variation in specific mRNAs and proteins, we shall discuss RNA polymerase II throughout this chapter.

RNA polymerase I: It is responsible for the transcription of the tandem arrays of genes encoding rRNA. RNA polymerases, cannot however initiate transcription and need other protein factors. The essential DNA sequence for recognition of polymerase is located within the 50 bases immediately upstream of the start site of transcription. The sequences around –50 are recognised by a protein transcription factor known as UBF (upstream binding factor). Subsequently, another regulatory protein known as SL1 (also known as TIFIB) is recruited via protein-protein interaction with UBF and in turn SL1 recruits the RNA polymerase I and its associated factors. Hence, the initiation of transcription by RNA polymerase I is regulated by transcription factors SL1 and UBF.

RNA polymerase II: The gene regulation mechanisms discussed in this chapter are, unless specifically mentioned, mediated by RNA polymerase II.

RNA polymerase III: The downstream promoter that is used for transcription of 5S RNA requires the complex

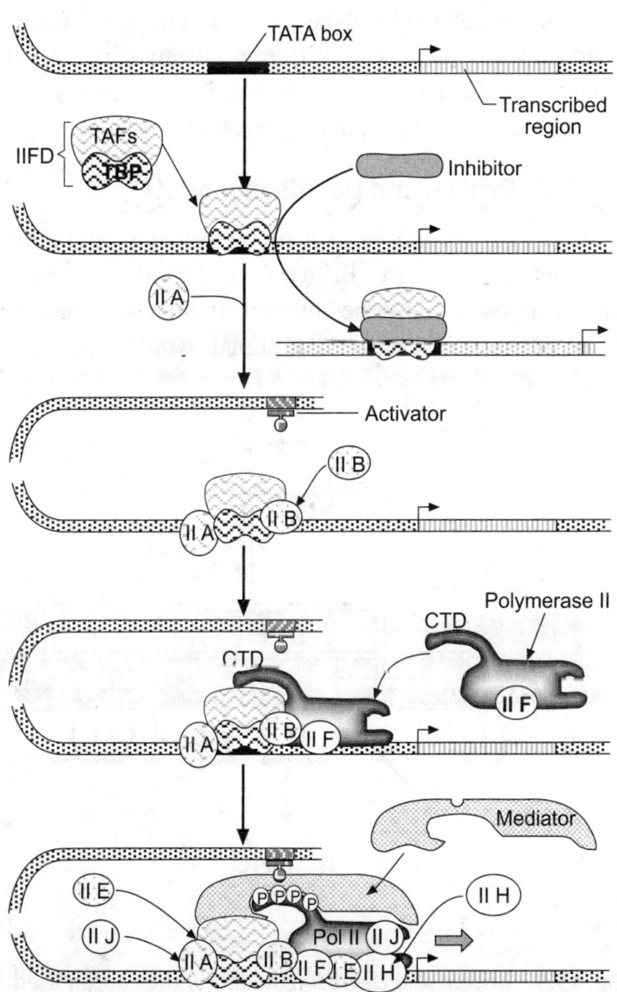

Fig. 15.8 Model of the assembly of the RNA polymerase II holoenzyme complex and initiation of transcription on a promoter with a TATA box

formed by three transcription factors (TFIIIA, TFIIIC and TFIIIB) to recruit RNA polymerase III. The genes encoding the tRNAs also have internal promoters. But TFIIIC and TFIIIB are the two transcription factors essential to recruit RNA polymerase III. It is clear therefore, that TFIIIB plays an essential role in the transcription of three different types of RNA polymerase III promoters.

15.9 MODELS FOR REGULATION

In higher eukaryotes, operon-type structures transcribed into polycistronic RNAs do not exist. Rather, individual genes are transcribed into single monocistronic RNAs encoding single proteins. Hence, genes whose protein products are required in parallel are present in the genome as individual genes whose expression must be coordinated. Moreover, such coordinately expressed genes are not even closely associated in the genome and often they are present on different chromosomes. This kind of genetic arrangement in eukaryotes demands far greater flexibility in regulating gene expression.

15.9.1 Britten and Davidson model

A model of the mechanism of coordinated regulation was put forward by Britten and Davidson (1969). They proposed that genes that are regulated in parallel in response to a particular signal would contain a common regulatory element, which would cause the activation of the gene in response to that signal (**Fig. 15.9**). According to this model, individual genes could contain more than one regulatory element, some of which would be shared with other genes, which in turn could possess elements not present in the first gene. Specific signals causing gene activation would act by stimulating a specific integrator gene whose product would activate all the genes containing one particular sequence element. This mechanism would allow the observed activation of distinct but overlapping sets of genes in response to specific signals via the activation of particular integrator genes.

15.9.2 Regulation of enhancer activity

Many enhancers and their corresponding proteins have been identified and it is now clear that several are common to the majority of promoters. Enhancer elements differ from promoters in many ways. They can exert their positive influence on transcription even when separated by thousands of base pairs from a promoter. They work when oriented in either directions and they can work upstream (5') or downstream (3') from the promoter (**Fig. 15.10**). Enhancers are promiscuous (not gene-specific) and they can stimulate any promoter in the vicinity and may act on more than one promoter. On the other hand, they often show species and cell type specificity. For example, the enhancer for mouse

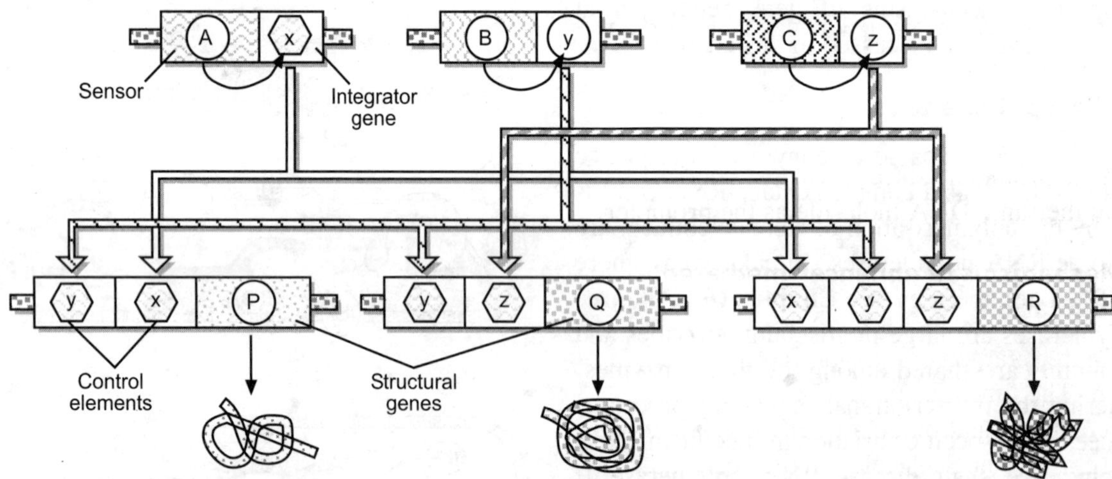

Fig. 15.9 Britten and Davidson model for the coordinated expression of unlinked genes in eukaryotes; sensor elements (A,B,C) detect changes requiring alterations in gene expression and switch on appropriate integrator proteins (X,Y,Z). Their products activate the structural genes (P,Q,R) containing the appropriate control elekents. According to this model, a particular structural gene can be activated with or without another structural gene by selectively activating the integrator gene.

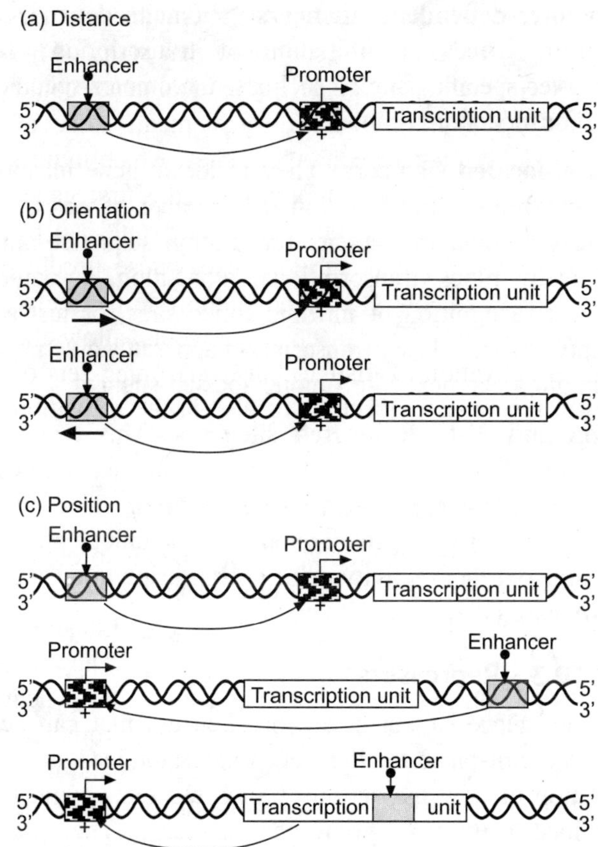

Fig. 15.10 Characteristics of enhancer elements (a) Can activate a promoter at a distance (b) In either orientation relative to the promoter (c) When positioned upstream, downstream or within a transcription unit

immunoglobulin H chains is active in myeloma cells (normally make antibody), but not in fibroblasts (do not make antibody).

An enhancer does not produce any product that in turn acts on the promoter, since it is active only when it exists within the same DNA molecule as the promoter.

15.9.3 Mechanisms of enhancer-mediated regulation

Enhancers are targets for proteins regulating transcription. The functions of these various sequences relate to recognition and binding of RNA polymerase, but there are as yet no molecular studies that indicate how this maneuver is performed, although there are some clues. A striking feature of the characterised enhancers is that they exhibit tissue specificity. The factors bound at the enhancer sequence and the promoter sequence can act in a cooperative manner, either positively or negatively. This indicates that these proteins are probably in contact, at least some of the time, and suggests that the intervening DNA folds or bends to permit this contact (**Fig. 15.11**).

Initially it was thought that enhancers carried out their function through physical interactions with RNAP II. Now it is believed that their effects are indirect, involving intermediary factors that convey signals from activators to the transcription proteins.

Early summaries of promoter structure and function suggested that upstream factors and response elements were restricted to well-defined areas of the gene and directed gene expression in well-defined patterns. Recent studies, however, have shown that both silencers and enhancers can be located 5' or 3' of the transcription initiation site, in introns, exons or even on the transcribed RNA itself.

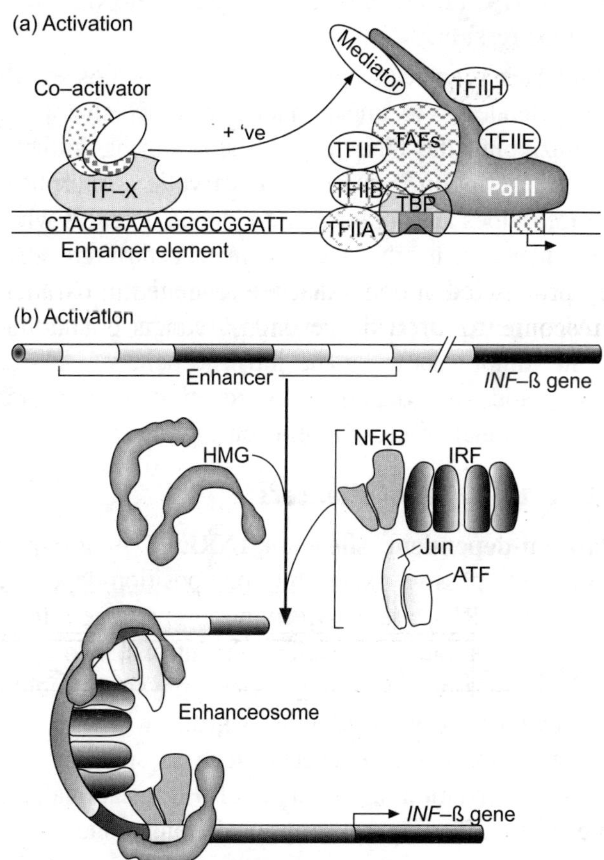

Fig. 15.11 (a) Active TF–X binds to its consensus sequence; interaction with flanking sequence produces a conformation which allows binding to a co-activator complex. The complex increases the rate of GTF assembly and subsequent expression. (b) The human β interferon enhanceosome (***Source:*** Redrawn from Ogbourne and Antalis 1998; Watson et al 2004)

15.10 NEGATIVELY ACTING SEQUENCE ELEMENTS

15.10.1 Silencers

Silencers were initially defined as sequences capable of repressing promoter activity in a manner independent of orientation and position for a native or a heterologous promoter. However, in practice, a silencer is now generally considered to be a short, specific sequence of nucleotides, which are located in the 5' upstream promoter region of a given gene. These sequences are capable of recruiting TF proteins to the promoter, which in turn carry out specific functions. More recently, it has become clear that many different types of silencers exist that are capable of affecting many aspects of gene regulation, such as cytoplasmic retention of TFs, activity of positive acting TFs, chromatin structure, intron splicing, 3'-upstream untranslated signal recognition, as well as GTF assembly, to ultimately downregulate gene expression.

AT-rich sequences have been shown to possess the strong tendency to unpair nucleotides, which may be by unzipping the DNA double helix during initiation of transcription. During eukaryotic interphase, chromosomes are organised into looped domains which are anchored to the nuclear matrix through DNA regions called matrix-attachment regions (MARs). MAR-binding proteins generally recognise and bind to the minor groove or the intrinsic bend of AT-rich DNA, and subsequently may function as important transcriptional control proteins during interphase.

15.10.2 Types of silencers

Position-dependent silencers (NREs): A group of silencers, passive silencers or position-dependent silencers (NREs) are now recognised in a large number of gene promoters, as well as in introns, exons and various flanking sequences. NREs physically inhibit the interaction with specific signals which control various transcriptional events, such as splicing sites, 5'-polyadenylation signals and ATG translational start sites or by affecting transcriptional elongation.

Orientation-dependent silencers: A small group of silencers have been described as orientation-dependent. The activity of various TF-binding sites depends heavily on the sequences directly adjacent to the binding motifs and, in a large number of orientation-dependent silencers, changes in silencer properties due to an anti-sense orientation is probably a result of this fact.

Promoter-dependent silencers: Promoter-dependent silencers affect the initiation of transcription via promoter-specific factor, such as a promoter-specific TAF or mediator component.

Intron-located silencers: They reside in gene introns and can repress transcription in a number of ways. A repressor could bind to its recognition sequence and physically block transcriptional elongation. It could prevent recognition of intronic splice sites, or just as simply abrogate basal transcription apparatus assembly as might a classical 5'-promoter-located silencer.

Exon and 3'-UTR located silencers: This type of silencers may behave similar to 5'-promoter located silencers. However, it has been demonstrated that silencers located within an exon have the unique ability to carry out their activity prior to transcription or as an RNA-based element.

15.10.3 Repressors

The existence of enhancers or silencers that can act over long distance to alter gene expression leads to the question of how the action of such elements could be confined to the appropriate gene or genes. Therefore, some mechanism must exist to prevent the effect of enhancers or silencers spreading to the other genes in adjacent regions of the chromosome and producing an inappropriate pattern of gene expression. This problem is solved by the presence of repressors that block the spread of enhancer or silencer activity (**Fig. 15.12**)

NRSF/NRSE

NRSF (or 'REST') and NRSE [or 'repressor element-1' ('RE-1')], which play a key role in directing neuron-specific expression represent one of the more understood systems. NRSF was one of the first transcriptional regulators to be implicated in negative gene regulation during neuronal development in vertebrates. NRSF/REST was originally identified as a silencer-binding protein for the neuron-specific sodium channel gene (*scg10*), which binds to a conserved element known as NRSE and was simultaneously cloned in separate studies. NRSF/REST is known to be a 116-kDa protein containing eight non-canonical zinc fingers, only one of which has been shown to be crucial for repression of the type II sodium channel gene. The rat m4 muscarine acetylcholine receptor gene (*mAChR*), the human synapsin I gene and the rat dopamine ß-hydroxylase gene have recently revealed the presence of cell-specific silencers that were identified as NRSEs by sequence. In

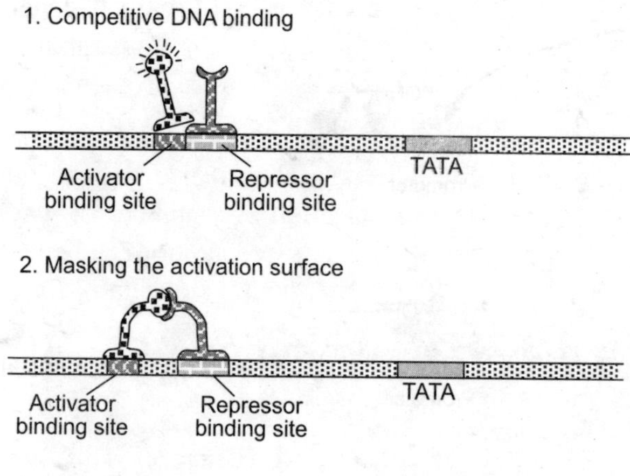

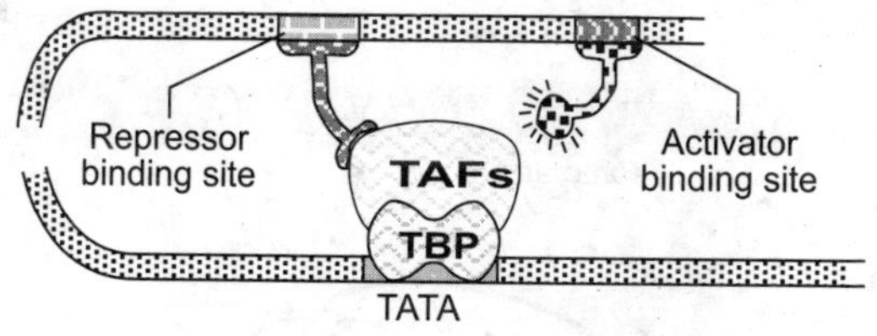

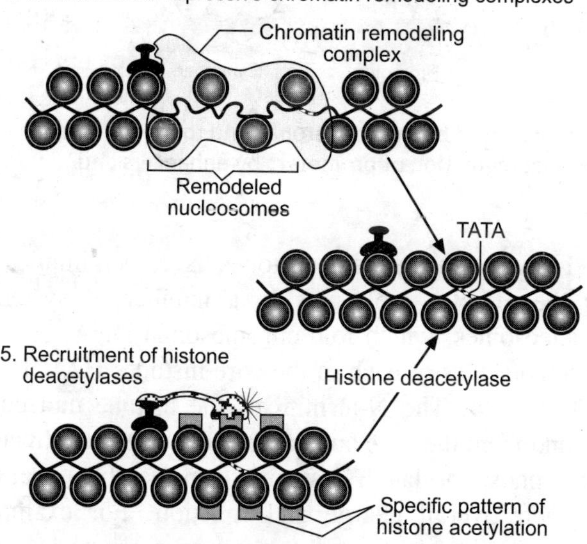

Fig. 15.12 Five possible ways in which eukaryotic gene repressor proteins can operate (***Source:*** B Alberts, A Johnson, J Lewis et al 2002 *Molecular Biology of the Cell*, 4th ed.)

all three systems, NRSE was shown to silence promoter expression in several non-neuronal cells, but not in neuronal cells.

15.10.4 Polycomb group proteins

In general, TFs play a transient role in regulation of transcription, and as such are required in continuous supply to carry out their function. However, the identification of the polycomb group (Pc-G) of repressors from *Drosophila*, and the functionally homologous yeast Rme1p, *Xenopus* XPc and human HPH1 proteins has revealed a mechanism which overcomes this requirement. Pc-G proteins have been shown to keep homeotic genes stably and heritably silenced during development in *Drosophila.* These proteins show no sequence-specific binding, suggesting that they anchor to the DNA by other pre-bound proteins. In fact, several Pc-G proteins have been shown to act as transcriptional repressors when tethered to DNA-binding domains.

15.10.5 Insulators

If enhancers can activate expression of a given gene several kilobases away from the targeted initiation site, why is the expression of closely linked genes not activated? Short-range repressors that function over 50 to 100bp have been shown to inhibit or quench closely linked enhancers, thereby allowing independent function of enhancers within a modular promoter. Several mechanisms have been proposed to account for this regulatory compartmentalisation, but the best understood rely on insulator elements, also called boundary elements. Insulator elements (insulators) are DNA sequences that bind specialised proteins and have two specific properties. First, they buffer genes from the repressing effects of heterochromatin. Secondly, they can block the action of enhancers. Insulators act as dominant repressors that function over long distances to block interactions between *cis*-elements and inappropriate promoters (**Fig. 15.13**). For example, a 340-bp insulator DNA within the *Drosophila* gypsy retrotransposons, disrupts gene activity by blocking the interactions of distal enhancers with target promoters. It is noteworthy that reduction of *mod*(*mdg*4)$^+$ causes the insulator to act as a promoter-specific repressor. It is understood that insulators might block distal enhancers via DNA bending, mimicking the structural changes mediated by TFIID binding to the promoter. Alternatively, the mechanism of 'insulation' may involve chromatin looping to separate genes into distinct nuclear compartments. Insulators appear to possess some regulatory specificity, as they selectively block enhancers but not repressors. Insulators are also found to be able to inhibit the spread of chromatin modifications.

15.10.6 Co-repressors

Recently, several groups have identified secondary repressor complexes, named CoRs, which appear to elicit

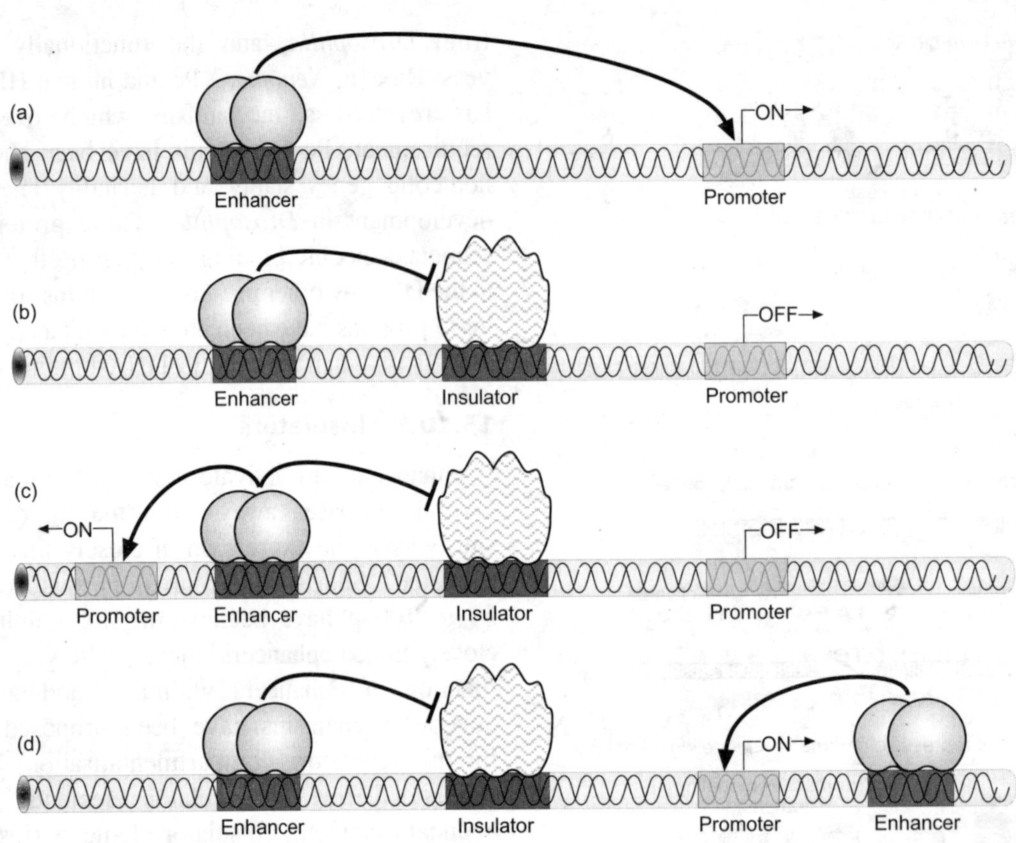

Fig. 15.13 Activation of insulator block by enhancer (a) Promoter is activated by the activator bound to the enhancer. (b) Activation of enhancer is blocked by insulator. (c) and (d) Differential regulation of promoters by enhancers and insulators on multiple genes.

repression from several previously identified repression systems. The repression activity of unliganded thyroid hormone and retinoic acid receptors is dependent on a conserved motif in the ligand-binding domain known as the CoR box. The CoR box has since been shown to bind to the 270 kDa protein nuclear-receptor co-repressor (N-CoR), which functions via a signal to the GTFs. The CoR complex that contains N-CoR has also been shown to contain members of the Sin3 family, as well as the histone deacetylase, RPD3. It is thought that this complex binds to unliganded hormone receptors and affects local histone acetylation, producing alterations in chromatin structure that block transcription. Upon ligand binding, the CoR complex disassociates and is replaced by a co-activating complex which contains the cAMP-response-element-binding protein, p300, and/or p/CAF, each of which has histone acetyl transferase activity.

15.10.7 Histones

Although histones are not strictly repressors or CoRs, they are capable of repressing transcription *in vitro* and *in vivo*, and recently histones have been implicated as the effector molecules in a number of systems. Nucleosomes, which fold chromosomal DNA, contain two molecules of each of the core histones H2A, H2B, H3 and H4. The N-termini of the histone molecules extend from the core and can be acetylated, methylated and phosphorylated, causing considerable changes to their overall charge and function. For example, acetylation neutralises the positively charged lysine residues of the histone molecules, decreasing their affinity for DNA. As a consequence, DNA can unfold, providing free access to transcription factors and, ultimately, the GTFs.

15.11 REGULATION OF EUKARYOTIC TRANSCRIPTION FACTORS

Transcription of eukaryotic genes is regulated by a combination of activators and repressors that bind to specific DNA regulatory sequences. Whether or not a specific gene in a multicellular organism is expressed in a particular cell at a particular time is largely a consequence of the binding and activity of the transcription factors that interact with the regulatory sequences of that gene. Therefore, it is necessary that the activity of transcription factors must be controlled.

15.11.1 Transcriptional control by hormones

Among the known regulators of eukaryotic genes, hormones are the most extensively studied transcription activation factors. Hormones are extracellular substrates, small molecules, polypeptides or small proteins that are carried from hormone-producing cells to target cells. Many of the steroid hormones (e.g., estrogen, progesterone, aldosterone, glucocorticoids and androgens) and general metabolic hormones (e.g., insulin, glucagon and so on) frequently act by turning on transcription (**Fig. 15.14**). However, the penetration of a target cell by a hormone and its transport to the nucleus is a complex process.

The basic steps involved in transcriptional activation by hormones are:

- A hormone binds directly to the DNA and facilitates binding of RNA polymerase or transcription factors.
- Hormone binds to an effector protein (such as the CRP protein in the *lac* operon) that must bind to a site on the DNA prior to RNA polymerase binding.
- A DNA-bound protein is activated by the hormone and the activated form is needed for RNA polymerase binding.
- A hormone is an inducer that inactivates a repressor.

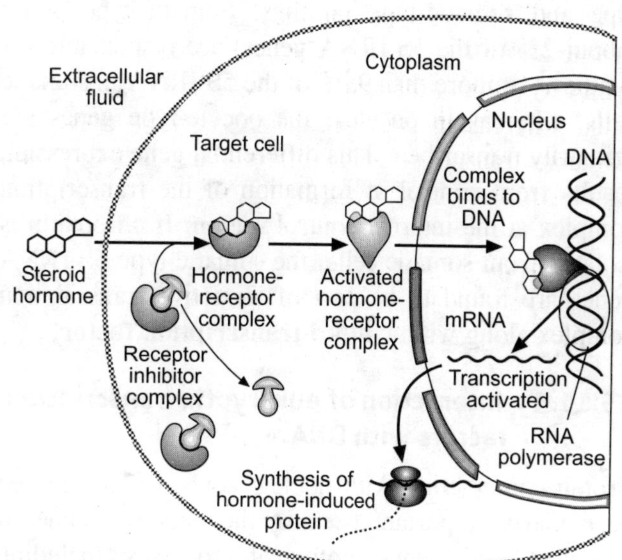

Fig. 15.14 Diagrammatic representation of cortisol-induced gene expression

- A hormone modifies chromatin (structural change in chromatin) that makes the DNA available to RNA polymerase. At present, only a few hormonal-induced systems have been studied.

15.11.2 Formation of transcription complex

Transcription factors form a transcription complex to exert their effect on initiation of transcription. This process can be explained with suitable examples (**Fig. 15.15**). The *Xenopus* genome contains two multigene families encoding 5S rRNA, called oocyte-

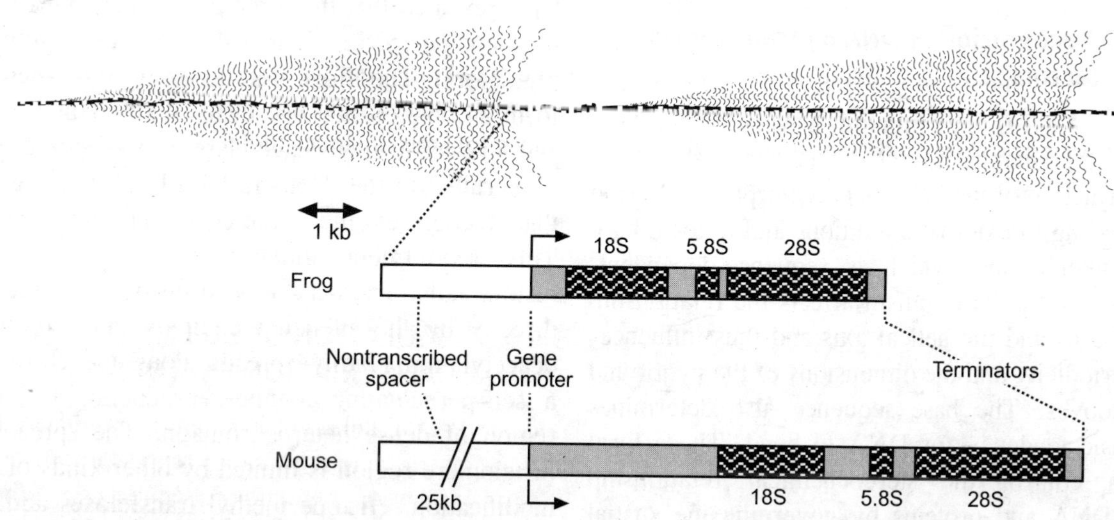

Fig. 15.15 Synthesis of rRNA and the rRNA transcription unit

type and somatic-type families. Somatic-type genes (about 2% of the 5S rRNA genes) are responsible for synthesis of more than 95% of the 5S rRNA in somatic cells, whereas in oocytes, the oocyte-type genes are primarily transcribed. This differential gene expression results from control of formation of the transcription complex at the internal control region. If chromatin is isolated from somatic cells, the somatic-type 5S rRNA genes are found to be part of an active transcription complex along with a **40-kd transcription factor**.

15.11.3 Interaction of eukaryotic transcription factors with DNA

Proteins interacting with nucleic acids are considered particularly important because they control some of the most fundamental biological processes, including replication, recombination, DNA repair, transcription, RNA processing and protein synthesis. DNA–protein interactions can fulfil many roles in the cell and are grouped into four major categories.

- Structural and packaging roles (e.g., histones, capsid proteins)
- Transport and localisation roles including DNA segregation and localisation, RNA export and localisation and plasmid transfer
- Metabolism and rearrangement roles (e.g., DNA and RNA polymerases, nucleases, helicases, DNA repair enzymes, recombinases, topoisomerases)
- Gene expression roles (e.g., RNA polymerases, transcription factors, ribosomes and initiation factors, the RNA splicing apparatus, amino acyl-tRNA synthetases).

15.11.4 Recognition of DNA by transcription factors

Cellular DNA exists predominantly as very long dsDNA whose overall structure is relatively constant. However, the conformation of the helix is polymorphic, differing both according to external conditions and to local base stacking interactions, which is sequence-dependent. Local structural polymorphism affects the relationship between bases and the helical axis and thus influences helical periodicity and the dimensions of the major and minor grooves. The base sequence also determines the intrinsic tendency for DNA to bend. These local variations control the stereochemical relationship between DNA and proteins by governing the spatial organisation of atoms and the ability of DNA to change conformation upon protein binding. Proteins binding to dsDNA may be divided into three categories:

- Those interacting with DNA ends (e.g., DNA ligases, exonucleases)
- Those enclosing DNA or binding it in a deep crevice (e.g., DNA polymerases, topoisomerases)
- Those interacting with the face of the helix. The first two categories consist of DNA-processing enzymes, whereas the third category is the largest and includes most transcription factors, restriction enzymes, DNA-packaging proteins, site-specific recombinases and DNA repair enzymes.

There are also proteins, which interact with ssDNA, example, RecA, SSB and capsid proteins. Penetration of the major groove is important for sequence-specific proteins, which facilitates sequence readout by direct binding to bases. DNA-binding proteins are assigned to families according to the structural motif used for DNA recognition (**Table 15.2**).

15.12 GENE SILENCING BY MODIFICATION OF HISTONES AND DNA

Silencing means a gene is silenced because of its location and not in response to a specific environmental signal. The most common form of silencing is associated with a dense form of chromatin called heterochromatin. The heterochromatin regions do not contain any protein coding sequences. But, if a coding gene is moved experimentally to this region, it will be switched off, particularly if they are weakly expressed in their usual location. Mutations in which silencing is relieved (gene is expressed when placed in telomere region) have revealed that the three genes encoding regulators of silencing, SIR2, 3, and 4 (SIR, silent information regulators). The three proteins encoded by these genes form a complex that associates with silent chromatin, and Sir2 is a histone deacetylase. The silencing complex is recruited to the telomere by a DNA-binding protein that recognises the telomere's repeated sequences. This recruitment initiates local deacetylation of histone tails. The deacetylated histones are recognised directly by the silencing complex, and so the local deacetylation readily spreads along the chromatin in a self-perpetuating manner, producing an extended region of dense heterochromatin. The spreading out of telomere region is limited by other kinds of histone modification. Histone methyl transferases add methyl groups to specific lysine residues in the tails of histone H3 and H4.

Table 15.2 Principal recognition structures in DNA and RNA binding proteins

Structures	*Recognition*	*Distribution*	*Examples of resolved structures*
DNA-binding structures			
Helix-turn-helix	α-helix	All	*E. coli lac* repressor, *Drosophila* Antennapedia
HMG domain	α-helix	Eukaryotes	Rat HMG1
Zinc finger	α-helix	Predominantly eukaryotes	Mouse Zif268 *Drosophila* Tramtrack
Steroid receptor family	α-helix	Eukaryotes	Human retinoic acid receptor, rat glucocorticoid receptor
Binuclear cluster	α-helix	Yeast	*S. cerevisiae* BAL4
Basic domain	α-helix	Eukaryotes	*S. cerevisiae* GCN4, human MyoD, mouse Max
BPV E2 motif	α-helix	Papillomaviruses	BPV E2 protein
Ribbon-helix-helix and other sheet structures	ß-sheet	All	*E. coli* MetJ repressor, *Arabidopsis* TATA-binding protein
Histone-core	Electrostatic	Eukaryotes	Human $TAF_{II}31$, $TAF_{II}80$
-linker	α-helix	Eukaryotes	Human H5
Rel homology	Loop	Eukaryotes	NF-kB p50 homodimer
RNA-binding structures			
RNP domain	ß-sheet	Eukaryotes	U1A snRNP
K-homology	Loop	Eukaryotes	HnRNP K
DsRBD	ß-sheet	All	*Drosophila* Staufen

15.13 REGULATION OF ELONGATION

Regulation of elongation is one of the methods of control of transcription. In such cases, a block in elongation of the nascent transcript is released. In some genes, there are sequences downstream of the promoter that cause pausing or stalling of the polymerase soon after initiation. At those regions, the presence or absence of certain elongation factors greatly influences the level at which the gene is expressed. For example, the *hsp70* gene of *Drosophila* is normally activated by heat shock and is controlled by two activators working together. The GAGA binding factor recruits enough of the transcription machinery to the gene for initiation of transcription, but in the absence of a second activator, HSF, the initiated polymerase stalls 100 bp downstream of the promoter. In response to heat shock, HSF binds to specific sites at the promoter and recruits a kinase, P-TEF, to the stalled initiated machinery. The kinase phosphorylates the C-terminal domain of the largest subunit of RNA polymerase freeing the enzyme from the stall and allowing transcription to proceed through the gene. Regulation of elongation is also responsible, for example, for the ten-fold decline in mRNA levels encoding the cellular oncogene c-*myc* which occurs when the human pro-myeloid cell line HL-60 is induced to differentiate into a granulocyte type cell.

Termination of elongation

Once initiation is complete, nuclear RNA (nRNA) synthesis occurs *via* one of the three RNA polymerases (RNAPs), and finally elongation of the nRNA transcript is terminated. Termination of elongation with RNAP I occurs at a discrete 18-bp recognition sequence found between 100 and 4,000 bp downstream of the mature 3'-end, while RNAP III terminates at the second, third or fourth uridine in a run of four uridine bases surrounded by GC pairs.

15.14 POST-TRANSCRIPTIONAL REGULATION

Following transcription, a number of processes modify the primary transcript of genes which are transcribed by RNA polymerase II before it can be translated into protein.

In eukaryotic cells all mRNA transcripts, with the exception of those for histone genes, contain non-coding introns. Transfer of these precursor mRNAs from nucleus to cytoplasm involves 'RNA processing', that is excision of introns, as well as the 5' and 3' end modifications (maturation of hnRNA). These post-transcriptional modifications are important aspects of gene regulation since they enable control over the number, kind and stability of mRNA molecules that enter the cytoplasm of particular tissues for translation. The RNA splicing system that removes introns, for example, by recognising specific sequences at each intron junction is apparently common to different mRNA precursors in different eukaryotes. However, in theory, an RNA species transcribed in several tissues might be correctly spliced to yield functional RNA in one tissue and remain unspliced in another tissue. An unspliced RNA would either be degraded within the nucleus or, if transported to the cytoplasm, would be unable to produce a functional protein due to the interruption of the protein-coding regions. Several cases of such processing versus discard decisions have now been described in *Drosophila* and a similar pathway in which unspliced RNAs are specifically degraded within the nucleus has been characterised in yeast.

15.14.1 Splicing and polyadenylation of mRNA

Although such processing versus discard decisions are not widespread in mammals, numerous cases of alternative RNA splicing have been described, both in mammals and other organisms. The splicing and polyadenylation of pre-mRNA depends on the precise identification of splice sites and polyadenylation sites, respectively. In many cases, however, a single pre-mRNA transcript is processed into several different mRNAs of differing functions. Such differential processing may be cell-, tissue- or developmental-specific. It is known, for instance, to control the production of foetal or adult forms of myosin, the synthesis of membrane-bound and secreted forms of IgM heavy chain, the cellular and plasma forms of fibronectin, prohormones that include or exclude various neuropeptides, sex determination in *Drosophila* and multiple isoforms of more than 50 different known proteins. This kind of processing is called **alternative processing**. Despite considerable investigation much remains to be learned of the mechanisms whereby this different processing is controlled. There are different forms of alternative splicing in the control of gene expression.

Spliced or retained introns

An intron may be removed or retained and its retention may result in the addition of extra amino acids to a protein or may totally disrupt protein synthesis by, for instance, inserting a premature stop codon. A well-characterised example is the *Drosophila* P element, the transposition of which is dependent on a transposase that is only expressed in germ line cells. In some somatic cells, the third intron of the P element pre-mRNA is retained. The resulting 66-kDa protein lacks transposase activity and functions as a negative regulator of transposition.

Alternative 5'-donor and 3'-acceptor splice sites

The transcript of the SV40 early gene is a model of the use of alternative splice sites. It can be processed to produce the mRNA encoding large T antigen or that of small t antigen by the utilisation of alternative 5'-splice sites but a common 3'-splice site. Thus, what is part of an intron in the splicing of small t antigen mRNA becomes part of an exon during processing to produce a large T antigen. The polypyrimidine element of the 3'-splice site is involved in the selection. Mutations that increase the pyrimidine content result in an increased production of large T antigen whereas those that increase purine content have enhanced amounts of small t. *Trans*-acting factors have also been implicated in differential splicing.

Use of alternative promoters

In these cases, two alternative primary transcripts are produced by transcription from different promoter elements and these are then processed differentially. The use of alternative promoters may dictate a variant splicing pattern. This is well demonstrated in the alternative processing of chicken skeletal muscle protein, in which two forms of the muscle protein myosin called alkali light chains LC1 and LC3 are produced. The myosin gene has two different TATA sequences, which yield two different primary transcripts. These two transcripts are processed differently to form mRNA molecules encoding distinct forms of the protein (**Fig. 15.16**). The mechanism involved depends on whether the mRNA 5'-end comprises exons 1 and 4 or exons 2 and 3 to decide which of the two promoters is used during transcription. In such cases, it is assumed that the different primary transcripts fold into different secondary structures which favour the different splicing events.

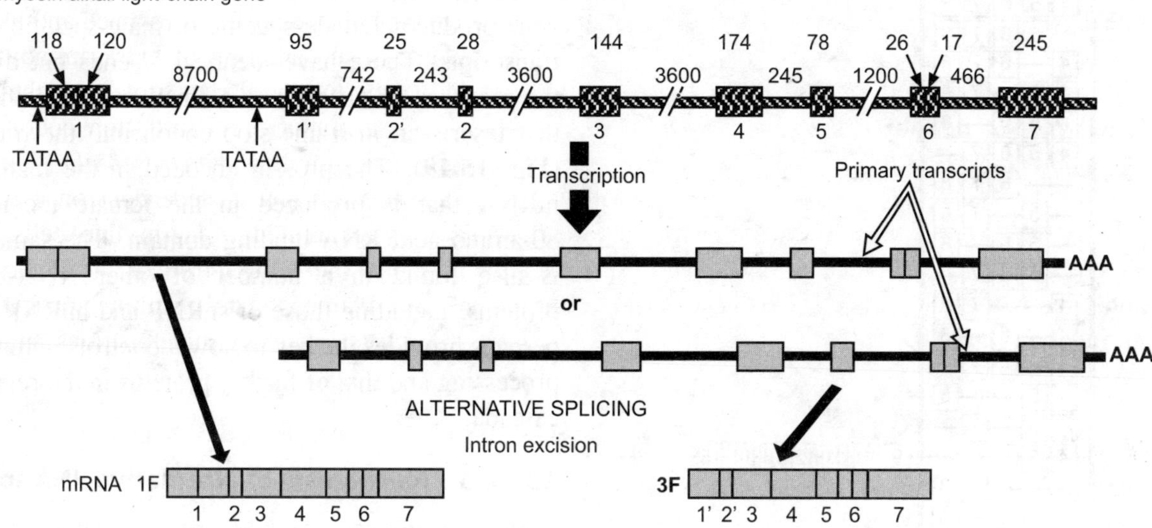

Fig. 15.16 Alternative splicing of the myosin alkali light chain transcripts in different muscle cell types produces two mRNAs (1F and 3F) differing at their 5′ ends. The two alternative start sites of transcription used to produce each of the RNAs are indicated (two distinct TATAA) together with the intron-exon structure of the gene

Alternative polyadenylation/cleavage sites

Many examples are known of the use of variant 3'-exons coupled with alternative polyadenylation sites. However, it is not always clear if the alternative cleavage/polyadenylation dictates the splicing pattern or whether the splicing pattern is the critical event controlling the 3'-processing. The best-defined example of this process occurs in the genes encoding the immunoglobulin heavy chain of the antibody molecule, which plays an important role in the regulation of the antibody response to infection. The production of membrane-bound and secreted immunoglobulin molecules is controlled by the alternative splicing of different RNA molecules differing in their 3'-ends (**Fig. 15.17**).

Mutually exclusive exons

In this variant of alternative splicing, one of a pair of exons is always spliced into an mRNA but never both. The rat ß-tropomyosin gene contains 11 exons and gives rise to two isoforms by differential splicing of the transcript. Exons 1–5, 8 and 9 are common to both mature mRNAs but exons 6 and 11 are only incorporated into the mRNA of fibroblast tropomyosin 1 whereas exons 7 and 10 are used in that of skeletal muscle ß-tropomyosin.

Cassette exons

Cassette exons can be included or excluded from the mature mRNA independently of others and usually with the maintenance of the reading frame. Where multiple cassette exons occur in a transcript, various patterns of exclusion or inclusion can give rise to a high degree of diversity in mRNA. The gene coding leukocytecommon antigen spans 130 kb and includes 44 exons. Various isoforms of the complete transcript are generated by the variable use of exons 4, 5 and 6. Extreme variants of the resultant proteins are the 180-kDa isoform that lacks all three cassette exons and a 220-kDa polypeptide that includes all the three. At least six of the eight possible isoforms have been detected. In another example like the skeletal muscle troponin T gene, the same RNA can be spliced in up to 64 different ways in different cell types (**Fig. 15.18**).

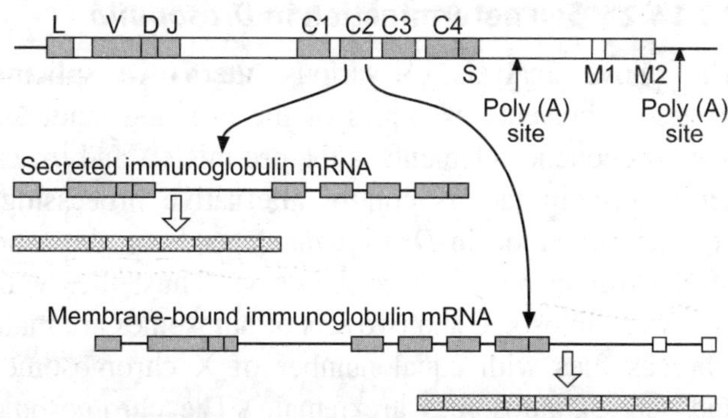

Fig. 15.17 Alternative splicing of the immunoglobulin heavy chain transcript at different stages of B-cell development; the two unspliced RNAs produced by the two alternative polyadenylation sites in the gene are shown together with the spliced mRNAs produced from them

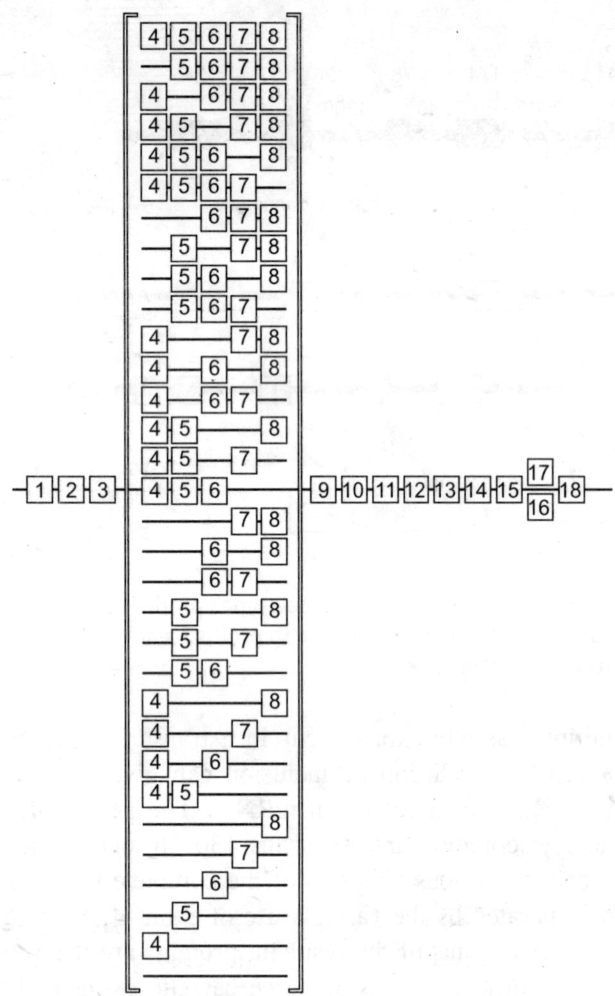

Fig. 15.18 Alternative splicing of the four combinatorial exons (4–8) and the two mutually exclusive exons (16 and 17) can result in up to 64 distinct mRNAs from the rat troponin T gene

15.14.2 Sex determination in *Drosophila*

The above analysis of various alternative splicing strategies presents examples of the growing evidence that *cis*-sequence elements in the pre-mRNA and *trans*-acting protein factors control alternative processing. Sex determination in *Drosophila* depends on the ratio of X chromosome to autosomes (A). Thus, flies with one X chromosome and two sets of autosomes are male whereas flies with equal number of X chromosomes and sets of autosomes are female. The chromosome ratio, a one to one X:A activates the so-called sex-lethal locus (Sxl) which positively controls its own expression and controls three other groups of genes. These are the genes involved in the sexual differentiation of somatic cells and germ-line cells and dosage compensation. Through most of the life of the *Drosophila*, the Sxl gene produces female-specific or male-specific spliced transcripts. These have identical 5'-ends but differ in the presence or absence of a small male-specific exon that inserts an in-frame stop codon into the transcript (**Fig. 15.19**). The protein encoded in the mature Sxl mRNA that is produced in the female includes an 80-amino acid RNA-binding domain. The same motif is also found in a number of other RNA-binding proteins, including those of snRNP and hnRNP, and it perhaps provides the key to how it controls both its own processing and that of further proteins in the regulatory cascade.

15.14.3 Mechanism of alternative RNA splicing

From the above examples of alternative splicing, it is clear that certain factors are necessary for each pattern of splicing in genes such as calcitonin/CGRP or troponin. It is likely that these factors recognise *cis*-acting sequences within the RNA transcript itself. The interaction of these factors with such sequences could produce alternative splicing either by promoting splicing at the site of the *cis*-acting sequence at the expense of the alternative splice site or by inhibiting splicing at the site of binding and thereby promoting the use of the alternative splice site.

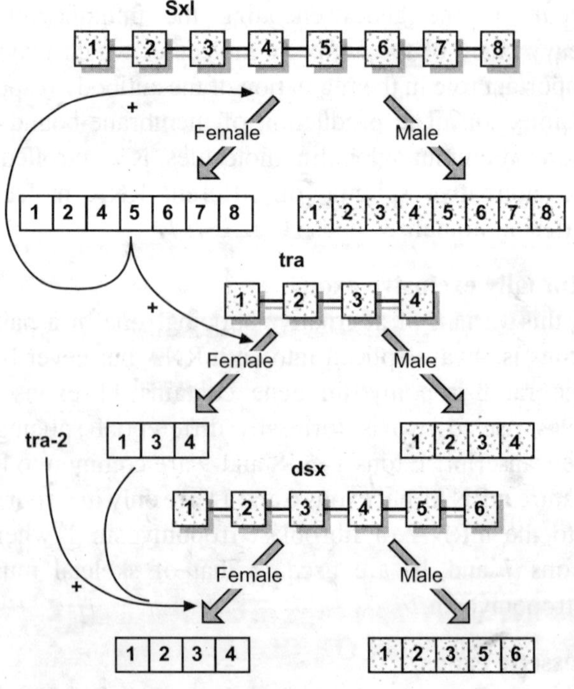

Fig. 15.19 Schematic diagram of the hierarchy of alternatively spliced genes which controls sex determination in *Drosophila*

Both these mechanisms appear to be used in different cases and can be seen in the hierarchy of alternatively spliced genes, which regulates sex determination in *Drosophila*. In this hierarchy, each gene product controls the alternative splicing of the next gene in the pathway resulting in different protein products in males and females. In turn, these products differentially regulate the splicing of the next gene in the hierarchy leading ultimately to the production of a male or female fly. The Sxl gene is differentially spliced in males and females, with the product of the female-specific mRNA not only controlling the splicing of the next gene in the hierarchy, *tra*, but also promote its own female-specific splicing. Mutations, which affect this autocatalytic function of Sxl on its own RNA map in the intron between exon 2 and the male-specific exon 3, indicate that the Sxl protein acts by inhibiting the male-specific splicing of exons 2 and 3. Similarly, Sxl prevents the use of the male-specific exon 2 in the *tra* gene by binding within intron 11 and preventing the binding of the constitutive splicing factor U2AF which in turn recruits the U2snRNP particle which is essential for removal of introns.

In contrast, the action of the *tra* and *tra*-2 gene products on the splicing of the dsx transcript appears to be mediated by promoting the use of the female-specific splice rather than the constitutive male-specific slicing event. Mutations in dsx which affect its splicing map in the intron between exon 3 and the female-specific exon 4, indicate that the alternative splicing factor promotes the splicing of exons 3 and 4. In addition to responding to the presence or absence of a specific factor, specific RNA sequences can also regulate the pattern of alternative splicing in response to cellular signaling pathways.

15.14.4 Gene regulation by RNA editing

Gene expression can be regulated by RNA editing. A change in the mRNA sequence as a result of RNA editing is well explained in the case of apolipoprotein B, which plays an important role in lipid transport (see **Chapter 10**) .

15.15 REGULATION OF CYTOPLASMIC TRANSPORT OF NUCLEAR RNA

In mammalian cells, about 1/20th of the total mass of RNA synthesised ever leaves the nucleus. But most of the RNA molecules undergo extensive processing and the 'leftover' RNA fragments are degraded in the nucleus. This degradation is carried out by the **exosome**, a large protein complex that contains several different RNA exonucleases.

After pre-mRNA has been processed (capping, polyadenylation and splicing) in the nucleus, each of the resulting mature mRNAs is transported into the cytoplasm through one of the ≈4,000 nuclear pores of typical mammalian cells. The transport of RNA to the cytoplasm is little understood. While it is not known, for instance, how transcription and/or processing is coupled to transport or about the nature of the transport mechanism, it has been suggested that both maturation and transport of mRNA occur in association with the nuclear matrix with the involvement of 5'-cap in transportation.

Nuclear pore complexes are composed of dozens of proteins with a total nuclear mass of ≈120 million Da. They are immense compared with the eukaryotic ribosome of ≈4 million Dalton. This nuclear pore complex functions as a gated channel through which ribonucleoproteins are selectively transported to the cytoplasm by a mechanism that requires ATP hydrolysis.

15.15.1 Recognition of 5'-cap structures

Studies of snRNA transport have demonstrated that the 5' cap of RNAs is somehow recognised by the nuclear transport mechanism, supporting the notion that mRNPs move through the nuclear pore complex in the 5'→3' direction. U1, U2, U4, and U5 snRNAs, which are synthesised by RNA polymerase II, initially are produced with a m^7G cap identical to that of pre-mRNAs at their 5' ends. Following their synthesis, these snRNAs are transported to the cytoplasm where they assemble with S_m protein and other proteins to form the corresponding snRNPs. The discovery that capped snRNAs are initially transported into the cytoplasm, but that uncapped U6 snRNA is not, suggested that the cap structure might be required for transport to the cytoplasm.

It is critical that only fully processed mRNAs be transported from the nucleus to the cytoplasm, because translation of incompletely processed pre-mRNAs containing introns would produce defective proteins, which might interfere with the function of the corresponding normal proteins. The direct evidence for existence of an mRNA-binding protein is the Gle1 protein found in yeast. As expected, Gle1 protein contains a short amino acid sequence known as a nuclear export signal (NES). A similar NES sequence has also been identified in the Rev protein of the human immunodeficiency virus (HIV) which is involved in regulating the transport of different HIV mRNAs from the nucleus to the cytoplasm. Rev is an RNA binding protein and binds to a specific site in the second intron of the HIV transcripts. This binding promotes the transport of these RNA to the cytoplasm (**Fig. 15.20**). Similarly, the TFIIIA transcription factor also contains a NES sequence.

15.15.2 Regulation of mRNA transport in the cytoplasm

In addition to controlling the transport of mRNA from the nucleus to the cytoplasm, it is also possible for regulatory processes to control the location of an mRNA species within the cytoplasm of a polarised cell, resulting in the corresponding protein being made only in certain parts of the cell.

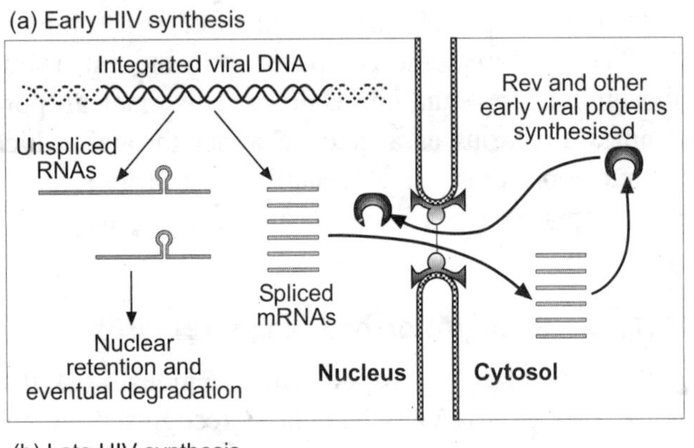

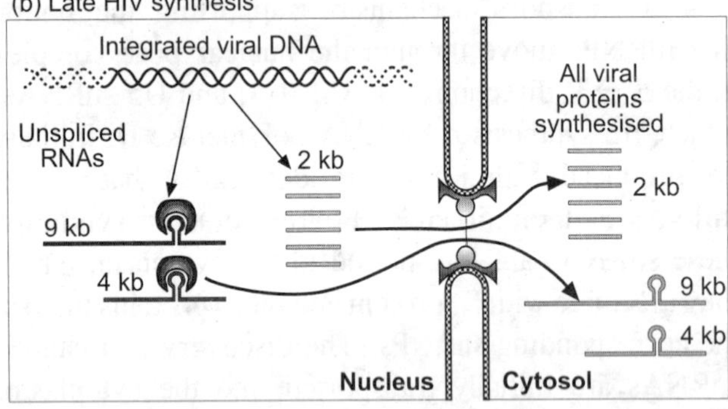

Fig. 15.20 Regulation of nuclear export by the HIV mRNAs by Rev protein (a) Early in HIV infection, only the fully spliced RNAs (~2 kb and ~4 kb alternately spliced transcripts) are exported from the nucleus and translated in the cytoplasm. (b) Once sufficient Rev protein has accumulated and transported into the nucleus, unspliced viral RNAs (~9 kb) can be exported from the nucleus. Most of these RNAs are translated into protein and the full length transcripts are packaged into new viral particles (***Source***: B Alberts, A Johnson, J Lewis et al 2002 *Molecular Biology of the Cell*, 4th ed.; Lodish H, A Berk, SL Zipursky et al 2004 *Molecular Cell Biology*, 4th ed.)

Regulation of cytoplasmic localisation of mRNA

Similar to the nuclear pre-mRNAs and mature mRNAs before transport, cytoplasmic mRNAs are associated with protein in ribonucleoprotein complexes. However, the ratio of protein to RNA is lower in cytoplasmic mRNPs than in nuclear mRNPs. The most abundant cytoplasmic mRNP is the cytoplasmic poly A-binding protein (PABP), which is tightly associated with the poly A tails of mRNAs. The translation of many RNAs in the cytoplasm does not take place free in solution. For example, polyribosomes can be intimately associated with membranes of the rough endoplasmic reticulum. The proteins synthesised at these sites are exported from the cell or become components of new membranes. The association of some RNAs with the cytoskeleton has been demonstrated by microscopic examination of individual cells extracted with a mild non-ionic detergent. Some experimental observations have led to the suggestion that the poly A-binding protein associated with mRNA poly A tails may associate with the portion of the cytoskeleton made of actin microfilaments.

Direction of some mRNAs to specific cytoplasmic sites

Some mRNAs are directed to specific cytoplasmic sites by sequences in their 3' untranslated regions of the cytoplasm. Such mRNA localisation can be visualised in mammalian myoblasts as they differentiate into myotubes—the fused, multinucleated cells that make up muscle fibres.

The most elaborate examples of localisation of mRNA involve certain mRNAs in the cytoplasm of *Drosophila* oocytes. The mRNAs encoded by the *bicoid* and *nanos* genes are synthesised in nurse cells, which surround the developing oocyte, and then move through cytoplasmic bridges into the oocyte. Analysis of *Drosophila* mutants in defective early embryonic development has identified three genes required for localisation of *bicoid* mRNA. One of these genes, *staufen*, encodes an RNA-binding protein that is believed to bind to sequences in the 3' untranslated region of *bicoid* mRNA required for its localisation. Localisation of *nanos* mRNA may be even more complicated, since eight genes have been identified that are required for its localisation.

15.16 REGULATION OF TRANSLATION

Translational control refers to the regulation of the number of times a processed mRNA molecule is translated. Eukaryotes generally control the amount of specific proteins by varying the level of transcription of the gene and/or by RNA processing, but some controls occur in the cytoplasm. The presence of multiple copies of 5'–AUUUA–3', usually in the 3'-non-coding region, marks the mRNA for rapid degradation and thus limited translation. Another form of translational control involves proteins binding directly to the mRNA (called '**masked RNA**') and preventing translation. Some non-coding sequences can cause mRNA to be located in a specific part of the cytoplasm and, when translated, can give rise to a gradient of protein concentration across the cell. For example, the mammalian reticulocyte cell uses older mRNA molecules to synthesise α and ß-globins. Since these cells do not posses nuclei, transcriptional control can no longer occur, yet globin synthesis takes place only in the presence of haeme. When haeme is absent, protein synthesis stops. The agent responsible for translational control in this case has been identified as a protein that becomes activated when the concentration of haeme falls. In its activated inhibitory form, hemin-controlled repressor (HCR) phosphorylates one of the translation initiation factors (eIF-2), thereby preventing synthesis. Thus, the following types of controls are possible:

- Extension of the lifespan of mRNA.
- The presence of appropriate initiation, elongation and termination factors as well as the ability of ribosomes to attach to particular mRNA molecules or to particular sections of such molecules.
- The formation of stems and loops can inhibit exonucleases and increase the longevity of the mRNA templates.
- The kind and amount of tRNA molecules carrying the necessary anticodons to enable translation.
- Prokaryotes sometimes make short antisense RNA molecules that form duplexes near the ribosome-binding site of certain mRNAs, thus inhibiting translation.

15.16.1 Regulation of RNA stability

It is now clear that the rate of turnover of an mRNA plays an important role in determining its level in the cell. The concentration of mRNA is a function of both its rates of synthesis and degradation. The same rate of transcription produces a much higher steady-state concentration of an mRNA with a longer lifespan than of an mRNA that is degraded rapidly once it reaches the cytoplasm. The stability of an mRNA also determines how rapidly the synthesis of the encoded protein can be shut down. For a stable mRNA, synthesis of the encoded RNA persists long after transcription of the gene is repressed. Most bacterial mRNAs are unstable and they decay exponentially with a half-life, typically of a few minutes. This allows the cell to rapidly readjust the synthesis of proteins to accommodate changes in the cellular environment. In contrast, most mRNAs of higher eukaryotes have half-lives of many hours. However, some proteins in eukaryotic cells are required in bursts, for example, hormones called lymphokines and transcription factors. The mRNAs encoding these proteins have unusually short half-lives, and their expression is controlled by rapidly turning gene transcription on and off. Many of these short-lived mRNAs contain multiple copies of the sequence AUUA, often overlapping, in their 3' untranslated region. The mechanism by which these sequences destabilise mRNAs is not yet understood.

Mechanisms of regulation of stability

Stability of RNA is regulated by a number of mechanisms (**Fig. 15.21**). In many cases, sequences associated with RNA stability are located in the 3' untranslated region of the mRNA, downstream of the stop codon that terminates production of the protein. For example, the cell-cycle dependent regulation of histone H3 mRNA stability is controlled by a 30-nt sequence at the extreme 3' end of the molecule. Similarly, the sequence responsible for the destabilisation of the mRNA encoding the transferrin receptor in response to the presence of iron can be traced to the 60-nt long 3' untranslated region (**Fig. 15.22**). Both of these sequences have the potential to form stem-loop structures by intramolecular base pairing, suggesting that changes in stability might be brought about by alterations in the folding of this region of RNA in response to a specific signal. The sequences involved in the regulated degradation of specific mRNA species located to the 3' untranslated region also play important role in the RNA stability. Such sequences may act either by promoting endonucleolytic cleavage within the RNA transcript or by promoting loss of the poly A tail, opening of the RNA to exonucleolytic attack via its free 3'- end.

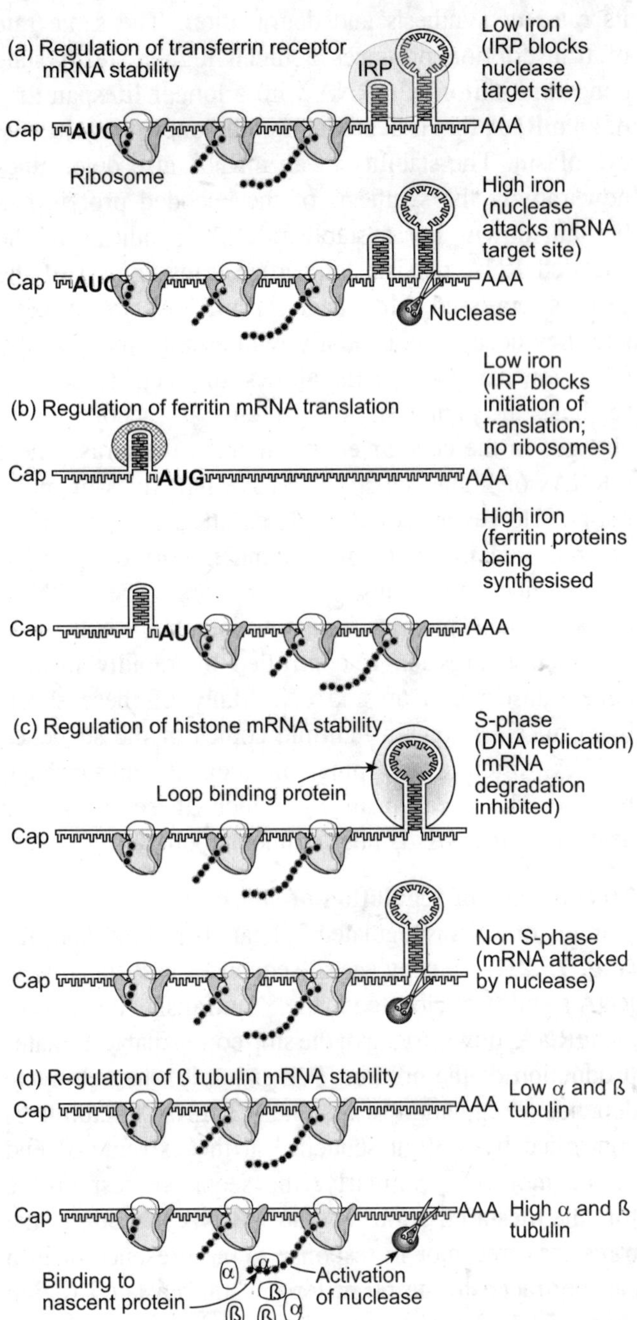

Fig. 15.21 Examples for the regulation of stability of mRNA

15.16.2 Control of eukaryotic translation by IRE-BP

The best-known example of translational control in eukaryotes mediated by a specific RNA-binding protein involves the mRNAs encoding several proteins that function in iron metabolism. IRE-BP, the protein that regulates degradation of TfR mRNA, also regulates the translation of the mRNAs encoding the two chains of ferritin and the enzyme **5-aminolevulinate (ALA)** synthase in erythrocytes (**Fig. 15.23**). Ferritin is an intracellular protein that binds iron ions, thereby preventing accumulation of toxic levels of free ions. When intracellular stores of iron are low, translation of ferritin mRNAs is repressed, so that the iron transported into the cell by the transferrin receptor is available for iron-requiring enzymes; when iron is in excess, ferritin synthesis is derepressed so that iron ions are bound by newly synthesised ferritin. ALA synthase catalyses the first step in synthesis of haeme, the iron-containing prosthetic group in haemoglobin. When iron concentrations are low, production of ALA synthase is repressed. As a result, iron is shunted away from haeme synthesis and is made available for iron-requiring enzymes essential for cell viability. When iron concentrations are adequate, production of ALA synthase is derepressed, permitting synthesis of haem.

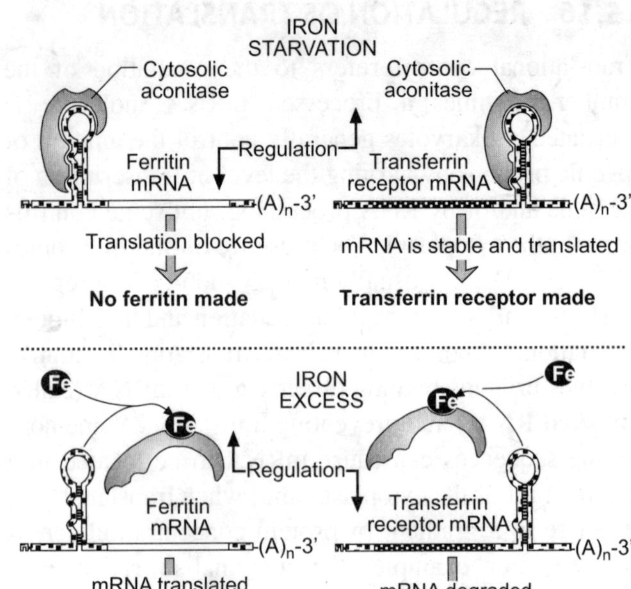

Fig. 15.22 Two post-translational controls mediated by iron; when iron is in excess in the cytosol, a cell increases its synthesis of ferritin in order to bind the extra iron and decreases its synthesis of transferring receptors in order to import less iron across the plasma membrane. When iron is absent in the cytosol, the two genes respond otherwise. (***Source***: B Alberts, A Johnson, J Lewis et al 2002 *Molecular Biology of the Cell*, 4th ed.)

Heat shock proteins

There is also evidence that a number of proteins made by eukaryotes in response to elevated temperatures, called heat shock proteins, can be controlled at both

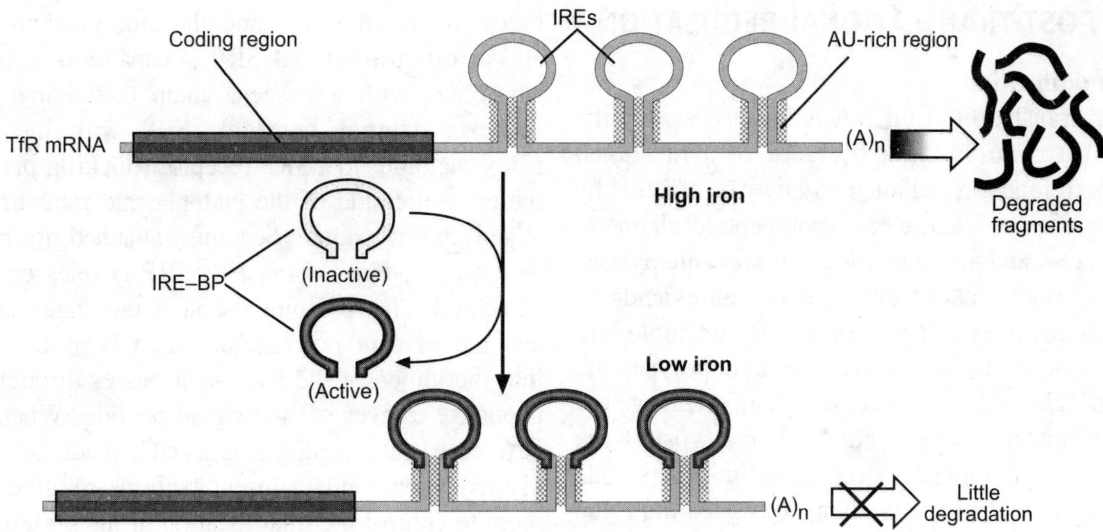

Fig. 15.23 A model for iron-dependent regulation of stability of transferring receptor (Tfr) mRNA; IREs: Iron response elements

transcriptional and translational levels. In *Drosophila*, for example, heat shock protein mRNA is first transcribed on exposure to heat, but is then also preferentially translated compared to other mRNAs.

Polyproteins

In prokaryotes, coordinated regulation of the synthesis of several gene products is accomplished by regulating the synthesis of a single polycistronic mRNA molecule encoding all of the products. The analogue to this arrangement in eukaryotes is the synthesis of a polyprotein. Many mRNAs for hormones in eukaryotes are translated to give a long single polypeptide chain (polyprotein) that is cleaved subsequently by specific proteases to produce multiple mature proteins from one translation product. In this kind of system, the coding sequences of each gene in the polyprotein unit are not separated by stop and start codons but instead by specific amino acid sequences that are recognised as cleavage sites by particular protein-cutting enzymes. Polyproteins with up to eight cleavage sites (e.g., polio virus) have been studied. The cleavage sites are not cut simultaneously, but are cut in a specific order. This type of processing serves to maintain an equal molar ratio of the constituent proteins, as is seen in the production of *Xenopus* rRNA.

Regulation of gene expression by antisense RNA

Evidence for the control of eukaryotic genes by antisense sequences includes the detection of antisense RNA to the dihydrofolate reductase-thymidylate synthase mRNA in the protozoan parasite, *Leishmania*; sequences complementary to the mRNA encoding the P53 protein that is associated with the transformation of murine erythroleukemia cells and the detection of myelin basic protein antisense sequences in mouse. Antisense RNA appears to function by hybridising to the target RNA, thus masking the ribosome binding site and rendering the message functionally inactive (**Fig. 15.24**).

"Sense" construct

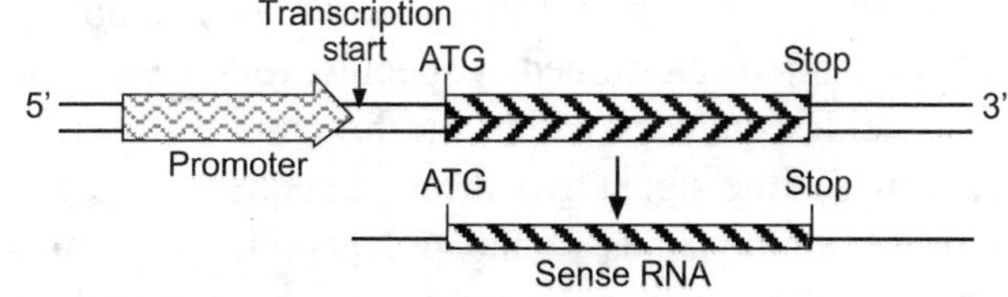

"Antisense" construct

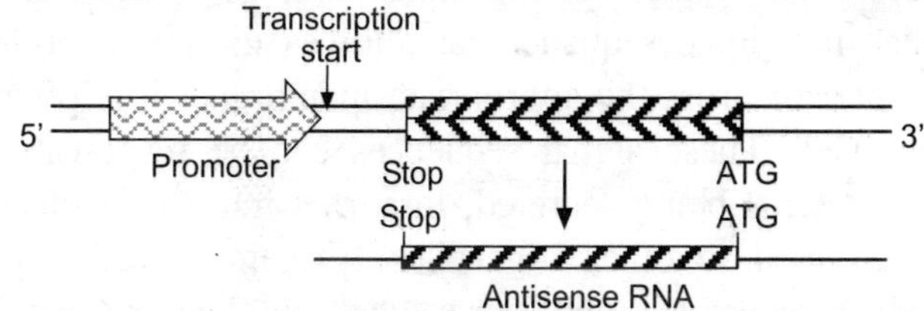

Fig. 15.24 Antisense gene technology; the transcribed RNA sequence is reversed with respect to the promoter so that the other strand of DNA is transcribed. The resulting RNA is, thus, complementary to the normal 'sense' RNA. Consequently, the latter will not be available for translation by ribosomes.

15.17 POST-TRANSLATIONAL REGULATION

Protein modification

Ribosomal translation of mRNA is not necessarily the last step that controls the appearance of a functional protein. Both function and longevity may be affected by proteolytic enzymes that cleave polypeptide chains at specific places, and also by various kinases, methylases and other enzymes that modify specific amino acids on specific polypeptides. Many proteins, for example, are known to be activated, when their precursor polypeptide chains are cleaved by proteases. The cleavage of chymotrypsinogen yields active chymotrypsin; the partial degradation of the preproinsulin polypeptide yields proinsulin, which is then converted into the final active insulin molecule; the fibrinogen precursor molecule is converted into blood-clotting fibrin by the proteolytic action of thrombin. Ubiquitin is made as a polyprotein containing multiple copies linked end-to-end, and this must be cleaved to generate the individual ubiquitin molecules.

Many chemical modifications have been shown to take place on the N and C termini, as well as on most of the 20 amino acid side chains, with the exception of Ala, gly, ILe, Leu, Met and Val. The modifications include acetylation (e.g., histone), hydroxylation (e.g., collagen), phosphorylation (e.g., transcription factor activity), methylation, glycosylation and even addition of nucleotides.

Protein targeting

It has been discovered that the cellular location of proteins is often determined by specific, relatively short, amino acid sequences within the proteins themselves. According to the signal hypothesis, a sequence of about 20 or so amino acids at the polypeptide N-terminal serves to orbit the polypeptide chain to particular membranes either during translation or afterwards. Once the signal sequence has allowed the polypeptide to enter or cross the appropriate membrane, it is often removed. These signal sequences can be responsible for proteins being secreted, imported into the nucleus or targeted to other organelles. Protein secretion in both prokaryotes and eukaryotes involves a signal sequence in the nascent protein and specific proteins or, in the latter, a RNP particle, signal recognition particle (SRP), that recognises it (see **Chapter 13**).

As soon as a cytosolic ribosome begins to translate mRNA encoding a protein that is to be secreted, SRP binds to the ribosome and the emerging polypeptide and arrests translation. SRP is capable of recognising ribosomes with a nascent chain containing a signal sequence (**signal peptide**). SRP with the arrested ribosome binds to a SRP receptor (**docking protein**) on the cytosolic side of the endoplasmic reticulum (ER). When the ribosome becomes attached to ribosome receptor proteins on the ER, SRP is released and can be reused. The ribosome is able to continue translation, and the nascent polypeptide chain is pushed through into the lumen of the ER. As it passes through, signal peptidase cleaves off the signal peptide. When protein is released into the ER it is usually modified, often by glycosylation, and different patterns of glycosylation seem to control the final location of the protein.

Protein degradation

Half-lives of proteins varies depending upon the protein type. For example, regulatory proteins tend to turn over rapidly and cells must be able to dispose off faulty and damaged proteins. In eukaryotes, the N-terminal residue plays a critical role in inherent stability. A protein that is damaged, modified, or has an inherently destabilising N-terminal residue becomes ubiquitinated by the covalent linkage of molecules of the small, highly conserved protein, ubiquitin, via its C-terminal Gly, to lysine residues in the protein. The ubiquitinylated protein is digested by the ATP-dependent 26S protease complex (proteasome) and releases intact ubiquitin for reuse.

15.18 GENE EXPRESSION MECHANISMS AND SPECIALISED CELLS

Regulation of gene expression need not be restricted to switching the genes on or off. It may be related to the developmental stages of multicellular organisms and can even extend to creation of a set of specialised cells. These specialised cells generally maintained through many subsequent generations. This phenomenon of cell memory is essential for the creation of organised tissues and for the maintenance of stably differentiated cell types. In this section, a few well-understood examples are discussed.

Mating type

The common baker's yeast, *Saccharomyces cerevisiae*, is one of the well-studied single cellular eukaryote, because of its ability to differentiate into three distinct cell types. It can exist either in a haploid (a and α cells) or diploid state (a/α cells). Diploid cells form by a process known as **mating**, in which two haploid cells fuse. These haploid cells differ in mating type (sex) and are specialised for mating. Each produces a specific diffusible signalling molecule (mating factors, i.e., express different a-specific and α-specific genes) and a specific cell surface receptor protein. The resulting diploid cells after mating, called a/α, can under stress conditions, form spores giving rise to haploid cells by meiosis.

The mating type of the haploid cell is determined by a single locus, the **mating type (MAT) locus**, which in a-type cell, encodes a single gene regulatory protein, a1 and in an α cell encodes two gene regulatory proteins, MATα1 and Matα2. The a1 protein has no effect in the a-type haploid cell that produces. It will be functional later in the diploid cell. However, the α-type haploid cell produces the proteins specific to its mating by default. In contrast, the α2 protein acts in the α cell as a transcriptional repressor that turns off an 'a-specific' gene, while the α1 protein acts as a transcriptional activator that turns on the α-specific genes. Once the two mating types have fused, the combination of the a1 and α2 regulatory proteins generates a completely new pattern of gene expression. However, they can switch occasionally between a and α cell types by a mechanism of gene rearrangement (**Fig. 15.25**). Since the change involves the removal of one gene from the active 'slot' and its replacement by another, this mechanism is called the **cassette mechanism** (**Fig. 15.25**).

A regulatory gene for organ formation

For example, eye development in *Drosophila*, mice and humans is controlled by a gene regulatory protein

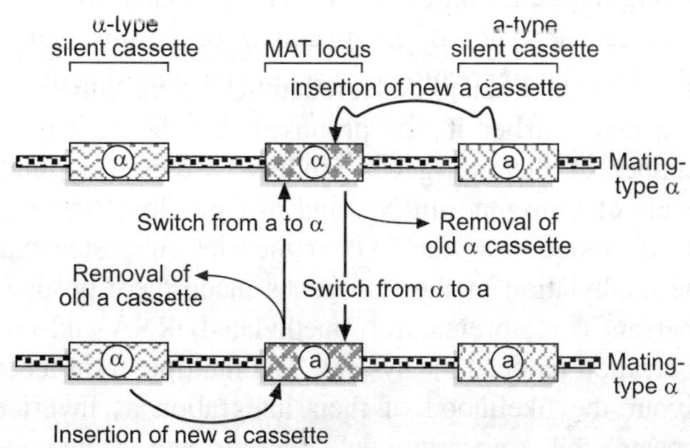

Fig. 15.25 Cassette model of yeast mating type gene switching

(called Ey in flies and Pax-6 in vertebrates). When expressed normally, it can trigger the formation of not just a single cell type but a whole organ (an eye), composed of different types of cells organised in three-dimensional space.

The classic example for this type of regulation comes from experiments in fruit flies in which the *ey* gene is artificially expressed early in development in groups of cells that normally will go on to form leg parts. This abnormal gene expression causes eyes to develop in the middle of the legs (**Fig. 15.26**). The Ey directly controls the expression of many other genes by binding to their regulatory regions. Some of the genes controlled by Ey are themselves gene regulatory proteins that, in turn, control the expression of other genes. In this way, action of just one regulatory protein can turn on a cascade of gene regulatory proteins and cell–cell interaction mechanisms which result in an organised group of many different types of cells.

Other regulatory mechanisms

Besides the above mechanisms, gene inactivations and reactivations caused by the insertion and excision of transposable elements are a significant example of specialised gene expression. Examples are genes encoding proteins involved in pigment biosynthesis such as the chalcone synthase (CHS) gene of petunia and *Antirrhinum*. The phenotype is the well-known 'streaky' flowers, with zones of pigmentation corresponding to excision of the element from an inactivated CHS gene in some cells during floral development. This gives rise to an active CHS gene, and hence pigmentation, in the progeny of these cells produced by cell division. The effect of a transposable element on the expression of the *bronze* gene in maize, giving spots or zones of pigmentation in the seed coat, has also been extensively characterised (see **Chapter 18**).

15.19 REGULATION OF GENE EXPRESSION BY RNA INTERFERENCE

Genomes of complex organisms are particularly potential targets of invasion by viruses and transposable elements (TEs). Thus, organisms have evolved a mechanism to fight off such invasions to prevent the genome from being completely taken over by molecular invaders. The mechanisms evolved to protect the integrity of genomes are based almost similar to those of vertebrate immune system, which is based on two fundamental

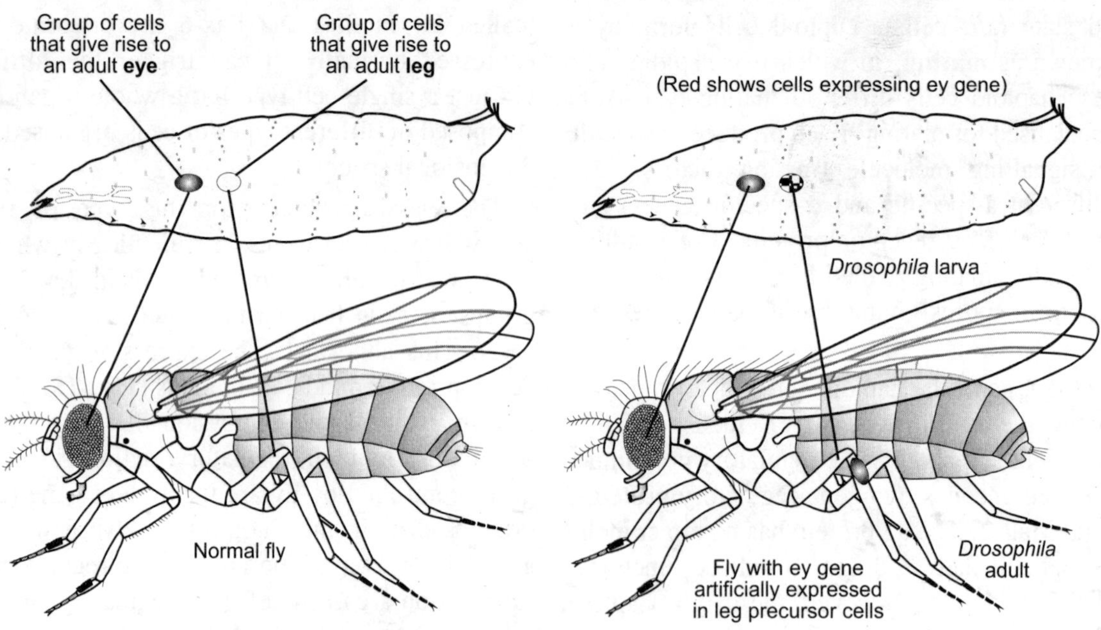

Fig. 15.26 Expression of the *Drosophila ey* gene in precursor cells of the leg triggers the development of an eye in the leg

mechanisms, i.e., how to recognise self from non-self and how to amplify an initial response in a sequence-specific fashion.

Recently, an RNA-based silencing mechanism has emerged, conserved among species from different kingdoms (fungi, animals, and plants) and very likely acts as the 'immune system' of the genome. Gene silencing in plants is a genetic control mechanism implicated in virus resistance, genome maintenance and developmental control. Post-transcriptional gene silencing (PTGS) and co-suppression as well as RNA-mediated virus resistance in plants, RNA interference in animals and silencing in fungi (quelling in *Neurospora*) and algae are all based on the same core mechanism. This conclusion is based on the discovery of common mechanistic elements (such as the **small interfering RNAs**, **siRNAs**) and of homology between genes required for this mechanism in plants, animals and fungi and algae. The precise mechanism of this group of phenomena is now referred to as RNA silencing.

15.19.1 Functions of RNA silencing

An organism will not normally encounter highly concentrated dsRNAs of identical sequence to that of their endogenous genes. This can be clearly seen in plants, which show post-transcriptional gene silencing (PTGS) and virus-induced gene silencing, recognised as mechanisms that protect against frequently occurring viral infections. The advantage of this defence system is that the signal can spread in such a way that inoculation in one area of a leaf can confer immunity on surrounding cells. Protection against viruses and transposons may be the natural function of the core of the RNAi pathway. One of the most striking features of RNAi in *C. elegans* is the systemic effect. Injection of naked dsRNA into one region of the animal may affect gene expression elsewhere, and dsRNA present in the lumen of the gut as part of the food is apparently taken up and affects gene expression in progeny that arises in the gonads. In plants, grafting experiments have shown immunity travelling over 30 cm of stem tissue. This ability may add to the protection in case of repeated infection by a virus.

Many theories have been attempted to explain how sequence-specific degradation system might be activated. Earlier it was proposed that the high copy number of the transgenes produced excessively high levels of transgene mRNA and that this level induced the degradation system. Other theories suggested that the methylation of the transgenes made them produce aberrant (i.e., prematurely methylated) RNA and this aberration induced the system. The multiple transgenes favour the likelihood of their integration as inverted repeats, by transcriptional read-through from one transgene into the other to produce duplex-forming, self-complementary RNA.

Post-transcriptional gene silencing (PTGS)

Homology-based gene silencing was first observed in plants that overexpressed either endogenous or foreign genes. Introduction of transcribed sense transgenes leads to downregulating the expression of homologous endogenous genes, a phenomenon called co-suppression. It results in the degradation of endogenous gene and transgene RNA after transcription. Post-transcriptional RNA degradation can affect a wide range of transgenes expressing plant, bacterial or viral sequences. PTGS can be classified according to the nature of the silencing source, which can be a sense transgene, an antisense transgene, simultaneously expressed sense/antisense transgenes, or viruses. PTGS, like TGS, can occur in *cis* (RNA from the silencing source is degraded), simultaneously in *cis* and *trans* (RNA from both silencing source and all homologous RNA are degraded), or in *trans* (only RNA transcribed from the silencing source and all homologous RNA are degraded), or in *trans* (only RNA homologous to RNA transcribed from the silencing source is degraded).

Methylation and chromatin remodelling in PTGS

The coding sequences of PTGS-inducing transgenes are frequently found to be methylated. PTGS can be established in plants with a mutant methyltransferase (met1), but during growth, the silencing becomes impaired, reactivating the silenced gene in some sectors of the plant. Furthermore, PTGS can fail to establish in mutant plants lacking the chromatin remodelling protein DDM1. These results suggest a role for DNA methylation and/or chromatin structure in both establishment and maintenance of PTGS.

It seems unlikely that the DNA methylation mechanism associated with PTGS and TGS is involved directly in protecting plants against most RNA viruses. The vast majority of these viruses have exclusively cytoplasmic lifecycles and no homologous DNA sequences in plant genomes. It is possible that dsDNA-directed methylation is involved in inhibiting the handful of known plant retroviruses or pararetroviruses during their DNA phases within the nucleus. It is even more likely that the mechanism is primarily for defence against TEs.

Potential applications of RNA interference

Small amounts of dsRNA are able to silence a large surplus of target mRNA in *C. elegans*. This is due to the following reasons:

- The dicer enzyme cuts the long dsRNA molecules into short 'primary' siRNAs (**Fig. 15.27**). Because each siRNA can potentially target a homologous mRNA, this provides a level of amplification that, depending on the length of the dsRNA, could easily measure 10- to 20-fold.
- A catalytic mechanism, in which siRNAs are used multiple times, can provide further amplification.
- Short RNAs may be able to serve as primers on target mRNA and subsequent generation of 'secondary siRNAs' (target-directed amplification) and thus initiate an RNA-directed RNA polymerisation reaction.

Clinical applications of RNA interference

Viral infections and cancer are the most important targets for the development of new types of molecular drugs. Viruses ravage an infected cell by synthesising mRNA that code for viral proteins, which disrupt cellular activities. But most cells that become cancerous contain mutations in certain genes (called oncogenes), causing affected cells to produce mutant mRNAs that are translated into abnormal versions of cellular proteins. In both the situations, a drug that destroys specific mRNAs, ignoring other mRNA molecules, would be most effective. RNA interference is a significant phenomenon in this respect. Mammalian cells can be subjected to RNAi-a process that leads to the degradation of specific mRNA. When cells are incubated with a 21–23 nt synthetic siRNA, the molecules are taken up by the cells, incorporated into a mRNA-cleaving ribonucleoprotein complex, which attacks the complementary mRNA. A wide variety of cultured cells carrying disease-causing genes have been subjected to RNA interference technology.

A type of leukemia, for example, is caused by a gene called the fusion form of *bcr-abl*. A siRNA against the mRNA produced by the *bcr-abl* fusion gene has proven capable of converting the cells to a normal phenotype in culture. In the case of AIDS, it is hoped that stem cells can be isolated from a patient, transfected with a vector carrying a siRNA that targets a viral mRNA, and then reinfused back into the patients' bloodstream. Such cells have the potential to produce the siRNA more or less continuously, causing the cell and its descendants to be resistant to destruction by the virus.

One of the methods to deliver the oligonucleotide-based drug agents to tissue targets was described by Song and colleagues in 2005. It consists of antibody-

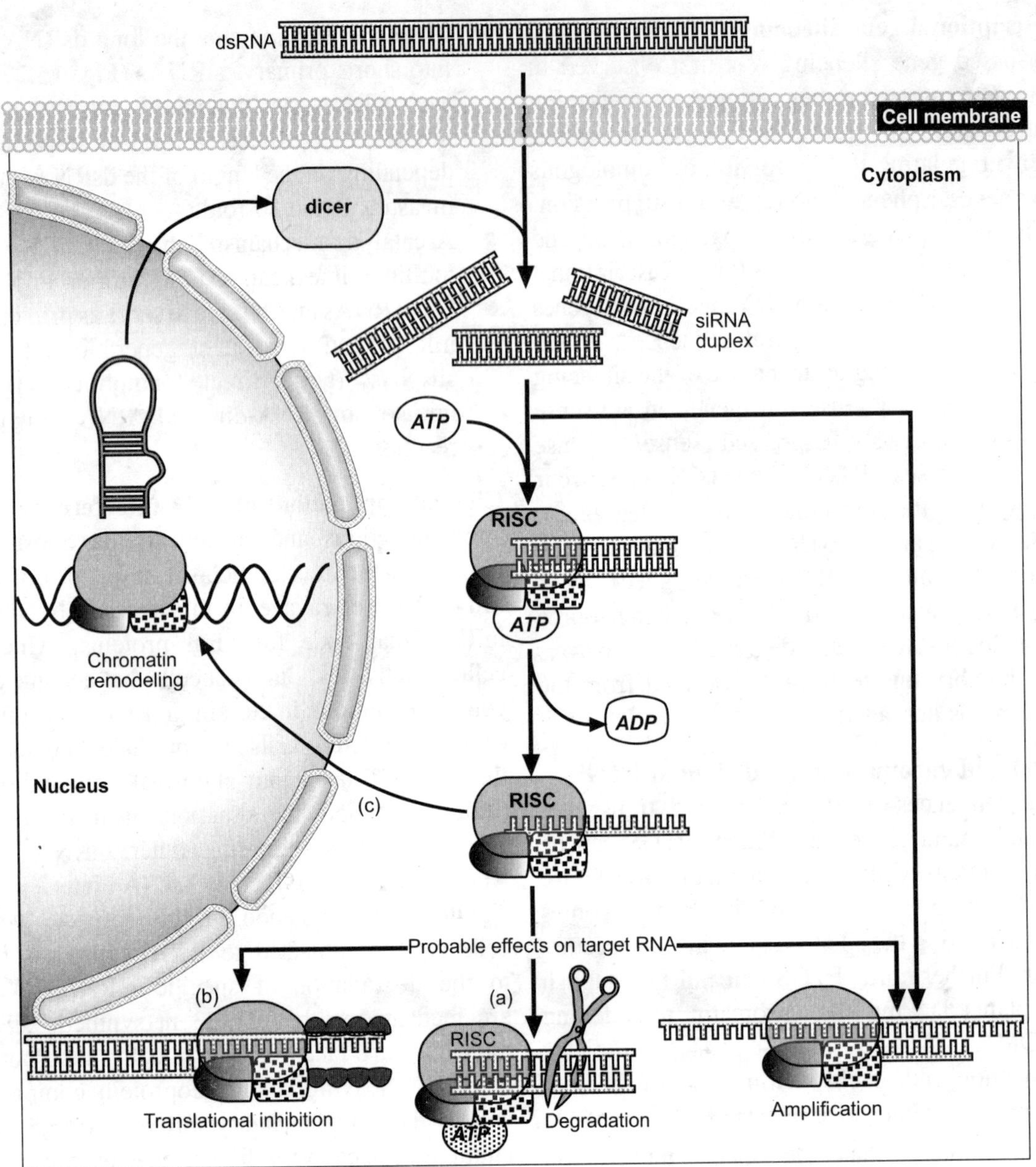

Fig. 15.27 RNA*i* silence the expression of a given gene when dsRNA with homology to that gene are introduced or made in that cell. This involves processing of the dsRNA to make short interfering RNAs (siRNAs) by the enzyme Dicer. These siRNAs then direct a complex RISC (RNA-induced silencing complex) to repress genes in at least three ways. (a) It attacks and digests mRNA with homology to the siRNA (b) It interferes with translation of those mRNAs and (c) It directs chromatin modifying enzymes to the promoters that expression of those mRNAs.

mediated *in vivo* delivery of siRNAs via cell surface receptors (**Fig. 15.28**). Song and coworkers used the information already known, i.e., coupling protamine (a protein that binds DNA) to the Fab protein of an antibody enables the complex to deliver small pieces of DNA to a cell expressing the appropriate antigen on its surface. In this case, they substituted siRNA (targeted to the HIV-1 gag protein) for DNA and fused protamine to a Fab directed against the HIV-1 envelope protein. When the mixture was added to the target cells, the complex honed in on cells expressing the HIV-1 envelope antigen, which caused suppression of viral replication in infected primary T-cells.

Limitations of RNAi technique

The unique feature of RNAi technology is its extraordinary sequence specificity, which can be both advantageous and otherwise.

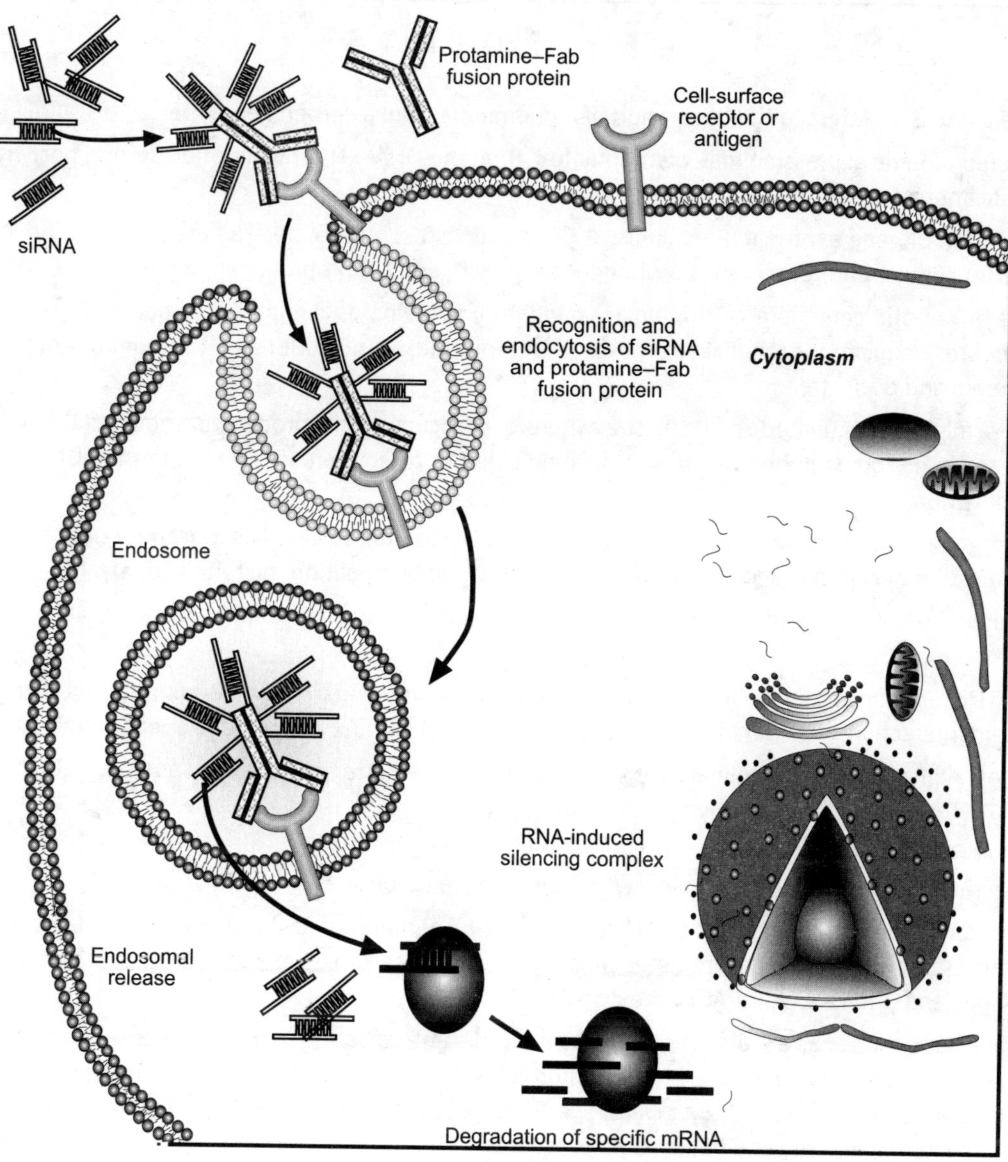

Fig. 15.28 Cell-specific targeting of silencing RNA specifically against HIV1-infected melanoma cells

- Since viruses mutate rapidly, their genome sequences also change enormously, which leads to the production of mRNAs that are no longer fully complementary to the therapeutic siRNA.
- Delivering RNAi (or the vectors that encode them) to target cells is another major drawback with this technology.

KEY POINTS

- Gene expression refers to the generation of a phenotype from a certain genotype.
- Any step of gene expression may be modulated, from the DNA–RNA transcription step to post-translational modification of a protein.
- Regulation of gene expression facilitates cells to exercise control over structure and function. It forms the basis for cellular differentiation, morphogenesis, growth and adaptation of an organism.
- More eukaryotic genes are controlled by regulatory proteins as compared to bacterial genes. In higher eukaryotes, regulatory sequences can extend to thousands of nucleotides away from the promoter—both upstream and downstream.
- Genes might be activated in the tissue where a particular gene product is required through selective increase in its copy number, via an amplification event or by some rearrangement of the DNA necessary for its activation.
- Three general ways in which chromatin structure can be altered are: Nucleosome remodelling, covalent modification of histones and replacement of core histone by one more histone variants.
- It is postulated that the attachment of the supercoiled DNA loops to the matrix plays a role in the control of gene expression.
- Many DNA-binding proteins bend the DNA to which they bind and that the bending causes changes in other protein interactions. The bending can serve as a switch for turning on and off a specific biological activity.
- In most higher plants and animals, DNAs are often modified after synthesis by an enzymatic conversion of many cytosine bases to 5-methylcytosine bases.
- Acetylation of the arms of histones is an important process in transcription and has been correlated with transcriptionally active chromatin in a wide variety of cell types.
- Promoter sequences in eukaryotic DNA are used for recognition of RNA polymerase , binding and initiation of transcription. The proximal promoter consists of up to three known regions: the start of transcription (TCS); TATA box and often a CAAT or AGGA box.
- Many eukaryotic genes are also regulated by more distant *cis*-acting regulatory elements called enhancers. A silencer is generally considered to be a short, specific sequence of nucleotides, which are located in the 5′ upstream promoter region of a given gene. These sequences are capable of recruiting TF proteins to the promoter, which in turn carry out specific functions.
- Insulator elements (insulators) are DNA sequences that bind specialised proteins and have two specific properties; they buffer genes from the repressing effects of heterochromatin and they can block the action of enhancers.
- Among the known regulators of eukaryotic genes, hormones are the most extensively studied transcription activation factors.
- Proteins interacting with nucleic acids are considered particularly important because they control some of the most fundamental biological processes, including replication, recombination, DNA repair, transcription, RNA processing and protein synthesis.
- The post-transcriptional modifications enable control over the number, kind and stability of mRNA molecules that enter the cytoplasm of particular tissues for translation.
- In mutually exclusive alternative splicing, one of a pair of exons is always spliced into an mRNA but never both.
- Regulation of gene expression by RNA editing result in sequence change in mRNA.
- The final stage in the expression of a gene is the translation of its messenger RNA into protein.

- A number of proteins made by eukaryotes in response to elevated temperatures, heat shock proteins, are controlled at both transcriptional and translational levels.
- Antisense RNA function by hybridising to the target RNA, thus masking the ribosome binding site and rendering the message functionally inactive.
- Post-transcriptional regulation can be achieved by very small RNA molecules of between 19 and 25 bases in length, which also interact with specific mRNAs to inhibit their activity called RNA interference (RNAi).

Self-assessment Questions

1. Explain with the help of a diagram the principal stages of regulation of gene expression.
2. How is gene expression regulated by genome rearrangements?
3. Explain the different types of regulation by genome uncoating.
4. Describe the topological changes in genome as a mechanism of regulation.
5. Describe how histone modification is an important mechanism of regulation.
6. Which features of gene organisation are involved in regulation?
7. Explain the role of regulatory sequences in regulation.
8. Give a brief account of the mechanism of negative regulation.
9. How do repressors regulate gene expression?
10. Explain how transcription factors play an important role in regulation.
11. Describe briefly the different mechanisms of post-transcriptional gene silencing.
12. Explain the mechanism of mating type regulation.
13. How does RNAi regulate gene expression?
14. Explain in detail the different mechanisms of post-translational regulation.

FURTHER READING

Textbooks

Carey M and ST Smale (2000). *Transcriptional Regulation in Eukaryotes: Concepts, Strategies and Techniques.* New York: Cold Spring Harbor Laboratory Press.

Hannon GJ (2003). RNAi: A Guide to Gene Silencing. New York: Cold Spring Harbor Laboratory Press.

Hartl DL and EW Jones (1998). *Genetics: Principles and Analysis.* Jones and Bartlett Publishers.

Latchman DS (2005). *Gene Regulation*, 5th ed. Taylor and Francis Group.

Krebs JE, ES Goldstein and ST Kilpatrick (2011). Lewin's *Genes X.* Jones and Bartlett Publishers.

Twyman RM (1998). *Advanced Molecular Biology.* New Delhi: Viva Books Private Limited.

References

Bell AC, AG West and G Felsenfeld (2001). Insulators and boundaries: Versatile regulatory elements in the eukaryotic genome. *Science* 291:447–50.

Berger SL (2002). Histone modifications in transcriptional regulation. *Curr. Opin. Genet. Dev.* 12:142–48.

Bird AP and AP Wolffe (1999). Methylation-induced repression: Belts, braces and chromatin. *Cell* 99:451.

Butler JE and JT Kadonaga (2002). The RNA polymerase II core promoter: A key component in the regulation of gene expression. *Gene Dev.* 16:2583–92.

Brivanlou AH and JE Darnell Jr (2002). Signal transduction and the control of gene expression. *Science* 295:813–18.

Courey AJ and S Jia (2001). Transcriptional repression: The long and the short of it. *Gene Dev.* 15:2786–96.

Cowell IG (1994). Repression versus activation in the control of gene transcription. *Trends Biochem Sci.* 1:38.

Englesberg E and G Wilcox (1974). Regulation: Positive control. *Ann. Rev. Genet.* 8:219–42.

Gehring WJ, M Affolter and T Burglin (1994). Homeodomain proteins. *Annu. Rev. Biochem.* 63:487–526.

Grunstein M (1990). Histone function in transcription. *Ann. Rev. Cell Biol.* 6:643–78.

Hannon GJ (2002). RNA interference: *Nature* 418: 244–51.

Harrison SC (1991). A structural taxonomy of DNA-binding domains. *Nature* 353:715–16.

Hill CS and R Treisman (1995). Transcriptional regulation by extracellular signals: Mechanism and specificity. *Cell* 80: 199–11.

Kim DK, Y Yamaguchi, T Wada and H Handa (2001). The regulation of elongation by eukaryotic RNA polymerase II: A recent view. *Mol. Cells* 11:267–74.

Malik S and RG Roeder (2000). Transcriptional regulation through mediator-like coactivators in yeast and metazoan cells. *Trends Biochem. Sci.* 25:277–83.

Ogbourne S and TM Antalis (1998). Transcriptional control and the role of silencers in transcriptional regulation in eukaryotes. *Biochem. J.* 331:1–16.

Pabo CT and RT Sauer (1992). Transcription factors: structural families and principles of DNA recognition. *Ann. Rev. Biochem.* 61:1053–95.

Plasterk RHA (2002). RNA silencing: The genome's immune system. *Science* 296:1263–69.

Ptashne M (1988). How eukaryotic transcriptional activators work. *Nature* 335:683–89.

Riechmann JL, J Heard, G Martin, L Reuber, CZ Jiang et al (2002). *Arabidopsis* transcription factors: genome-wide comparative analysis among eukaryotes. *Science* 290:2105–10.

Singh KB (1998). Transcriptional regulation in plants: The importance of combinatorial control *Plant Physiol.* 118: 1111–20.

Song E, P Zhu, SK Lee SK et al (2005). Antibody-mediated *in vivo* delivery of small interfering RNAs via cell surface receptors. *Nat. Biotechnol.* 23:709–17.

Sterner DE and SL Berger (2000). Acetylation of histones and transcription-related factors. *Microbiol .Mol. Biol. Rev.* 64:435.

Waterhouse PM, MB Wang and T Lough (2001). Gene silencing as an adaptive defense against viruses. *Nature* 411:834–42.

Wilusz CJ, M Wormington and SW Peltz (2001). The cap-to-tail guide to mRNA turnover. *Nat. Rev. Mol. Cell Biol.* 2:237–46.

Zamore PD (2002). Ancient pathways programmed by small RNAs. *Science* 296:1265–69.

CHAPTER 16

Epigenetic Regulation

LEARNING OBJECTIVES

After reading this chapter, you will be able to:

- Understand the mechanism of epigenetics
- Appreciate the biological significance of epigenetics
- Describe the mechanism of extrasomatic evolution and apogenetics

16.1 INTRODUCTION

Traditional genetics explains phenotypic traits in an organism through the presence or absence of, or size and position of a specific nucleotide sequence in its genomic DNA. The nucleotide changes in genes have been used in characterising the biological features in organisms. However, the primary DNA sequences alone do not adequately explain the expression or otherwise of all genes. In fact, transcription of some genes is controlled by non-genetic mechanisms. Gene expression can be influenced by changes in the genome that do not involve mutations in the DNA sequence. These heritable yet reversible changes are said to be epigenetic. Thus, **epigenetics** can be defined as the study of heritable changes in gene expression that occur without a change in the DNA sequence. Control mechanisms such as DNA methylation and chromatin modifications, produce 'on/off' states of gene expression. Epigenetic changes may be induced spontaneously, in response to environmental factors, or in response to the presence of a particular allele, even if absent from subsequent generations. Breakdown of such control mechanisms is associated with a variety of diseases related to behavioural or neurological manifestations and disorders of tissue growth, including cancer. Current interests are focussed on understanding heterosis in plants, developmental regulation, inter-individual variations in epigenetic markings in human populations and interaction between epigenetics and environment. These would help in developing approaches to correct adverse epigenetic states.

There are three dimensions to the inheritance patterns in organisms. The first dimension is the DNA-based genetic systems for somatic evolution and the second dimension is the epigenetic inheritance for both somatic and extrasomatic evolution. The third dimension which relates mostly to extrasomatic evolution is that of non-genetic inheritance. All the three evolutionary processes are highly interdependent and have contributed enormously in the evolution of most life forms, including human beings. In this chapter, the epigenetic (second dimension) and the apogenetic (third dimension) inheritance systems will be discussed.

16.2 EPIGENETICS

Holliday (2002) suggested three definitions for epigenetics:

- Nuclear inheritance that is not based on differences in DNA sequence
- The transfer of stable and heritable trait to progeny cells that is not encoded in nucleic acid sequences
- The inheritance of gene expression patterns, in the absence of either mutation or the initiating signal.

Therefore, epigenetic inheritance refers to a set of reversible heritable changes in gene function or other cell phenotype that occur without a change in genotype. The term **epigenotype** means the different patterns of gene expression that give rise to the phenotypes of all the specialised cells in the organism even though their somatic cells have the same genotype. In contrast to classical gene regulation by transcription factors, which regulate the production of RNA over a broad range, epigenetic modifications produce more or less binary on/off transcriptional states. Epigenetic regulation involves two steps:

- Establishment of epigenetic markings on the DNA and chromatin
- Maintenance of these marks during subsequent development and tissue growth.

Cellular differentiation processes rely almost entirely on epigenetic rather than genetic inheritance from one generation to the next. If this were not so, then somatic cell cloning would be impossible and a normal organism could not be recovered from a differentiated cell nucleus. In fact, a somatic cell can be reprogrammed to become totipotent. Therefore, epigenetics enhances our understanding of the inheritance patterns in embryonic morphogenesis, regeneration, turnover of cells, tumours, cell culture and replication of single-celled organisms. Epigenetic inheritance might be maintained even through the production of germ cells (meiosis) and therefore, move from one generation to the next.

Nuclear and cytoplasmic interactions are fundamental to early differentiation. As cells move to specific sites during embryonic development, neighbouring cells influence the cytoplasmic composition. During development there are many branching processes, in which a given cell gives rise to two independent cell types. Such a process can be regarded as a developmental switch in which one daughter gains a new phenotype, or potential phenotype, whereas the other retains its original phenotype. This is a stem cell situation, but in other cases both daughter cells may have phenotypes different from each other and also different from the parent cell. There is also the stability of differentiated cells, so that lymphocytes, fibroblasts and keratinocytes retain a stable karyotype through many divisions. These cells contain the same chromosomes, the same genes, and the same DNA sequences, but their patterns of gene expression are different. Epigenetic inheritance explains the ability of different states, which may have different phenotypic consequences, to be inherited without any change in the sequence of DNA.

16.2.1 Epigenetic markings

Epigenetic mechanisms involve physical and biochemical alterations that are outside the primary DNA sequence, but affect the ability of the genes to be expressed. A number of molecular markings on both nucleic acids and proteins have been well characterised. The important epigenetic markings are DNA methylation, histone modification and non-coding RNA. The two important types of epigenetic markings that have been thoroughly studied are: DNA methylation and histone modifications.

DNA methylation

Many studies have established that methylation is layered on to the primary nucleotide sequence to produce on or off states of gene expression. Genomic imprinting must involve a marking process. A specific gene or chromosome must be marked dissimilarly during spermatogenesis versus oogenesis. Immediately after fertilisation, this differential marking affects the expression of specific genes. In mammals, DNA methylation is a common modification of making imprinted genes. DNA methylation involves the attachment of a methyl group ($-CH_3$) to the bases in DNA. The most common forms of methylated DNA are shown in **Fig. 16.1**. Imprinted genes methylate differentially during gametogenesis. In this example, both the offspring inherit an imprinted allele from their mother and a non-imprinted allele from their father. This pattern of imprinting is maintained in the somatic cells of both offspring. However, when the male offspring makes gametes, the imprinting is erased during early spermatogenesis so that the male can pass on both active alleles to its offspring. In the female also, the imprint is erased during early oogenesis, but then *de novo* methylation occurs in both genes. Therefore, the female offspring transmits inactive alleles of this gene to her offspring.

Most DNA methylation observed in eukaryotic genomes is constitutive and maintenance methylation is sufficient for its propagation. However, global changes

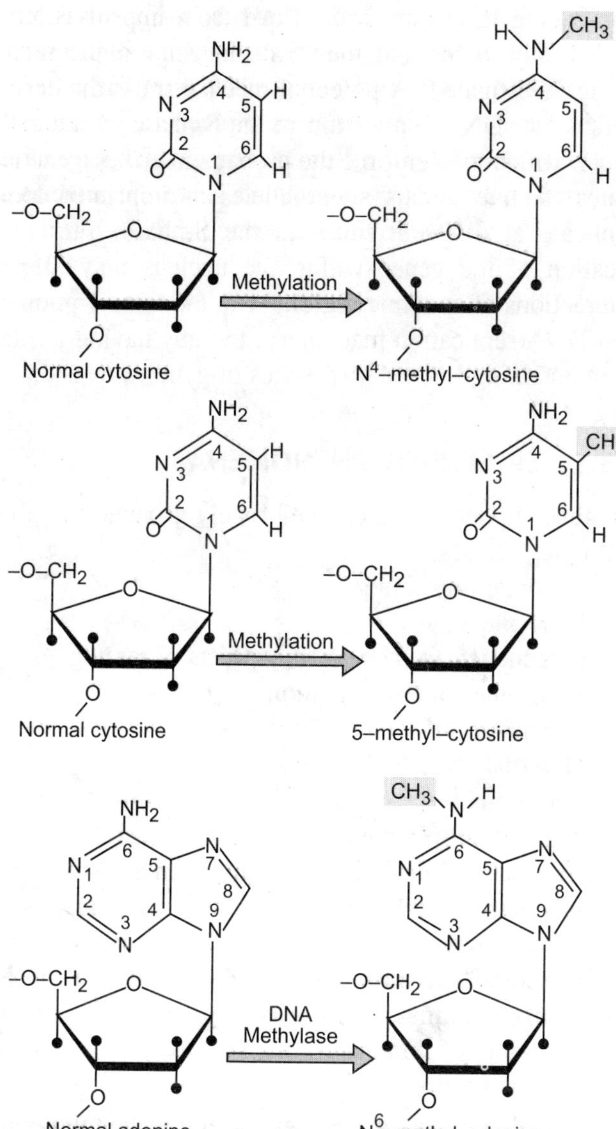

Fig. 16.1 Classification of DNA methylases; C5-methylcytosine and N6 methyladenine are two common forms of methylated DNA bases

in methylation states have been observed during both animal and plant development, which predicts the existence of *de novo* methylases and demethylases that are yet to be characterised. Moreover, additional methylated cytosines occur at asymmetric sites and are implicated in the control of DNA replication.

16.2.2 Methylation and gene regulation in mammals

Now it is clear that genes could be deactivated not only by mutation, but also DNA methylation. This was demonstrated by the use of the potent demethylating agent 5-azacytidine, which re-activated at high frequency the genes silenced by DNA methylation.

Many lines of evidence suggest a link between DNA methylation and transcription, specifically between undermethylation or hypomethylation (lack of methylation at certain sites) and transcriptional activity (**Fig. 16.2**). Evidence arises from the manipulation of **CpG islands** (non-methylated, CG-rich regions, 1–2 kbp in size predominantly associated with the 5' ends of some genes) and abolition of transgene expression where the transgene is methylated before introduction into the cell.

Methylation can change DNA either to activated or silenced form. DNA methylation frequently occurs in repeated sequences, and may help to suppress 'junk DNA'. It could directly interfere with the binding of transcription factors with cytosine in their recognition site. Alternatively, it could recruit proteins which specifically recognise methylated DNA and hence block or displace transcription factors; two methyl-CpG-binding proteins of this nature have been isolated. Finally, it could regulate chromatin structure by influencing either the positioning of the nucleosome or interactions with other chromatin proteins.

The widespread methylation observed in higher eukaryotic genomes suggests that it may reduce 'genetic noise'. A recent observation is that most methylated cytidine residues in mammalian genomes lie within transposable elements. Two classes of retroelements, LINEs and SINEs, are particularly abundant in mammalian DNA and comprise much of the middle repetitive DNA of the mammalian genome.

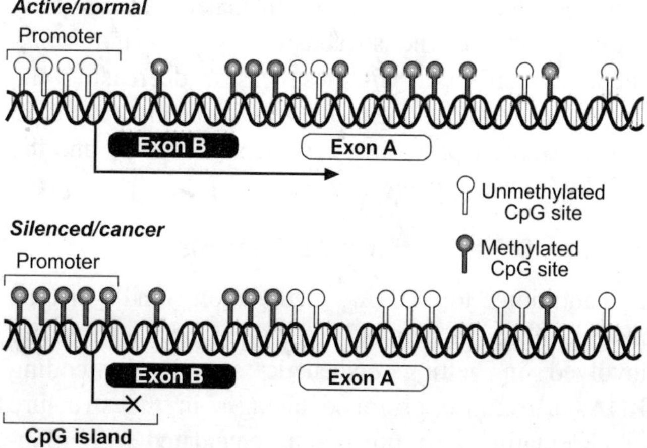

Fig. 16.2 DNA methylation by DNA methyltransferases at CpG islands

It is well established that DNA methylation represses the transposition of plant and bacterial transposable elements; so it is likely to be used in the same way in vertebrates. To establish this, recombinant retroviruses introduced into the mouse genome are often silenced by hypermethylation. Interestingly, the incidence of mutations and chromosome rearrangements caused by mobile elements is low in mammals but high in *Drosophila*, where DNA is unmethylated.

Global regulation of methylation patterns

In mammals, the pattern of methylation characteristic of the adult is established just prior to gastrulation. The level of methylation in the zygote is high, but global demethylation occurs during cleavage, and subsequent *de novo* methylation occurs only in the epiblast lineage, which gives rise to the somatic tissues of the adult. The extra-embryonic lineage, which includes the extra-embryonic membranes and germ cell primordia, remains undermethylated until gametogenesis, when the germ cells are also methylated *de novo*.

16.2.3 Methylation and gene regulation in plants

Much information about DNA methylation and the epigenetic control of gene activity is now available in plants. The integration of extra copies of an endogenous gene or multiple copies of a transgene into a plant genome often results in epigenetic silencing of both endogenous and exogenous genes. The more the copies of transgene present at a given locus, the greater the chances that expression from those *cis*-linked transgenes is silenced and greater their ability to act in *trans* to silence related transgenes elsewhere in the genome. Endogenous loci with repeated gene sequences in maize, *Arabidopsis*, petunia, tobacco and soyabean show *cis*- and *trans*-silencing behaviour. In these cases, decreasing the transgene or the endogenous gene typically results in re-activation of previously *cis*-silenced regions and the inability to cause *trans*-silencing.

16.2.4 Other marking mechanisms

In addition to DNA methylation and histone modifications, a number of other mechanisms are involved in setting epigenetic states. Non-coding RNAs can cause profound changes in the structure of chromatin. This function is elucidated by a non-translated RNA called XIST that is essential for deactivating the X chromosome. A variety of other non-coding RNAs are transcribed from imprinted loci on the autosomes and their role in gene silencing is being investigated. A potential mechanism is the early versus late DNA replication in the S-phase of the cell cycle, which may enforce the propagation of epigenetic states. Active versus silent alleles at imprinted loci replicate at different times in the S-phase. Further, location of the genes within the nucleus may affect interactions of genomic regions with the transcriptional and DNA replication machinery, thereby having a role in maintaining 'on' or 'off' states of gene expression.

16.3 EPIGENETIC PHENOMENA

Epigenetic phenomena can be broadly grouped into the following classes:

- Genomic imprinting
- Gene silencing
- Epigenetic carcinogens and effects of teratogens
- X-chromosome inactivation
- Paramutation
- Maternal effect
- Centromere
- Alternative splicing
- Histone modifications by methylation and acetylation
- Prion diseases
- Structural inheritance
- Reprogramming
- Assembly of immunoglobulin genes
- Regulatory proteins

We shall discuss each of these briefly in the following sections.

16.3.1 Genomic imprinting

The earlier notion that the genotype could be predicted by its alleles regardless of the contributing parent has been undermined by the emergence of a group of genes whose phenotypic effects are determined by the parent that donated the particular allele. This phenomenon is called genomic imprinting (or **molecular imprinting** or **parental imprinting**). It means that the expression of certain alleles differs, depending on the gamete from which it originates. Therefore, a small subset of genes is expressed according to the parent of origin. That is genes (~100) are expressed from only one parental allele, maternal or paternal, depending on the gene and such monoallelically expressed genes are said to be **'imprinted'**. Imprinted genes expressed from maternally inherited chromosomes are silenced on the paternal chromosome; while other imprinted genes show the opposite expression pattern and are only expressed from a paternally inherited chromosome. Interestingly, 'imprints' can act as a silencer or an activator for imprinted genes.

Parental imprinting

In somatic cells, most genes are expressed on both maternal and parental chromosomes, but imprinted genes are expressed in a parent chromosome-specific manner. Conventionally, an allele is said to be imprinted if it is repressed (e.g., at maternally imprinted loci, only the paternal allele is expressed), but this is only an arbitrary definition. At the molecular level, the imprint is DNA methylation, but in some loci, methylation corresponds to gene activity and in others it corresponds to inactivity. The consequences of parental imprinting in a genetic cross are shown in **Fig. 16.3**. Although both parents contribute equally in terms of chromosomes, the locus is functionally hemizygous and the results of reciprocal crosses are therefore non-equivalent. The first evidence for parental imprinting was the failure of parthenogenetic development in mammals. Nuclear transplantation experiments then showed that both parental genomes are required for normal development. Uniparental mouse embryos, i.e., those from zygotes containing either two male nuclei (androgenous) or two female nuclei (gynogenous), are abnormal and die before birth. Gynogenous embryos undergo normal early development but the extra-embryonic membranes do not develop. Androgenous embryos are growth-retarded although the membranes develop normally. Such aberrant concepts are termed as **complete hydatidiform moles**; **partial moles** contain a triploid foetus with two paternal and one maternal genome.

The role of imprinting

Although parthenogenetic mice fail to develop, chimeric mouse and normal cells are viable. The distribution of parthenogenetic cells in chimeric mouse embryos suggests that differentially imprinted cells play an important role in the development of the brain, e.g., parthenogenetic cells are excluded from the hypothalamus but are abundant in the forebrain. On the other hand, the **parental conflict model** suggests that imprinting arose as a result of competition between maternal and paternal genomes in polygamous and promiscuous mammals. In promiscuous species, females gestate offspring from multiple fathers who would try to maximise the ability of their offspring to survive their half-siblings. Thus, paternal alleles would be selected for increased growth, whereas maternal alleles would be selected for reduced growth.

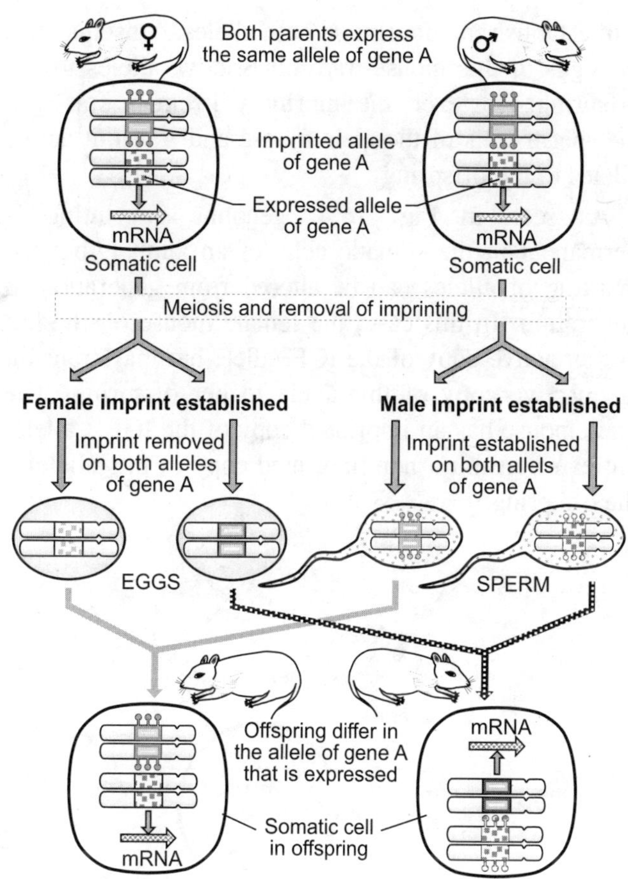

Fig. 16.3 Parental imprinting in the mouse

At the cellular level, imprinting can be divided into three stages:

- Establishment of the imprint during gametogenesis
- Maintenance of the imprint during embryogenesis and in adult somatic cells
- Erasure and re-establishment of the imprint in the germ cells.

The two mice shown in the example have inherited the IGF_2 allele from their father and the IGF_{2m} allele from their mother. Because of imprinting, both mice express the IGF_2 allele in their somatic cells. In other words, the imprint that deactivates the IGF_{2m} allele is maintained as each mouse grows from a fertilised egg into an adult. However, the imprint is maintained only in the somatic cells. In the germ cells (sperm and eggs), the imprint is erased, but it may be re-established depending on the gender. The female mouse on the left will establish the imprint on both alleles; consequently, the eggs of this mouse transmit inactive alleles to her offspring. However, the imprint will remain erased in the sperm cells of the male mouse and transmit active alleles to the offspring.

As seen in **Fig. 16.4,** genomic imprinting is permanent in the somatic cells of an animal, but the marking of alleles can be altered from generation to generation. In this case, the female mouse on the left has an active copy of the IGF_2 allele but may transmit an inactive copy of this allele to her offspring. The male mouse has an imprinted copy of the IGF_{2m} allele, but can transmit a non-imprinted copy of this allele to the offspring.

Diseases related to imprinting

Recent studies have revealed that some epigenetic marks are quite uniform in human populations, while others can vary among individuals. Further characterisation of the nature and extent of such inter-individual epigenetic variation may reveal pervasive influences of epigenetics on human traits and possibly in complex diseases. Now it is known that certain imprinted genes are involved in the etiology of several well-defined genetic disorders. Some human diseases associated with imprinting include ovarian and placental tumours, overgrowth syndrome like Beckwith–Wiedemann syndrome (BWS) and its associated paediatric embryonal tumours, neurodevelopmental disorders like Prader–Willi

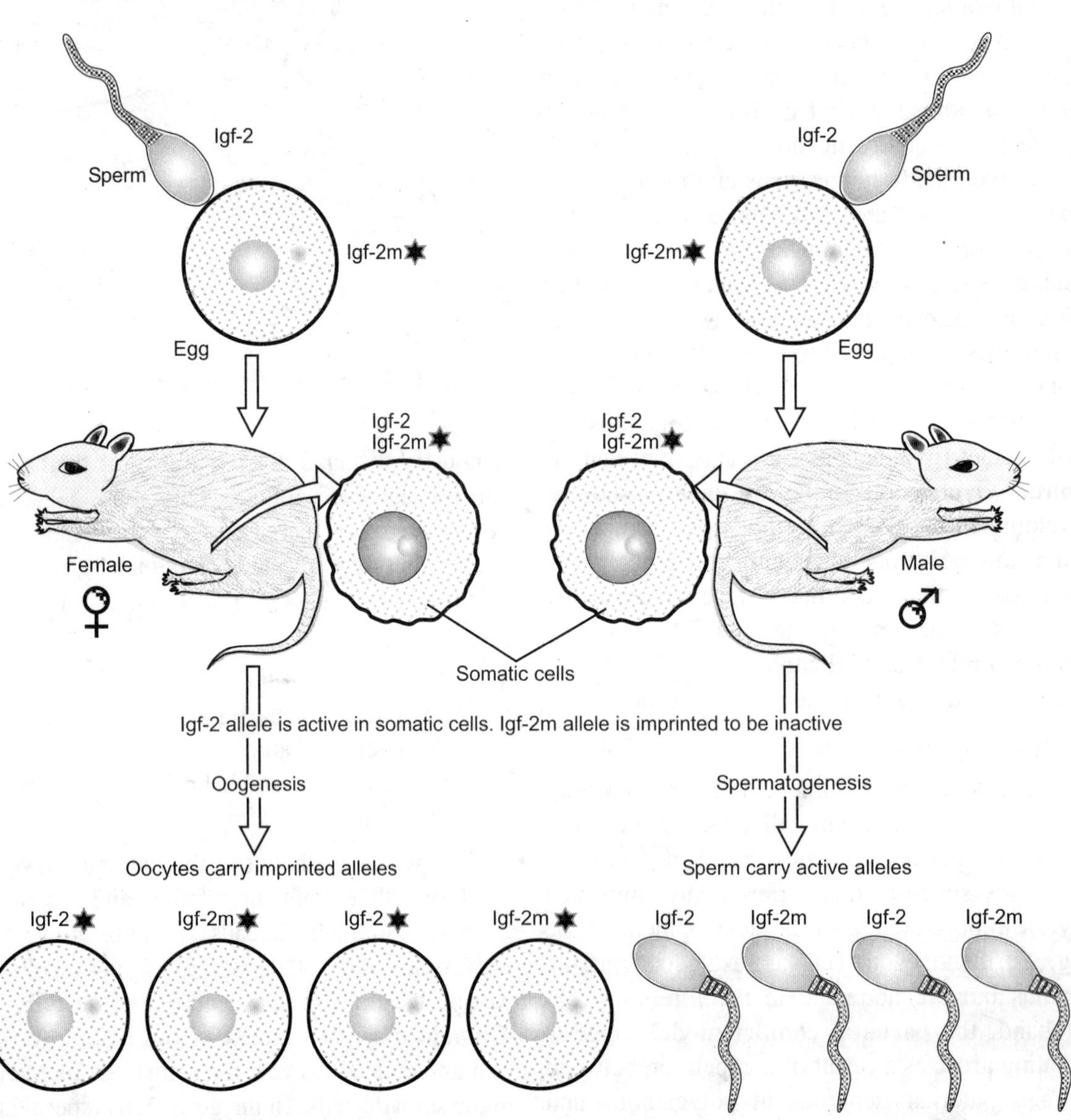

Fig. 16.4 Genomic imprinting during gametogenesis; the example involves a mouse gene IGF_2, which is found in two alleles designated IGF_2 and IGF_{2m}

Table 16.1 Mammalian genes and inherited human diseases associated with imprinted genes

Gene	Allele expressed	Function
WT1	Maternal	Wilms tumour suppressor gene;suppresses uncontrollable cell growth
INS	Paternal	Insulin involved in cell growth and metabolism
Igf2	Paternal	Insulin-like growth factor II; similar to insulin
Igf2r	Maternal	Receptor for insulin-like growth factor II
H19	Maternal	Unknown
Snrpn	Paternal	Splicing factor
Gabrb	Maternal	Neurotransmitter receptor

syndrome (PWS) and Angelman syndrome (AS), and growth retardation disorders like Russell–Silver syndrome (RSS) (**Table 16.1**). These are caused by different mechanisms, including chromosomal, epigenetic or genetic abnormalities. The **syntenic** regions in the human genome are associated with classic imprinting disorders. Prader–Willi syndrome (hypotonia, obesity, hypogonadism) is due to two copies of chromosome 15 being maternally imprinted, and thus lacking in paternal genes; Angelman syndrome (happy puppets) is due to two copies of chromosome 15 being paternally imprinted, and thus lacking the maternal genes. The individual who inherits an imprinting pattern of chromosome 15 from one parent (called uniparnetal disomy) has Prader–Willi or Angelman syndrome, depending on which parent it comes from (**Fig. 16.5**).

16.3.2 Epigenetic gene silencing in cancer

Many genes required for cell proliferation, differentiation and apoptosis have been shown to undergo genetic alterations such as mutation, deletion or amplification in tumours. This classical theory of recessive oncogenesis postulated a mutational mechanism for the inactivation of tumour suppressor (TS) genes. This prediction has been amply confirmed, but an alternative non-mutational pathway for loss of TS gene activity has also been discovered. For some TS genes, the epigenetic pathway is more frequent than the mutational one.

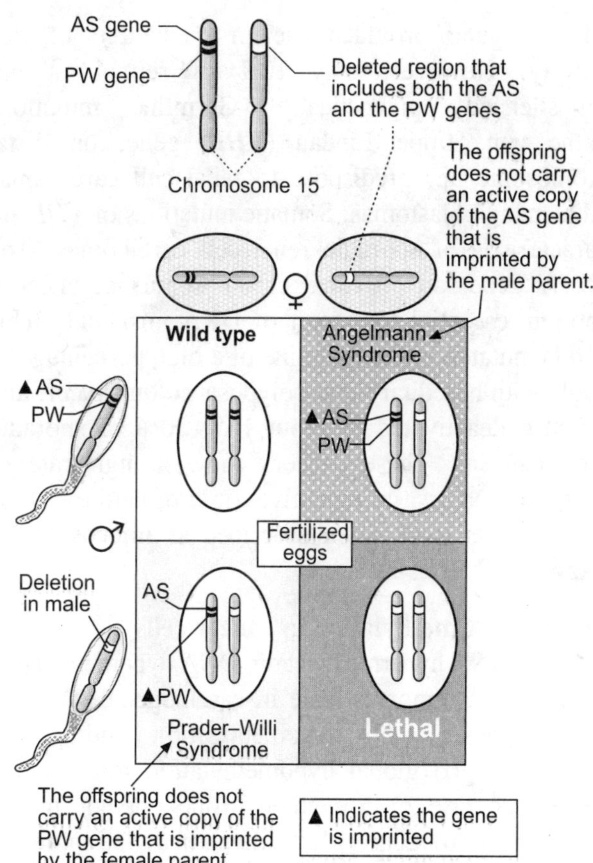

Fig. 16.5 The role and inheritance of imprinting in the development of Angelman syndrome and Prader–Willi syndrome

DNA hypermethylation in cancer cells

The best-studied DNA modification that correlates with epigenetic gene silencing is methylation of cytosine residues in CpG sequences. CpG methylation has recently been linked to a general mechanism of epigenetic silencing, i.e., histone deacetylation. Some potential mechanisms for *de novo* methylation of TS genes in cancer cells include: spreading of DNA methylation from repetitive sequences into promoter-associated CpG islands secondary to loss of transcriptional activator proteins, gain of methylation secondary to hyperexpression of transcriptional repressors, primary hypermethylation due to hyperexpression of methyltransferases and inter-allelic transfer of methylation via gene pairing. In some cancers, environmental pressures that select for a hypermethylating cellular phenotype may also be driving these processes.

The *Rbi* gene is silenced by promoter hypermethylation in some sporadic retinoblastomas, and this gene provided the first example of this pathway. However, only 10% of retinoblastomas are silenced epigenetically. Germline mutations in the von Hippel–Lindau (***VHL***) gene, on distal chromosome 3p, predispose to renal cell carcinomas and hemangioblastomas. Somatic mutations of *VHL* are characteristic of sporadic renal cell carcinomas. The *hMLH1* gene, on proximal chromosome 3p, encodes a protein essential for repair of DNA mismatch. This gene is mutated in the germline of a high percentage of people with hereditary non-polyposis colon cancer, and it is also deactivated in about 15%–20% of sporadic colon cancers. These cancers show a high rate of replicative errors in repetitive (microsatellite) DNA tracts, a phenotype often described as **microsatellite instability** (MIN).

Loss of DNA methylation in cancer cells

Similar to DNA hypermethylation, DNA hypomethylation also plays an important role in carcinogenesis. There are two types of DNA hypomethylation conditions in cancer cells: (i) global hypomethylation refers to an overall decrease of 5-methylcytosine content in the genome and (ii) gene-specific hypomethylation refers to a decrease in cytosine methylation relative to the normal methylation level of that gene. Net decrease in the content of methylcytosines in cancer cells often exceeds the localised increase in DNA methylation. In case of colon adenocarcinomas, the genomic 5-methylcytosine content is reduced by an average of 10%. Global DNA methylation has been found in the pre-malignant or early stages of some neoplasms and is implicated as an important factor for tumour progression. DNA methylation in mammalian genomes is viewed as a defence mechanism by which repetitive DNA is transcriptionally silenced to prevent it from propagating. Hypomethylation of normally methylated DNA can disrupt such a defence mechanism, leading to structural and functional alterations of the genome. The significant loss of methylation, especially at repetitive DNA sequences, is associated with genomic instability. Gene-specific hypomenthylation is often seen in cancer cells but not in normal ones.

16.3.3 Deactivation of the X chromosome

In heterosexual mammals, sex determination is genetically identified by the presence of either two X chromosomes for females or an X and a Y chromosome in males. In contrast to the large X chromosome, the Y chromosome is very small and rudimentary, although it contains unique male-determining sequences. This imbalance in the DNA content poses the problem of equalising the dosage of X-linked genes in males and females. This problem, called 'dosage compensation', has been overcome in various ways during evolution. It is a well-studied example of epigenetic inheritance.

Dosage compensation

The purpose of dosage compensation is to offset differences in the number of active sex chromosomes. One of the sex chromosomes is altered so that males and females will have similar levels of gene expression. In mammals, dosage compensation adjusts the level of expression of many genes on the sex chromosomes (X chromosome) such that their expression is similar in both genders even though males and females have different complement of sex chromosomes. The term was coined by Müller (1932) to explain the effects of mutations in eye colour in *Drosophila*. Müller observed that female flies homozygous for certain X-linked eye colour alleles had the phenotype similar to hemizygous males. For example, an X-linked gene conferring an apricot colour to the eye produces a very similar phenotype in homozygous females and hemizygous males. In contrast, a female that has one copy of the apricot allele and a deletion of the apricot gene on the other X chromosome has eyes of paler colour. Therefore, one copy of the allele in the female is not equivalent to one copy of the allele in the male. Depending on the species, dosage compensation occurs through different mechanisms.

Dosage compensation occurs in somatic cells of female mammals by the random inactivation of one X chromosome. Epigenetic change causes inactivation of the X chromosome. The change that occurs with X-inactivation is epigenetic. It is not restricted to females. It also occurs in males with Klinefelter syndrome who have more than one X chromosome. In 1961, Marry Lyon proposed that dosage compensation in mammals occurs by the inactivation of a single X chromosome in females. In 1949, Murray Barr and Ewart Bertram identified a highly condensed structure in the interphase nuclei of somatic cells in female cats that was not found in males and this structure became known as the **Barr body**. In 1960, Susumu Ohno proposed that the Barr body is actually a highly condensed X chromosome (condensed into heterochromatin). The mechanism of X-inactivation, also known as the **Lyon hypothesis**, (**Fig. 16.6**.)

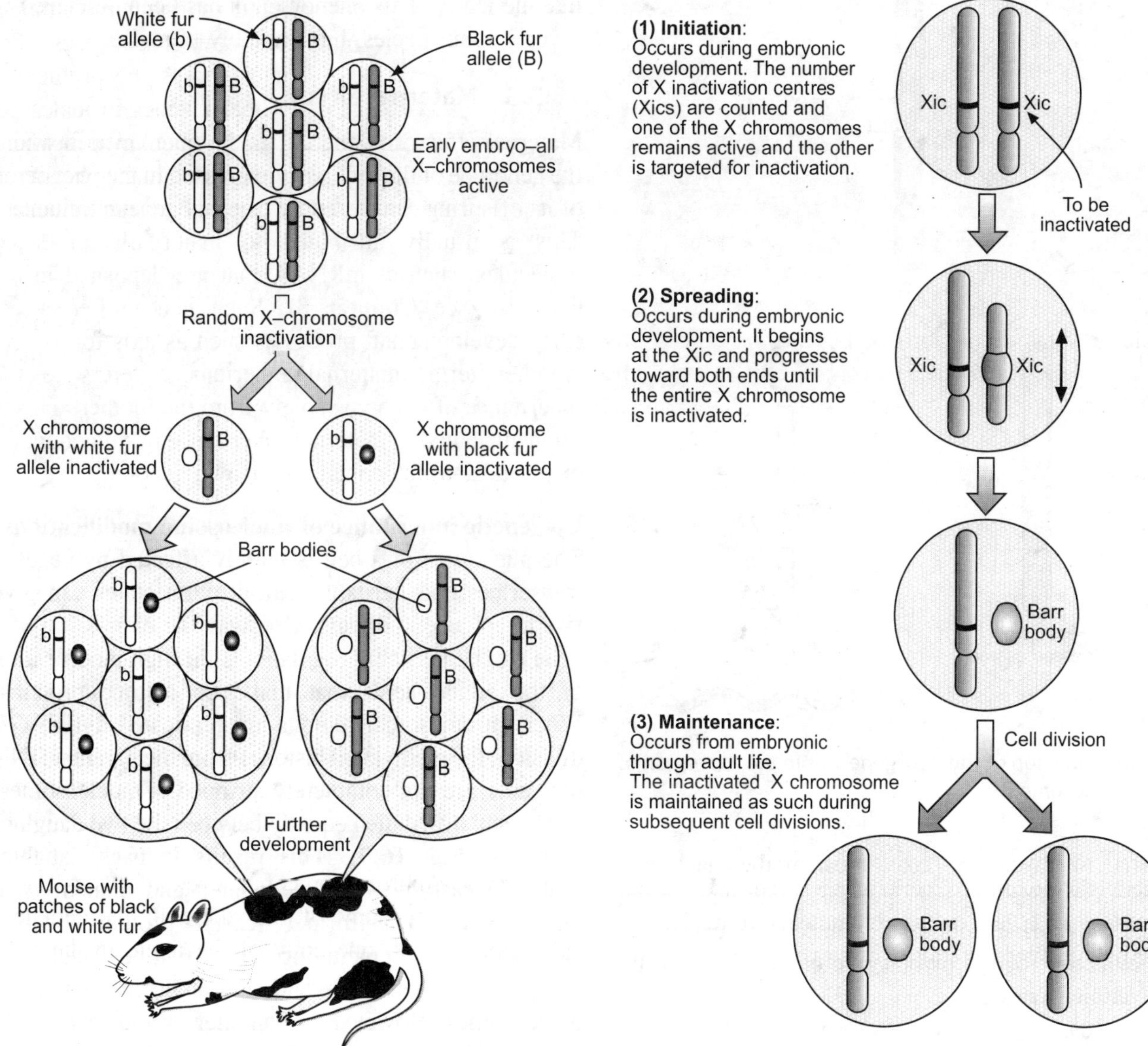

Fig. 16.6 The mechanism of X chromosome inactivation

Fig. 16.7 The function of the Xic locus during X chromosome Inactivation

The molecular basis of X-chromosome inactivation

Since the 1960s, researchers have been interested in the genetic control of X-inactivation. They have found that mammalian cells possess the ability to count their X chromosomes and allow only one of them to remain active. X-chromosome inactivation can be divided into four stages: counting, initiation, propagation (spread) and maintenance (**Fig. 16.7**). An X-linked *cis*-acting site, apparently involved in all four stages, is the **X-inactivation centre (Xic)**, which has been mapped by studying deletions and translocations of the X chromosome resulting in the disruption of inactivation. If the region containing Xic is deleted, the mutant X remains active. If the X chromosome is involved in a translocation, the fragment containing Xic is inactivated (as a part of the autosome to which it is attached) while the other fragment remains active.

Fine mapping of Xic originally revealed a single locus, **Xist** or **X-inactivation specific transcript**, which is expressed specifically on Xi (**Fig. 16.8**). Xist encodes a large untranslated RNA, which associates with Xi in the nucleus. This RNA transcript has some important features that have a role in the process of X-inactivation: (i) transcription of Xist was solely from the inactive X-chromosome (ii) the Xist transcript had no evolutionary conserved translational reading frames, indicating a lack of protein-coding functions and (iii) when examined by *in situ* hybridisation, the Xist RNA

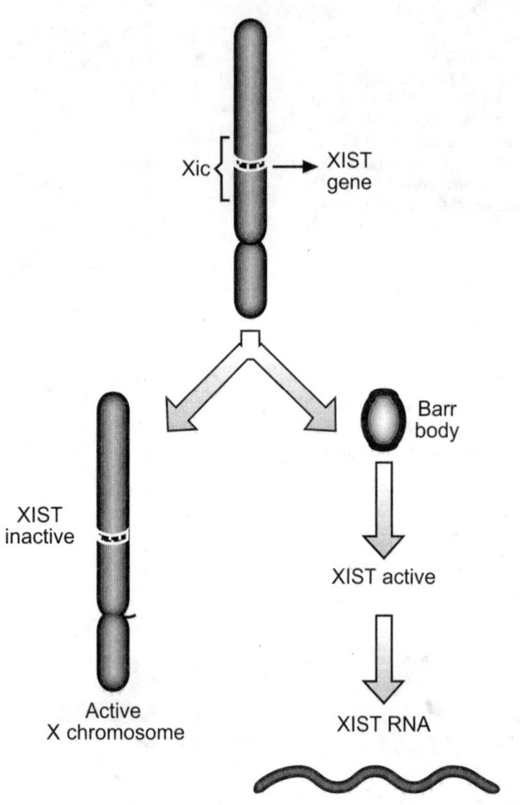

Fig. 16.8 Location of the XIST gene within the Xic region of the X chromosome

appeared to physically 'coat' most of the inactive X chromosome. The knockout and transgenic experiments in mice and murine embryonal stem cells have provided clear evidence for the pivotal role of Xist RNA in the inactivation process.

16.3.4 Paramutation

Epigenetic paramutation is an interaction between two alleles of a single locus, resulting in a heritable change of one allele that is induced by the other allele. It is an example of meiotically heritable epigenetic change. Paramutation violates Mendel's first law, which states that during the formation of gametes the allelic pairs separate and that each gene remains completely uninfluenced by the other. In paramutation, an allele in one generation heritably affects the other allele in future generations, even if the former is not transmitted. What may be transmitted in such a case are RNAs such as piRNAs, siRNAs, miRNAs or other regulatory RNAs. These are packaged into the egg or sperm and cause paramutation upon transmission to the next generation. This means that RNA is a molecule of inheritance, just like the DNA. This phenomenon has been described in several plant species, most notably in maize.

16.3.5 Maternal effect

Maternal effect, in genetics, is the phenomenon where the genotype of the mother is expressed in the phenotype of its offspring, unaltered by paternal genetic influence. This is usually attributed to maternally produced molecules, such as mRNAs, that are deposited in the egg cell (see **Chapter 8**). Such genes often affect early developmental processes such as axis formation. Another term, maternal inheritance, refers to the inheritance of genotype solely from the mother. This is often attributed to maternal inheritance of mitochondria or plastids, which contain their own genome.

Epigenetic inheritance of nucleosome modifications

The phenotype of a cell is mainly affected by genes it transcribes and heritable transcription states can give rise to epigenetic effects. One way the expression of a gene can be heritably regulated is through modification of the amino acids that make up histone proteins. When chromosomal regions are replicated during cell division, the methylated histones from the parental DNA molecule are not completely stripped of nucleosomes, but end up distributed equally between the two daughter duplexes (**Fig. 16.9**). This results in each daughter molecule carrying both methylated and unmethylated nucleosomes. The former recruit proteins bearing chromodomains, including the histone methylase itself, which then methylates the adjacent unmodified nucleosomes. However, a daughter strand that lacked methylated histones altogether, that is, the one from an unmethylated parent, would not recruit the methylase. This implies that the methylated nucleosome provides the information needed to methylate newly synthesised nucleosomes. In this way, modified chromatin can be maintained through generations. However, acetylation, methylation and ubiquitylation of the unstructured termini of histones makes them highly modified. Therefore, these are not inherited from one generation to another.

Other mechanisms

Structural inheritance systems: Both unicellular and multicellular organisms show inheritance of cell structures. Ciliates such as *Tetrahymena* and *Paramoecium*, show genetically identical cells with heritable differences in the patterns of ciliary rows

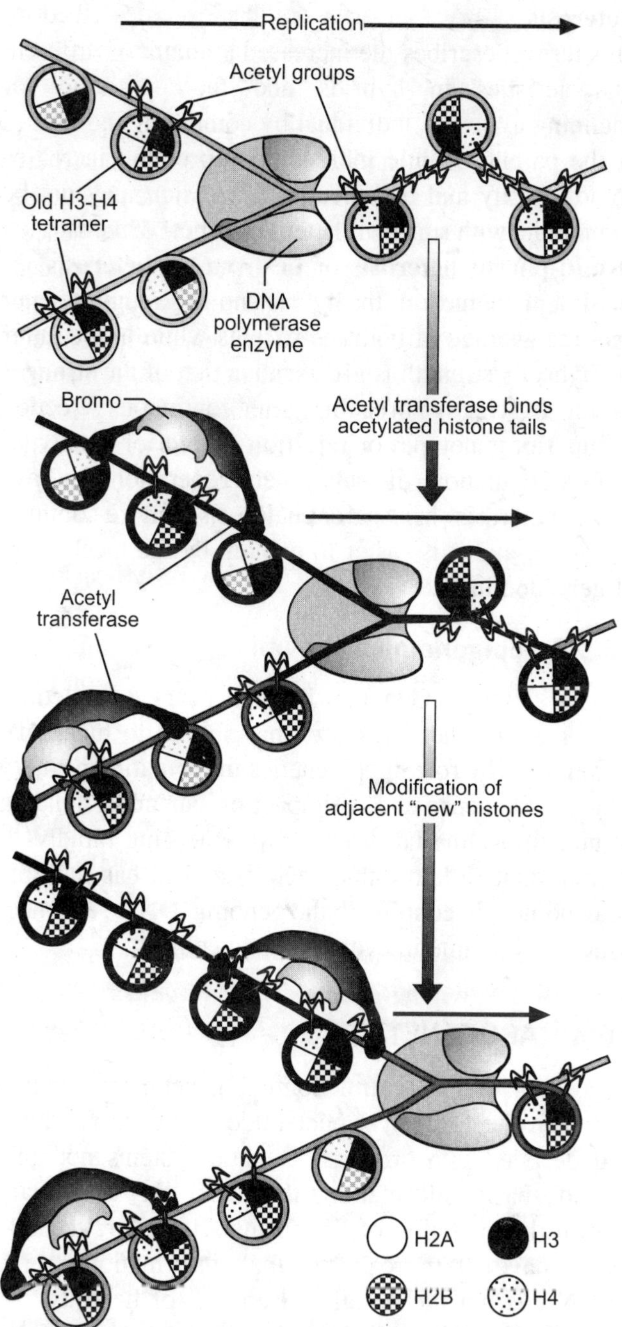

Fig. 16.9 Inheritance of parental H3–H4 tetramers facilitates the epigenetic inheritance of chromatin states

on their cell surface. Experimentally altered parental patterns can be transmitted to daughter cells. It appears that existing structures act as templates for new structures.

Assembly of immunoglobulin genes: The changes in DNA that occur in the assembly of immunoglobulin genes in cell lineages of the immune system should be regarded as an epigenetic mechanism. It depends on the creation of new DNA sequences.

Regulatory proteins: There are regulatory proteins that stably repress or induce particular genes without any involvement of DNA methylation. But, it is not clear whether such gene activation or deactivation are stably inherited through mitosis.

16.3.6 Environmental influences on epigenetic traits

The influence of environment on epigenetic marks is complex and could depend on a number of factors like the stability or strength of the epigenetic mark, the critical period for the modification and exposure to specific environmental elements. Some of these elements are diet, early embryonic environment and maternal behaviour that could have effect until adulthood.

Diet: There is evidence that environmental exposure of compounds such as heavy metal cadmium and arsenic and reduced intake of folate, for some individuals, may be a predisposing factor that leads to epigenetic instability, ageing and possibly cancer. Both folate and the enzyme methyltetrahydrofolate reductase (MTHFR) are important for DNA synthesis and methylation.

Early embryonic environment: Imprinting is known to cause problems in cloning, with clones having DNA that is not methylated in the right places. This could be due to not enough time being available for proper reprogramming. When a nucleus is added to an egg during somatic cell nuclear transfer, the egg starts dividing in minutes, as compared to the days or months it takes for reprogramming during embryonic development. An association has been reported between human imprinting disorders and intervention with assisted reproductive technology (ART) following a history of infertility. In murine experiments, it has been shown that embryo culture conditions can affect the reprogramming of imprints during the preimplantation phase of early embryonic development.

Maternal behaviour: In mice, maternal behaviour in the neonatal period correlates with epigenetic programming in adulthood. A recent study indicates that mothers showing strong nurturing behaviour towards their pups, by frequently licking and grooming their offspring, produce alterations in the patterns of DNA methylation in the promoter of the glucocorticoid receptor gene in the hippocampus of the pups, which persist throughout adulthood. There is evidence that these changes in chromatin structure result in decreased anxiety and a strong maternal nurturing instinct in the offspring during their adulthood. Further, this epigenomic state could be reversed by administering a histone deacetylase inhibitor that removes the epigenetic mark and correspondingly alters the behaviour of the mice.

16.3.7 Commercial applications of epigenetic processes

Stem cell therapy

It is hoped that stem cell-based regeneration to help repair diseased tissue could be used widely as a treatment method in the future. We need to understand that epigenetic signals give stem cells their unique ability to self-renew and transform into different cell types.

A human embryonic stem cell is prevented from giving up its unique characteristics of self-renewal and pluripotency by a protein modification that stifles any genes that would prematurely instruct the cell to develop into heart or other specialised tissue. But, the simultaneous presence of different protein modifications prime and poise the stem cells to develop into specialised body tissue. Thus, epigenetic research opens up the possibility of reprogramming cells. By manipulating epigenetic marks, cells can be transformed into other cell types without changing their DNA. As a stem cell develops, chemical tags on the DNA or its surrounding histone proteins switch the genes on or off, controlling a cell's fate.

Inducing grafting vigour in plants

In plants also, imprinting may be the reason for early flowering in the grafted plants. The mango plants grown from seeds, generally take 6–9 years to bring forth flowers for the first time. However, when a scion from matured plant is grafted to stock derived from a seedling, the plant flowers within three years. This shows that the flowering in the plant is epigenetically programmed in matured plants and this property could be maintained and expressed in the grafted plants. This property has great commercial application in many horticultural plants. The grafting vigour due to a particular of stock and scion combination clearly explains the epigenetic regulation of many metabolic genes.

Heterosis

This terms describes the increased strength of different characteristics in hybrids and the possibility of obtaining a 'better' individual by combining the virtues of the parents. While inbreeding depression increases homozygosity and outbreeding depression may not be compatible with survival, heterosis is best demonstrated in **mid-parent heterosis** or **best-parent heterosis**. In the first phenomenon, the hybrid shows strength greater than the average of both the parents while in the latter the hybrid's strength is greater than that of the stronger parent. In selective breeding, after continuous selection of superior genotypes or rejection of inferior genotypes from a population, the subsequent generation can have higher heterosis. Mid-parent heterosis is more common in nature, and it is easier to explain by the mechanism of gene dominance.

16.3.8 Epigenomics

As the Human Genome Project nears completion, the focus of genetic research has shifted to functional genomics. The role of epigenetics in regulating primary genomic sequence and its impact on our understanding of human health and diseases are emerging rapidly. It is important to know that each organism has multiple epigenomes, in contrast to the genomic DNA sequence, which is the same in every somatic cell.

16.4 APOGENETICS

According to the central dogma, genetic information stored in the form of a nucleotide sequence is copied and decoded into proteins. Genetic systems not only provide basic information, they also play a primary role in the evolution of life forms. However, recently many have start realising that information stored in DNA is not the be all and end all of heredity. As mentioned previously, the role of epigenetics is now clearly understood in development and inheritance of epigenetic regulations. Epigenetic inheritance is the second order in evolution and the patterns of gene expression are governed by the cellular material, the epigenome. These epigenetic 'marks' tell the genes to switch on or off. The evolution of DNA based systems and epigenetic regulation has fundamentally contributed to somatic evolution, which is the major aspect of most books. However, the evolution of inheritance systems are not restricted to somatic evolution alone, organisms have extended their evolutionary process beyond the somatic (extrasomatic). Extrasomatic evolution, utilises non-biological and non-cellular elements to aid the organism in continuation of its existence. Most of the human activities after birth are dominated by non-somatic behaviour. The aspects related to human behaviour, communication abilities, social evolution and so on, are beyond somatic evolution and are acquired in nature, though they are heritable across the generations. Extrasomatic evolution is very complex and it is directly or indirectly related to the somatic evolution.

16.4.1 Apogenetics

In 1990, a new field 'meme theory', was established to explain replicators that behave like genes. Meme is an information pattern, held in an individual's memory, which is capable of being copied to another individual's memory. In general, the characteristics of memes are similar to genes and to other replicators, such as computer viruses or crystals. Supporters of the meme concept regard memes as cultural analogues to genes, in that they self-replicate and respond to selective pressures. Meme-theorists contend that memes evolve by natural selection (similar to that of biological evolution) through the processes of variation, mutation, competition and inheritance influencing an individual entity's reproductive success. Historically, memes have spread through exclusively by direct demonstration or by word-of-mouth transmission. This method was commonly used for parental transmission; since children generally inherited the trades of their parents, they could learn their parents' skills by imitation. The second method, word-of-mouth transmission, is memetic spread via direct verbal communication. This is common, of course, in the spread of religious or political memes. Though 'Mimeme' comes from a suitable Greek root, there was a need for a monosyllable that sounded a bit like 'gene' and explained all the events associated with extra-somatic inheritance. Here, a new term 'Apogenetics' has been introduced to discuss all the aspects related to extra-somatic evolution under a single heading. Apogenetics is a combination of two words 'apo' (meaning, away) and 'genetics' (meaning, different from). The word 'apogenetics' here refers to 'above' or 'beyond' the genetics and epigenetic systems, though they are interconnected. In fact the strength of somatics will be assessed based on the fitness of the extrasomatic performance. Apogenetics is very essential to understand the life forms in a third dimension. The

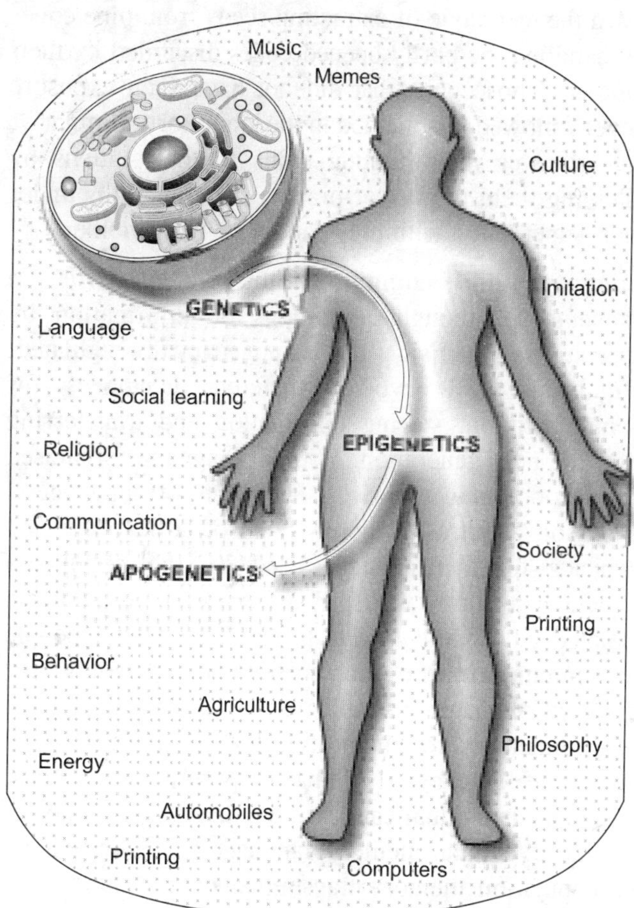

Fig. 16.10 The relationship between Genetics --> Epigenetics --> Apogenetics. The magnitude of environmental influence on each system is lowest to highest in the order: Genetics to Apogenetics.

concept of apogenetics interlinks all the aspects of extrasomatic activities with somatic evolution (**Fig. 16.10**). Since the apogenetic manifestation is highly influenced by the environmental factors, explaining it on an empirical basis may be difficult.

16.4.2 Components of Apogenetics

To understand how extrasomatic information might be an advantage compared to somatic evolution, it helps to envisage how the various information stores of a biological system interact.

Behaviour: It has been reported in many organisms that an organism's behaviour is transmitted to the next generation without the DNA being involved. It encompasses most forms of social learning in animals. For example, the development of a dog's instinct to bark and the imitation of tearing the bones and flesh of a carcass by biting furniture or old cloth. Rat pups learn the technique of extracting seeds from pine cones by handling the partly opened cones discarded by their mother. Those pups that did not have such exposure never followed this method in the laboratory. Similarly, cats transfer their hunting skills to their kittens by allowing them to play or practice hunting skills under their supervision with prey caught by their mother.

Language and communication: Language is an exchange of thoughts, opinions or information by speech, writing or signs and is apogenetically regulated. Organisms have evolved different languages for communication, storing information and transferring it to the next generation. Without language, organised civilisation cannot thrive. The type of language(s) a person acquires is purely via social learning and depends on the language used by the population in which a person lives. If a person is raised in an isolated environment with all facilities except people to communicate with, such person may not acquire any civilised language. Instead, he or she may purely depend on rudimentary communication through gestures. The ability of language learning is purely apogenetics, since any individual can learn any language of choice. Similar to cellular memory, language acts as individual or population memory transmitted over a number of generations. The terminology used in a language not only helps to communicate with others, it also helps to store the information mentally and literally.

Evolutionary psychology: The field of evolutionary psychology addresses the question of human brain and behaviour from an evolutionary perspective. Animals, in particular humans, have evolved a unique ability to think and develop a concept (by thought processing) based on the information acquired or by mere imagination, which enhance the individual. The information stored in the brain is modular in nature and can assist in recognitiion of existing things or conceptualise new forms.

Cultural evolution: Human culture has long been recognised as a highly specialised, extrasomatic means of adapting to a rapidly changing environment. Culture is an evolutionary, progressive and extrasomatic means of adaptation. Cultural evolution also has properties similar to biological evolution: cultural innovation (similar to genetic variation), cultural transmission (similar to heredity) and differential multiplication and survival of culture (similar to natural selection). Human populations are genetically structured through biological reproduction but reproduction is socially structured through culturally expressed systems of marriage. Thus, there is a need to take into account the cultural dimension of human societies when considering the genetic structure of modern human populations. Socially structured reproduction benefits the population, and the population at large will have better adaptability and be more sustainable.

Energy and human evolution: The most remarkable human innovation is the use of extrasomatic energy. Here, energy is made to accomplish human ends outside the bodies of its users. The control of fire and the exploitation of fossil fuels have made it possible for humans to release extrasomatic energy to modify the environment to suit their needs. This has allowed an expansion of human population that can be sustained and continues to grow. The abundant, cheap energy provided by fossil fuels has made it possible for humans to exploit a staggering variety of resources. In particular, the development of mechanised agriculture has allowed relatively few farmers to work vast tracts of land, producing an abundance of food and making possible an uncontrolled growth of population. The exploitation of animal power has also played an important role in the densification of population. But, fire has remained the most important source of extrasomatic energy. The development of charcoal improved on the energy density of untreated wood, and made a substantial contribution to metallurgy. The development of coke and its suitability for steelmaking set off the Industrial Revolution.

Computers are extrasomatic organs: Computers are considered as extrasomatic organs of our brain. They help us to maximise our thinking capabilities, driving us to carry out our daily processes faster and more efficiently than the human brain. To envisage how extrasomatic information is nested with somatic evolution, it is essential to envisage how the various information stores of a biological system interact. In general, information flow can be divided into three domains—the genetic, epigenetic and the apogenetic. The domains may be pictured as a nested hierarchy, with the genetic inside the epigenetic and the epigenetic inside the apogenetic. Each sphere gives rise to the next outer, and co-evolves with it. The accumulated apogenetic information of a society can serve as an extended memory and it possess a nearly limitless capacity without the metabolic costs. The understanding of the apogenetic process can be

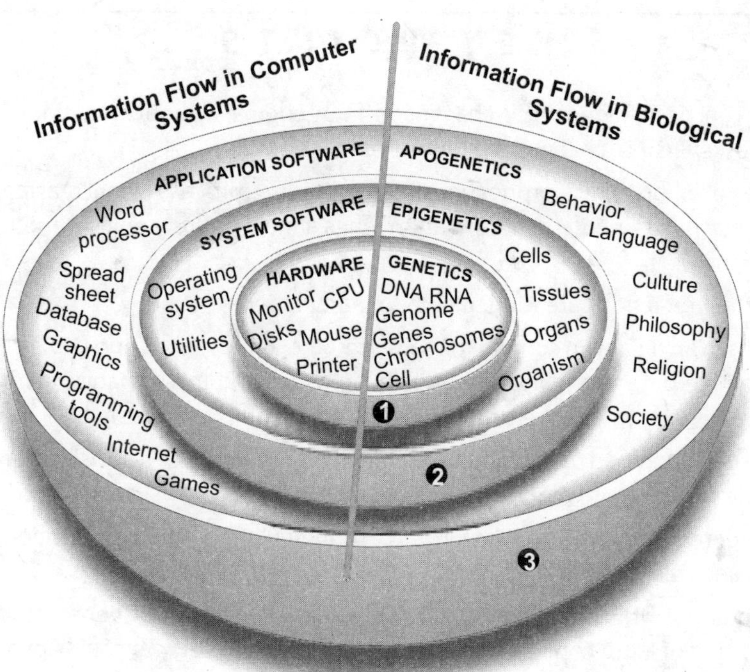

Fig. 16.11 Parallelism between information flow in computer systems and genetic information flow in biological systems.

further simplified by using the analogy of organisation of information in computer systems in comparison with biological systems (**Fig. 16.11**). The diagram explains the existence of parallelism between the information flow in the computer and in biological systems. The three globes represent the kind and level of information organised in both the systems. The inner globe represents the first level of information organisation and represents the system development. It involves organisation of biomolecules, cell organelles and functioning of cell in biological systems (genetics), similarly it represents the assembly of all the hardware parts in the computer systems. The second globe represents the establishment of communication between various components of a cell and between cells, tissues and organs of an organism in biological system (epigenetic regulation) and different components of hardware in computer systems (system software). However, its utility and application lies in the third level. In a biological system, it can be explained under apogenetics, which is acquired in nature and operates at the population level and like genetic information, it is also heritable from generation to generation among the populations or communities. Similarly, in case of computer systems, the application software makes computers useable for various purposes by the end users.

Thus, extrasomatic inheritance plays a significant role in human evolution. Inheritance of all the extrasomatic activities can be discussed conveniently under a single term 'apogenetics' along the lines of genetics and epigenetics. The evolution of apogenetics is highly influenced by the environment and is better explained with an analogy of evolution of computer systems.

KEY POINTS

- Gene expression can be influenced by changes in the genome without mutations in the DNA sequence. These heritable yet reversible changes are termed epigenetic.
- Epigenesis refers to embryonic morphogenesis as a gradual process of increasing complexity in which organs are formed *de novo*. Cellular differentiation relies almost entirely on epigenetic inheritance from one cell generation to the next.
- Nucleus–cytoplasm interactions are fundamental to early differentiation.
- Though all the cells in an organism contain the same chromosomes, genes and DNA sequences, their patterns of gene expression are different. Switching of gene activities and the somatic inheritance of a given set of active and inactive genes, provide the basis for a new epigenetic mechanism.
- In eukaryotes, the only modified base commonly found in DNA is 5-methylcytosine (5-meC) and this probably represents the sole programmed modification.
- Many lines of evidence suggest a link between DNA methylation and transcription, specifically between hypomethylation and transcriptional activity.
- Much information about DNA methylation and the epigenetic control of gene activity is now available in plants. The integration of extra copies of an endogenous gene or multiple copies of a transgene into a plant genome often results in epigenetic silencing of both endogenous and exogenous genes.
- In addition to DNA methylation and histone modifications, a number of other mechanisms are also involved in setting epigenetic states.
- Genomic imprinting is the phenomenon whereby a small subset of genes in the genome are expressed according to the parent of origin. That is, genes are expressed from only one parental allele, maternal or paternal, depending on the gene and such monoallelically expressed genes are said to be 'imprinted'. An imprint is a form of epigenetic information.
- It is now known that certain imprinted genes are involved in the etiology of several well-defined genetic/epigenetic disorders.
- Inactivation of one X chromosome out of the two in females is due to the random deactivation of one X chromosome, which is epigenetically regulated.
- In plants, heterosis or vigour of the hybrids can also be attributed to epigenetic regulation.
- Extrasomatic evolution process is called apogenetic and it is said to play a major role in the development of abilities of language, thought and cultural values as well as adaptation to geographical conditions.

Self-assessment Questions

1. Define epigenetics.
2. Differentiate between genetics, epigenesis and epigenetic regulation.
3. What are the mechanisms that contribute to epigenetic behaviour?
4. Explain the mechanism of epigenetic gene silencing.
5. What is maternal and paternal imprinting?
6. What are the epigenetic problems associated with cloning?
7. Give a brief account on the commercial exploitation of epigenetic phenomenon.
8. List some of the important features of human development that have been associated with apogenetics.

FURTHER READING

Textbooks

Holiday R (1984). The significance of DNA methylation in cellular ageing. In *Molecular Biology of Ageing* (ed.) Woodhead AD, AD Blackett and A Hollaender), New York: Plenum.

Jablonka E and MJ Lamb (1995). *Epigenetic Inheritance and Evolution*. Oxford: Oxford University Press,.

Krebs JE, ES Goldstein and ST Kilpatrick (2011). Lewin's *Genes X*. Jones and Bartlett Publishers, LLC.

Turner BM (2001). Chromatin and Gene Regulation: Molecular Mechanisms in Epigenetics, Oxford: Blackwell Science.

References

Alleman M et al (2006). An RNA-dependent RNA polymerase is required for paramutation in maize. *Nature* 442:295–98.

Baylin SB, JG Herman, JR Graff, PM Vertino and JP Issa (1998). Alterations in DNA methylation: A fundamental aspect of neoplasia. *Adv. Cancer Res.* 72:141–96.

Beck S, A Olek and J Walter (1999). From genomics to epigenomics: A loftier view of life. *Nat. Biotech.* 17:1144.

Cezar GG, MS Bartolomei, EJ Forsberg, NL First et al (2003). Genome-wide epigenetic alterations in cloned bovine fetuses. *Biology of Reproduction* 68:1009–14.

Constanica M, G Kelsey and W Reik (2004). Resourceful imprinting. *Nature* 432:53–57.

Dawkins, Richard (1989). Memes: The new replicators, *The Selfish Gene*, 2nd ed. Oxford: Oxford University Press.

Feinberg AP and B Tycko (2004). The history of cancer epigenetics. *Nature Reviews Cancer* 4:143–53.

Ferguson-Smith AC and MA Surani (2001). Imprinting and the epigenetic symmetry between parental genomes. *Science* 293:1086–89.

Graham, Gordon (2002). *Genes: A philosophical inquiry*. New York: Routledge.

Grewal SI and D Moazed (2003). Heterochromatin and epigenetic control of gene expression. *Science* 301:798–802.

Hake SB, A Xiao and CD Allis (2004). Linking the epigenetic 'language" of covalent histone modifications to cancer. *British Journal of Cancer* 90:761–69.

Hollick JB, JE Dorweiler and VL Chandler (1997). Paramutation and related allelic interactions. *Trends Genet.* 13:302–08.

Holliday R (1994). Epigenetics: An overview. *Dev. Genet.* 15:453–457.

Holliday R (2002). Epigenetics comes of age in the twenty-first century. *J. Genet.* 81:1–4.

Holliday R and T Ho (1998). Evidence for gene silencing by endogenous DNA methylation. *Proc. Natl. Acad. Sci.* USA 95:8727–32.

Horwich AL and JS Weissman (1997). Deadly conformations: protein misfolding in prion disease. *Cell* 89:499–510.

Hsiao K et al. (1989). Linkage of a prion protein missense variant to Gerstmann-Straussler syndrome. *Nature* 338:342–45.

Jaenisch R and A Bird (2003). Epigenetic regulation of gene expression: How the genome integrates intrinsic and environmental signals. *Nature Genetics Supplement* 33:245–254.

Jouvenot Y, V Ginjala, L Zhang, PQ Liu, et al. (2003). Targeted regulation of imprinted genes by synthetic zinc-finger transcription factors. *Gene Therapy* 10:513–22.

Lalande M (1996). Parental imprinting and human disease. *Anu. Rev. Genet.* 30:173–95.

Lee JT (2002). Is X chromosome inactivation a homology effect? *Adv. Genet.* 46:25–50.

Lund AH and van Lohuizen M (2004). Epigenetics and cancer. *Genes and Development* 18:2315–35.

Martienssen RA and V Colot (2001). DNA methylation and epigenetic inheritance in plants and fungi. *Science* 293: 1070–74.

Monk M (1995). Epigenetic programming of differential gene expression in development and evolution. *Developmental Genetics* 17:188.

Polychronakos C and A Kukuvitis (2002). Parental genomic imprinting in endocrinopathies. *European Journal of Endocrinology* 147:561–69.

Prusiner S (1982). Novel proteinaceous infectious particles cause scrapie. *Science* 216:136–44.

Reik W and ER Maher (1997). Imprinting in clusters: Lessons from Beckwith–Wiedemann syndrome. *Trends Genet.* 13: 330–34.

Richards EJ (1997). DNA methylation and plant development. *Trends Genet.* 13:319–23.

Richards EJ and SC Elgin (2002), Epigenetic codes for heterochromatin formation and silencing: Rounding up the usual suspects. *Cell* 108:489–500.

Roygeulle C and E Heard (2002). Antisense RNA in imprinting: Spreading silence through Air. *Trends in Genetics* 18:434–37.

Serio TR and SL Lindquist (1999). [psi+]: An epigenetic modulator of translation termination efficiency. *Ann. Rev. Cell Dev. Biol.* 15:661–703.

Siegfried Z and H Cedar (1997). DNA methylation: A molecular lock. *Curr. Biol.* 7:R305–R307.

Sleutels F and DP Barlow (2002). The origins of genomic imprinting in mammals. *Adv. Genet.* 46:119–63.

Stewart, Julian (1956). Cultural Evolution. *Scientific American.* May 1956. Pp. 65–80.

Tilghman SM (1999). The sins of the fathers and mothers: Genomic imprinting in mammalian development. *Cell* 96:185–193.

Weksberg R, P Sadowsk, AC Smith and B Tycko (2007). Epigenetics. Rimoin DL, JM Connor, R Pyeritz and BR Korf. (eds.) In *Principles and practices of medical genetics.* Vol 1: 81–100. Elsevier Ltd.

Wolffe AP (2000). Transcriptional control: Imprinting insulation. *Curr. Biol.* 10:R463–R465.

Yoder JA, CP Walsh and TH Bestor (1997). Cytosine methylation and the ecology of intragenomic parasites. *Trends Genet.* 13:335–40.

CHAPTER 17

Genetic Recombination

LEARNING OBJECTIVES

After reading this chapter, you will be able to:

- Understand the principle and mechanism of genetic recombination
- List out the different types and models of genetic recombination
- Appreciate the biological significance of genetic recombination

17.1 INTRODUCTION

One of the first discoveries in genetics was the observation that, before transferring to the offspring, the different parental genes are shuffled. Such shuffling can be due to recombination within (intrachromosomal—crossing over) or between (interchromosomal—independent assortment) the chromosomes. Genetic recombination is a fundamental process in all cells and is responsible for rearranging or altering the sequences between different pieces of DNA, limiting the divergence of repeated DNA sequences, pairing the chromosomes before cell division and promoting repair of DNA.

Movement of large blocks of DNA leads to new arrangements of genes. This process is mediated by genetic processes, such as **recombination** and **transposition** (see **Chapter 18**). Genetic recombination is the production of one or more new DNA molecules by rearrangement of the two parental ones. It usually occurs by either of the two methods:

- exchange of genetic material (reciprocal recombination) so that the new DNA molecules carry genetic information derived from both parental molecules,
- transfer from one parent molecule to the other (**gene conversion**).

Depending on the requirements for the DNA molecules and the proteins involved, three types of recombination have been described: **homologous** (also called as **general**), **site-specific recombination** and **illegitimate**.

In molecular biology, recombination generally refers to the molecular process by which genetic variation found associated at two different places in a continuous piece of DNA becomes disassociated (shuffled). One mechanism leading to such molecular recombination is chromosomal crossing over. Such shuffling of variation is also possible between duplicated loci within the same DNA molecule. If the shuffling process changes the number of loci in each of the recombinant molecules, it is referred to as 'unbalanced' recombination or unequal crossing over. Enzymes called recombinases catalyse this reaction. Molecular mechanism of recombination involves a physical rearrangement of the parental DNA and can be observed during:

- A mixed infection of two viruses or plasmids carrying different genetic markers
- Integration of bacteriophage, viral or plasmid DNA into a host cell chromosome

- Transformation of cells with a foreign DNA
- Crossing over at meiosis in eukaryotic cells
- Gene conversion
- Post-replication repair and so on.

Genetic exchange (i.e., recombination due to blending and rearrangement of DNA sequences), occurs constantly in all organisms, especially during meiosis of higher organisms. In populations, offspring frequently have different kinds and combinations of alleles than their parents. In addition to mutations, this can happen in two major ways: by independent assortment of chromosomes; and by crossing over between homologous chromosomes. However, at the molecular level, only crossing over contributes towards genetic recombination. **Figure 17.1** illustrates several variations of recombination. All are characterised by exchange of a strand or crossing over events that join DNA segments that were previously separated. Recombination can be intermolecular or bimolecular, which involves crossing over between two independent DNA molecules. Generally, recombination is reciprocal—in which the two participants trade DNA segments. But it can be one crossing over event, or two or more, and the number of events strongly influences the nature of the final products. In both prokaryotes and eukaryotes, genetic recombination involves the breakage and rejoining of homologous DNA double helices. Recombination and transposition generate new combinations of genes. These processes are significant for evolution because they markedly enlarge the genetic repertoire from which natural selection can occur. Without recombination, genes will remain linked forever on the same chromosome.

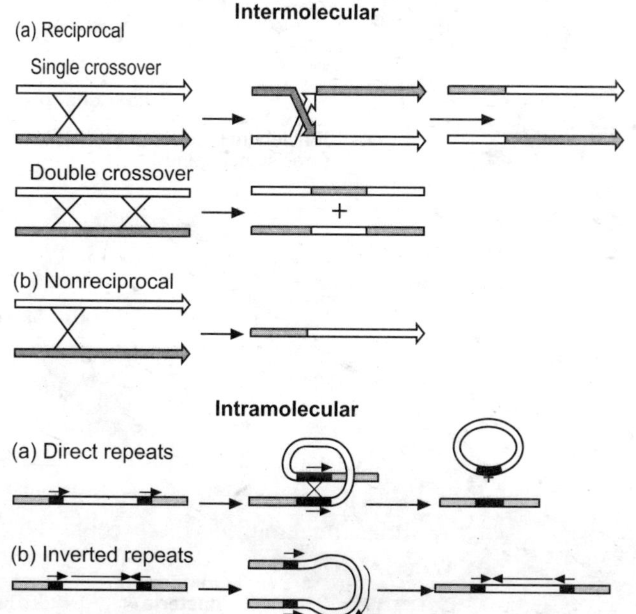

Fig. 17.1 Examples of different types of recombination; the Xs indicate cross over events between the two chromosomes or parts of the same chromosome

Genetic maps derived from frequency of crossing over between different genes have given information about the structure of chromosomes. It was earlier believed that genes were arranged in a fixed, linear order. However, recent discoveries have suggested that the gene order does change, although rarely, and it is becoming clear that such changes are a critical feature of the evolution of chromosomes. Now it is well accepted that recombination is not accidental, but instead an essential cellular process catalysed by enzymes. The enzymes allow cells to retrieve sequences lost when DNA is damaged by mutagens by providing the new strand from a homologous chromosome. Moreover, special types of recombination regulate gene expression. Molecular models of recombination are quite complex because heteroduplexes may or may not lead to recombination between flanking genetic markers.

17.2 TYPES OF RECOMBINATION

Genetic recombination can be broadly grouped into the following categories. They share common features as well as distinguishing characteristics.

Homologous recombination: It is also known as generalised recombination. It occurs in meiosis and depends on long regions of DNA homology between recombining partners. Many enzymes are involved in this process (e.g., recA in *E. coli*). They are not sequence-specific but are homology-dependent.

Site-specific recombination: This depends on the similarity of a few sequences between the recombining DNAs and almost always involves the same DNA regions. For example, it is observed during the insertion of λ phage DNA into the host DNA, or excision of the λ DNA. The enzymes mediating this process (**site-specific recombinases**) require a short specific DNA sequence in the partner recombinant molecules.

Illegitimate recombination/transposition: The movement of a DNA element from one locus to another generally requires a little similarity of sequence between the recombining DNAs, and it is therefore not site-specific. Therefore, it is also called 'illegitimate recombination'. The enzymes (transposases and integrases) recognise specific, short DNA sequences in one of the recombining partners (i.e., on transposable element) only. The recipient sites are usually non-specific in sequence and recombination integrates the transposable element into the host DNA (see **Chapter 18**).

Artificial recombination: This refers to recombination carried out by artificial ligation *in vitro* between the two free ends using purified enzymes and substrates.

17.3 GENETIC RECOMBINATION IN BACTERIA

Bacteria have no sexual reproduction and therefore, do not have alternation of diploid and haploid generations, formation of gametes and meiosis. But, bacteria have three mechanisms for genetic recombination: **transformation**, **conjugation** and **transduction (Fig. 17.2**). These three mechanisms enable bacteria to acquire DNA from other cells and incorporate it into their cellular DNA.

17.3.1 Transformation

It is the process in which the DNA released into the environment by the death of a donor bacterium enters a recipient cell. Many bacteria can acquire new genes by taking up DNA molecules from their surroundings. Certain bacteria (e.g., bacillus, haemophilus, *Neisseria*, *Pneumococcus*) can take up DNA from the environment and incorporate it into the recipient's chromosome by recombination. Some bacteria are able to take up DNA naturally during a particular time in their growth cycle when they produce a specific protein called a **competence factor**. At this stage, the bacteria are said to be competent. The first demonstration of bacterial transformation was done with *Streptococcus pneumoniae* and led to the discovery that DNA is the substance of the genes. The path leading to this epoch-making discovery began in 1928 with the work of a bacteriologist, Fred Griffith (see **Chapter 5**).

Uptake of DNA by Gram-positive and Gram-negative bacteria is different. The former take up DNA as a single strand while the latter do as a double-stranded DNA. After the donor DNA is taken up, a **reciprocal recombinant** event occurs between the chromosome and the donor DNA. This recombination requires homology between the donor DNA and the chromosome and results in the substitution of DNA between the recipient and the donor as shown in **Fig. 17.2**. Recombination requires the bacterial recombination genes **(recA, B and C)** and homology between the DNAs involved. This type of recombination is called legitimate or homologous or general recombination, since it requires DNA homology.

17.3.2 Transduction

Transduction is the process often seen in phages. A bacteriophage carrying DNA is picked up by growth in a donor bacterium. In the process of assembling new

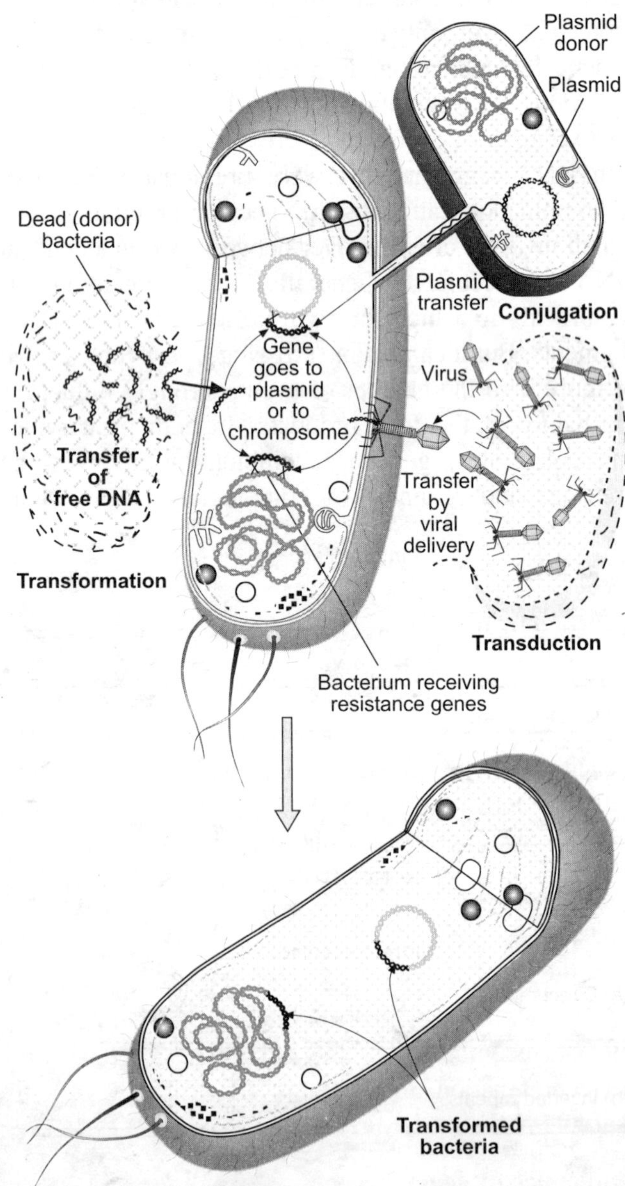

Fig 17.2 Genetic recombination in bacteria; they are able to exchange genes in nature by three processes: transformation, conjugation and transduction

phages, some host DNA may be incorporated in them. When these phages infect new bacteria, the 'hitchhiker' bacterial gene(s) may be inserted into the DNA of the new host giving it an altered phenotype. This phenomenon is called **transduction**. The ability of a phage to mediate transduction is related to its lifecycle. There are two types of transduction: generalised and specialised. In the former, potentially any bacterial gene from the donor can be transferred to the recipient. Phages that mediate generalised transduction break down the host DNA into smaller pieces. One of the pieces of the host DNA is randomly packaged into a phage and upon infection, the donor DNA enters the recipient. The new DNA substitutes the host's DNA by generalised recombination. On the other hand, in specialised transduction, only certain donor genes are transferred to the recipient. Different phages may transfer different genes but an individual phage can only transfer certain genes. During excision of the prophage, occasionally an error occurs and some of the host DNA is also excised along with the phage DNA. Only the host DNA on either side of the location where the prophage has inserted can be transferred (i.e., specialised transduction).

17.3.3 Conjugation

Some bacteria (e.g., *Escherichia coli*) can transfer a portion of their chromosome to a recipient with which they are in direct contact through the sex pilus. As the donor (male) replicates its chromosome, the copy is injected into the recipient (female). When the donor and recipient are separated, the transfer of genes is disrupted. Those genes that successfully find a new home replace their equivalent in the recipient's chromosome. In each type, the transferred DNA is not capable of continual replication, and its maintenance requires that the DNA be inserted into the host DNA, usually by a substitution mechanism. Conjugation can only occur between cells of opposite mating types (the donor or 'male') with a fertility factor (F^+) and the recipient ('female') without the factor (F^-). This resembles, in many ways, the exchange between chromatids during meiotic recombination in eukaryotes. However, the recombination mechanisms in prokaryotes and eukaryotes have many differences.

Significance of genetic recombination in bacteria

The importance of genetic recombination through transformation, conjugation and transduction in nature is not well known. But in the laboratory, it has been well exploited for gene transfer experiments. Recent studies on completion of the sequence of the entire genome of a variety of different bacteria suggest that genes have in the past moved from one species to another. The transfer of resistant genes within populations and even between the species may have aided the spread of resistance genes in response to multiple antibiotics. Interestingly, many bacteria have enzymes that enable them to destroy foreign DNA that gets into their cells. It seems unlikely that these would be needed if that did not occur in nature.

17.4 RECOMBINATION IN EUKARYOTES

Recombination in eukaryotes is usually assessed at the genetic level by analysing the products of meiosis in terms of the expression of concerned alleles. In specially designed situations, it can be followed at the cytological level by analysing the morphological features of the concerned chromosome pair in the products of meiosis. A detailed work by Stern in *Drosophila* and McClintock in maize brilliantly demonstrated the involvement of recombination between linked genes to facilitate a physical exchange of homologous chromosome segments (crossing over). It also became clear that crossing over most likely involves a breakage and reunion between the homologous chromosomes. These observations were based on the analysis of the products of meiosis.

Broadly, the process of genetic recombination is of two types: reciprocal (transfer in both directions) and non-reciprocal (unidirectional transfer)

17.4.1 Reciprocal recombination

The most common type of reciprocal recombination is general or homologous recombination. It is guided by base pairing and exchange of homologous regions (reciprocal but not sequence-dependent) between two DNA molecules (**Fig. 17.3**). Complementary base pairing between strands unwound from two different chromosomes puts them in exact register so that crossing over generates **homologous recombination**. Usually, these are sequences on two equivalent regions of homologous chromosomes, but recombination can also be detected between homologous segments in non-equivalent regions. Such unequal crossing over gives rise to duplications and deletions in chromosomes (**Fig. 17.4**). The general recombination reaction is essential for every proliferating cell,

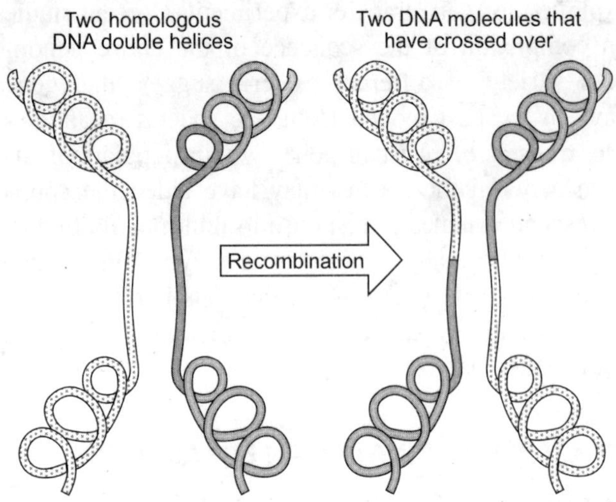

Fig. 17.3 General recombination; the breaking and rejoining of two homologous DNA double helices create two DNA molecules that have crossed over

because accidents occur early in every round of DNA replication that interrupt the replication fork and require general recombination mechanisms to repair. General recombination is also essential for the accurate chromosome segregation that occurs during meiosis and generation of genetic diversity.

In diploid eukaryotes, the homologous recombination commonly occurs during meiosis when the homologous duplicated chromosomes line up in parallel in metaphase I. After chromosomal replication during gametogenesis, the four chromatids are tightly meshed and form a **tetrad**. The homologous sites on each pair of chromatids exchange genetic information. Thus, after replication, the tetrad contains two pairs of two identical chromatids, but after crossing over, each of the four chromatids carries a unique set of genetic information.

Synapsis: During pairing of the homologous DNAs, they are well aligned. The common stretch of bases that will be recombined is known as the homology between the sequences and can be as few as ~50–100 bp or as much as the whole chromosome. During synapsis, base pairs form between complementary strands from the two DNA molecules. This base pairing then extends to guide the general recombination process, allowing it to occur only between DNA molecules that contain long regions of matching. The correlation between recombination and formation of synaptonemal complex is well established and recent work has shown that all mutations that abolish chromosome pairing in *Drosophila* or in yeast also prevent recombination.

Process of crossing over

Crossing over involves the breaking and rejoining of DNA at the same position in two homologous DNA molecules, which is initiated by breaks in the dsDNA. The cellular enzymes (**Table 17.1**), known as recombinases catalyses the reactions that allow for cross over to occur. A recombinase creates a nick in one strand of a DNA double helix, allowing the nicked strand to pull apart from its complementary strand and anneal to one strand of the double helix on the opposite chromatid. A second nick allows the unannealed strand in the second double helix to pull apart and anneal to the remaining strand in the first, forming a structure known as a **cross-strand exchange** or **Holliday junction**. It is a tetrahedral structure which can be 'pulled' by other recombinases, moving it along the four-stranded

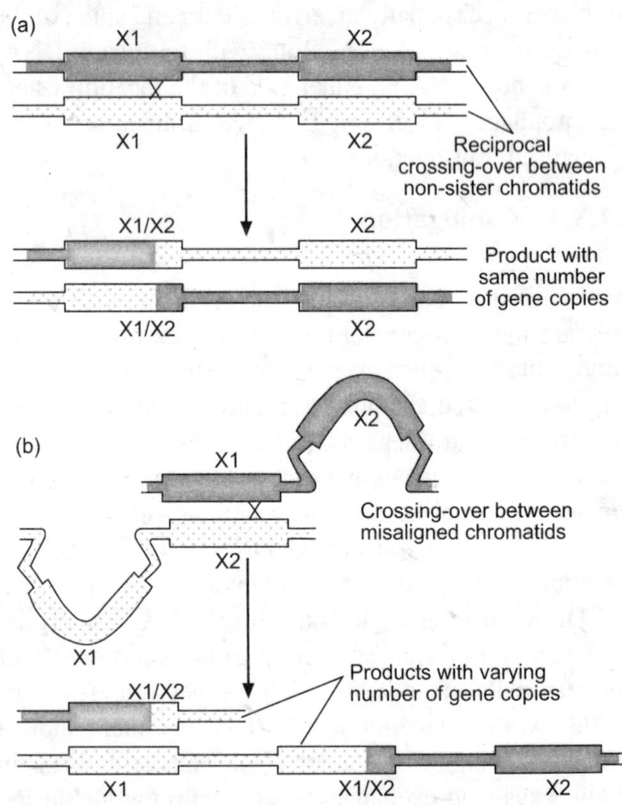

Fig. 17.4 Increase in gene number by homologous recombination involving unequal crossing over; the diagram shows two tandem copies, X1 and X2, of a gene on two recombining non-sister chromatids at meiosis. Reciprocal crossing over between properly aligned chromatids (a) does not affect gene number, whereas (b) unequal crossing over on misaligned chromatids generates different products with low or high copies of the gene.

Table 17.1 Enzymes involved in homologous recombination

Name (gene or protein)	*Function*
*rec*A/RecA	Binds single-stranded DNA, DNA-dependent ATPase, promotes base pairing (assimilation) and control of SOS system
*rec*BC/RecBC	ssDNA and dsDNA exonuclease (part of ExoV), chi(ζ)-specific endonuclease, DNA helicase, ATPase, initiator of recombination by unwinding a dsDNA molecule and creating a region of single strandedness in an area containing a ζ (chi) site
*rec*D/RecD	Exonuclease, part of ExoV
*rec*E/RecE	5'→3' dsDNA exonuclease, ExoVIII
*rec*F/RecF	Binds ssDNA and ATP, can substitute for recBCD
*rec*G/RecG	Binds to Holliday junctions, ATPase, involved in branch migration and resolution of Holliday junctions
*rec*J/RecJ	5'→3' ssDNA exonuclease, can substitute for recBCD
*rec*N/RecN	Binds ATP, can substitute for recBCD
*rec*O/RecO	Binds ssDNA and promotes renaturation, can substitute for recBCD
*rec*Q/RecQ	dsDNA helicase can substitute for recBCD
*rec*R/RecR	ssDNA binding and renaturation
*rec*T/RecT	Endonuclease, cleaves Holliday junctions
*rus*A/RusA	Binds to Holliday junctions, involved in branch migration
*ruv*B/RuvB	Binds to ruvA, ATPase, involved in branch migration
*ruv*C/RuvC	Endonuclease specific for Holliday junction, involved in resolution

structure. No base pairs are added to or deleted from the DNA as a result of the breakage and rejoining events. Thus, a very early step in recombination is the generation of a break in each double helix involved along a pair of synapsed chromosomes in meiosis, thereby stimulating the recombination process.

Strand transfer

Breaking of the homologous DNA strands: During strand transfer, a single DNA strand is transferred from one duplex to the other. The first transfer marks the initiation of recombination as it invades the homologous duplex and (if the recipient duplex is intact) displaces a resident strand. This process may generate a short region of heteroduplex DNA which consists of strands from different parental molecules which may contain base mismatches reflecting sequence differences (different alleles) in the parental duplexes. If the recipient duplex is intact, the displaced resident strand is able to pair with the free strand of the initiating duplex. The two transferred strands cross each other, forming a Holliday junction. The site of the Holliday junction may move in relation to its original position by progressive strand exchange between duplexes. This process is called **branch migration**, and may increase or decrease the amount of heteroduplex DNA. It can take place at any point where two single DNA strands with the same sequence begin to pair with the same complementary strand.

Repair and resolution

Repair involves re-forming the phosphodiester bonds to join the two homologous strands. They do not occur in a fixed order as this depends upon the recombining partners and the availability of appropriate enzymes. It consists of three different processes.

Conservative recombination: In the simplest case, the recombining duplexes are intact (i.e., there is no genetic information missing from either duplex) and repair involves re ligation of the broken strands. This is conservative recombination. The Holliday junction can then be resolved in either of the two planes to generate one of the two products—a **patch** of heteroduplex DNA in a non-recombinant background, or a **splice** of heteroduplex DNA with recombination of flanking markers.

Non-conservative recombination: If genetic information is missing from either duplex (i.e., if there is a single-strand gap or a break) DNA repair replaces the missing information from the homologous duplex as a template. In the extreme case, where an entire chromosome segment is missing, resolution of the Holliday junction yields a replication fork which can duplicate the missing segment.

Post-replication repair: The third type of DNA repair, that is repair of mismatch in the heteroduplex DNA, is generally random in direction and may cause gene conversions.

Several models have been proposed to explain how these events take place. In copy choice models, recombination occurs during DNA replication: the nascent strand switches onto a new template as it elongates. In breakage and reunion models, recombination occurs in the absence of DNA replication: DNA strands are broken, exchanged between duplexes and re-ligated. There are also hybrid models combining features from both of the above. All these types of recombination occur under different circumstances.

Consequences of cross over

The resulting haploid gametes, after meiosis, contain information derived from both maternally and paternally inherited chromosomes, thus ensuring that an individual will inherit genes from all the four grandparents. However, the exact molecular details of the initiation of recombination are not clear.

Somatic cells of eukaryotes carry two copies of each gene, each referred to as an allele. Both the parents pass on one allele to each offspring. Thus, an individual gamete inherits a complete haploid complement of alleles on chromosomes that are independently selected from each pair of chromatids lined up on the metaphase plate. Without recombination, all alleles for those genes linked together on the same chromosome would be inherited together. Meiotic recombination allows a more independent selection between the two alleles that occupy the positions of single genes, as recombination shuffles the allele content between the sister chromatids. However, recombination does not have any influence on another offspring having the same combination. This theory of 'independent assortment' of alleles is fundamental to genetic inheritance. However, the frequency of recombination is actually not the same for all gene combinations. This leads to the notion of 'genetic distance', which is a measure of recombination frequency averaged over a sample of pedigrees. This means recombination is greatly influenced by the proximity of the genes. The chances of separating the closely placed genes by recombination are very low than that of the genes placed farther apart. Genetic linkage describes the probability of genes to be inherited together as a result of their location on the same chromosome. Genetic disequilibrium describes a situation in which some combinations of genes occur more or less frequently in a population than would be expected from their distances.

Problems of cross over: Crossover recombination can occur between any two double helices of DNA which are very close in sequence and come into contact with each other. Thus, cross over may occur between *Alu* repeats on the same chromatid, or between similar sequences on two completely different chromosomes. These processes are called unbalanced recombination. Even though fairly rare compared to normal recombination, it can create severe problems if a gamete containing unbalanced recombinants becomes a part of a zygote.

17.4.2 Models for homologous recombination

As mentioned earlier, there are two primary roles for homologous recombination, viz., **genetic mixing** and **DNA repair**, although it is not clear which role the process first evolved to fulfil. Homologous recombination is linked to DNA replication, but does not occur at replication forks. Rather, it takes place between intact double helices. Homologous recombination is precisely accurate to one base pair. The current models for the mechanism of homologous recombination share many essential steps.

The recombination among non-sister chromatids must be based on breakage and reunion of DNA molecules. In eukaryotes, each chromatid contains a single DNA double helix and each bivalent during meiosis would, therefore, have four DNA double helices. But, the recombination event occurs between two of the four DNA double helices present in a bivalent. The models of breakage and reunion of DNA molecules are grouped into two broad categories: those involving single-strand breaks and those with double breaks.

Models involving single-strand breaks

The earliest models were similar in their requirements: two homologous chromosomes each break at exactly the same point along their length and then rejoin in a new combination. This was referred to as the **breakage and rejoining model**. The **copy choice model** was first proposed by John Belling in 1931 and later revived by Alfred Sturtevant in 1949 and Joshua Lederberg in 1955 in an attempt to explain recombination between bacteriophage DNA and a bacterial DNA. This model proposed that DNA recombination was the result of DNA replication in which more than one parental chromosome acted as a template for the newly synthesised daughter-chromosome. In 1961, Meselson and Weigle reported the recombination of mutations in genes that were distant from each other and physically located on two different radioactive tags and genetically marked by different mutations. The two chemically and genetically marked λ phages were mixed, used to infect *E. coli*, and allowed to recombine. The products from this experiment proved that DNA molecules were breaking and rejoining during recombination. A few years later, studies in eukaryotes led Harold Whitehouse (1963) and Robin Holliday (1964) to independently propose the first molecular models for recombination.

These two models differ only in their statement of the nature of single-strand breaks. The Whitehouse model envisages breaks in the two DNA duplexes to occur in strands of opposite polarity (i.e., in complementary strands), whereas in the Holliday model they affect the strands of same polarity (i.e., in homologous strands that have the same sequence). As a result, the Holliday model is relatively simpler.

Holliday model for homologous recombination

Robin Holliday's model for reciprocal recombination does not mention DNA synthesis. But it clearly illustrates central features like invasion of the DNA strand, branch migration and the process of resolution of the Holliday junction (**Fig. 17.5**). The different stages of his model are as follows:

Pairing of homologous DNA duplexes: The first step begins with the pairing of two homologous DNA

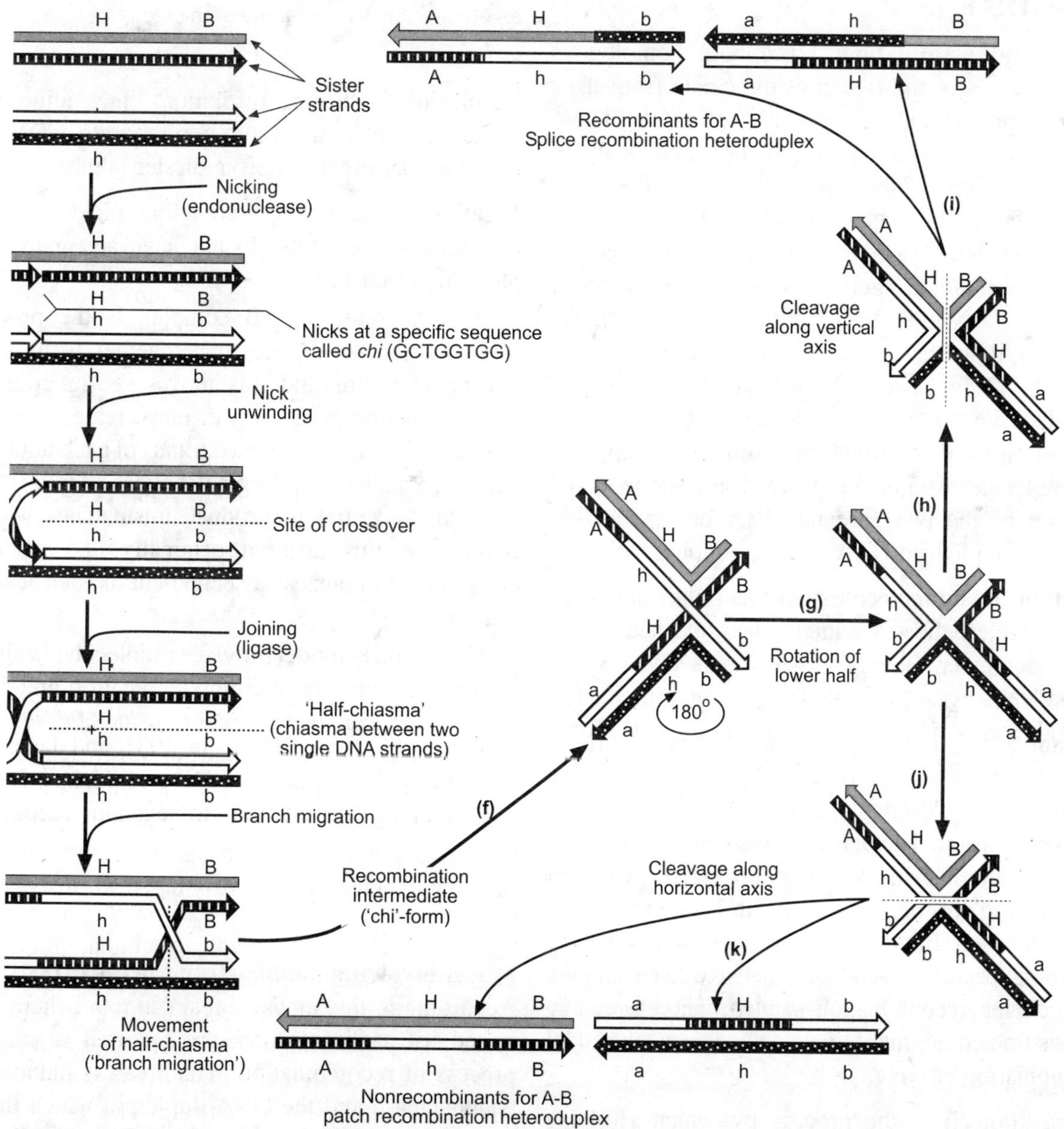

Fig. 17.5 The Holliday model for homologous recombination between two DNA duplexes involving single-strand breaks (SSB)

duplexes. The duplex DNA from two non-sister strands of a tetrad align themselves in register so that the subsequent exchange does not delete or duplicate any genetic information.

Strand breakage or nicking: Strands of the same polarity are nicked at equivalent positions. This process is initiated by an endonuclease in corresponding positions in two homologous DNA strands. In eukaryotes, the protein spo11 accomplishes this.

Unwinding: The break is followed by 5'→3' exonuclease activity to widen the gaps formed in the double helix (unwind) and create 3' single-polynucleotide-strand tails (**Fig. 17.5 b, c**).

Invasion and assimilation: Once both homologous strands are broken, the free ends dissociate from their complementary strands (non-nicked strands). These tails are coated with recA protein which then catalyses the invasion of one of the single strands into the intact double helix in the opposition (**Fig. 17.5d**). Then, hydrogen bonds are formed with the complementary sequence to create a short segment of heteroduplex DNA. The ssDNA by DNA polymerase I and DNA ligase then replaces sections of previously digested DNA (**Fig. 17.5e**). The point at which the exchanged strands in a joint molecule cross each other is called **recombinant joint** or **branching point**. Each duplex in a joint molecule has a region in which it has one strand from each of the two parental DNA duplexes. This region is called hybrid DNA or heteroduplex DNA.

Separation of the molecules: At this point, however, the two double helices are interlocked and need to be freed of each other. When the two strands have crossed over and DNA ligase has sealed the new intermolecular phosphodiester bonds, a Holliday junction—also known as a half-chiasma or *chi* structure or cross-strand exchange—is formed as shown in **Fig. 17.6**.

Branch migration: The branch in the Holliday junction can migrate in either direction simply by breaking old base pairs and forming new ones, in a process called branch migration. But the rate at which it occurs *in vitro* cannot be expected to generate extensive heteroduplex regions during recombination *in vivo*. Therefore, any extensive branch migration *in vivo* must be catalysed by a recombination enzyme.

Isomerisation: It is the process by which Holliday junctions rearrange without breaking the hydrogen bonds between bases.

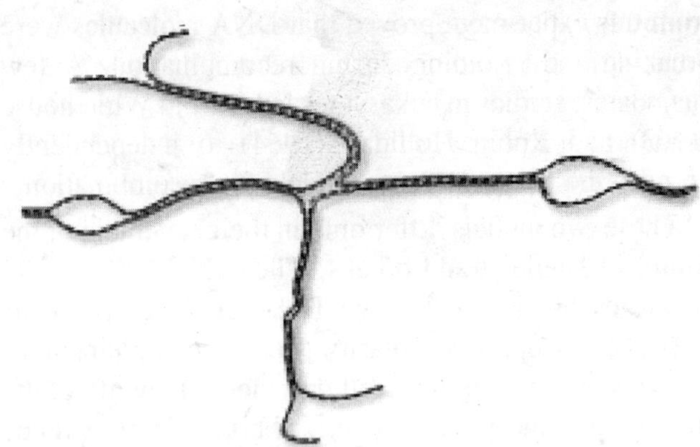

Fig. 17.6 A *chi* structure with four duplex arms held together at the site of strand exchange

Resolution: During resolution, the joined DNA sequences are resolved into two separate dsDNAs by a nuclease that breaks phosphodiester bonds.

Ligation: It seals any remaining nicks so that the two separate dsDNAs do not have any gaps in their phosphodiester backbone.

In *E.coli*, the ruvAB complex is the product of the *ruvA* and *ruvB* genes that together form an ATP-dependent motor and this moves the junction point. As the junction points move, they create heteroduplex DNA, places where the two strands of each double helix come from different original helices. These stretches have the potential to produce mismatches where the two chromatids differed originally. To resolve the cross-linked duplexes, a second cut at each junction is required.

The Holliday model provides a molecular basis for the association between aberrant segregation and crossing over. Gene conversion and post-meiotic segregation may be the result of repair of mismatches, on one or both the chromosomes in a symmetric structure containing heteroduplex DNA. It also explains how 5:3 and 6:2 segregations can occur without recombination of flanking markers, in crossing over in *Neurospora asci*.

Model involving double-strand breaks (DSB)

According to this model, breaks in two helices (double strand) of a chromosome are required to initiate the process of recombination. The break is induced by an endonuclease and the DNA duplex in which the break is induced is called the recipient. This is followed by exonuclease action that nibbles away one strand of the duplex on either side of the break. This generates 3' single-stranded termini on both sides of the break, and the break is enlarged into a gap. One of the single-stranded 3' ends now invades the homologous region of the other DNA duplex called the donor. The homologous strand in the donor duplex is replaced by the invading strand. This generates a D-loop in the donor duplex (**Fig. 17.7**).

The D-loop is then extended using the 3' end of the invading strand as primer and the complementary strand of the donor duplex as template. The D-loop spans the entire length of the gap in the recipient duplex and pairs with the single-stranded region of this duplex, which is complementary to the displaced strand. Now the donor has a heteroduplex region and the recipient has a single-stranded region corresponding to the gap generated at the site of original cleavage. This region is occupied by the displaced strand of the donor duplex (**Fig. 17.7**). The repair synthesis fills single-stranded region of the recipient duplex using the other 3' end at the gap as the primer using the displaced strand of the donor as the template. The nicks remaining in the two duplexes are sealed by DNA ligase. Branch migration converts the two DNA duplexes into a joint DNA molecule having two recombinant joints. Both the joints have to be resolved to recover two DNA duplexes from this structure. The location of the nicks in the two joints determines whether the two duplexes are parental or recombinant. If the two joints are resolved in the same manner, the original non-cross over DNA duplexes will be obtained. However, the donor would have a heteroduplex region, while the recipient duplex would contain a sequence from the donor. If, however, the two joints are resolved in the opposite manner, a genetic cross over or recombination occurs.

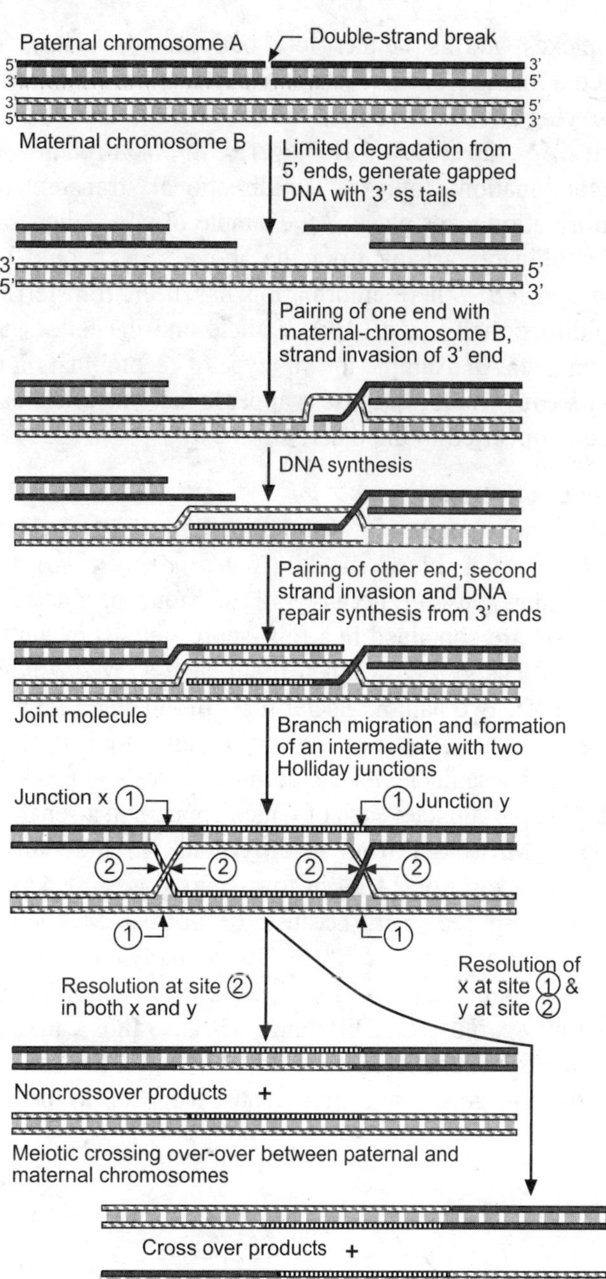

Fig. 17.7 General recombination in meiosis; double strand break (DSB) repair model for homologous recombination

Asymmetric strand transfer model

This is an alternative to the Holliday model. The main drawback with Holliday's model is that it assumes two breaks in exactly corresponding positions in homologous DNA strands (i.e., perfectly reciprocal). As single-strand breaks can occur frequently, it is not difficult to imagine a break in any given position in a DNA molecule. However, it seems unlikely that two breaks, conveniently located right across from one another in two paired molecules can occur spontaneously. Therefore, this model suggests that initial pairing will be asymmetric. That is, in the initial interaction, one molecule donates a single polynucleotide strand to a recipient molecule and a heteroduplex region is formed only in the invaded molecule. By replicative extension of the invading strand and by branch migration, the heteroduplex DNA is enlarged and by a novel rearrangement (isomerisation), a heteroduplex region is formed in both the molecules.

17.4.3 Non-reciprocal recombination

Non-reciprocal recombination is both symmetrical and conservative (i.e., information is exchanged between duplexes and is neither lost nor gained). Thus, if recombination occurs between parental chromosomes carrying alleles *AB* and *ab*, the reciprocal recombinants are *Ab* and *aB*. However, in non-reciprocal recombination, only a unidirectional transfer of information takes place. An example of non-reciprocal recombinants arising from the above cross would be *Ab* and *AB*, where information has been transferred unidirectionally from *A* → *a* allele and the latter has been lost. An example for this type of recombination is **gene conversion**, because it appears that one allele has been converted to the other.

Gene conversion

For genetic studies of recombination, fungi such as yeast or pink bread mould (*Neurospora crassa*) are particularly useful because all the four products of meiosis are contained in a four-spore (tetrad) or eight-spore (octads) ascus. When fungi such as *Neurospora* sporulate, two haploid nuclei fuse, producing a diploid nucleus that undergoes meiosis to give four haploid nuclei. These nuclei then experience mitosis to produce eight haploid nuclei each of which appears in a separate spore. Most asci from heterozygous *Aa* individuals contain a ratio of 2*A*:2*a* in four-spored asci, or 4*A*:4*a* in eight-spored asci, because of normal Mendelian segregation (**Fig. 17.8**). Occasionally, however, aberrant ratios are also found, such as 3*A*:1*a* or 1*A*:3*a* in four-spored asci. Different aberrant ratios can also occur (for example, 5*A* and 3*a*). Such asci are said to result from gene conversion because it appears as though one allele has 'converted' the other allele into a form like itself. Gene conversion is frequently accompanied by recombination (general recombination) between genetic markers on either side of the conversion event, even when the flanking markers are tightly linked. The accepted explanation of gene conversion is that it results from mismatch repair or gap repair.

Mismatch repair occurs in a small segment of duplex DNA containing one strand (*a* being converted to *A*) from each parental molecule (**Fig. 17.9**). The region of hybrid DNA is called a heteroduplex region. Repair systems recognise mispaired bases in heteroduplex DNA, and then may excise and replace one of the strands to restore the complementary status. This leaves three strands with the *A* allele, and only one with the *a* allele. Now DNA replication will produce three *A* duplexes and only one *a* duplex.

During gap repair, one DNA duplex may act as a donor of genetic information that directly replaces the corresponding sequences in the recipient by a process of gap generation, strand exchange and gap filling.

Gene conversion is not confined to meiotic events (crossing over) alone. Various biases can be seen when recombination is examined at the molecular level. Either direction of gene conversion may be equally likely, or allele-specific effects may create a preference for one direction. Now it is clear that **recombination hot spots** represent sites at which double-strand breaks are initiated, and that gaps are enlarged and converted

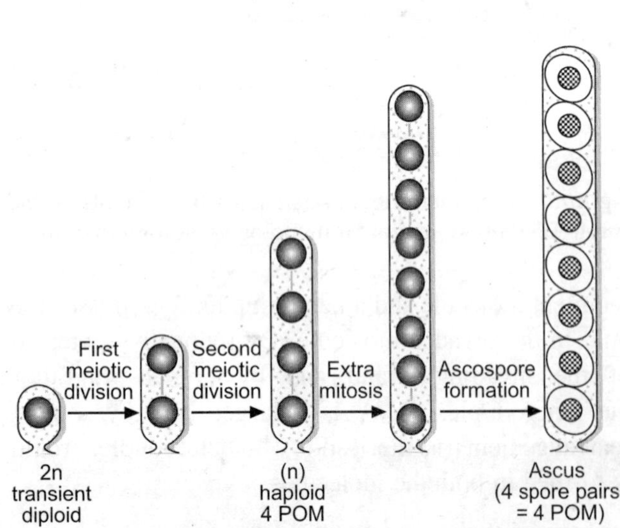

Fig. 17.8 Spore formation in *Neurospora*; production of paired tetrad in an ascus, due to linear meiosis and subsequent mitosis (POM = products of meiosis)

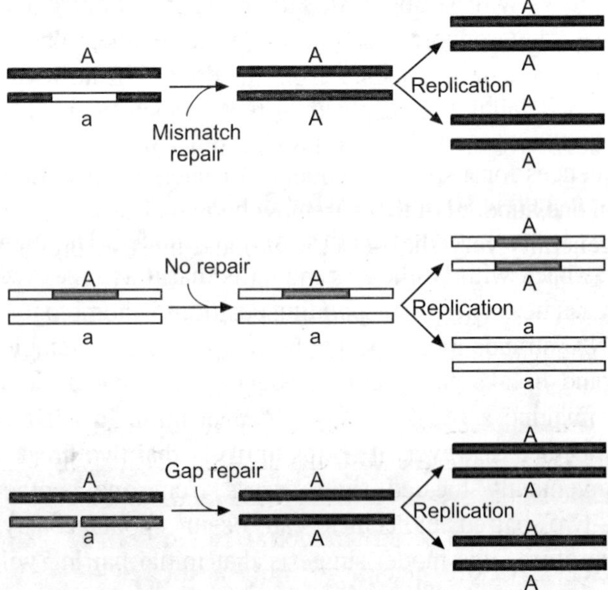

Fig. 17.9 A model for gene conversion in sporulating *Neurospora*

to long single-stranded ends. The gene conversion event may involve the transient pairing of two similar, but non-identical genes, followed by conversion of one gene sequence to that of the other. It is also the mechanism that switches the mating type of baker's yeast and antigenic variation in trypanosomes. In fact, the latter is a unidirectional recombination in which one form always replaces the other.

Programmed non-reciprocal recombination

Non-reciprocal recombination can be programmed to vary information present at a specific gene locus. This mechanism of gene regulation, termed **information cassette switching**, requires an active locus where information can be expressed and silent loci where alternative forms of the information, termed **cassettes**, are stored in a constitutively unexpressed state. Non-reciprocal recombination between a silent locus and the active locus is induced by a double-strand break at the active locus, which initiates repair and gene conversion. Cassette switching is a form of passive transcription. It differs from active transposition and other forms of information exchange, such as the use of alternative antibiotic resistance cassettes in bacterial integrations. In both active transposition and integron modulation (which involves site-specific recombination), the locus where the cassette is integrated is initially empty. In the case of switching, there is always information present at the active locus and switching is always a replacement process rather than an integration process.

17.5 SITE-SPECIFIC RECOMBINATION

In site-specific recombination, a type of protein called a recombinase acts on both the participating DNA sequences for a specific sequence of nucleotides. More than one hundred of them are now known. The sequence is generally very short (14–50 bp) and must be present on both the participating segments. Every site-specific recombinase catalyses a unique sequence. Site-specific recombination can take place between two separate DNA molecules as long as both have the same specific site (**intermolecular**). It can also take place within the same DNA molecule (**intramolecular**) as long as the specific site is present at least twice on the molecule.

Site-specific recombination is described as **conservative site-specific recombination**, (**CSSR**) for the following reasons.

- No nucleotides are added or lost as a result of a recombinational event.
- DNA replication is not needed, and thus it typifies a breakage and rejoining mechanism rather than a copy choice or break copy mechanism.
- The thermodynamics of site-specific recombinational events are extremely energy-efficient and no ATP is required.

Site-specific recombination gives precise DNA rearrangements and requires only limited homologies between the recombining species. This involves the exchange of non-homologous but specific pieces of DNA, and often with the result that different genes or sets of genes are then expressed. Site-specific recombination have been found in organisms ranging from bacteria to vertebrates.

17.5.1 Biological roles of site-specific recombination

Most biological systems including cells and viruses use CSSR for a wide variety of biological functions. Some of these functions are discussed in the following sections.

Site-specific recombination in bacteria

The intestinal bacterium *Salmonella typhimurium* can spontaneously change the protein in its flagella. A colony of bacteria derived from a cell show the same kind of flagellar protein. If this colony grows large enough, most of these cells generally will have the original type of flagellin, but some may have a new variety. Clearly, some of the bacteria had replaced one kind of protein in their flagella with another. This is referred to as **phase variation**. Further, a colony with the new type of flagellin, or phase, could give rise to subcolonies with the original phase.

Salmonella contains two genes (*H1* and *H2*) for flagellins and expresses only one of them at a given time. A shift in phase occurs about once every thousand cell divisions. This switching between flagellins *H1* and *H2*, helps the bacterium evade the immune response of its host. To be effective, the change must be complete. In phase 2, for example, the presence in the flagellum of even a few H1 molecules amidst many H2 would render this motile appendage vulnerable to deactivation by antibodies directed against *H1* which is the type produced in phase 1. The switch must be both absolute and reversible.

Surface glycoprotein genes in trypanosomes

Antigen switching describes a defence mechanism used by trypanosomes and *Borrelia* to avoid the immune responses of vertebrate hosts. Trypanosomes are protozoan (a single-celled) parasites responsible for sleeping sickness. The immune response after infection is directed against the variant surface glycoprotein (VSG) of the protozoan. However, the parasite can escape from the host's defence by changing the VSG. There are about a thousand VSG genes in each trypanosome making up almost 10% of the genome. However, only one type—the **expression linked copy (ELC)**—is active at any time. Most of the genes are arranged in tandem, separated by several copies of incompletely homologous 70-bp repeats and all the genes have a common sequence of 6–8 bp at their 3' ends. In addition to the normal chromosomes, trypanosomes have about 100 minichromosomes, each one of which has a VSG gene towards each end. These subtelomeric VSG genes have copies of the 70-bp repeats in their upstream region and towards the end of the chromosomes, they have tandem repeats of the telomeric sequence 5'-CCCTAA-3'. Up to 20 of these subtelomeric copies have the potential to be active (have functional promoter), but only one is active at a time. Which of the expression-linked copies is transcribed appears to be determined at random, but the outcome is that parasites with a series of different surface glycoproteins are present in the bloodstream early in infection. VSG switch rate (recombination) ranges from 10^{-2} to 10^{-3} switch events per cell generation.

Expression-linked copies are always subtelomeric and differ from the other genes by the presence of a 5'-exon of 35 bp. This miniexon, which is absent from all non-functional copies of the gene, is present in 200 or more copies as part of a tandem repeat which may form part of a complex promoter. As most VSG genes are not present in expression-linked sites, their expression requires the transposition of an inactive VSG gene to the expression-linked site to replace the resident gene. This takes place most readily between subtelomeric genes, and although reciprocal recombination events are known, transfer more commonly requires the basic copy gene to be duplicated and for one copy to replace the ELC.

Recombination involves the 70-bp repeats on the 5' side of the gene and the common 6–8 bp homologous region at the 3' ends of the genes. The 5' miniexons are not lost from the ELC. Transposition of a subtelomeric gene into the EL site may involve recombination only at the 5' end of the gene so that the whole of the end of the chromosome is replaced. Alternatively, activation of a subtelomeric VSG may occur by a non-duplicative event involving reciprocal recombination between the ends of two chromosomes. Occasionally, recombination occurs at sites within the gene to generate chimeric VSGs. Recombination involving the ends of chromosomes is also common on the malarial parasite.

Nitrogen-fixing genes

Loss of DNA also occurs during the development of the cyanobacterium, *Anabaena*, from a vegetative cell to a heterocyst when there is a rearrangement of the genes required for nitrogen fixation (*nif*). There are two nitrogen fixation operons, known as *nifH* (containing the genes *nifH*, *nifD* and *nifK*) and *nifB* (containing the genes *nifB*, *fdxN*, *nifS* and *nifU*) together with the genes *nifQ* and *nifV*. In the vegetative cell, the *nifK* gene is separated by 11kb from the *nifD* and *nifH* genes. But during formation of the heterocyst, the intervening DNA is excised to form a circular molecule which remains in the heterocyst. Excision occurs as a result of site-specific recombination between two 11-bp, directly repeated sequences and requires the products of the *xisA* gene. The rearrangement allows the three genes to function as a single transcriptional unit.

A second excision event, occurring independently as a result of recombination between two 5-bp directly repeated sequences, deletes a 55-kb element from the *fdxN* gene, thereby juxtaposing the *nifB*, *fdxN*, *nifS* and *nifU* genes. These two excisions bring together the two operons in the heterocyst. The heterocysts develop at regular intervals along a filament of vegetative cells and, as with the antibody-producing cells, such differentiation is irreversible. Here, we have an example of a colony of bacteria in which some cells sacrifice their potential for multiplication in order to supply fixed nitrogen compounds to the remaining cells.

Since recombination results in a loss of the intervening sequence, such rearrangements differ from the situation where the inversion of an intervening stretch of DNA can bring together two previously separated regions of DNA such as flagellar phase variation in *Salmonella* and *E. coli* and genome isomerisation in the herpes simplex virus, which is a potentially reversible reaction in which no DNA is lost).

Phase variation in pilus

This is an example in eukaryotes of the transfer of a gene from a site in which it is inactive into an expression-linked site. In *Neisseria gonorrhoeae* (causes gonorrhoea) and *N. meningitidis* (causes meningococcal meningitis), the presence and nature of the pilus (required in bacterial conjugation) is variable. Thus, some cells lack a pilus (i.e., fail to produce a monomer—pilin—molecule) and others have pili containing one of a series of different pilin molecules. The active pilin gene is expressed from the expression-linked site *pilE,* whereas other genes are available as a result of efficient transformation. Exchange of pilin genes at the *piliE* site depends on *recA* and involves recombination between the incumbent *pilE* gene and one of the other available *pil* genes. The loss of an effective pilin (P^+ to P^-) occurs at a frequency of about 10^{-3}, whereas reversion to P^+ which requires production of effective pilin is only one-tenth as frequent.

Rearrangement of immunoglobulin genes

Humoral system in mammals consists of antibody-producing B lymphocytes. These lymphocytes have the ability to recognise a wide range of foreign antigens in solution and bind them by means of a cell surface antibody (immunoglobulin) molecule. Site-specific recombination can generate different antibodies to novel antigenic stimuli.

During the production of immunoglobulins in B lymphocytes, the DNA sequence within the antibody

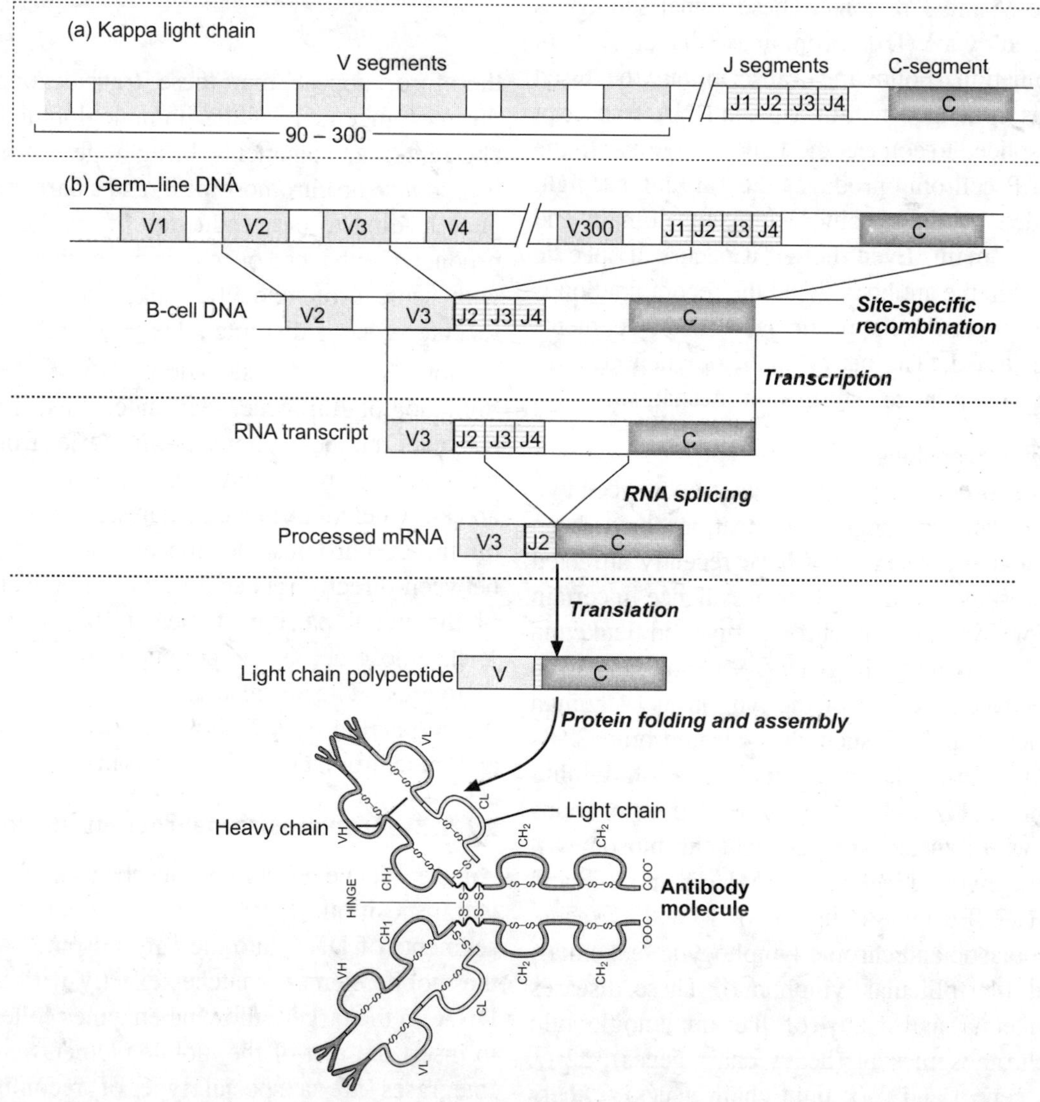

Fig. 17.10 Site-specific recombination within the gene that encodes the kappa κ light chain for immunoglobulin G (IgG) proteins (a) Kappa light chain (b) Germ line DNA

genes is rearranged by enzymes that recognise specific sites within those genes and catalyse the breakage and reunion of DNA segments. **Figure 17.10** shows the site-specific recombination of an antibody precursor gene. One type of light chain is the κ (kappa) light chain, which is a component of a class of antibodies known as immunoglobulin G (IgG). The sequences of the variable regions of this chain in the germ cell are encoded by three gene segments: V, D and J. During the differentiation of B cells, the κ light precursor gene is cut and rejoined so that one variable sequence becomes adjacent to a joining sequence. This recombination event is catalysed by an enzyme known as **V(D)J recombinase**. At the 3' end of every V domain and the 5' end of the J domains, there is a recombination signal sequence that functions as a location for site-specific recombination between the V and J regions. These signal sequences are recognised by a V(D)J recombinase that carries out the recombination. Following transcription, the fused VJ region is contained within a pre-mRNA transcript that is also spliced to connect the J and C regions. In the next step, a B cell only produces the particular κ light chain encoded by the specific VJ fusion domain and the constant domain. Even though it occurs at specific junctions within the antibody gene, the recombination is random with regard to the particular V and J segments that can be joined. The heavy-chain polypeptides are produced by a similar recombination mechanism.

Chromosomal translocations

This is the reciprocal exchange of material between two chromosomes and although translocations have been observed for many years, they have recently attracted greater interest as a result of their prevalence in certain cancers. For example, in chronic myeloid leukemia (CML), there has been (in at least 90% of cases), an exchange between the end of the long arms of human chromosomes 9 and 22 such that chromosome 22 is shortened to form the characteristic **Philadelphia chromosome** (**Fig. 17.11**). Other translocations associated with cancers are between chromosomes 8 and 21 (acute myeloid leukemia, AML); 8 and 2, 8 and 14 or 8 and 22 (Burkitt's lymphoma); 15 and 17 (acute non-lymphoblastic and chronic lymphocytic leukemia); and 18 and 14 (follicular lymphoma). These diseases are all leukemia and many of the immunoglobulin genes on chromosomes 14 (heavy chain genes), 22 (λ light chain genes) and 2 (κ light chain genes). Others occur nearer to the T cell receptor genes. It would,

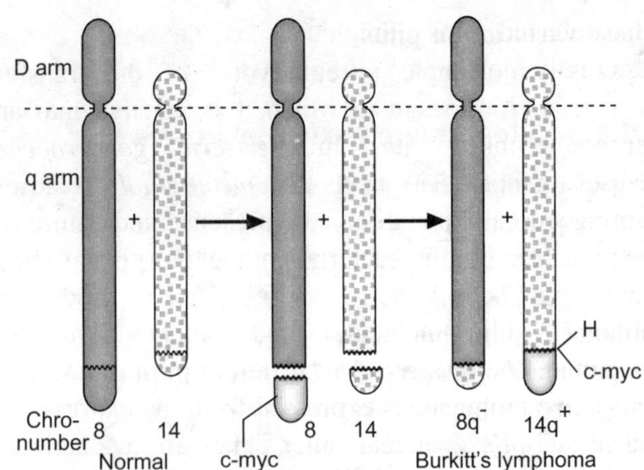

Fig. 17.11 Chromosomal translocation involved in the initiation of Burkitt's lymphoma brings the *c-myc* oncogene from chromosome 8 under the control of the immunoglobulin heavy chain promoter on chromosome 14.

therefore, appear that these translocations occur in the rearrangement of the immunoglobulin genes. The aberrant rearrangements bring together an oncogene e.g., *c-myc* on chromosome 8, and a part of the antibody gene leading to the production of an altered oncogene product or enhanced oncogene transcription. In a patient with acute promyelocytic leukemia, the retinoic acid receptor gene is disrupted due to the rearrangement.

Not all translocations involve aberrant immunoglobulin gene rearrangements. For instance, the small number of females who suffer from Duchenne muscular dystrophy have an X autosome translocation and large deletions in the human steroid sulphatase locus on the X chromosome appear due to recombination between directly repeated sequence elements. The site of the initial pairing at meiosis between the X- and Y-chromosome is a common sequence near the telomeres. This is also the site at which obligatory recombination occurs between one chromatid of the X and one chromatid of the Y chromosome.

17.5.2 Transpositional recombination

Another unique but complex mechanism of recombination is transposition. It also involves the integration of small segments of DNA into the chromosome. A segment of the mobile element matches exactly with a segment of DNA on the target, allowing enzymes called **integrases** to insert the rest of the mobile element into the target. Integrases are a special type of recombinases. The DNA segments that transpose themselves are known as **transposable elements** (**TEs**). They have been also referred to as 'jumping genes', because they are inherently mobile (see **Chapter 18**).

17.5.3 Illegitimate recombination

It is also called aberrant replication and recombination and it refers to the promiscuous recombination mechanisms, requiring little or no homology. However, on further analysis, it has become clear that, in many instances, short homologous regions are present and recombination occurs between these regions. In other cases, recombination occurs as a consequence of breakage of DNA molecules or as a result of replication errors. Illegitimate recombination takes place in the absence of recA; it is frequently associated with very small repeated sequences. The **short sequence repeats (SSR)** can be as small as 3–4 bp. The frequency of illegitimate recombination decreases as the SSR are located further apart and it increases exponentially as the SSR get longer i.e., at SSR of 8 bp, illegitimate recombination can be quite frequent.

Generally, this type of recombination is due to normal cellular processes using incorrect substrates and carried by end-joining, aberrant replication and recombination. When a double-stranded break occurs somewhere between repeats, resection of the intervening DNA can allow annealing of the complementary strands of the adjacent repeats. Ligation then results in the joining of the two fragments at a common sequence. Alternatively, a copy choice mechanism involves errors in replication in which the polymerase jumps from one repeat to the other. The repeats may even be present on different chromosomes, and the resulting errors are similar to those involved in the generation of frameshift mutations. Illegitimate recombination often causes mutation by gene disruption, and is prominent in human genetic diseases and cancer.

Manipulation of DNA *in vitro* by using recombination techniques is very recent, but well established and it is known as artificial recombination.

KEY POINTS

- Genetic recombination is a fundamental process in all cells and is responsible for rearranging sequences between different pieces of DNA, shaping the genome by altering the sequences that are present, limiting the divergence of repeated DNA sequences, pairing chromosomes before cell division and promoting DNA repair.
- There are three types of recombination that have been described: homologous , site-specific and illegitimate.
- Genetic exchange is a process that occurs constantly in all organisms, especially during meiosis of higher organisms. It works to blend and rearrange the chromosomes in most organisms.
- Genetic recombination can be broadly viewed at two levels, reciprocal and non-reciprocal. The former involves transfer of information in both directions, whereas in the latter there is only unidirectional transfer.
- General recombination is also known as homologous recombination. It is guided by base pairing and the exchange of homologous regions between two DNA molecules. In diploid eukaryotes, the homologous recombination commonly occurs during meiosis when the homologous duplicated chromosomes line up in parallel in metaphase I.
- At the end of meiosis, the resulting haploid gametes contain information derived from both maternally and paternally inherited chromosomes, thus ensuring that an individual inherits genes from all the four grandparents.
- The primary roles of homologous recombination are genetic mixing and DNA repair.
- The events involved in the exchange of sequences between DNA molecules were first described in bacterial systems during bacterial conjugation and in phage transduction.

- Non-reciprocal recombination is both symmetrical and conservative. An example for this type of recombination is gene conversion, because in this process one allele has been converted to the other.
- In site-specific recombination, a specific type of protein called a recombinase acts on both participating DNA sequences at a very specific sequence of nucleotides. The sequence is generally very short and present on both participating segments and the sequences are different for different site-specific recombinases. Lambda phage integration and excision is an example of site-specific recombination (CSSR).
- Genetic recombination, in addition to promoting DNA pairing, DNA repair and genetic exchange can also serve to change the DNA sequence at a specific chromosomal location. This type of recombination is sometimes used to regulate gene expression as seen in controlling the mating type of the budding yeast, antigen switching, nitrogen-fixing genes, pilus phase variation and rearrangement of immunoglobulin genes.
- A unique but complex mechanism of recombination is transposition. In some ways, transposition resembles site-specific recombination. It involves integration of small segments of DNA into the chromosome.
- Artificial recombination involves the manipulation of DNA *in vitro*.

Self-assessment Questions

1. What is genetic recombination?
2. Explain the mechanism of recombinational events in bacteria.
3. Briefly explain the different types of recombination.
4. What is crossing over?
5. Explain in detail the mechanism of homologous recombination.
6. What is Holliday junction and explain how it is resolved.
7. What is gene conversion?
8. Explain briefly about mismatch and gap repair mechanism.
9. Explain the model for site-specific recombination of phage.
10. Discuss the biological role of site-specific recombination.

FURTHER READING

Textbooks

Berg DE and MM Howe (ed.) (1989). *Mobile DNA.* Washington DC: American Society for Microbiology.

Brown TA (2002). *Genomes,* 2nd ed. New York: John Wiley and Oxford: BIOS Scientific Publishers Limited.

Gardner EJ, MJ Simmons and DP Snustad (1991). *Principles of Genetics.* 8th ed., New York: John Wiley and Sons, Inc.

Hartl DL and EW Jones (1998). *Genetics Principles and Analysis.* Jones and Bartlett Publishers.

Lewin B (2000). *Genes VII.* 7th ed. New York: Wiley.

Old RW and SB Primrose (1985). *Principles of Gene Manipulation.* Oxford: Blackwell.

Shapiro JA (ed.) (1983). *Mobile Genetic Elements.* New York: Academic Press.

Weaver RF (1999). *Molecular Biology.* New York: The McGraw-Hill Companies.

References

Cameriniotero RD and P Hsieh (1995). Homologous recombination proteins in prokaryotes and eukaryotes. *Annu. Rev. Genet.* 29:509–52.

Chen Y and PA Rice (2003). New insight into site specific recombination from FLP recombinases-DNA structures. *Annu. Rev. Biophys. Biomol. Struct.* 32:135–59.

Court DL, JA Sawitzke and LC Thomason (2002). Genetic engineering using homologous recombination. *Ann. Rev. Genet.* 36:361–88.

Cox MM (2000). Recombinational DNA repair of damaged replication forks in *Escherichia coli*; Questions *Annu. Rev. Genet.* 34:53.

Craig NL, R Craigie, M Gellert and AM Lambowitz (eds.) (2002). *Mobile DNA II.* American Society for Microbiology.

Derbyshire KM and NDF Grindley (1986). Replicative and conservative transpositions in bacteria. *Cell* 47:325–27.

Finnegan DJ (1985). Transposable elements in eukaryotes. *Int. Rev. Cytol.*, 93:281–25.

Gellert M (2002). V(D)J recombination: RAG proteins, repair factors, and regulation. *Annu. Rev. Biochem.* 71:101–32.

Grindley NDF and RR Reed (1985). Transpositional recombination in prokaryotes. *Annu. Rev. Biochem.*, 54: 863–96.

Haber JE (2002). Switching of *Saccharomyces cerevisiae* mating–type genes, In *Mobile DNA II* (ed. Craig NL, R Craigie, M Gellert and AM Lambowitz).

Holliday R (1964). A mechanism for gene conversion in fungi. *Genet. Res.* 5:282–304.

Kowalczykowski SC, DA Dixon, AK Eggleston, SD Lauder and WM Rehrauer (1994). Biochemistry of homologous recombination in *Escherichia coli. Microbiol. Rev.* 58:401–65.

Lichten M and ASH Goldman (1995). Meiotic recombination hotspots. *Annu. Rev. Genet.* 29:423–44.

Lusetti SL and MM Cox (2002). The bacterial recA protein forks. *Annu. Rev. Biochem.* 71:71–100.

Mayerhofer R, Z Koncz-Kalman, C Nawrath, G Bakkeren, A Crameri, K Angelis, GP Redei, J Schell, B Hohn, C Koncz (1991). T-DNA integration: A mode of illegitimate recombination in plants. *EMBO J.* 10(3):697–704.

McKim KS, JK Jang and EA Manheim (2002). Meiotic recombination and chromosome segregation in *Drosophila* females. *Annu. Rev. Genet.* 36:205–32.

Paques F and JE Haber (1999). Multiple pathways of recombination induced by double-stranded breaks in *Saccharomyces cerevisiae. Microbiol. Mol. Biol. Rev.* 63: 349–04.

Pastink A, JC Eeken and PH Lohman (2001). Genomic integrity and the repair of double-strand DNA breaks. *Mutat. Res.* 480–481:37–50.

Saedler H and A Gierl (1996). Transposable elements. *Curr. Top. Microbiol. Immunol.* 204.

Shapiro JA (1979). Molecular model for the transposition and replication of bacteriophage Mu and other transposable elements. *Proc. Natl. Acad. Sci.*, USA, 76:1933–37.

Shinagawa H and H Iwasaki (1996). Processing the Holliday junction in homologous recombination. *Trends Biochem. Sci.* 21:107–11.

Sigal N and B Alberts (1972). Genetic recombination: The nature of crossed strand exchange between two homologous DNA molecules. *J. Molec. Biol.* 71:789–93.

Smith GR (1988). Homologous recombination in prokaryotes. *Microbiological reviews* 52:1–28.

Smith GR (2001). Homologous recombination near and far from DNA breaks: Alternative roles and contrasting views. *Annu. Rev. Genet.* 35:243–74.

Stavnezer J (1996). Antibody class switching. *Adv. Immunol.* 61:79–146.

Stark WM, MR Boocock and DJ Sherratt (1992). Catalysis by site-specific recombinases. *Trends Genet.* 8:432–39.

Symington LS (2002). Role of RAD52 epistasis group genes in homologous recombination and double-strand break repair. *Microbiol. Mol. Biol. Rev.* 66:630–70.

Watson JD, NH Hopkins, JW Roberts, J Steitz and AM Weiner (ed.) (1987). The molecular biology of the gene, Vol. 1. Menlo Park: Benjamin Cummings.

CHAPTER 18

Mobile Elements: Drivers of Genome Evolution

LEARNING OBJECTIVES

After reading this chapter, you will be able to:

- Learn about the different types of transposons
- Understand the molecular mechanism of transposition
- Appreciate the biological significance of transposons in genomic evolution

18.1 INTRODUCTION

The concept of chromosomes as invariable entities of genetic information transmitted unchanged from generation to generation has now been changed in favour of a more dynamic view (i.e., **'genome in flux'**). Rearrangements (for example, inversion and duplication of genes) occur at low frequency in both bacteria and eukaryotes. Recent investigation has found that the formation of new arrangements of genes is due to the movement of large blocks of DNA. Genetic determinants, often identified as specific nucleic acid sequences repeatedly change their positions within or between cellular genomes. Because their genomic positions are not fixed, these determinants may conveniently be studied under the heading **mobile genetic elements**. In higher eukaryotes, a significant portion of the genome is made up of functional and 'ghost' transposable elements (genes inactivated by mutations). Mobile genetic elements are segments of DNA that can move from site to site in the genome, or between genomes in the same cell. They are a diverse group differing in structure, mechanism of mobilisation, distribution, freedom of movement and level of autonomy. The mobility of some elements has a cellular function, but most are believed to be selfish DNA.

Movements of DNA segments are generally mediated by two genetic processes: **recombination** and **transposition**. Transposition is a specific form of genetic recombination that moves certain genetic elements from one DNA site to another. The frequency of transposition can vary widely, depending on the transposon. Transposition is generally low in both bacteria and eukaryotes (10^{-7} to 10^{-2} per generation) and it should not significantly harm the organism. Movement of mobile elements requires DNA exchange and can occur with or without duplication of the element and in some cases the recombination reaction involves a transient RNA intermediate.

Transposons or transposable elements (TE) can move virtually to any position in a cell's genome without the benefit of homology. Thus, transposable elements may be defined as segments of DNA that are bounded by specific terminal DNA sequences and have the ability to move to new DNA sites with little or no specificity for the latter. Transposition has also been called illegitimate recombination because it requires no homology between sequences nor is it site-specific. Consequently, it is relatively inefficient, but upon insertion it can completely disrupt gene function. Therefore, transposable elements have been considered the most common source of new mutations in many organisms and are important causes of mutations leading to genetic diseases in humans. The flexibility of transposable elements to insert so promiscuously in DNA has also led to their modification and use as mutagens and DNA delivery vectors in experimental biology. Mobile elements are present in the genomes of all life forms. More than 50% of human and maize genomes are composed of transposon-related DNA sequences and are highly variable across genomes of different organisms. Mobile elements of prokaryotes and eukaryotes exhibit differences as well as similarities, and will, therefore be discussed separately.

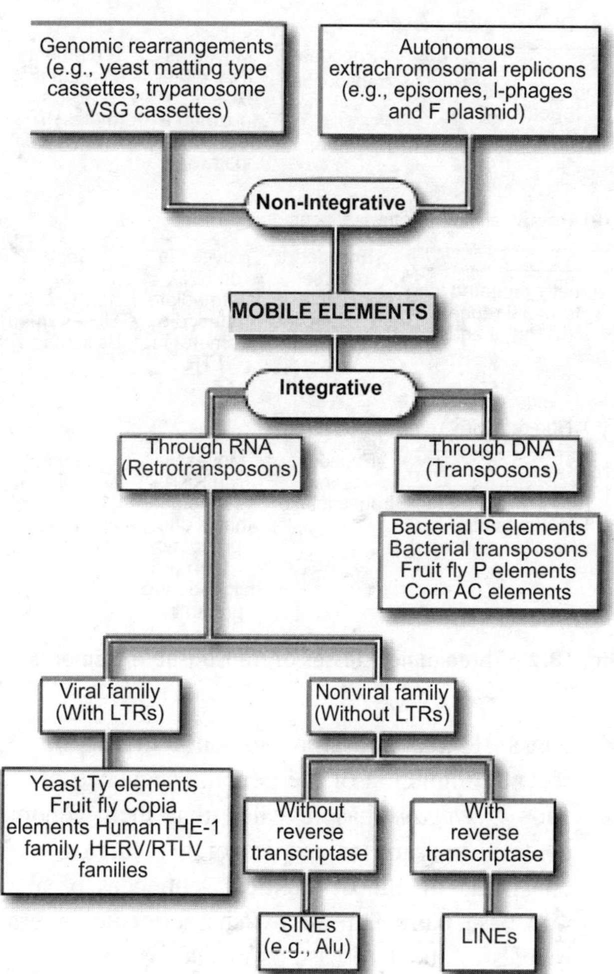

Fig. 18.1 Classification of mobile elements

18.2 CLASSIFICATION OF MOBILE ELEMENTS

Mobile elements have been classified variously. According to their level of independence, they can be grouped into two broad categories (**Fig. 18.1**): either **integrated** into the host genome or as **autonomous extrachromosomal replicons** (termed as **episomes**). Also the other elements whose mobility is controlled by the cell to mediate specific **genomic rearrangements** are termed **cassettes** (e.g., trypanosome VSG and yeast mating type). The mobility of episomes and cassettes is facilitated by homologous or site-specific recombinations, which have target site choice. But those elements, which cannot replicate outside the host genome, but control their own mobility within it, are termed **transposable elements (TE)**. Thus, the mobile elements that transpose through DNA are called **transposons** and those *via* RNA are referred to as **retrotransposons**. Since the pioneering studies of Barbara McClintock, various types of transposable elements have been found in bacteria, fungi, plants and animals. Transposable elements fall into different classes based on their organisation and active transition mechanisms (**Fig. 18.2**).

- **Class I transposable elements (retroelements)** utilise an RNA intermediate during the transposition process. These are known as retroelements, retrotransposons or retroposons. These increase in number during retrotranspositional events. Retrotransposons work by copying themselves and pasting copies back into the genome in multiple places. They behave very similarly to retroviruses, such as HIV, giving a clue to the evolutionary origins of such viruses.
- **Class II transposable elements (transposons)** are DNA transposons; their transposition mechanism does not involve an RNA intermediate. Different types of transposases work in different ways. Some can bind to any part of the DNA molecule and the target site can therefore be anywhere, while others bind to specific sequences. In **simple** or **conservative transposition**, the TE is removed from its original site and transferred to a new target site (cut and paste mechanism). However, in case of replicative type, the transposon replicates itself to a new target site.

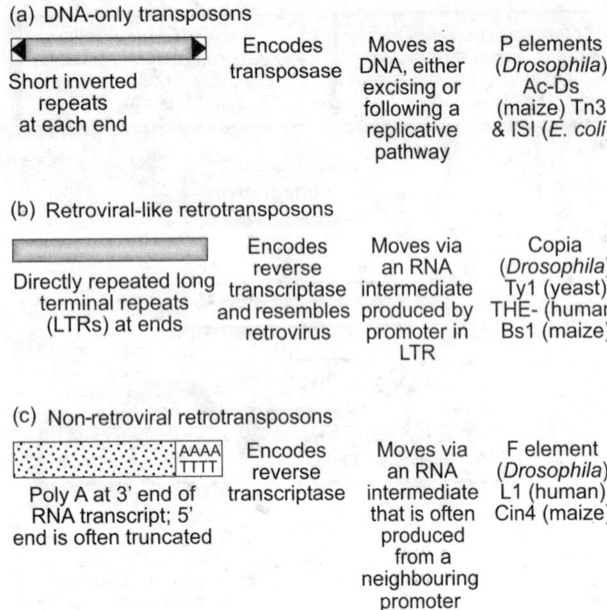

Fig. 18.2 Three major classes of transposable elements

- **Class III transposable elements (MITEs):** The recent completion of the genome sequence of rice and *C. elegans* has revealed that their genomes contain thousands of copies of a recurring motif consisting of almost identical sequences of about 400 base pairs flanked by characteristic inverted repeats of about 15 base pairs such as

5' GGCCAGTCACAATGG.. ~400 nt.. CCATTGTGACTGGCC 3'
3' CCGGTCAGTGTTACC.. ~400 nt.. GGTAACACTGACCGG 5'

MITEs are too small to encode any protein. Just how they are copied and moved to new locations is still uncertain. There are over 100,000 MITEs in the rice genome (representing some 6% of the total genome). Some of the mutations found in certain strains of rice are caused by insertion of a MITE in the gene. MITEs have also been found in the genome of humans, *Xenopus* and apples.

- Other transposable elements, termed as **defective**, lack enzyme functions but retain the *cis*-acting sites and may be mobilised in *trans* by an active element (also called as **non-autonomous** or **passive**, which are derived from related families of active sequences). A large and structurally diverse family of non-autonomous retroelements represents sequences with no similarity to known autonomous elements (also called as **processed pseudogenes**), and appears to be cellular RNAs transposed adventitiously when reverse transcriptase and integrase are supplied in *trans*.

18.3 GENERAL MECHANISMS OF TRANSPOSITION

The end result of the transposition process is the insertion of a transposon between two nucleotide pairs in a recipient DNA molecule. Base sequence studies of many transposons and their insertion sites reveals that there is no sequence homology; however, the analysis of sequences of the regions in which the inserted element joins the recipient DNA yielded several clues needed for understanding the mechanism of transposition. Within each family (transposons and retroelements), several specific types of transposition mechanisms can be used. The following are some general mechanisms applicable to all transposable elements.

Proliferation: All transposition events have the potential to increase the number of copies of the TE in the genome; this is accomplished either actively or passively.

Stages of transposition: Transposons actively mobilise in three stages (**Fig. 18.3**): (i) the **donor cut stage**, an endonucleolytic cleavage reaction catalysed by transposase, which separates the TE from the host DNA (ii) the **strand exchange stage**, a pair of transestrification reactions also catalysed by transposase, which join the 3' ends of the TE to host DNA at the target site (iii) the **repair stage**, DNA synthesis undertaken by the

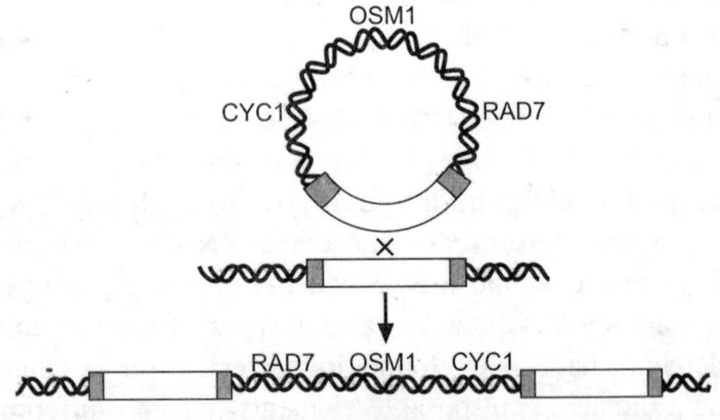

Fig. 18.3 Possible mechanism of transposition of the CYM1–OSM1-RAD7 gene cluster. Recombination between the Tys flanking the three gene cluster has generated a circle containing one Ty and the CYC1, OSM1 and RAD7 genes. The CYC1-OSM1-RAD7 gene cluster is re-integrated into the yeast genome.

cell, which fills any remaining gaps. Retrotransposons precede these three stages by transcription and reverse transcription to generate a cDNA for processing and integration.

Target site duplication: For most TEs, the strand exchange stage of transposition occurs at a staggered break in the target DNA, so that the freshly integrated element is initially flanked by ssDNA tail. Repair synthesis over the gaps generates direct repeats of the sequence at the target site, so-called **target site duplications (TSDs)**. Most TEs generate TSDs of specific length, but the heterogeneous class 1.2 retroelements generate variable sized TSDs even within the same genome, probably because they utilise broken DNA ends adventitiously.

Target preference: Transposable elements vary in their target site preference, some integrating at specific sequences, others appearing to integrate at random. For example, the *E. coli* transposon Tn7 usually integrates at a specific site, but can switch to random transposition if the site is unavailable. Most elements avoid overlapping the integration site i.e., they display *cis* immunity (**twintrons**). Where individual transposition events can be studied, many elements demonstrate topological preference. Some eukaryotic transposons, move only a short distance (1–10 kbp) from their donor site, this is **regional re-integration**. Bacterial replicative transposons prefer to move to plasmids rather than from plasmids to the bacterial chromosome.

18.3.1 The transposition machinery

During transposition, three DNA substrates interact with the two inverted repeats at the ends of the transposon and the target DNA. The inverted repeats are the sites of breaking and joining of the transposon DNA to the target DNA. They also bind transposition proteins to bring them into close proximity. For most transposons, there is little or no apparent sequence specificity for the target, but some hotspots may be present on the target DNA for some transposons. The major enzyme responsible for transposition is **transposase**. It mediates specific recognition of the transposon ends, as well as breaking and joining of the transposon to the target DNA.

18.3.2 Accessory proteins involved in transposition

Accessory proteins encoded by the transposon: Some transposons encode additional proteins that play a role in moving or regulating the movement of the element. Tn7, for example, has one protein that mediates recognition of the ends of the element and another protein that is responsible for breaking and joining the DNA molecules. Several other Tn7 proteins control the end-binding activity and the breaking and joining activities of these proteins.

Accessory proteins encoded by the host: Some accessory proteins synthesised by the host can regulate transposition, directly affecting the breaking and joining reactions, or help clean up the DNA after transposition has taken place. The *E. coli* protein, *Dam* methylase is responsible for methylating newly replicated DNA. Fully methylated DNA represses the transposition of several elements (e.g., IS10, IS50 and IS903). In *dam* mutants, the DNA is unmethylated, which results in increased transposition frequency. *E. coli* contains several, small DNA-binding proteins that are involved in constructing and maintaining DNA-protein complexes. DNA-binding proteins named IHF (integration host factor), HNS (histone-like nucleoid structuring protein), HU (histone-like protein) and FIS (regulator of transcription) aid in the transposition reactions of several different elements including Mu, IS1, IS10 and IS50. Other types of proteins involved in transposition are DNA replication and repair proteins.

18.4 TYPES OF TRANSPOSITION REACTIONS

Because of their ability to move from one place to another, transposons are sometimes called 'jumping genes'. There are two types of mechanisms for transposition: **conservative** (cut-and-paste) and **replicative** transposition (**Fig. 18.4**). If the transposon leaves one place and jumps to the other, it is called **conservative transposition**, because both strands of the original DNA are conserved as they move together from one place to the other. If transposition involves DNA replication, so one copy of the transposon remains at its original site as another copy inserts at the new site, it is called **replicative transposition**.

18.4.1 Replicative transposition

Some DNA transposons move using a mechanism called replicative transposition (also known as **copy and paste**). Replicative transposition involves duplication of the element, one copy remaining at the donor site and one copy integrating at the target site. Replicative

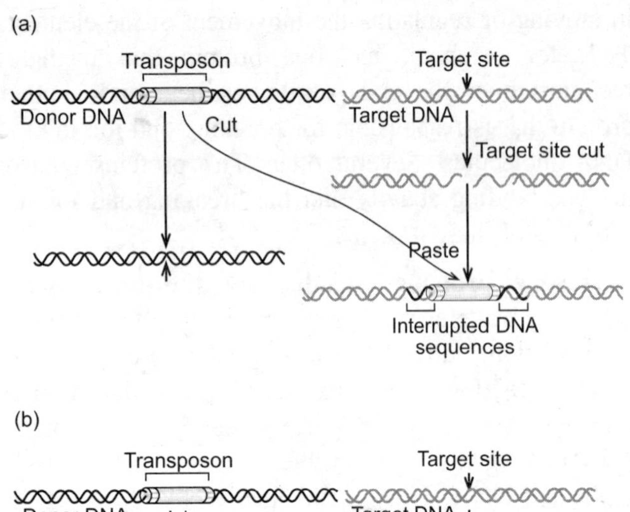

Fig. 18.4 A unified model of transposition (a) Conservative (b) Replicative

transposition is exhibited by class II transposons and bacteriophage Mu. A model of replicative transposition is shown in **Fig. 18.5**. Its key feature is the formation of an intermediate, evidence for the existence of which has been obtained by Mizuuchi and coworkers using bacteriophage Mu. As shown in the figure, the first step involves the assembly of the transposase protein on the two ends of the transposon DNA to generate a transpososome. The next step is DNA cleavage at the ends of the transposon DNA: staggered endonucleolytic break in the target DNA that produces 5'-protruding ends, and a single nick at each end of the transposable element leaving a free 3'-OH on the sides adjacent to the element. This reaction is catalysed by the transposase within the transpososome. The 5'-ends of the target is then ligated to the 3'-OH groups on the element, resulting in forks. It is assumed that host enzymes attach to the forks of this intermediate and replicate through the transposable element, without necessarily replicating any other DNA. The product of this, containing two copies of the transposon and both donor and acceptor chromosomes, is termed a **'co-integrate'**. In class II transposons, the resolvase encoded by the *tnpR* gene binds to the *res* site to promote homologous recombination regenerating host and target

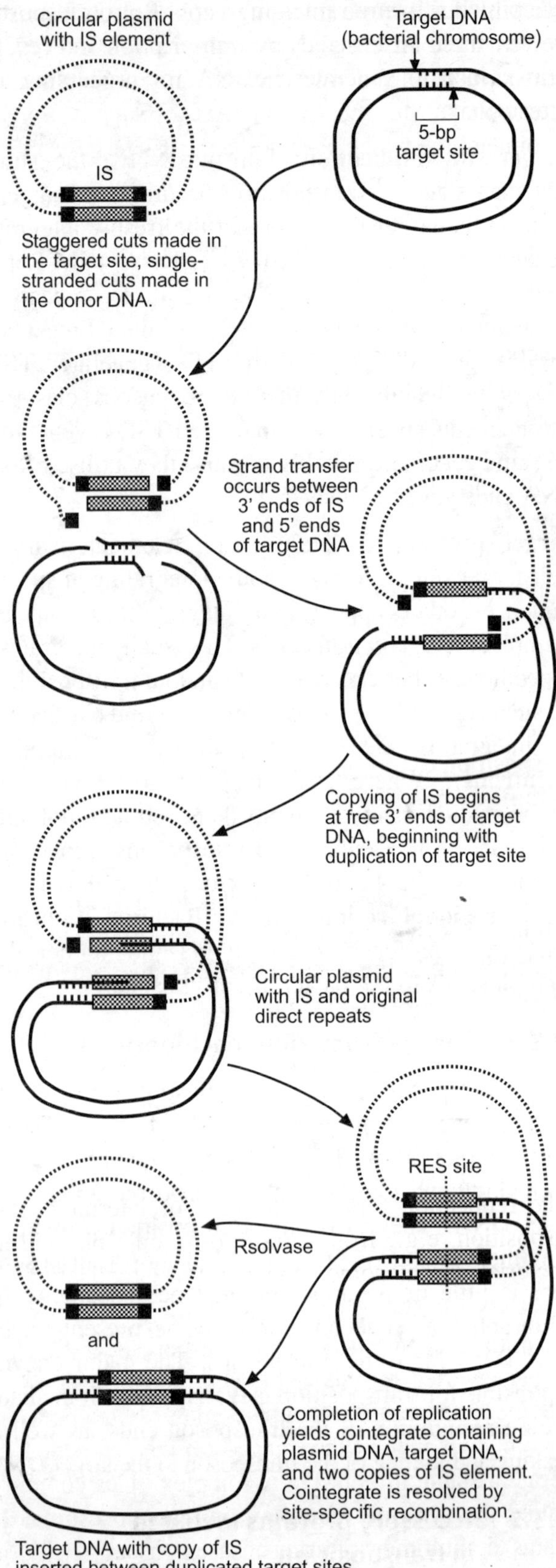

Fig. 18.5 Model for replicative transposition of certain bacterial insertion sequences (e.g., gd)

replicons. An enzyme to resolve cointegrates seems to have emerged in class II transposons because of the intrinsic instability of two different fused genomes. In bacteriophage Mu, there is no need for such resolution of cointegrate, which is merely the bacterial genome containing Mu as prophage.

Following replicative transposition between, for example two plasmids, the circles become composite structure, with copies of the transposable element marking the boundary between each contributing replicon. Transposons that move in this manner also carry functions that separate and then cointegrate into the two constituent replicons. This process is termed **resolution**. Each transposon encodes **resolvase**, a site-specific recombinase, which acts on a short sequence within the transposon, the internal resolution site (*res*). Recombination between two *res* elements in cointegrates catalysed by resolvase separates the two molecules. The active form of the resolvase is a dimer.

18.4.2 Conservative transposition

Conservative transposition (or **non-replicative transposition**), also known as **cut and paste mechanism**, is exhibited by class I transposons (e.g., Tn5 and Tn10), and is also an option available to certain class II and class III transposons. The transposition is lost from its original location in the donor DNA during the transfer (moves conservatively), and as the double-stranded break in the donor DNA on either side of the transposon is not repaired, the donor DNA is, in fact, destroyed (**Fig. 18.6**). Despite the loss of the donor, it is possible that there is net increase in the number of transposons. This can happen intramolecularly when transposition occurs from one of the two copies of the newly replicated transposon to a position ahead of the replication fork. In the case of intermolecular transposition, e.g., from a transposon on a plasmid to a bacterial chromosome, compensatory replication of another copy of the plasmid can occur.

To initiate recombination, the transposase binds to the terminal inverted repeats at the end of the transposon. Once the transposase recognises these sequences, it brings the two ends of the transposon DNA together to generate a stable protein DNA complex. This complex is called the **synaptic complex** or **transpososome**. It contains a multimer of transposases (usually two or four subunits) and the two DNA ends. This complex ensures that the DNA cleavage and joining reactions needed to move the transposon occur simultaneously on the two ends of the element's DNA and also protects the DNA ends from cellular enzymes during recombination. The next step, the transposase subunits within the transpososome, first cleave one DNA strand at each end of the transposon, at the junction between the transposon DNA and the host sequence in which it is inserted (called **flanking host DNA**). Upon cleavage, the transposon sequence terminates with free 3'-OH groups at each end of the element's DNA. At the same time, the other DNA strand at each end of the element will also be cleaved. After excision of the transposon, the 3'-OH ends of the transposon DNA attach the DNA phosphodiester bonds at the site of the new insertion. This DNA segment is called the **target DNA**. As a result of this attack, the transposon DNA is covalently joined to the DNA at the target site. This DNA joining reaction occurs by a one-step transesterification reaction that is called **DNA strand transfer** (this reaction is similar to RNA splicing).

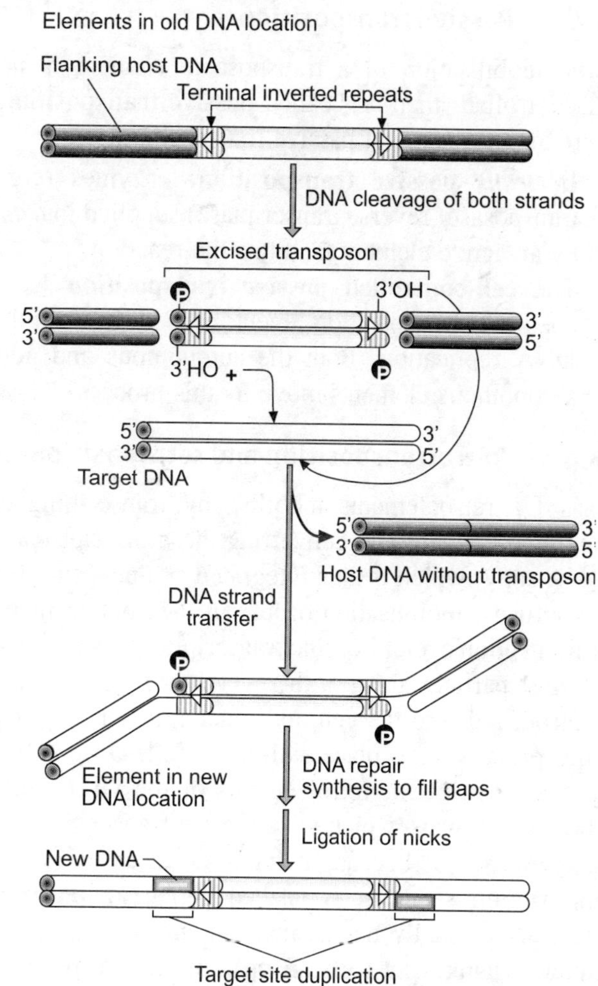

Fig. 18.6 The cut and paste mechanism of transposition

18.4.3 Passive transposition

If the mobilisation of a transposable element is not self-controlled, then it is called **passive transposition**. There are two types of passive transposition:

- **In *trans* passive transposition**, enzymes (e.g., transposase, reverse transcriptase) supplied in *trans* by an active element facilitate the reaction.
- The **cell-controlled passive transposition** is an apparent transposition event caused by cellular DNA replication. Both the autonomous and non-autonomous elements move by this process.

18.4.4 Retrotransposition and retroposition

Class I.1 retroelements mobilise by transcribing an integrated element and converting the transcript into a cDNA copy, which is then integrated at a new site. For retroviruses, mobilisation occurs in the context of the viral infection cycle, i.e., the transcript is packaged into the viral particle along with reverse transcriptase and is introduced into the cell and other retroviruses. The entire process takes place within the cell. Once cDNA has been generated, retroelements of the viral family follow a pathway, which is similar to transposition by transposons, but because it includes an extra reverse transcription stage, it is termed **retrotransposition**. The cDNA usually has a small amount of host DNA sequence flanking it (often 2 bp), and this is removed from the 3' end of each strand by integrase in a reaction analogous to the 3' donor cut stage of transposition. The integrase then facilitates a transesterification reaction between the free 3' ends of the cDNA and the host DNA at the target site. Integration is followed by repair synthesis, which generates 2-bp target site duplications and presumably removes the remaining 22-bp of the 'old' donor site DNA from the 5' ends of the element. Class I.2 elements mobilise using a mechanism unrelated to retroviral replication. The cDNA is generated either by self-priming or by DNA-priming at the target site. Integration is achieved by cDNA 'repair synthesis' across the target site using RNA as the template. Because this pathway lacks most of the steps in classic transposition but involves a reverse transcription step, it is termed **retroposition**.

Priming reverse transcriptase

The most common mechanism for priming reverse transcription of the LTR retroelements is by the annealing of the 3' end of a specific tRNA species to the primer-binding site (PBS) adjacent to the upstream LTR. This 3'-end then serves as a primer for the reverse transcriptase that copies the RNA in a complex series of events into a duplex DNA, which occurs in the cytoplasm. A few elements have developed an alternate priming strategy in which the 3' end of the element transcript can prime reverse transcription from an internal priming site in the transcript. Priming reverse transcription of non-LTR elements uses the poly A stretch at their 3'-end as the PBS. The most likely primer for this process is the 3'-end of a nick in the genomic integration target site to prime reverse transcription from the poly A end of the RNA and is referred to as **target-primed reverse transcription (TPRT)**. The use of a poly A tract as the priming target for these non-stringent elements is likely to be a critical factor in allowing the non-autnomous elements to take advantage of this mechanism and have the same type of priming occur on their A-rich regions. In the case of SINEs, their A-rich region is not at the extreme 3' end of the transcript.

Integration

The LTR retroelements differ from the non-LTR elements in their integration as they do not utilise their RNA in the integration process. Thus, the LTR elements have to convert their RNA into a dsDNA molecule that is transported to the nucleus and integrated. This integration process is similar to the integration of the DNA transposons, with an element-encoded nuclease making specific nicks in both the element DNA and the integration site to catalyse the integration process. Different transport mechanisms to the nucleus have been found for different elements in yeast. Non-LTR elements usually have two ORFs that encode proteins that are essential for the retrotransposition activity. The first ORF encodes an RNA-binding protein that shows some specificity for binding L1 RNA. The second ORF encodes the reverse transcriptase and an endonuclease that appears to nick the chromosomal insertion site. The L1 endonuclease cleaves at a consensus sequence of 3'-AATTTT-5', which potentially allows the T's at the 3'-terminus of the nick to prime reverse transcription from the polyadenylated L1 RNA. Some non-mammalian LINE elements, such as R2Bm, have been shown to have an endonuclease with restriction enzyme-like specificity for cleavage at a specific site in the rRNA gene cluster. Others, such as TRAS1 and SART1 in *Bombyx mori*, insert preferentially in the telomeric repeat and their specific site preferences can be swapped by swapping their endonuclease domains.

18.4.5 Poly A retrotransposons

The major characteristic of poly-A retrotransposons is that they do not have the terminal inverted repeats present in the other transposon classes. Instead, the two ends of the element have distinct sequences (**Fig. 18.7**). One end of this retrotransposon is called 5'-UTR end and the other end has the region called the 3'-UTR followed by a stretch of AT base pairs called the **poly A sequence**. These elements are also flanked by short target site duplications. The transposition activity of this class of retrotransposons mainly depends on two genes, which they carry and are known as ORF1 and ORF2. ORF1 encodes an RNA-binding protein. ORF2 encodes a protein with both reverse transcriptase activity and an endonuclease activity. These proteins play essential roles during recombination. The poly A retrotransposons exist commonly in both autonomous and non-autonomous forms.

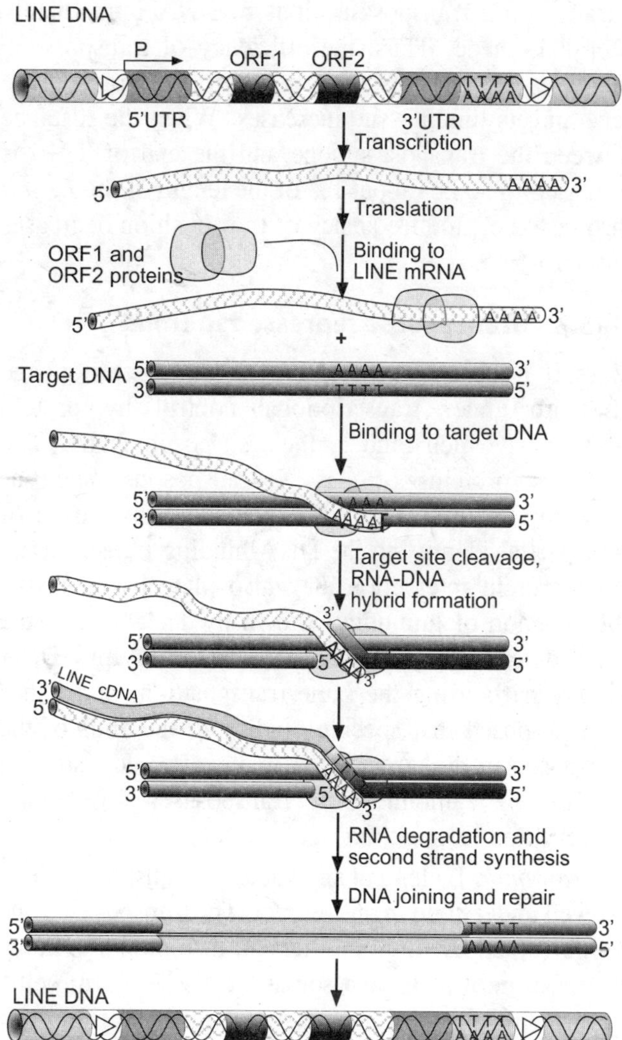

Fig. 18.7 Transcription of a poly A retrotransposon by target site-primed reverse transcription (***Source***: Watson et al 2004)

18.5 REGULATION OF TRANSPOSITION

The successful colonisation of transposons in all life forms is mainly attributed to their regulation of transposition, which helps to establish a harmonious coexistence with the host cell. This coexistence is very essential for survival of transposons, since without the host DNA, they cannot survive. On the other hand, transposons can cause various kinds of mutations, altering gene expression and promoting large-scale DNA rearrangements. Therefore, excessive transposition would be lethal. Transposons are regulated mainly for their copy number and choice of target site. Such regulatory mechanisms control not only the frequency of the transposon, but also its timing in relation to the cell cycle and nature of target site. Typically, the transposition frequency is enhanced when a transposon enters a naive host or immediately after DNA replication. There are several different ways in which transposition is subject to regulation. These are discussed briefly in the following section.

18.5.1 Regulation by *Dam* methylation

D-adenine methylase (the *dam* gene product) adds a methyl group to the adenine residue in the sequence 5' GATC 3' ('*dam* sites'). Because the sequence is present in both strands of dsDNA, both strands of DNA are methylated within the paired GATC sequences. However, when the sequence is copied during DNA replication initially, only the DNA strand derived from the parent is methylated. When one of the strands is methylated and the other is not, the site is hemimethylated. Hemimethylated *dam* sites provide a signal that DNA replication has occurred. This signal is recognised by some DNA-binding proteins. In transposon Tn10, two *dam* sites play an important role in regulation of transposition.

18.5.2 Transposase in *cis*

Increasing the concentration of transposase increases the rate of transposition. However, cells with multiple copies of a transposon may not show a higher frequency of transposition. Transposase is very unstable, so it is not normally accumulated to high concentrations. Rather than diffuse through the cell to find the end of a transposon, transposase binds to DNA immediately after it is made. Thus, the efficiency of transposition decreases as the distance between the transposase gene and its binding site increases. When the distance between the transposase gene and the end of Tn10 is increased to 50 kb (about 1% of the length of the *E. coli* chromosome), the frequency of transposition decreases about 15-fold.

18.5.3 Transposase repressed in *trans*

Class II transposons, such as Tn3 and bacteriophage Mu, are under transcriptional control by protein repressors, which bind to the transposon DNA. The repressor proteins of class I transposons, such as those for IS1 and Tn5, also bind DNA by virtue of their partial identity to the DNA-binding transposases. DNA-protein interaction may also play a role in the phenomenon of immunity, shown by Tn3. This is the reduced tendency for transposon to occur into DNA already harbouring the same transposition, and it has been proposed that specific binding to the ends of the transposon in the target DNA may cause sequestration of limiting amounts of transposase, preventing transcription from occurring.

Drosophila P element and bacterial transposon Tn10 are well understood in this respect. The transposase gene is interrupted by three introns and differential splicing occurs in germ cells and somatic cells. In germ cells, all three introns are removed and the mature transcript encodes a polypeptide of 87 kDa with transposase activity. In somatic cells, intron 3 is retained and a truncated polypeptide is generated due to an in-frame stop codon within the intron. This truncated polypeptide may act as a repressor of transposition.

18.6 CONSEQUENCES OF TRANSPOSITION

Although mobile element insertions cause a significant level of damage to the human genome, unequal recombination events between dispersed elements cause even more damage. An estimate suggests that at least 0.3% of all human genetic diseases are caused by unequal *Alu* or *Alu* homologous recombination events that cause moderate-size deletions or duplications.

18.6.1 Genetic consequences of transposition

Transposable elements were first recognised by their ability to cause mutations and insertional mutation is the most direct consequence of transposition. Insertional mutation can affect single genes or many genes simultaneously. Transposons can cause mutations in the following ways:

Mutation: If a transposon inserts itself into a functional gene, it will probably disable that gene. Insertion into exons, introns and even into DNA flanking the genes (which may contain promoters and enhancers) can destroy or alter the gene's activity. Faulty repair of the gap left at the old site (in cut and paste transposition) can lead to mutation there.

Modulation of gene expression: Transposons also influence gene expression by interfering with the relationship between the gene and its regulatory elements or by modulating DNA structure. The modulation of gene expression without deactivation by direct insertion occurs if the element carries an endogenous gene whose expression is controlled within the element, or if it carries an endogenous regulatory element which influences the expression of adjacent genes.

Alteration in genomic organisation: Gene expression may be influenced by modulation of the structural properties of DNA. The insertion of IS1 can activate cryptic operons by altering the topology of the local DNA.

Gene silencing: DNA methylation is used by some eukaryotes to restrict proliferation of transposons, but if a transposon inserts near a repetitive endogenous sequence, this can be targeted by the same system, resulting in epigenetic silencing.

Unequal crossing over: The presence of a string of identical repeated sequences presents a problem for precise pairing during meiosis. A string of five *Alu* sequences on the 'invading strand' of one chromatid ensures that it pairs with the third sequence in the other strand. If it accidentally pairs with one of the other *Alu* sequences, the result will be an unequal crossover—one of the most common causes of duplications.

Genomic expansion and the C-value paradox: The genome of *Arabidopsis thaliana* contains 1.2×10^8 base pairs (bp) of DNA. About 14% of this consists of transposons; the rest are functional genes (about 25,000 of them). The maize (corn) genome contains 20 times more DNA (2.4×10^9 bp) but surely has no need for 20 times as many genes. In fact, 60% of the corn genome is made up of transposons. The figure for humans is 42%. Most of the 2.5×10^{11} bp of DNA in the genome of *Psilotum nudum* is presumably 'junk' DNA. So, it seems likely that the lack of an association between size of the genome and the number of functional genes, the C-value paradox, is caused by the amount of transposon DNA accumulated in the genome.

Human genetic diseases

Transposons often cause a number of genetic diseases in humans. SINEs (mostly *Alu* sequences) and LINEs cause only a small percentage of human mutations, however, they have been found to be the cause of the mutations responsible for some cases of human genetic diseases, including:

- Haemophilia A (factor VIII gene) and Haemophilia B (factor IX gene)
- X-linked severe combined immunodeficiency (SCID) gene for part of the IL-2 receptor
- Porphyria
- Predisposition to colon polyps and cancer (APC gene)
- Duchenne muscular dystrophy (dystrophin gene)
- Additionally, many transposons contain promoters, which drive transcription of their own transposase. These promoters can cause aberrant expression of linked genes, causing disease of mutant phenotypes.

18.7 BACTERIAL TRANSPOSONS

Prokaryotes contain several classes of transposable elements as natural constituents of the chromosome or of extrachromosomal plasmids and bacteriophage genomes. The earliest molecular understanding of mobile DNA elements came from studies with *E. coli* mutants. James Shapiro and others provided the experimental background for the discovery of bacterial transposons (**Fig. 18.8**).

18.7.1 Structure and classification of transposons

Recently, the term transposon has been adapted to encompass all transposable elements, which mobilise directly as DNA, i.e., all class II transposable elements. All transposons have a conserved structure comprising one or more open reading frames flanked by **inverted terminal repeats (ITRs)**. The terminal repeats are necessary for transposition and are the substrates recognised by the transposase. However, they are often not sufficient, and further internal sequences are required.

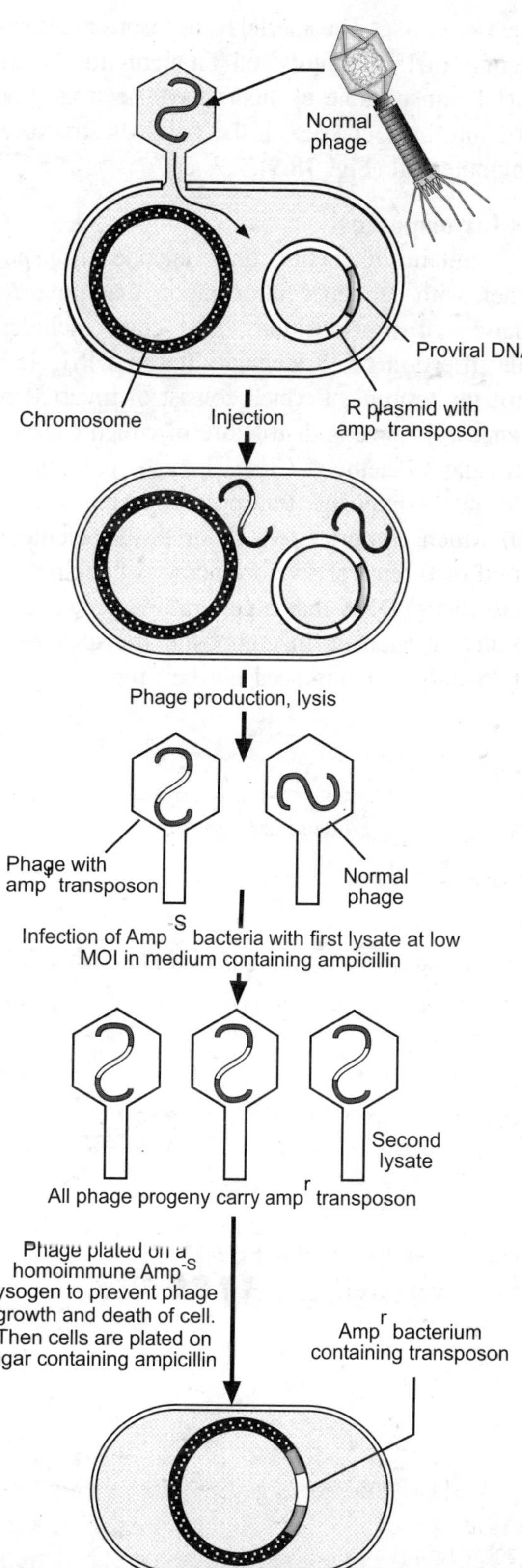

Fig. 18.8 Experiment demonstrating transposition by moving ampicilin resistance gene by generalised transduction using bacteriophage; the antibiotic resistance determinant from the transposon is used as the selectable marker to track the movement of amphicilin resistance gene

The two classes of bacterial transposons are **insertion sequences** or **IS elements** and **Tn elements**. However, bacterial transposable elements have been assigned to one of the three classes on the basis of structure and coding potential (**Fig. 18.9**).

Class I transposons

They contain elements that encode transposase, together with sequence information that governs the regulation of transposition. This class includes the simple insertion or IS elements (termed IS1, IS5 and so on), the termini of which consist of inverted repeat sequences, the size and structure of which characterise a particular IS element. Class I also contains what are termed composite transposons (e.g., Tn5, Tn9, Tn10) which comprise two IS or IS-like elements (in inverted or direct relative orientation) flanking a piece of 'passenger' DNA that is generally a gene specifying antibiotic resistance. In composite transposons, both IS or IS-like elements need not be functional, and the extreme terminal repeat contributed by each member of the pair may be regarded as defining the limits of the transposon. Nevertheless, individual functional modules can transpose independently.

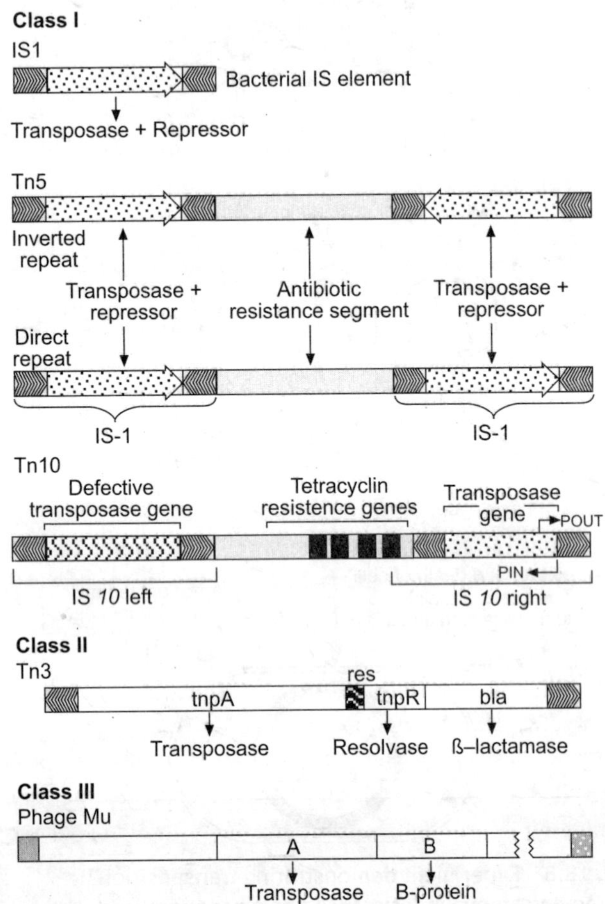

Fig 18.9 The structure and genetic organisation of some bacterial transposable elements

Class II transposons

They (e.g., Tn3, $\gamma\delta$) are complex transposons. They also carry various 'passenger' genes (e.g., the *bla* gene, encoding ß-lactamase), but have only single terminal repeats (38 bp) at their extremities. In addition to a gene coding for a transposase (*tnpA*), they are typified by the fact that they contain a gene (*tnpR*) coding for a site-specific recombinase, the resolvase, and an internal site (*res*), which the latter protein recognises. The resolvase can also act as a repressor of the transcription of the *trpA* and *tnpR* genes.

Class III transposons

This group includes the bacteriophages such as Mu, which replicate by a transpositional mechanism. In addition to encoding a considerable number of bacteriophage functions (coat proteins etc), Mu possesses two genes, A and B, involved in transposition. Gene A is the transposase and gene B greatly enhances the efficiency of transposition. It is, in fact, the high efficiency of transposition of Mu (100 transpositions per lytic infection) compared with that of other transposons (10^{-7} to 10^{-5} per generation) that has made it particularly suitable for studying the mechanism of transposition.

18.8 IS-MEDIATED DNA REARRANGEMENTS

Besides being discovered through their ability to transpose (i.e., to insert at new locations on DNA and to excise subsequently), IS elements were originally revealed as mediators of integration of conjugative F and R plasmids into the bacterial chromosome, as well as of mediators of the fusion between two F-plasmid derivatives (also called **co-integration**). The fused replicon can resolve to yield two replicons, each carrying one copy of the IS element that had mediated the process (**Fig. 18.10**). IS element also mediates a deletion during segregation of cointegrates. The IS1-mediated deletions, start precisely at the end of IS1 and terminate at various non-random sites. **Figure 18.11** summarises elementary DNA rearrangements mediated by IS elements. Combination of these DNA rearrangements would lead to more complex DNA restructuring.

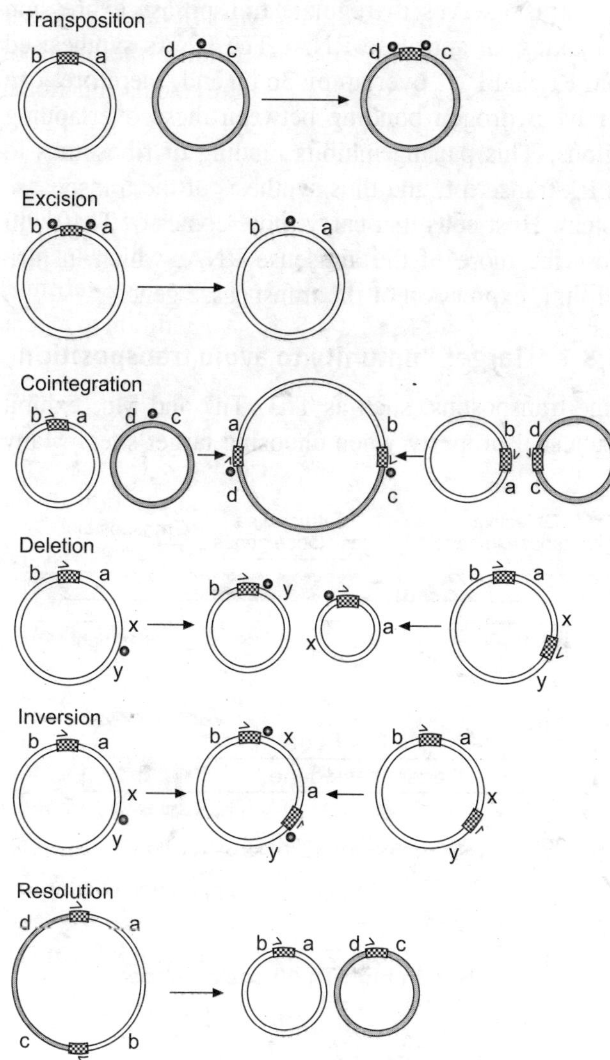

Fig. 18.10 Schematic representation of DNA rearrangements mediated by IS elements; the donor and target DNA molecules represent before (left and right) and after (centre) rearrangement events. Boxes represent IS elements, small arrows symbolise DNA polarity, small dots indicate the position of the target sequences undergoing duplication. Letters are either genetic or physical markers.

[*Source*: Iida et al (edited by) Shapiro JA 1983]

The arrows indicate the orientations of the Tys. The Xs indicate the sites of crossing over. Reciprocal crossing over between Tys that lie in the same orientation of the same chromosome results in the deletion of the Ty element. Reciprocal crossing-over between two Tys that lie in opposite orientations on the same chromosome result in inversion.

Gene amplification: Tandem duplication and further amplification to partially oligomeric states of genomes have been observed in many systems. They could have resulted from recombination between small repeats of about 10 bp or repeats of two copies of the same IS element. IS elements, provide homology for the rec pathways or a site-specific recombination systems (e.g., $\gamma\delta$).

Recombinational switching by inversion: Recombination between two inverted repeats causes the inversion of the DNA segment between them. Such an inversion can lead to an altered expression of genes located on the inverted segment immediately

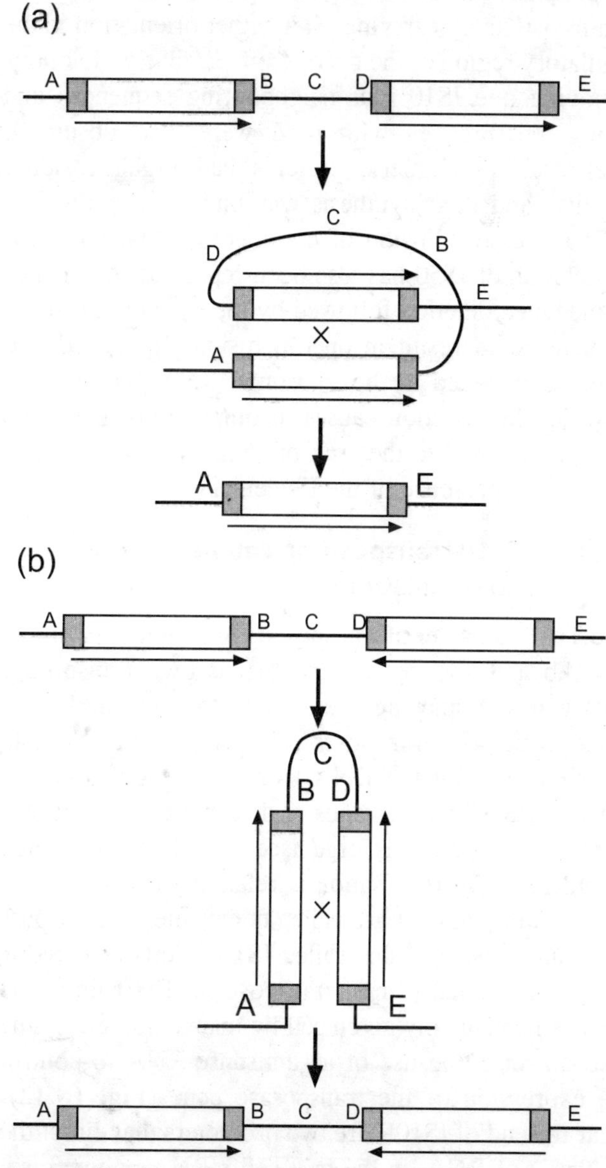

Fig. 18.11 Deletion and inversion due to δ–δ recombination. A, B, C, D and E indicate genetic markers in the unique sequence DNA flanking the Ty elements

outside. Inversions may be mediated by a site-specific recombination system, as was discovered to regulate the synthesis of alternative flagellar antigens in *Salmonella* and to determine two alternative host ranges in phage Mu and probably phage P1. Recombination between inversely repeated IS elements can also act as a genetic switch for gene expression.

Operon fusion: In principle, the strong polar effect exerted by many IS elements could be reverted not only by precise or nearly precise excision but also by deletion of a part of the element together with adjacent regions, thereby fusing a new promoter to the operon. This type of deletion was observed among Gal$^+$ revertants of strains originally carrying ISl in either orientation in the regulatory region of the *gal* operon. Similarly, deletions removing one IS10 plus neighbouring sequences and fusing Tn10 genes with lacZ were also observed. The resulting chimeric proteins had ß-galactosidase activity, and their synthesis was under the control of a Tc repressor. Fusion of *his* operon to Tn5 operons by deletion of IS50 has also been reported. In general, IS-mediated deletion followed by the excision of the IS element could result in operon fusion. If IS-mediated deletion extended to the ribosomal RNA operon *rrnX* in which IS insertion causes incomplete polar effects, the operon fused to the *rrnX* operon could be activated even in the presence of the IS element.

18.8.1 Tn10 transposition coupled to replication

The Tn10 (a bacterial transposon) is a compact element of 9 kb and encodes a gene for its own transposase and genes imparting resistance to the antibiotic, tetracycline. It transposes *via* the cut and paste mechanism using the DNA hairpin strategy to cleave the non-transferred strands and also has the site for IHF binding. Tn10 is organised into three functional modules. This organisation is relatively common, and elements that have it are called composite transposons. The outermost modules called IS10L (left) and IS10R (right) are actually mini transposons. Tn10 limits its copy number in any given cell by many strategies. One mechanism is the use of an antisense RNA to control the expression of the transposase gene (**Fig. 18.12**). Near the end of IS10R are two promoters that direct the synthesis of RNA by the host cell's RNA polymerase. The promoter that directs RNA synthesis inward (P_{IN}) is responsible for the expression of the transposase gene. The promoter that directs transcription outward (P_{OUT}), in contrast, serves to regulate transposase expression by making an antisense RNA. The RNAs synthesised from P_{IN} and P_{OUT} overlap by 36 bp and, therefore, can pair by hydrogen bonding between these overlapping regions. This pairing inhibits binding of ribosomes to the P_{IN} transcript, and thus synthesis of the transposase protein. Host cells that carry more copies of Tn10 will transcribe more of the antisense RNA, which in turn will limit expression of the transposase gene.

18.8.2 Target immunity to avoid transposition

Some transposons, such as Tn3, Tn7 and Mu, exhibit an unusual property when choosing target sites. Many

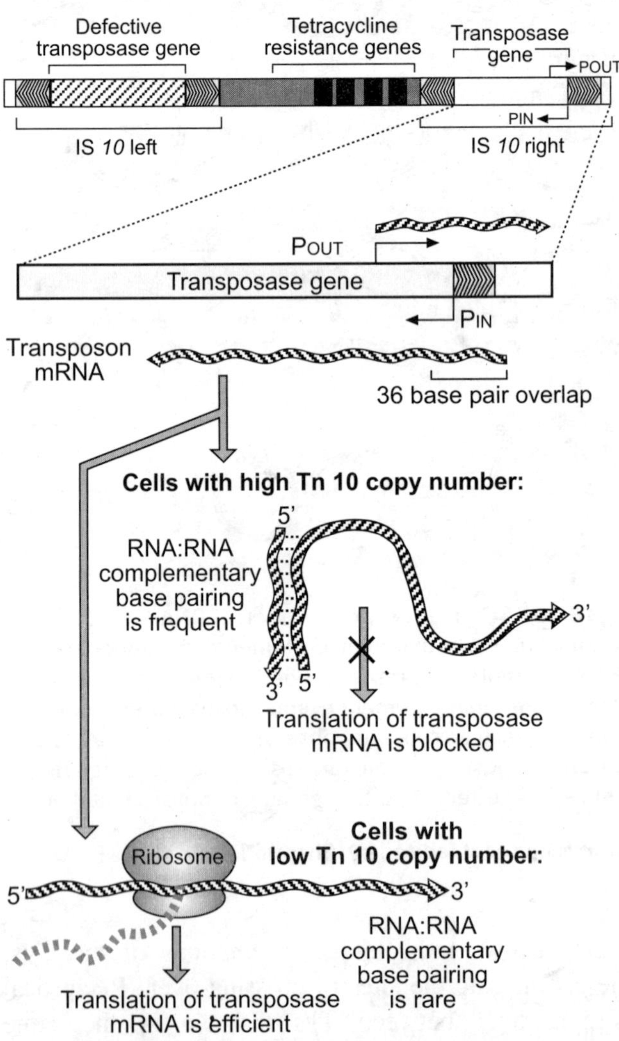

Fig. 18.12 Antisense regulation of Tn 10 expression; a map of the overlapping promoter regions is shown. The leftward promoter (PIN) promotes expression of the transposase gene; the rightward promoter (POUT), a 36 bp to the left of PIN promotes expression of an antisense RNA.

transposons show very little sequence preference at the target sites. But if transposition is regulated by a process called **transposition target immunity**, DNA sites surrounding a copy of the Mu element, including the element's own DNA, are rendered very poor targets for new transposition. If a target piece of DNA already contains a transposon, then a second copy of that same transposon cannot insert into that target DNA in the vicinity of the first element. This phenomenon is called transposition immunity or target immunity. For example, MuA-MuB interactions prevent MuB from binding to the DNA near the point where MuA is bound. Target immunity can be effective over very long distances. For example, Tn7 can provide immunity over distances of approximately 175 kb. Target immunity is specific to a particular element. For example, Tn3 provides immunity only to another Tn3 and not to any other elements, even if they are closely related.

18.9 EUKARYOTIC TRANSPOSABLE ELEMENTS

Transposable elements were, in fact, first identified in eukaryotes. Barbara McClintock discovered the first mobile elements while doing classical genetic experiments in maize during the 1940s. She characterised genetic entities that could move into and back out of genes, changing the phenotype of corn kernels. Her theories were very controversial until similar mobile elements were discovered in bacteria, where they were characterised as specific DNA sequences, and the molecular basis of their transposition was deciphered.

Eukaryotic transposons are termed class II transposable elements. Most eukaryotic transposons are similar in structure to bacterial IS elements, comprising a central region encoding a putative transposase and other functions required for transposition, flanked by short inverted terminal regions. On the basis of their structure, size and nature of terminal repeats transposons in eukaryotes can be classified into four types.

- Long terminal direct repeats (e.g., Copia in *Drosophila*, Ty in yeast and IAP in mice)
- Long terminal inverted repeats (e.g., FB, TE in *Drosophila*).
- Short terminal inverted repeats (e.g., P and I in *Drosophila*; Ac/Ds in maize; Taml in snapdragon and Tcl in *Caenorhabditis*)
- Without terminal repeats (e.g., Alu sequences in mammals).

However, the classification on the basis of mechanism of transposition contains two classes: those which transpose by reverse transcription of an RNA intermediate—**class I** (e.g., *copia* like, Ty, IAP, L1, F) and those, which transpose directly, DNA to DNA—**class II** (e.g., P, Ac, Spm/En, Tam, Tcl). The two best-characterised eukaryotic transposons in this category are the P factors of *Drosophila* and the controlling element, **Activator**, of maize. These are bounded by short (31 bp and 11 bp, respectively) inverted repeat sequences, and in their central portion contain genetic information for a transposase. Thus, structurally they bear some resemblance to the Tn3 family of bacterial transposons, although mechanistically they are more like the non-replicative class I bacterial transposons.

18.9.1 Transposable elements in yeast

Transposable elements in yeast, called Ty elements are a family of dispersed, repetitive DNA sequences. As in maize and *E. coli*, yeast Ty elements were first identified by the genetic properties of the mutations they caused. In addition to extinguishing gene function by insertion into coding or regulatory sequences, Ty insertions had the unusual property of sometimes activating genes that were previously silent. Ty elements were originally described by Cameron et al (1979). They characterised a Ty sequence present at the yeast SUP4 locus. The demonstration that this element occurred at the SUP4 locus of some yeast strains but not others and the presence of terminally repeated sequences suggested that the sequence was transposable. As shown in **Fig. 18.13**, each Ty element consists of a central region of about 5.6 kbp of DNA flanked by direct repeats of an approximately 330 bp sequence called delta (δ). There are 30–35 Ty elements per haploid yeast genome. In addition to the δ sequences associated with Ty elements, the yeast genome contains at least 100 solo δ sequences.

Ty transposition results in a duplication of DNA present at the target site. A 5-bp sequence that occurs only once in the sequence of the wild type gene is found at both junctions between the inserted Ty and the wild type gene sequence. The duplication of target DNA at the insertion site is a feature of transposition for bacterial transposons, *Drosophila* mobile elements and retroviruses. Ty insertions leading to loss of HIS4 function occur spontaneously at a frequency of approximately 2×10^{-8} cells. All of the Ty insertion mutations characterised thus far occurred in the

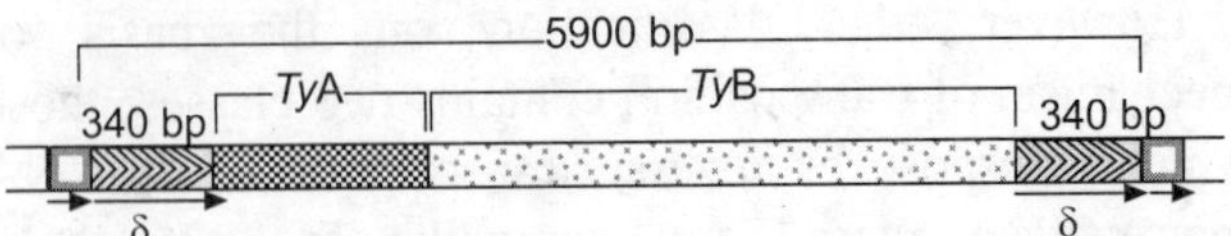

Fig. 18.13 Genetic organisation of yeast Ty elements. The long terminal repeat (LTR) sequences bound each side are denoted by the Greek letter (δ). Ty elements apparently contain two genes TyA and TyB. The small arrows indicate the direct repeats at the site of integration. Sequence length is shown in base pairs

regulatory regions adjacent to genes and not in the structural genes themselves indicating some kind of sequence or structural specificity during Ty insertion.

Yeast Ty elements display considerable sequence divergence, by simple base pair changes and also by insertion, deletion and substitution mutations.

18.9.2 Recombination between Ty and δ sequences

Recombination between Ty elements at dispersed sites in the yeast genome results in a variety of DNA rearrangements. The Ty simply acts as a portable region of homology, making it possible for the normal recombination system to promote rearrangements. Recombination events involving Ty elements fall into two main classes—those that lead to a recombinant configuration of markers outside the Tys and those that leave the flanking markers in the parental configuration. This exchange leads to recombination between markers within the Tys and also to recombination between genetic markers that lie outside the Ty elements. Because there are more than 30 Ty elements distributed throughout 17 chromosomes, the potential for DNA rearrangements due to Ty–Ty recombination is high. Recombination between Ty and δ sequences has been detected in several different experimental systems. In addition, spontaneous mutations at the SUP4 and CYC1 loci sometimes result from recombination between Ty or δ sequences in the chromosomal regions.

18.9.3 Excision

One of the most frequent recombination events involving Tys is reciprocal crossing over between the homologous δ sequences at opposite ends of a single Ty element. This recombination results in deletion of the internal region of the Ty and leaves behind a single copy of the δ sequence. The haploid genome of *S. cerevisiae* contains approximately 100 solo δ sequences that are unassociated with Ty element. It is likely that most of these 'solo δs' are the result of Ty excision.

18.9.4 Deletions and inversions

Recombination between δs or Ty elements that are in the same orientation on the same chromosome results in deletion of the DNA between the recombining δ or Ty sequences (**Fig. 18.14**). Recombination between δs that is separated by non-Ty DNA results in deletion of chromosomal DNA of unique sequence. Deletion of the *SUP4* genes resulting from δ–δ recombination occurs spontaneously at a frequency of 3 × 10^{-7} cells. Tandem duplications of Ty elements occur at a number of sites in the yeast genome. Deletions promoted by Ty–Ty recombination occur at a high frequency

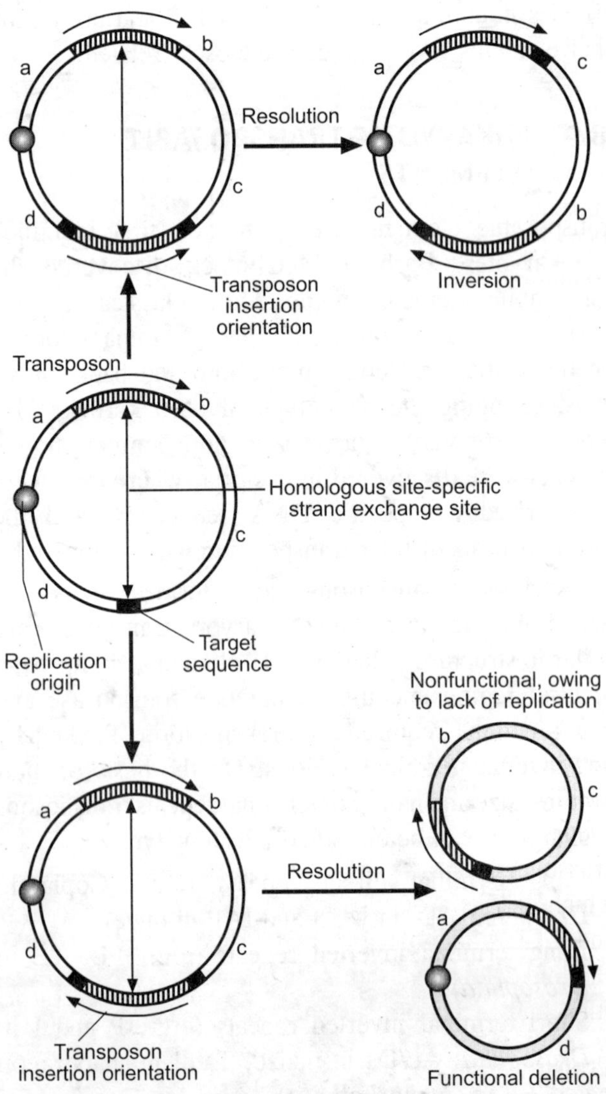

Fig. 18.14 Model for production of genetic deletions and inversions

in certain yeast strains known as *DEL1*. However, reciprocal recombination between Ty elements in opposite orientations on the same chromosome results in inversion of the DNA between the two Tys. One revertant of the his4-912 mutant carries an inversion of most of the left arm of chromosome III. There are intact Ty elements at both inversion breakpoints, but the restriction maps of these Tys are slightly different from that of Ty912. These data suggest that the inversion occurred by reciprocal recombination between Ty912 and a Ty element near the left end of chromosome III.

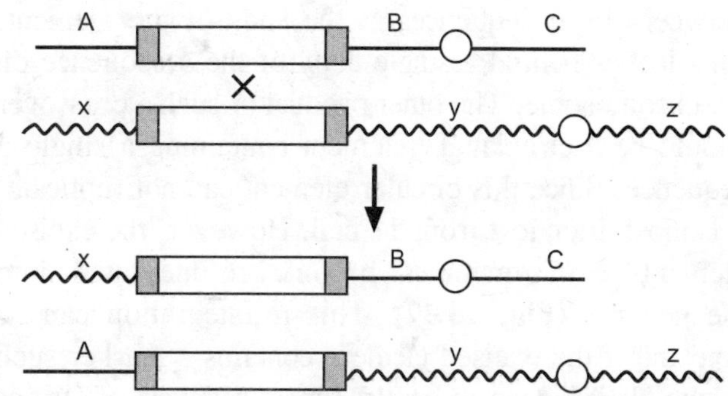

Fig. 18.15 Reciprocal chromosome translocation due to Ty–Ty recombination. Genetic markers represented by A, B and C on the chromosome (solid line) and x, y and z on the chromosome (wavy line). The open circles represent centromeres. A reciprocal crossover between two Tys on different chromosomes results in the generation of two new genetic linkage groups.

Translocations

Recombination between two Ty elements on different chromosomes, result either in the generation of a dicentric chromosome and a chromosome fragment (lethal) or in a reciprocal chromosome translocation (**Fig. 18.15**). Chromosome translocations have been detected in his4$^+$ revertants of the his4-912 mutant. Two Ty912-promoted translocations have been described; one between chromosomes I and III and the other between chromosomes III and XII. In both cases, the chromosome III translocation breakpoint occurs within Ty912. Cloning and characterisation of DNA segments containing the translocation breakpoints indicate that intact Ty elements are present at both the junctions between chromosome III DNA and chromosome I (or XII) DNA. These results suggest that both translocations are the result of reciprocal recombination between homologous Tys on different chromosomes.

Unequal crossing over

Recombination between Ty elements at different sites on homologous chromosomes results in deletion on one chromosome and duplication on the other (**Fig. 18.16**). Unequal crossover events of this sort have been observed between Ty elements on chromosome III. In diploid cells that are undergoing meiosis, reciprocal recombination between the Ty917 element on one chromosome III and the Ty1-17 element on the homologous chromosome occurs in about 1% of all meiotic events. This unequal crossover generates two aberrant chromosomes, one of which carries a duplication of all DNA between HIS4 and LEU2 and one of which lacks this information. Haploid cells that are missing the 40 kbp DNA between HIS4 and LEU2 are still viable.

Transposition

In addition to being able to promote their own transposition, Ty elements can sometimes mediate the transposition of unique sequence DNA. Stiles et al (1981) have characterised a yeast strain in which this CYC1-OSM1-RAD7 (13 kbp) DNA segment has undergone transposition to chromosome VII. This transposition is thought to have involved a circular DNA molecule resulting from recombination between the Tys flanking the CYC1 region. Transposition would result when the circular chromosomal fragment reintegrated into the genome, possibly by recombination with another Ty element. In two of the his$^+$ revertants derived from the his4-912 mutant, the *his4* gene is transposed to a new chromosomal location.

Excision and re-integration

As described before, Ty elements can become excised from the yeast genome by reciprocal recombination

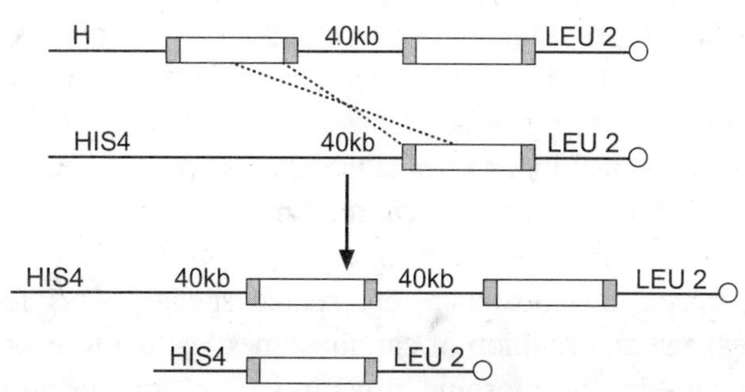

Fig. 18.16 Unequal crossing-over due to recombination between the Ty elements

[***Source***: Redrawn from Roeder GS and Fink GR (edited by) Shapiro JA (1983)]

between the δ sequences at the ends of the element. This leaves behind a single copy of the δ sequence on the chromosome. The other product of such a crossover should be a circular Ty element containing a single δ sequence. Since this circular element can not replicate, it is most often lost from the cell. However, the excised element can sometimes become re-integrated into the genome (**Fig. 18.17**). This re-integration can be detected if the excised element contains a marker such as the URA3 gene in Ty912 (ura). Analysis of strains containing Ty912 (ura) indicates that approximately 1% of the excised Ty elements are re-integrated.

Replacement of Ty sequences

Ty elements can also engage in recombination events that do not alter the arrangement of flanking markers. Several *ura3*⁻ derivatives of Ty912(ura) and Ty917(ura) retain an intact Ty element at the *HIS4* locus. The Ty elements are in the same position and in the same orientation as the original elements; however, they differ from the parental Ty by the loss of URA3 DNA and by changes in the restriction endonuclease cleavage patterns. Several observations indicate that these replacement events result from gene conservation rather than double crossovers.

18.10 CONTROLLING ELEMENTS IN CORN

Controlling elements are transposable elements capable of causing unstable insertion mutations as well as a variety of chromosomal rearrangements, including deletions, duplications, inversions and translocations. Mutations due to insertion of unstable controlling elements give a characteristic pattern of gene expression in the organism. The variegation in colour observed in the kernels of Indian corn was caused by an unstable mutation. This is due to a factor encoded by the maize C locus. These genetic elements in maize were found to be responsible for turning the expression of genes on or off; hence the name '**controlling elements**'. It was also shown that these controlling elements when located on maize chromosomes represented specific sites for breakage and reunion of chromosomes leading to gross changes in chromosome structures. Perhaps the most striking feature of controlling elements is their ability to sense developmental parameters. Controlling elements transpose and activate other types of genetic change at a predetermined time and with a characteristic frequency

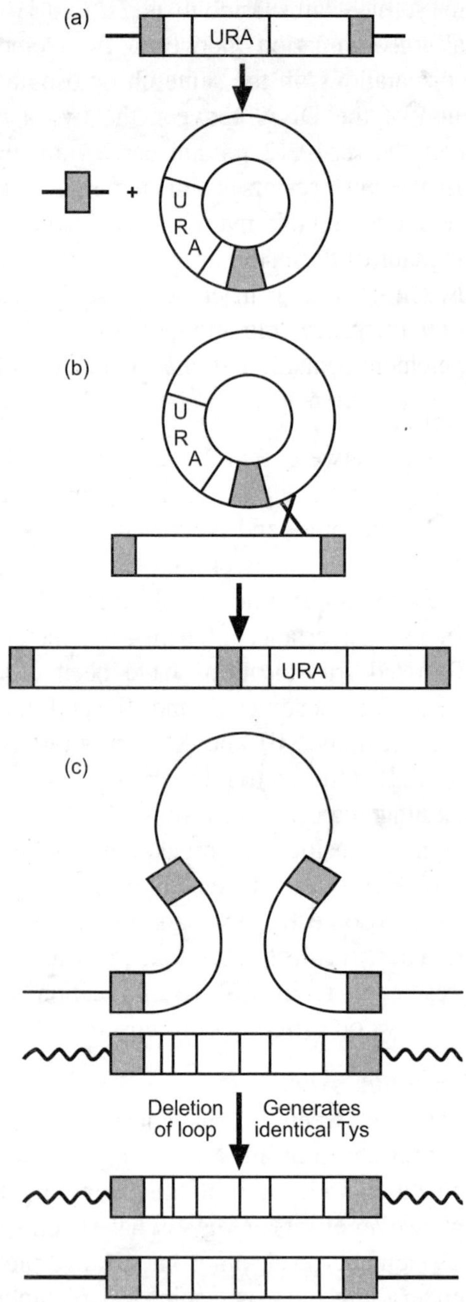

Fig. 18.17 Excision and re-integration of a Ty element. (a) A Ty element containing the URA3 gene is excised from the yeast genome by δ–δ recombination (b) The circular Ty engages in reciprocal recombination with a Ty in a yeast chromosome (c) Recombination between a Ty element and a DNA segment containing two Tys

[***Source:*** Redrawn from Roeder and Fink (edited by) Shapiro JA (1983)]

in the development of the organism. The nature, frequency and timing of element-associated changes are genetically specified and can be altered by mutation.

There is some evidence that controlling elements can be pre-programmed to alter the timing and frequency of element-associated genetic events at a precise time in development.

The genetic behaviour of the controlling elements in maize is the best understood amongst eukaryotic transposable elements. The main factor in the analysis of controlling elements is that there are many genes that are expressed in maize and inactivation has little or no deleterious effect on viability and fertility. Insertion of a controlling element can give rise to an unstable allele of a locus. Some of these are **one-element system** (e.g., Dt or Dotted, Mp or modulator, Uq or ubiquitous, Bg or Bergamo, Mu or mutator and so on) and others are **two-element system** (e.g., Ac-Ds system, Spm-dSpm system and so on). The number, type and location of these elements are characteristic of each strain of maize.

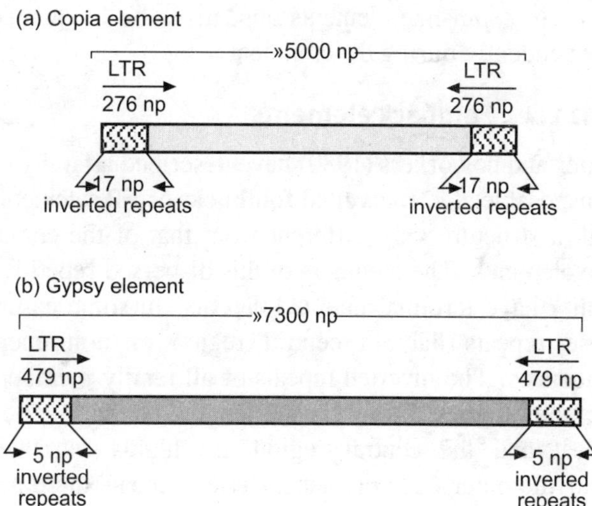

Fig. 18.18 Structures of retrotransposons in *Drosophila*: (a) Copia element (b) Gypsy element

18.11 TRANSPOSONS IN DROSOPHILA

Since McClintock's early work on mobile elements in corn, transposons have been identified in other eukaryotes. Most well studied eukaryotic transposons are of yeast and the fruit fly. In *Drosophila*, as much as 10% of the genome may consist of transposons, most important of these are copia-like elements, the fold back (FB) and P (and I) elements. Although most of the mobile elements in *Drosophila* function as retrotransposons, at least one—the **P element**—functions as a DNA transposon, moving by a cut-and-paste mechanism similar to that used by bacterial insertion sequences. Current methods for constructing transgenic *Drosophila* depend on engineered, high-level expression of the P-element transposase and use of the P-element inverted terminal repeats for transposition.

18.11.1 Copia-like elements

The prototype *Drosophila* transposon is called **copia** (**Fig. 18.18**), because it is present in copious quantity (*Drosophila* middle repetitive sequences). In fact, copia and related transposons called copia-like elements account for about 1% of the total genome. The copia-like elements are strikingly similar in structure to the yeast Ty1 element and to the integrated proviruses of RNA tumour viruses. Their size may range from 5 kb to 8.7 kb and their number may range from 10 to 80. Each carries (i) a long direct terminal repeat (LTR) ranging from 268 to 512 bp and (ii) a short imperfect inverted repeat, ranging from zero to 18 bp. There is now direct evidence that several *Drosophila* mutations result from the insertion of copia-like elements. Insertion of these elements causes mutations. A classical example is white apricot (W^a) mutation for eye colour caused by insertion of a copia-like element. FB elements range from a few hundred to a few thousand base pairs and carry long inverted repeats at the termini. The repeats may be adjoining (FB1, FB5) or may be separated (FB3, FB4, FB5, FB7, FB8, FB9), but can always fold back to form stem and loop structures. They cause mutations or alter the gene expression.

Insertion of a copia-like element is accompanied by the duplication of a few base pairs of DNA present once in the target DNA at the site of insertion and found immediately adjacent to the ends of the element after insertion. The integration of such duplication appears to be a universal property of transposable elements in prokaryotes and eukaryotes, and probably results from a staggered cut in the initial cleavage of the target DNA. Other two structural features include the direct repeats of copia-like elements conserved within an element but have diverged between elements and circular forms.

The RNA transcripts of the copia-like elements are unusual in several respects. These transcripts are often abundant. Six of the seven, characterised copia-like elements are known to be transcribed. The transcription of mdg1 and mdg3 is not fully asymmetric; about 95% of the transcripts are complementary to one strand and 5% to the other. The amount of RNAs encoded by different *copia*-like elements appears to be modulated independently during development.

18.11.2 Foldback elements

Potter and coworkers (1980) have described a family of transposable elements called **foldback**, or FB, elements with a structure very different from that of the copia-like elements. The members of this dispersed repetitive family have terminal inverted repeats. In some cases, these repeats flank a central region of non-repeat sequences. The inverted repeats of all family members are homologous; however, the lengths of the inverted repeat and the central region are highly variable. Near the outer ends of each inverted repeat, there are multiple copies of a 10 bp sequence; however, these 10 bp repeats are generally separated by large stretches of more diverse DNA. At about 300 bp in from the end of the element, the 10 bp repeat expands to 20 bp, and the stretches of DNA between repeats are shorter and less diverse. At 500 bp into the element, the repeat unit again expands to 31 bp, and these repeats are now contiguous and may extend for hundreds of base pairs.

The genetic properties of mutations caused by the insertion of FB elements are revealed by the behaviour of two such mutations—white-crimson (w^c) and white-dominant zest-like (w^{DZL}). Both are highly unstable mutations at the white locus. The white locus is a complex genetic locus with alleles affecting both the amount and pattern of pigmentation of the eye. Flies homozygous for deletions of this locus have bleached white eyes but are viable and fertile. A wide variety of perturbations in the normal expression of white are thus easily recognised phenotypically.

18.11.3 P elements

A family of P factors (and defective derivatives called P elements) is one of the elements involved in a complex genetic phenomenon called **hybrid dysgenesis**, with structures clearly distinct from those of the copia-like or FB elements (**Fig. 18.19**). A peculiarity of the transposition is that it only occurs in the germ line, and thus there must be a mechanism for the repression of transcription in somatic cells. These small transposons terminate in 31 np inverted repeats and are flanked by 8 np target site duplications. The complete sequence of a 2.9 kb π factor shows the presence of four potential open reading frames (they are, in fact, exons) all of which are required for transposition. This is because in the germ line the portions of the transcript corresponding to these four reading frames are spliced to produce a single open reading frame in the mRNA encoding the 87 kDa transposase. In somatic cells however, only the first two introns are removed, and the open reading frame in the mRNA terminates before the RNA corresponding to the final open reading frame of the element. The smaller protein encoded by this mRNA lacks transposase activity but it is a repressor of transposition, acting by preventing the removal of the third intron.

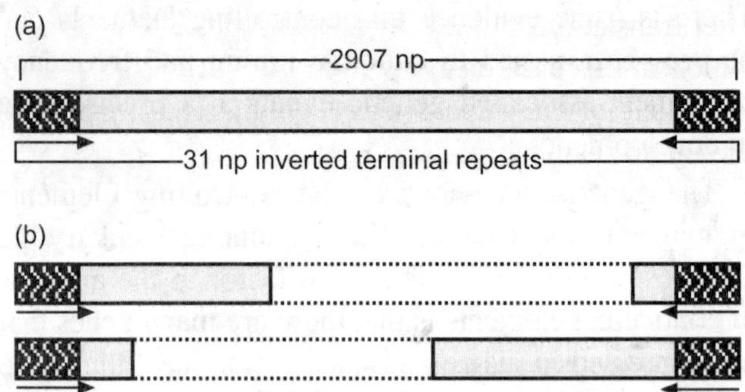

Fig. 18.19 Structure of P elements in *Drosophila*: (a) A complete, or autonomous, P element (b) Incomplete, or non-autonomous P elements that were derived from complete elements by the deletion of internal sequences

18.12 MOBILE INTRONS AND INTEINS

Mobile introns are a class of transposable elements found in yeast mitochondria. These introns contain an ORF encoding a site-specific endonuclease, similar to bacterial restriction endonucleases except that the target sequence is much longer, typically >20 bp. The size of the target sequence means that it is rare, and in fact it is only found once in the genome, at the site where the intron is inserted. The endonuclease, therefore, cannot cleave host DNA unless there is an intronless allele of the same gene present in the cell. If such an allele is encountered, a double-strand break is introduced which may be repaired by homologous recombination with the intron-containing allele—a process which results in the intron being copied and inserted into the intronless allele by gene conversion. These remarkable elements may be termed **homing introns** because of their ability to home in on a particular site within the target gene. Even more intriguing, however, are the **homing inteins**, which are polypeptides spliced out of nascent proteins after translation. These may also encode endonucleases, which in this case cleave within the ORF of an intein-less allele of the same gene and initiate the transfer of the intein segment by gene conversion.

18.13 BACTERIAL INTEGRONS

Integrons are bacterial operons incorporating various structural genes. The integron consists of a regulatory region, an ORF encoding a site-specific recombinase and a target site for the recombinase, attl, which can integrate a variety of antibiotic resistance gene cassettes carrying a 59-base element, which is also recognised by the recombinase. Unlike other types of cassette systems, where the target locus is never 'empty' and one cassette replaces another by non-reciprocal recombination, integrons may be empty or may integrate one or more cassettes in tandem. In this case, recombination can occur between 59-base elements of different cassettes to eliminate particular genes or establish stable junctions. Transcription through attl facilitates the expression of all inserted genes in a common operon. Like bacteriophage λ, the integron cassettes mobilise by site-specific recombination rather than transposition. They are not true transposons, and the term site-specific transposon is used to describe such elements.

18.14 RETROELEMENTS

In the virus lifecycle, the initiating events of provirus includes virus entry, DNA synthesis, integration, expression of viral genes, virus production and transformation, which can be viewed as a mechanism of DNA transposition. A single provirus in a single cell can engineer the production of thousands of infectious virus particles, and these particles can introduce the viral genome into new cells in which new viral DNA can then be synthesised and integrated. These elements can persist, either as a component of the host chromosome or as the genome of an extracellular particle capable of resisting environmental factors and infecting new host cells. The simple structural features that assign retroviruses to the class of mobile elements include Tn9 in bacteria, Ty1 in yeast and copia, 412 several others in *Drosophila*. All of these share an internal coding domain of one to several kilobase pairs; large terminal direct repeat units of several hundred base pairs, usually with short inverted terminal repeats and short flanking direct repeats of host DNA duplicated during insertion.

Retroelements (Class I elements) are genetic elements, which mobilise by reverse transcribing an RNA intermediate and then integrating the cDNA into the genome. Retroelements can be divided into two major families. The **viral family (class I.1 transposable elements)** includes the retroviruses and non-viral elements, which resemble them in structure and mechanism of transposition (termed **retrotransposons**). The non-viral family (**class I.2 transposable elements**) includes autonomous transposable elements, which do not resemble retroviruses (termed **retroposons**) and DNA sequences, which appear to have been mobilised by accident, due to the adventitious action of reverse transcriptase obtained in *trans*. These are known as **retrosequences** and those corresponding to cellular mRNA are **retrogenes**.

Retroviruses compose a large class of animal viruses united by a single principle: the single stranded RNA genomes are replicated through dsDNA intermediates. These elements are 5–15 kbp long and resemble the integrated forms of retroviruses, much like the Ty elements of yeast. Retroviruses, although identified as agents involved in the onset of cancer but the recent epidemic of AIDS has shown to be due to the HIV retrovirus. A striking feature of retroviruses is the high frequency with which genetic recombinants (10%–50%) form after dual infection of closely related viruses. Retroviruses can be considered transposons in that their genomes are found inserted randomly into host DNA (**Fig. 18.20**).

18.14.1 Integrating retroelements

The second category of eukaryotic transposons exemplified by copia and Ty has a replicative mode of transposition. This differs from that in prokaryotes as it involves a reverse transcriptase similar to that found in retroviruses. In order to appreciate this, one may consider how the double-stranded proviral DNAs of retroviruses function as a transposable element.

Sequencing of the human genome has revealed that the most common LTR retrotransposons-related sequences in humans are derived from endogenous retroviruses (ERVs). Most of the 443,000 ERV-related DNA sequences in the human genome consist only of isolated LTRs. These are derived from full-length proviral DNA by homologous recombination between the two LTRs, resulting in deletion of the internal retroviral sequences.

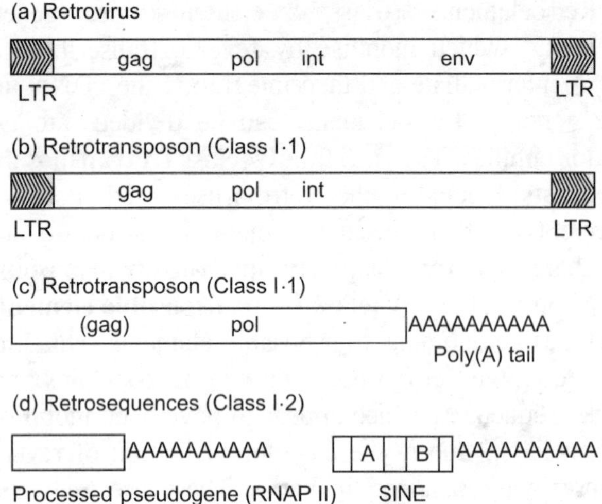

Fig. 18.20 Structures of eukaryotic retroelements LTR: Long terminal repeat and gag, pol, int, and env are virus ORFs

18.15 LINEs AND SINEs

The most abundant mobile elements in mammals are retrotransposons that lack LTRs. These are sometimes called non-viral retrotransposons. These moderately repeated DNA sequences fall into two categories in mammalian genomes: **long interspersed elements (LINEs)** and **short interspersed elements (SINEs)**. The full length LINEs in humans consists of repeats about 6–7 kb long and SINEs are less than 500 bp. A large number of LINEs and SINEs in higher eukaryotes have accumulated over evolutionary time by repeated copying of sequences at a few positions in the genome and insertion of the copies into new positions (representing 20.4% of the human genome). LINEs have been observed in most eukaryotes but SINEs are found primarily in mammalian DNA.

LINEs were first detected in rodents and in man. The main difficulty in studying the vertebrate LINEs is in identifying a functional member in the large genome, most of which is defective. For example, many copies lack varying portions of the 5'-end present in the longest versions, consistent with their generation by different extents of reverse transcription. Human DNA contains three major families of LINE sequences that are similar in their mechanism of transposition, but differ in their sequences: L1, L2 and L3. LINE sequences are present at ~900,000 sites in the human genome accounting for a staggering 21% of total human DNA. The general structure of a complete LINE is shown in **Fig. 18.21.** LINEs are usually flanked by short direct repeats, the hallmark of mobile elements and contain two long open reading frames (ORFs). The most common LINE elements constitute the L1 LINE family. Some 50,000 copies of L1 elements occur in the human genome, accounting for ≈5% of total human DNA. The full length L1Md and L1Hs elements are 6–7 kb long and contain two ORFs (≈1 kb and ≈4 kb long). They both terminate in a polyadenylation/processing signal followed by a polyA tail at their 3'-end, but differ in their 5'-untranslated regions, which are nevertheless thought to contain the promoters for transcription by cellular RNA polymerase II.

Since LINEs do not contain LTRs, their mechanism of transposition through an RNA intermediate differs from that of LTR retrotransposons. ORF1 and ORF2 proteins are translated from LINE RNA. *In vitro* studies indicate that transcription by RNA polymerase II is directed by promoter sequence at the left end of the integrated LINE DNA. LINE RNA is polyadenylated by the same post-transpositional mechanism that polyadenylates other mRNAs. The LINE RNA is then transported into the cytoplasm, where it is translated into ORF1 and ORF2 proteins. Multiple copies of ORF1 protein then bind to the LINE RNA and ORF2 protein binds to the polyA tail.

The LINE RNA is then transported back into the nucleus as a complex with ORF1 and ORF2. ORF2 then makes staggered nicks in chromosomal DNA on either side of any A/T-rich sequence in the genome. Reverse transcription of LINE RNA by ORF2 is primed

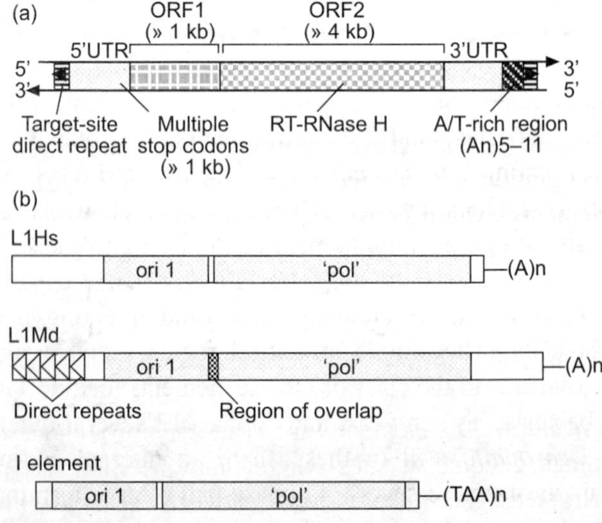

Fig. 18.21 (a) Diagram of the general structure of a typical LINE element (b) Structure of some members of the LINE family showing open reading frames

by the single-stranded T-rich sequence generated by the nick in the bottom strand, which hybridises to the LINE polyA tail (step 2). ORF2 then reverse transcribes the LINE RNA (step 3) and then continues this new DNA strand, switching to the single-stranded region of the upper chromosomal strand as a template (steps 4 and 5). Cellular enzymes then hydrolyse RNA and extend the 3' end of the chromosomal DNA top strand, replacing the LINE RNA strand with DNA (step 6). Finally, the 5' and 3' ends of DNA strands are ligated, completing the insertion (step 7). These last steps (steps 5 and 7) probably are catalysed by the same cellular enzymes that remove RNA primers and ligate Okazaki fragments during DNA replication. The complete process results in insertion of a copy of the original LINE retrotransposons into a new site in chromosomal DNA. A short direct repeat is generated at the insertion site because of the initial staggered cleavage of the two chromosomal DNA strands.

18.15.1 SINEs

The second most abundant class of mobile elements in the human genome, SINEs constitute ~13% of total human DNA. Varying in length from about 100–400 bp, these retrotransposons do not encode protein, but most contain a 3' A/T-rich sequence similar to that in LINEs. SINEs are transcribed by RNA polymerase III, the same nuclear RNA polymerase that transcribes genes encoding tRNAs, 5S rRNAs and other small stable RNAs. Most likely, the ORF1 and ORF2 proteins expressed from full length LINEs mediate transposition of SINEs by the retrotransposition mechanism depicted in **Fig. 18.22**.

SINEs occur at about 1.6 million sites in the human genome. Of these, ~1.1 million are *Alu* family of elements, so named because most of them contain a single recognition site for the restriction enzyme Alu1. *Alu* elements exhibit considerable sequence homology and may have evolved from 7SL RNA (294 nt cytoplasmic RNA), a component of the signal recognition particle. This abundant cytosolic ribonucleoprotein particle aids in targeting certain polypeptides, as they are being synthesised, to the membranes of the endoplasmic reticulum. *Alu* elements are scattered throughout the human genome at sites where their insertion has not disrupted gene expression: between genes, within introns, and in the 3' untranslated regions of some mRNAs. For instance, nine *Alu* elements are located within the human ß-globin gene cluster.

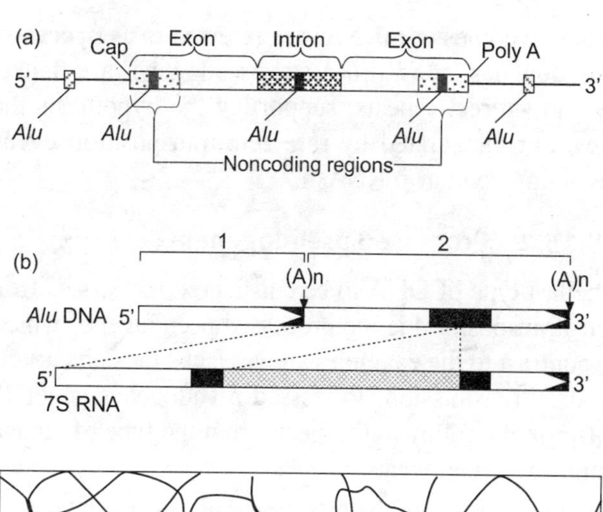

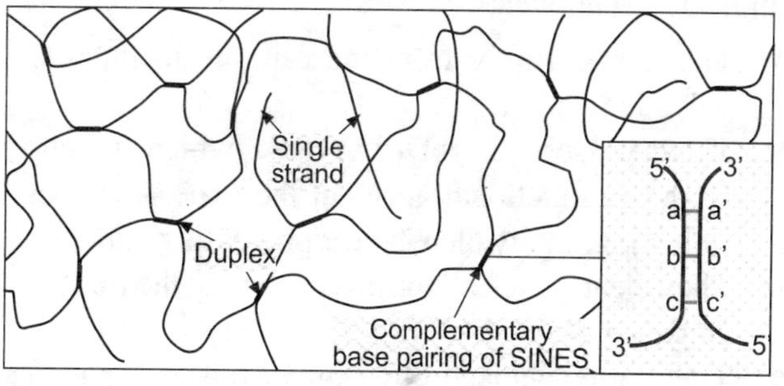

Fig. 18.22 Genetic organisation of SINEs (a) Positions of *Alu* sequences in and around a typical protein-coding transcription unit (b) The structural features of *Alu* DNA (SINEs) and relationship to 7S RNA (c) A diagrammatic representation of electron micrograph of human DNA fragments renatured under conditions that permit reassociation only of repetitive sequences. Insert illustrates base pairing of complementary strands of SINEs.

Similar to other mobile elements, most SINEs have accumulated mutations from the time of their insertion in the germ line of an ancient ancestor of modern humans. Like LINEs many SINEs also are truncated at their 5' end. In addition to the mobile elements discussed above, DNA copies of a wide variety of mRNAs appear to have integrated into chromosomal DNA. Since these sequences lack introns and do not have flanking sequences similar to those of the functional gene copies, they clearly are not simply duplicated genes that have drifted into non-functionality and become pseudogenes. Instead, these DNA segments appear to be retrotransposed copies of spliced and polyadenylated (processed) mRNA. Compared to normal genes encoding mRNA, these inserted segments generally contain multiple mutations, which are thought to have accumulated since their mRNAs were first reverse-transcribed and randomly integrated into the genome of a germ cell in an ancient ancestor. These functional genomic copies of mRNAs are referred to as **processed pseudogenes**. Most processed pseudogenes are flanked by short direct repeats, supporting the hypothesis that they were generated by rare retrotransposition events involving cellular mRNAs.

18.15.2 Processed pseudogenes

The last type of DNA to be considered that arises from retrotransposition is the processed pseudogene, which, in contrast to the examples above, lacks the capacity for further transmission. Processed pseudogenes generally differ in the following respects from the type of tandem duplicate pseudogene:

- They extend only from the cap site to the site of polyadenylation.
- They possess a poly(dA) tail (which is absent from the functional gene) in the expected position following a polyadenylation/processing signal.
- They clearly lack all introns present in the functional gene.
- They are flanked by direct-repeat sequences (typically 11–15bp) immediately preceding the transcriptional start site and immediately following the poly(dA) tail; and they reside at chromosomal locations distant from the functional gene (**Fig. 18.23**).

These characteristics clearly indicate that these pseudogenes originated from spliced polyadenylated mRNAs, DNA copies of which were inserted at sites in the chromosome created by a staggered endonucleolytic break. Most of such processed pseudogenes have been inserted at sites at which the lack of a promoter renders transcription impossible, and most have accumulated mutations that would render any transcript non-functional. However, in the case of lacking frameshift or nonsense mutations there is evidence for expression. A similar situation is obtained for the functional rat insulin I gene in which, due to its location on the same chromosome as the rat insulin II gene and the fact that it contains one of the three introns of the latter, had been thought to be derived from gene II by a tandem duplication. Subsequent analysis demonstrated that the gene I has poly A tail, is flanked by direct repeats and is 100,000 kb from gene II. It appears that the insulin I gene arose from the reverse transcription of a partially processed transcript of gene II that had initiated upstream from the cap site and included the promoter.

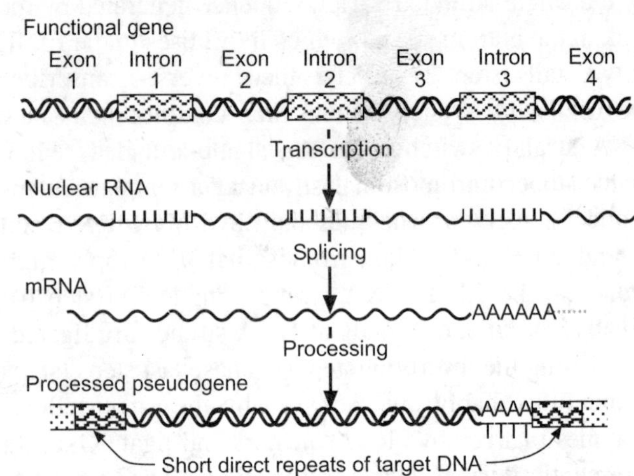

Fig. 18.23 Diagrammatic representation of the origin and features of processed pseudogenes

Vertical transmission of processed pseudogenes can only occur if they arise in germ cells and the most numerous processed pseudogenes correspond to mRNAs that one would expect to be abundant in undifferentiated egg or sperm cells. These include multiple processed pseudogenes corresponding to glycolytic enzymes and to cytoskeletal proteins such as actin. In actin, it is mRNAs corresponding to the cytoplasmic (ß and γ) isotypes, rather than the muscle-specific isotypes, that have given rise to processed pseudogenes. Nevertheless, processed pseudogenes corresponding to α-globin and immunoglobulin λ, normally the products of differentiated cells have been described. These, like the insulin I gene, are atypical in corresponding to transcripts extending upstream beyond the normal 5'-cap site. The possibility that such mRNAs are expressed in the germ line (rather than being somehow transmitted horizontally from differentiated cells) is interesting in relation to the gene conversion thought to be operating on globin and immunoglobulin genes. It has been suggested that this conversion, which would, of course, have to occur in the germ line, might possibly involve reverse transcripts of such transiently expressed mRNAs.

18.16 USE OF TRANSPOSONS AS MOLECULAR TOOLS

Transposons are valuable tools for molecular biologists because they can get inserted randomly around the chromosome and bring about many changes which can be used in the molecular studies. Some of their uses are discussed in the following paragraphs.

Gene conversion

New genes can evolve from the mobile elements by repeated insertions that lead to higher copy number, followed by the accumulation of random mutations and gradual divergence of sequence. There have been sporadic examples of instances of *Alu* elements that get inserted in the ancestral primate genome, but have recently undergone gene conversion to a new subfamily of *Alu* members. Recent studies have shown that almost 20% of the newer *Alu* subfamily members have undergone partial gene conversion to a different subfamily. These partial gene conversions only involved alterations in short stretches of the Alu element, typically on the order of 50 bp. similar observations were also made for L1 elements.

Evolutionary issues

The widespread distribution of transposable elements suggests that they have played a role in evolution. One hypothesis is that these elements are nature's tools for genetic engineering. Their ability to copy, transpose and rearrange other DNA sequences can be construed as a benefit for the organisms which carry them. Since excessive transposon activity can destroy a genome, many organisms seem to have developed mechanisms to reduce transposition to a manageable level. Bacteria may undergo high rates of gene deletion as part of a mechanism to remove transposons and viruses from their genomes. Eukaryotic organisms have developed the RNA interference (RNAi) mechanism as a way of reducing transposon activity. Transposons may have been co-opted by the vertebrate immune system as a means of producing diversity in antibodies. For example, the V(D)J recombination system operates by a mechanism similar to that of transposons.

Research tools

The remarkable properties of transposable elements have made them most versatile tools for the molecular biologist. Their ability to jump between genomes has enabled them to be used as gene delivery vectors, and their tendency to disrupt genes has allowed them to be exploited as mutagens. They have been used to introduce genome rearrangements, as mobile reporter systems and as tags to clone nearby genes. Several transposons, including Mu, Tn5 and Tn10, have been developed into sophisticated genetic tools, because a transposon moves any DNA fragment that is between its two ends. A number of ways in which transposable elements can be exploited are summarised in **Table 18.1**.

Table 18.1 Applications of transposable elements

Use	Examples
Gene delivery vector	Introduction of *in vitro* modified genes into host cells (bacteria, yeast, animal and plant cells) by cloning in a transposon and introducing by transformation. Transposons have shown that genes can be transferred artificially if carried by a transposon in a plasmid. For instance, a fragment of DNA carrying rosy gene within a P element was transferred in a plasmid, which was injected into embryos of a strain of *Drosophila*, homozygous deficient for the rosy gene. It was found that 50% of the progeny did carry rosy+ progeny indicating successful transfer.
Gene mutation	Generation of mutants by random insertional mutagenesis.
Structural rearrangements	Generation of nested deletions in bacteria by intermolecular transposition and specific genome rearrangements by delivering FLP/FRP recombinase system into genome can be studied.
Gene expression	Overexpression of neighbouring genes can be induced using strong promoter and endogenous regulatory functions using entrapment vectors can be assayed.
Cloning	Cloning genes directly from mutant organisms using transposition as a flag-transposon tagging. Inclusion of plasmid origin of replication in transposon facilitates plasmid rescue. In vivo cloning with bacteriophage Mu, isolate genes directly from mutant cells by packaging into phage heads.
Gene mapping	By insertional mutagenesis plasmids can be mapped and introducing portable sequencing primer template facilitates large-scale sequencing experiments.
Gene tagging and isolation	Insertion of transposons usually causes mutations due to insertion in structural or regulatory region. Therefore, these mutations can be used for the study of structural and regulatory regions of a gene. Transposable elements also produce chromosome breakage. In case of Ds of maize and P of *Drosophila*, breaks can lead to the loss or rearrangement of chromosomal material. Further, a probe carrying a transposon can be used to screen the restriction fragments containing the mutant gene. This feature of transposons has also been utilised for isolation of genes though transposon tagging.
Genetic markers	Mobile elements that were inserted in the *Drosophila* genome demonstrate a geographic gradient in copy number. Since mobile elements are multicopy in a genome, they increase the probability of being passed on to progeny relative to any other genomic allele. The multicopy nature of the elements provides an increased probability that some copies of the element will be propagated. The fixation of a specific mobile element insertion sites in a species can be used as a distinct character for phylogenetic analysis. If two species share a common retroelement insertion at a given location and a third species does not, the first two species are likely to be more closely related.

KEY POINTS

- Recent investigation has found that the formation of new arrangements of genes is due to the movement of large blocks of DNA. Because their genomic positions are not fixed, these determinants have been studied under the heading 'mobile genetic elements'.
- The movements of DNA segments are mediated by two genetic processes, which include recombination and transposition.
- Transposition is a specific form of genetic recombination that moves certain genetic elements from one DNA site to another.
- Transposons or transposable elements (TE) can move to virtually any position in a cell's genome without benefit of homology.
- During transposition, the two inverted repeats at the ends of the transposon and the target DNA interact.

- Some DNA transposons move using a mechanism called replicative transposition and others as conservative transposition (or non-replicative transposition).
- After a transposon leaves a gene, the resulting gap is not repaired correctly, which leads to the altered site.
- As a result of the insertion of IS element, gene expression may also be affected by DNA rearrangements involving longer segments. Tandem duplication and further amplification to partially oligomeric states of genomes have been observed in many systems. They could have resulted from recombination between small repeats of about 10 bp or repeats of two copies of the same IS element.
- Transposable elements were found to fall into two basic categories: those that transpose directly as DNA and those that transpose *via* an RNA intermediate from the mobile element by an RNA polymerase and then converted back into dsDNA by a reverse transcriptase.
- Transposable elements in yeast, called Ty elements are a family of dispersed, repetitive DNA sequences. Yeast Ty elements display considerable sequence divergence.
- McClintock's discovery of mobile elements came from observation of certain spontaneous mutations in maize kernels. Mutant kernels are white and wild-type kernels are purple. McClintock called the agent responsible for the first class of mutations the activator (Ac) element and those responsible for the second class dissociation (Ds) elements because they also tended to be associated with chromosome breaks.
- The prototype *Drosophila* transposon is called copia and can cause mutations or alter the gene expression.
- A family of P factors, one of the elements is involved in a complex genetic phenomenon called hybrid dysgenesis.
- Retroelements (Class I elements) mobilise by reverse transcribing an RNA intermediate and then integrating the cDNA into the genome.
- Ty elements of *Saccharomyces cerevisiae* are a family of dispersed, repetitive DNA sequences. An example of Ty element in yeast is Ty-1, which is present in 35 copies in yeast genome.
- The virus-like particles associated with copia and Ty differ from those of retroviruses in not being infectious as they lack extracellular phase. In these cases they are, in fact, acting as transposable elements analogous to copia and Ty.
- The most abundant mobile elements in mammals are retrotransposons that lack LTRs, sometimes called non-viral retrotransposons. These moderately repeated DNA sequences fall into two categories in mammalian genomes: long interspersed elements (LINEs) and short interspersed elements (SINEs).
- The last type of DNA to be considered that arises from retrotransposition is the processed pseudogene, which lacks the capacity for further transmission.
- Transposons are valuable tools for molecular biologists, because they can insert DNA randomly around the chromosome. Thus, new genes can evolve from the mobile elements by repeated insertions that lead to higher copy number. Mobile elements may be used as genetic markers because they change the pattern of restriction fragment lengths.

Self-assessment Questions

1. Define mobile elements.
2. Discuss briefly the general mechanism of transposition.
3. Give a brief account of the different types of transposons.
4. Describe the process of regulation of transposition.
5. Explain the mechanism of retrotransposition.
6. Discuss the structure and functions of bacterial transposition.
7. What is gene amplification?
8. Explain the different yeast transposons in brief.
9. Give a brief history of controlling elements in corn.
10. What are the P elements?
11. Explain the mechanism of gene conversion with the help of a diagram.
12. Explain the evolutionary issues of transposons.
13. How are transposons used in genetics and molecular biology studies?

FURTHER READING

Textbooks

Berg DE and MM Howe (ed.) (1989). *Mobile DNA*. Washington DC, American Society for Microbiology.

Hartl DL and EW Jones (1998). *Genetics Principles and Analysis*. Jones and Bartlett Publishers, Inc.

Lewin B (2000). *Genes VII*. New York: Wiley.

Malacinski GM (2003). *Essentials of Molecular Biology* 4th ed. Jones and Bartlett Publishers.

Old RW and SB Primrose (1985). *Principles of Gene Manipulation*. Oxford: Blackwell.

Sambrook J, EF Fritsch and T Maniatis (ed.) (1989). *Molecular Cloning: A laboratory manual*, 2nd ed. Cold Spring Harbor Laboratory Press.

Shapiro JA (ed.) (1983). *Mobile Genetic Elements*. Academic Press.

Watson JD, NH Hopkins, JW Roberts, J Steitz and AM Weiner (ed.) (1987). *The Molecular Biology of the Gene*, Vol. 1. Benjamin Cummings.

References

Cameron JR, EY Loh and RW Davis (1979). Evidence for transposition of dispersed repetitive DNA families in yeast. *Cell* 16:739–51.

Campbell A, DE Berg, D Botstein, EM Lederberg, RP Novick, P Starlinger and W Szybalski (1979). Nomenclature of transposable elements in prokaryotes. *Plasmid* 2:466–73, and *Gene* 5:197–206.

Craig NL (1997). Target site selection in transposition. *Annu. Rev. Biochem.* 66:437–74.

Derbyshire KM and NDF Grindley (1986). Replicative and conservative transpositions in bacteria. *Cell* 47:325–27.

Finnegan DJ (1985). Transposable elements in eukaryotes. *Int. Rev. Cytol.* 93:281–325.

Finnegan DJ (1990). Transposable elements and DNA transposition in eukaryotes. *Curr. Opinion Cell Biol.* 2:475–80.

Gierl A, H Saedler and PA Peterson (1989). Maize transposable elements. *Annu. Rev. Genet.* 23:71–85.

Goff SP (1992). Genetics of retroviral integration. *Annu. Rev. Genet.* 26:527–44.

Haren L, B Ton–Hoang and M Chandler (1999), Integrating DNA: Transposases and viral integrases. *Annu. Rev. Microbiol.* 53:245–81.

Kleckner N (1990). Regulation of transposition in bacteria. *Annu. Rev. Cell Biol.* 6:297–327.

Lambowitz AM and M Belfort (1993). Introns as mobile genetic elements. *Annu. Rev. Biochem.* 62:587–622.

Leplae R, A Hebrant, SJ Wodak and A Toussaint (2004). ACLAME: A classification of mobile genetic elements. *Nucleic Acids Research* 32:D45–D49.

Potter SS, M Truett, M Phillips and A Maher (1980). Eukaryotic transposable genetics elements with inverted terminal repeats. *Cell* 20:639–47.

Prak ETL and HH Kazazian Jr. (2000). Mobile elements in the human genome. *Nat. Rev. Genet.* 1:134–44.

Rice PA and TA Baker (2001). Comparative architecture of transposases and integrative complexes. *Nat. Struct. Biol.* 8:302–307.

Saedler H and A Gierl (1996). Transposable elements. *Curr. Top. Microbiol. Immunol.* 204.

Stavnezer J (1996). Antibody class switching. *Adv. Immunol.* 61:79–146.

Stiles JI, LR Friedman, C Helms, S Consaul and F Sherman (1981). Transposition of the gene cluster *CYC1–OSM1–RAD7* in yeast. *J. Mol. Biol.* 148:331–36.

Yoder JA, CP Walsh and TH Bestor (1997). Cytosine methylation and the ecology of intragenomic parasites. *Trends Genet.* 13:335–40.

CHAPTER 19

Mutagenesis and DNA Repair

LEARNING OBJECTIVES

After reading this chapter, you will be able to:

- Identify mutagenic agents and the types of mutations
- Understand mechanisms of mutagenesis and their biological significance
- Appreciate the different DNA repair systems

19.1 INTRODUCTION

DNA is the ultimate repository for genetic information and any damage to it can lead to permanent changes. Therefore, the perpetuation of genetic material from generation to generation depends on maintaining rates of mutation at low levels. If progeny have to survive better and maintain species identity, the DNA sequences must be passed on largely unchanged in the germ line. If the rate of mutations in somatic cells were high, the incidence of cancer would increase. Recent advances have made localisation and isolation of numerous disease-related genes and direct detection of the underlying pathologic lesions possible. Point mutations are the most frequently encountered changes in the human genome; duplications, inversions, expansions and complex rearrangements are other important types of mutations. Now it is well confirmed that there is a relationship between DNA damage and ageing. Evidence is also accumulating that DNA damage may contribute to several diseases (e.g., Alzheimer's disease, mutation in APP gene). These and many other examples serve to illustrate that DNA damage, whether a primary cause, or secondary effect, is associated with a wide range of health-related issues.

Biomolecules in a cell are highly susceptible to both chemical and physical alterations that arise from environmental damage or errors during biosynthesis. It is very important to protect the genome from such damage in macromolecules including RNAs, proteins, or other cellular molecules. The damage can be avoided by replacement of these molecules through normal turnover. Cellular integrity is vital for survival and reproduction of an organism and its information must be protected and preserved over the lifespan of the cell and also generation after generation. To maintain the fidelity and integrity of genetic information, several types of enzymatic DNA repair processes have developed during evolution. Some of the safeguards include high-fidelity replication systems and repair systems that correct damage that might alter information in the DNA.

19.2 BASIC TYPES OF MUTATIONS

Mutations have been classified in a number of ways. At the level of the gene, mutations may fit into more than one category.

On the basis of origin

Spontaneous mutation: It occurs during normal cellular activities, primarily DNA replication and repair.

Induced mutations: This type of mutation occurs as a result of exposure to a mutagenic agent or polluted environment. Mutation rate is generally higher than background levels (i.e., spontaneous). Ionising radiation, non-ionising radiation (UV light) and chemicals can cause induced mutations.

Mutator mutations: Such mutations influence the mutability of other genes. They are further classified as: (i) Specific mutators—limited to one locus and (ii) Non-specific mutators—that do not affect just one locus; that is, they occur in genes that control DNA repair.

Based on the magnitude of phenotypic effect

Change in mutation rate: Alleles mutate at different rates; some can be distinguished based on their rate of mutation.

Isoalleles: They produce identical phenotypes in homozygous or heterozygous combinations with each other, but prove to be distinguishable when in combination with other alleles.

Mutants effecting viability

Subvitals—relative viability is greater than 10% but less than 100% compared with wild type.

Semilethals—cause more than 90% but less than 100% mortality.

Lethals—kill all individuals before adult stage; albino mutations in plants are examples of these.

Beneficial mutation—enhance the survival of the organism.

Based on direction

Forward mutation: A forward mutation changes the wild-type genotype into some new variation (generally abnormal phenotype).

Reverse or back mutation: It has the opposite effect. It changes the altered nucleotide sequence back to its original sequence.

Suppressor mutation: It produces a change from abnormal (i.e., mutated) phenotype back to wild type. Alternatively, a second mutation will sometimes affect the phenotypic expression of a first mutation. Such mutations are generally called **suppressors** or **suppressor mutations**. There are two types of suppressor mutations: (i) **Intragenic suppressor**—a mutation in the same gene as was originally mutated, but at a different site, that results in restoration of wild type function (ii) **Intergenic suppressor**—a mutation in another gene that results in restoration of wild type function (i.e., a nonsense mutation may be suppressed by a mutation in the tRNA for that codon so that it now inserts an amino acid).

Polar mutations: Some mutations have an effect on the genes downstream. Polar mutations are unique in bacteria, as they frequently contain operons or a single promoter followed by several genes. If a mutation occurs in an upstream gene, it can affect the expression of the downstream genes.

Based on the cell type

Somatic cell mutation: It occurs in any cell of the body, excluding the sex cells, often producing a mutant phenotype in only a sector of the organism (mosaic or chimera), and not a heritable change. Somatic mutations frequently arise in plants as mutant branches or sprouts, due to the mutation being limited to a part of a plant meristem. This means that the mutation will not be sexually transmitted through the gametes.

Germ cell (gametic) mutations: They occur in cells that will ultimately form the gametes and therefore, are sexually transmitted (i.e., heritable change).

Based on size

Point mutation: It is a change in a very small segment of DNA, usually considered involving a single nucleotide or nucleotide pair.

Gross (large) mutations: These are changes involving more than one or a few nucleotides. They may involve megabases of DNA that may encompass the entire gene, the entire chromosome, or sets of chromosomes (polyploidy).

Deletions: These mutations involve the loss of a part of the normal gene sequence. The size of the deletion can vary greatly from just a few bases to the entire gene sequence.

Insertions: These mutations involve the insertion of a segment of DNA into the normal gene sequence from elsewhere in the genome.

Rearrangements: These are reciprocal movements between segments of a gene sequence and sequences elsewhere in the genome.

Effect on protein (codons)

Silent mutation, same sense or synonymous mutation: This is a point mutation, which does not affect the sense of a codon and thus has no effect on polypeptide structure. This type of mutation is possible because a point mutation at the third nucleotide position of a codon (code is wobble), does not alter the amino acid coded for because of the degeneracy of the code.

Mis-sense mutation or non-synonymous mutation: Mis-sense mutations are point mutations that result in changes in the amino acid. A codon is mutated to result in incorporation of a different amino acid, making an inactive or unstable enzyme.

Nonsense mutation: It is a point mutation, that changes a codon specifying an amino acid into a termination codon, which will prematurely terminate the polypeptide produced, changing the activity of the protein and altering the phenotype.

Neutral mutation: Enormous variation is manifest in the DNA sequences of any two randomly chosen human haploid genomes. Clearly, not all variations within a gene cause abnormal expression of protein products. Indeed, single nucleotide substitutions/polymorphisms (SNP) occur in 1 in 600–1200 nucleotides in the non-coding region of DNA. Similarly, some single nucleotide substitutions in the coding regions of genes may also be normal polymorphic variants even if they alter the polypeptide product. Changes are such that the new amino acid behaves similar to the original one (e.g., has a similar functional group) and does not alter the function of the protein.

Frameshift mutation: It often results in insertions or deletions of $3n \pm 1$ and not in multiples of three (code). These can cause changes in the amino acids downstream from the mutation, resulting in novel polypeptide sequence distal to the mutation and consequently, a mutant phenotype.

Non-frameshift indel: This represents a small insertion or deletion of $3n$ nucleotides (indel is generic for insertion or deletion).

Based on the effect on DNA

Based on changes in the nucleotide content of the gene, mutations are classified as follows.

Transition: It is a point mutation in which a purine is changed to a purine (A → G) or a pyrimidine to a pyrimidine (T → A).

Transversion: It is a point mutation in which the change is purine to pyrimidine or pyrimidine to purine (A or G → T or C).

Inversion: It refers to the exclusion of a part of the DNA followed by its insertion into the same position with a reverse orientation.

Addition mutation: It means insertion (or addition) of one or more nucleotides; may cause frameshift mutations if inserted nucleotides are not triplet.

Deletion: Deletion of one or more nucleotides causes frameshift mutations, if inserted nucleotides are not triplets.

Based on the effect on gene function

Loss of function mutation: It is a mutation that results in a lack of gene function; this can result from a number of different types of mutations and is recessive in nature.

Gain of function mutation: A mutation that results in a new or different gene function; it can result from a number of different types of mutations and it will be dominant in nature.

Splice site mutations: These mutations alter the signal sequences for splicing that occur at the 5'- and 3'-ends of exons (**splice sites**) leading to a failure of RNA transcribed from the mutated gene to be spliced properly. Mutations in introns also result in the creation of splice sites causing splicing to occur abnormally.

Biochemical mutations: They may not always be observed visually, but can be identified only by biochemical analysis.

Chromosome rearrangements

Changing the location of a piece of DNA within the genome can result in large structural changes (translocations or inversions) in genes or may change the expression of a gene by placing it under the control of a different promoter (called a **position effect**).

Translocations: Movement of DNA to a non-homologous chromosome; usually an exchange occurs between two non-homologous chromosomes.

Inversions: Movement of DNA within the same chromosome; an 180° rotation or 'flip'.

Transposition: It refers to the transposition of a transposable element (TE) into a gene can also create mutations in genes called intergenic suppression.

Conditional mutations

Among the mutations that are most useful for genetic analysis are those whose effects can be turned on or off at will. These are called conditional mutations because they produce changes in one set of environmental conditions (called the **restrictive conditions**) but not in another (called the **permissive conditions**). A temperature-sensitive mutation (or Ts mutation), or drug resistance, for example, is a conditional mutation.

Regulatory mutations: Mutations may be regulatory in nature. This would mean that their expression is detected by the inability to control another gene. Constitutive mutants in the lactose operon are such phenotypes.

Mutations in promoter regions

Mutations can occur outside the coding sequence and still influence gene expression. For example, a mutation may alter the sequence within the core promoter of a gene. If the mutant promoter sequence becomes more like the consensus sequence, the mutation may increase the rate of transcription. This is called an **up-promoter mutation**. In contrast, a **down-promoter mutation** occurs when a mutation causes the promoter to become less like the consensus sequence, decreasing its affinity for regulatory factors and decreasing the transcription rate.

Cytologically visible effects

These are detectable through microscopic analysis of chromosomes (**macrolesions**), and non-visible changes which occur at the nucleotide level (**microlesions**). Macrolesions can be subdivided into changes in chromosome number (gain or loss of single chromosomes or sets of chromosomes) and changes in chromosome structure (breaks, deletions, rearrangements). Microlesions consist of base-pair substitutions or base-pair addition/deletion changes. A strain carrying such changes is called a mutant. A mutant by definition differs from its parental strain in genotype, the precise sequence of nucleotides in the DNA of a genome. But in addition, the observable properties of the mutant, its phenotype, may also be altered relative to the parental strain. It is common to refer to a strain isolated from nature as a wild type strain.

Phenotypic effects

The phenotypic effects can range from minor alterations that are detectable only by biochemical methods to drastic changes that produce clearly defined effects, such as unrestrained cell proliferation or cell death. The effect of a mutation is determined by the type of cell containing the mutant allele, by the stage of the lifecycle or development of the organism that the mutation affects, and in diploid organisms, by the dominance or recessiveness of the mutant allele. They arise, either through spontaneous errors in DNA replication or meiotic recombination, or as a consequence of the damaging effects of physical or chemical agents on the DNA (**induced mutation**).

Genetic mutations

These range from the involvement of a single base pair to an entire chromosome and are broadly divided into two classes: **chromosomal mutations**, involving modifications of large sections of chromosomes or even the entire chromosome and **gene mutation**, involving only a single gene or a small fragment of a chromosome. In chromosomal mutations, although the number of copies of a gene may change, the actual genes may be unaffected. There are however, gene mutations, which change the base pair sequence of individual genes. When these mutations involve only one base pair, they are called **point mutations**. Occasional mistakes in the normal duplication of DNA give rise to spontaneous mutations.

19.3 ORIGIN OF SPONTANEOUS MUTATIONS

Mutations occur *in vivo* spontaneously or after induction with mutagenic agents. Spontaneous mutations are those that arise randomly without any readily apparent cause. The possible sources of spontaneous mutations are normal cellular events and mutagens in the environment. The normal cellular events include errors in DNA replication (before proofreading and mismatch correction) and spontaneous alterations of nucleotides (tautomerism of the nucleotide bases, methylation followed by deamination of cytosine and activities of transposable elements). The mutagenic chemicals or radiation present in the environment (also called premutagenic damage) can cause mutations.

Errors during DNA replication (incorporation errors)

It is estimated that DNA polymerase incorporates wrong bases during DNA replication with a frequency of 10^{-5}. Mutations have been isolated in DNA polymerase which make it more efficient; however, these mutations significantly slow down the speed of the enzyme. Conversely, mutations in DNA polymerase that increase its speed also increase the misincorporation rate.

Errors due to tautomerism

A tautomer is a structural isomer of each base. Tautomers form as a result of spontaneous, transient rearrangement of bonds through the shifting of hydrogen atoms in each of the bases. This arrangement of hydrogen atoms alters the hydrogen bonding properties of the bases. If the rare tautomer is present when the template is being replicated, a wrong base may be incorporated by DNA pol III into the daughter strand.

Alteration by depurination

Depurination means the removal of the purine from the 2-deoxyribose-phosphate backbone. In order to depurinate, the N, N-glycosidic bond holding the purine to the 2-deoxyribose has to be broken. This can be accomplished naturally if environmental temperature increases or the pH decreases. Hydrolysis of the N,N-glycosidic bond between a base and its 2-deoxyribose in a DNA molecule results in an apurinic site or less frequently, an apyrimidinic site.

Alteration by deamination

Cytosine, adenine and guanine contain exocyclic amino groups (NH_2) that can be lost spontaneously or when exposed to base modifiers such as nitrous oxide. Deamination means the loss of NH_2– groups. Deamination of C, A or G changes their hydrogen bonding properties, creating a potential for mismatched base pairs. Deamination of cytosine, in addition to DNA pol III incorporation errors, explains the occurrence of uracil in a DNA molecule. Uracil pairs with adenine creating a C-G to A T transition.

Alterations by spontaneous genetic rearrangement

Deletions, duplication and amplifications can occur spontaneously in DNA. In many cases, small regions (4–5 bp) of homology are found at the endpoints of the rearrangements, which are termed as **microhomology**. Several mechanisms have been proposed to account for these rearrangements. DNA strands can slip relative to each other. The microhomology is able to stabilise the slipped DNA. Replication through the slipped DNA results in a deletion. Alternatively, low-level recombination at the microhomologies by recombination system can result in rearrangements.

Free radicals: Spontaneous damage to the DNA occurs frequently in most cells due to the free radicals of oxygen. These arise in cells as a result of oxidative metabolism and are also formed by physical agents such as radiation. An important oxidation product is 8-hydroxyguanine, which mispairs with adenine, resulting in G:C to T:A transversions.

Alterations caused by transposition

Transpositions are mobile elements of DNA that move around the chromosome (see **Chapter 18**). Every time a transposon moves, it creates an insertion mutation. If a transposon moves into a gene it will usually lead to a loss of function. If the gene is in an operon, frequently, the insertion will be polar on the downstream genes. When a transposon comes out of the chromosome to move to another place, several alterations can occur. These can include deletions, duplications and rearrangements. Transposon-induced mutations and rearrangements of chromosomes lead to alterations of the genome structure and genome expansion, which acts as a substrate for evolution.

19.4 RATES OF MUTATION

The mutation rate is the number of mutations that arise per cell per division in single-celled organisms, or the number of mutations that arise per gamete in higher organisms. Each organism has a characteristic spontaneous mutation rate that depends on the growth conditions of the organism. The mutation rates vary tremendously depending upon the length of the genetic material, the kind of mutation and other factors (**Table 19.1**).

Spontaneous and induced mutants arise as a result of structural changes in the genome which includes:

Table 19.1 Representative spontaneous mutation rates for various organisms

Organism	Rate
E. coli	2×10^{-6} to 1×10^{-9} per cell division
Chlamydomonas reindhardi	1×10^{-6} per cell division
Neurospora crassa	6×10^{-8} per asexual spore
Zea mays	1×10^{-6} per gamete/generation
Drosophila melanogaster	4×10^{-5} per gamete/generation
Homo sapiens	3×10^{-5} per gamete/generation

- Genome mutation may cause changes in the chromosome number.
- Chromosome mutation may change the order of the genes within the chromosome (e.g., deficiency, deletion, inversion, duplication or translocation).
- Gene or point mutations may result from changes in the base sequence in a gene.

19.5 MOLECULAR BASIS OF MUTATIONS

19.5.1 Point mutations

Alterations in the base pattern of DNA may arise in various ways. The following are the few important types of point mutations (**Fig. 19.1**) classified based on the nature of the change in nucleotides.

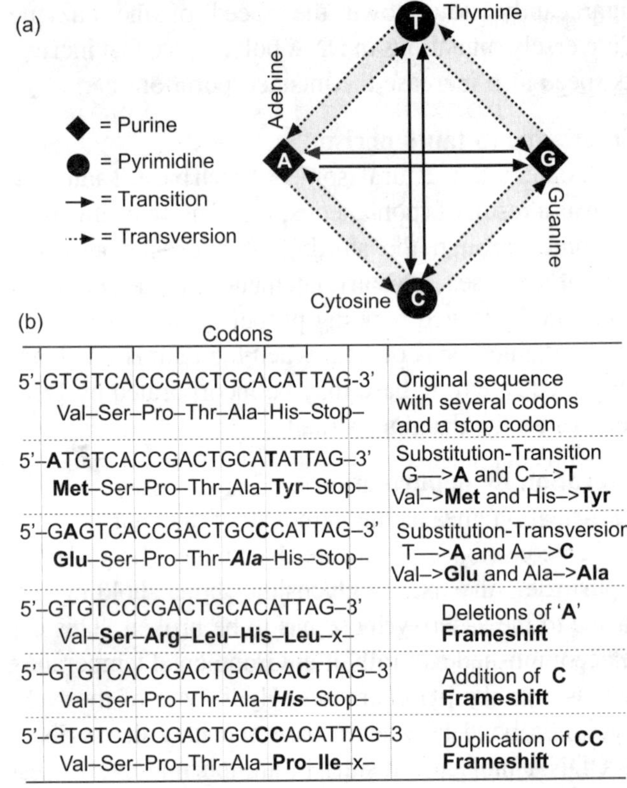

Sequence	Mutation
5'-GTG TCA CCG ACT GCA CAT TAG-3' Val–Ser–Pro–Thr–Ala–His–Stop–	Original sequence with several codons and a stop codon
5'-**A**TGTCACCGACTGCA**T**ATTAG-3' **Met**–Ser–Pro–Thr–Ala–**Tyr**–Stop–	Substitution-Transition G—>**A** and C—>**T** Val->**Met** and His->**Tyr**
5'-G**A**GTCACCGACTGC**C**CATTAG-3' **Glu**–Ser–Pro–Thr–***Ala***–His–Stop–	Substitution-Transversion T—>**A** and A—>**C** Val->**Glu** and Ala->**Ala**
5'-GTGTCCCGACTGCACATTAG-3' Val–**Ser–Arg–Leu–His–Leu**–x–	Deletions of '**A**' **Frameshift**
5'-GTGTCACCGACTGCACA**C**TTAG-3' Val–Ser–Pro–Thr–Ala–***His***–Stop–	Addition of **C** **Frameshift**
5'-GTGTCACCGACTGC**CC**ACATTAG-3 Val–Ser–Pro–Thr–Ala–**Pro–Ile**–x–	Duplication of **CC** **Frameshift**

Fig. 19.1 (a) Diagram illustrating the base substitutions that are possible in DNA. These include four transitions (purine for purine or pyrimidine for pyrimidine; solid arrows) and eight transversions (purine for pyrimidine and pyrimidine for purine; dashed arrows). (b) Diagrammatic representation of various types of point mutations. Letters with bold face indicates the point mutation in nucleotide and respective amino acid change.

Substitution mutation

The simplest mutation is a **base substitution**, in which a nucleotide pair in a DNA duplex is replaced with different nucleotide pairs. For example, in a single base change, A→G substitution, A is replaced with a G in one of the DNA strands. This can be either a **transition** in which one purine (or pyrimidine) is replaced by the other (e.g., T→C, C→T or A→G, G→A); or a **transversion** where a purine replaces a pyrimidine or vice versa (e.g., T→A, T→G, C→A, or C→G or A→T, A→C, G→T, or G→C). The phenotypic effects of such a mutation can be various. If it is in a non-coding or non-regulatory piece of DNA or in the third position of a codon, which often has no effect on the amino acid incorporated into a protein, then it may be silent (or **silent substitutions**). If it results in an altered amino acid in a gene product then it is a mis-sense mutation whose effect can vary from none to lethality, depending on the amino acid affected. A base substitution that creates a new stop codon is called nonsense mutation and gives rise to truncated protein products. There are three kinds of nonsense mutations, each representing one of the three particular base sequences that do not correspond to any amino acid. They are called **amber** (Am), **ochre** (Oc) and **opal**.

Insertion and deletion

Insertions or deletions involve the addition or loss of one or more bases. These can produce frameshift mutations in genes, where the translated protein sequence to the C-terminal side of the mutation is completely changed. Transitions are routinely produced by base analogues. Two of the most widely used base analogues are 5-bromouracil (5BU) and the purine analogue 2-aminopurine (2AP). Many analogues are found to be incorporated into DNA. Ethyl methane sulphonate (EMS) and ethyl ethane sulphonate (EES) are agents that cause the removal of purine rings from DNA. The multistep process begins with the ethylation of a purine ring and ends with the hydrolysis of the glycosidic bond, causing the loss of the base.

Frameshift mutation

A point mutation may consist of replacement, addition, or deletion of a base. Point mutations that add or subtract a base are, potentially, the most devastating in their effects on the cell or organism because they change the reading frame of a gene from the site of mutation onward. It generates small insertions or deletions. In coding regions, slipping causes frameshift. But the term has been adopted to describe all mutations of this nature

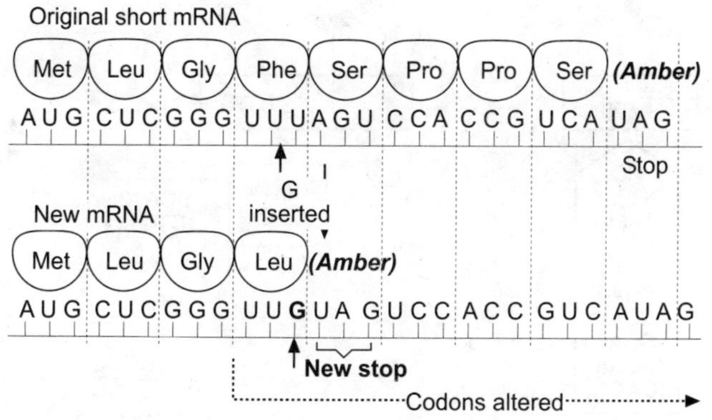

Fig. 19.2 Possible effects of a frameshift mutation; the insertion of a single base results in the creation of a new stop sequence (amber). The result will be premature termination of translation

regardless of where they occur. A frameshift mutation causes two problems (**Fig. 19.2**). First, all the codons from the frameshift on will be different and thus yield a useless protein. Second, the stop signal information will be misread. One of the new codons may be a nonsense codon, which causes translation to stop permanently or translation may continue beyond the end of the gene.

19.5.2 Misalignment mutagenesis

Insertions and deletions of nucleotide sequences can also occur due to the replication slippage or **unequal crossing over**. Most of these are less than 20 base pairs long. In the Human Gene Mutation Database (HGMD), the deletion of one base pair accounts for 48% of small deletions and an additional 30% involve two or three nucleotides. The majority of microdeletions (78%) result in an alteration of the reading frame. Most microdeletions occur in regions that contain direct repeats of 2 bp or more. The most common length of direct repeat is 3 bp (48% of direct repeats associated with short deletions). The most plausible mechanism for small deletions mediated by the presence of direct repeats is the slipped mispairing model (**Fig. 19.3**).

The genomes of most eukaryotes contain, at a very large number of locations, tandem repeats of any of a number of short nucleotide sequences. A repeat of this type is called a **simple tandem repeat polymorphism (STRP)**. The number of copies of the repeat often differs from one chromosome to the next. Hence, populations are usually highly polymorphic for the number of repeating units. STRPs are usually polymorphic because they are susceptible to errors in replication or recombination that change the number of repeats or the length of the run. Any short sequences repeated in tandem a number of times, may gain or lose a few copies because of these types of errors. An increase in the number of repeating units in STRPs is genetically equivalent to an insertion, and a decrease in the number of copies is equivalent to a deletion.

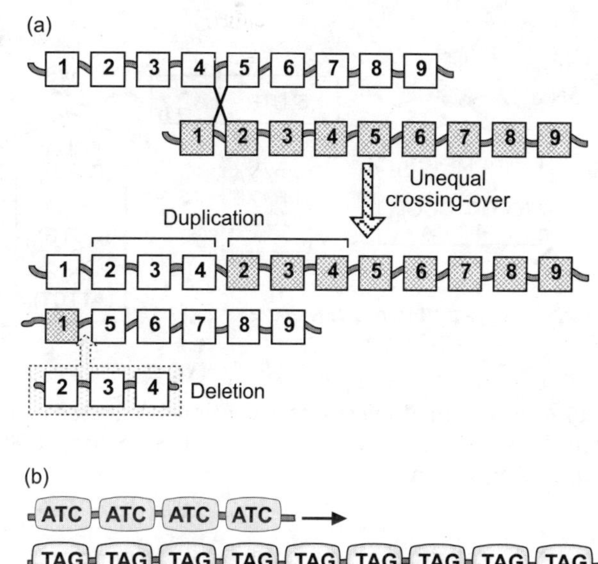

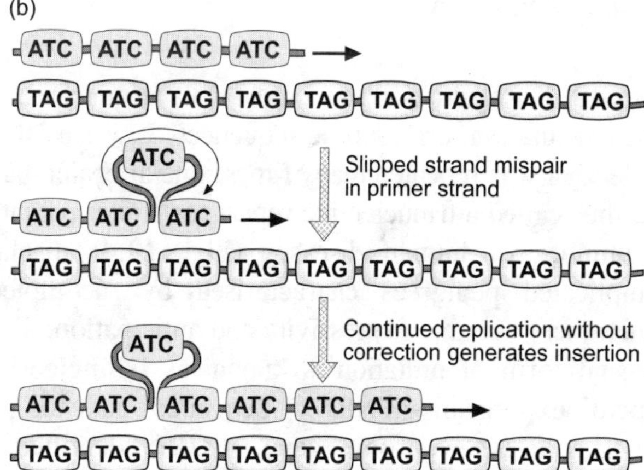

Fig. 19.3 Deletions and insertions are generated by direct repeats in DNA. (a) Direct repeats are hotspots for unequal exchange; two chromatids can misalign and consequently crossing over generates reciprocal duplication and deletion products. (b) Short direct repeats are hotspots for replication slipping; backward slipping of primer strand in this case generates an insertion.

19.5.3 Triplet repeat syndromes

Another mechanism of human gene mutation causing hereditary disease is the instability of certain trinucleotide repeats and their expansion in affected genes. More than 140 disorders recorded in the HGMD and the majority of which involve neuromuscular tissues, have been found to be due to or associated with the expansion of repeat sequences, and of these 23 are expansions of triplet repeats.

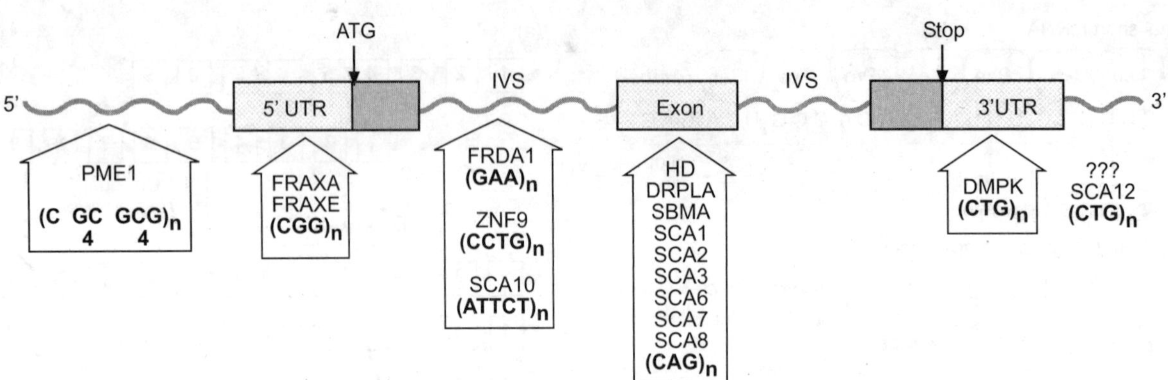

Fig. 19.4 Location of the repeat expansion in selected human disorders SCA: spinocerebellar ataxia UTR: untranslated region FRDA: Friedreich ataxia HD: Huntington disease FRAX: fragile X DMPK: myotonicdystrophy DRPLA: dentatorubral pallidoluysian atrophy

In 1991, scientists discovered a totally unexpected type of mutation, error-prone sequences. It was a DNA triplet that was repeated many times, one after another, and thus called a trinucleotide repeat (or triplet repeat). A number of human diseases (**Fig. 19.4**) display complicated pedigrees characterised by incomplete penetrance, variable expressivity and anticipation. This unusual form of mutation is known as **trinucleotide repeat expansion** (**TNRE**). Sequence analysis of the fragile X syndrome gene, *FMR1* identified a polymorphic region of tandemly repeated triplets $(CGG)_n$, in the 5' untranslated region. In normal individuals, the repeat copy was less than 50, whereas in affected persons, the region had expanded and contained over 1,000 copies. Since then, more than 10 additional triplet repeat syndromes have been identified. Three classes of triplet repeat syndrome, have been identified.

Hotspots of mutations

The frequencies of mutation at any given base pair can be determined by statistical analysis, except for hotspots, where the frequency is increased by higher order of magnitude. Some sites gain far more than the number of mutations expected from a random distribution; these sites are called hotspots (**Fig. 19.5**). Spontaneous mutations may occur at hotspots, and different mutagens may have different hotspots. Many hotspots result from modified bases, for example, 5-methylcytosine spontaneously deaminated to thymine.

19.6 GROSS VARIATIONS OF CHROMOSOMES

Gross variation can occur in both chromosomal structure and number. Variations in chromosome number include aneuploidy, euploidy and mosaicism. Variations in chromosome structure include a number of DNA alterations, such as single-strand, double-strand and centromeric breaks, deletions, insertions, duplications, inversions and translocations. In the following sections, only the heritable structural variations will be discussed.

19.6.1 Effect on structure of chromosome

Variation in chromosome structures that are relatively stable are deletions, insertions, duplications, inversions and translocations (**Fig. 19.6**).

Gross deletions

Deletions or deficiencies represent a loss in chromosomal material. The classic example for gross deletion is the syndrome known as *cri du chat,* so called because of the cat-like cry. Microcephaly, congenital heart disease and severe mental retardation are common symptoms. This disorder arises from a deletion in chromosome 5. Most other deletions studied (4p-, 13q-, 18p-, 18q-) also result in microcephaly and severe mental retardation.

At nucleotide sequence level, gross deletions are common causes of certain disorders and rare in others. In most of the X-linked disorders, for example, large deletions account for about 5% of molecular defects. In case of steroid sulphatase deficiency, large deletions of the STS gene account for 84% of patients and the deletion can be as large as 1 Mb. The same is true for disorders such as Duchenne muscular dystrophy, growth hormone deficiency and thalassemia. In Kallmann

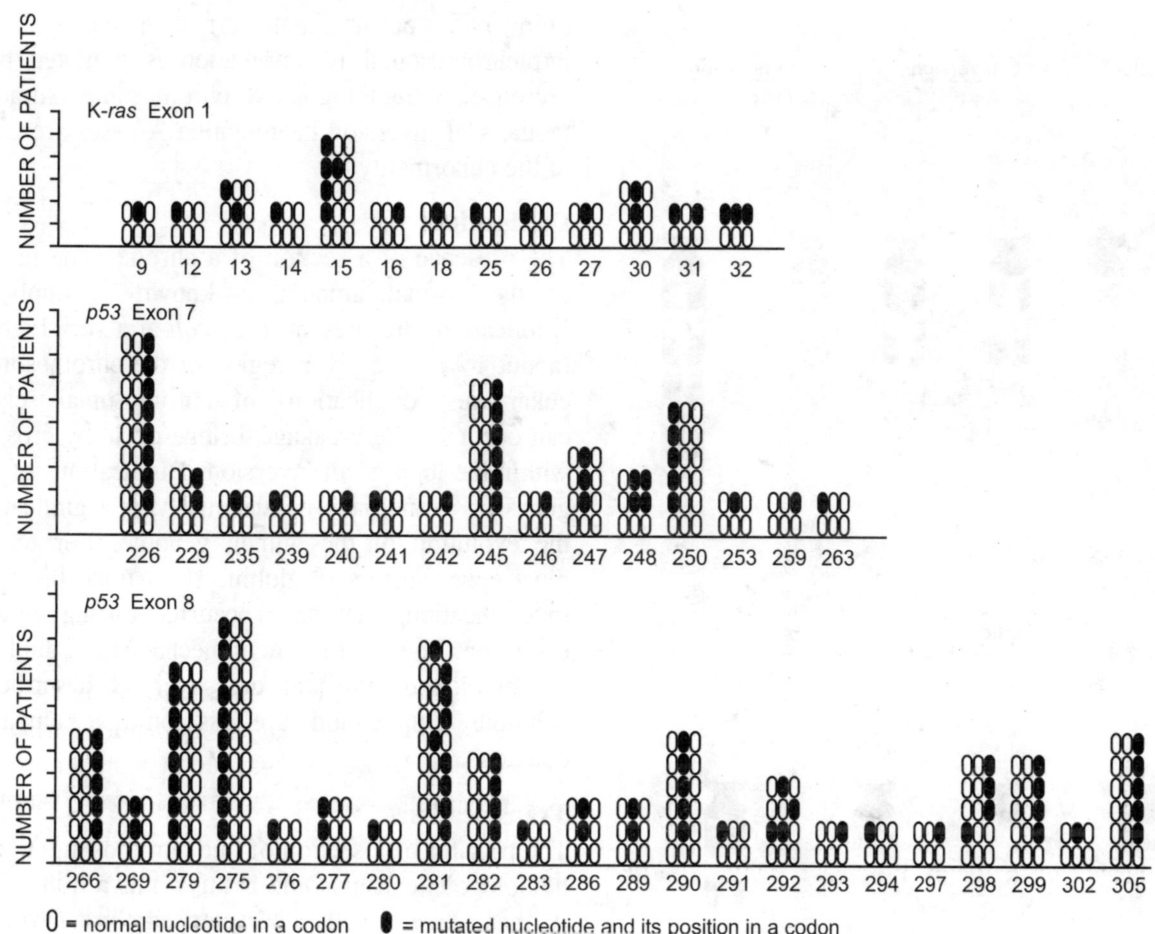

Fig. 19.5 The mutation hot-spots in (a) exon 1 of *K-ras* (b and c) Exons 7 and 8 of the *p53* genes from 27 lung cancer tissues. Each triplet codon (a group of three circles) which is located above the normal codon represents the mutated codon of one individual. [***Source***: Gao HG, Chen JK, Stewart J, Song B, Channarayappa, Whong WZ, and T Ong 1997. Distribution of *p53* and K-*ras* mutations in human lung cancer tissues, *Carcinogenesis*]

syndrome, translocation can occur as a result of unequal mispairing of X and Y homologous sequences. A number of common genetic disorders are due to large deletions (or duplications) caused by unequal crossing over of homologous sequences. Approximately 5% of human genome is duplicated either intra- or inter-chromosomally. The large deletions or duplication due to duplication crossover are also termed genomic disorders.

Insertions

A less common but very interesting mechanism of human gene mutation is the *de novo* insertion of repetitive elements via retrotransposons. The phenomenon was first observed in humans in the *F8* gene in two unrelated *de novo* cases of severe haemophilia A. Truncated LINE repetitive elements were introduced into exon 14 of the *F8* gene, where they caused disruption of the reading frame. The inserted elements contained a poly A tract and caused a target site duplication of more than 12 nucleotides. LINEs probably integrate into genomic DNA by a process called target-primed reverse transcription. Similar retrotransposons that involve members of the *Alu* sequence family have also been reported in several genes.

Inversions

Two breaks in the same chromosome can also lead to inversion, in which the middle section is reattached but in the inverted configuration without changing chromosomal material. Inversions have many interesting features such as linkage analysis, position effect etc. They have been well characterised in *Drosophila*. They are the most common mutations. Most inversions, which are high-frequency-independent recurring events, involve the distal sequence. The vast majority

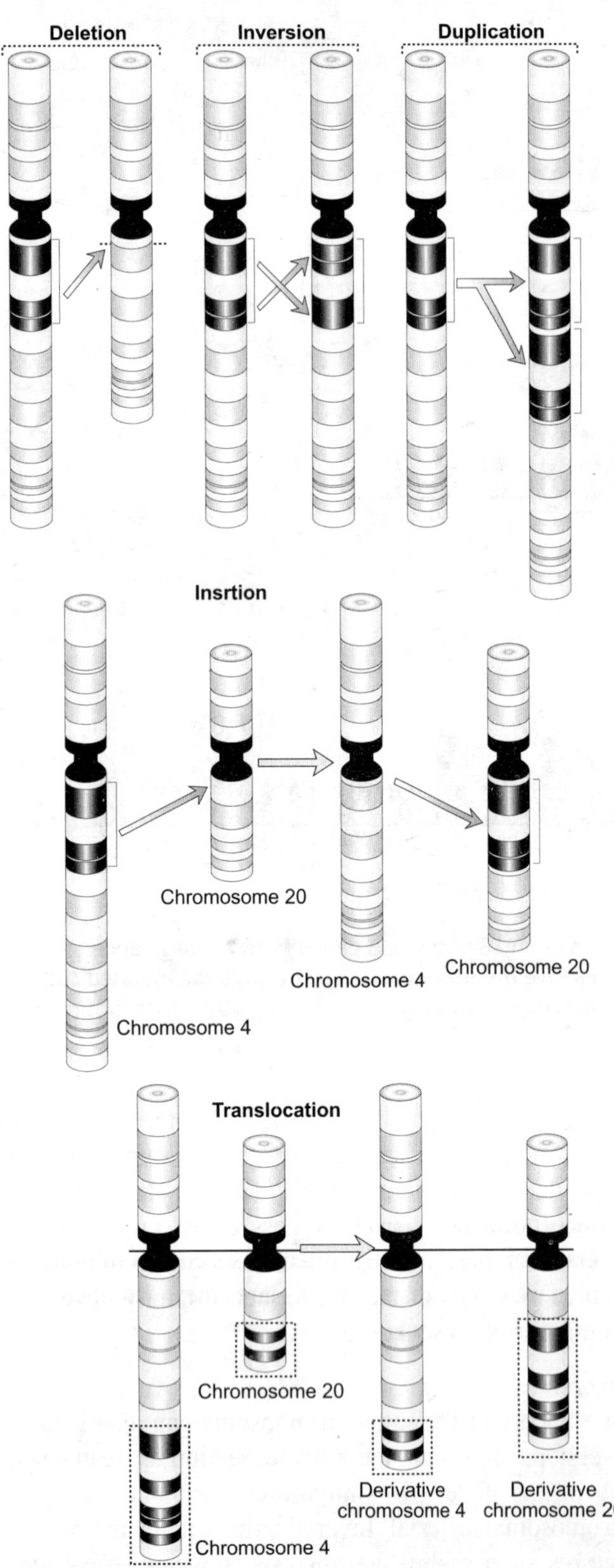

Fig. 19.6 Illustration of five types of chromosomal mutations and effect on structure

of inversions occur in male germ cells, perhaps because intrachromosomal recombination is inhibited by the presence of homologous X chromosome. Almost all mothers of inversion haemophilia A cases are carriers of the abnormality.

Duplications

The presence of a section of a chromosome in excess of the normal amount is known as duplication. Duplications are present in *E. coli* at a very high level (about 1%) for a given region of the chromosome. In eukaryotes, duplications of chromosomal segments can occur by the breakage–bridge cycle by crossovers within the loop of an inversion. Duplications of whole genes or exons have contributed very significantly to the evolution of the human genome. For example, most gene clusters (ß-globin, Hox) trace their origin to duplications that have occurred during vertebrate evolution. The most frequent mechanism of duplication is homologous unequal crossover as described for deletions. Duplications are less common compared to deletions.

Translocations

The transfer of a section of one chromosome to a non-homologous chromosome is known as a translocation. It has been well characterised in *Drosophila*. Translocations may be simple or shifts and or reciprocal types. In humans, when one of the parents is a translocation heterozygote in whom the short arm of chromosome 5 is translocated to a chromosome of another group (e.g., 15 in the D group), the parent is normal, but he or she will nevertheless produce some unbalanced gametes.

19.6.2 Amplifications

Amplification may be defined as excessive duplication. Specifically, it has been shown for a variety of *Streptomycetes* wherein after various intense treatments, one finds massive tandem duplications of small regions of the chromosome. While it is not clear at this point what the function of these is, it is clear that a system for generating such an amplified region does exist in at least some prokaryotes. A number of examples for gene amplification in eukaryotes have also been reported. When an organism is exposed to high concentrations of chemicals (drugs, heavy metals, toxins), drug resistance genes will be overexpressed, often due to gene amplification.

19.6.3 Gene conversions

Gene conversion is the modification of one of the two alleles by the other. It involves the non-reciprocal correction of an 'acceptor' gene or DNA sequence by a 'donor' sequence that remains physically unchanged. In most known instances of gene conversion as a cause of human genetic disease, the functional gene has been wholly or partially converted to the sequence of a highly homologous and closely linked pseudogene, which therefore acts as the donor sequence. Probable examples include the genes for steroid 21-hydroxylase (CYP21), polycystic kidney disease (PKD1), neutrophil cytosolic factor p47-*phax* (NCF1), immunoglobulin λ-like polypeptide 1 (IGLL1), glucocerebroidase (GBA), von Willebrand factor (VWF) and phosphomannomutase (PMM2). Most of these gene–pseudogene pairs are closely linked.

19.6.4 Insertion–deletion (indel)

A relatively rare type of mutation causing human genetic disease is the indel, a complex lesion that appears to represent a combination of microdeletion and microinsertion. An example is provided by the nine deleted base pairs encoding codons 39 through 41 of the α_2-globin gene (HBA2) that were replaced by eight inserted bases that served to duplicate the adjacent downstream sequence. Some 0.67% of lesions in the HGMD belong to this category.

19.6.5 Molecular misreading

Long runs of adenines (also other nucleotides) promote a phenomenon termed molecular misreading by which DNA replication/RNA transcription and/or translation result in erroneous products with the number of as different from that derived from the original DNA sequence. In a family with hypobetalipoproteinemia, a deletion of one C in the A_5CA_3 coding sequence of the apolipoprotein B gene (ApoB) results in a run of A_8. The patient, however, did not have the disease, because some apoB protein was made. This was the result of molecular misreading in which approximately 10% of the resulting mRNAs contained A_9 instead of the expected A_8. This partially restored the reading frame, thereby templating the synthesis of low amounts of normal apoB.

19.7 INDUCED MUTATIONS

Any mutation that occurs in response to the externally applied agent is an induced mutation. The agents capable of inducing mutations are known as mutagenic agents (mutagens) and their capacity for inducing mutations is termed as mutagenic property. The process of inducing mutations through treatment with a mutagen is known as **mutagenesis**, while the exploitation of induced mutations for improvement of organisms is called **mutation breeding**.

19.7.1 Mutagenic agents

A mutagenic agent can be simply defined as a natural or human-made agent which can alter the structure or sequence of DNA. It is now well established that a wide variety of chemical, physical and biological agents can induce mutations (**Table 19.2**). Many mutagens induce more than one type of potentially mutagenic lesions. Thus, they frequently cause mutation directly as a result of pairing errors and indirectly as a result of errors during the repair process. The alterations induced by mutagenic agents are listed below:

- Spontaneous deamination of 5-methyl cytosine to thymidine
- Chemical alterations of bases by deaminating or monofunctional alkylating agents
- Dimerisation of two adjacent pyrimidines on the same strand under the action of UV light
- Cross-links between the two strands, caused mainly by UV and by bifunctional alkylating agents such as di-(2-chloroethyl) methylamine (nitrogen mustard), psoralen in the presence of light and mitomycin
- Breaks of one or both strands mostly by ionising radiation through the formation of highly reactive radicals in the water surrounding the DNA by the decay of radioactive atoms (e.g., ^{32}P, ^{3}H) incorporated in the DNA
- Gaps in the growing strand caused by inhibition of DNA synthesis, which are then widened by exonuclease V, the recBCD product.

19.7.2 Physical mutagens

Exposure to low pH

At low pH (below 4), some bases become positively charged and strand separation occurs, which may be lethal. At pH 4 or above, this effect is negligible. But low pH can also affect depurination by removing adenine or guanine and cause disintegration of the backbone.

Table 19.2 Major types of mutagens and their mode of action

Mutagens	*Mode of action*
Physical agents	
Ionising radiation	
Particulate radiation:	
α-rays, ß-rays, fast neutrons, thermal neutrons	Ionise and produce free radicals
Non-particulate radiation:	
X-rays, γ-rays	Free-radical attack on DNA, breaking the chain
Non-ionising radiation	
Ultraviolet radiation:	
Short-wavelength UV (200–300 nm)	Pyrimidine dimer formation
Long-wavelength UV (300–400 nm)	Mutagenic effect in the presence of various dyes
Heat	Strand separation, deamination and depurination of DNA
Chemical agents	
Base analogues	
5-Bromouracil (5-BU)	Incorporated like T; occasional faulty pairing with G, has higher likelihood for tautomerisation to the enol form
2-Aminopurine (2-AP)	Incorporated like A; faulty pairing with C
Alkylating agents:	
Monofunctional (e.g., Ethyl methane sulphonate, EMS and Methylmethane sulphonate, MMS)	Put methyl on G; faulty pairing with T
Bifunctional (e.g., *cis*-platinum, benzopyrene metabolites, nitrogen mustard, mitomycin, nitrosoguanidine)	Cross-link DNA strands; faulty region excised by DNase
Anticancer drugs (cyclophosphamide, chlorambucil, nitrosourea) 1-acetyl-imidazole and dimethylcarbamyl chloride	DNA strand breaks, sister-chromatid exchanges (SCEs), chromosomal aberrations
Procarcinogens that require metabolic activation Polycyclic and heterocyclic aromatic hydrocarbons Benz(a)anthracene Benzo(a)pyrene Dibenz(a,h)anthracene 7,12-Dinethylbenz(a)anthracene	Multiple DNA damage
Aromatic amines, amides, azo dyes 2-Naphthylamine (B-naphthylamine), 2-Acetylaminofluorene, Dimethylaminoazobenzene (butter yellow)	
Chemicals reacting with DNA	

Nitrous acid (HNO_3)	Deaminates A and C
Hydroxylamine (NH_2OH), HA	Reacts with C
Intercalating agents	
Acriflavin, proflavin, acridine orange, ethidium bromide	Insert between two base pairs and cause frameshift mutations; DNA intercalators induce DNA breaks.
Natural plant and microbial products	
Aflatoxin B1, griseofulvin, betel nuts	
Other chemical mutagens	
Sodium azide, nitrosamine and amides, vinyl chloride, nickel, chromium, insecticides, fungicides, polychlorinated biphenyls (PCBs), arsenic, asbestos.	Causes various types of DNA mutations
Caffeine	Causes mutations by inhibiting DNA repair enzymes
Biological agents	
Mutator genes	Increase of error-prone DNA polymerase
IS-elements, transposons and bacteriophages	Insertion of DNA sequences
Replication and repair enzymes	Error-prone repair, mis-incorporation of bases
Hormones	
Viruses	Can cause cancer, either by inactivating the gene by viral insertions or by activating the oncogenes
None	Spontaneous deamination of 5-methylcytosine (MeC). G:MeC→A:T

Exposure to high temperatures

High temperature causes strand separation and depurination of DNA. Heat also produces GC–AT transitions by deamination of C→U.

Radiation

Based on the effect they cause on the atom, radiations are grouped into two classes: **Ionising** and **non-ionising**.

Non-ionising radiation: They cause molecular vibrations or promotion of electrons to higher energy levels within the target molecules. This can lead to the formation of new chemical bonds. The most important form causing DNA damage is ultraviolet (UV) light (present in solar radiation or produced by mercury vapour lamps). Their wavelength ranges 100–3900 Å (10–390 nm) and specifically absorbed by purines and pyrimidines present in DNA. UV is normally classified in terms of its wavelength: UV-C (180–290 nm) is the most energetic and lethal. It is absorbed by the ozone layer. UV-B (290–320 nm) is a major lethal/mutagenic fraction of sunlight. UV-A (320 nm to visible), also called 'near UV' has deleterious effects (primarily because it creates oxygen radicals) but it produces very few pyrimidine dimers. Thymidine dimers are produced maximally at 254 nm, which is also the most mutagenic wavelength of UV.

The mutagenic action of UV is the consequence of both its direct and indirect effects on DNA. The direct effect of UV on DNA is of two types: formation of **pyrimidine dimers** and **pyrimidine hydrates**. Absorption of UV light by pyrimidines produces pyrimidine dimers from adjacent pyrimidine bases, primarily thymine. Thymine dimers are the most commonly formed, but cytosine dimers and thymine-cytosine dimers are also produced. This chemical linkage brings the bases closer, causing a distortion of the helix, which blocks transcription and transiently, DNA replication. The major product of UV irradiation is the cyclobutane thymine dimer shown in **Fig. 19.7** in which adjacent thymine bases are linked at the fifth and sixth atoms.

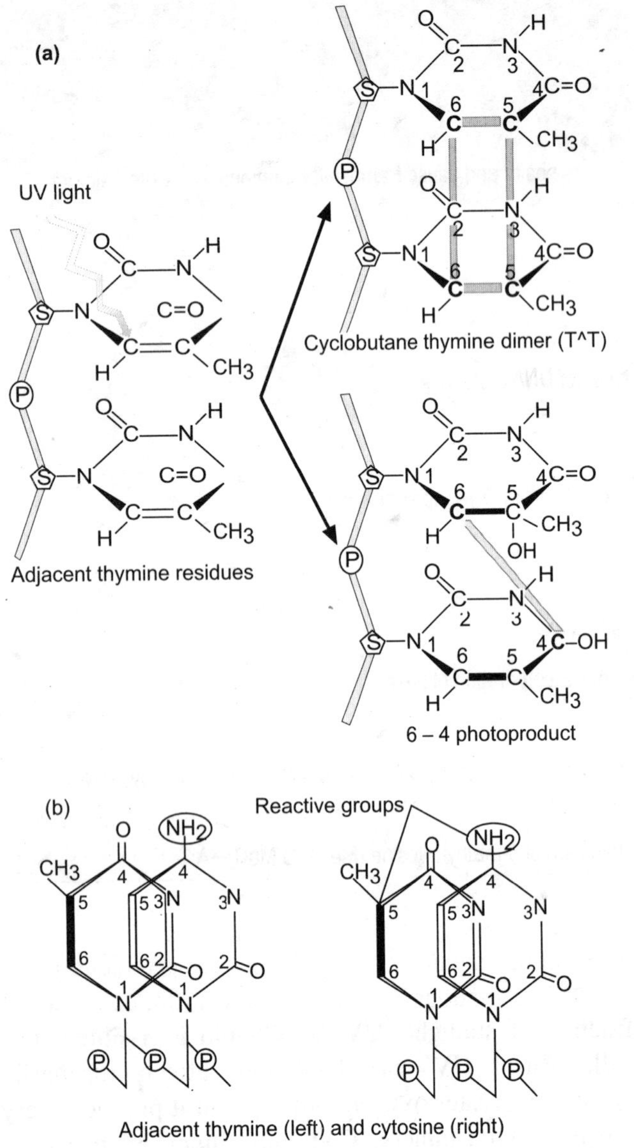

Fig. 19.7 Pyrimidine dimers (a) Two important pyrimidine dimer photoproducts produced by light (b) Dimer formation between thymidine and cytosine

The indirect effect of UV is due to the organic peroxides and free radicals produced in the medium, but its effect in the medium is shortlived.

Ionising radiation: Ionising radiation is a more powerful form of radiation. There are two types of ionising radiation: particulate and non-particulate. Particulate radiation consists of high-energy atomic particles generated due to radioactive decay like β-rays, α-rays and fast neutrons. Non-particulate ionising radiation result from loss of electrons by the target molecules on absorption of high-energy ionising radiation (e.g., X-rays and γ-rays). These electrons can cause extensive chemical alterations to DNA, including breaks in the strand and destruction of base sugar. These radiation cause water and other substances to ionise (forming highly reactive free radicals). These free radicals react with macromolecules (e.g., DNA) in the cell and alter its functions (result in carcinogenic and mutagenic effects). In contrast to UV radiation, ionising radiation penetrates readily through glass and other materials. Because of this, ionising radiation is used frequently to induce mutations in animals and plants.

19.7.3 Chemical and biological mutagens

A large number of chemicals have been found to be mutagenic, carcinogenic or teratogenic. Since mutagenic chemicals (**Table 19.2**) greatly increase the probability that a mutation will occur, it is clear they interfere in maintaining fidelity of DNA replication. Mutagens are studied not only because they are useful for modifying genes, but also for their adverse effect on the genetic material of organisms.

Base and nucleotide analogues

Base analogues are derivatives of the normal bases (artificially produced) with altered base pairing properties and can cause direct mutagenesis when incorporated into a DNA duplex in the course of normal replication. Among the most important analogues are the halogenated pyrimidines and those bases where nitrogen has been substituted for a $-CH_3$ group. The action of these unnatural bases seems, at least in some cases, to be two-fold: (1) They may generally block some stage in the biosynthesis of the normal purine and pyrimidine nucleotides. For example, 5-azacytidine is incorporated largely into RNA and this rapidly interferes with protein synthesis. 2-aminopurine is incorporated into DNA where it normally pairs with thymine, although it can also pair with cytosine. This capacity leads to frequent transitions both of A:T to G:C and G:C to A:T. The base 5-bromouracil is an analogue of thymine and the bromine atom is about the same size as the methyl group of thymine. The rare shift in the configuration of the 5-bromouracil base is due to the influence of bromine group and it pairs preferentially with guanine. Consequently, it induces AT→GC transition. DNA containing 5-bromouracil instead of thymine is very susceptible to breakage at light-induced bromouracil dimers. 2-aminopurine (2-AP) can be incorporated similar to adenine yet it also occasionally pairs with cytosine, resulting in

A:T to G:C transitions and occasionally G:C → T:A transitions.

Agents that modify DNA

A wide range of natural and synthetic organic and inorganic chemicals can react directly with DNA and alter its properties. These mutagens are active on both replicating and non-replicating DNA, in contrast to the base analogues, which are mutagenic only when DNA replicates.

Alkylating agents: Ethyl methane sulphonate (**EMS**) and ethylnitrosourea (**ENU**) are potent mutagens, which produce lesions that usually have to be repaired to prevent serious disruption to the processes of transcription and replication (**Fig. 19.8**). Alkylating agents (e.g., $-CH_3$ and $-CH_2-CH_3$) carry one, two, or more alkyl groups in reactive form and include the well-known compounds sulphur mustard or di-(2-chloroethyl) sulphide and nitrogen mustard or methyl-di-(2-chlroethyl)amine. They are electrophilic reagents and react with nucleophilic centres, particularly with guanine at the N7 atom.

Deamination agents: Deamination of nucleotide bases can happen by enzymatic or by chemical modifications. For example, nitrous acid (HNO_2) is a potent mutagen, which acts both on replicating and non-replicating DNA. It leads to oxidative deamination of cytosine (removal of 6 NH_2), adenine (removal of 6 NH_2) and guanine (removal of 2 NH_2) in which the amino groups ($-NH_2$) are replaced by keto (=O) groups yielding uracil, hypoxanthine and xanthine, respectively. When nitrous acid deaminates cytosine to produce uracil in a DNA, it base-pairs with adenine and causes G:C→A:T transitions upon subsequent replication. Similarly, deamination of adenine to the guanine analogue hypoxanthine results in A:T→G:C transitions. However, xanthine (similar to guanine) pairs with cytosine; hence, deamination of guanine does not yield mutations.

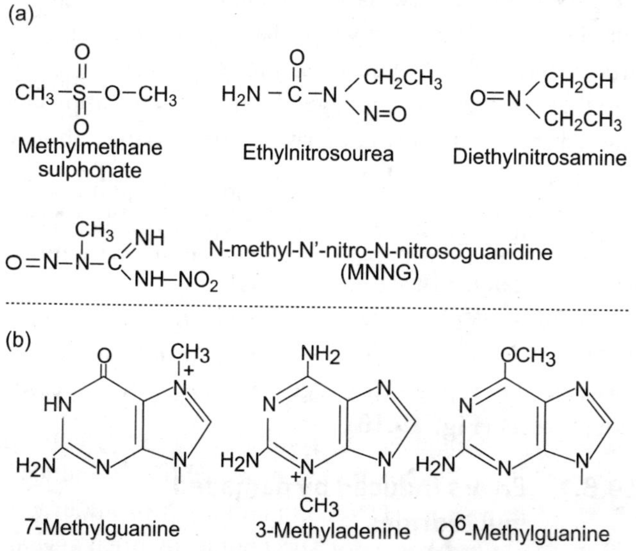

(a)

Methylmethane sulphonate | Ethylnitrosourea | Diethylnitrosamine

N-methyl-N'-nitro-N-nitrosoguanidine (MNNG)

(b)

7-Methylguanine | 3-Methyladenine | O^6-Methylguanine

Fig. 19.8 (a) Alkylating agents (b) Alkylated bases

Intercalators: These insert themselves between and parallel to the plane of adjacent base pairs, resulting in structural changes in DNA. Intercalating agents have been widely used to probe the DNA structure inside chromatin. Ethidium bromide (EB) has been reported to bind preferentially to such distinctive structures as bulges and guanine tetramers are found in telomeres.

Intercalators generate insertion and deletion mutations (**Fig. 19.9**). The acridine dyes (e.g., proflavine, ethidium bromide and propidium di-iodide) molecules, whose dimensions are roughly the same as those of a purine–pyrimidine pair can cause mutations. These substances get inserted between adjacent base pairs of DNA. This process is called intercalation. This brings about some unwinding of the DNA duplex which interferes with transcription.

By partially unwinding the DNA double helix, intercalating agents can lead to the production of frameshift mutations in which at replication, extra bases are inserted or deleted opposite to the intercalating molecule.

Is caffeine a mutagen?

Caffeine is a purine base, very similar in structure to adenine and guanine, but different enough not to be substituted for the normal purines in DNA. It produces mutations in bacteria and fungi by interfering with the ability of cells to repair damage to DNA from other causes. In plants, hamster and human cell cultures, high concentrations of caffeine have been shown to cause chromosome breaks. In humans, ingested caffeine is rapidly excreted, although it does reach all body fluids and tissues.

Mutator genes

Other than chemical and physical mutagens, mutations can also be introduced into genetic material through other processes, such as mutator genes, transposon mutagenesis and bacteriophages.

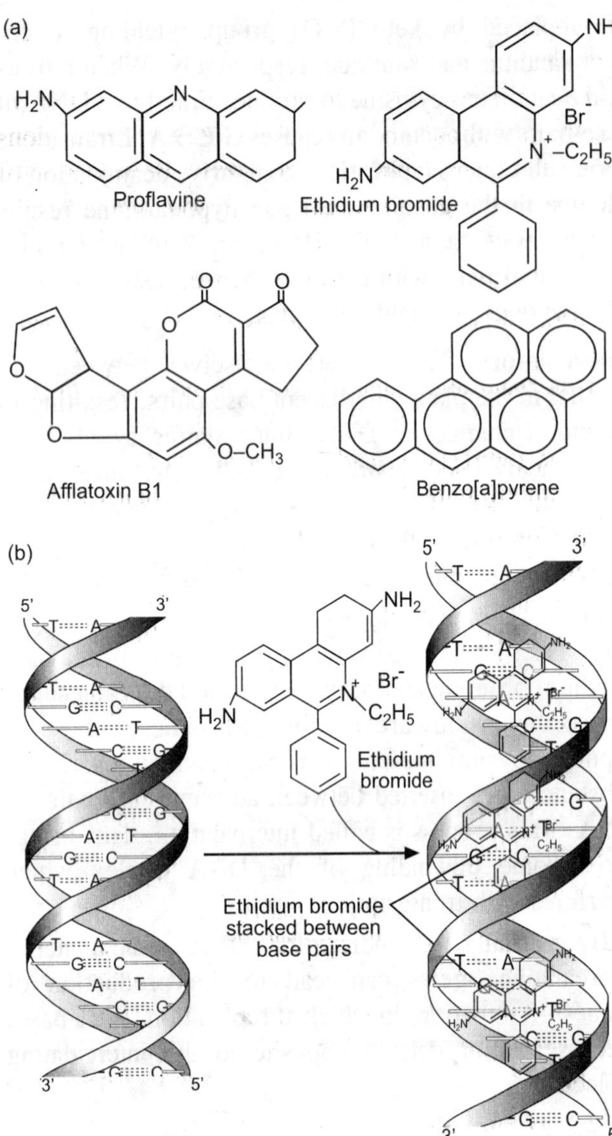

Fig. 19.9 (a) Chemical structures of intercalating agents (b) The process of intercalating by ethidium bromide results in lengthening and untwisting of the DNA double helix

Mutations in certain genes will raise the general mutation rate of the cell and can also cause other mutations to appear frequently in other genes throughout the chromosome. This is called mutator phenotype, and the corresponding genes are called *mut*. The normal function of *mut* genes is to keep the mutation frequency low; that is, it is only when the product of a mutator gene is itself defective that the frequency of spontaneous mutations increases. For example, *mut*D is a mutation in the proofreading function of DNA polymerase III (dnaQ).

19.8 MUTATIONS FROM DNA REPLICATION AND REPAIR

19.8.1 Fidelity of replication

The error rate of DNA replication (misincorporation of nucleotides during DNA synthesis) is much lower than that of transcription because of the need to preserve the meaning of the genetic message from one generation to the next. Of the various types of replication errors caused by erroneous action of the replicative apparatus, three types have been well characterised: a **single-base mispair** leading to a base substitution, a **single-base bulge** leading to a single-base frameshift, and a **multiple-base mismatch** leading to a sequence substitution. Since the replication errors consist only of undamaged nucleotides, premutagenic damage is referred to as **error of native replication**. At the next round of DNA replication, the native replication error would readily be fixed into a mutation. Replication errors by DNA polymerase are not limited to single base pair changes. Deletions or additions of one or a few base pairs are thought to result from displacement (looping out) of bases from either the template strand (giving a deletion) or the growing strand (giving an addition). These mistakes occur particularly at runs of identical bases in the DNA, where the looped-out structure can be stabilised by normal base pairing beyond the unpaired bases. Spontaneous breathing of the terminus can facilitate simple slippage events. Since duplex DNA undergoes such breathing only when the DNA polymerase is dissociated from the primer-template, misalignment would hardly occur during the processive DNA synthesis of chromosomes. Another possible source of slippage events is the improper reattachment of the primer terminus back to the template during the course of the proofreading process. Template switching, another type of erroneous action by the replicative apparatus, potentially generates multiple-base mismatch. Based on the definitive requirement of pseudo-inverted repeats for sequence substitution mutagenesis, **intra-** or **inter-strand template switching** between the repeats has been proposed for specific mutagenesis (**Fig. 19.10**).

19.8.2 Errors induced by damaged nucleotides

There are two additional pathways in which replication errors are induced during DNA synthesis by the replicative apparatus. One is the misinsertion of a

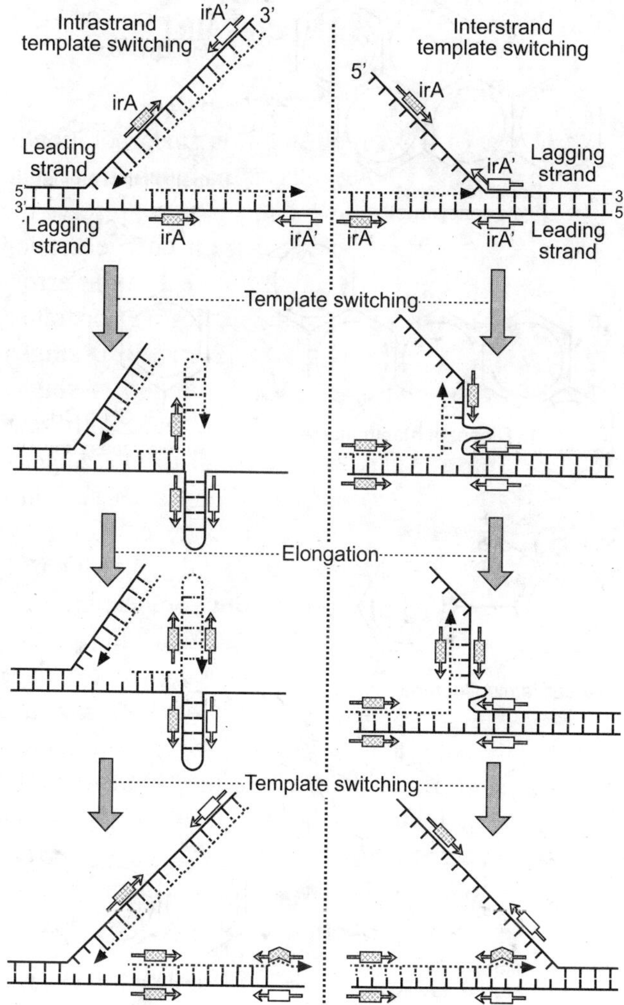

Fig. 19.10 Two models for sequence substitution mutagenesis. Arrows with boxes indicate pseudo-inverted repeats, irA and irA'. Intrastrand template switching occurs when the lagging strand synthesis pauses within the irA', while interstrand template switching occurs when the leading strand synthesis pauses within the irA

mutagenic nucleotide. For example, 8-oxodGTP is incorporated opposite either C or A residues of template DNA with almost equal efficiency by DNA polymerase III holoenzyme of *E. coli*. In this case, the induced replication error is a mispair containing the damaged nucleotide and would be converted to a pre-mutation at the next round of replication. Fixation of the pre-mutation into a base substitution requires a further round of DNA replication. The other pathway is the misinsertion of a normal nucleotide opposite a miscoding type of spontaneous DNA lesion. Spontaneous DNA lesions caused by hydrolytic deamination of C or 5-MetC are classified into a special class of miscoding lesions.

19.8.3 Translesional DNA synthesis

Many spontaneous DNA lesions have inhibitory effects on DNA synthesis and generally block the chain elongation reaction by replicative DNA polymerases. In addition, miscoding-type base-damaged DNA lesions can be processed into replication-blocking type by the action of DNA glycosylases. Replication-blocking type DNA lesions are subjected to TLS and produce pre-mutations (**Fig. 19.11**). Various DNA polymerases participating in TLS have been identified in prokaryotes and eukaryotes. Genetic and biochemical analyses of these TLS polymerases suggest that they differ in lesion specificity and in the accuracy of TLS.

19.8.4 Direct and indirect mutagenesis

Direct mutagenesis results from the presence of a stable, unrepaired base with altered base pairing properties in the DNA. All that is required for this lesion (non-permanent) to be fixed as a mutation (permanent and

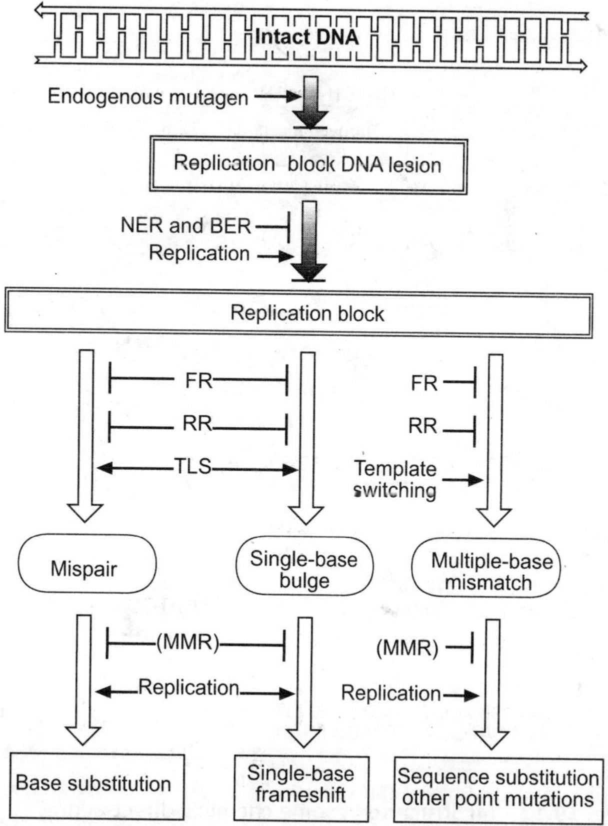

Fig.19.11 Translesional DNA synthesis leading to various types of spontaneous mutations. NER: nucleotide excision repair BER: base excision repair TLS: translesion DNA synthesis MMR: mismatch repair FR: fork regression RR: recombination repair (Redrawn from Maki H 2002)

heritable) is DNA replication. For example, the *keto* tautomer of the thymine analogue 5-bromouracil base-pairs with adenine as expected, but the frequently occurring *enol* form (due to electronegativity of Br) pairs with guanine (**Fig. 19.12**).

Indirect mutagenesis

Many chemical compounds such as progesterones, steroidal and hormonal compounds do not react with DNA. They are classified as non-genotoxic mutagens (**Fig. 19.13**). Many chemicals become activated to interact with DNA after enzymatic conversion in a cell or in tissues. The liver microsomal enzymes play a major role in conversion of non-mutagens into mutagens.

(a)

ß-Propiolactone

Ethyl methanesulphonate (EMS)

Dimethyl sulphate (DMS)

Methyl nitrosourea (MNU)

Nitrogen mustard

Dimethyl nitrosamine

(b)

5-Bromouracil → 5-BrU Guanine

(c)

AGTGTACGT / TCACATGCA —(i)→ AGTGBACGT / TCACATGCA

(ii) → AGTGBACGT / TCACGTGCA and AGTGTACGT / TCACATGCA

(iii) → AGTGBACGT / TCACATGCA and AGTGCACGT / TCACGTGCA

Fig. 19.12 (a) Structure of some chemical direct-acting carcinogens (b) Structure of common nucleotide base analogue, 5-bromouracil (c) (i) the 'T' analogue 'B' can be readily incorporated in place of 'T' (ii) After one round of replication, 'G' may be incorporated opposite the 'B' (iii) After a second round, a 'C' will be incorporated opposite the 'G'

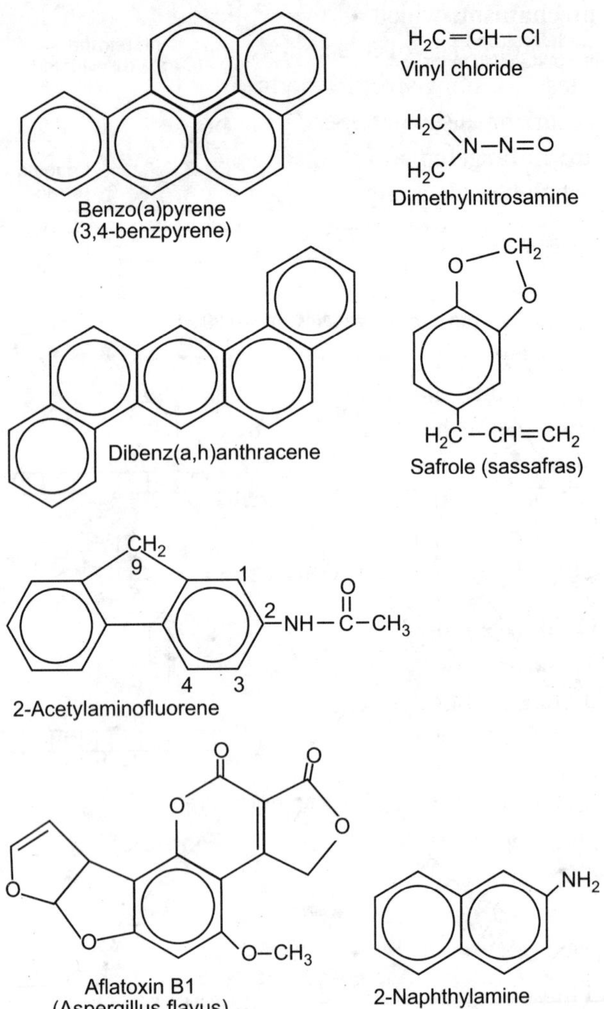

Fig. 19.13 Structures of some indirect-acting chemical carcinogenes; must be metabolised before they can react with DNA

19.9 TYPES OF DNA DAMAGE

Lesions

A lesion is an alteration of the normal chemical or physical structure of the DNA, i.e., a damaged region of DNA. The effects of DNA damage may be two-fold. If a lesion alters the specificity of base pairing, it is known as **mis-instructional lesion** and may generate a mutation during the next round of replication by tautomeric shift. If a lesion renders the DNA unable to specify a complementary base, it is known as **non-instructional lesion** and may cause a **replication block**, which is lethal. DNA damage may thus cause mutation or killing. To avoid the lethal effect of replication blocks, cells have evolved a **damage tolerance**

mechanism, which allows replication to continue even in the presence of a lesion. Mutations occurring at the sites of lesions are **targeted mutations**, whereas those occurring elsewhere, e.g., due to replication errors, are **untargeted mutations** (gene targeting). A lesion, which heralds a mutation, is a **pre-mutagenic lesion**.

Apurinic or apyrimidinic (AP) site

Depurination is another spontaneous hydrolytic reaction that involves cleavage of the N-glycosylic bond between N-9 of the purine bases A and G and C-1' of the deoxyribose sugar and hence loss of purine bases from the DNA. The sugar–phosphate backbone of the DNA remains intact. The resulting apurinic site is a non-coding lesion. Depurination occurs at the rate of 10,000 purines lost per human cell per hour at 37°C. Though less frequent, depyrimidination can also occur.

Oxidative damage

This occurs under normal conditions due to the presence of reactive oxygen species (ROS) in all aerobic cells. usually hydrogen peroxide and the hydroxyl radical (OH^-). This radical can attack DNA at a number of points, producing a range of oxidation products with altered properties, such as 8-oxoguanine, 2-oxoadenine and 5-hydroxymethyluracil. The levels of these can be increased by hydroxyl radicals from the radiolysis of water caused by ionising radiation.

Alkylation

Alkylating agents are electrophilic chemicals, which readily add alkyl (e.g., methyl) groups to various positions on nucleic acids distinct from those methylated by normal methylating enzymes. Common examples are methylmethane sulphonate (MMS) and ethylnitrosourea (ENU). Typical examples of methylated bases are 7 methylguanine, 3-methyladenine, 3-methylguanine and O^6-methylguanine. Some of these lesions are potentially lethal as they can interfere with the unwinding of DNA during replication and transcription. Most are also indirectly mutagenic.

Genotoxic effects

DNA damage consists of two broad categories: visible effects detectable through cytological analysis of chromosomes (macrolesions) and non-visible changes, which occur at the nucleotide level (microlesions). Chromosome aberrations (**Fig. 19.14**) from structural changes are classified according to their time of formation relative to the cell cycle. Chromatid breaks exhibit one normal and one broken chromatid of a pair, whereas in chromosome breaks, both chromatids are broken at identical locations. The production of either chromatid and/or chromosome aberrations by an agent depends on the nature of the clastogen (chromosome breaking agent) and the cell cycle stage the target cell was in at the time of exposure (**Fig. 19.15**). Cytogenetic evaluations alone cannot be used to define mutagens since not all mutagens produce DNA lesions that lead to the formation of chromosome aberrations. The cytogenetic technique that is currently in wide use is the analysis for SECs. This mechanism proposed to explain the staining technique and formation of SECs (**Fig. 19.16**). Another screening technique developed to assess the induction of chromosome damage is the test for production of micronuclei (**Fig. 19.17**).

19.10 DIRECTED MUTAGENESIS

The methods of mutagenesis which have been discussed up to this point, are completely undirected and the sites of these changes in the DNA are random. The development of gene technology has led to revolutionary new methods, which make it possible to isolate mutants of specific genes of interest. Deletions can be induced near the restriction site in circular chromosomes by removing a few nucleotides at one end after restriction digestion and re-ligation. Insertions may be produced using transposons; Tn5 and Tn7 are commonly used, which inactivate the genes upon insertion. Synthetic oligonucleotides of known base sequence, may be used as primers for replication of a single-stranded DNA molecule (template).

Deletion mutagenesis

In organisms with small genomes, and/or rapid generation times, it is possible to create and analyse mutants produced *in vivo*, but this is not easy or acceptable in higher organisms such as humans. However, if cloned genes have been isolated, it is quite feasible to mutate them *in vitro* and then assay for the effects by expressing the mutant gene *in vitro* or *in vivo*. For both cDNA clones, it is common to delete progressively from the ends of the coding region to produce either N-terminally or C-terminally truncated proteins to discover which parts of the whole protein have particular properties. **Figure 19.18** shows a cDNA cloned into a plasmid vector in the multiple cloning sites (MCS). There are several restriction sites in the MCS at each side of the insert, and by choosing two that are unique and that give a 5'- and a 3'-recessed end on the vector and insert respectively, unidirectional deletions can be created. The enzyme exonuclease III can remove one strand of nucleotides in a 3'→5' direction from a recessed 3'-end, but not from a 3'-protruding end. Therefore, it will remove the lower strand of the insert only progressively with time.

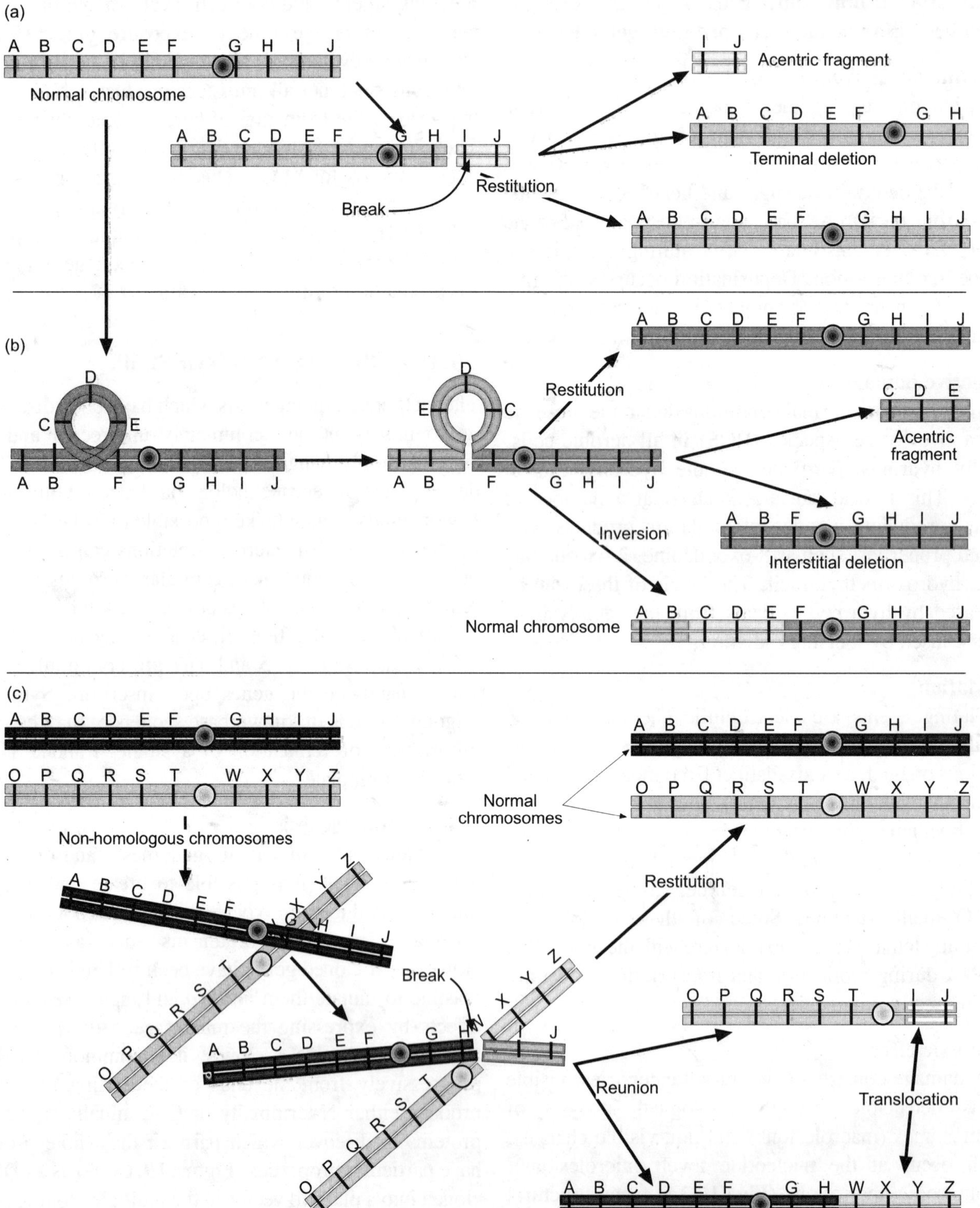

Fig. 19.14 A diagrammatic illustration of the possible modes of origin of structural aberrations in chromosomes (a) Single chromosome breaks (b) Two chromosome breaks (c) Breaks in non-homologous chromosomes

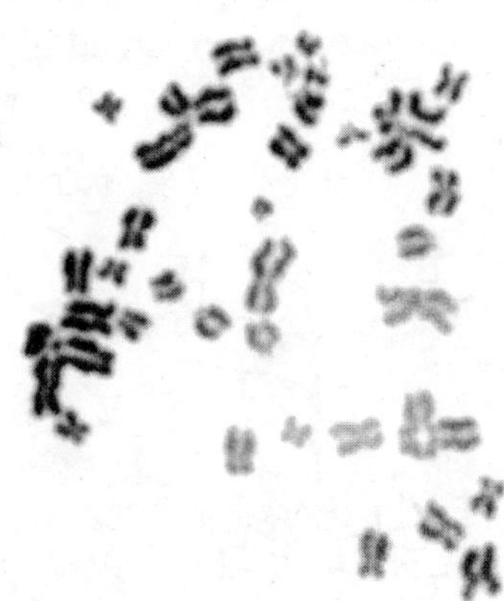

Fig. 19.15 Metaphase chromosomes showing different types of chromosomal aberrations (***Source:*** Channarayappa, 1990)

(a)

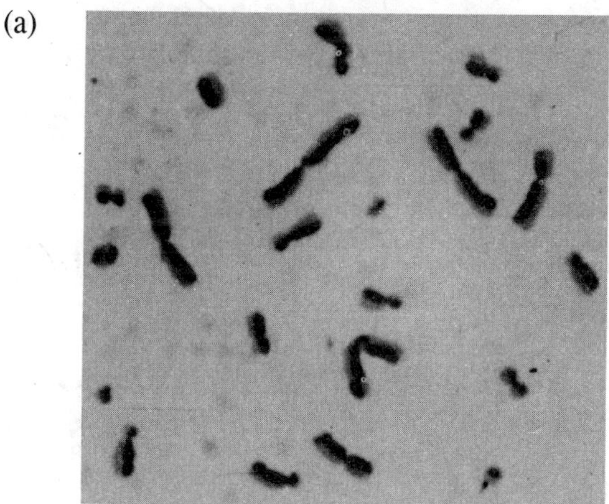

(b)

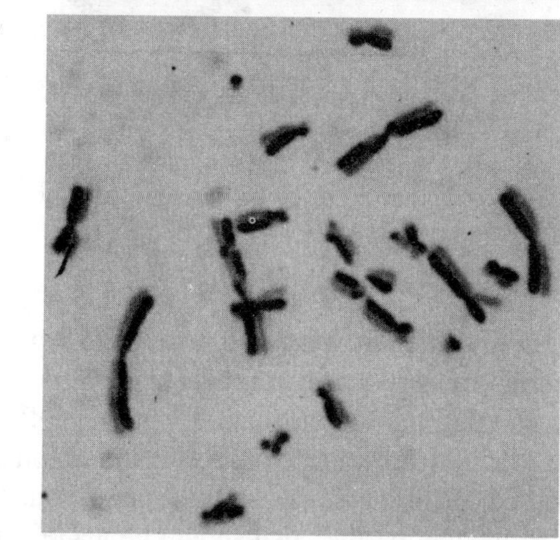

Fig. 19.16 Sister chromatid exchanges: mice bone marrow cells (a) Untreated (b) Treated with mitomycin a clastogen and harlequin-stained (***Source:*** Channarayappa 1990)

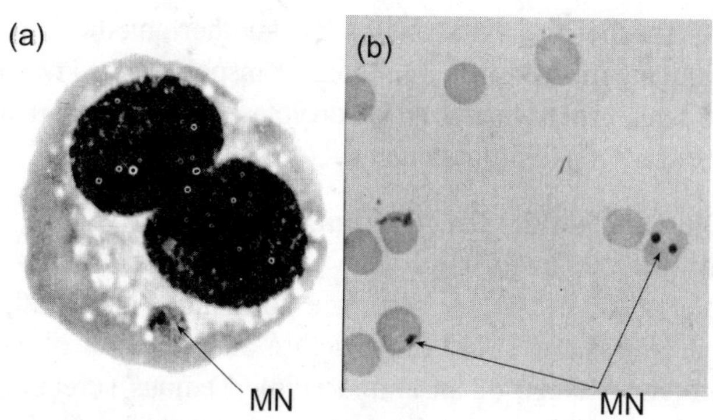

Fig. 19.17 Micronucleus; photomicrographs of micronucleus containing (a) Binucleated cell (b) Mice bone marrow cells (***Source:*** Channarayappa 1990)

Insertion mutagenesis

By using a linker–adaptor approach, an adaptor DNA or a linker DNA can be inserted at the recognition site where the restriction enzyme has acted. An adaptor molecule is a chemically synthesised dsDNA that can be used to connect the ends of two DNA molecules. A linker molecule is similar but possesses a recognition site for a restriction enzyme.

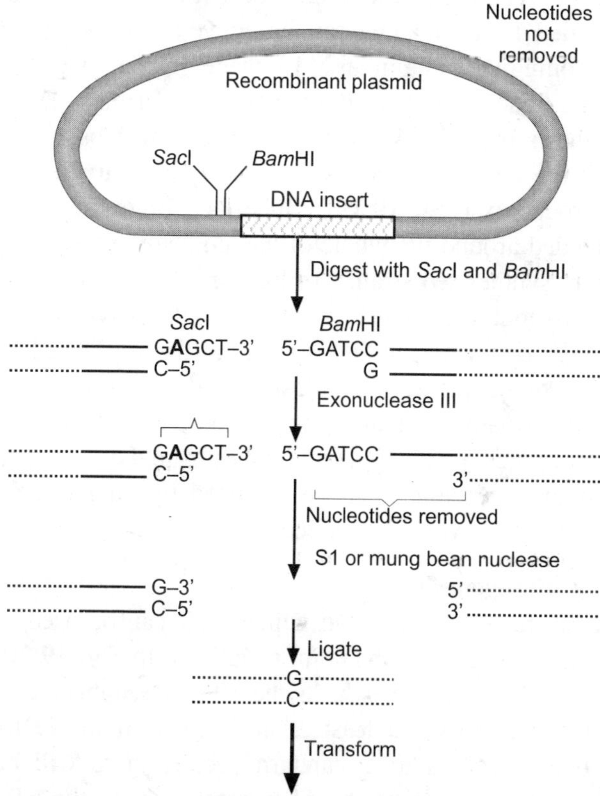

Fig. 19.18 Unidirectional deletion mutagenesis

Transposon mutagenesis is another method for inducing mutation via insertion. Transposons are known in both prokaryotes and eukaryotes and can insert at arbitrary sites in the genome.

Oligonucleotide-directed mutagenesis

In order to understand gene function, it is important to change the DNA sequence (or single nucleotide) and assess the effect on structure and function of the protein (change of an amino acid). Tedious screening of randomly generated mutations can only be applied effectively to relatively simple organisms such as fruit fly and bacteria. Only broad functional domains can be identified and the structure–function relationships at molecular level cannot be defined. Specific and site-directed mutagenesis allows the gene to be isolated, modified and then returned to the organism where the function can be examined. Site-directed mutation can be used to precisely insert a mutation into a specific region of DNA. Oligonucleotide-directed mutagenesis begins with the synthesis of an oligonucleotide of about 7–20 nucleotides, which contains a base mismatch with the complementary wild-type sequence. The procedure involves using a cloned gene whose base sequence is known. Then the DNA must be available in single-stranded form and this can be conveniently achieved by cloning the gene in an M13 phage or in a phagemid clone. The synthetic oligonucleotide primes DNA synthesis by a DNA polymerase and it will base pair with the complementary sequence template to form a heteroduplex (**Fig. 19.19**). Once the primer has been extended around the whole of the template, ends of the newly synthesised strand are ligated to form a circular dsDNA molecule. The heteroduplex is then transformed into *E. coli* where either strand can be replicated to form homoduplexes. The plasmid that grows up in the colony usually contains only one type of plasmid. In order to pick out the mutant colonies from the wild-type colonies, the clones can be screened by nucleic acid hybridisation with ^{32}P-labelled probe.

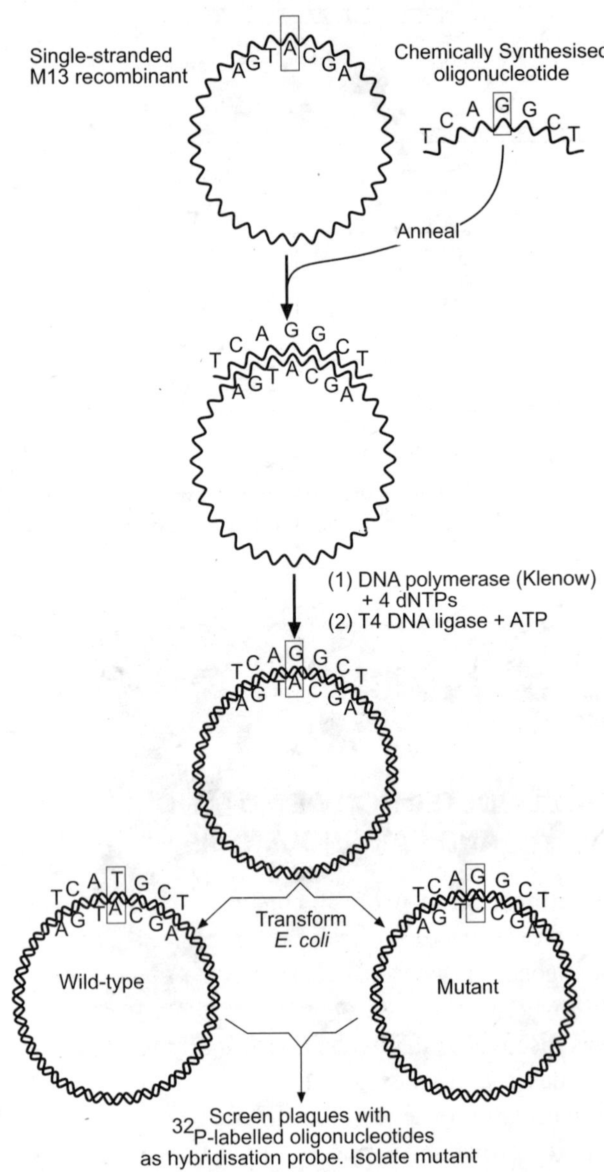

Fig. 19.19 Oligonucleotide-directed mutagenesis (the bracket indicates the mismatched bases)

PCR mutagenesis

There are several ways in which PCR can be used to create both deletion and point mutations. In **Fig. 19.20**, a pair of primers that have the altered sequence and which overlap by at least 20 nt is used. If the DNA to be mutated is in a standard vector, there will be standard primer sites in the vector such as the SP6 and T7 promoter sites of pGEM, which can be used in combination with the mutagenic primers. Two separate PCR reactions are performed, amplifying the 3'-portion of the insert using the forward and T7 primers. If the two PCR products are purified, mixed and amplified using SP6 and T7 primers, then a full length, mutated molecule is the only product that should be made. This will happen when a 5'-sense strand anneals via the 20 nt overlap to a 3'-antisense strand or vice versa, and is extended. This method is nearly 100% efficient and very quick.

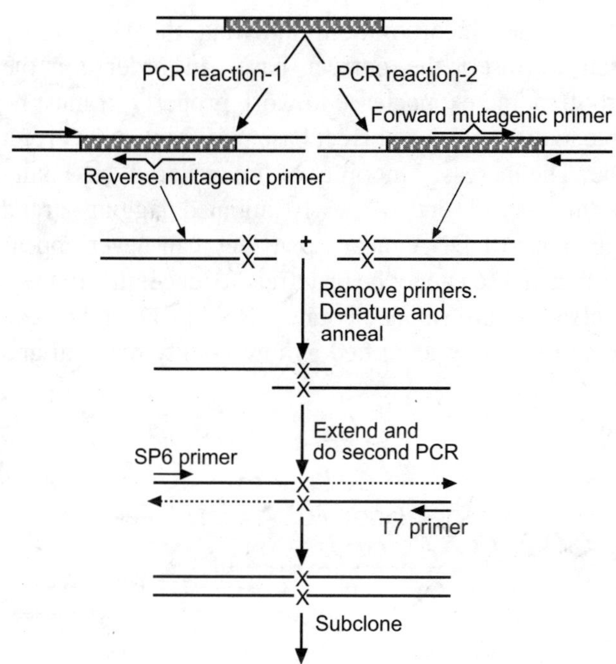

Fig. 19.20 PCR mutagenesis; X is the mutated site in the PCR product

19.11 DETECTION OF MUTATIONS AND CARCINOGENESIS

Identification of proto-oncogenes

Most exciting discoveries of recent years are that of oncogenes or genes that cause cancer. These genes were discovered in a class of viruses called retroviruses. It was discovered that when a virus carrying an oncogene invades a cell, the growth pattern of the cell can be converted into one identified as a cancerous (neoplastic) type of growth. As long as the virus remains, the cell shows tumour growth. Two different possibilities are now being explored to explain how these oncogenes of viruses are pathogenic. First, when transferred to the host cell, these genes are controlled by viral promoters and not cellular promoters. This may allow the genes to be expressed constantly and at high levels. Second, oncogenes acquire mutations in the process of becoming oncogenes from proto-oncogenes. This genetic damage prevents the gene from being controlled in the normal manner, and thus it becomes pathogenic. A common feature of most types of cancer is that of uncontrolled cellular proliferation.

Determination of genotoxic compounds

Much of the data derived from genotoxicity testing is used for development of compounds as well as for assessing environmental risk. The International Commission for Protection against Environmental Mutagens and Carcinogens (ICPEMC) has developed an operational definition for a genotoxic material (**Table 19.3**) as 'an agent that produces a positive response (e.g., mutation, UDS, chromosome breakage) in any bioassay measuring and genetic endpoint'. A non-genotoxin is also defined by the ICPEMC as a chemical that is inactive in a series of tests covering five types of endpoints considered critical to the production of all genetic damage. The five main types of tests are as follows:

- Test for alteration of DNA base sequence (nucleotide substitutions, additions, deletions)
- Test for alteration of DNA integrity (adducts/breaks/cross-links)
- Test for DNA rearrangement (recombinogenic effects including SCE)
- Test for altered chromosome segregation (numerical)
- Test for altered chromosome integrity (structural)

Table 19.3 Classification of genotoxic agents

Genotoxicant	*Definition*
Carcinogen	An agent promoting neoplastic transformation of eukaryotic cells; many carcinogens are also mutagens, but the converse is not necessarily true.
Clastogen	An agent inducing chromosome fragmentation, i.e., double-strand breaks
Mutagen	An agent promoting mutagenesis either by directly interacting with DNA or by generating metabolic products which do so
Oncogen	An agent inducing tumour formation
Supermutagen	An extremely potent mutagen, such as ethylmethane sulphonate (EMS)
Telogen	An agent which causes premature termination of replication, e.g., 2',3'-dideoxynucleotides
Teratogen	An agent inducing developmental abnormalities; not necessarily a genotoxic agent, i.e., includes molecules which mediate their effects without altering DNA structure
Spindle poison	An agent which damages spindle fibres; damage to the spindle fibres during cell division may cause aneuploidys.

19.12 DNA REPAIR

Repairing damaged DNA ensures an intact chromosome for DNA replication, mRNA transcription and protein translation. Thus, in the course of evolution, efficient systems for correcting the occasional mistakes that occur in the genetic material have developed. DNA repair refers to a collection of processes by which a cell identifies and corrects the damage to the DNA molecules that encode its genome. The number and types of DNA repair systems reflect both the importance of DNA repair to cell survival and the diverse sources of DNA damage. Some repair systems are extraordinarily accurate, whereas many are error-prone.

Repair systems are generally placed in five broad categories:

- Direct repair or damage reversal
- Indirect repair or excision repair
- Double-strand repair
- Post-replicative repair
- Reverse mutation and suppression mutation.

Different processes that are important in the repair of aberrant or damaged DNA are discussed below.

19.12.1 Direct repair mechanisms

Many classes of lesions are subject to direct reversal of DNA damage. Many types of DNA damage can be corrected without removing a base or nucleotide. Simple nicks can be directly re-ligated, certain radiation damaged products can be repaired by photoreactivation, and certain alkylated bases can be repaired by removal of adducts. In addition, in both eukaryotes and prokaryotes, there are enzyme systems, which repair damaged nucleotides before they are inserted into the DNA.

Multiple steps of mutation avoidance mechanisms

In *E. coli*, for example, the spontaneous mutation rate is about one error per 10^{10} bases incorporated during replication. This is primarily due to the presence of the minor tautomeric forms of the bases, which have altered base pairing properties. The error rate is minimised by a variety of mechanisms. DNA polymerases only incorporate an incoming nucleotide if it forms the correct Watson–Crick base pair with the template nucleotide in its active site. The occasional error (**Fig. 19.21**) is detected by the 3'→5' proofreading exonuclease associated with the polymerase. This removes the incorrect nucleotide from the 3'-end before any further incorporation, allowing the polymerase then to insert the correct base. In order for the proofreading exonuclease to work properly, it must be able to distinguish a correct base pair from an incorrect one. The increased mobility of 'unanchored' base pairs at the very 5'-end of newly initiated lagging strand fragments of DNA means that they can never appear correct and so cannot be proofread. Hence, the first few nucleotides are ribonucleotides (RNA) so that they can be subsequently identified as low fidelity material and

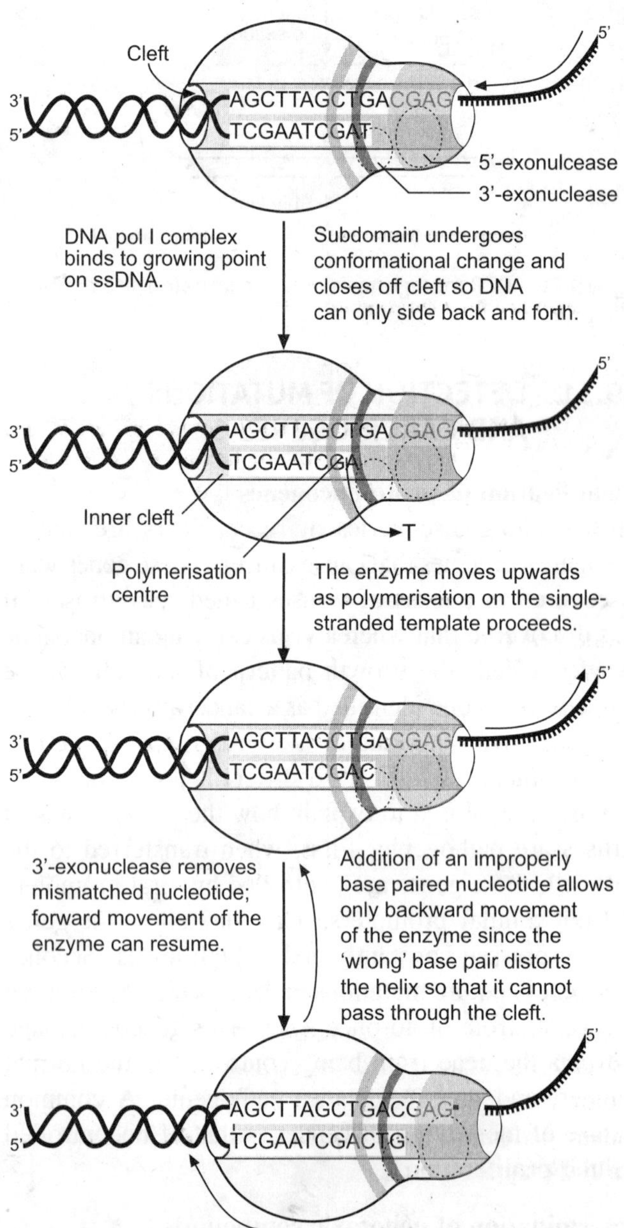

Fig. 19.21 Diagrammatic view of DNA proofreading after incorporation of mismatched base into DNA (***Source:*** Redrawn from Watson et al. 1987)

replaced with DNA elongated (and proofread) from the adjacent fragment. Errors which escape proofreading are corrected by a mismatch repair mechanism.

Photoreactivation

Exposure of organisms to visible light, particularly to blue light, after UV irradiation greatly reduces the yield of mutations. In 1949, Albert Kelner reported that when the bacterium, *Streptomyces griseus* was irradiated with a large dose of UV light, its chance for survival increased several hundred folds if it was also exposed to visible light. This phenomenon is called **photoreactivation** or **photorestoration** (**PR**). Photoreactivation is a light-dependent DNA repair mechanism in which certain types of pyrimidine dimers are cleaved (monomerised). Short-wavelength ultraviolet (UV) irradiation affects DNA in a number of ways, but a well-established action is the formation of thymine dimers, a state in which two adjacent thymine molecules are chemically joined, so that replication of the DNA cannot occur. When bacteria damaged by UV light are exposed to an intense source of visible light, a large proportion of the damaged cells recover. Various enzymes can recognise and catalyse the direct reversal of specific DNA damage (**Fig. 19.22**).

Repair of damaged single strand

In the DNA double helix, when only one of the two strands has a defect, the other strand can be used as a template to guide the correction of the damaged strand. In order to repair damage to one of the two paired molecules of DNA, there exist a number of excision repair mechanisms that remove the damaged nucleotide and replace it with an undamaged nucleotide complementary to that found in the undamaged DNA strand.

Purine insertion: Purine is inserted into apurinic sites. The enzyme does not work at random but is reported to re-form the correct base pairing.

Excision repair

It is a ubiquitous multistep enzymatic process by which a stretch of a damaged DNA strand is removed from a duplex molecule and replaced by resynthesis using the undamaged strand as a template. Various enzymes can sense damage or distortion in the DNA double helix. In contrast to photoreactivation, which is possible with ssDNA, the complementary strand is required for excision repair. It operates on a wide variety of lesions (UV-induced lesions, alkylation or deamination) and is essentially error-free. There are two forms: **base excision repair (BER)** and **nucleotide excision repair (NER)**.

Base excision repair: A base can be removed from a nucleotide within DNA in several ways: by direct action of an agent such as radiation, spontaneous hydrolysis, an attack of oxygen-free radical, or by DNA glycosylases. Modified bases are recognised by relatively specific DNA glycosylases (**Fig. 19.23**). The N-glycosylic bond is recognised and cleaved by a second class of enzymes termed AP endonucleases, which introduces a nick at the AP site.

Following the 5' excision to the AP site, the enzyme deoxyribophosphodiesterase (dRpase), is required to hydrolyse the 5' residue by base flipping between the altered base and the sugar, leaving an apurinic or apyrimidinic (AP) site (**Fig. 19.23**, step 1). AP sites are also produced by spontaneous loss of base. An AP endonuclease then cleaves the DNA at this site and a gap may be created by further exonuclease activity (**Fig. 19.23**, step 2). The gap is generally larger in NER and can be as small as one nucleotide in BER. From this point on, both forms of excision repair are essentially the same. In *E. coli*, DNA polymerase I (**Fig. 19.23**, step 3) fills the gap; DNA polymerase uses the intact complementary strand to fill the gap. The repair is completed by a phosphodiester bond made by DNA ligase (**Fig. 19.23**, step 4), much as in the final stages of processing of the lagging strand fragments during DNA replication. The replacement of just one base takes place 80%–90% of the time. In the remaining 10%–20% of cases, several nucleotides may be removed depending probably on, which DNA polymerase first repairs the site. In eukaryotes, excision repair is coupled to transcription so that the transcribed (genetically active) regions of the DNA are repaired more rapidly than non-transcribed ones. Thus, it helps to limit the production of defective gene products.

Nucleotide excision repair: It was first reported by Robert Setlow in 1964. NER was referred to as 'dark repair' because light was needed for irradiated cells to repair their DNA. This repair mechanism is the most intensively studied and most complicated, involving the products of over 30 genes. NER is efficient, ubiquitous to most organisms, and repair is non-specific in that it recognises several types of lesions. This system works by recognising major distortions to the shape of the double helix, such as those caused by a thymine dimer

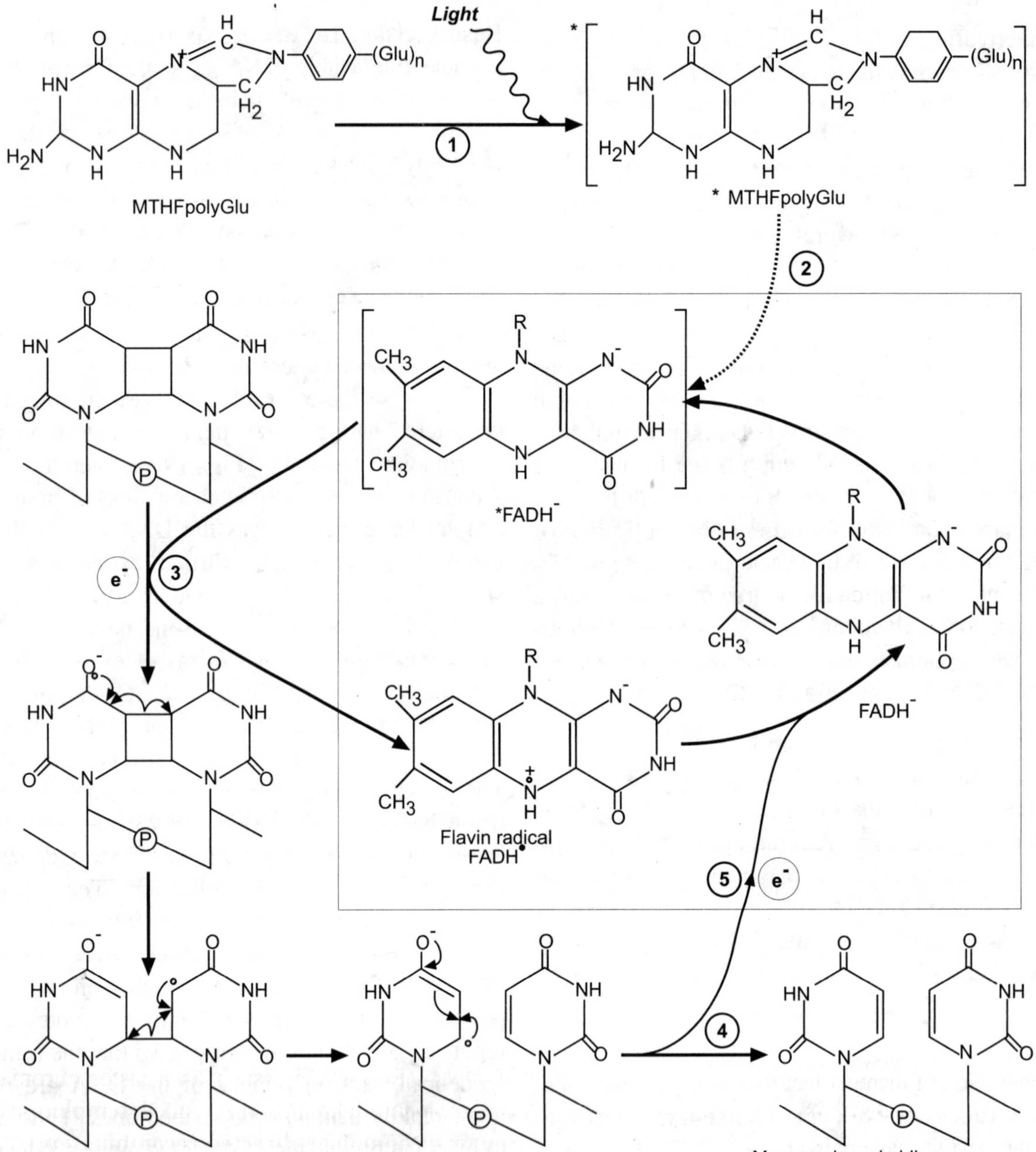

Fig. 19.22 Direct repair of damaged bases in DNA: Repair of a thymine dimer by photolyase; the two chromophores in *E. coli* photolyase, N5, N10-methenyltetrahydro-polyglutamate (MTHF polyGlu) and $FADH^-$, perform complementary functions. (1) A blue-light photon (300–500nm) is absorbed by MTHFpolyGlu (functions as photoantenna). (2) The excitation energy is transferred to $FADH^-$ in the active site of the enzyme. (3) The excited flavin (*FADH) donates an electron to the pyrimidine dimer. (4) Electronic rearrangement restores the monomeric pyrimidine. (5) The electron is transferred back to the flavin radical to regenerate $FADH^-$. (***Source:*** Nelson and Cox et al 2000)

or the presence of a bulky chemical adduct on a base. Such distortions trigger a chain of events that lead to the removal of a short, single segment (or patch) that includes the lesion.

Mismatch repair (MMR): It is an excision repair process, which corrects base pair mismatches in duplex DNA and covers about 99% of all DNA repairs. Base substitution mutations usually start as a single mismatched base in a dsDNA molecule. Mismatches can arise in a number of ways:

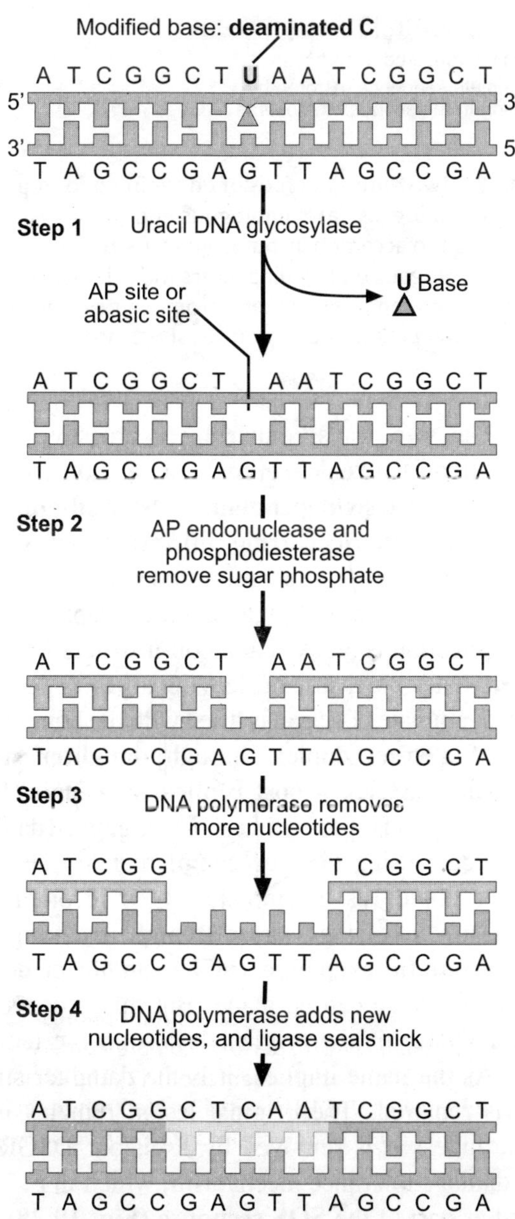

Fig. 19.23 Base excision repair of DNA

- By misincorporation during replication.
- By formation of heteroduplex DNA during recombination, for example, during meiotic recombination
- Spontaneous or induced base modifications for example, by the deamination of 5-methylcytosine, to generate thymine
- In addition, short insertions or deletions can arise by strand slipping during DNA replication and the formation of a heteroduplex with unequal numbers of residues.

Mismatch repair differs from other forms of DNA repair in that the excised material is the undamaged base normally found in DNA. Therefore, there must be a mechanism to discriminate between the correct and incorrect strands so that the mutagenic nucleotide and not the correct nucleotide on the template strand is removed. The main pathway for correction of mismatches arising during replication is the *MutHLS* pathway, described in *E. coli* (**Fig. 19.24**).

19.12.2 Repair of breaks in double strand

There are two mechanisms to repair double strand breaks (DSBs): non-homologous end joining (NHEJ) and recombinational repair. DSBs, in which both strands in the double helix are severed, are particularly hazardous to the cell because they can lead to genome rearrangements. Some damage to DNA, such as that caused by ionising radiation, is capable of breaking both strands of the double helix. When that happens, the cell uses one of the two mechanisms to repair the broken ends. It can simply bring the ends back together, by a process called **non-homologous end joining** (**NHEJ**) or it can use a mechanism that relies on the nucleotide sequences of a homologous region of DNA (**Fig. 19.25**), such as a sister chromatid or a homologous chromosome (called **template-assisted repair** or **homology-directed recombination**).

19.13 RECOMBINATIONAL REPAIR

Genetic recombination is a major part of molecular genetics and now it is clear that homologous genetic recombination systems did not evolve to generate genetic diversity. Rather, a primary role in the non-mutagenic recombinational DNA repair of stalled or collapsed replication forks has been the major alternative hypothesis, i.e., coping with DNA damage without repairing it. Recombination repair (also called

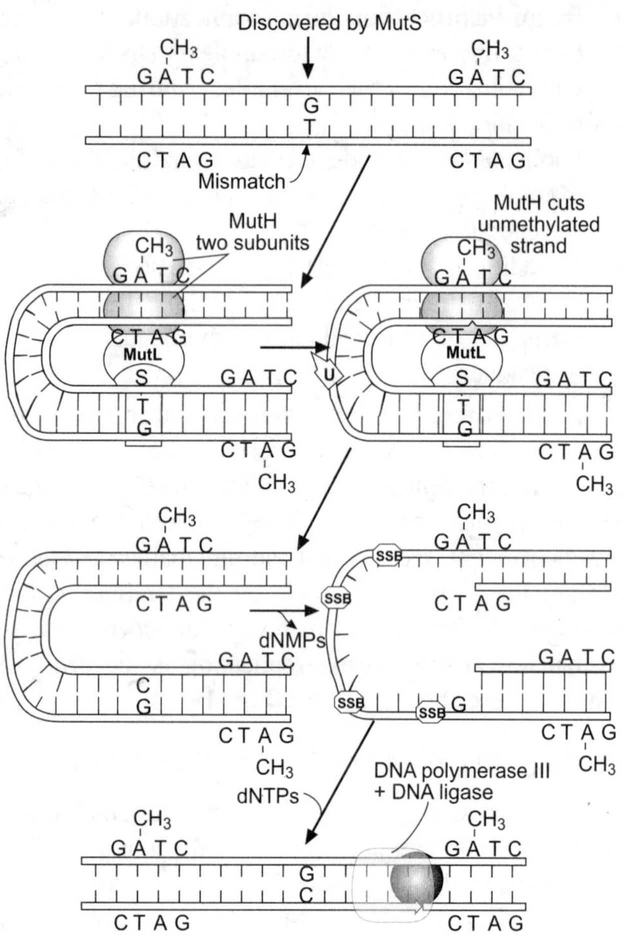

Fig. 19.24 DNA mismatch repair in *E. coli* is initiated by the interplay of the MutS, MutL, MutH and MutU proteins. G:T mismatch is recognised by MutS, MutH cleaves the backbone in the vicinity of the mismatch. MutU helps in unwinding the single strands stabilised by SSB. A segment of the DNA strand containing the erroneous T is removed by exonuclease I and synthesised anew by DNA polymerase III.

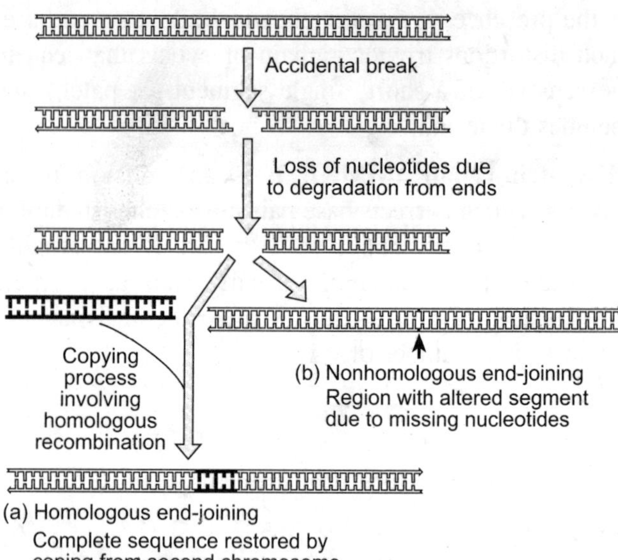

Fig. 19.25 Two different types of end-joining for repairing double-strand breaks. (a) Homologous end-joining is more difficult to accomplish, but is much more precise. (b) Nonhomologous end-joining alters the original DNA sequence when repairing broken chromosomes. These alterations can be either deletions or short insertions

post-replication repair is any DNA repair or damage tolerance strategy involving genetic recombination. It requires the presence of an identical or nearly identical sequence to be used as a template for repair of the break. The existence of such repair mechanisms is indicated by differences in sensitivity to ionising radiation demonstrated by haploid yeast cells arrested in G_1 and G_2. More direct evidence for recombination repair comes from studies, *E. coli* and *S. cerevisiae*, which are also radiation-sensitive.

19.13.1 Mechanism of recombination repair

There are two substrates for recombination repair: double-stranded breaks (**Fig. 19.26**) and single-stranded gaps (**Fig. 20.32**). Double-stranded breaks may be repaired in two ways, either by a crossover repair (**double-strand break repair**) or a non-crossover repair (**synthesis-dependent, strand-annealing repair**), the latter occurring in specialised systems such as P-element transposition and the switching of mating types in yeast. Single-stranded gaps can arise when replication is blocked at a non-instructional lesion and re-initiates downstream, leaving a gap opposite the damaged template. The gap is filled with the homologous region of a sister duplex through **daughter strand gap repair (DSGR)** or **post-replicative repair (PRR)** (**Fig. 19.27**). And repair synthesis of the gapped daughter strand occurs using the free complementary strand of the sister duplex as a template. By a recombination mechanism involving **sister-strand exchange** or **sister-chromatid-exchange (SCE)**, an intact double-stranded molecule can be made. But this process does not actually repair the original lesion on the template strand. As the name implies, it is the daughter strand, which is repaired. The original lesion remains in the genome unless it is removed by excision. DSGR is a major damage tolerance mechanism, which in *E. coli* is induced as part of the **SOS response (Fig. 19.28)**. The five key reactions in the recombinational DNA repair of replication forks are as follows.

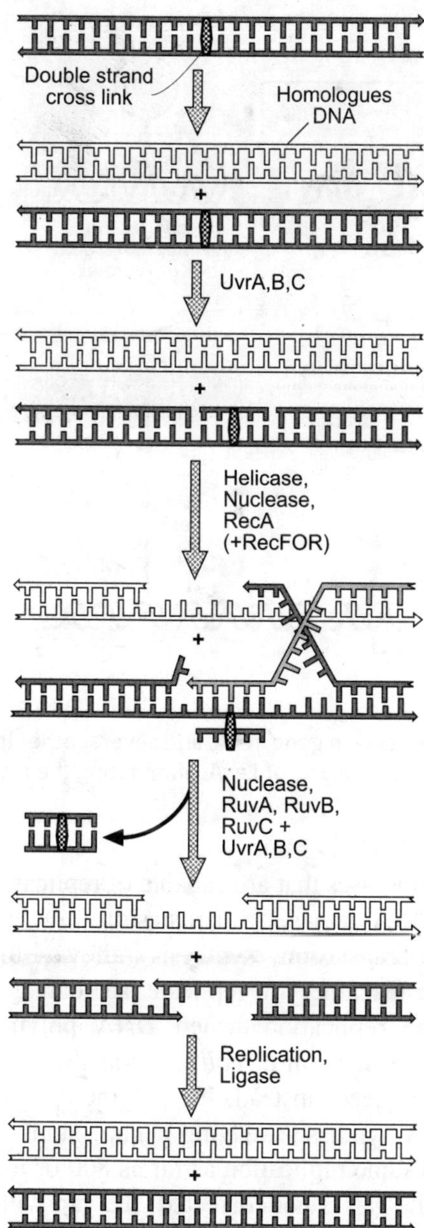

Fig. 19.26 Models for recombinational DNA repair: Double-strand cross-link repair

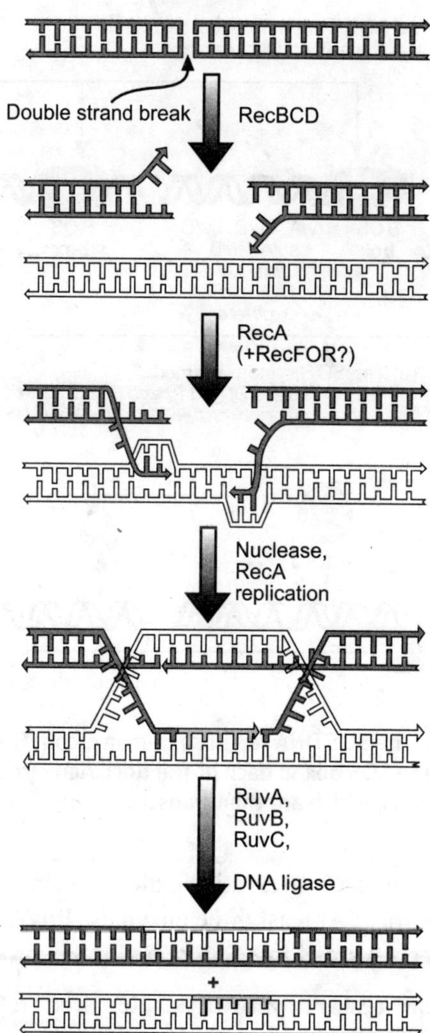

Fig. 19.27 Models for recombinational DNA repair: Postreplication repair

Replication fork regression: Fork regression (fork reversal) involves the re-pairing of the recently replicated template strands at a staled fork, in such a way that the fork is moved backward. The newly synthesised strands are displaced and then paired as the fork is regressed to create a four-branched structure, the **chicken foot**.

Holliday intermediate cleavage: The four-branched Holliday structure is a signature recombination intermediate, associated with a wide range of recombination processes. The enzymes involved in resolving the Holliday structure have been identified in many organisms. In *E. coli*, there are at least two such enzymes, RuvC and RusA. The reaction generally involves symmetric cleavage of two opposing strands of the intermediate so that two branches remain with each product (**Fig. 19.29**). The resolution of Holliday intermediate formed behind the replication fork during fork repair and cleavage of that intermediate can lead to the formation of chromosomal dimers (**Fig. 19.29**).

Branch migration: The migration of DNA branches is closely related to fork regression. In the fork regression process, branch migration proceeds in one direction creating a Holliday intermediate. Migration of the Holliday intermediate opposite to regression (or fork reversal) can reset the replication fork. If the chicken foot structure is not cleaved, then this branch

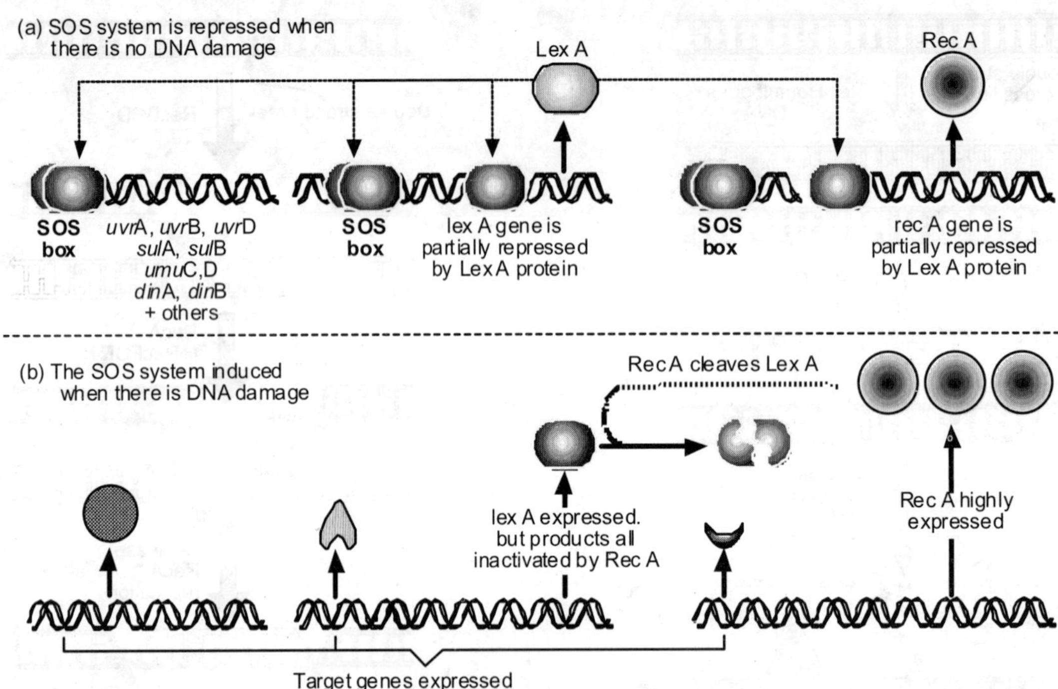

Fig. 19.28 The SOS DNA repair system in E.coli. The LexA protein represses its own gene, recA, and several other loci by binding at the SOS box in each of the tloci. Activated RecA protein causes auto catalysis of LexA, eliminating the repression of all these loci, which are then transcribed and translated

migration offers a conservative path to restoration of the replication fork. At least three enzymes (RuvAB, RecG and PriA) are known to be involved in this process.

Replication restart pathways: Repaired replication forks go through a restart process to resume replication (**Fig. 19.29**). This process is origin-independent and replication restart requires the activity of a seven-protein complex, which is known as **restart primosome**. The seven proteins of restart primosome are PriA (protein n, factor Y), PriB (protein n), PriC (protein n), DnaT (protein I), DnaC, the DnaB helicase, and the DnaG primase.

The 3'-end invasion: The invasion of 3'-ends into a homologous duplex DNA has been a key component and the process is central to the late stages of bacteriophage T4 replication. One or a few RecA monomers form a nucleus on a ssDNA molecule. The filaments are then extended 5' to 3' from that point.

19.13.2 Translesion synthesis

When pre-replicative repair systems (e.g., direct reversal and excision repair) fail, the cell has alternative mechanisms for dealing with damaged DNA at the replication fork. Cells have evolved several specialised DNA polymerases that are capable of replicating across a site of DNA damage (a **translesion**, e.g., thymidine dimers). Translesion synthesis allows the DNA replication machinery to replicate past damaged DNA. During the replication, when DNA polymerase III encounters damage in *E. coli*, such as thymine dimers, it cannot proceed; instead, the polymerase pauses and stops DNA synthesis, leaving a gap, skips down the DNA to resume replication as far as 800 or more bases away. If the gap is not corrected, the gap will result in deficient and broken DNA. Since a part of one strand is absent and the other has damage, there appears to be no viable template for replicating new DNA. However, the cell has two mechanisms to repair this gap: one uses polymerases that can replicate these lesions, and the other is a repair process that uses homologous DNA.

19.13.3 Post-replicative repair

In addition to using repair polymerases, the cell can use a second repair mechanism to replicate damaged DNA when the polymerase leaves a gap. A replication fork creates two DNA duplexes. Thus, an undamaged copy of the region with the lesion exists on the other daughter duplex. A group of enzymes, with one being

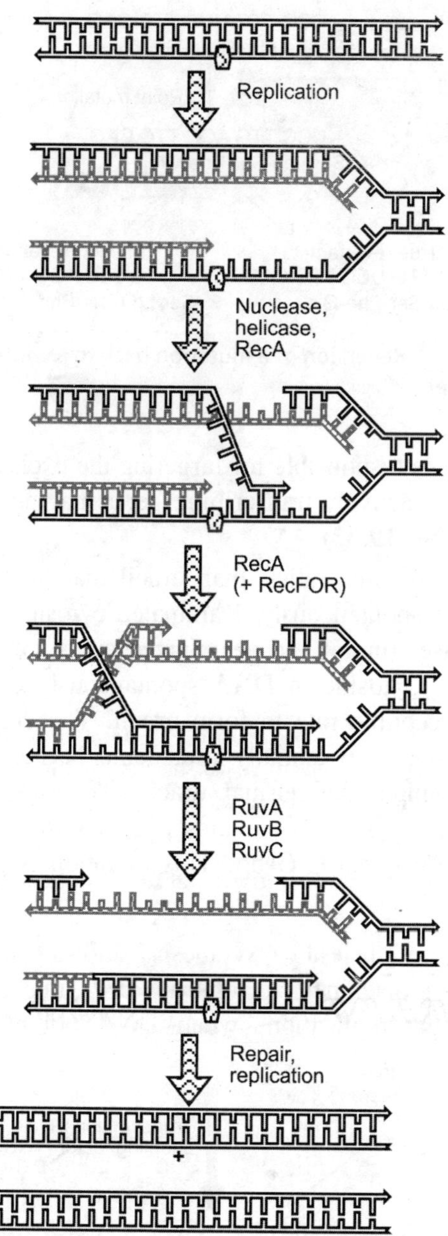

Fig. 19.29 Models for recombinational DNA repair for replicative type

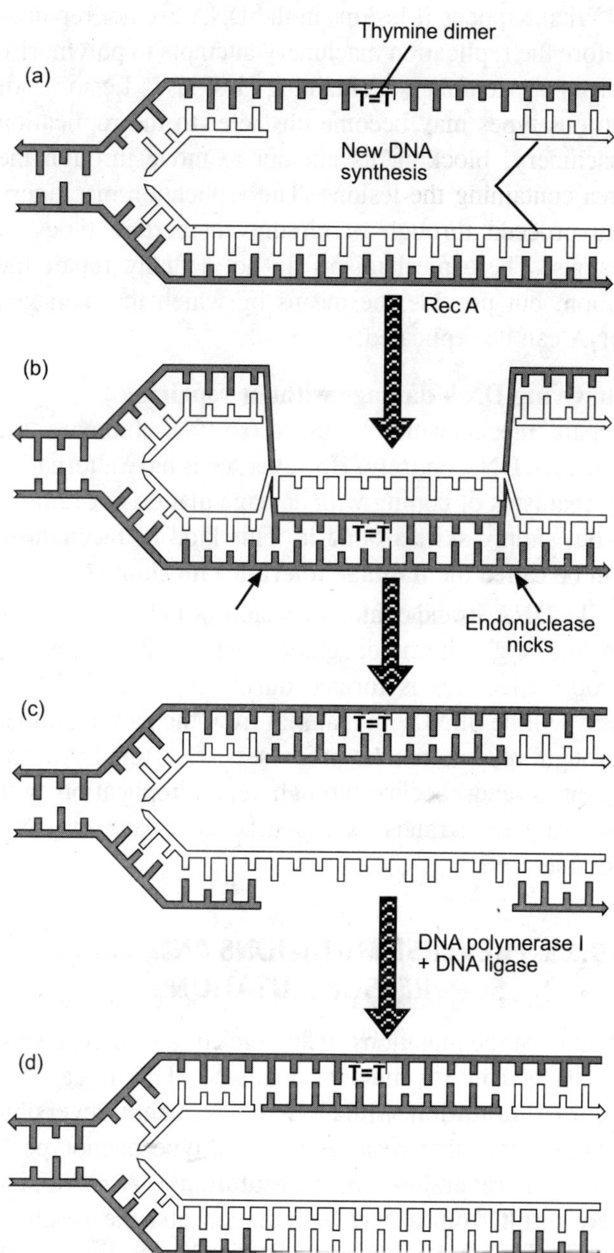

Fig. 19.30 Post-replicative repair: RecA-dependent post-replicative DNA repair

specified by the recA locus, repairs the gap. This process is called **post-replicative repair (Fig. 19.30)**.

19.14 GLOBAL RESPONSE TO DNA DAMAGE

Cells exposed to potent DNA damaging agents like ionising radiation, UV-light or chemicals are prone to acquire multiple sites of bulky DNA lesions and double-strand breaks. Moreover, these agents can damage other biomolecules such as proteins, carbohydrates, lipids and RNA. The accumulation of damage, specifically double strand breaks or adducts stalling the replication forks, is among the known stimulation signals for a global response to DNA damage. The global response to damage is an act directed towards the cell's own preservation and triggers multiple pathways of macromolecular repair, lesion, tolerance or apoptosis. The common feature of global response is induction of multiple genes, cell cycle arrest and inhibition of cell division.

What happens if lesions in the DNA are not repaired before the replication machinery attempts to polymerise across a template containing lesions? Lesions of various types may become obstacles to the replication machinery, blocking its attempt to move through the area containing the lesions. The replication machinery can proceed through or circumvent certain types of lesions. These mechanisms do not actually repair the lesion, but provide the means by which the damaged DNA can be replicated.

Surviving DNA damage without repairing

Repair mechanisms discussed so far eliminate the defective DNA entirely. However, cells have alternative mechanisms of coping with damage that do not remove it but simply skirt around it. This kind of mechanism can be called the **damage tolerance mechanism**.

If a DNA strand contains a lesion, which hinders base pairing, a gap in the daughter strand of approximately 1000 nucleotides is formed during replication. These gaps will be filled with material from the parent strands through recombination processes. The repair of the parent strands occurs through repair replication with the daughter strands as a matrix by means of DNA polymerase I.

19.15 REVERSE MUTATIONS AND SUPPRESSOR MUTATIONS

In most of the mutations so far studied, a wild-type gene is changed into a mutant phenotype. This is called a **forward mutation**. Mutations are frequently reversible and an event that restores the wild-type phenotype is called a **reversion** (**back mutation**). A reversion may result from a reverse mutation and the resultant mutant is called a revertant. Reversion may occur in two ways: a **true reversion** due to a back mutation that exactly restores the original DNA sequence and a **pseudoreversion** or **suppressor mutation** due to an additional mutation at a second site that restores the original phenotype (**Fig. 19.31**).

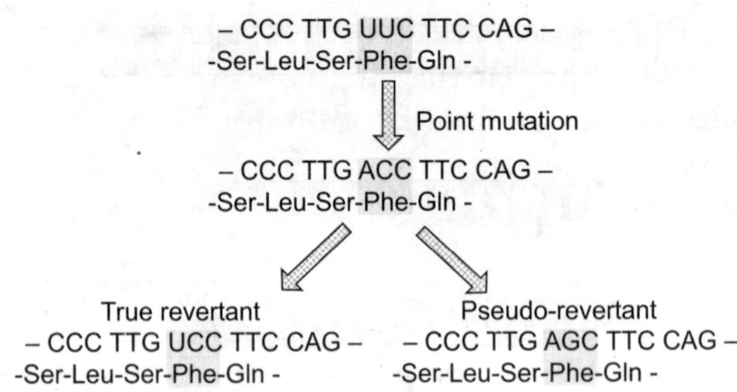

Fig. 19.31 Reversion of a mutation back to a wild-type phenotype

It is very interesting to mention that not all DNA is repaired at the same rate. Actively transcribing genes are more efficiently repaired than non-transcribed DNA. In eukaryotes, the mechanism that couples DNA repair and transcription is not completely understood. Certain general factors, such as TFII H, play a role in both transcription and excision repair. In *E. coli*, a protein known as **transcription repair coupling factor** (**TRCF**) is responsible for targeting the excision repair system to actively transcribing genes having damaged DNA (**Fig. 19.32**).

It is thymine rather than uracil that facilitates the repair of spontaneously deaminated cytosine in DNA. Both bases (uracil and deaminated cytosine) pair with adenine. Cytosine in DNA spontaneously deaminates at a perceptible rate to form uracil. The deamination of cytosine is potentially mutagenic because it pairs with adenine like normal uracil. Therefore, one of the daughter strands contains a U-A base pair rather than the original C-G bp. This mutation is prevented

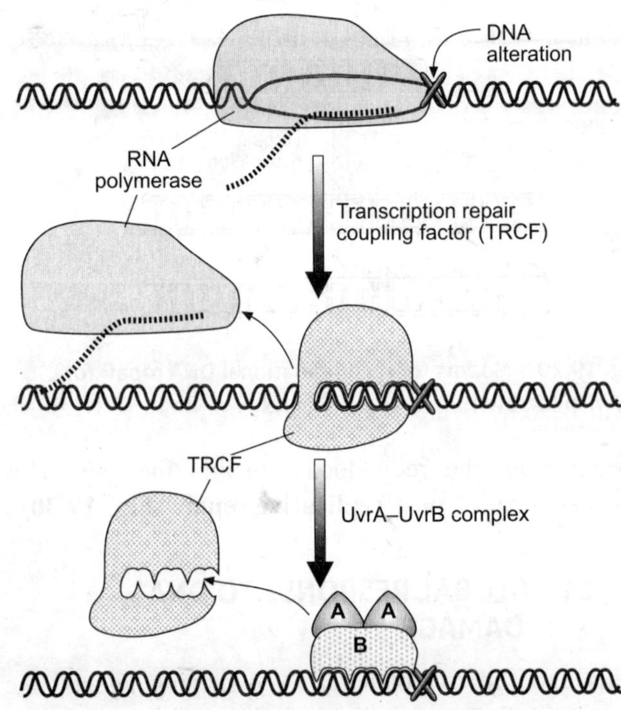

Fig. 19.32 Recruiting of the nucleotide excision repair system to an actively transcribing gene containing a DNA alteration

by a repair system that recognises uracil to be foreign to DNA. This enzyme, **uracil DNA glycosylase**, hydrolyses the glycosidic bond between the uracil and deoxyribose moieties but does not attack thymine-containing nucleotides.

19.16 RNA DAMAGE AND REPAIR

Damage to DNA in cells caused by chemical or physical agents are corrected by various repair systems. However, the discovery that damaged RNA can also be repaired is very recent. Some of the enzymes involved in the repair of damaged DNA are also found to be involved in repair of damaged RNA. DNAs, RNAs and proteins have all been reported to be chemically damaged by aberrant methylation. These methyl groups might come from cellular methyl donors (e.g., S-adenosylmethionine) or from environmental pollutants such as tobacco smoke.

19.17 PATHOLOGICAL EFFECTS OF DEFECTIVE DNA REPAIR

Defective repair of DNA causes cancers

Many cancers are caused by mutations in genes associated with growth control. Exposure to mutagens and/or defects in DNA repair systems are expected to increase the overall frequency of mutations and hence, the likelihood of a cancer-causing mutation.

Hereditary repair defects

Defects in the NER mechanism are responsible for several genetic disorders, including xeroderma pigmentosum, Cockayne syndrome and trichothiodystrophy. **Xeroderma pigmentosum (XP)** is an autosomal recessive disorder characterised phenotypically by extreme sensitivity to sunlight and a high incidence of skin tumours. In infancy, severe changes in the skin become evident and worsen with time. The skin becomes dry, and there is a marked atrophy of the dermis. Keratoses appear, the eyelids become scarred and the cornea ulcerates. Skin cancer usually develops at several sites. Many patients die before age 30 from metastases of these malignant skin tumours. Studies of skin fibroblasts from patients with XP have revealed a biochemical defect in one form of this disease. In normal fibroblasts, half the pyrimidine dimers produced by UV radiation are excised in less than 24 hours. In contrast, almost no dimers are excised in this time interval in fibroblasts derived from patients with XP. The results of these studies show that XP can be produced by a defect in the exonuclease that hydrolyses the DNA backbone near a pyrimidine dimer. Defects in at least seven different genes can cause XP, indicating the complexity of excision repair in mammalian cells.

In **Cockayne syndrome** too, patients are sensitive to sunlight, have short stature and suffer progressive neurologic degeneration. Unlike XP, Cockayne syndrome is not associated with cancer. Cultured cells from patients are hypersensitive to UV damage and have defective DNA repair of actively transcribing genes, i.e., defective in transcription coupled excision repair. Patients with **trichothiodystrophy (TTD)** have photosensitivity, short stature, mental retardation and brittle hair deficient in sulphur. TTD is also not associated with cancer.

Some other DNA repair disorders include, Werner syndrome characterised by premature ageing and retarded growth, Bloom syndrome characterised by photohypersensitivity, high incidence of malignancies, and ataxia telangiectasia which involves sensitivity to ionising radiation and some chemical agents.

Defects in other repair systems can increase the frequency of different tumours. For example, HNPCC or Lynch syndrome, results from a defective DNA mismatch repair. The frequency of HNPCC is about 1 in 200 people. The known cause for this cancer is mutations in two genes, called *hMSH2* and *hMLH1*, account for most cases of this hereditary predisposition to cancer.

KEY POINTS

- DNA provides the ultimate repository for genetic information and damage to DNA can lead to permanent changes. Therefore, perpetuation of the genetic material from generation to generation depends on maintaining rates of mutation at low levels.
- One of the most important requirements for the genetic material is its stability and integrity. However, potential for change is also necessary.
- Biomolecules in a cell are highly susceptible to both chemical and physical alterations that arise from environmental damage or errors during biosynthesis.
- DNA replication and repair are not perfect processes and in rare instances mutations can occur spontaneously.
- Alterations in the base pattern of DNA may arise in various ways. For example, bases may be replaced or they may be deleted or new bases may be inserted in the DNA. Mutations that add or subtract a base are the most devastating in their effects on the cell because they change the reading frame of a gene from the site of mutation onward (frameshift mutation).
- Gross variations can occur in both chromosomal structure and number. Variations in chromosome number include aneuploidy, euploidy and mosaicism and variations in chromosome structure includes a number of DNA alterations, deletions, insertions, duplications, inversions and translocations.
- Mutations can be induced by externally applied agents and the agents capable of inducing mutations are known as mutagens. A wide variety of chemical, physical and biological agents can induce mutations.
- Radiation was the first mutagenic agent known and its effects on genes were first reported in many organisms.
- A large number of chemicals have been found to be mutagenic, carcinogenic or teratogenic. A wide range of natural and synthetic organic and inorganic chemicals can react directly with DNA and alter its properties.
- Of the various types of replication errors caused by erroneous action of the replicative apparatus, three types have been well characterised: a single-base mispair leading to a base substitution, a single-base bulge leading to a single-base frameshift and a multiple-base mismatch leading to a sequence substitution.
- Direct mutagenesis results from the presence of a stable, unrepaired base with altered base pairing properties in the DNA. However, some chemicals, can induce mutations by complex mechanisms without interacting with DNA.
- Other than chemical and physical mutagens, mutations can also be introduced into genetic material through biological agents, such as mutator genes, transposon mutagenesis and bacteriophages.
- Damage to DNA by environmental, physical and chemical agents are diverse and classified broadly into different groups: lesions, AP sites, oxidative damage, alkylation, bulky adduct, thymidine dimers, SCEs and chromosomal aberrations.
- The development of gene technology has led to revolutionary new methods, which make it possible to introduce various types of mutations at will.
- The occurrence of a mutation event at various levels can be detected by using a number of techniques: at the gene level can be detected by biochemical, cytological or alterations in the phenotypic expression, identification of oncogene activation, from genotoxicity testing it and many other genetic end points.
- All organisms have evolved efficient systems for correcting the occasional mistakes in the genetic material (DNA repair). Repair systems are generally placed in five broad categories: damage reversal, excision repair, double-strand repair, post-replicative repair, reverse mutation and suppression mutation and recombination repair.
- Agents that damage DNA also activate important defensive pathways by inducing multiple proteins involved in DNA repair, cell cycle checkpoint control, protein trafficking and degradation.

- Many cancers are caused by mutations in genes associated with growth control. Defects in the NER mechanism are responsible for several genetic disorders, including xeroderma pigmentosum, Cockayne syndrome and trichothiodystrophy. Defects in other repair systems can increase the frequency of different tumours.

Self-assessment Questions

1. List the different types of mutations.
2. Explain briefly the different kinds of chemical mutagens.
3. How does radiation affect DNA and induce lesions?
4. Distinguish between macromutations and point mutations.
5. Discuss the different biological agents that can cause cancer.
6. How do replication errors contribute to mutagenesis?
7. What are the different repair mechanisms that operate in humans?
8. How does the SOS repair system work?
9. Differentiate between base excision and nucleotide excision repair.

REFERENCES

Textbooks

Brusick D (1987). *Principles of Genetic Toxicology.* New York Plenum Press.

Cooper DN and M Krawczak (1993), *Human Gene Mutation.* Oxford: BIOS Scientific Publishers.

Hollaender A (1971). *Chemical Mutagens.* New York: Plenum Press.

Kornberg A and T Baker (1992). *DNA Replication.* New York: WH Freeman and Company.

References

Aas PA. m Otterlei, PO Folnes et al (2003). Human and bacterial oxidative demethylases repair alkylation damage in both RNA and DNA. *Nature* 421:859–63.

Batty DP and RD Wood (2000). Damage recognition in nucleotide excision repair of DNA. *Gene* 241:193–204.

Begley TJ and LD Samson (2003). A fix for RNA. *Nature* 421:795–96.

Berneburg M and AR Lehmann (2001). Xeroderma pigmentosum and related disorders: Defects in DNA repair and transcription. *Adv. Genet.* 43:71–102.

Bishop JM (1987), The molecular genetics of cancer. *Science* 235:305–11.

Bridges BA (1999). DNA repair: Polymerases for passing lesions. *Curr. Biol.* 9:R475–R477.

Channarayappa (1990). *Cytogenetic analysis of clastogens and aneuploidogens by immunofluorescent and micronucleus methods.* PhD thesis submitted to the West Virginia University, Morgantown, West Virginia, USA. **W Microfilm 3365.**

Channarayappa, J Nath and T Ong (1990). Micronuclei assay in cytokinesis-blocked binucleated and conventional mononucleated methods in human peripheral lymphocytes. *Teratogenesis, Carcinogenesis and Mutagenesis* 10:273–79.

Channarayappa, J Nath and T Ong (1992). Clastogenic and aneuploidogenic effects of cigarette smoke condensate, mitomycin C and vincristine sulfate. *Mutagenesis.* 7(6): 457–60.

Channarayappa, J Nath and T Ong (1992). Immunofluorescent staining of kinetochores in micronucleus for detection of aneuploidy inducing agents. *J. of Tiss. Cult. Meth.* 14:125–32.

Channarayappa, T Ong and J Nath (1992). Cytogenetic effects of vincristine sulfate and ethylene dibromide in human peripheral lymphocytes: Micronucleus analysis. *Environmental and Molecular Mutagenesis,* 20:117–26.

Chu G and L Mayne (1996). Xeroderma pigmentosum, Cockayne syndrome and trichothiodystrophy: Do the genes explain the disease? *Trends Genet.* 12:187–92.

Cox MM (2001). Recombinational DNA repair of damaged replication forks in *Escherichia coli*: Questions. *Ann. Rev. Genet.* 35:53–82.

De Laat WL, NG Jaspers and JH Hoeijmakers (1999). Molecular mechanism of excision nucleotide repair. *Genes Dev.* 13: 768–85.

De Serres FJ and A Hollaender (1977–1986)(eds.) *Chemical Mutagens: Principles and Methods for their Detection.* Vols. 6–14, New York: Plenum Press.

Drake JW (1991). Spontaneous mutation. *Annu. Rev. Genet.* 25:125–46.

Echols H and MF Goodman (1991). Fidelity mechanisms in DNA replication. *Ann. Rev. Biochem.* 60:477–511.

Featherstone C and SP Jackson (1999). Double strand break repair. *Current Biology* 9:R759–61.

Friedberg EC, GC Walker and W Siede (1995). *DNA Repair and Mutagenesis.* Washington, DC: ASM Press.

Gao HG, JK Chen, J Stewart, B Song, Channarayappa, WZ Whong and T Ong (1997). Distribution of p53 and K-*ras* mutations in human lung cancer tissues. *Carcinogenesis* 18(3):473–478.

Goodman MF (2002). Error-prone repair DNA polymerases in prokaryotes and eukaryotes. *Ann. Rev. Biochem.* 71:17–50.

Kolodner RD (1995). Mismatch repair: mechanisms and relationship to cancer susceptibility. *TIBS* 20:397–401.

Lambert MW and WC Lambert (1999). DNA repair and chromatin structure in genetic diseases. *Prog. Nucleic Acid Res. Mol. Biol.* 63:257–310.

Lindahl T, P Karran and RD Wood (1997). DNA excision repair pathways. *Curr. Opin. Genet. Dev.* 7:158–69.

Lindahl T and RD Wood (1999). Quality control by DNA repair. *Science* 286:1897–1905.

Maki H (2002). Origins of spontaneous mutations: specificity and directionality of base-substitution, frameshift, and sequence-substitution mutageneses. *Annu. Rev. Genet.* 36:279–303.

Modrich P and R Lahue (1996). Mismatch repair in replication fidelity, genetic recombination and cancer biology. *Annu. Rev. Biochem.* 65:101–33.

Prolla TA (1998). DNA mismatch repair and cancer. Current *Opin. Cell Bio.* 10:311.

Sancar A (1996). DNA excision repair. *Annu. Rev. Biochem.* 65:43–81.

Seeberg E, L Eide and M Bjoras (1995). The base excision repair pathway. *TIBS* 20:391–97.

Singer B and JT Kusmierek (1982). Chemical mutagenesis. *Annu. Rev. Biochem.* 52:655–93.

Smith GR (2001). Homologous recombination near and far from DNA breaks: alternative roles and contrasting views. *Annu. Rev. Genet.* 35:243–374.

Taylor JH (1958). Sister chromatid exchanges in tritium-labeled chromosomes. *Genetics* 43:515-29.

Walker GC (1995). SOS-regulated proteins in translesion DNA synthesis and mutagenesis. *Trends Biochem. Sci.* 20:416–20.

Weinberg RA (1983). A molecular basis of cancer. *Sci. American* 249:126–42.

Wood RD (1996). DNA repair in eukaryotes. *Annu. Rev. Biochem.* 65:135–67.

Wood RD (1997). Nucleotide excision repair in mammalian cells. *J. Biol. Chem.* 272:23465–68.

CHAPTER 20

Molecular Biology of Cancer

LEARNING OBJECTIVES

After reading this chapter, you will be able to:

- List the different types and characteristics of cancers
- Understand the mechanism of carcinogenesis
- State the action of carcinogenic agents
- Discuss the different techniques used for diagnosis and treatment of cancer

20.1 INTRODUCTION

Cancer is a genetic disease that manifests due to abnormal cell proliferation, blockage of differentiation or defective cell turnover in multicellular organisms. The body of a multicellular organism functions as a community or society and each cell is a faithful member, multiplying by cell division and organised into collaborative assemblies (tissues or organs). In this cellular society, self-sacrifice—as opposed to survival of the fittest—is the rule. In multicellular organisms, cell division is carefully regulated and responsive to specific needs of the body. Normally, in a young animal, cell multiplication exceeds cell death and so the animal increases in size. In an adult, the processes of cell birth and death are balanced to produce a steady state. However, occasionally, the controls that regulate cell multiplication break down and, although the body has no need for further cells of its type, a cell begins to grow and divide (**neoplasia**, Greek, *neo* = new; *plastein* = to form; new growth). When such a cell has descendants that inherit the propensity to grow without responding to regulation, the result is a clone of cells able to expand indefinitely. Ultimately, a cellular mass called **tumour** may be formed from the clone of unwanted cells. The study of tumours is called **oncology**. Since the tumours can devastate the cellular integrity of an organism, much research has gone into understanding how they form and spread.

Cancer is not a single disease, but a group of diseases, arising from the progressive accumulation of mutations or other metabolic changes that promote the clonal expansion of cells. Cancer is characterised by uncontrolled division of cells and the ability of these cells to spread, either by direct growth into adjacent tissue through invasion or by implantation into distant sites by metastasis. Cancer may affect people at all ages, but risk tends to increase with age. Cancer cells grow without appropriate signals from other cells (**loss of contact inhibition**) or with a diminished requirement for them (**Fig. 20.1**). The fundamental feature of cancer cells is that when they divide, both daughter cells are also cancerous. Most known cancer genes fall into three broad classes:

- **Proto-oncogenes** (Greek, *oncos* = mass *genein* = to generate) are positive regulators of cell proliferation and survival. Cancer occurs due to their inappropriate activity (i.e.,) gain of function mutations.

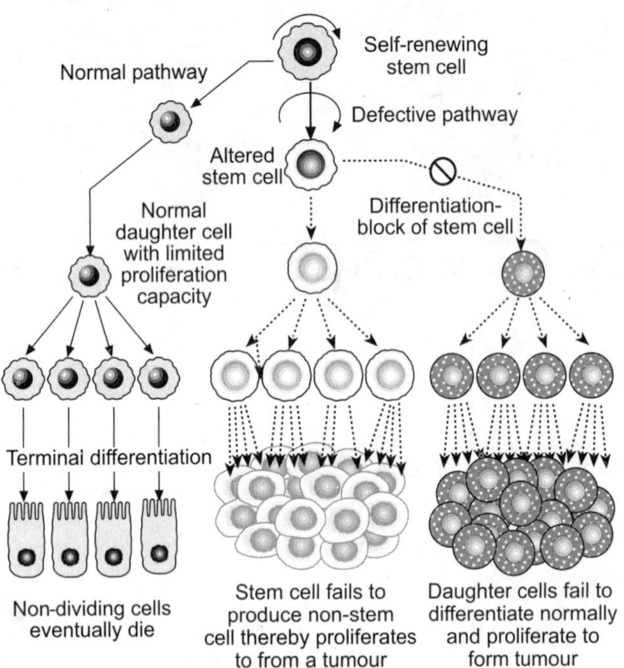

Fig. 20.1 Normal and genetically altered (transformed or defective differentiation) regulation of cell proliferation from stem cells

- **Tumour suppressor genes (TSGs)**, which are negative regulators, inhibit cell proliferation and cancer results from their reduced activity, (i.e.,) loss of function mutations.
- **DNA repair genes**, which are responsible for the detection and repair of genetic damage may fail or repair with defects.

The coordinated regulation of these gene classes is responsible for maintaining tissue homeostasis. Cancer reflects a fundamental breakdown in this control.

20.2 TYPES OF CANCER

A tumour is any abnormal proliferation of cells, which may be benign, progressive or aggressive (malignant). There are more than one hundred distinct types of cancers, which can vary substantially in their complexity and severity.

Leukemias, for example, are diseases that cause excessive production of leucocytes, which originate in the bone marrow (from haemopoietic cells). Two types of leukemias cause overproduction of white blood cells: **lymphomas** cause excessive production of lymphocytes, whereas **myelomas** cause excessive production of myelocytes. Both types of cancer originate in the bone marrow, lymph nodes and spleen, which are the origin of the embryological endoderm.

Sarcomas are tumours of tissue such as muscle, bone and cartilage that arise from the embryological mesoderm.

About 85% of cancers are **carcinomas** or tumours arising from epithelial tissue such as glands, breast, skin, and the linings of the urogenital, digestive and respiratory systems (origin of embryological ectoderm or endoderm).

Mesothelioma are tumours derived from the mesothelial cells lining the peritoneum and the pleura.

Glioma are tumours derived from glia, the most common type of brain cell.

Germinoma are tumours derived from germ cells, normally found in the testicles and ovaries.

Choriocarcinoma are malignant tumours derived from the placenta. All cancers are genetic in that they come about from alterations in genes that control cell growth. Cancer is stable and inherited during cell division. The clonal nature of cancer suggests that they arise from a single aberrant cell which then proliferates. The important issue in cancer biology is the detection and differentiation of benign and malignant tumours.

20.3 TUMOUR CLONALITY: DERIVATIVES OF A SINGLE ABNORMAL CELL

One of the unique characteristic features of cancer development is tumour clonality—the expansion of tumours from single cells that begin to proliferate abnormally. Even at advanced stage (metastasised), its origins can be traced to a single **primary tumour**. Chromosomal aberrations and rearrangements are common features in the cells of most cancers. For example, in almost all patients with **chronic myelogenous leukemia (CML)**, leukemic white blood cells can be distinguished from normal ones by a specific chromosomal abnormality: the so called **Philadelphia chromosome**, created by a translocation between the long arms of chromosome 9 and 22. When the DNA at the site of translocation is cloned and sequenced, it is found that the site of breakage and rejoining of the translocated fragments is identical in all the CMLs, but differs slightly (a few or thousand base pairs) from one patient to another. The single cell origin of many tumours can also be demonstrated by the analysis of X chromosome inactivation (**Fig. 20.2**). Normal tissues

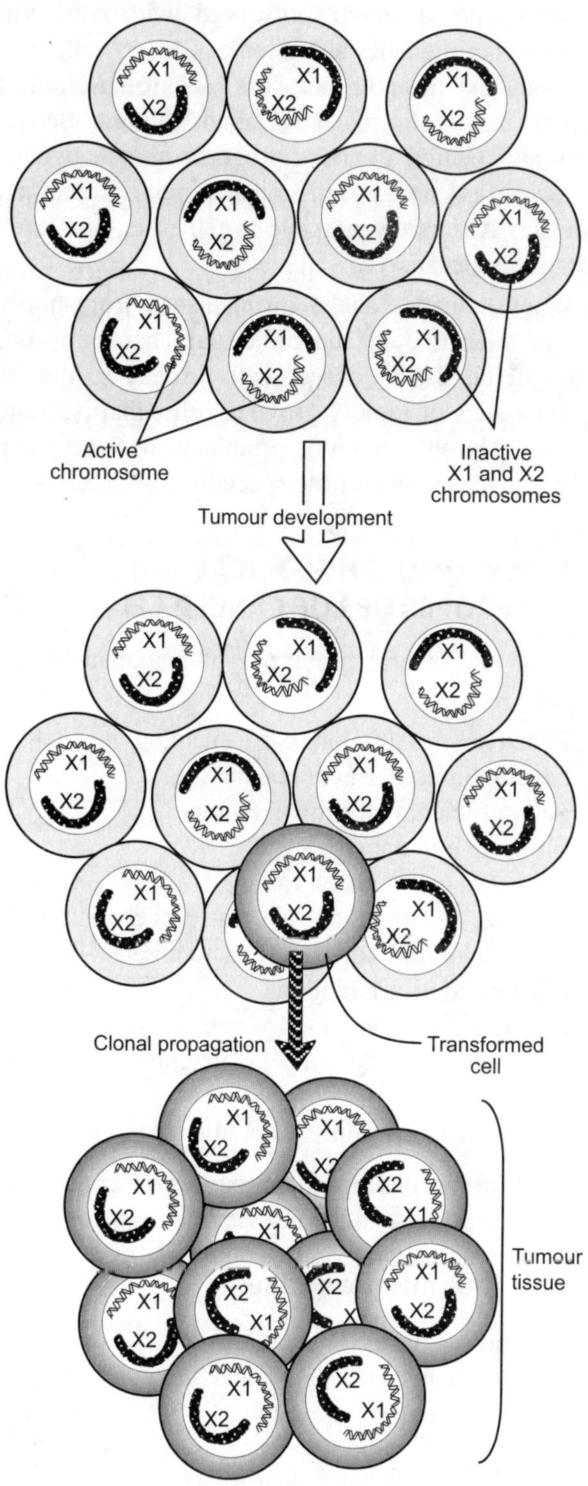

Fig. 20.2 Tumour clonality; normal tissue of females is a mosaic of cells in which different X chromosomes (X1 and X2) have been functionally inactivated. All the tumour cells developed from a single abnormal cell display the same pattern of X inactivation. In the above diagram, X1 and X2 are randomly inactivated in a normal cell but in tumours, all the cells contain active X1 and inactive X2.

of heterozygote females contain mixtures of cells with different X chromosomes, expressing both the alleles. In contrast, tumour tissues generally express only one allele of a heterozygous X chromosome gene, implying that all of the cells constituting such a tumour were derived from a single cell of origin, in which the pattern of X inactivation was fixed before the tumour began to develop.

20.4 MOLECULAR BASIS OF CARCINOGENESIS

Carcinogenesis is the process that induces cancer in normal tissues. The various types of agents that are responsible for the initiation of neoplasia are called **carcinogens**. Some other factors predisposed to or help in the action of these carcinogens are called **co-carcinogens**. Some of the fundamental changes at the molecular level that lead to carcinogenesis are:

- Non-lethal genetic damage (mutations) acquired by the action of many agents including chemicals, radiation, viruses in the environment or metabolic factors and germline defects
- Damage of two normal regulatory genes such as the growth-promoting proto-oncogenes and the growth-inhibiting cancer suppressor genes
- Multiple and step-wise acquisition of tumour progression.

20.4.1 Cancer: A multistep process

The clonal origin of tumours does not, however, imply that the original cell had acquired all of the characteristics of a cancer cell. On the contrary, the development of cancer is a multistep process in which cells gradually become malignant through a progressive series of alterations. Berenblum and Shublik (1949) suggested the two phases of carcinogenesis. The first phase is the **initiation** of the tumour. This is associated with alkylation of DNA and a transient inhibition of DNA synthesis. At this level, no morphological markers are apparent, but some biochemical markers characteristic of altered DNA are detectable. Changes at the molecular level during initiation may remain latent for the lifespan of the animal without showing any morphological features. The second phase is the **promotion** of the tumour. This is mainly associated with the alteration in the gene activity, causing a reversal of cell differentiation. After initiation, however, a promoting factor is essential to allow expression of their neoplastic phenotype and growth to form a clinically diagnosable neoplasm. Cell proliferation and hyperplasia are the most conspicuous changes induced by the promoters. Armitage and Doll proposed that carcinogenesis consisted of at least six (1954) or seven stages.

At the cellular level, the development of cancer occurs in several stages, usually starting with a mild growth (benign) disorder (dysplasia), which gradually becomes more serious (tumour progression). Primary cultured cells are initially like cells *in vivo* i.e., they show serum dependence, contact inhibition, finite lifespan etc. They can turn malignant lacking these inhibitions by a multistep process of **growth transformation**.

20.5 TUMOUR CELLS

Tumour cells have two heritable properties: they reproduce in defiance of the normal restraints on cell division and invade and colonise territories normally reserved for other cells. Tumours arise with great frequency in multicellular organisms, especially in older animals, humans and also in plants, but most pose little risk to their host because they are localised. The mild and localised tumours are generally known as **benign.** An example is warts. They are rarely life-threatening, grow within a well-defined capsule which limits their size and maintain the characteristics of the tissue of origin and are thus usually well differentiated. In contrast, malignant cells lack differentiation and have atypical structure. The major characteristics that differentiate malignant tumours from benign ones are their invasiveness and spread. Malignant tumours do not remain localised and encapsulated; instead they invade surrounding tissues, get into the body's circulatory system, and set up areas of proliferation away from the site of their original appearance. This property, known as **colonisation** in other tissues, spreads the tumour cells and leads to establishment of secondary areas of growth. This process is called **metastasis** (Gr., *meta* = subsequent; *stasis* = location), which is often the most life-threatening (**Fig. 20.3**).

Malignant cells are usually less well differentiated and their properties often vary over time and different clones within a tumour. For example, lung cancer cells may lack certain enzymes or factors characteristic of lung cells and may ultimately evolve to a state in which they lack most lung-specific functions. This variability of phenotype is often correlated with variability of genotype (genetic alteration). Cancer cells have abnormal and unstable numbers of chromosomes, as well as many chromosomal abnormalities. Because apparently benign tumours progress into malignancy and the earliest stages of malignant tumours are difficult to detect, it is rarely possible to tell when and how a malignancy began.

Malignant cells can be distinguished from normal ones by microscopic and molecular characterisation. In a specific tissue, cancer cells usually exhibit the characteristics of rapidly growing cells, that is, a high nucleus-to-cytoplasm ratio, prominent nucleoli, many mitoses, and relatively little specialised structure.

20.6 CYTOMORPHOLOGICAL PROPERTIES OF CANCER CELLS

Some of the general properties of malignant tumour cells are:

- Reactivation of telomerase enabling the cells to have unlimited growth potential
- Self-sufficiency of growth factors
- Insensitive to apoptosis
- Increased cell division rate
- Blocked differentiation so that cancer cells would not develop into terminally mature cells
- Lack of contact inhibition
- Ability to promote blood vessel growth
- Gain function to invade neighbouring tissues (metastasis)
- Ability to build metastasis at distant sites.

The cyto-morphological properties of cancer cells are discussed in the following section.

20.6.1 Intermitotic nucleus

Shape and volume: The nucleus of the cancer cell sometimes changes it shape and contents. It has a markedly lobulated appearance. Many neoplastic nuclei appeared to be sprouting, deformed by bulby protrusions, constricted at their base or even completely separated from the central nuclear mass. The presence of deep indentation in the nucleus increases the surface area. In cancer cells, the nuclear and nucleolar volume increases. Nuclear hypertrophy, chromatin lumps and the amount of DNA may also increase. The cancer cells are, thus, generally polyploids.

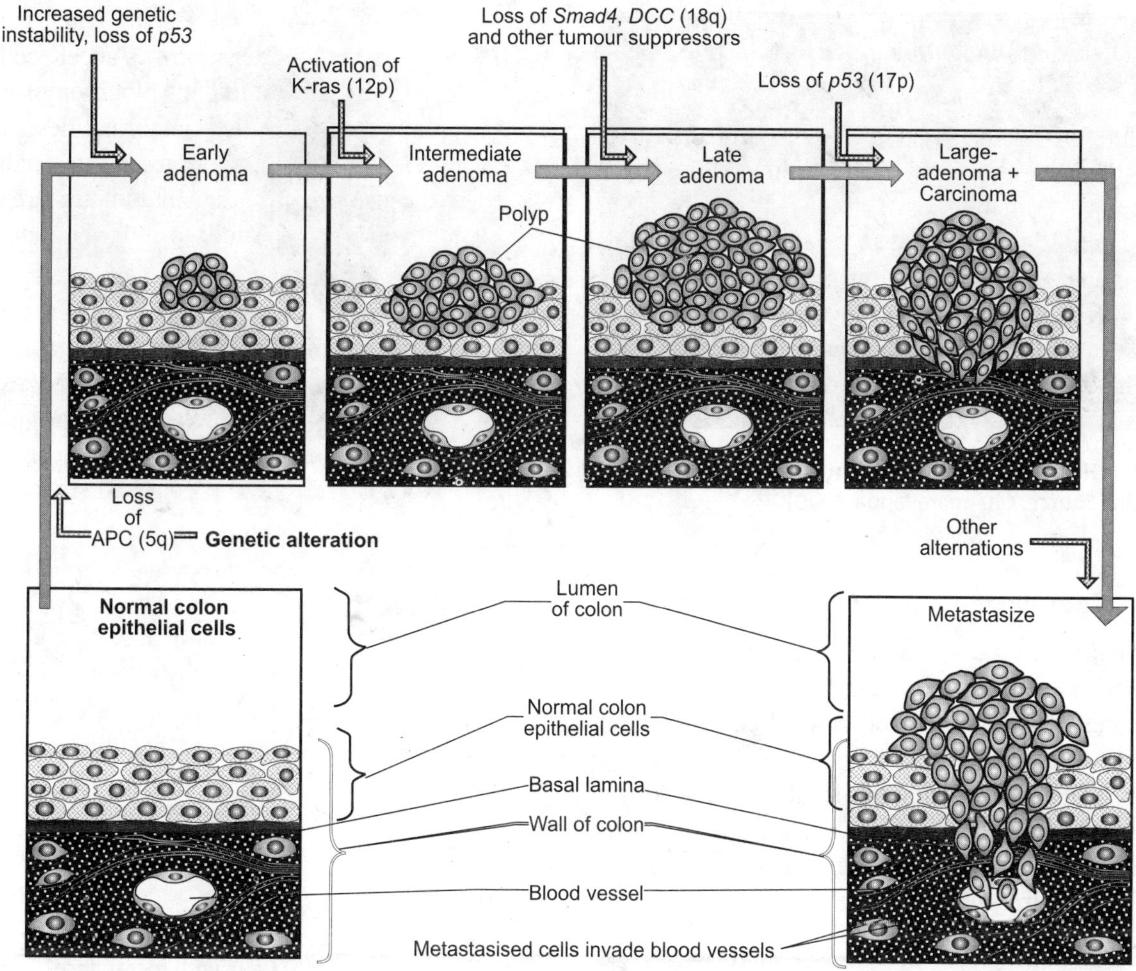

Fig. 20.3 A model of sequential genetic alterations leading to metastatic colon cancer (colerectal carcinoma); the information in the brackets are gene position in the chromosome; smaller arrows indicate the possible genetic alterations and the bigger arrows indicate different stages and the direction of cancer development

20.6.2 Mitotic nucleus

As the tumour strain evolves, genetic instability increases and the behaviour of chromosomes becomes abnormal. Generally, the range of chromosome variation is narrower in tumours evolved from single cell clones. In tumours of diploid origin, the number of chromosomes shifts to polyploidy. The form of chromosomes is extremely variable in cancer cells. They may be contracted, elongated, filamentous or thin. Dicentric (**Fig. 20.4**) and annular chromosomes are also commonly observed. Some commonly found anomalies in cancer cells are:

- The chromosomes of the equatorial plate spread away from the centre, which results in **hollow metaphase**. At anaphase, certain chromosomes fail to move to the poles due to chromosomal stickiness and the two adherent chromosomes form a chromosomal bridge (**Fig. 20.5**).
- The lack of movement of chromosomes may be duc to the absence of spindle fibres, and arrangement of chromosomes at the equatorial plate is irregular. The chromosomes divide normally but do not move towards the opposite poles.

Normal mitosis may be interrupted at any stage. This is simply a sequel to cytoplasmic disturbance and leads to the formation of multinucleated giant cells (**Fig. 20.6**). The chromosomes are scattered, no equatorial plate is formed and the nucleus has double the normal number of chromosomes. Lack of chromosomal separation involves duplication without separation of chromosomes (endoreduplication) and results in nuclear hypertrophy.

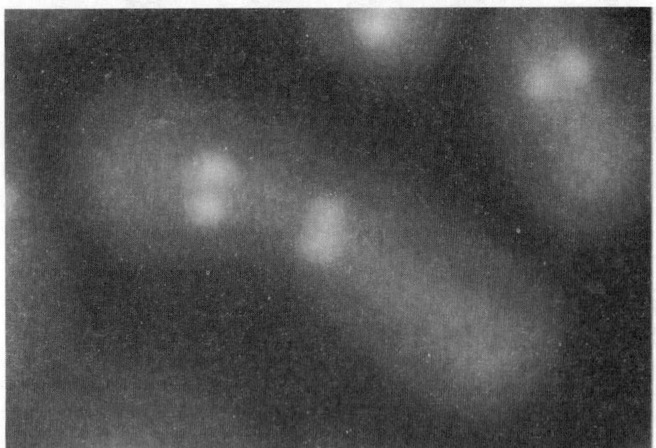

Fig. 20.4 Dicentric chromosome: a type of chromosomal aberration (***Source:*** Channarayappa 1990)

20.6.3 Cytoplasm

The gross structural changes of cytoplasm include shrinking and decrease in cytoplasmic–nuclear ratio. Tumour cells rarely show well-defined cell membrane and edge-to-edge contacts between cells always overlaps. They often contain smaller mitochondria than normal cells. The mitochondria become round and swollen and eventually transform into empty sacs. The cristae become irregular in shape.

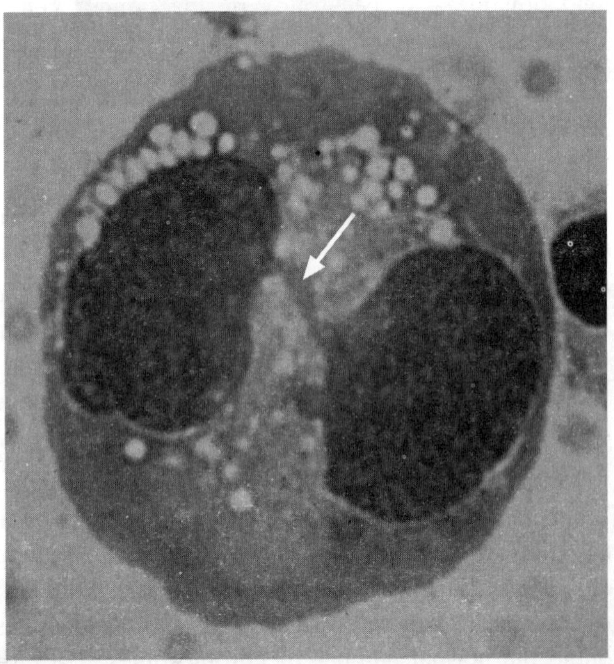

Fig. 20.5 Sticky or dicentric chromosomes can form chromosomal bridges (arrow) which can lead to abnormal distribution of chromosomes into daughter cells during mitosis (***Source:*** Channarayappa 1990)

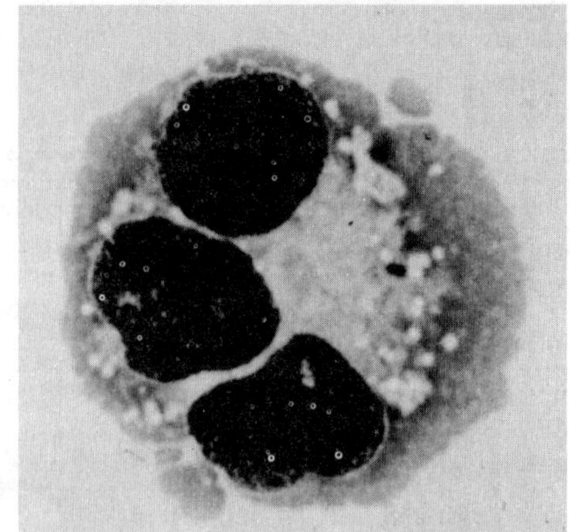

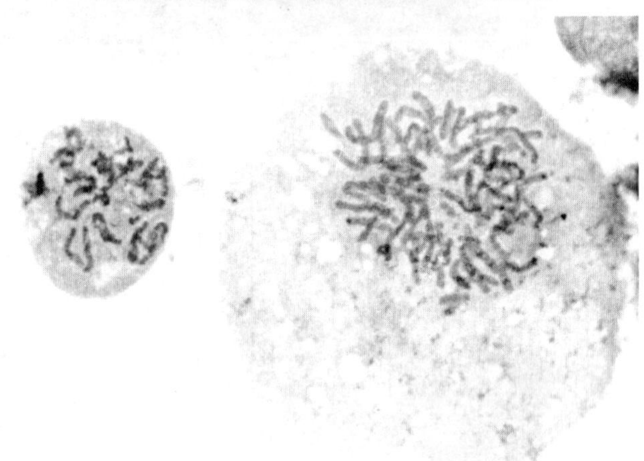

Fig. 20.6 Light microscopic photographs showing (a) Multinucleated cell (b) Cells with metaphase chromosomes of normal (left) and giant cell (right) (***Source:*** Channarayappa, 1990)

20.6.4 Lack of normal control on cell growth

Normal cells, generally, either grow or are **quiescent**. Cell growth involves an increase in cellular mass that leads a cell to divide. Cell growth and cell division are thus interrelated phenomena. Quiescent cells carry out the characteristic functions of tissues, such as the synthesis of export proteins in the liver or the transmission of impulses in a nerve. Cell growth is a regulated process, so the fraction of growing cells in a given tissue is a function of both the age of the organism and the properties of the tissue. A myriad of cell-specific cytokines signal individual cell types whether to divide or not. The control of cell division is one of the most important aspects of an animal's physiology. Cells of an adult must divide frequently enough to allow tissues to remain in a steady state and division must be stimulated at wounds or when special requirements are placed on a tissue.

20.6.5 Immortalisation

Immortalisation describes the process of indefinite cell growth (without any other gross alterations). Except some blood cells and cells of the immune system, all other animal cells grow in culture only on the surface. When normal animal cells are cultured *in vitro*, they grow for about 50 cell generations after which they start dying. If a piece of a typical tissue is placed on a glass or specially treated plastic surface, various types of fibroblast cells grow from the tissue mass. These dividing cells are called **primary culture**, and divide only for a few generations. Then most of the cells die and disintegrate (called **crisis**). The number of divisions that occur before this happens is sometimes called the **Hayflick limit**. If any cells survive after this crisis, they have gained the ability of indefinite growth and are said to be immortalised. These cells can constitute a **permanent** or **established cell line**. Immortalised cells grow only as a thin layer of single-cell thickness on a substratum (called **monolayer**). But these cells are **non-tumorigenic**; they do not form tumours when introduced in test animals. In cultures, these cells produce various kinds of aneuploids.

20.6.6 Metastasis

Metastasis (Greek, *meta-* subsequent; *stasis-* location) describes the stage at which the cancer cell gains the ability to invade normal tissue, so that it can move to new locations from the tissue of origin through the blood stream and establish a new colony elsewhere in the body.

Patterns of spread: The following patterns are commonly observed.

Direct invasion: The tumour leaves the capsule and invades and destroys the adjacent tissue.

Seeding of body cavities: The tumour seeds cavities. It is critical to avoid this especially during surgical resection.

Lymphatic spread: It is the commonest pathway, with less extensive basement in membrane than in vessels.

Haematogenous spread: It usually follows the pattern of drainage.

20.6.7 Altered gene transcription

In transformed cells, when compared to normal cells, mRNAs of some specific genes have decreased expression while expression of some other genes increases. Only 3% of mRNA population is specific to transformed cells. These transformation-sensitive genes, although low in concentration, have severe effects on cell growth and morphology. Such specific mRNAs have also been found in embryonic cells, suggesting that transformation may alter the genes that are involved in embryonic development.

20.6.8 Angiogenesis

Diffusion limits of blood supply mean that no cell can be more than approximately 2–5 mm from a blood vessel. Once the tumour mass reaches about 2 mm in diameter, it must choose either to stop growing or find blood supply. Angiogenesis (new vessel formation) is activated through the elaboration of angiogenic factors and the suppression of angiostatic factors.

Cancer cells secrete growth factors (**Table 20.1**) that promote the formation of new blood vessels (called **angiogenesis**). It is needed to support the growth of a tumour beyond the size of about a million cells (**Fig. 20.7**). At this stage, new blood vessels are required to supply oxygen and nutrients to the rapidly growing tumour cells. The formation of such new blood vessels is important not only in supporting tumour growth, but also in metastasis. The actively growing new capillaries developed in response to angiogenic stimulation is easily penetrated by the tumour cells, providing a ready opportunity for cancer cells to enter the circulatory system and begin the metastatic process.

Table 20.1 Factors that promote or inhibit angiogenesis

Pro-angiogenic factors	*Anti-angiogenic factors*
Fibroblast growth factors	Fumagillin
Vascular endothelial growth factors	Thrombospondin
Platelet-derived growth factors	Angiostatin
Transforming growth factors	Endostatin
Angiogenin	
Leptin	
Angiopoietins	
Matrix metalloproteinases	
Integrin αVß3	

20.6.9 Heredity of cancer

Most forms of cancers are 'sporadic' in nature, and have no set hereditary rules. There are however, a number of recognised syndromes of cancer with a hereditary component: often a defective tumour suppressor allele. A few examples are:

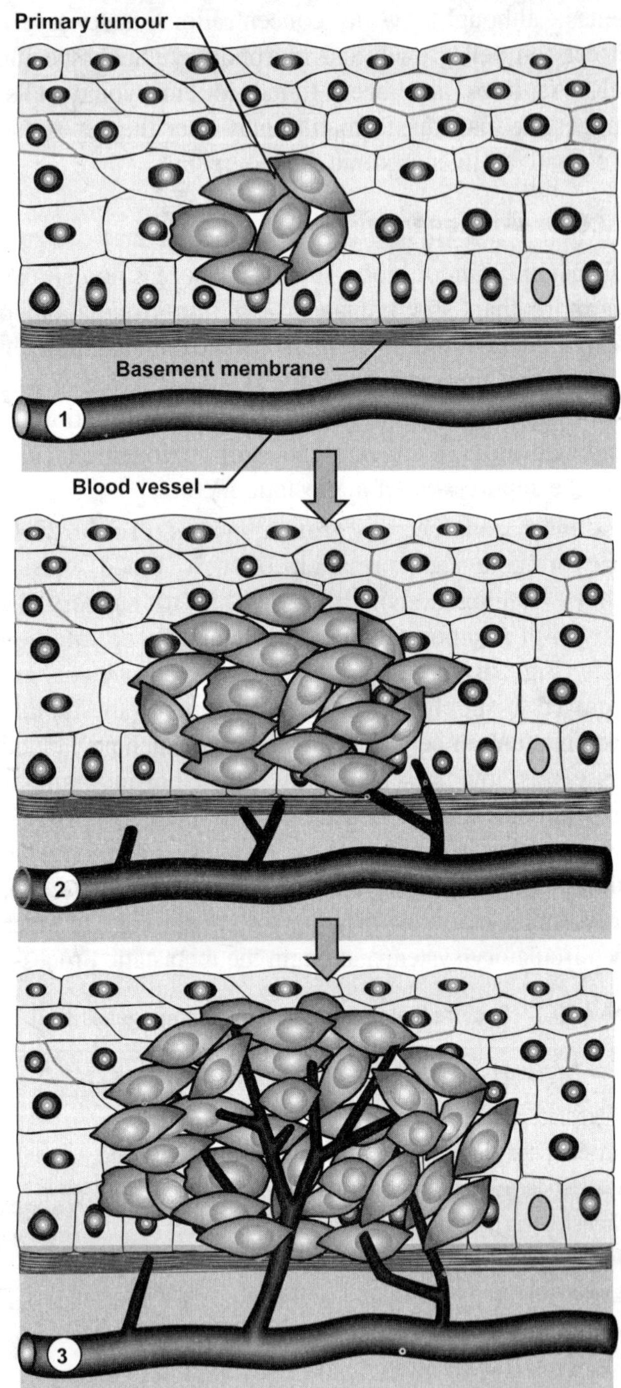

Fig. 20.7 Angiogenesis and tumour growth; steps in the vascularisation of a primary tumour (1) The tumour proliferates to form a small mass of cells (2) The tumour mass start producing angiogenic factors that stimulate the endothelial cells of nearby vessels to grow out toward tumour cells (3) The tumour has become vascularised and is now capable of unlimited growth

- Certain inherited mutations in the genes *BRCA1* and *BRCA2* are associated with an elevated risk of breast cancer and ovarian cancer.
- Tumours of various endocrine organs in multiple endocrine neoplasia (MEN types 1, 2a, 2b)
- Li-Fraumeni syndrome (e.g., osteosarcoma, breast cancer, soft-tissue sarcoma, brain tumours) due to mutations of p53
- Turcot syndrome (brain tumours and colonic polyposis)
- Familial adenomatous polyposis due to an inherited mutation of the *APC* gene that leads to early onset of colon carcinoma
- Retinoblastoma in young children is an inherited cancer.

20.6.10 Importance of apoptosis

Multicellular organisms have evolved one of the most important protection mechanisms against cancer—death of cells called apoptosis. It is induced when cells are damaged, stressed, unwanted or life-threatening. Apoptosis is also an integral part of the differentiation program of many cell types. It is a key mechanism in developing organisms, used to rid the body of excess cells. Most cancer cells fail to undergo apoptosis, consequently their lifespan increases when compared to their normal counterparts. This failure of cancer cells to undergo programmed cell death contributes substantially to tumour development. For example, the survival of many normal cells is dependent on signals from growth factors or from the extracellular matrix that prevent apoptosis. However, tumour cells are often able to survive in the absence of growth factors required by their normal counterparts. The failure of apoptosis in cancer cells leads to the development of resistance to irradiation and many chemotherapeutic drugs, which act by damaging the DNA.

20.7 ONCOGENES

As discussed earlier, cancer is caused by the alterations in critical regulatory genes that control cell proliferation, differentiation and survival. Alterations can lead to disruption of cell proliferation control mechanisms, either through the activity of a virus (or occasionally bacterial signalling) at the level of protein function, or by mutation of critical growth regulation genes. In 1989, Michael Bishop and Harold Varmus received the Nobel prize for their pioneering contributions in the field of proto-oncogenes and their potential to transform into oncogenes causing cancer. Proto-oncogenes are cellular genes that encode proteins having a specific function in the normal cell. But molecular hybridisation also revealed that the viral oncogenic sequences were almost identical to sequences found in the normal cellular DNA. Thus, proto-oncogenes can be defined as the normal cell genes from which the viral oncogenes originated. **Oncoproteins** and their corresponding **proto-oncoproteins** are distributed throughout the cell including nucleus, cytoplasm and the nuclear membrane (**Fig. 20.8**). On the nuclear membrane, they may be found on the cytoplasmic face or may be found as transmembrane proteins. Based on their function they are classified into the following four groups:

- Tyrosine kinases (subdivided into cytoplasmic proteins and membrane-bound growth factor receptors)
- Growth factors
- GTP-binding proteins
- Nuclear proteins.

The oncogenes in a virus are denoted with the prefix v (e.g., v-*src*). The corresponding cellular proto-oncogene is given the prefix c (e.g., c-*src*). Most

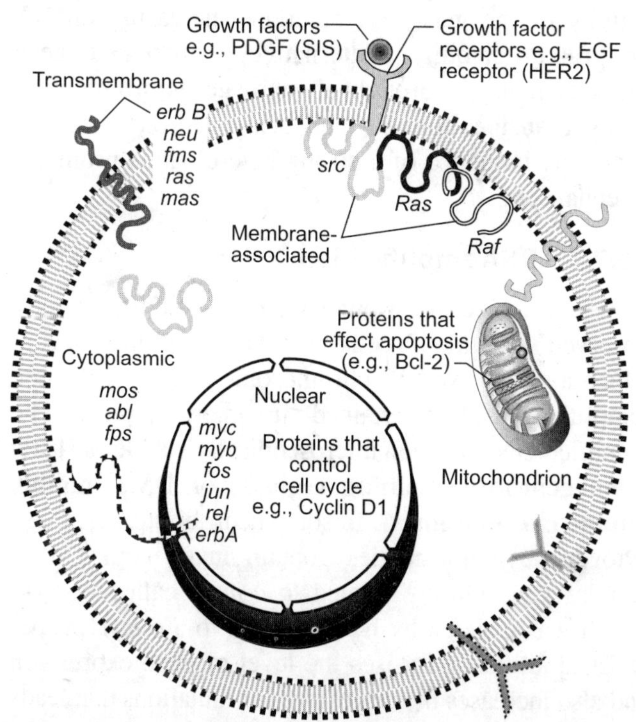

Fig. 20.8 A schematic diagram summarising the types of proteins encoded by proto-oncogenes and their location

known proto-oncogenes play a role in the control of cell growth. So far ~100 different *c-onc* genes have been identified by detection of homology between *v-onc* genes and cellular genomes and looking directly for cancer-causing genes in the genomes of cancer cells by transfection experiments. The viral oncogene is under the regulation of viral promoter and enhancer sequences. Consequently, oncogenes are expressed at much higher levels than the proto-oncogenes and are sometimes transcribed in inappropriate cell types. In some cases, such abnormalities of gene expression are sufficient to convert a normally functioning proto-oncogene into an oncogene that drives cell transformation. In addition to such alterations in gene expression, oncogenes frequently encode proteins that differ in structure and function from those encoded by their normal homologues.

There are many signal transduction systems and regulatory mechanisms which control cell proliferation and differentiation. In principle, disruptions to any of the pathways could cause cancer. The oncogene mutated most frequently in human tumours is *FAS*, which encodes a GTP-binding protein (Ras) that functions as an on-off switch for a key signalling pathway controlling cell proliferation. Oncogenic *RAS* mutants typically encode a protein whose GTPase activity cannot be stimulated, which leaves the molecule in an active GTP-bound form, sending continuous proliferation signals along the pathway. Other genes often found to be involved in cancer are those which function in DNA repair or cellular responses to DNA damage. Although mutations in these genes are not tumourigenic *per se*, they cause an increase in the mutation rate, making a subsequent oncogene or TSG mutations more likely.

Most proto-oncogenes have the potential to promote tumour development. Of the many oncogenes studied, most are derivatives of normal cellular genes. They are important for normal cellular process, but they can be altered, often in very simple ways, to become oncogenes.

20.7.1 Basis of oncogene activity

Oncogenes were originally discovered through the analysis of tumour viruses, whose infection can result in the induction of tumour growth (**Fig. 20.9**). Using the ubiquitous *Alu* human-specific sequences as a handle, biologists cloned the active genes from the DNA of the tumour cells. Many of these genes turned out to be

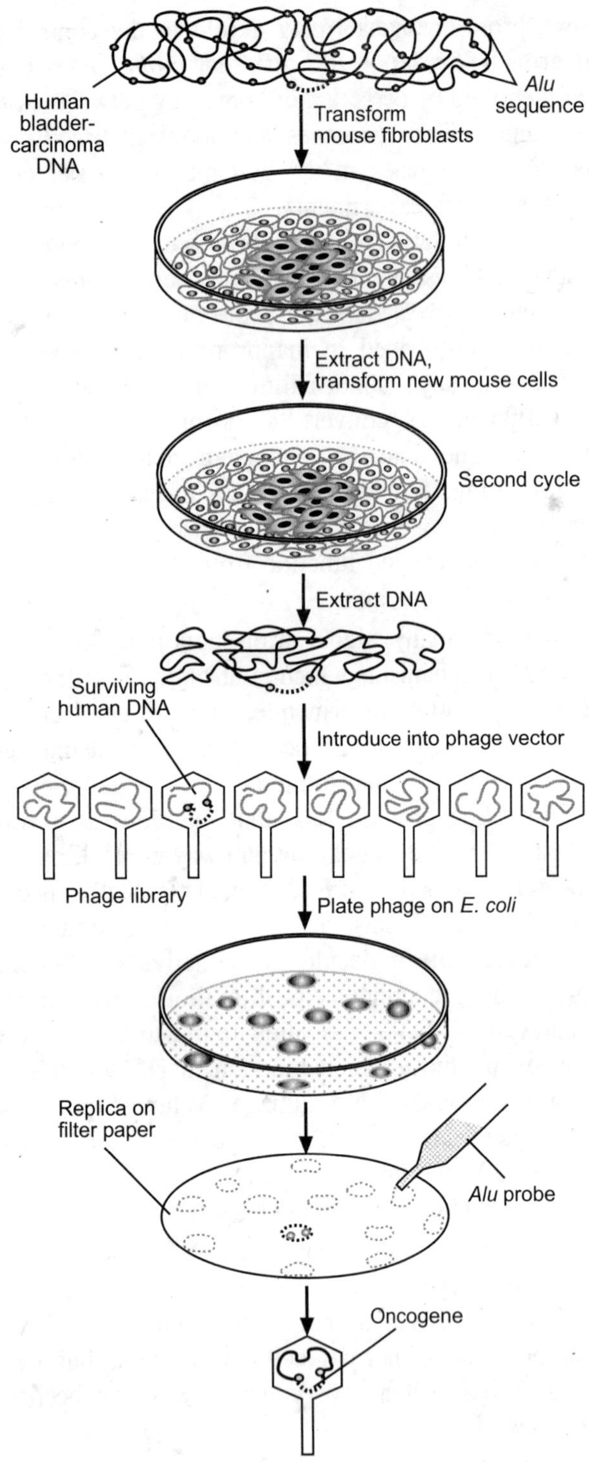

Fig. 20.9 The isolation of a cellular oncogene from a human tumour by molecular cloning

the same ones found in viruses. In particular, the *ras* family of genes was identical as common oncogenes in human and animal tumours.

20.7.2 Activation by mutation

Although many oncogenes have been identified by the analysis of acute transforming retroviruses, most cancers occur in the absence of viral infection. Another approach to identifying oncogenes is the direct analysis of DNA, either structurally, by looking for differences between the DNA of normal and neoplastic cells, or functionally by transfecting tumour cell DNA into cultured cells and assaying for growth transformation.

Comparisons of corresponding *v-onc* and *c-onc* sequences, and of *c-onc* sequences from normal and transformed cells, have shown that the oncogenic potential of a proto-oncogene arises either from a change in structure (qualitative change) or increased expression (quantitative change). Qualitative changes have the most severe effects where the normal oncoprotein is constrained by its structure (e.g., receptor which needs to be activated by ligand), whereas quantitative changes have the most severe effects when the activity of the normal oncoprotein is regulated by its abundance (e.g., a transcription factor). The simplest qualitative changes are gene mutations, either point mutations or truncations, which alter the structure of the oncoproteins and make it hyperactive even when produced in normal amounts. Quantitative changes can involve mutations in regulatory elements increasing the transcriptional activity of the gene, or mutations increasing mRNA or protein stability. Alternatively, increased gene expression may be brought about by gene **amplification** or by **translocations**, which bring a proto-oncogene under the influence of a strong heterologous promoter or enhancer.

20.7.3 DNA amplification

A different type of chromosomal alteration frequently observed in tumour cells is DNA amplification, in which a segment of the chromosome is over-replicated (amplified) and maintained in many copies. The amplified DNA is often microscopically visible as HSRs (**homogeneously staining regions**) or DMs (**double minute chromosomes**). In some neuroblastomas, these cytogenetic abnormalities contain many copies of a cellular gene closely related to c-*myc*, called N-*myc*, which was isolated by hybridisation to a c-*myc* probe. DNA amplification raises the level of gene expression and also increases the target size for mutations that leads to activation of proto-oncogenes. Molecular analysis

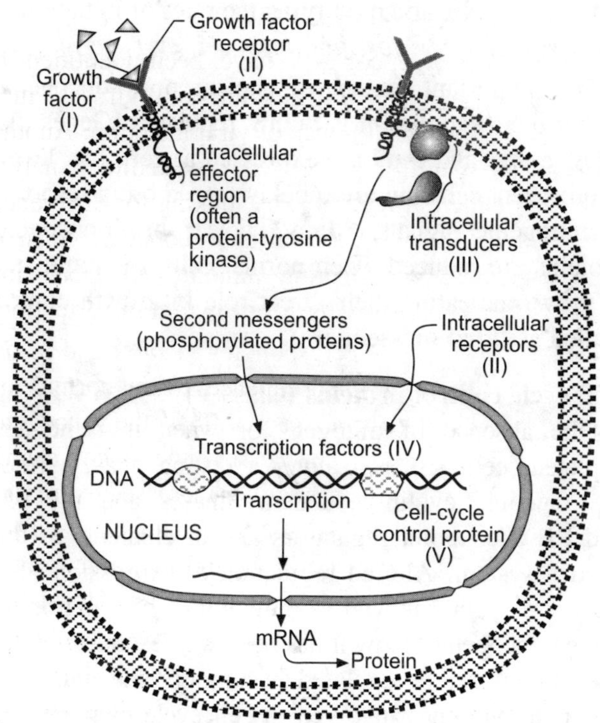

Fig. 20.10 Location of oncoproteins in a cell; control of cell growth involves five types of proteins whose mutant forms lead to cancer

of proto-oncogene structure in tumours now routinely reveals amplification of proto-oncogene sequences.

20.7.4 Oncogenes and control of cell growth

Many studies have supported the hypothesis that oncogenes would be involved in the essential functions of the cell related to control of cell proliferation and differentiation. The proteins produced by proto-oncogenes, which are suspected to be involved in controlling cell growth can be grouped into six types:

- Growth factors
- Growth factor receptors
- Intracellular signal transducers
- Nuclear transcription factors
- Cell cycle control proteins, and
- Products that affect apoptosis (**Fig. 20.10**).

Growth factors (Class I)

Growth factors constitute the products of the first group of proto-oncogenes and exist as polypeptides, oligopeptides or steroid hormones (PDGF, EGF, FGF1-7, IGF1 and 2 and TGFα and ß). Understanding the control of growth starts from the observation that cells in culture can be induced to grow by adding specific growth factors to the medium, epidermal growth factor (EGF) being a paradigm. Certain hormones also induce the cultures to grow. Such growth factors and hormones induce many responses in cells, such as mobilisation of energy stores and differentiation, as well as entry into the growth cycle. The responding cell has a specific receptor (e.g., the EGF receptor) that is sensitive to the signal. Receptors for many growth factors have kinase activity. Because different cells have different receptors, specific signals can produce a response in some cell types but not with others. For example, EGF stimulates growth of epithelial cells, but not other cells (**Fig. 20.11**). The extracellular N-terminal region binds the ligand that activates the receptor. The intracellular C-terminal region includes a domain that has tyrosine kinase activity.

Growth factor and hormone receptors (Class II)

A second group of proto-oncogenes (*erbb1, trk, bek, fgfr1-4, erbb2 her2/neu, igf-1r, met, fig, fms, ret, sea, eph, ros, eek, kit, elk*) encode either the growth factor receptors themselves or their functional homologues. A receptor recognises and binds a specific factor

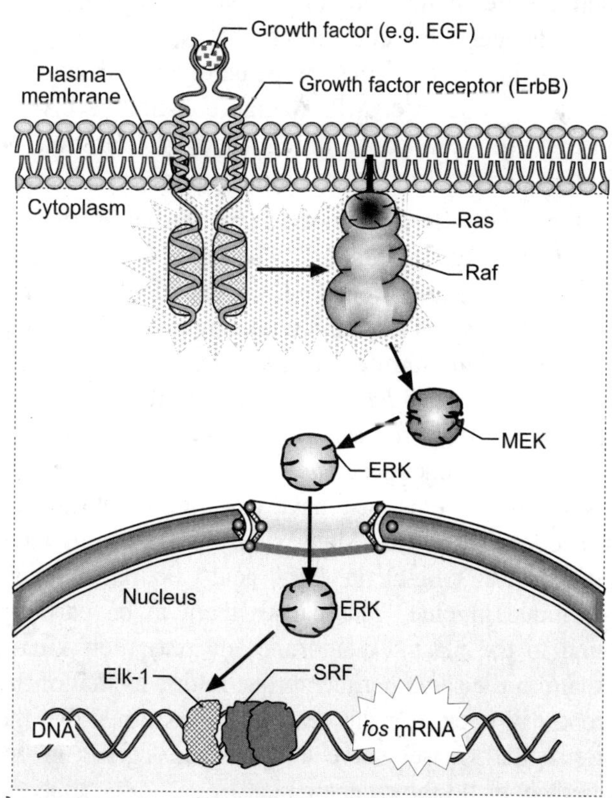

Fig. 20.11 Oncogenes proteins act as growth factors (e.g., EGF), growth factor receptors (e.g., ErbB) and intracellular signalling molecules (Ras and Raf). ERK: Extracellular signal-regulated kinase: MEK, MAP kinase/ERK kinase

and sends a growth signal to the cell through one or more signal transduction pathways. Some cell surface receptors have integral protein-tyrosine kinases in their cytoplasmic domains. These receptors transmit the growth signal by phosphorylating tyrosine residues on themselves as well as on one or more target proteins, thus initiating a cascade of events. The genes for such receptors become oncogenes when they are mutated in such a way that the receptor remains active even in the absence of its bound ligand.

Signal transducers (Class III)

The largest class of oncogenes consists of proto-oncogenes that encode proteins that occupy the area just inside the plasma membrane and act as intracellular transducers (*ras, gsp, gip, abl, src, fes, fgr, syn, lck, raf, mos and cot*)—proteins that transmit signals from a receptor to a cellular target. Among the best-understood transducers are the G proteins, one of which, G_s, controls cAMP synthesis. G_s protein senses that a ligand has occupied a cell-surface receptor, binds GTP, and activates the adenyl cyclase enzyme that forms cAMP. It then hydrolyses the GTP and returns to an inactive state. A mutation in the gene for G_s can make it an oncogene by eliminating the GTPase activity, thus causing it to signal perpetually and produce a constitutive rise in cAMP. A high constitutive cAMP concentration can cause unregulated proliferation of pituitary cells and therefore pituitary tumours.

Other class III oncoproteins are variant forms of different transducing proteins. The single largest family of oncogenes consists of those that encode the protein-tyrosine kinase (PTK) found on the inner surface of the membrane. This family, includes class III oncogenes, such as *src*, *abl* and *lck* that encode non-receptor PTK. These differ from the cell surface receptors produced by the class II oncogenes in that they are cytoplasmic or nuclear proteins that have no transmembrane or extracellular domains. Many such PTK proteins have myristate, a long-chain fatty acid, bound to their N-terminal glycine. This causes them to be partially bound to the plasma membrane and puts their kinase domain in the same perimembrane region as that of the receptor kinases. For this reason, it is assumed that the cytoplasmic kinases serve to transduce signals in the same way as do the receptor kinases.

Nuclear transcription factors (Class IV)

The proteins encoded by the class IV oncogenes exert direct effect on transcriptional rates by actually binding to DNA and modifying the rates of initiation of transcription (*myc, fos, jun, myb, ski, evi1, rel*).

One important feature of nuclear proto-oncogenes is that their expression is highly regulated. Genes like c-*fos*, c-*myc* and c-*rel* are expressed at very low levels in quiescent cells but are quickly turned on in response to mitogenic signals. Many nuclear proto-oncogene proteins are induced when normal cells are stimulated to grow, indicating their direct role in growth control (called as **third messengers**).

Cell cycle control proteins (Class V)

In the absence of mitogens or when nutrients are depleted, cells can also enter a resting phase termed G_0. Cyclins, cyclin-dependent kinases and the p53 and RB checkpoint regulators are all elements of this control system. Altered regulation of expression of at least one cyclin, as well as mutation of *p53* and/or *rb*, can be oncogenic. Now it has been well established that checkpoints are deregulated in tumours and mutations occur in the genes involved in cell cycle components, leading to genomic instability and cancer.

Products that affect apoptosis (Class VI)

Apoptosis plays an important role in tumorigenesis as well as in many degenerative and autoimmune diseases. In addition to the well-established morphological features, the genetic events leading to apoptosis have been recently studied in detail and a number of genes and proteins, some acting within the cells and others externally, have now been identified.

The *myc* gene has a dual role in cells.When it complexes with its protein partner Max, it either can drive cells to proliferate or it can make cells commit to enter into the apoptotic pathway. The choice of which pathway to follow appears to depend on the cell's microenvironment, particularly the availability of growth factors. Overexpression or deregulation of *myc* in tumours, therefore, has important implications for the control of apoptosis. Another oncogene that is closely linked to apoptosis is *bcl*-2, which encodes a membrane-bound protein that inhibits apoptosis. The role of *bcl*-2 in apoptosis is most clearly revealed in the phenotypes of knockout mice lacking *bcl*-2 genes. Once formed, the lymphoid tissues of these mice undergo dramatic regression as the result of widespread apoptosis. Like *myc*, the product of the *bck*-2 gene becomes oncogenic when it is expressed at higher-than-normal levels, as can occur when the gene is translocated to an abnormal site on the chromosome. Certain human lymphoid cancers (follicular B-cell lymphomas) are correlated with the translocation of the *bcl*-2 genes next to a gene that codes for heavy chain of antibody molecules.

20.7.5 Viral oncogenes and cell transformation

For a long time it has been known that DNA viruses (e.g., SV40) and RNA viruses (e.g., retrovirus, Rous sarcoma virus), are capable of transforming the cells that they infect. These viruses also became important study materials since they have provided useful clues for understanding the molecular basis for transformation and have led to the identification of cellular oncogenes. Peyton Rous (1911) demonstrated that sarcomas (a type of tumour) could be induced in chicken by a virus now called Rous sarcoma virus (RSV). Since then many tumour-inducing (oncogenic) viruses have been found in animals. Oncogenic viruses are found in almost all families of DNA viruses. Among the RNA viruses, only some of the retroviruses produce tumours. Therefore, both DNA and RNA (retroviruses) viruses can be transforming agents and are called tumour viruses.

Tumour viruses cause transformation as a consequence of their ability to integrate their genetic information into the host cell's DNA. They carry discrete genetic elements that can transform cells to become tumorigenic (or oncogenic). These genes, for example, include the T antigen genes of SV40 and polyoma virus, E1A and E1B of adenovirus and E6 and E7 of papilloma viruses. Viral oncogenes encode oncoproteins, which most often cause the chronic production of one or more oncoproteins that maintain an infected cell in the transformed state.

Transformation by DNA viruses

Members of several families of animal viruses, called tumour viruses, are capable of directly causing cancers in both animals and in humans. Some important viruses that cause human cancer include hepatitis B and C viruses (liver cancer), papilloma viruses (cervical and other anogenital cancers), Epstein–Barr virus (Burkitt's lymphoma and nasopharyngeal carcinoma), Kaposi sarcoma-associated herpes virus (Kaposi sarcoma), and human T-cell lymphotropic virus (adult T-cell leukemia). In addition, HIV is a indirectly responsible for the cancers that develop in AIDS patients as result of immunodeficiency.

Although all types of DNA viruses can be tumour viruses, one group of very small DNA viruses, the papovaviruses (e.g., SV 40) is best-studied. Infection of cells by a papova virus is divided into an early phase, during which it makes one set of viral proteins and induces cellular proteins, and a late phase, during which viral DNA is replicated, coat protein is made, and new virions mature. At the end of the early phase, a quiescent infected cell is activated to synthesise cellular DNA. The small papova viral genome encodes a few key proteins that induce the cell to make the enzymes necessary for DNA replication. The virus accomplishes this by causing the cell to progress from the G_0 to G_1 phase into the S phase. Thus, both cellular and viral DNA is replicated during the infection cycle (**Fig. 20.12**).

Transformation by RNA viruses

Many different oncogenes, derived from cellular proto-oncogenes, have been identified in retroviruses. During evolution, retroviruses captured cellular genes that gave them their dramatic growth-transforming ability. The isolation of mutant strains of RSV that were deficient in transformation provided the first evidence that the virus carried a specific genetic locus, termed ***src*** (for ***sarc***oma-producing), that encoded its ability to transform cells. The presence of *src*-related sequences was later identified in many natural hosts for RSV. The

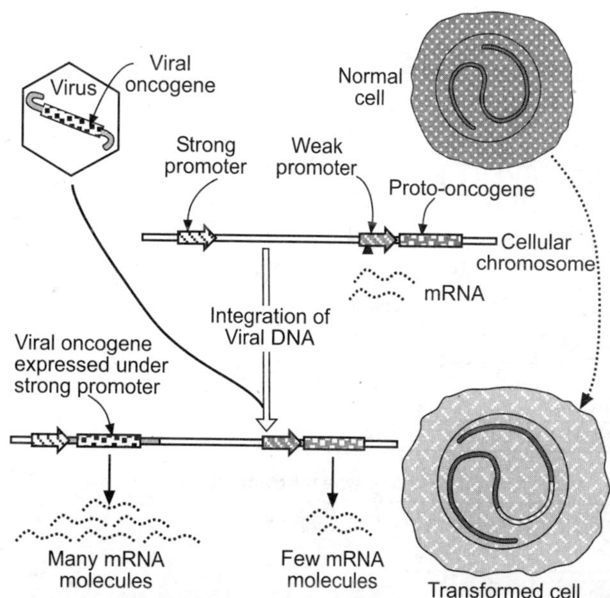

Fig. 20.12 A method by which a viral oncogene having similar coding sequence as its cellular counterpart produces excess oncogene product

characteristics of different transducing retroviruses are determined in part by which oncogenes they contain. For instance, some transform only adherent cells, whereas others transform both adherent and non-adherent cells. In animals, the ones that transform non-adherent cells cause leukemia while the others cause sarcomas or carcinomas.

The demonstration in transducing retroviruses (**Fig. 20.13**) that cellular genetic information can be an agent of transformation clearly implies that human genomes harbour genes that can, in certain circumstances, cause cancer.

Human tumour viruses

Identification of many RNA and DNA tumour viruses in animals has raised the question of whether humans also have tumour viruses; if so, what fraction of human cancer might be caused by them. After years of investigation, the first plausible candidate, **human T-cell leukemia virus (HTLV)**, was reported. Infection with this virus is clearly associated with a

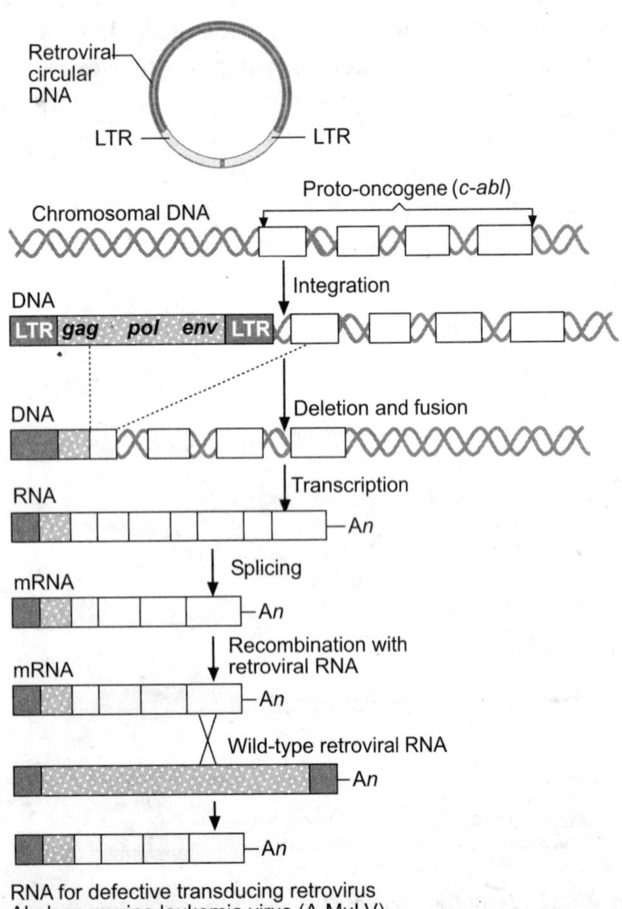

Fig. 20.13 Formation of a transducing retrovirus

type of T-lymphocytic human tumour. Epstein–Barr virus, herpes-like DNA virus, probably plays a role in at least two human tumours: Burkitt's lymphoma and nasopharyngeal carcinoma.

Infection with **hepatitis B virus**, a very small DNA virus that is mainly responsible for human serum hepatitis, is correlated with liver cancers, especially in underdeveloped areas of the world. Although this virus has a DNA genome, it multiplies through an RNA intermediate like a retrovirus. Its genome has been found integrated into the DNA of liver tumour cells. The **papilloma viruses**, which are small DNA viruses, not only are responsible for warts and other benign human tumours but also for malignancies. Out of the 30 currently identified, two types of papilloma virus are associated with human cervical carcinoma and appear to be the cause of this sexually transmitted disease. Papilloma viruses, which also cause a variety of warts, have multiple oncogenes that together transform cells. They are probably the most significant viruses for inducing cancers in developed countries.

Acquired immunodeficiency syndrome (AIDS) is a retrovirus induced disease that, among other symptoms, makes its victims susceptible to numerous cancers. AIDS is caused by the **human immunodeficiency virus (HIV)**, a retrovirus different from all of the others discussed thus far (**Fig. 20.14a**). The HIV genome includes five or six genes in addition to the *gag*, *pol* and *env* genes of other retroviruses (**Fig. 20.14b**). These extra genes, which are all small and expressed from a battery of spliced mRNAs, allow HIV to grow more vigorously than other retroviruses in infected cells. The exact function of HIV-specific genes is still debated but involves activation of transcription, regulation of splicing or transport of mRNA, and possibly other alterations of macromolecular metabolism. HIV kills cells rather than transforming them. Its sole target are the cells that mount the immune response and it causes a profound immunodeficiency. HIV infection has two consequences: short term and long term. During the first six weeks or so following exposure, the HIV-infected person develops flu-like symptoms and there is abundant virus in the blood. Thereafter, the virus largely, but entirely, disappears from the blood, probably because of the actions of T-killer cells. Over years, there is chronic low-level virus production that finally causes the immune system to collapse and AIDS to develop. One type of immune system cells, T helper cells, slowly disappear from the blood and the AIDS

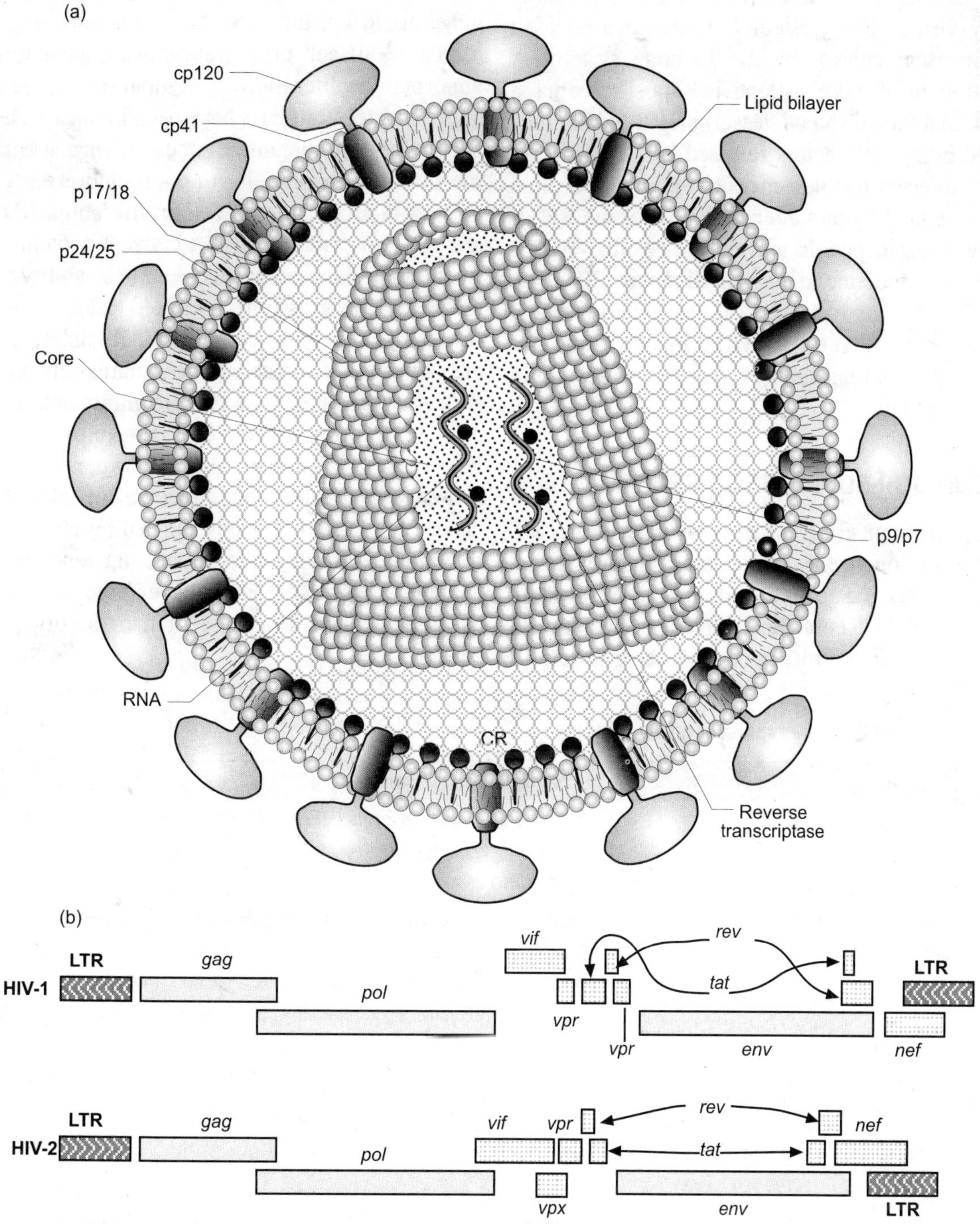

Fig. 20.14 (a) Schematic structure of the human immunodeficiency virus (HIV) particle showing the location of various components. (b) Structures of the genomes of two strains, HIV-1 and HIV-2 which cause AIDS.

patient begin to suffer from infectious diseases that his or her body cannot fight off.

AIDS patients often suffer from cancers, probably caused by viruses like Epstein–Barr that are released from their usual control by the immune system. A common tumour of AIDS patients is Kaposi sarcoma, a tumour of endothelial (blood vessel) origin, which may be induced by growth factors released by HIV-infected cells. HIV infected people generally die of AIDS at an average of about 10 years after infection. Some 5% or so of HIV-infected people may never develop AIDS, but since the epidemic only started about 20 years ago, insufficient time has passed to be certain that any infected persons will survive. AIDS appears to be one of the most devastating diseases ever to spread within the human population.

20.8 TUMOUR SUPPRESSOR GENES

The formation of cancerous tissues is regulated by growth promoting proto-oncogenes, which are counterbalanced by growth suppressing genes called **tumour suppressor genes (Table 20.2)**. These genes represent the opposite side of cell growth control, normally acting to inhibit cell proliferation and tumour development. These genes encode proteins that restrain cell growth and prevent cells from becoming malignant. In many tumours, these genes are lost or inactivated, thereby removing negative regulators of cell proliferation and contributing to the abnormal proliferation of tumour cells. At present, approximately 24 genes have been implicated as tumour suppressors in humans. In one way or another most of the proteins encoded by tumour suppressor genes act as negative regulators of cell proliferation. RB and p53 are the two prototypes of a class of proteins encoded by tumour suppressor genes. Their ability to check cell cycle progression, and hold cells in quiescence or even lead cells to commit suicide unless conditions are appropriate for cell cycle progression, means that they can prevent cells from becoming cancerous. This provides a strong selective advantage to cells that lose both copies of a tumour suppressor gene and therefore can grow even when conditions are not optimal. Certain cancer cells lack both copies of RB because of multiple mutations, chromosome losses or gene conversion. Tumour suppressor genes have often been described as **anti-oncogenes** because of their ability to prevent cancer.

Table 20.2 Important tumour suppressor genes

Gene	*Location*	*Gene product*	*Function and type of cancer*
RB	13q14	PRB (1055 kD)	Primary tumour is retinal, binds E2F. It has DNA-binding ability; inactivated by forming complex with E1A. Involved in retinoblastoma, sarcoma, bladder, breast and lung carcinoma
P53	17p	P53	A transcription factor (cell cycle and apoptosis regulation). Suppresses effect of *ras*, forms complex with *hsc70*. Brain tumours, breast, lung tumours, neurofibrosarcomas, bladder cancer, cervical tumours, testicular tumours, ovarian, colon/rectum, esophageal, liver and lung carcinomas, sarcomas, leukemias and lymphomas
WT-1	11p13	WT-1 (345 amino acids)	A transcription factor, contains zinc finger domains. Wilms' tumour of the kidneys
DCC	18q	Phosphoprotein (190 kD)	—
NF-1	17q11.2	NF-1	Stimulates GTPase activity of *ras*; downregulates of $p21^{ras}$. Neurofibrosarcoma. Syndrome, Neurofibromatosis type 1.
NF-2	—	NF-2	—
APC	5	—	Colorectal, familial colon cancer, binds ß-catenin acting as transcription factor; its inheritance suggests that it is a tumour suppressor gene. Example: familial adenomatous polyposis

BRCA-1	—	BRCA-1 protein	Acts as a transcription factor, contains a zinc finger domain and involved in DNA repair. Example: familial breast cancer
BRCA-2	—	BRCA-2 protein	Familial breast cancer
MSH2, MLH1	—	—	Colorectal, involved in mismatch repair, shows HNPCC syndrome
E-Cadherin	—	Cell adhesion molecule	Breast, colon, familial gastric cancer
p16 (MTS1)	—	MTS1 protein	Acts as an inhibitor of cyclin-dependent protein kinases; causes familial melanoma
INK4a	—	p16: Cdk inhibitor $p14^{ARF}$: stabilises p53	Melanoma, pancreatic, lung carcinoma, brain tumours, leukemias, lymphomas, Example: familial melanoma
PTCH	—	–	Basal cell carcinoma
PTEN	—	PIP_3 phosphatase	Breast, thyroid, brain tumours, melanoma, prostate, endometrial, kidney and lung carcinomas. Example: Cowden disease
Smad2	—	—	Colon/rectum carcinoma
Smad4	—	—	Colon/rectum carcinoma, pancreatic carcinoma
TßRII	—	—	Colon/rectum carcinoma
VHL	—	RNA poll II elongation	Renal cell carcinoma, involved in protein degradation. von Hippel-Lindau syndrome

20.8.1 Retinoblastoma gene

Loss or inactivation of the retinoblastoma (RB) locus involved in a childhood disease (paediatric tumour) in which a tumour forms at a the retina and other forms of cancer, including osteosarcomas and small cell lung cancers are seen. It occurs both as a sporadic and heritable trait (by somatic mutation) in a pattern consistent with autosomal dominant inheritance. In the inherited form of disease, one parental chromosome carries an alteration in this gene. A somatic event in retinal cells that causes loss of the other copy of the RB gene causes a tumour. In the sporadic form, however, the parental chromosomes are normal and both RB alleles are lost by somatic mutations. Alfred Knudson (1971) was the first to explain the genetic basis of retinoblastoma. He proposed that the development of retinoblastoma requires that both copies of the RB gene of a retinal cell be either eliminated or mutated before the cell can give rise to a tumour.

The role of pRb in regulating the cell cycle

Molecular cloning of the *rb*1 gene was greatly facilitated by the identification of a chromosome 13q14 DNA marker that detected rearrangements of retinoblastoma alleles by Southern blot. Following its identification, it was established that *rb*1 is a large gene with 27 exons, spanning more than 200 kbp of genomic DNA. The penetrance of *rb*1 germline mutations is high, and it has been estimated that 90% of individuals with a germline *rb*1 mutation will develop retinoblastoma. The *rb* gene is located in band q14 of human chromosome 13, and encodes RB protein and the disease is associated with deletion of band q14. RB is a nuclear protein that interacts with a variety of cellular proteins (E2F and cyclins D and E) and tumour antigens (SV40 T antigen, adenovirus E1A and HPV E7). It is specifically phosphorylated at the end of G_1 phase by cyclin–CDK complexes and dephosphorylated during mitosis, and remains in this state till it is phosphorylated again at G_1/S boundary. Unphosphorylated RB specifically binds several proteins, including E2F and cyclins D and E. E2F is a group of transcription factors and the binding of E2F with RB inhibits its ability to activate transcription of some genes. The consequences of RB binding cyclins D and E are, however, not clear.

The products of E2F are essential for S phase initiation. The unphosphorylated RB binds to E2F and prevents entry of cells into the S phase. In addition, RB–E2F complex directly represses transcription of some target genes. Phosphorylation of RB at G_1/S boundary abolishes its ability to sequester E2F, which now can activate its target genes. Further, the repression of genes that are targets of RB–E2F complex is also relieved allowing the cell to enter S phase. SV40 T

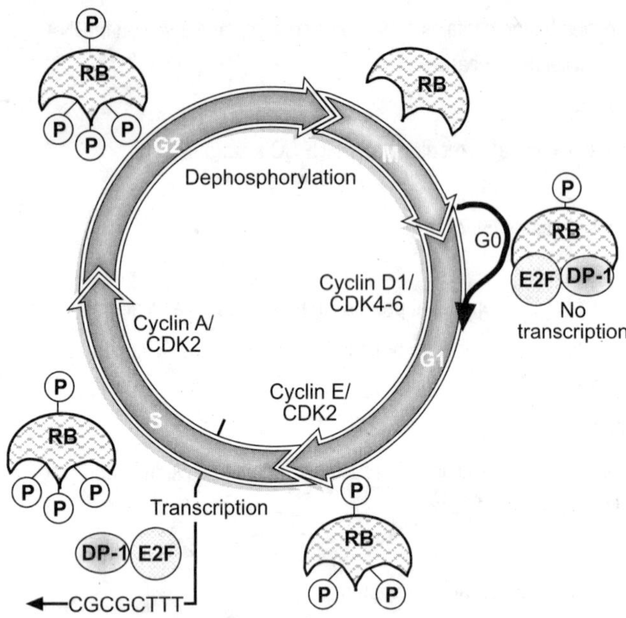

Fig. 20.15 Mode of action of RB; during G_0 and G_1 phases of cell cycle retinoblastoma protein (RB) is underphosphorylated and is bound to the E2F-1 transcription factor complexed with DP-1. During G_1, cyclin D1/CDK4–6 and cyclin E/CDK2 phosphorylate RB; consequently E2F-1 is released to interact with and promote transcription from genes necessary for S phase. Phosphorylation status of RB is maintained by cyclin A/CDK2 until mitosis. Dephosphorylated RB is ready either to re-enter G_1 or to go into the stationary phase.

antigen and adenovirus E1A protein specifically bind the unphosphorylated from of RB. This suggests that tumour antigens sequester unphosphorylated RB and leaves the cellular proteins like E2F free to activate cell division (**Fig. 20.15**). A loss of both the alleles of *rb* gene will have the same effect. It may be added that proteins p107 and p130 have sequences related to RB protein, and have similar properties. Similarly, small proteins p16, p21 and p27 contribute to tumour suppression by ensuring that cells remain in G_0 state and also ensure that CDK4, 6-cyclin complexes remain inactive: p16 acts on CDK-cyclin D, while p21 and p27 act on CDK2-cyclin E complexes as well. CDK-cyclin activity is necessary to phosphorylate RB protein.

20.8.2 The *p53* gene: Guardian of the genome

The vertebrate *p53* protein was first identified in the late 1970s. It is a master tumour suppressor gene that senses DNA damage and is the most important and extensively studied. About 50% of all human cancers are associated with defects in this gene. The *p53* protein is involved in the control of cell cycle, apoptosis and in the maintenance of genetic stability. This gene encodes a nuclear phosphoprotein. Mutant *p53* gene is associated with Li-Fraumeni syndrome inherited as a dominant autosomal gene. Affected individuals show cancers in a variety of tissues, and they have mis-sense mutations in the *p53* gene. The mutations in *p53* gene act as dominant negatives and function by overwhelming the normal protein and preventing it from functioning. It is also called the guardian of the cell.

Roles of *p53* as transcription factor

Several cellular functions for *p53* have been identified. It activates various pathways through its role as a transcription factor. These various roles of *p53* may be divided into the following three groups:

Arrest of cell cycle in G_1: *p53* binds to single-stranded region of <40 bases and to mismatches produced by deletions/additions of 1–3 bases. This binding somehow activates the DNA binding and transcription activation of *p53* domains, which now leaves the site of damage and activates the expression of target genes. In cells that lack *p53*, the loss of the ability to check cell cycle progression (G_1/S) leads to increased rates of DNA alteration (mutations and translocations) speeding up the evolution of new oncogenes.

Apoptosis: It is most likely to play an important role in the inhibition of tumorigenesis. The ability of *p53* to trigger apoptosis is less well understood. Following its activation by DNA damage, *p53* possibly activates a set of genes that leads to apoptosis. It can activate apoptosis pathways in two ways. One is to cause the production of proteins that act on the mitochondrion to trigger its apoptotic functions or produce or activate the cell surface receptors that trigger apoptosis.

Maintenance of genomic stability: This pathway is initiated by activation of transcription of *gadd45*, which encodes a DNA repair protein *gadd45*. Other repair pathways that respond to irradiation damage also activate this repair protein.

The level of *p53* in a healthy G_1 cell is very low. However, if a G_1 cell gets injured, the concentration of *p53* rises rapidly. The increase in *p53* levels is not due to increased expression of the gene but to a decrease in the protein's degradation. The level of *p53* expression is regulated by an interaction between *p53* and cellular onco-protein Mdm2 (**Fig. 20.16**).

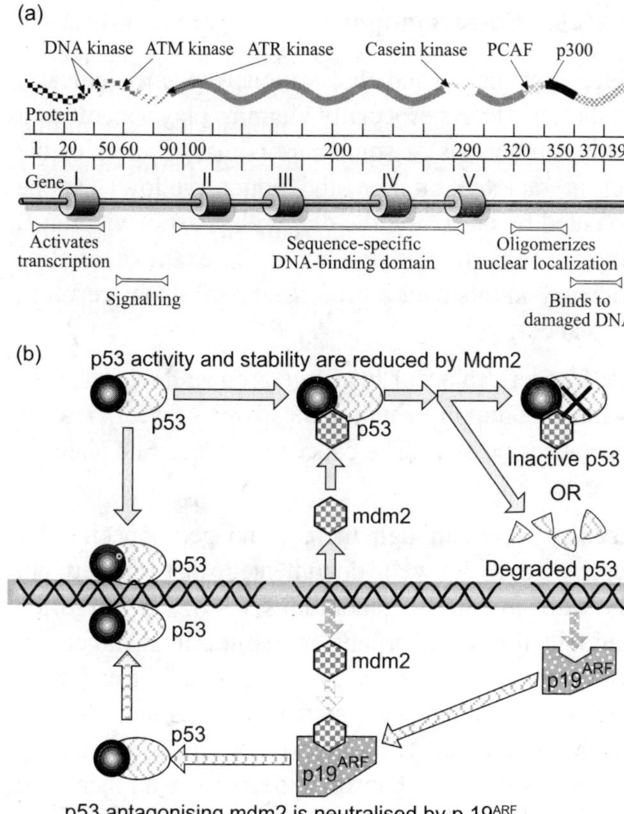

Fig. 20.16 Regulation of p53 activity by mdm2 and p19ARF

20.9 CANCER: A MULTI-CAUSAL PROCESS

Cancer is induced by a variety of agents, which include physical, chemical and biological agents of endogenous or environmental origin, called as mutagens. There are many non-mutagenic agents, which initiate cancers. These carcinogens contribute to cancer development by stimulating cell proliferation, rather than by inducing mutations, and are referred to as tumour promoters. Hormones, particularly estrogens, are important as tumour promoters in the development of some human cancers. The proliferation of cells of the uterine endometrium, for example, is stimulated by estrogen, and exposure to excess estrogen significantly increases the likelihood that a woman will develop endometrial cancer.

20.10 CHEMICAL CARCINOGENS

Chemicals of different origin have been implicated to be the cause of mutations and also many human cancers. These were generally identified through experimental studies in animals.

20.10.1 Metabolic conversion of chemicals into active carcinogens

Chemical carcinogens have a very broad range of structures with no obvious unifying chemistry. Because many carcinogens have structures that appear to be highly unreactive, they need to be activated before they become cancer inducers. There are two categories of chemical carcinogens: **direct-acting** (e.g., ß-propiolactone, EMS, DMS, MNU) and **indirect-acting** (e.g., benzo(a)pyrene, dibenz(a,h)anthracene, 2-naphthylamine, dimethylnitrosamine, vinyl chloride, safrole, aflatoxin B_1). The indirect acting chemical carcinogens require metabolic activation to become carcinogens. The direct acting carcinogens, of which there are only a few, are reactive electrophiles that can directly interact with negatively charged centres on the DNA. Indirect carcinogens are converted to ultimate carcinogens by the introduction of electrophilic centres.

The metabolic activation of carcinogens is carried out by enzymes that are normal body constituents. Animals have such enzymes, especially in the liver, because they are a part of the system that detoxifies noxious chemicals that make their way into the body. Therapeutic drugs, insecticides, polycyclic hydrocarbons and some natural products are often fat-soluble and water-insoluble that they would accumulate continually in fat cells and lipid membranes if there were no way for the animal to excrete them. The detoxification system works by solubilisation, i.e., adding hydrophilic groups to water-insoluble compounds, thus allowing the body to rid itself of noxious or insoluble materials.

Once a carcinogenic chemical is inside the cell, electrophiles can react with negatively charged centres on many different molecules (the most obvious ones being DNA, RNA and proteins). It is only the reaction with DNA, however, that leads to cancer. Depending on their size and structure, electrophiles react with particular positions on individual DNA bases, causing changes in the DNA sequence.

The mutagenicity of chemicals is easily assayed in bacteria and most compounds identified as carcinogens for experimental animals have been shown to be mutagenic. Importantly, the mutagenic potential of compounds is roughly proportional to their carcinogenic potential. For this reason, bacterial mutagenesis has become a potential chemical screening test and the most popular one is the Ames test.

20.10.2 Epigenetic alterations in teratocarcinomas

A most significant class of non-genetic cancer-inducing mechanisms involves epigenetic inheritance. These mechanisms may underlie formation of malignant mouse teratocarcinomas. These tumours of very early embryonic cells maintain the ability to differentiate into the full range of body cell types. In fact, if certain teratocarcinoma cells are incorporated into a preimplantation embryo, they contribute as normal cells to all of the tissues of the adult, including the germ line cells, without forming tumours. Conversely, if teratocarcinoma cells are injected into adult mice, they form lethal tumours. Thus, their malignant state is conditional on their environment. Mouse teratocarcinomas can be formed by the simple procedure of transforming early embryos into older mice at sites such as the testis.

20.10.3 Hormones as carcinogens

Hormones are the key regulators of gene expression and known to exert a trophic influence on their target organs by controlling their proliferative and secretory activities. In the endometrium and the breast, phases of cell proliferation alternate with secretory activity due to a balanced secretion of estrogens and progesterone, in turn controlled by pituitary gonadotrophins. Such hyperplasia involution cycles also occur in the thyroid gland. Testosterone has a trophic action on the prostate. Examples include nodular swellings in the breast in mammary dysplasia; adenomas and nodular enlargements of the thyroid, prostate etc. Such tumours are solely induced by excess hormonal stimulation and are called as **hormone-dependent tumours**.

20.10.4 Endogenous carcinogens

It has been well studied that many endogenously synthesised biomolecules are converted into carcinogens. The carbon ring structure of sex hormones, sterols, bile acids and cholesterol is similar to that of carcinogenic hydrocarbons. The endogenous carcinogens may originate through some error in synthesis or metabolism of sex hormones or bile acids; such errors may be genetic based and appear at various age periods or be governed by mutations. Methyl cholanthrone (a known carcinogen) is synthesised from deoxycholic acid, a normal constituent of bile.

20.10.5 Co-carcinogens

Diet: Experimental and observational evidence indicates that dietary factors especially vitamins play a significant role in the genesis of some carcinomas. Cancer of the liver in rats fed on rice and butter (yellow) can be prevented by enriching the diet with yeast or vitamin B complex, in particular riboflavin. For example, choline deficiency in rats over a prolonged period may result in liver cancer.

Age: Cancer can occur in all age groups; however, they are more common in individuals over 45 years (called cancer-bearing age). The cancer incidence rate plateaus above age 80.

Heredity: Even though there is no gene specific for neoplasms, it is well documented that cancer not only runs in family generations ($\sim 10\%$), but often occurs in the same organs or tissues in members of various generations in the family group. For example, cancers of the breast, stomach, intestine, cervix, neuroblastomas etc run commonly in some family generations. The mechanism appears to be an increased potential of certain tissues or organs to give off mutant cells more regularly and in progressive earlier years. In these cases, heredity plays its part in producing some anatomical congenital error of development, which in later years gets superimposed by malignancy, e.g., intestinal multiple polyposis.

Trauma: The effect on cancer induction is indirect. It just aggravates already existing neoplasm at the site of the trauma, which was unnoticed earlier. Trauma by inducing haemorrhage can cause an existing tumour to enlarge suddenly, thus making it more apparent.

Chronic irritation: Chronic irritation can induce inflammation with repeating occurrence of epithelial necrosis, regeneration and so on. Consequently, continued regenerative activity in due course of time generates mutant cells. Examples for appearance of malignant cells at the site of chronic irritation are: gall stones and cancer gall bladders, renal stones and carcinoma of renal pelvis, squamous cell or basal cell carcinomas, margins of chronic ulcers and sinuses not responding to routine therapeutic measures.

Precancerous conditions: Certain diseases or malformations are superimposed by malignant changes with great frequency so that detection of such diseases or malformations increases the chances of malignant tumours appearing in the patient. These include leukoplakia, syphilitic glossitis, longstanding mammary dysplasia, xeroderma pigmentosum, intestinal multiple polyposis, senile keratosis, kraurosis vulvae and so on.

20.10.6 Environmental causes of cancer

The environment plays a major role in an organism's survival as well as well-being. Many environmental factors (both biotic and abiotic) are found to be mutagenic or teratogenic or carcinogenic. Avoidable substances in the environment and the diet are estimated to cause 80% to 90% of all cancers, although the exact mechanisms by which these agents induce transformation are generally unknown. The environmental agents that bring about tumorigenic mutations in somatic cells are called carcinogens, and most of these apparently act by causing mutations (>90% of carcinogens are mutagens). Identifying environmental causes of cancer in humans is usually difficult, because years or even decades may need to express the action of carcinogens and the clinical detection of tumours.

20.11 HUMAN CANCERS

A central aspect of basic cancer research is to understand human cancer well enough to be able to alter its course. Naturally occurring cancer is most often a consequence of multiple factors that interact over long periods of time. Because each of these factors increases the possibility that a cancer will occur, they are called risk factors. For example, Epstein–Barr virus is a risk factor for Burkitt's lymphoma, although it is clearly not the sole cause. Identification of cigarette smoking is a crucial risk factor in lung cancer. A risk factor of this potency gives a clear indication of how to act to avoid lung cancer: avoid smoking of cigarettes. A diet high in animal fat is thought to be a risk factor for colon and breast cancer, and many viruses and chemicals have been correlated with minor cancers.

20.11.1 Telomere shortening and onset of cancer

Most human cells appear to have a built-in limit to their proliferation: they show replicative senescence when they are grown in culture. Replicative cell senescence in humans is thought to be caused by changes in the structure of **telomeres**—the replicative DNA sequences and associated proteins. These telomeric DNA sequences are synthesised and maintained by a special mechanism that requires the enzyme **telomerase**. In most human cells, other than those of the germ line and some stem cells, expression of the gene coding for the catalytic subunit of telomerase is switched off, or at least not fully activated. As a result, the telomeres in these cells tend to become a little shorter with each round of cell division. Eventually, the telomeric cap on the chromosome end can become shortened to the point where a danger signal is generated, arresting the cell cycle. Therefore, replicative senescence might be expected to favour the development of cancer.

When compared with normal cells in humans, most human cancer cells express telomerase. This is thought to be the reason why, unlike normal cells, they tend to divide without limit in culture and it is added evidence that maintenance of telomere has a significant role in cancer (**Fig. 20.17**).

20.12 PREVENTION AND TREATMENT OF CANCER

Translating the growing understanding of cancer into practical applications in cancer prevention and treatment is a challenging task for molecular biologists. It is increasingly being felt that conventional approaches to combating cancer, namely surgery, chemotherapy and radiation, are often not sufficient to rid a patient of all cancer cells. A large number of clinical trials are being conducted to test a variety of new anticancer strategies.

Prevention and early detection

Many cancers can indeed be prevented—first and foremost, by avoiding the use or exposure to environmental as well as occupational hazardous material. Much of the promise for cancer comes from observational epidemiological studies that show association between modifiable lifestyle factors or environmental exposures and specific cancers. Examples of modifiable factors include alcohol consumption (associated with increased risk of oral, esophageal, breast and other cancers), smoking, physical inactivity and being overweight. Based on epidemiological evidence, it is now thought that avoiding excessive alcohol consumption, being physically active, and maintaining the recommended body weight may all contribute to reduction in the risk of certain cancers. Other lifestyle and environmental factors known to affect risk include certain sexually transmitted diseases, the use of exogenous hormones, exposure to ionising

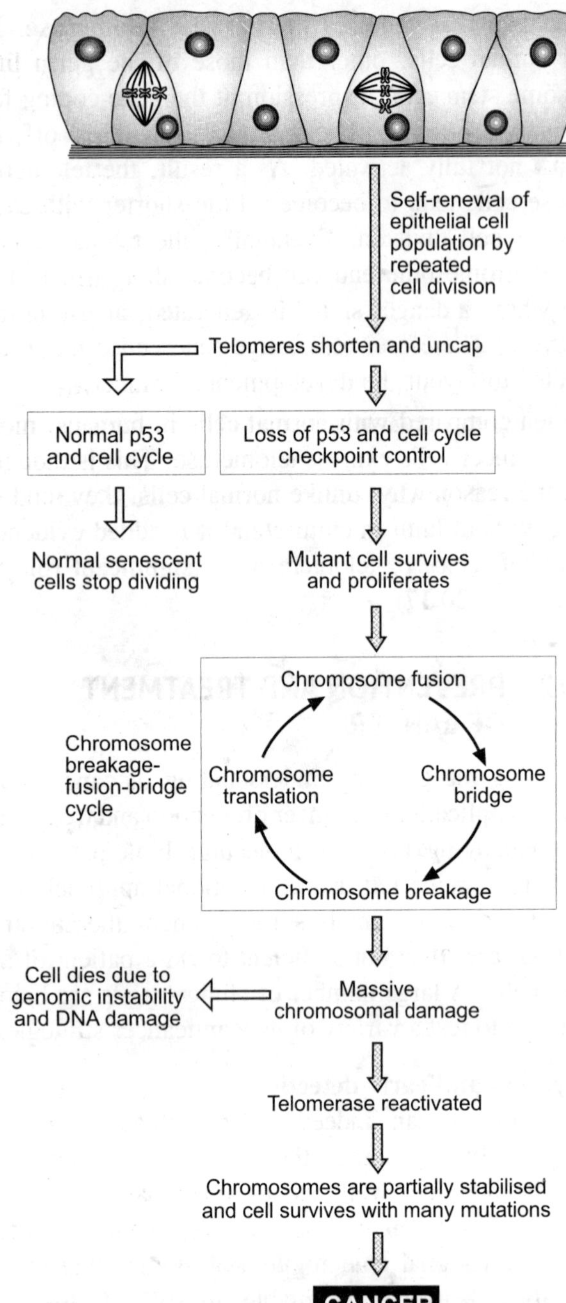

Fig. 20.17 Shortened telomeres may lead to chromosomal instability and cancer (***Source:*** Alberts et al 2002)

radiation and UV light, certain occupational and chemical exposure and infectious agents.

Diagnosis

The next best approach would be to reliably detect early pre-malignant stages of tumour development that can be effectively treated. Most cancers are initially recognised either because signs or symptoms appear or through screening. Roughly, cancer symptoms can be divided into three groups:

Local symptoms: Unusual lumps or swelling, haemorrhage, pain and/or ulceration

Symptoms of metastasis: Enlarged lymph nodes, cough and haemoptysis, hepatomegaly, bone pain, fracture of affected bones and neurological symptoms

Systemic symptoms: Weight loss, poor appetite and cachexia (wasting), excessive sweating, anaemia and specific paraneoplastic phenomena

The effective application of molecular biology to prevention and early detection may lie in the identification of individuals with inherited susceptibilities and carefully monitoring them to allow early detection and more effective treatments. Mutations in tumour suppressor and oncogenes can be detected by genetic testing and identifying high-risk individuals before disease develops. Isolation of cancer susceptibility genes has clear practical applications. Individuals with such inherited susceptibility genes could be appropriately advised to avoid exposure to relevant carcinogens and carefully monitored to detect tumours at early stages for effective treatment. Detection and molecular analysis of the oncogenes and tumour suppressor genes involved in particular types of tumours have the potential of providing information needed for diagnosis of cancer and in monitoring the effects of treatment. For example, translocation of *abl* (fusion of *abl* with the *bcr* gene generates Bcr/Abl oncogene protein) in chronic myeoloid leukemia is a good example of molecular marker for monitoring the mutations in oncogenesis.

20.13 TREATMENT OF CANCER

Our growing understanding of cancer cell biology and tumour progression is gradually leading to better methods for treating the disease. Cancers can be often treated by surgery, chemotherapy, radiation therapy, immunotherapy, monoclonal antibody therapy or other methods. The choice of therapy depends upon the location and grade of the tumour and the stage of the disease, as well as the general state of the patient. Complete elimination of the cancer cells without damage to the body is the goal of treatment. In theory, cancer can be cured if entirely removed by surgery, but this is not always possible. When cancer has metastasised to other sites in the body prior to surgery, complete surgical excision is usually impossible.

Current non-surgical approaches to the treatment of cancer suffer from the major limitation that they are not specific for malignant cells. Chemotherapy, even though significantly improves survival rates in certain tumours (e.g., ALL and testicular cancer), for most 'solid tumours', which comprise the majority of the cancer incidence and mortality rate (e.g., lung, breast, gastric, colon, ovary), survival figures have hardly improved over three decades of research in this area. The future hopes lie, however, in finding more powerful and selective ways to directly exterminate cancer cells. Pinpointing genetic lesions of cancer and using knowledge of cell biology, we can develop effective treatment regimens that can specifically kill them. Potential targets for therapy include the oncogenes themselves, their RNA transcripts, and their protein products.

20.13.1 Immunotherapy

A good number of cases for persons with metastatic cancer who were told they had only months to live, defy the prognosis and remain alive and cancer-free years later. The best-studied cases of such 'spontaneous remissions' were recorded in the late 1800s by a physician named William Coley. A patient with an inoperable tumour of the neck had gone into remission after contracting a streptococcal infection beneath the skin. His immune system activated against bacterial toxin also killed the tumour cells. In recent years, two broad treatment strategies involving the immune system have been followed: passive immunotherapy and active immunotherapy.

Passive immunotheraphy is an approach that attempts to treat cancer patients by administering antibodies as therapeutic agents. Antibodies recognise and bind to specific proteins that play a key role in the activities of the tumour that is being targeted. At least three basic components are necessary for targeting cell systems:

- A target should present either exclusively on the tumour cell or at least be expressed in higher amounts on the tumour compared to a normal tissue.
- A good carrier or delivery system is essential.
- A toxic agent or set of molecules, which will cause damage to the tumour after they have directed the target cell.

Ideally, the target would be present on the tumour and not on any normal cells or tissues. Since such tumour-specific target molecules are not frequently found, most antibody-mediated targeting has been directed at tumour-associated antigens, mostly differentiation antigens such as **carcinoembryonic antigen (CEA)**, or antigens present on particular cell types such as the melanoma-associated antigens. Of the entire carrier systems tried so far, antibodies have shown the greatest potential and have been the most intensively investigated to date. The use of antibodies to target 'warheads' such as radioisotopes, drugs, or toxins has great potential due to the unique specificity of the antibody for its target antigens. Monoclonal antibodies linked to radioisotopes have been used in diagnostics and also in studies aimed at therapeutic benefits (**Fig. 20.18**). Toxins linked to monoclonal antibodies have displayed good cytotoxic potency.

20.13.2 Gene therapy

Gene therapy is a treatment in which a patient's genotype is modified by the addition, deletion or alteration of a specific gene. An example of a successful approach in an animal model is the liposomal delivery of the human *apc* gene to rodent colonic epithelium where 100% of the epithelial cells which contacted the liposome-gene

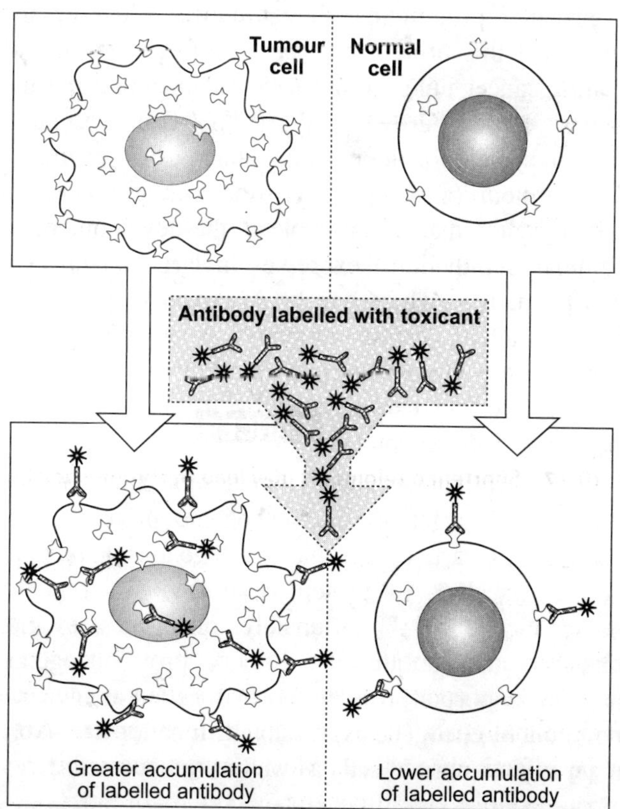

Fig. 20.18 Immunotargeting by using drug-linked antibodies

mixture were shown to have taken up the gene. The gene expression was, however, only transient and did not persist beyond a few days. Despite many problems, a number of clinical trials of gene therapy for the treatment of cancer have been approved or are already underway.

One of the most promising approaches has been the use of retroviral vectors containing a herpes simplex virus thymidine kinase (HSVtk) gene in patients with brain tumours. Results of the clinical studies using this approach are awaited with great interest. As discussed earlier, tumour cells lack functional tumour suppressor genes whose products restrain the growth of normal cells. The therapeutic possibilities of introducing extra copies of the wild type *p*53 and *rb*1 genes into the DNA of tumour cells are readily apparent. The *p53*-bearing viruses have been shown to restore normal growth to malignant cells both in culture and in laboratory animals. Research continues to be in the area of development of efficient, multiple therapeutic gene delivery systems.

20.13.3 Inhibiting the activity of cancer-promoting proteins

Many cancer cells are known to depend heavily on a particular protein that they produce. Blocking the activity of this protein may be an effective means of treating cancer unless it has severe damage on normal tissues. Hence, growth factors and their receptors are candidates for therapeutic attack. For example, Her2 protein (a receptor tyrosine kinase related to EGF receptor that plays a role in the development of mammary epithelium) expresses in large quantity in breast tumours, which can be inactivated by using a blocking agent, or an antibody. Another approach to destroying tumours targets the delivery of a toxin directly into the cancer cells by using cancer cell-specific surface receptors.

By a different approach, small molecules can be designed to target specific oncogenic proteins. For example, in case of chronic myelogenous leukemia, the chimeric Bcr-Abl protein is an obvious target for therapeutic attack. Unfortunately, oncogenes are not unique to just tumour cells, because proto-oncogenes play an important role in normal cells, as general inhibitors of oncogene expression or function are likely to act against normal cells as well as tumour cells.

The first therapeutic regimen targeted against a specific oncogene was used for the treatment of acute promyelocytic leukemia (PML). PML is due to the chromosomal translocation in which the gene that encodes the retinoic acid receptor (RARα) is fused to another gene (PML) to form the PML/RARα oncogene. The PML/RARα protein is thought to function, at least in part, by blocking cell differentiation. However, treatment with high doses of retinoic acid appears to overcome (but temporarily) the inhibitory effect of the PML/RARα oncogene protein. Later, combined treatment with retinoic acid and standard chemotherapeutic agents was found to significantly reduce the incidence of relapse.

20.13.4 Inhibition of angiogenesis

One promising approach to cancer therapy is the use of drugs that inhibit tumour growth by interfering with angiogenesis. Because tumour cells require the formation of new blood vessels to grow to more than a millimetre or so in size, treatments that block angiogenesis should block tumour growth in many different types of cancer. It has led to the development of new drugs that inhibit angiogenesis by blocking the proliferation of endothelial cells.

20.13.5 Use of antisense RNA or oligonucleotides

Following the early discovery of natural antisense RNAs in prokaryotes, numerous applications of antisense RNA-mediated regulation have been demonstrated in a variety of experimental systems. These oligonucleotide RNAs, directly repress gene expression by hybridising to a target RNA, rendering it functionally inactive.

20.13.6 Chemotherapy

Chemotherapy refers to treatment of diseases (infections or cancer) by administration of chemical drug(s), under the supervision of a qualified physician. Various chemical drugs (natural and synthetic) have been used to treat the patients for different types of cancer and infections, either singly or in combination with other agents including radiation.

Chronotherapy

Chronotherapy refers to the administration of drugs at a certain time of the day—when efficacy is highest and side effects are lowest. An example of successful chronotherapy is the use of the cholesterol-lowering drugs known as statins. Statins inhibit HMG-CoA reductase, the rate-limiting enzyme in cholesterol biosynthesis. The activity of HMG-CoA reductase has circadian rhythmicity, as it is highest at night. Therefore, statins are most effective when administered before bedtime. It is widely accepted that cells enter various phases of the cell cycle in a circadian manner. For example, chronotherapy using doxorubicin and cisplatin showed significant improvement in the survival rate of patients with ovarian cancer when doxorubicin was administered in the morning, followed by cisplatin 12 hours later. The circadian clock shares several conceptual and molecular similarities with the cell cycle. Both consist of interlocked autoregulatory loops, and they both rely on sequential phases of transcription, translation, post-transcriptional modification and degradation.

KEY POINTS

- Cancer is an abnormality of multicellular organisms, arising from the progressive accumulation of mutations or other metabolic changes that promote the clonal expansion of cells.
- There are many types of cancer and are generally classified depending on tissue and cell types originally involved.
- One of the characteristic features of cancer development is tumour clonality: the expansion of tumours from single cells that begin to proliferate abnormally.
- Carcinogenesis is a process which induces cancer in normal tissues. The various types of agents that are responsible for the initiation of neoplasia are called carcinogens.
- The development of cancer is a multistep process in which cells gradually become malignant through a progressive series of alterations.
- Tumours may be either benign or malignant. Benign tumours are rarely life-threatening, grow within a well-defined capsule which limits their size. In contrast, malignant cells lack differentiation and structure is atypical.
- The nucleus of the cancer cell sometimes changes it shape and contents. It may become hypertrophic.
- Immortalisation describes the process of indefinite cell growth.
- Transformation describes the process of acquiring the new abnormal features that facilitate the cell to overcome normal constraints of growth.
- Cancer cells secrete growth factors that promote the formation of new blood vessels (called angiogenesis).
- Cancer cells fail to differentiate normally. Such defective differentiation is closely coupled to abnormal proliferation, since, most fully differentiated cells cease cell division.
- Most cancer cells fail to undergo apoptosis, consequently their lifespan increases when compared to their normal counterparts.
- Proto-oncogenes are cellular genes that encode proteins having a function in the normal cell.
- In human tumours, oncogenes have been found to be activated by different cellular activation mechanisms such as mutations, translocations, amplification and DNA rearrangements. Many oncogenes become tumorigenic due to their increased expression.
- Activation of oncogenes and inactivation of tumour suppressor genes are involved in induction of cancer.
- Proto-oncogenes are counterbalanced by growth suppressing genes called tumour suppressor genes. The *p53* gene is a master tumour suppressor gene that senses DNA damage and is the most important and extensively studied tumour suppressor gene.
- Cancer is induced by a variety of agents, which include physical, chemical and biological agents of endogenous or environmental origin, called as mutagens.

- Most human cancer cells express telomerase. This is thought to be the reason why, unlike normal cells, they tend to divide without limit in culture and it is added evidence that telomere maintenance has a significant role in cancer.
- Gene therapy is a treatment in which a patient's genotype is modified by the addition, deletion or alteration of a specific gene.
- Many cancer cells are known to depend heavily on a particular protein that they produce. Blocking the activity of this protein may be an effective means of treating cancer unless it has severe damage on normal tissues.
- A promising approach to cancer therapy is the use of drugs that inhibit tumour growth by interfering with angiogenesis. Depriving blood supply to the cancer cells can block tumour growth.
- Mutations can also affect gene regulation at post-transcriptional stages. A number of such mutations which affect the processes of RNA splicing or alternative splicing have been described.
- Detailed study of the role of aberrant gene regulation in cancer and many other diseases may help to develop good control strategies for various diseases.

Self-assessment Questions

1. Explain briefly the characteristics of different types of cancer.
2. Discuss with an example the multistep process of cancers.
3. What are the major differences between benign and malignant tumours?
4. Explain briefly the cytomorphological characteristics of cancer cells.
5. What is meant by immortalisation of cells?
6. What are oncogenes? What is their influence on cell transformation?
7. Explain the role of viral oncogenes in cancers.
8. What are tumour suppressor genes and explain their mechanism of action.
9. List a few chemical carcinogens and their mode of action.
10. Explain how epigenetic alterations influence carcinogenesis.
11. Explain the role and mechanism of telomerase in cancer.
12. What are the different therapies available for treatment of cancers?

FURTHER READING

Textbooks

Cooper G (1990). *Oncogenes*. Boston: Jones and Bartlett.

Franks LM and NM Teich (1991). *Introduction to Cellular and Molecular Biology of Cancer*, 2nd ed. New York: Oxford University Press.

Macdonald F and CHJ Ford (1997). *Molecular Biology of Cancer*. Bios Scientific Publishers.

Tannock IF and RP Hall (1987). *The Basic Science of Oncology*. New York: Pergamon Press.

Varmus H and RA Weinberg (1993). *Genes and the Biology of Cancer*. Scientific American Library.

References

Adams JM and S Cory (1991). Transgenic models of tumour development. *Science* 254:1161–67.

Balmain A and IB Pragnell (1983). Mouse skin carcinomas induced *in vivo* by chemical carcinogens have a transforming *Harvey-ras* oncogene. *Nature* 303:72–74.

Behbens J et al (1992). Cell adhesion in invasion and metastasis. *Semin. Cell Biol.* 3:169–78.

Bergers G and LE Benjamin (2003). Tumorigenesis and the angiogenic switch. *Nature Revs. Cancer* 3:401–10.

Bishop JM (1983). Cellular oncogenes and retroviruses. *Ann. Rev. Biochem.* 52:301–54.

Bishop JM (1985). Viral oncogenes. *Cell* 42:23–38.

Brehm A and T Kouzrides (1999). Retinoblastoma protein meets chromatin. *Trends in Biochem. Sci* 24:142–45.

Carpenter G and S Cohen (1975). Human epidermal growth factor and the proliferation of human fibroblasts. *J. Cell Physiol.* 88:227–38.

Cavanee WK and RL White (1995). The genetic basis of cancer. *Sci. Amer.* 272:50–57.

Channarayappa, J Nath and T Ong (1992). Clastogenic and aneuploidogenic effects of cigarette smoke condensate, mitomycin C and vincristine sulfate. *Mutagenesis* 7(6): 457–60.

Channarayappa, J Nath and T Ong (1992). Immunofluorescent staining of kinetochores in micronucleus for detection of aneuploidy inducing agents. *J. of Tiss. Cult. Meth.* 14:125–32.

Channarayappa, T Ong and J Nath (1992). Cytogenetic effects of vincristine sulfate and ethylene dibromide in human peripheral lymphocytes: Micronucleus analysis. *Envi. Molec. Mutagen.* 20:117–26.

Channarayappa (1990). Cytogenetic analysis of clastogens and aneuploidogens by immunofluorescent and micronucleus methods. PhD thesis submitted to the West Virginia University, Morgantown, West Virginia, USA **(W Microfilm 3365)**.

Couzin J (2003). Tracing the steps of metastasis: Cancer's menacing ballet. *Science* 299:1002–06.

Diffley JFX and G Evan (1999). Oncogenes and cell proliferation. Cell cycle, genome integrity and cancer–a millennial view. *Curr. Opin. Genet. Devel.*, 10:13–16.

Druker BJ (2002). Inhibition of the Bcr-Abl tyrosine kinase as a therapeutic strategy for CML *Oncogene* 21:8542–46.

Elledge RM and WH Lee (1995). Life and death by *p53*. *BioEssays*, 17:923–30.

Fidler IJ (1990). Critical factors in the biology of human cancer metastasis. *Cancer Res* 50:6130.

Gao HG, JK Chen, J Stewart, B Song, Channarayappa, WZ Whong and T Ong (1997). Distribution of p*53* and K-*ras* mutations in human lung cancer tissues. *Carcinogenesis* 18(3):473–78.

Hanahan D (1988). Dissecting multistep tumorigenesis in transgenic mice. *Ann. Rev. Genet.* 22:479–521.

Hermans A et al (1987). Unique fusion of *bcr* and c-*abl* genes in Philadelphia chromosome positive acute lymphoblastic leukemia. *Cell* 51:33.

Hunter T (1991). Cooperation between oncogenes. *Cell* 64:249.

Jacks T (1996). Tumour suppressor gene mutations in mice. *Annu. Rev. Genet.* 30:603–36.

Lees EM and E Harlow (1995). Cancer and the cell cycle. In: *Cell Cycle Control: Frontiers in Molecular Biology*, (eds) C Hutchinson and DM Glowver pp.228–63.

Levine AJ, J Momand and CA Finlay (1991). The *p53* tumour suppressor gene. *Nature* 351:453–56.

Lozano G and SJ Elledge (2000). *p53* sends nucleotides to repair DNA. *Nature* 404:24–25.

Marshall CJ (1991). Tumour suppressor genes. *Cell* 64: 313–26.

Payne GS, JM Bishop and HE Varmus (1982). Multiple arrangements of viral DNA and an activated host oncogene in basal lymphomas. *Nature* 295–314.

Prolla TA (1998). DNA mismatch repair and cancer. *Current Opinion in Cell Biology* 10:311–16.

Rous P (1911). A sarcoma of the fowl transmissible by an agent separable from the tumour cells. *J. Exp. Med.* 13: 397–411.

Weinberg RA (1989). Oncogenes, anti-oncogenes and the molecular basis of multistep carcinogenesis. *Cancer Res.* 49:3713–21.

Index

D

E

F

G

T